Engineering Statistics

Engineering Statistics

ALBERT H. BOWKER
Professor of Mathematics and Statistics
Stanford University

GERALD J. LIEBERMAN
Professor of Statistics and Industrial Engineering
Stanford University

PRENTICE-HALL, INC.
Englewood Cliffs, N. J.

Library of Congress Catalog Card Number: 59-11116

Fourth printing......... May, 1961

PRINTED IN THE UNITED STATES OF AMERICA

27948-C

Preface

This book is intended as a text for a first course in statistics for students in engineering and the physical sciences. At Stanford it is the basis for a one-quarter course entitled "Statistical Methods in Engineering and the Physical Sciences," which takes up the material in the first ten chapters. This course meets for approximately forty periods (four times a week) for one quarter and would be roughly the equivalent, in classroom hours, of a semester course meeting three times a week. A course limited to the first ten chapters concentrates primarily on basic probability and distribution theory and experimental statistics. Additional topics in experimental statistics, and the important subjects of quality control and acceptance sampling, could be incorporated in a two-quarter or two-semester course. This would include the material in chapters XI, XII, and XIII. Chapter XIII, in particular, is a very long and detailed presentation of sampling inspection, including the basic tables now in practical use for both attribute and variables inspection.

The book attempts to present statistics as a science, rather than an art, for making decisions. After a thorough grounding in the necessary fundamentals of probability and distribution theory, the text turns to a discussion of the statistical techniques commonly used by engineers and physical scientists. Ideally in statistical work, decision procedures should be specified before the data are available and the experiment is run. The book emphasizes the notion of operating characteristic curves and the considerations in the choice of a particular experimental procedure and sample size. Each technique is illustrated by fully worked examples drawn from engineering and science. A complete set of tables, both for the design of the experiment and the final test of significance, is included.

The book is not intended to be a cookbook or a treatise in mathematical statistics. It is our hope to provide a logical presentation of the statistical techniques used in engineering and physical sciences. The only prerequisite for the book is an elementary knowledge of calculus. Sections which can be deleted without losing the continuity of the text are starred. As a rule these are sections which have detailed or difficult mathematical arguments.

We are grateful to Professors E. L. Grant and W. G. Ireson, editors of the *Handbook of Industrial Engineering and Management*, for many helpful comments and for their permission to reprint much of the material from our chapter in that handbook. We are indebted to many members of the Statistics Department and the Applied Mathematics and Statistics Laboratory at Stanford, and to several classes of students who have helped to eradicate some of the errors and clarify some of the explanations. Particular mention should be made of Mrs. Peggy Hathaway, who typed the manuscript, of Miss Barbara Bonesteele, who undertook the tedious chore of proofreading, of Professor Donald Guthrie and Mr. Peter Zehna who assisted in the course and helped check much of the material, of Mrs. Gladys Garabedian for invaluable assistance in the preparation of the charts, of Miss Shirley Smith for her general assistance, and of Professor Herbert Solomon, without whose aid this book might never have been completed.

Finally, we are indebted to Professor Sir Ronald A. Fisher, Cambridge, and to Dr. Frank Yates, Rothanstad, and to Messrs. Oliver and Boyd Ltd., Edinburgh, for permission to abridge Tables III, IV, and V from their book, *Statistical Tables for Biological, Agricultural, and Medical Research*.

A. H. B.

G. J. L.

Table of Contents

CHAPTER I

Histograms and Empirical Distributions

1.1 Introduction

The science of statistics deals with drawing conclusions from observed data. However, the popular conception of statistics is that it involves large masses of data and concerns itself with percentages, averages, or presentation of data in tables or charts. These, in fact, represent only a small part of the field today and are of less interest to engineers than are other aspects of statistics, e.g., quality control, sampling inspection, and the design and analysis of experiments. The latter are treated in this book.

Most scientific investigations, whether concerned with the effect of a new drug on polio, methods of allaying traffic congestion in a big city, consumer reaction to a new product, or the quality of manufactured products, depend on observations, even if they are of a rudimentary sort. In scientific and industrial experimentation, these observations are taken to study the effect of variation of certain factors or the relation between certain factors. One may wish to study the quality of a raw material from a new supplier, the relation between tensile strength and hardness for a particular alloy, or the optimum combination of conditions for a manufacturing process. Ultimately, these observations are to be used for making decisions, and the main subject matter of this book will deal with providing procedures for making decisions with preassigned risks on the basis of the limited information in samples. These procedures will be illustrated by examples.

The raw material of statistics consists of data or observations. Usually these observations are physical measurements: length, decimeter temperature, or resistance, and may be any number on a continuous scale. But these continuous measurements are not the only kind of observations. In the inspection of manufactured products, items are often classified as defective or non-defective and the observation for each item is not a number but simply the category into which it falls. In studying the grading system in a statistics course, each observation may be a letter, say A or B or C, etc.—the grades given to each student. Whether the observations are numbers, letters, categories, or whatever, the same general methods of statistical analysis are applicable. In most of the book, the concern is with small numbers of observations, but before turning to this topic a few of the techniques useful in arranging, presenting, and summarizing large numbers of observations will be discussed.

1.2 Empirical Distributions

A basic notion of statistics is the notion of variation. There is no real meaning in speaking of the life of an incandescent lamp produced under certain conditions; some bulbs may last many times as long as others. Consider, for example, the data in Table 1.1 on the lifetimes in hours of 417 forty watt, 110 volt, internally frosted incandescent lamps taken from forced life tests.[1] Many factors including variations in raw materials, workmanship, function of automatic machinery, and, in fact, test conditions may account for these differences; some of these factors may be controlled carefully but some pattern of variation is inherent in all observational data. Perhaps no one would expect all light bulbs to have exactly the same life, but even the results of the most easily controlled experiments, such as those designed to measure the velocity of light, exhibit variation from experimental error.

Returning to the life of bulbs, the data in Table 1.1 are listed in the order in which the life tests were made; arranged serially in this way the numbers present a meaningless jumble. The best way to arrange the data depends, of course, on the question to be answered. In a study of the life of light bulbs, there are many questions which might arise. How long do the bulbs last on the average? What fraction of light bulbs manufactured in this way can be expected to last 1,000 hours? What value of life will 95% of the bulbs outlast? What range of life can we expect? What interval around the average will include 50% of the observations? A careful investigation of the table will reveal that the life varies from a low of 225 hours to a maximum of 1,690, but the other questions cannot be answered from this table.

[1] D. J. Davis, "An Analysis of Some Failure Data," *Journal of the American Statistical Association*, Vol. 47, 1952.

Table 1.1. Life in Hours of 417 Incandescent Lamps

Date	Item Lifetimes										Average of Sample
1- 2-47	1,067	919	1,196	785	1,126	936	918	1,156	920	948	997
1- 9-47	855	1,092	1,162	1,170	929	950	905	972	1,035	1,045	1,012
1-16-47	1,157	1,195	1,195	1,340	1,122	938	970	1,237	956	1,102	1,121
1-23-47	1,022	978	832	1,009	1,157	1,151	1,009	765	958	902	978
1-30-47	923	1,333	811	1,217	1,085	896	958	1,311	1,037	702	1,027
2- 6-47	521	933	928	1,153	946	858	1,071	1,069	830	1,063	937
2-13-47	930	807	954	1,063	1,002	909	1,077	1,021	1,062	1,157	998
2-20-47	999	932	1,035	944	1,049	940	1,122	1,115	833	1,320	1,029
2-27-47	901	1,324	818	1,250	1,203	1,078	890	1,303	1,011	1,102	1,088
3- 6-47	996	780	900	1,106	704	621	854	1,178	1,138	951	923
3-13-47	1,187	1,067	1,118	1,037	958	760	1,101	949	992	966	1,014
3-20-47	824	653	980	935	878	934	910	1,058	730	980	888
3-27-47	844	814	1,103	1,000	788	1,143	935	1,069	1,170	1,067	993
4- 3-47	1,037	1,151	863	990	1,035	1,112	931	970	932	904	993
4-10-47	1,026	1,147	883	867	990	1,258	1,192	922	1,150	1,091	1,053
4-17-47	1,039	1,083	1,040	1,289	699	1,083	880	1,029	658	912	971
4-23-47	1,023	984	856	924	801	1,122	1,292	1,116	880	1,173	1,017
5- 1-47	1,134	932	938	1,078	1,180	1,106	1,184	954	824	529	986
5- 8-47	998	996	1,133	765	775	1,105	1,081	1,171	705	1,425	1,015
5-15-47	610	916	1,001	895	709	860	1,110	1,149	972	1,002	922
5-22-47	990	1,141	1,127	1,181	856	716	1,308	943	1,272	917	1,045
5-29-47	1,069	976	1,187	1,107	1,230	836	1,034	1,248	1,061	1,550	1,130
6- 5-47	1,240	932	1,165	1,303	1,085	813	1,340	1,137	773	787	1,058
6-12-47	1,438	1,009	1,002	1,061	1,277	892	900	1,384	1,148		1,123
6-19-47	1,117	1,225	1,176	709	1,485	1,225	1,011	1,028	1,227	1,277	1,148
6-26-47	1,222	912	885	1,562	1,118	1,197	976	1,080	924	1,233	1,111
7- 3-47	1,135	623	983	883	1,088	1,029	1,201	898	970	1,058	987
7-10-47	1,160	831	1,023	1,354	1,218	1,121	1,172	1,169	1,113	1,308	1,147
7-17-47	1,166	1,470	1,635	1,141	1,555	1,054	1,461	1,057	1,228	1,187	1,295
8- 7-47	1,016	744	1,197	1,122	666	1,022	964	1,085	612	1,003	943
8-14-47	1,235	942	1,055	893	1,235	1,056	968	1,056	1,014	1,096	1,055
8-21-47	1,013	889	1,430	926	1,297	1,033	1,024	1,103	1,385		1,122
8-28-47	1,077	813	1,121	960	1,156	1,033	1,255	225	525	675	884
9- 4-47	1,211	995	924	732	935	1,173	1,024	1,254	1,014		1,029
9-11-47	798	1,080	862	1,220	1,024	1,170	1,120	898	918	1,086	1,018
9-18-47	1,028	1,122	872	826	1,337	965	1,297	1,096	1,068	943	1,055
9-25-47	1,490	918	609	985	1,233	985	985	1,075	1,240	985	1,051
10- 2-47	1,105	1,243	1,204	1,203	1,310	1,262	1,234	1,104	1,303	1,185	1,215
10- 9-47	759	1,404	944	1,343	932	1,055	1,381	816	1.067	1,252	1,095
10-16-47	1,248	1,324	1,000	984	1,220	972	1,022	956	1,093	1,358	1,118
10-23-47	1,024	1,240	1,157	1,415	1,385	824	1,690	1,302	1,233	1,331	1,260
10-30-47	1,109	827	1,209	1,202	1,229	1,079	1,176	1,173	769	905	1,068

The first step in analyzing data of this sort is to arrange them into a frequency table, grouping adjacent observations into classes which are usually called class intervals or cells. If this is done by hand, the class intervals are listed on a ruled sheet and a tally mark for each observation is placed opposite the appropriate interval. A convenient way to keep track is to make the fifth tally mark diagonally through the preceding four. A worksheet is presented in Table 1.2;

Table 1.2. Tally Sheet for Length of Life of Incandescent Lamps

Class Interval	Frequency
201- 300	|
301- 400	
401- 500	
501- 600	|||
601- 700	𝍸 𝍸
701- 800	𝍸 𝍸 𝍸 𝍸 |
801- 900	𝍸 𝍸 𝍸 𝍸 𝍸 𝍸 𝍸 𝍸 𝍸
901-1,000	𝍸 𝍸 𝍸 𝍸 𝍸 𝍸 𝍸 𝍸 𝍸 𝍸 𝍸 𝍸 𝍸 𝍸 𝍸 𝍸 𝍸 𝍸 |
1,001-1,100	𝍸 𝍸 𝍸 𝍸 𝍸 𝍸 𝍸 𝍸 𝍸 𝍸 𝍸 𝍸 𝍸 𝍸 𝍸 𝍸 𝍸
1,101-1,200	𝍸 𝍸 𝍸 𝍸 𝍸 𝍸 𝍸 𝍸 𝍸 𝍸 𝍸 𝍸 𝍸 𝍸 𝍸 𝍸
1,201-1,300	𝍸 𝍸 𝍸 𝍸 𝍸 𝍸 𝍸 𝍸 ||||
1,301-1,400	𝍸 𝍸 𝍸 𝍸 |||
1,401-1,500	𝍸 ||||
1,501-1,600	|||
1,601-1,700	||

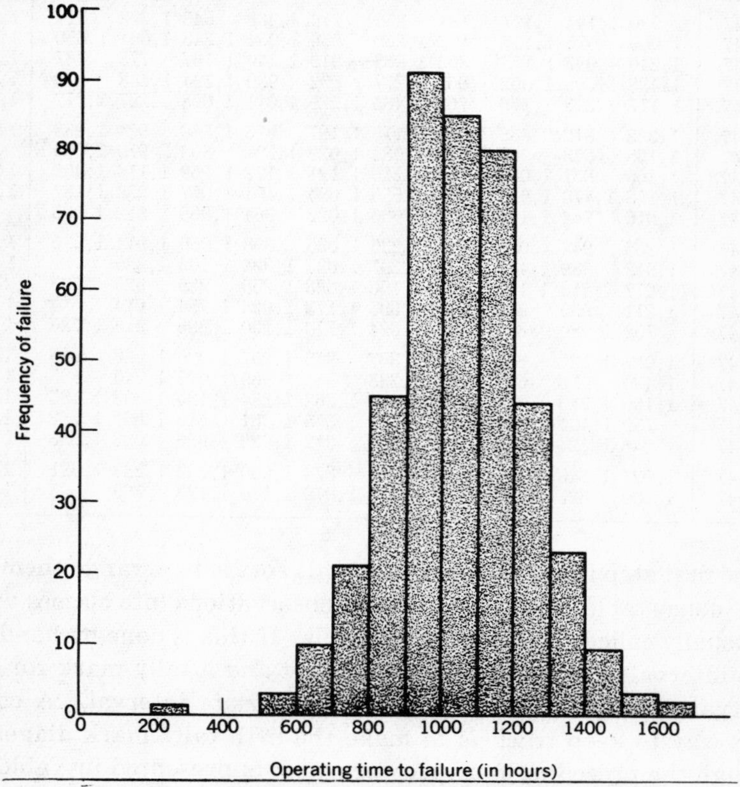

Fig. 1.1. Life length histogram for incandescent lamps.

the completed frequency table is given in Table 1.3. This table indicates that there is one lamp with a life between 201 and 300, none between 301-400 and 401-500, three with lives between 501 and 600, etc.

The frequency table is often displayed in graphical form by drawing a series of rectangles, each with the class interval as base and height equal to the number of items in the interval (see Figure 1.1). Charts of this form are called histograms. Both the histogram and the frequency table provide at a glance much more information than the original data. There is a high concentration of lives between 900 and 1,300 hours, and all but a handful lie between 600 and 1,500 hours.

We can answer questions about life of bulbs much more precisely if, instead of presenting the frequency for a given class interval, we present the number that will achieve a specified value. This is particularly useful with life data; a common question is: how many bulbs will fail by 1,000 hours ? From the frequency table, we can calculate the number of bulbs with lives that fail before 1,000 hours

Table 1.3. Frequency Table for Length of Life of Incandescent Lamps

Class Interval (100 hr)	Frequency f
201- 300	1
301- 400	—
401- 500	—
501- 600	3
601- 700	10
701- 800	21
801- 900	45
901-1,000	91
1,001-1,100	85
1,101-1,200	80
1,201-1-300	44
1,301-1,400	23
1,401-1,500	9
1,501-1,600	3
1,601-1,700	2
	417

by adding the frequencies in cells below 1,000. We would get $1+3+10+21+45+91=171$. If we do this for every cell in the table, we obtain a cumulative frequency table as in Table 1.4. If we divide each of the cumulative frequencies by the total number of observations, we obtain the fraction cumulative frequency distribution. A plot of this cumulative frequency function is presented in Figure 1.2.

Table 1.4. Cumulative Frequency Table

Class Interval (100 hr)	Frequency f	Cumulative Frequency	Fraction Cumulative Frequency
201- 300	1	1	0.002
301- 400	—	1	0.002
401- 500	—	1	0.002
501- 600	3	4	0.009
601- 700	10	14	0.034
701- 800	21	35	0.084
801- 900	45	80	0.192
901-1,000	91	171	0.410
1,001-1,100	85	256	0.614
1,101-1,200	80	336	0.806
1,201-1,300	44	380	0.913
1,301-1,400	23	403	0.966
1,401-1,500	9	412	0.988
1,501-1,600	3	415	0.995
1,601-1,700	2	417	1.000

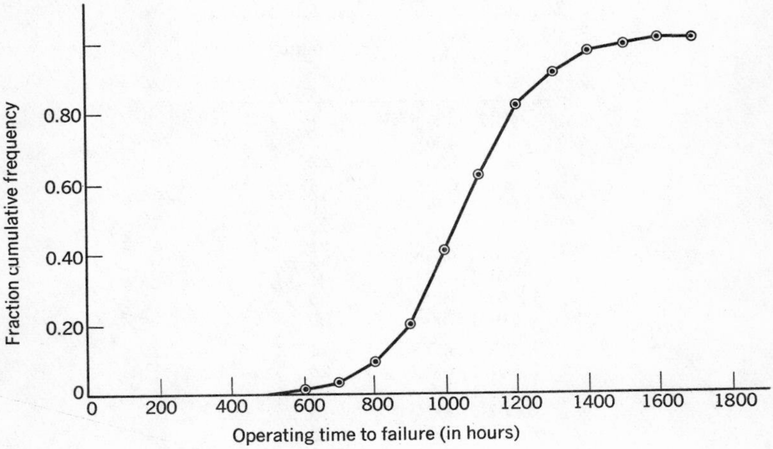

Fig. 1.2. Cumulative frequency function.

From this curve we can easily see that 90% of the bulbs will exceed 820 or that half of them will last 1,040 hours. From such a curve we can schedule replacement of bulbs so as to minimize total cost or forecast our future demand for bulbs.

1.3. Measures of Central Tendency

The preceding sections have been concerned with presenting a mass of data in useful form. Often this is not enough and it is necessary to replace a collection of data by a single number. Someone may ask, "How long do these light bulbs really last? How long do they last on the average? Can you give me some idea about length of life in one number?" The questioner may not be satisfied with the evasive answer that it all depends, and that some bulbs last longer than others. Reluctantly, the statistician must agree that there are certain questions which can be answered by a single number.

The most common and most useful single description measure is the mean or arithmetic average. If the observations are denoted by $x_1, x_2, x_3, \cdots, x_N$, the arithmetic mean is

$$\bar{x} = \frac{\sum_{i=1}^{N} x_i}{N} .$$

If the data as grouped in a histogram present a well-behaved picture which rises and falls smoothly, the mean is a reasonably typical value in the sense that it will occur where the observations cluster. For the light bulb data,

$$\bar{x} = \frac{435{,}921}{417} = 1{,}045,$$

which is a value close to the intuitive center of the frequency curve.

A word of caution must be introduced. Most data in engineering and science follow reasonably smooth curves but occasionally one may observe a pathological frequency function—for example, one which might look like Figure 1.3:

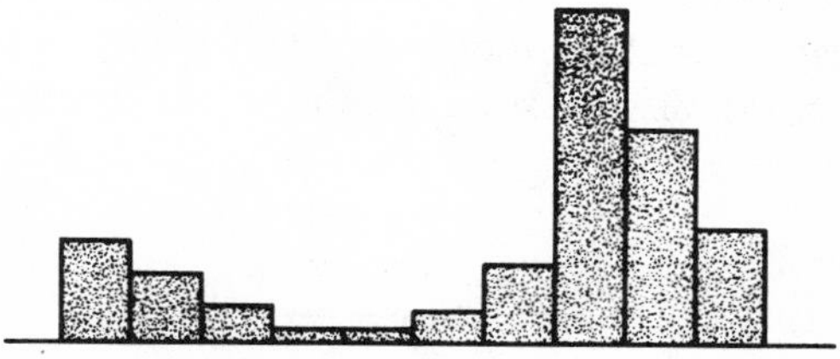

Fig. 1.3. Histogram of a pathological frequency function.

The mean of such a set of data is still the middle in some sense but does not represent a point of high concentration of the data. Actually, if we consider each observation to have an associated unit mass, the mean is simply the center of gravity.

Another common measure of central tendency is the median, defined as the middle observation when the numbers are arranged in order of magnitude. If there are an odd number, the median is uniquely defined; if there are an even number, the average of the neighboring numbers is usually taken: that is, the median of the numbers 7, 11, 21, 24, 31, 92, 1,017 is the middle one, or 24. The median of the numbers 7, 11, 21, 24, 31, 92 is $(21+24)/2$ or 22.5. The median is easy to calculate if the data are arranged properly or can be sorted, say, on punch cards. If a cumulative frequency function is given, the median can be read off easily, at least approximately.

A third measure of central tendency sometimes used is the mode, defined as the observation of maximum frequency. This measure is defined for grouped data and is taken to be the midpoint of this class interval with maximum frequency, 950 for the light bulb data given in Table 1.3.

This text does not make much use of the median and the mode as they do not play an important role in the field of analytical statistics. However, they are so widely used in descriptive work that it seems inappropriate to follow our inclination to write a statistics book without mentioning them.

1.4. Measures of Variation

Two methods of summarizing data have been presented: one which groups the data but still exhibits the patterns of variation, and one which replaces the data by a single measure of central tendency. A compromise is to characterize data by two statistics: one a measure of central tendency, and one a measure of variation. One simple measure of variation is the range of the observation, the largest minus the smallest. Another more common measure is the standard deviation, s, defined as the square root of the average squared deviation from the mean; i.e.,

$$s = \sqrt{\frac{\sum_{i=1}^{N} (x_i - \bar{x})^2}{N-1}}.$$

Some writers divide by N instead of $N-1$. For small numbers of observations, there are important reasons for dividing by $N-1$ rather than N; for large samples it doesn't make any appreciable numerical difference.

A detailed discussion of the relation between s and the pattern of variation will not be presented here because this can be illustrated very quickly after we introduce the normal curve. Let it suffice to say that small values of s are associated with high concentration of the observations around the mean and that this is, strictly speaking, valid only for reasonably symmetric bell-shaped histograms. s^2 is really the moment of inertia of the observations.

The algebraic identity

$$\sum_{i=1}^{N}(x_i - \bar{x})^2 = \sum_{i=1}^{N} x_i^2 - N\bar{x}^2 = \sum_{i=1}^{N} x_i^2 - \frac{\left(\sum_{i=1}^{N} x_i\right)^2}{N}$$

provides a useful computation formula for s. It is not efficient to subtract the mean from numbers and square the difference, but much better to calculate the sum and sum of squares of the observations. For the life data, we find

$$\sum_{i=1}^{417} x_i = 435{,}921 \quad \text{and} \quad \sum_{i=1}^{417} x_i^2 = 470{,}808{,}333.$$

Hence

$$s^2 = \frac{470{,}808{,}333 - \dfrac{190{,}027{,}118{,}241}{417}}{416} = 36{,}316.85$$

and

$$s = 190.57\ .$$

1.5. Computation of the Mean and Standard Deviation from the Frequency Table

Besides providing a useful tabulation of a large quantity of data, the frequency table may be used to compute the mean and standard deviation. There is some approximation involved in the method; essentially it replaces each observation by the midpoint of the class interval in which it falls, but the error is usually negligible. The computing method is illustrated in Table 1.5; it involves a new scale with an arbitrary origin and the width of the class interval taken as unity. The mean and standard deviation are computed in terms of the new scale and then transformed to the original units. For our life data, the mean computed from grouped data is 1,048 compared with 1,045 from raw data; the standard deviations are 190.7 and 190.6.

Note that the midpoint of the class interval 901–1,000 is taken as 950. This assumes that a bulb which expires with a life of say 900.2 hours is recorded as 901; the test rack is checked every hour. If the data were measurements of some dimension, the interpretation would be different. It might be that 901 actually represented not any number between 900 and 901 but any number which would round to 901, say any number between 900.5 and 901.5. In this case the class interval 901 to 1,000 would actually represent 900.5 to 1,000.5 and its midpoint would be 950.5. The precise definition of endpoints of class intervals affects both computation and graphical presentations and deserves consideration when class intervals are chosen.

Table 1.5. Computation of Mean and Standard Deviation from Grouped Data

Class Interval (100 hr)	Frequency f	Midpoint of Class Interval	Deviation from Arbitrary Origin d	fd	fd^2
201- 300	1	250	−8	−8	64
301- 400	—	350	−7	0	0
401- 500	—	450	−6	0	0
501- 600	3	550	−5	−15	75
601- 700	10	650	−4	−40	160
701- 800	21	750	−3	−63	189
801- 900	45	850	−2	−90	180
901-1,000	91	950	−1	−91	91
1,001-1,100	85	1,050	0	0	0
1,101-1,200	80	1,150	1	80	80
1,201-1,300	44	1,250	2	88	176
1,301-1,400	23	1,350	3	69	207
1,401-1,500	9	1,450	4	36	144
1,501-1,600	3	1,550	5	15	75
1,601-1,700	2	1,650	6	12	72
TOTALS	417			−7	1,513

Mean:

$$\bar{x} \text{ in class interval from arbitrary origin of } 1{,}050 = \frac{\sum fd}{N} = \frac{-7}{417} = -0.0167.$$

$$\bar{x} \text{ in original units} = \text{arbitrary origin} + \frac{\sum fd}{N}(\text{class interval})$$

$$= 1{,}050 - (0.0167)(100) = 1{,}048.$$

Standard Deviation:

$$s \text{ class interval units} = \sqrt{\frac{\sum fd^2 - \frac{(\sum fd)^2}{N}}{N-1}}$$

$$= \sqrt{\frac{1{,}513 - \frac{49}{417}}{416}} = 1.907.$$

s in original units $= s$ in class interval units $\times\ 100 = 190.7$.

(CLASS INTERVAL)

PROBLEMS

1. For the data shown in the table (a) compute a frequency table with class intervals of 0.02 starting with 6.60–6.61, 6.62–6.63, etc., (b) draw a histogram, and (c) a cumulative frequency diagram.

Diameters of Rivet Heads in Hundredths of an Inch

6.72	6.77	6.82	6.70	6.78	6.70	6.62
6.75	6.66	6.66	6.64	6.76	6.73	6.80
6.72	6.76	6.76	6.68	6.66	6.62	6.72
6.76	6.70	6.78	6.76	6.67	6.70	6.72
6.74	6.81	6.79	6.78	6.66	6.76	6.72
6.74	6.70	6.78	6.76	6.70	6.76	6.76
6.67	6.62	6.68	6.74	6.74	6.81	6.66
6.68	6.72	6.74	6.64	6.79	6.72	6.82
6.80	6.74	6.73	6.81	6.77	6.60	6.72
6.68	6.78	6.76	6.74	6.70	6.64	6.78
6.72	6.71	6.64	6.70	6.70	6.75	6.79
6.67	6.72	6.76	6.64	6.69	6.73	6.74
6.67	6.66	6.84	6.73	6.66	6.66	6.64
6.62	6.72	6.80	6.72	6.76	6.72	6.80

2. From the raw data of Problem 1, compute the mean and standard deviation. From the grouped data of Problem 1, compute the mean and standard deviation.

3. In an experiment measuring the percent shrinkage on drying, 40 plastic clay test specimens produced the following results:

19.3	20.5	17.9	17.3
15.8	16.9	17.1	19.5
20.7	18.5	22.5	19.1
18.4	18.7	18.8	17.5
14.9	12.3	19.4	16.8
17.3	19.5	17.4	16.3
21.3	23.4	18.5	19.0
16.1	18.8	17.5	18.2
18.6	18.3	16.5	17.4
20.5	16.9	17.5	18.2

(a) Group these percentages into a frequency distribution with class intervals of 1% starting with 12.0–12.9. (b) Draw a histogram, and (c) a cumulative frequency diagram.

4. In Problem 3, compute the mean and standard deviation from the raw data and from the grouped data.

5. Take the first 42 observations (Column 1) of Table 1.1 and (a) compute a frequency table with class intervals of 100 hours starting with 201–300, (b) draw a histogram, and (c) a cumulative frequency diagram.

6. From the raw data of Problem 5 compute the mean and standard deviation. From the grouped data of Problem 5 compute the mean and standard deviation.

7. Repeat Problem 5 using the 43–84 observations (Column 2).

8. From the raw data of Problem 7 compute the mean and standard deviation. From the grouped data of Problem 7 compute the mean and standard deviation.

9. Repeat Problem 5 using the 85–126 observations (Column 3).

10. From the raw data of Problem 9 compute the mean and standard deviation. From the grouped data of Problem 9 compute the mean and standard deviation.

11. Repeat Problem 5 using the 127–168 observations (Column 4).

12. From the raw data of Problem 11 compute the mean and standard deviation. From the grouped data of Problem 11 compute the mean and standard deviation.

13. Show that $\sum_{i=1}^{N} (x_i - \bar{x}) = 0$.

14. Prove the algebraic identity $\sum_{i=1}^{N} (x_i - \bar{x})^2 = \sum_{i=1}^{N} x_i^2 - N\bar{x}^2$.

15. If all the observations in Table 1.1 are changed by subtracting 200 hours from each, what effect will this have on the mean and standard deviation?

16. If all the observations in Table 1.1 are changed by dividing each by 1,000 hours, what effect will this have on the mean and standard deviation?

CHAPTER II

Random Variables and Probability Distributions

2.1 Introduction

Chapter I dealt with the aspect of statistics related to the processing of data. It was assumed that a large amount of data was amassed and had to be processed. In this chapter and the ensuing chapters, that aspect of statistics dealing with making decisions in the face of uncertainties will be considered.

In most engineering problems decisions must be made on the basis of experimentation. Experiments or observations usually are repeated several times under uniform or constant conditions. Even though great care is taken to keep the conditions of the experiment as uniform as possible, the individual observations exhibit an intrinsic variability that cannot be eliminated. For example, steel ingots may be produced and placed in a Rockwell hardness testing machine[1] for the purpose of measuring the hardness of a particular type of steel. Even if the ingots are taken from the same batch and the experiment performed under similar conditions, the readings for the different ingots will generally differ. This inherent variability is often referred to as the experimental error, which is a convenient name for a source of variation that eludes control. Thus, in all types of repeated experiments performed under " controlled " conditions, the outcomes of the individual repetitions vary, and hence the results of any given repetition usually cannot be predicted exactly.

[1] The principle of a Rockwell hardness testing machine consists of impressing a hardened steel or diamond point into the surface to be tested, and measuring the depth of the penetration—the resistance of the metal indicating the degree of hardness. The depth measurement is recorded on a dial, the scale of which depends upon the amount of pressure used in making the indentation.

2.2 Set of All Possible Outcomes of the Experiment

Although the results of any given experiment cannot be predicted exactly, it is possible to characterize the set of all possible outcomes of the experiment. Denote this set by the letter S. For example, suppose a single steel ingot is to be placed in the Rockwell hardness testing machine. The Rockwell hardness of this ingot on the B scale must fall between 0 and the upper limit of the scale U. Prior to doing the experiment, the value of the actual observation to be obtained will be unknown, but will lie within the interval 0 and U. Thus, the set S consists of all points in the interval 0 and U. If two steel ingots are measured, each ingot can have a hardness between 0 and U. Denote the Rockwell hardness of the first ingot by x_1 and the Rockwell hardness of the second ingot by x_2. Thus, the set S consists of all points x_1, x_2 such that

$$0 \leqq x_1 \leqq U \quad \text{and} \quad 0 \leqq x_2 \leqq U.$$

If $U = 70$, one possible outcome of the experiment is $x_1, x_2 = 43{,}56$. This would result if the Rockwell hardness of the first specimen were 43 and the second specimen 56. Thus 43,56 is a single point in S. S consists of all such points x_1, x_2 such that $0 \leqq x_1 \leqq U$ and $0 \leqq x_2 \leqq U$. Similarly, if a sample of n ingots are measured and x_i represents the Rockwell hardness of the ith ingot, the set of all possible outcomes of the experiment S consists of all points $x_1, x_2, \cdots, x_n$ such that

$$0 \leqq x_1 \leqq U, \quad 0 \leqq x_2 \leqq U, \cdots, \quad 0 \leqq x_n \leqq U.$$

A second example concerns the tossing of a die. The experiment consists of tossing a die and observing the upturned face. Prior to the toss, the value of the upturned face cannot be predicted with certainty; but the set of all possible outcomes of the experiment, S, can be recorded. The upturned face can be 1, 2, 3, 4, 5, or 6. Hence, S consists of these 6 values. If two dice are tossed, each die can take on a number from 1 to 6 inclusive. Denote the value of the upturned face of die number 1 by x_1 and the value of the upturned face of die number 2 by x_2. The set S consists of all point x_1, x_2 such that x_1 and x_2 lie between 1 and 6 inclusive.

An enumeration of the 36 points in the set S is

$$\begin{array}{llllll} x_1, x_2 = 1{,}1; & 1{,}2; & 1{,}3; & 1{,}4; & 1{,}5; & 1{,}6 \\ 2{,}1; & 2{,}2; & 2{,}3; & 2{,}4; & 2{,}5; & 2{,}6 \\ 3{,}1; & 3{,}2; & 3{,}3; & 3{,}4; & 3{,}5; & 3{,}6 \\ 4{,}1; & 4{,}2; & 4{,}3; & 4{,}4; & 4{,}5; & 4{,}6 \\ 5{,}1; & 5{,}2; & 5{,}3; & 5{,}4; & 5{,}5; & 5{,}6 \\ 6{,}1; & 6{,}2; & 6{,}3; & 6{,}4; & 6{,}5; & 6{,}6 \end{array}$$

Any one of the 36 points above can be obtained as the result of experimentation. $x_1, x_2 = 3,4$ means that the result of the first toss is a 3 and the result of the second toss is a 4. If a single die is tossed twice, the set S is the same as that above where x_1 now denotes the value of the upturned face of the first toss and x_2 denotes the value of the upturned face of the second toss. If the same die is tossed n times, the set S consists of all points $x_1, x_2, \cdots, x_n$ such that x_i, the value of the upturned face of the ith toss, lies between 1 and 6 inclusive.

The final example concerns tossing a coin and is given to illustrate the fact that S need not consist of a set of numbers. The experiment consists of tossing a coin and observing the upturned face. The face can be either a head or a tail (ruling out the possibility of standing on edge) and hence, the set S consists of two values, a head or a tail. If the coin is tossed twice, the first toss can be either a head or a tail and similarly for the second toss. If x_1 denotes the outcome of the first toss and x_2 denotes the outcome of the second toss, the set S consists of all points x_1, x_2 such that x_1 represents a head or a tail and x_2 represents a head or a tail. An enumeration of the four points of S is given by $x_1, x_2 = H, H$; H, T; T, H; and T, T. Thus, $x_1, x_2 = H, H$ means that the result of the first toss is a head and the result of the second toss is a head.

This example is a special case of a more general experiment in which the outcome can be characterized by one of two values. Instead of flipping a single coin, the experiment can consist of observing the sex of a newborn child. The set S consists of the two points, male and female. Another type of experiment can consist of measuring whether an item is defective or non-defective. The set S consists of the two points, defective and non-defective.

The experiments described in the examples above were actually much more complex than indicated. In addition to measuring the Rockwell hardness of a steel ingot, the outcome of the experiment consisted of wear on the testing equipment, stress set up in the floor, energy consumed by the operator, generation of heat, etc. There are many factors resulting from the experiment which could be included in the set of all possible outcomes of the experiment but which are unimportant from the point of view of making decisions based upon the outcome of the experiment. Hence, these irrelevant factors will be excluded from the formal enumeration of the set S, and S will be the set of all possible outcomes *of the relevant factors* of the experiment.

2.3 Random Variables

Experiments are usually performed in order to obtain information on which to base decisions. Sometimes such decisions can be made directly by referring to the outcome of the experiment. For example, a manufacturer wants to decide whether to buy a certain type of steel. He will make his decision on the basis of measuring the hardness of a single steel ingot. Prior to performing the experiment, he decides that if the Rockwell hardness exceeds 60 on the B scale he will buy the steel. In this example, the manufacturer is partitioning the set of all possible outcomes into two parts: those points which are below 60 and those points which are above 60. If the actual outcome of the experiment lies in the partition with points below 60 he will not buy the steel. If the actual outcome of the experiment lies in the partition with points above 60 he will purchase the steel.

On the other hand, it is conceivable that the manufacturer may not want to make his decision by referring directly to the outcome of the experiment. For example, suppose a sample of 5 ingots are to be measured. If the average Rockwell hardness of the five specimens exceeds 60 on the B scale he will buy the steel. The average of the five specimens is not a point in the set of all possible outcomes of the experiment, and consequently the manufacturer cannot make his decision by referring directly to S. Instead the manufacturer makes his decision according to the results of a rule which assigns to each point in the set of all possible outcomes the average of the five specimens. The resulting set of points (averages of 5) is the set of all possible values of averages and will be denoted by R. This latter set of all possible values of averages is then partitioned into two parts: those points which are below 60 and those points which are above 60. If the actual result of the "rule" (average of the 5) lies in the partition with points below 60 the steel is not purchased. If the actual result of the "rule" lies in the partition with points above 60 the steel is bought.

The particular "rule" chosen is evidently of importance and hence is given the special name of "random variable." Thus, a random variable is a rule which assigns to each point in S a value in a set R, R consisting of all possible values assigned by the random variable. R is often referred to as the set of all possible values taken on by the random variable. In the first example, where the result of one steel ingot is used, the Rockwell hardness of this ingot is a random variable. The rule assigns each point in S to the same point in R. Hence R and S coincide and according to the definition the Rockwell hardness is a random variable.

In later chapters it will become clear that decisions are usually based upon the results of random variables. It must be emphasized that a random variable is a rule. The rule assigns points in S to points in R. Experimentation is equivalent to obtaining a point in the set of all possible values of the random variable. This point is the actual value that the random variable takes on. In the example where five ingots are measured, the average of the five specimens is a random variable. Whereas the set, S, of all possible outcomes of the experiment consists of all points

$$x_1, x_2, x_3, x_4, x_5$$

such that $0 \leqq x_i \leqq U$; the set, R, of all possible values of the random variable consists of all points in the interval 0 and U. This is seen by recognizing that the smallest value of the average occurs when each ingot is at its lowest value. The point 0, 0, 0, 0, 0 in S is assigned the point 0 (the average of five zeros) in R. The largest value of the average occurs when each ingot is at its highest value. The point U, U, U, U, U in S is assigned the point U (the average of five U's) in R. The other points are obtained in a similar manner. Thus, experimentation is equivalent to obtaining a point in the interval 0 and U. This is the value that the random variable actually takes on. If this quantity exceeds 60 the steel is purchased; otherwise it is not.

Other examples of random variables are as follows: The experiment consists of measuring the Rockwell hardness of five specimens. The quantity

$$s^2 = \tfrac{1}{4}[(x_1 - \bar{x})^2 + (x_2 - \bar{x})^2 + (x_3 - \bar{x})^2 + (x_4 - \bar{x})^2 + (x_5 - \bar{x})^2]$$

is a random variable. s^2 is a rule which assigns points in S as follows: For each point in S the average, $\bar{x}$, is computed. $\bar{x}$ is subtracted from each Rockwell hardness. These differences are squared and then summed. The resultant quantity is then divided by 4. R is the set of all possible values of s^2. Another random variable is x_3^2. For each point in S, the rule disregards all values of the Rockwell hardness except for the third specimen. The value for the third specimen is then squared. R is the set of all possible values of x_3^2. Another random variable is $x_5 - x_1$. For each point in S, the rule disregards all values of the Rockwell hardness except for the first and fifth specimens. The rule subtracts x_1 from x_5 for each point in S. R is the set of all possible values of $x_5 - x_1$. It is evident that there are an infinite number of random variables that can be obtained from this simple experiment. From a practical point of view, usually one, or just a few, are of interest.

In particular, a sequence of random variables of importance is generated as follows: An experiment consists of n trials. Denote the measurable quantity associated with the first trial by x_1, with the second trial by x_2, etc. $x_1, x_2, \cdots, x_n$ are then random variables.[1] This is verified by showing that each x_i is a rule which disregards the results of the other $(n - 1)$ outcomes and assigns to each point in S its identity. Thus, $x_1, x_2, \cdots, x_n$ is a particular sequence of random variables. It has already been indicated that certain functions of these random variables such as $\bar{x}$, s^2, x_3^2, and $x_5 - x_1$ are also random variables. In fact, *any* function of a random variable is also a random variable since a mathematical function is itself a rule; or equivalently, a function of a rule must also be a rule. Thus, such quantities as $\bar{x}$, s^2, x_3^2, $x_5 - x_1$, etc. can be verified to be random variables either by referring to the definition of a random variable (and the set S) or by recognizing them as functions of random variables, in particular, functions of the random variables $x_1, x_2, \cdots, x_n$.

If the set of all values the random variable takes on is continuous, the random variable is known as a continuous random variable. Rockwell hardness is an example of a continuous random variable.[2] On the other hand, if the values taken on by the random variable are discrete, the random variable is known as a discrete random variable. For example, consider the toss of two dice. A random variable is the sum of the upturned faces. For each point in S, this rule adds the value of the upturned face of the first die to the value of the upturned face of the second die. R is the set which consists of the points 2, 3, 4, 5, 6, 7, 8, 9, 10, 11, and 12. The random variable must take on one of these values; and consequently it is a discrete random variable. Similarly, another random variable is the square of the value of the upturned face of the first die. For each point of S, the values of the upturned face of the first die is squared. R is the set which consists of the points 1, 4, 9, 16, 25, and 36, and hence this

[1] The notation used is somewhat ambiguous. Here, x_i represents a random variable. In an earlier paragraph x_i represented an outcome of the experiment, or a particular value taken on by the random variable. The content of the paragraph should enable the reader to distinguish between the use of x_i as a random variable or a particular value taken on by the random variable.

[2] Naturally, every measuring device is really discrete so that Rockwell hardness is not a continuous random variable. However, from a practical point of view the Rockwell hardness scale can be considered as continuous.

random variable is discrete. In the coin tossing experiment, the face of the coin is a random variable. The rule assigns the point heads in S to heads in R, and tails in S to tails in R. Thus R and S coincide. This is not a numerical-valued random variable but can be made into one as follows. Let the rule assign the value 0 to the point head in S and 1 to the point tail in S. As a consequence R now consists of the points 0 and 1. This is now a numerical-valued random variable, the kind with which we shall usually concern ourselves.

2.4 Probability and Probability Distributions

The term "probability" has several meanings. It is used in every day speech in a qualitative manner, and almost everyone has some feeling for the concept. For the purposes of this text, however, a more quantitative definition is required. There are several alternatives available, and the frequency definition will be used here. Denote by m the number of successful occurrences of an event E in n trials. Let $P(E)$ denote the probability of a successful occurrence of this event. For large values of n, the ratio m/n will approximately equal $P(E)$. Four basic properties of probability are given here:

1. $P(E)$ is a non-negative number, i.e., $P(E) \geqq 0$.
2. The probability of an event that is certain is unity.
3. If the events E and F are mutually exclusive, $P(E + F) = P(E) + P(F)$ where $E + F$ indicates the event E or F. Events are mutually exclusive if the occurrence of any one of them makes the simultaneous occurrence of all the others impossible. This is the addition rule.
4. If the events E and F are independent, $P(EF) = P(E)P(F)$ where EF indicates the event E *and* F. This property is known as the multiplication theorem and can be used as a definition of two independent events.

With these definitions and properties it is possible to talk about probabilities which are associated with events defined over the set, S, of all possible outcomes of the experiment. For example, if the Rockwell hardness of two steel ingots is to be measured, the set S has been shown to consist of all points x_1, x_2 such that

$$0 \leqq x_1 \leqq U \quad \text{and} \quad 0 \leqq x_2 \leqq U.$$

We can speak of the probability of the event that x_1 is smaller than 50 and x_2 smaller than 45 (assuming that U is greater than or equal to 50). Similarly, for the toss of two dice example we can speak of the probability of the event which consists of obtaining the point 4,5.

In general then, probabilities are associated with events defined over the set S. However, it has been shown in the previous sections that decisions are not made by explicit reference to the set S which describes all possible outcomes of the experiment, but decisions are usually made on the basis of a random variable. Hence the random variable and the set R (which is derived from S for a particular random variable) require further study.

Associated with every random variable is a probability distribution. Before proceeding further it is necessary to define a probability distribution, and hence, the following notation is introduced: $P(x=a)$ denotes the probability of the event that the random variable x assumes the given value a; e.g., if the random variable x is the sum of the upturned faces of two dice, $P(x = 7)$ is the probability that the sum equals 7. Similarly $P(a < x \leqq b)$ denotes the probability that the random variable x belongs to the interval $(a, b]$. If $P(a < x \leqq b)$ is known for all values of a and b in the set R, a knowledge of the probabilities of all events is obtained. This is equivalent to specifying the probability distribution of the random variable x.

The probability distribution associated with a random variable is not arbitrary, but is induced by probabilities associated with events in the set, S, of all possible outcomes of the experiment. In the Rockwell hardness example, the probability distribution of the random variable, the average of two specimens, is induced by the probabilities associated with events of the type previously described. In the dice example, the probability distribution of the random variable, the sum of the upturned faces, depends upon the probabilities associated with obtaining each of the points in the set S.

Since decisions are usually based on the outcome of random variables, we usually concern ourselves only with probability distributions of the random variable of interest. Rarely is there need to be concerned with probabilities defined over S. In fact, assumptions are often made concerning the form of the probability distribution of the random variable, assumptions which can often be justified on theoretical grounds (see Section 3.9). Hence, whereas no subsequent mention will be made of the probabilities defined over S, and probability distributions of random variables will be studied in detail, readers should be aware of the relation between the two concepts.

The probabilities of events of the form $(x \leqq b)$ play an important role in probability theory: Let $F_x(b) = \mathrm{P}(x \leqq b)$ be defined as a cumulative distribution function (C. D. F.) of the random variable x. When there is no ambiguity, $F_x(b)$ will be written as $F(b)$. This function is defined for all values of b in the interval minus infinity to plus infinity even when R is a small subset of this interval. For example, in the die tossing situation, R consists of the points 1 through 6, yet the following are defined: $F_x(-2)$ is $\mathrm{P}(x \leqq -2)$ and is equal to 0; $F_x(3\frac{1}{2})$ is $\mathrm{P}(x \leqq 3\frac{1}{2})$ and is equivalent to $\mathrm{P}(x \leqq 3)$; and $F_x(7)$ is $\mathrm{P}(x \leqq 7)$ and is equal to 1. Further let a and b be any real numbers such that $a < b$. The events $(x \leqq a)$ and $(a < x \leqq b)$ are mutually exclusive as seen in Figure 2.1.

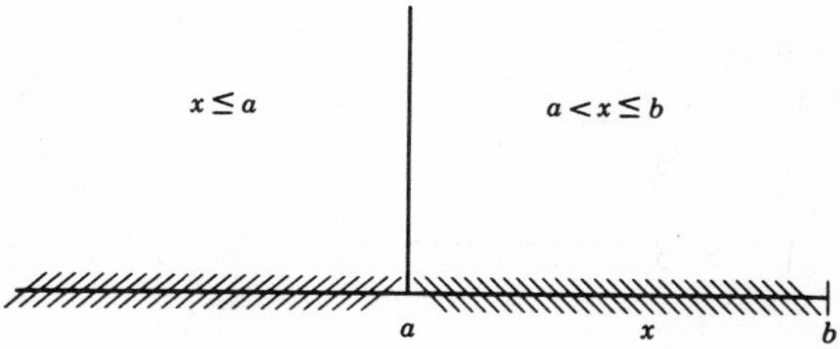

Fig. 2.1. Events $(x \leqq a)$ and $(a < x \leqq b)$.

The sum of the events $(x \leqq a)$ and $(a < x \leqq b)$ is $(x \leqq b)$, so that by means of the addition rule of probability

$$\mathrm{P}(x \leqq b) = \mathrm{P}(x \leqq a) + \mathrm{P}(a < x \leqq b).$$

It follows that

CDF to B. - CDF to A.

$$\mathrm{P}(a < x \leqq b) = F(b) - F(a).$$

Thus, it is clear that a knowledge of the C.D.F. is equivalent to the knowledge of the probability distribution function. The C.D.F. is useful if there is interest in probabilities of events of the form $(x \leqq b)$.

Every cumulative distribution function has the following properties:

1. $F(b)$ is a non-decreasing function of b. This results from the fact that every probability is a non-negative number.
2. $F(+\infty) = \lim\limits_{b \to \infty} F(b) = 1$. This statement says that the probability that the random variable will assume some value in R is 1.
3. $F(-\infty) = \lim\limits_{b \to -\infty} F(b) = 0$. This statement says that the probability that the random variable assumes no value in R is 0.

2.5 Discrete Probability Distributions

For a discrete random variable, a probability distribution is characterized by all the values the random variable x takes on in the set R, together with the probabilities associated with obtaining each value (p_i). Thus, $P(x = x_i) = p_i$ for all x_i in R. Since the probability of assuming some value in R is one, it follows that

$$\sum_{\substack{\text{over all } i \\ \text{for } x_i \text{ in } R}} p_i = 1.$$

The probability distribution for a random variable which takes on $k + 1$ values can be represented graphically as shown in Figure 2.2.

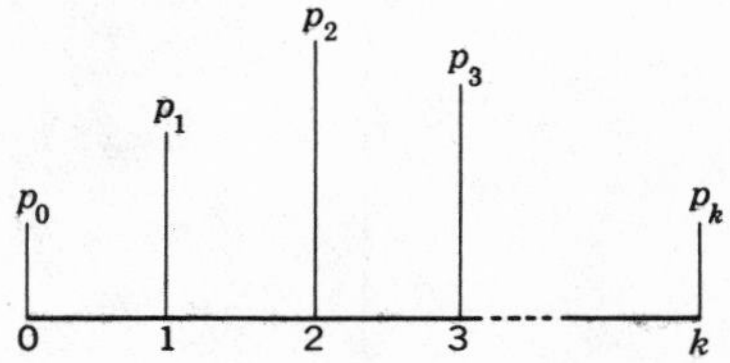

Fig. 2.2. Probability distribution of a discrete random variable.

The cumulative distribution function $F(b)$ of a discrete random variable is given by

$$F(b) = \sum_{x_i \leq b} p_i.$$

Thus a C.D.F. for a discrete random variable is a step function which is constant over every interval not containing any of the points x_i. This is shown in Figure 2.3.

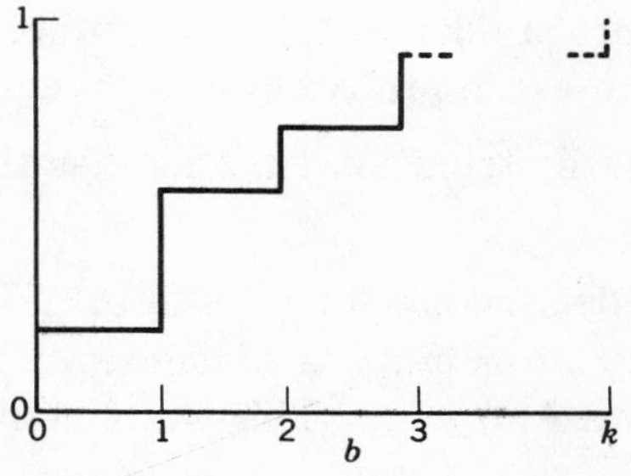

Fig. 2.3. Cumulative distribution function of a discrete random variable.

Some simple examples of the probability distribution and C.D.F. of a discrete random variable follow. If a fair die is tossed, each face is equally likely and the probability distribution would be as shown in Figure 2.4. The C.D.F. for this fair die is shown in Figure 2.5. An analogy can be drawn between a discrete probability distribution

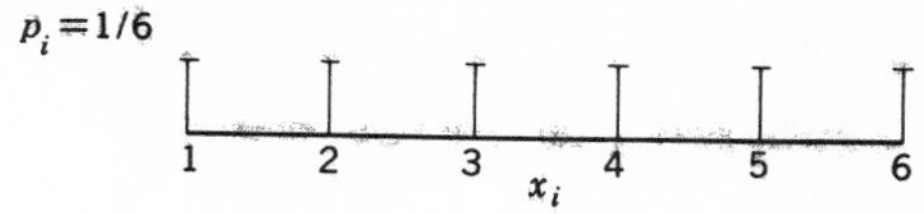

Fig. 2.4. Probability distribution of the random variable which is the upturned face of a fair die.

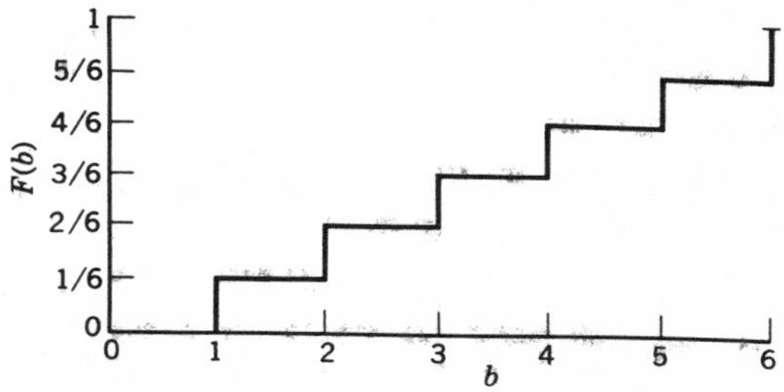

Fig. 2.5. C.D.F. of the random variable which is the upturned face of a fair die.

and a mass of unit weight distributed over a weightless bar. The unit mass is concentrated only at values that the random variable can take on. This is depicted in Figure 2.6 where each circle represents a mass of weight 1/6.

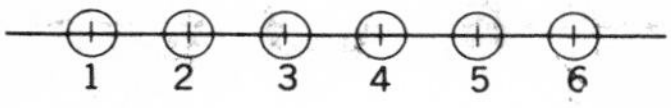

Fig. 2.6. Mass analogy to the probability distribution of the face of a fair die.

If the probability distribution is as shown above, i.e., the die is fair, the actual outcomes of a random variable should follow the laws of probability. In the long run each number should appear about 1/6 of the time. However, if the fairness of the die is in question, and its probability distribution unknown, as is the case in most statistical problems, the outcomes of the random variable will be used to make decisions about the unknown probability distribution.

A second example deals with the probability distribution of the random variable, the sum of two tosses of a fair die. The probability distribution of the random variable is given in Figure 2.7; the C.D.F. for this random variable is shown in Figure 2.8.

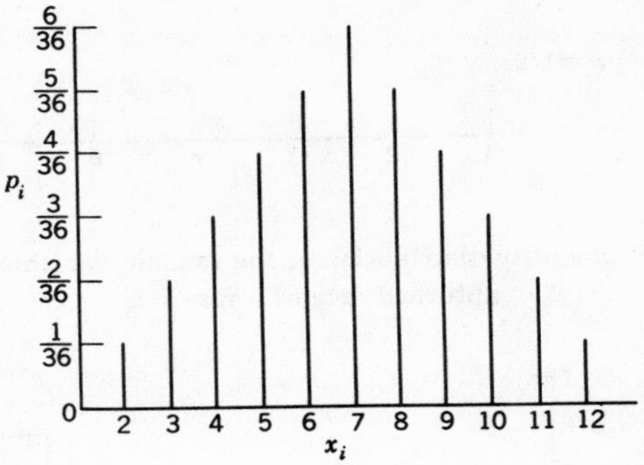

Fig. 2.7. Probability distribution of the random variable which is the sum of upturned faces of two fair dice.

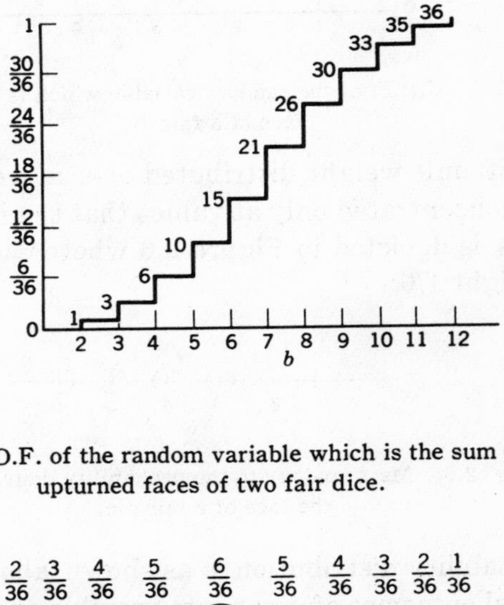

Fig. 2.8. C.D.F. of the random variable which is the sum of the upturned faces of two fair dice.

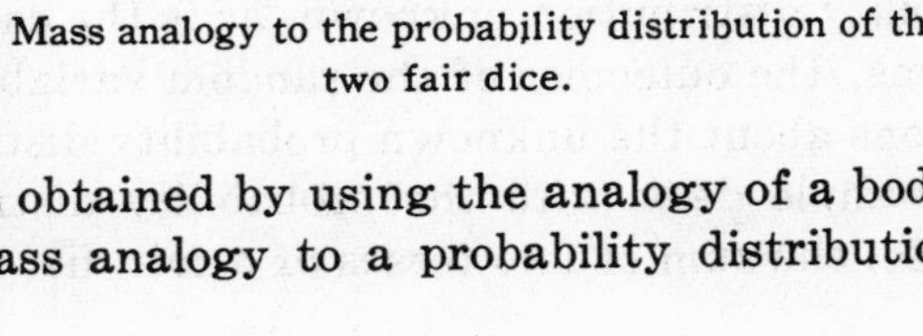

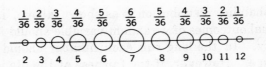

Fig. 2.9. Mass analogy to the probability distribution of the sum of two fair dice.

Figure 2.9 is obtained by using the analogy of a body having unit mass. The mass analogy to a probability distribution is a useful concept enabling one to use certain well known concepts belonging to the field of mechanics to illustrate definitions and properties of probability theory.

2.6 Continuous Probability Distributions

A continuous random variable has a probability distribution which is continuously distributed over the set of all possible outcomes of the random variable. Using the mass analogy of probability, the mass is distributed along a bar of unit mass according to some curve. This is illustrated in Figure 2.10.

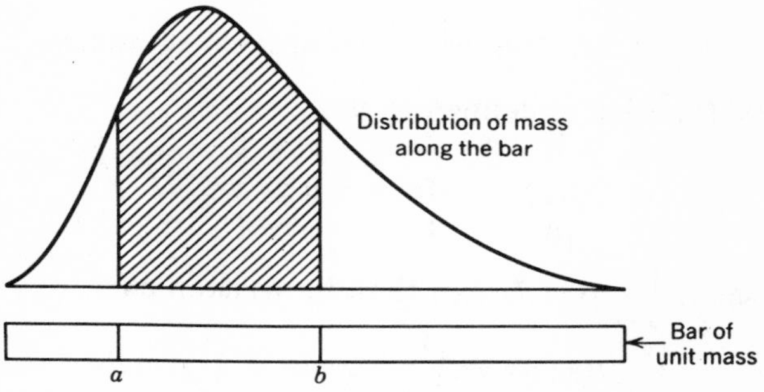

Fig. 2.10. Distribution of a continuous random variable.

The probability that the random variable is less than or equal to b but greater than a is given by the shaded area. The curve which gives the distribution of mass along the bar is known as a density function. Thus

$$P(a < x \leqq b) = F(b) - F(a) = \int_a^b f(x)\,dx$$

where $f(x)$ is the density function. If a is taken to be $-\infty$, the C.D.F. is seen to be

$$F(b) = \int_{-\infty}^b f(x)\,dx.$$

Furthermore if $a = -\infty$ and $b = +\infty$

$$\int_{-\infty}^{\infty} f(x)\,dx = 1.$$

If the random variable is defined over a finite interval, such as for Rockwell hardness which lies between 0 and some upper bound U, the meaning of $\int_{-\infty}^b f(x)\,dx$ needs some explanation. The set R of all possible values of the random variable can always be extended to the interval minus infinity to plus infinity. This is accomplished by extending the definition of $f(x)$ to cover the new points. $f(x)$ is taken to be zero over those points which were not originally in R. In the Rockwell hardness example, $f(x)$ is defined to be zero for $x < 0$ and $x > U$. Thus, if $f(x)$ in Figure 2.10 represents the density function of Rockwell hardness, it should really be drawn (with the range of x the interval minus infinity to plus infinity) as shown in Figure 2.10a.

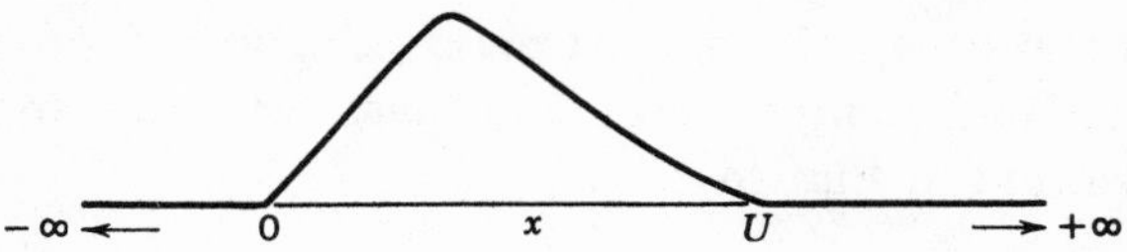

Fig. 2.10a. Density function of Rockwell hardness.

The density function is defined to be

$$\begin{array}{ccc} 0 & \text{for} & x < 0 \\ f(x) & \text{for} & 0 \leqq x \leqq U \\ 0 & \text{for} & x > U. \end{array}$$

The expression $\int_{-\infty}^{b} f(x)\,dx$ can then be written as

$$\int_{-\infty}^{b} f(x)\,dx = \int_{-\infty}^{0} f(x)\,dx + \int_{0}^{b} f(x)\,dx$$
$$= 0 + \int_{0}^{b} f(x)\,dx = \int_{0}^{b} f(x)\,dx.$$

Thus, extension of R to the interval minus infinity to plus infinity is done solely for the purpose of notation. However, it then follows that for continuous random variables, R can always be extended so as to consist of the entire real line.

The cumulative distribution function for a continuous random variable is illustrated in Figure 2.11.

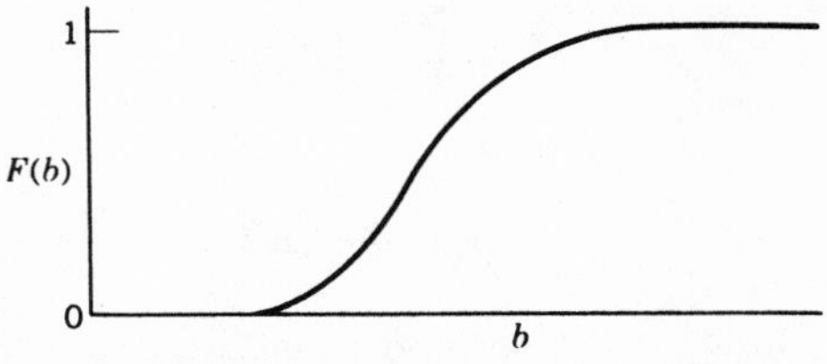

Fig. 2.11. C.D.F. of a continuous random variable.

The density function $f(x)$ has some interesting properties and interpretations. It is always non-negative and the area under the curve must equal one. The probability that a random variable will assume a value belonging to the infinitesimal interval $(x, x+dx)$ is given by $f(x)\,dx$. Thus $f(x)$ will be a measure of the frequency with which a random variable will tend to assume a value in the small interval dx. The probability that the continuous random variable equals b, i.e., $P(x = b) = 0$, for all b. Hence $P(x \leqq b)$ is equivalent to $P(x < b)$ so that $\int_{-\infty}^{b} f(x)\,dx$ can be given either interpretation. Finally, it is clear that the density function f is the derivative of the cumulative distribution function, i.e., $dF/dx = f$.

Some examples of a continuous random variable follow. In measuring the Rockwell hardness of a single specimen it will be assumed that the Rockwell hardness is known to vary from 50 to 70 on the B scale for the particular type steel used. In other words, the values the random variable takes on, the set R, extends from 50 to 70.[1] Suppose further that the probability is distributed among these values according to the density function

$$\begin{aligned} f(x) &= 0 && \text{for} \quad x < 50 \\ &= 1/20 && \text{for} \quad 50 \leqq x \leqq 70 \\ &= 0 && \text{for} \quad x > 70. \end{aligned}$$

This density function is known as a rectangular density and is illustrated in Figure 2.12. This function is always non-negative and its

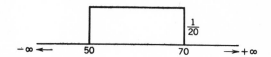

Fig. 2.12. The density function of a rectangularly distributed random variable.

total area is clearly one. The cumulative distribution function is given by

$$\begin{aligned} F(b) = P(x \leqq b) &= \int_{-\infty}^{b} f(x)\,dx \\ &= 0 && \text{for} \quad b < 50 \\ &= \int_{50}^{b} \frac{1}{20}\,dx = \frac{1}{20}(b - 50) && \text{for} \quad 50 \leqq b \leqq 70 \\ &= 1 && \text{for} \quad b > 70. \end{aligned}$$

[1] Although the set R is the interval from 50 to 70, it is extended to the interval minus infinity to plus infinity by defining $f(x)$ in the manner described below.

Thus,

$$P(x \leqq 60) = F(60) = \frac{1}{20}(60 - 50) = \frac{1}{2}.$$

This C.D.F. is shown in Figure 2.13.

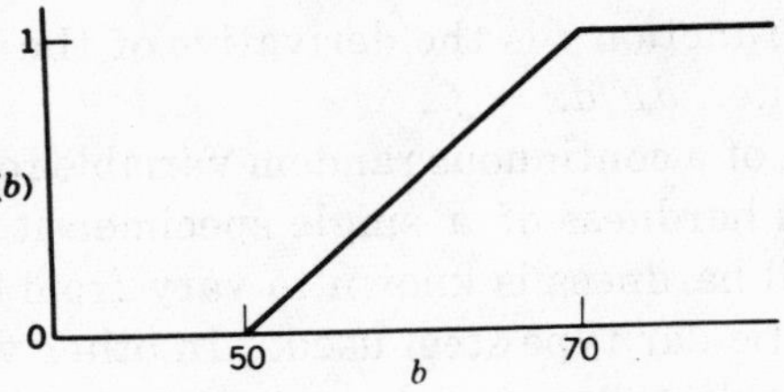

Fig. 2.13. The C.D.F. of a rectangular distribution.

A second example deals with the distribution of the life of electric light bulbs. A random variable x will represent the life of an electric light bulb. The density function for this random variable will be assumed to equal

$$\begin{aligned} f(x) &= 0 & x &< 0 \\ &= \frac{1}{1{,}000} e^{-x/1{,}000} & x &\geqq 0. \end{aligned}$$

This density function is known as an exponential density function and is illustrated in Figure 2.14.

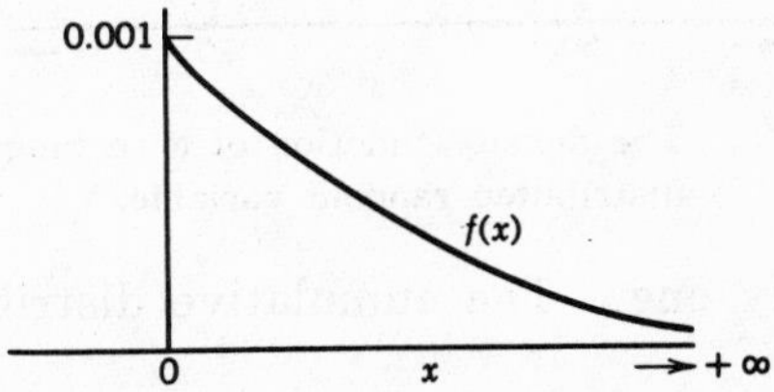

Fig. 2.14. The exponential density function.

This function is always non-negative and its total area can be verified to be one. The cumulative distribution function is given by

$$\begin{aligned} F(b) &= P(x \leqq b) = \int_{-\infty}^{b} f(x)\, dx \\ &= \int_{0}^{b} \frac{1}{1{,}000} e^{-x/1{,}000}\, dx = -e^{-x/1{,}000} \Big|_{0}^{b} \\ &= 1 - e^{-b/1{,}000}. \end{aligned}$$

Thus, the probability that an electric light bulb with the distribution above has a life of less than 1,000 hours is given by

$$F(1{,}000) = P(x < 1{,}000) = 1 - 1/e.$$

This C.D.F. is shown in Figure 2.15.

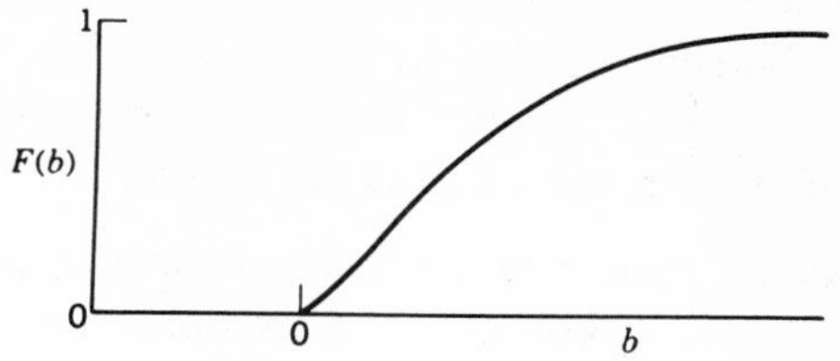

Fig. 2.15. C.D.F. of the exponential distribution.

2.7 Random Sample

A random sample is a series of n independent trials of some experiment leading to an observed outcome in the set S of all possible outcomes of the experiment. Alternatively, a random sample can be considered as a sequence of n independent random variables each having the same probability distribution, resulting in n observations. We say that n random variables, $x_1, x_2, \cdots, x_n$ are independent if $P(a_1 < x_1 \leqq b_1, a_2 < x_2 \leqq b_2, \cdots, a_n < x_n \leqq b_n) = P(a_1 < x_1 \leqq b_1) P(a_2 < x_2 \leqq b_2) \cdots P(a_n < x_n \leqq b_n)$ for all values $a_1, a_2, \cdots, a_n, b_1, b_2, \cdots, b_n$, i.e., if all the events of this type are independent. Heuristically, this definition implies that random variables are independent if knowledge of some of the outcomes has no effect on the probability distribution of those remaining. If the random variables (or trials) are not independent the sample is no longer a random sample.[1] This implies that in random sampling the result of a trial does not affect the probability distribution of any of the subsequent trials. Furthermore, in random sampling, the random variables each have the same probability distribution so that the entire sample, rather than a single trial, can be used to make decisions about the parameters (constants) of the probability distribution.

[1] The given definition of a random sample is somewhat restrictive in that it is inapplicable to situations where the probability distribution always depends upon the results of the previous trials. For example, if there are five red balls and five black balls in an urn and a sample of two is drawn, the probability of a black ball on the second trial depends upon the results of the first trial. For such cases, a different definition of a random sample is required. However, for the purposes of this book, the stated definition is sufficient.

2.8 Expectation

It is often necessary to describe a probability distribution by a "typical quantity." A quantity which is quite suggestive is the expected value of the random variable or, synonymously, the mean of the random variable or the average of the random variable.

Let x be a random variable and let the corresponding probability distribution of x be represented by p if the random variable is discrete. If x is continuous, let $f(x)$ denote the density function. The expected value of x (or mean of x) is defined as

$$E(x) = \sum_{\text{all } i} x_i p_i$$ if x is discrete, where x_i are the values taken on by a random variable and p_i are the corresponding probabilities

$$= \int_{-\infty}^{\infty} x f(x)\,dx$$ if x is continuous.

$E(x)$ is read as "the expected value of x" and will be denoted by μ. The expected value of the upturned face of the single die mentioned in Section 2.5 is given by

$$\begin{aligned}\mu = E(x) &= \sum_{\text{all } i} x_i p_i \\ &= (1)\left(\frac{1}{6}\right) + (2)\left(\frac{1}{6}\right) + (3)\left(\frac{1}{6}\right) + (4)\left(\frac{1}{6}\right) \\ &\quad + (5)\left(\frac{1}{6}\right) + (6)\left(\frac{1}{6}\right) \\ &= 3\frac{1}{2}.\end{aligned}$$

For the two dice,

$$\begin{aligned}\mu = E(x) &= (2)\left(\frac{1}{36}\right) + (3)\left(\frac{2}{36}\right) + (4)\left(\frac{3}{36}\right) \\ &\quad + (5)\left(\frac{4}{36}\right) + (6)\left(\frac{5}{36}\right) + (7)\left(\frac{6}{36}\right) + (8)\left(\frac{5}{36}\right) \\ &\quad + (9)\left(\frac{4}{36}\right) + (10)\left(\frac{3}{36}\right) + (11)\left(\frac{2}{36}\right) + (12)\left(\frac{1}{36}\right) \\ &= 7.\end{aligned}$$

For the example of the flip of a coin where the number 0 is assigned to heads and 1 to tails,

$$\mu = E(x) = (0)\left(\frac{1}{2}\right) + (1)\left(\frac{1}{2}\right) = \frac{1}{2}.$$

For the continuous random variable, Rockwell hardness, mentioned in Section 2.6,

$$\mu = E(x) = \int_{-\infty}^{\infty} xf(x)\,dx = \int_{50}^{70} x\,\frac{1}{20}\,dx = \left(\frac{1}{20}\right)\frac{x^2}{2}\bigg|_{50}^{70} = 60.$$

The expected value of the life of an electric light bulb having the density described in Section 2.6 is given by

$$\mu = E(x) = \int_{-\infty}^{\infty} xf(x)\,dx = \int_{0}^{\infty} \frac{x}{1{,}000}\,e^{-x/1{,}000}\,dx$$

$$= 1{,}000 \text{ hours.}$$

This result can be obtained by integration by parts or from any standard table of integrals.

The expectation is quite important because it follows the customary notion of an average. In repeated experiments, the sample average will tend to the expectation of the random variable as the number of trials gets large. For example, the expected life of an electric light bulb may be 1,000 hours. This is a parameter (a constant) of the probability distribution and must not be confused with a sample mean, which is a random variable. The sample mean tends towards a constant as the number of trials gets large, this constant being the expectation of the random variable. If the lives of a large number of light bulbs are investigated, the sample mean will tend towards 1,000 hours as the number of bulbs going into the computation of the average increases.

It is unnecessary to confine the discussion about expectation to the expectation of a random variable. According to the definition of a random variable any function of a random variable is itself a random variable. Thus, let $h(x)$ be an arbitrary function of x. The expected value of $h(x)$ is defined to be

$$E[h(x)] = \sum_{\text{all } i} h(x_i)p_i \qquad \text{if } x \text{ is discrete}$$

and

$$= \int_{-\infty}^{\infty} h(x)\,f(x)\,dx \quad \text{if } x \text{ is continuous.}$$

2.9 Moments

In elementary mechanics moments are associated with the physical properties of bodies of mass. The first moment about the origin is related to the center of gravity, and the second moment about the center of gravity is known as the moment of inertia. Similarly, for probability distributions, the concept of moments is extremely important. The analogy between bodies of unit mass and probability distributions leads to the usefulness of the analogy between the moment properties. Just as some bodies are completely characterized by their moments, some probability distributions are completely characterized by their moments.

The first moment about the origin is equivalent to the expected value (or the mean) of the random variable. Hence, the first moment about the origin is

$$\mu = E(x) = \sum_{\text{all } i} x_i p_i \quad \text{if } x \text{ is discrete}$$

$$= \int_{-\infty}^{\infty} x f(x)\, dx \quad \text{if } x \text{ is continuous.}$$

μ is also the center of gravity of the probability mass and can be interpreted as the point where the fulcrum is placed in order to balance the mass.[1]

The second moment about the mean, also known as the moment of inertia and variance, is defined as

$$E(x - \mu)^2 = \sum_{\text{all } i} (x_i - \mu)^2 p_i \quad \text{if } x \text{ is discrete}$$

$$= \int_{-\infty}^{\infty} (x - \mu)^2 f(x)\, dx \quad \text{if } x \text{ is continuous.}$$

$E(x - \mu)^2$ is read as the expected value of $(x-\mu)^2$ and will be denoted by σ_x^2 or by σ^2 when there is no ambiguity. The square root of the variance σ^2, is called the standard deviation and is denoted by σ. Similarly the second moment about the origin is defined as

$$E(x)^2 = \sum_{\text{all } i} x_i^2 p_i \quad \text{if } x \text{ is discrete}$$

$$= \int_{-\infty}^{\infty} x^2 f(x)\, dx \quad \text{if } x \text{ is continuous.}$$

$E(x)^2$ is read as the expected value of x^2.

The parallel axis theorem of elementary mechanics is, of course, applicable here. The second moment about the mean is equal to the second moment about any other point minus the square of the distance between the mean and this point, i.e.,

$$E(x - \mu)^2 = E(x - a)^2 - (\mu - a)^2.$$

In particular, the second moment about the origin can be written as

$$E(x)^2 = E(x - \mu)^2 + \mu^2 = \sigma^2 + \mu^2.$$

[1] The total probability mass is one. Hence it is unnecessary to divide this expression by the total mass as is usually done in mechanics to obtain the center of gravity.

This theorem is easily proven by reverting to the definitions. It will be presented here for the case of a continuous random variable, and the proof for the discrete case will be left to the reader. Using the definition of the variance, the following result is obtained:

$$E(x-\mu)^2 = \int_{-\infty}^{\infty} (x-\mu)^2 f(x)\,dx = \int_{-\infty}^{\infty} [(x-a)-(\mu-a)]^2 f(x)\,dx.$$

Squaring the terms in brackets, we find

$$E(x-\mu)^2 = \int_{-\infty}^{\infty} (x-a)^2 f(x)\,dx - 2(\mu-a)\int_{-\infty}^{\infty} (x-a)f(x)\,dx + (\mu-a)^2\int_{-\infty}^{\infty} f(x)\,dx.$$

Since

$$\int_{-\infty}^{\infty} f(x)\,dx = 1, \qquad \int_{-\infty}^{\infty} (x-a)^2 f(x)\,dx = E(x-a)^2 \qquad \text{and}$$

$$\int_{-\infty}^{\infty} (x-a)f(x)\,dx = \int_{-\infty}^{\infty} xf(x)\,dx - a\int_{-\infty}^{\infty} f(x)\,dx = \mu - a,$$

we have

$$\begin{aligned} E(x-\mu)^2 &= E(x-a)^2 - 2(\mu-a)^2 + (\mu-a)^2 \\ &= E(x-a)^2 - (\mu-a)^2. \end{aligned}$$

Returning to the die example of Section 2.5,

$$\begin{aligned} E(x)^2 &= \sum_{\text{all } i} x_i^2 p_i \\ &= (1)^2\frac{1}{6} + (2)^2\frac{1}{6} + (3)^2\frac{1}{6} + (4)^2\frac{1}{6} + (5)^2\frac{1}{6} + (6)^2\frac{1}{6} = 15\frac{1}{6}. \end{aligned}$$

$$\text{Variance of } x = \sigma^2 = E(x-\mu)^2 = E(x)^2 - \mu^2 = 15\frac{1}{6} - \left(3\frac{1}{2}\right)^2 = \frac{35}{12}.$$

The standard deviation of $x = \sigma = \sqrt{\dfrac{35}{12}}$.

For the two dice

$$\begin{aligned} E(x)^2 &= (2)^2\frac{1}{36} + (3)^2\frac{2}{36} + (4)^2\frac{3}{36} + (5)^2\frac{4}{36} + (6)^2\frac{5}{36} + (7)^2\frac{6}{36} \\ &\quad + (8)^2\frac{5}{36} + (9)^2\frac{4}{36} + (10)^2\frac{3}{36} + (11)^2\frac{2}{36} + (12)^2\frac{1}{36} \\ &= 54\frac{5}{6}; \end{aligned}$$

$$\sigma^2 = E(x)^2 - \mu^2 = 54\frac{5}{6} - (7)^2 = 5\frac{5}{6}.$$

For the example of the coin flip,

$$E(x)^2 = (0)^2\frac{1}{2} + (1)^2\frac{1}{2} = \frac{1}{2};$$

$$\sigma^2 = E(x)^2 - \mu^2 = \frac{1}{2} - \left(\frac{1}{2}\right)^2 = \frac{1}{4}.$$

For the continuous random variable, Rockwell hardness, mentioned in Section 2.6

$$E(x)^2 = \int_{-\infty}^{\infty} x^2 f(x)\,dx = \int_{50}^{70} x^2\frac{1}{20}\,dx = \frac{1}{20}\frac{x^3}{3}\Bigg|_{50}^{70}$$

$$= \frac{1}{60}(343{,}000 - 125{,}000) = 3{,}633\frac{1}{3};$$

$$\sigma^2 = E(x)^2 - \mu^2 = 3{,}633\frac{1}{3} - (60)^2 = 33\frac{1}{3}.$$

In general, the jth moment about the mean is denoted by

$$E(x - \mu)^j = \sum_{\text{all } i} (x_i - \mu)^j p_i \qquad \text{if } x \text{ is discrete}$$

and

$$= \int_{-\infty}^{\infty} (x - \mu)^j f(x)\,dx \qquad \text{if } x \text{ is continuous.}$$

$E(x - \mu)^j$ is read as "the expected value of $(x - \mu)^j$". Finally, the jth moment about the origin is denoted by

$$E(x)^j = \sum_{\text{all } i} (x_i)^j p_i \qquad \text{if } x \text{ is discrete}$$

$$= \int_{-\infty}^{\infty} x^j f(x)\,dx \quad \text{if } x \text{ is continuous.}$$

Thus, if the probability distribution is known, all its moments can be computed. Similarly, if all the moments of a probability distribution are known, the probability distribution is characterized. Unfortunately, in practice, a knowledge of moments beyond the second is rare. Hence, a reasonable question to formulate is "To what extent do the first two moments, μ and σ^2, characterize the probability distribution of the random variable?" The answer, known as Tchebycheff's Inequality, is as follows: For any real number k, the probability that the random variable, x, lies in the interval $(\mu - k\sigma, \mu + k\sigma)$ is larger than $1 - (1/k^2)$, i.e.,

$$\mathrm{P}(\mu - k\sigma \leqq x \leqq \mu + k\sigma) > 1 - \frac{1}{k^2}$$

where x is a random variable having *any* distribution with mean μ and variance σ^2. For example, if $k = 2$, we obtain

$$P(\mu - 2\sigma \leqq x \leqq \mu + 2\sigma) > \frac{3}{4};$$

if $k = 3$, we obtain

$$P(\mu - 3\sigma \leqq x \leqq \mu + 3\sigma) > \frac{8}{9}.$$

2.10 Some Properties of Functions of Random Variables

In Section 2.8, the expected value of the function $h(x)$ was defined as

$$E[h(x)] = \sum_{\text{all } i} h(x_i)p_i \qquad \text{if } x \text{ is a discrete random variable}$$

$$= \int_{-\infty}^{\infty} h(x)f(x)\,dx \quad \text{if } x \text{ is a continuous random variable.}$$

Some useful results about particular functions follow.

1. Let $h(x)$ be equal to ax, where a is a constant.

$$E(ax) = \left\{\begin{matrix} \sum_{\text{all } i} ax_ip_i & \text{if } x \text{ is discrete} \\ \int_{-\infty}^{\infty} axf(x)\,dx & \text{if } x \text{ is continuous} \end{matrix}\right\} = aE(x) = a\mu.$$

Thus, the expected value of a constant times a random variable is equal to the constant times the expected value of the random variable.

2. Let $h(x)$ be equal to $(ax - a\mu)^2$, where a is a constant.

$$E(ax - a\mu)^2 = \sigma_{ax}^2 = \left\{\begin{matrix} \sum_{\text{all } i} (ax_i - a\mu)^2p_i & \text{if } x \text{ is discrete} \\ \int_{-\infty}^{\infty} (ax - a\mu)^2f(x)\,dx & \text{if } x \text{ is continuous} \end{matrix}\right\} = a^2\sigma_x^2.$$

Thus, the variance of a constant times a random variable is equal to the square of the constant times the variance of the random variable.

3. Let $h(x)$ be a constant a.

$$E(a) = \left\{\begin{matrix} \sum_{\text{all } i} ap_i & \text{if } x \text{ is discrete} \\ \int_{-\infty}^{\infty} af(x)\,dx & \text{if } x \text{ is continuous} \end{matrix}\right\} = a.$$

Thus, the expected value of a constant is equal to the constant.

4. Finally, let $h(x)$ be $(a - a)^2 = 0$.

$$E(a - a)^2 = \sigma_a^2 = \left\{\begin{matrix} \sum_{\text{all } i} (a - a)^2p_i & \text{if } x \text{ is discrete} \\ \int_{-\infty}^{\infty} (a - a)^2f(x)\,dx & \text{if } x \text{ is continuous} \end{matrix}\right\} = 0.$$

Thus, the variance of a constant is zero.

PROBLEMS

1. Suppose two dice are thrown. Consider the random variable which is the total number of ones and twos that appear. Find the probability distribution for this random variable.

2. Let x be the random variable, the toss of a loaded die. The random variable takes on the values 1, 2, 3, 4, 5, and 6 with probabilities $\frac{1}{12}$, $\frac{1}{12}$, $\frac{1}{6}$, $\frac{1}{4}$, $\frac{1}{3}$, and d, respectively. (a) Find the value of d. (b) Find $P(2 \leq x < 4)$. (c) Find the C.D.F.

3. A die has its numbers removed and two of its sides painted black, two of its sides painted red, and two of its sides painted green. Describe the set of all possible outcomes of the experiment which consists of a single toss of the die.

4. In Problem 3 the random variable chosen assigns the number 0 to black, 1 to red, and 2 to green. If the original die is fair, describe a probability distribution which can be associated with this random variable.

5. During the course of a day, a machine turns out either 0, 1, or 2 defective items with probability $\frac{1}{6}$, $\frac{2}{3}$, and $\frac{1}{6}$, respectively. Calculate the mean value and the variance. In this problem, the random variable is the number of defective items produced, and the values the random variable takes on are 0, 1, or 2. Define briefly (in one or two sentences) the term *random variable*.

6. An airplane wing is assembled with a large number of rivets. The number of defective rivets is the factor of importance. Describe the set of all possible outcomes of the " experiment."

7. In Problem 6 the number of defective rivets x is also the random variable of interest. A probability distribution which closely approximates the probability distribution of this random variable is given by

$$P(x=d) = \frac{e^{-2}2^d}{d!} \quad \text{for} \quad d=0, 1, 2, \cdots$$

where $d! = (d)(d-1)(d-2) \cdots (1)$ and $0! = 1$. Find the probability that the number of defective rivets is less than or equal to 2.

8. If $f(x)=e^{-x}$, where $x \geq 0$ and zero otherwise, find the number a such that x is equally likely to be greater than or less than a. Find the number b such that the probability that x will exceed b is equal to 0.05.

9. Suppose the life in hours of a certain type tube has density $f(x) = a/x^2$, where $x \geq 750$, $f(x)=0$, where $x < 750$. Determine the expression for the C.D.F. What is the probability that a tube will last at least 1,250 hours ?

10. If $f(x)=k(2-x)$, where $0 < x < 1$ and zero otherwise, (a) determine k so that $f(x)$ is a density function and (b) find the number b such that the probability that x will exceed b is $\frac{1}{3}$.

11. If $f(x)=1$, where $1 < x < 2$ and zero otherwise, find the probability that $a < x < b$. Note that several cases arise.

12. The density function of coded measurements of pitch diameter of threads of a fitting is given by $1/(1+x)^2$ for $x>0$. Find the probability that x exceeds 1. What is the expression for the C.D.F.?

13. The density function of shear strength of test spot welds is given by

$$f(x)=x/160{,}000 \qquad 0\leq x\leq 400$$
$$=(800-x)/160{,}000 \qquad 400\leq x\leq 800\,.$$

Find the number a such that $P(x<a)=0.50$ and the number b such that $P(x<b)=0.90$.

14. The probability density of finished diameter on armored electric cable is given by

$$f(x)=\frac{x-0.750}{a} \qquad 0.750\leqq x\leqq 0.775$$
$$=\frac{0.800-x}{a} \qquad 0.775\leqq x\leqq 0.800\,.$$

Find the number a and the $P(x\leqq 0.740)$.

15. Let x be the random variable which denotes the life of electric light bulbs. The probability density function is given by

$$f(x)=a/x^3 \qquad \text{for} \quad 1000\leqq x\leqq 2000$$
$$=0 \qquad \text{otherwise}\,.$$

What is the probability of a bulb lasting less than 1,500 hours?

16. The density function of a certain random variable is described by the figure below.

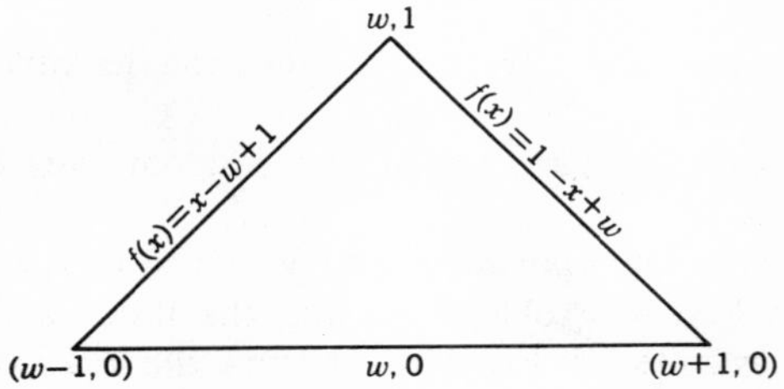

Fig. 2.16.

(a) Calculate the mean and variance. (b) The random variable above is an article which a manufacturer sells at a fixed price of \$1.00. He guarantees to refund the purchase money to any customer who finds the weight of his article is less than 8.25 ounces. His cost of production is C dollars per article (a returned article has no salvage value), and is related to the mean weight of the article by the relation

$$C=(0.05)(\text{mean weight})+0.30\,.$$

If the mean weight is set at $8\frac{1}{2}$ ounces, what is the manufacturer's expected profit per article?

17. A machine makes a product which is screened (inspected 100%) before being shipped. The measuring instrument is such that it is difficult to read between 1 and $1\frac{1}{3}$ (coded data). After the screening process takes place, the measured dimension has the density $f(x)=ax^2$ (a is a constant), where $0<x<1$, $f(x)=\frac{3}{2}$, where $1<x<1\frac{1}{3}$, and zero otherwise. (a) What fraction of the items will fall outside the twilight zone (between 0 and 1)? (b) Find the mean and variance of this random variable.

18. If $f(x)=2x$, where $0<x<1$ and zero otherwise, find the probability that $a<x<b$. Note that three cases must be treated: (a) $a<0$, (b) $b>1$, (c) $0<a<b<1$.

19. Prove the parallel axis theorem for a discrete random variable.

20. Find the mean and standard deviation of the random variable described in Problem 2.

21. Find the $E(x)$, $E(x)^2$, and $E(x-\mu)^2$ for the random variable described in Problem 4.

22. Show that the mean and variance of the random variable described in Problem 7 equal 2.

23. Determine the mean and variance of the random variable described in Problem 8.

24. Show that the mean of the random variable described in Problem 9 is not finite.

25. Find the $E(x)$ and $E(x-\mu)^2$ for the random variable given in Problem 10.

26. Find the mean and variance of the random variable given in Problem 11.

27. Find the mean and variance of the random variable described in Problem 12.

28. Find the mean and variance of the random variable described in Problem 13.

29. Find the mean and variance of the random variable described in Problem 14.

30. In Problem 15, find the mean and variance of the random variable.

31. Using the results of Problem 20, find the mean and variance of $4x$.

32. Using the results of Problem 21, find the $E(2x)$ and the variance of $2x$.

33. Using the results of Problem 25, find the mean and variance of $3x$.

34. Using the results of Problem 26, find the mean and variance of $\frac{1}{2}x$.

35. State whether the following statements are true or false. If x_1 and x_2 are independent random variables with means 3 and 4, respectively, and variances 16 and 25, respectively, then

(a) variance $2x_1=64$,
(b) variance of $(x_1+x_2)=41$,
(c) variance of $(x_1+x_2)=9$,
(d) $E(x_1-7)^2>16$,
(e) $E(x_2)^2-16=25$.

36. Let $x_1, x_2, \cdots, x_{10}$ be independent random variables each of which takes on the values 0, 1, 2, and 3 with probabilities $\frac{1}{16}$, $\frac{1}{8}$, $\frac{3}{16}$, and $\frac{5}{8}$, respectively. Determine whether the following are true or false.

(a) $P(\bar{x}>3)$ is a random variable.

(b) $\dfrac{E(x_1+x_2)}{2}-\dfrac{E(x_3+x_4+x_5)}{3}$ is a random variable.

(c) $\dfrac{x_1+x_2}{2}-\dfrac{x_3+x_4+x_5}{3}$ is a random variable.

(d) $\log(x_1+1)+\sin x_2+e^{x_3}$ is a random variable.

(e) $\Sigma(x_i-\bar{x})$ is a constant equal to zero.

(f) $E(x_7)=\frac{19}{32}$.

(g) The variance of $x_7=\frac{1}{2}$.

37. Let x be the face appearing on a cast die. Let $x_1, x_2, \cdots, x_n$ be n independent observations on x, and $\bar{x}$ their average. For each of the five expressions below, state whether it is: (a) a constant, not necessarily zero; (b) a random variable (which can take on more than one value); (c) neither of these.

1. $P(x<1.52)$
2. $E(\bar{x})$
3. $\bar{x}$
4. $\sum_{i=1}^{n}(x_i-\bar{x})^2$
5. $E[P(x<1.52)]$

38. Let $x_1, x_2, \cdots, x_N$ denote the results of throwing a fair die N times; let $y_1, y_2, \cdots, y_M$ denote the results of throwing a second fair die M times. Identify each of the following expressions as a random variable (R.V.), a constant, not necessarily zero (C), or as zero (0).

x_1-y_1	$x_N-E(x_N)$
$E(x_1-y_1)$	$\Sigma(y_i-\bar{y})$
$\bar{x}-\bar{y}$	$\Sigma(x_i-\bar{x})^2-(N-1)s_x^2$
$P(x_2\leq 3)$	$E(\bar{x})-E(x_2)$
$P(x_1<3)+P(x_1>3)-1$	$E(s_x^2)$

CHAPTER III

The Normal Distribution

3.1 Definitions

In Chapter I, histograms were drawn of experimental data. These histograms represented the outcomes of some random variables. One such random variable often encountered in practice is the continuous random variable whose probability distribution is the normal distribution. The density function of this distribution is defined by

$$f(x) = \frac{1}{\sqrt{2\pi D}} e^{-(x-C)^2/2D^2}, \qquad -\infty < x < \infty,$$

where $D > 0$ and C and D are parameters (constants).

The range of this density function is $-\infty$ to $+\infty$. This may appear to restrict the usefulness of this distribution. For example, measurement errors are often assumed to be normally distributed. However, it is evident that such errors are bounded. Measurements themselves are often assumed to be normally distributed yet are non-negative by the nature of the physical situation. In order to reconcile these apparent contradictions, it must be pointed out that the assumption of a random variable having a normal distribution is simply an assumption regarding the form of a mathematical model which, at best, is just an approximation to a real situation. In the examples above it is tacitly assumed that the probability of getting large measurement errors is very small and the probability of getting negative measurements is also very small. Although the physical phenomena require these probabilities to be zero, using the mathematical model approximates the real situation by assigning small probability to these events. This approximation is similar to the approximation made in assuming that random variables are continuous when measuring devices are all discrete.

The probability that a normally distributed random variable x is less than or equal to b is given by

$$F(b) = P(x \leqq b) = \int_{-\infty}^{b} \frac{1}{\sqrt{2\pi D}} e^{-(x-C)^2/2D^2} dx.$$

This density function cannot be integrated directly. However, the probability that x is less than or equal to b can be represented by the shaded area in Figure 3.1, and its magnitude is determined with the aid of tables as shown in Section 3.3.

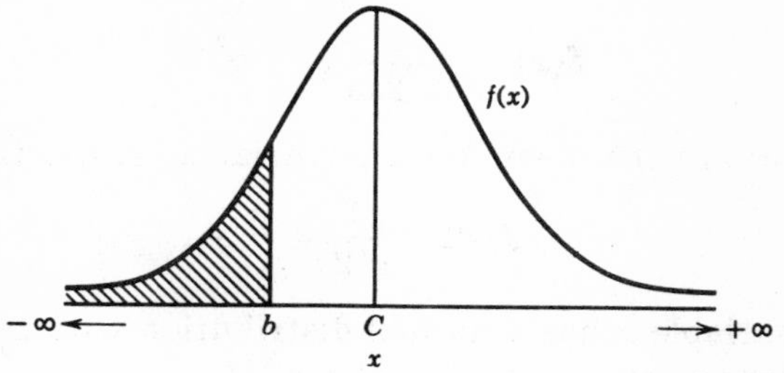

Fig. 3.1. The density function of the normal distribution.

A diagram of the cumulative distribution function is given in Figure 3.2.

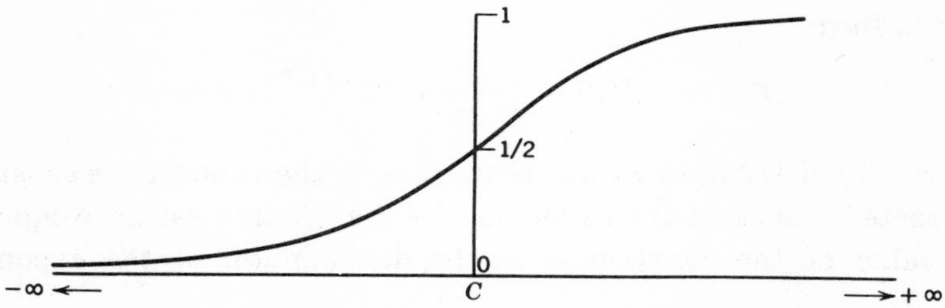

Fig. 3.2. The C.D.F. of the normal distribution.

It should be noted that the normal probability distribution is dependent only upon the two parameters C and D. If C and D are both specified, the probability of events of the above form can be obtained (using the method discussed in Section 3.3).

3.2 The Mean and Variance of the Normal Distribution

The constants C and D are related to the moments of the normal distribution. The expected value of x is equal to C, i.e.,

$$\mu = E(x) = \int_{-\infty}^{\infty} x \frac{1}{\sqrt{2\pi}D} e^{-(x-C)^2/2D^2}\, dx = C$$

and the variance of x is equal to D^2, i.e.,

$$\sigma^2 = E(x - \mu)^2 = \int_{-\infty}^{\infty} (x - \mu)^2 \frac{1}{\sqrt{2\pi}D} e^{-(x-\mu)^2/2D^2}\, dx = D^2.*$$

Thus, the two parameters which completely specify the normal distribution are its mean and its variance.

The density function of the normal distribution can then be written as

$$f(x) = \frac{1}{\sqrt{2\pi}\sigma} e^{-(x-\mu)^2/2\sigma^2},$$

where μ and σ^2 are the mean and the variance, respectively. If

$$f(x) = \frac{1}{\sqrt{2\pi}} e^{-x^2/2},$$

the random variable x has a normal distribution with mean equal to zero and variance equal to 1. Similarly, if

$$f(x) = \frac{1}{\sqrt{2\pi}(14.3)} e^{-(x-3.6)^2/2(14.3)^2},$$

x has a normal distribution with mean 3.6 and standard deviation 14.3. In other words, any random variable whose density function has the form

$$f(x) = \frac{1}{\sqrt{2\pi}\sigma} e^{-(x-\mu)^2/2\sigma^2}$$

is normally distributed with a mean equal to the value of the constant subtracted from x in the exponent of e, i.e., μ, and variance equal to the value of the constant σ^2 in the denominator of the exponent of e.

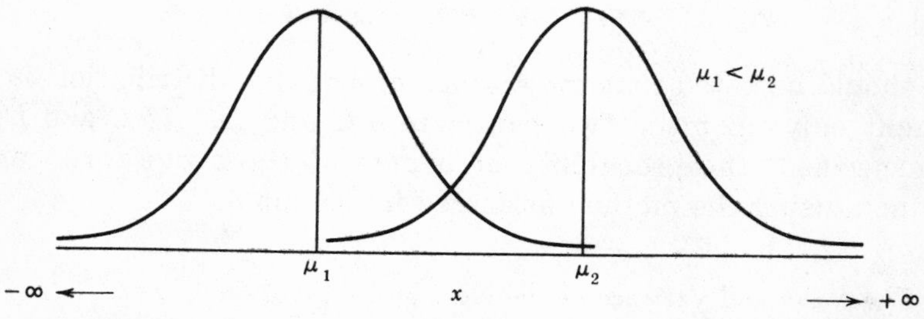

Fig. 3.3. Comparison of two normal distributions with different means.

*The proofs of these statements appear in Section 3.2.1.

Mean values can be represented on the graphs of the density functions as shown in Figure 3.3. It is evident that the normal density is symmetric about its mean.

The variance cannot be represented as easily, but it does have some intuitive meaning. It is a measure of the variability or dispersion of the random variable. The larger the variance, the larger is the variability. This can be shown graphically by Figure 3.4.

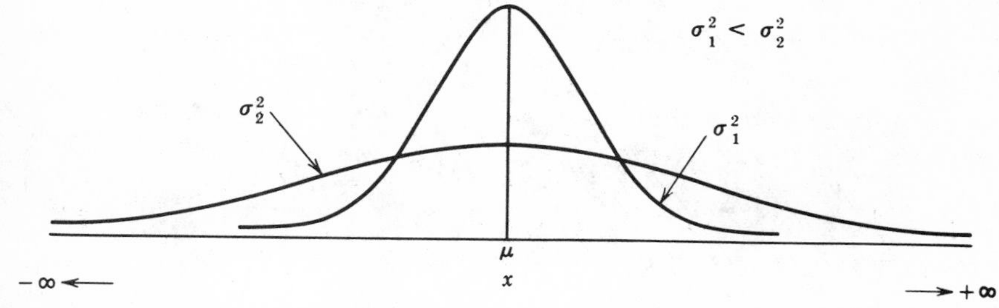

Fig. 3.4. Comparison of two normal distributions with different variances.

The standard deviation, σ, can be used to locate the points of inflection of the density function. The points of inflection appear at $\mu - \sigma$ and $\mu + \sigma$.

*3.2.1 Evaluation of the Mean and Variance of the Normal Distribution

This section will contain the proofs of the results that $E(x) = C$ and $\sigma_x^2 = D^2$. It will be shown first that

$$\mu = E(x) = C.$$

From the basic definition of the mean,

$$\mu = E(x) = \int_{-\infty}^{\infty} x \frac{1}{\sqrt{2\pi}D} e^{-(x-C)^2/2D^2}\, dx.$$

Let $\dfrac{x - C}{D} = y$ so that $dx = D\, dy$. Then

$$\mu = \int_{-\infty}^{\infty} \left(\frac{yD + C}{\sqrt{2\pi}}\right) e^{-y^2/2} dy = \frac{D}{\sqrt{2\pi}} \int_{-\infty}^{\infty} y e^{-y^2/2} dy + \frac{C}{\sqrt{2\pi}} \int_{-\infty}^{\infty} e^{-y^2/2}\, dy.$$

But

$$\frac{D}{\sqrt{2\pi}} \int_{-\infty}^{\infty} y e^{-y^2/2} dy = \frac{-D}{\sqrt{2\pi}} e^{-y^2/2} \Big|_{-\infty}^{\infty} = 0$$

and it is well known that

$$\int_{-\infty}^{\infty} \frac{1}{\sqrt{2\pi}} e^{-y^2/2}\, dy = 1.$$

Hence

$$\mu = C. \qquad \text{Q.E.D.}$$

The second result concerns the variance, i.e., it will be shown that

$$\sigma^2 = E(x - \mu)^2 = D^2.$$

From the basic definition of the variance

$$\sigma^2 = E(x - \mu)^2 = \int_{-\infty}^{\infty} \frac{(x - \mu)^2}{\sqrt{2\pi}D} e^{-(x-\mu)^2/2D^2}\, dx.$$

Let $(x - \mu)/D = y$ so that $dx = D\,dy$. Then

$$\sigma^2 = \int_{-\infty}^{\infty} \frac{D^2y^2}{\sqrt{2\pi}} e^{-y^2/2} dy = D^2 \int_{-\infty}^{\infty} \frac{1}{\sqrt{2\pi}} y^2 e^{-y^2/2}\, dy.$$

Integrating by parts, it can be shown that

$$\int_{-\infty}^{\infty} \frac{1}{\sqrt{2\pi}} y^2 e^{-y^2/2}\, dy = 1.$$

Hence

$$\sigma^2 = D^2. \qquad \text{Q.E.D.}$$

3.3 Tables of the Normal Integral

Given values of the mean and the variance of a random variable which has a normal distribution, it is only a matter of calculus to find such probabilities as

$$\mathrm{P}\,(x > b) = \int_{b}^{\infty} \frac{1}{\sqrt{2\pi}\sigma} e^{-(x-\mu)^2/2\sigma^2}\, dx$$

or

$$\mathrm{P}\,(x \leq b) = 1 - \mathrm{P}(x > b) = \int_{-\infty}^{b} \frac{1}{\sqrt{2\pi}\sigma} e^{-(x-\mu)^2/2\sigma^2}\, dx.$$

However, this is not an elementary function, and hence its integral cannot be written down in simple form. If it were tabulated in the form above, a separate table would be required for each pair μ and σ chosen. This is evidently an impossible task, but fortunately a simple transformation can be made to reduce the problem to the computation of a single table. Let

$$z = \frac{x - \mu}{\sigma}; \qquad dx = \sigma\, dz.$$

The probability that x is greater than b can then be written as

$$P(x > b) = \int_{(b-\mu)/\sigma}^{\infty} \frac{1}{\sqrt{2\pi}\sigma} e^{-z^2/2} (\sigma dz)$$

$$= \int_{(b-\mu)/\sigma}^{\infty} \frac{1}{\sqrt{2\pi}} e^{-z^2/2}\, dz = \int_{(b-\mu)/\sigma}^{\infty} f(z)\, dz.$$

It is noted that $f(z)$ has the form of a normal density function with mean 0 and variance 1. z is known as a standardized normal random variable. The probability that x is greater than b can always be written in the form of an integral of the normal density with mean 0 and variance 1, where the lower limit of integration depends on b, μ, and σ. Therefore, a tabulation of the normal integral when the random variable has mean 0 and variance 1 is sufficient. Appendix Table 1 is such a table. It presents the area of a standardized normal random variable from K_α to plus infinity, i.e., the shaded

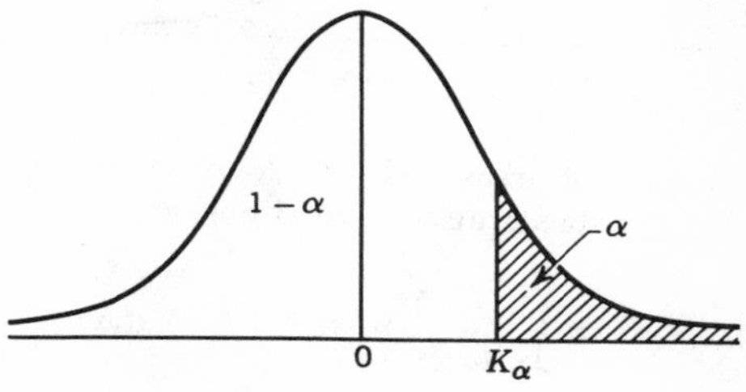

Fig. 3.5. The standardized normal distribution.

area in Figure 3.5. K_α is known as the upper α percentage point or the normal deviate corresponding to α and is defined by

$$P(z > K_\alpha) = \int_{K_\alpha}^{\infty} \frac{1}{\sqrt{2\pi}} e^{-z^2/2} dz = \alpha = \text{shaded area.}$$

Thus,

$$P(x > b) = \int_{b}^{\infty} \frac{1}{\sqrt{2\pi}\sigma} e^{-(x-\mu)^2/2\sigma^2}\, dx$$

$$= P(z > K_\alpha) = \int_{K_\alpha}^{\infty} \frac{1}{\sqrt{2\pi}} e^{-z^2/2} dz = \alpha$$

where $K_\alpha = (b - \mu)/\sigma$. This probability can be represented by the two equal shaded areas in Figure 3.6. In other words, in order to find this probability, subtract the mean from b and divide the result by the standard deviation, i.e., compute $K_\alpha = (b - \mu)/\sigma$. Enter Appendix Table 1 with this value and read out the probability. For example, let $\mu = 2$, $\sigma^2 = 9$, and $b = 8$. The probability that x is greater than 8 is given by

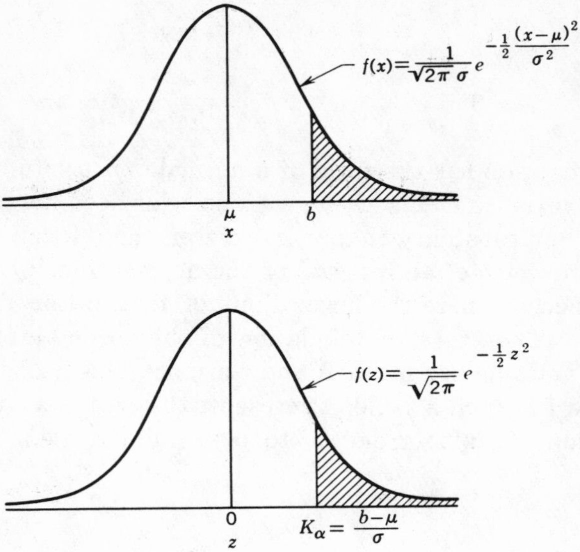

Fig. 3.6. Comparison of areas under the regular normal curve and the standardized normal curve.

$$P\,(x > 8) = \int_8^\infty \frac{1}{\sqrt{2\pi}\,(3)} e^{-(x-2)^2/2(9)}\,dx$$

$$= \int_{(8-2)/3=2}^\infty \frac{1}{\sqrt{2\pi}} e^{-z^2/2}\,dz = 0.0228$$

This result is obtained by entering Appendix Table 1 with $K_\alpha = 2$. Summarizing then, the probability that x is greater than 8 when the mean of x is 2 and the variance of x is 9 equals 0.0228 and is equivalent to the probability that z is greater than 2, where z is the standardized normal random variable. The value 0.0228 is the value of α corresponding to $K_\alpha = 2$.

Probabilities of events of the form $(x \leqq b)$ can be obtained by using the relation

$$P\,(x \leqq b) = 1 - P\,(x > b).$$

It is noted that Appendix Table 1 does not contain negative values of K_α. Such values of K_α can be obtained by recognizing that the standardized normal distribution is symmetric about 0. It is easily seen that the $P\,(z < -K_\alpha) = P\,(z > K_\alpha)$. This result can be verified by referring to Figure 3.7. Thus,

$$P\,(z < -2) = P\,(z > 2) = 0.0228.$$

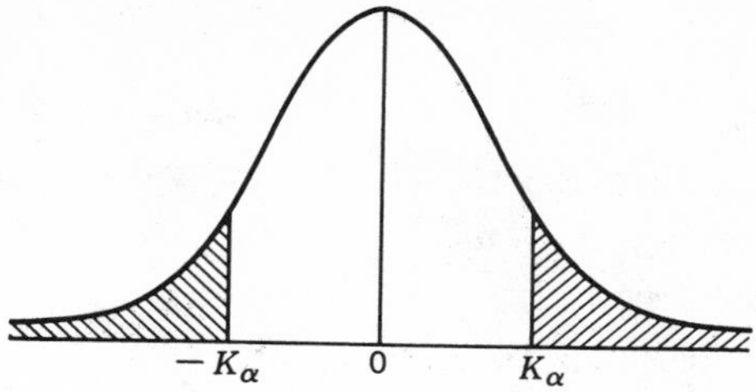

Fig. 3.7. The standardized normal distribution.

The probability of events of the form $(z \geqq -K_\alpha)$ is also easy to obtain.

$$P(z \geqq -K_\alpha) = 1 - P(z < -K_\alpha) = 1 - P(z > K_\alpha).$$

From this result we find:

$$\begin{aligned} P(z \geqq -2) &= (1 - 0.0228) \\ &= 0.9772. \end{aligned}$$

Events of the type $(a < x \leqq b)$ can also be obtained by a simple extension of the ideas above. This event is represented by the shaded area in Figure 3.8. This probability can be evaluated by considering the two events $x > a$ and $x > b$ shown in Figure 3.9.

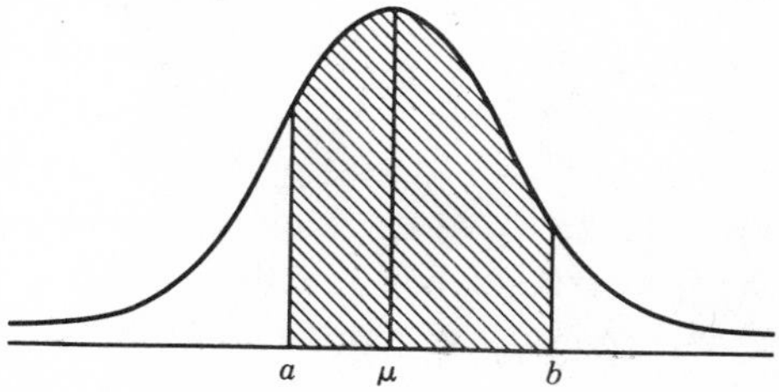

Fig. 3.8. Probability of the random variable falling in the finite interval $(a, b]$.

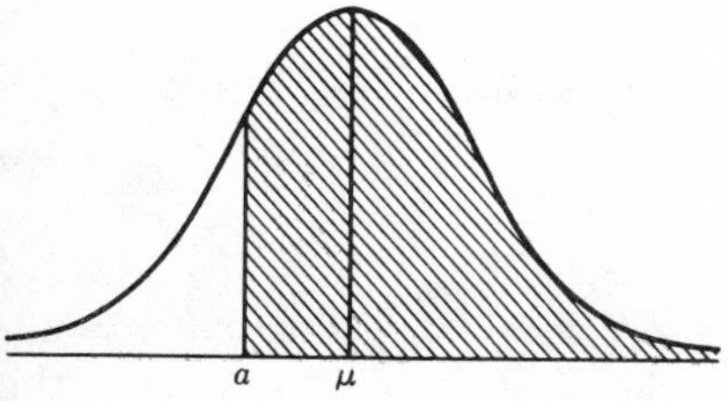

Fig. 3.9a. Probability of the event $x > a$.

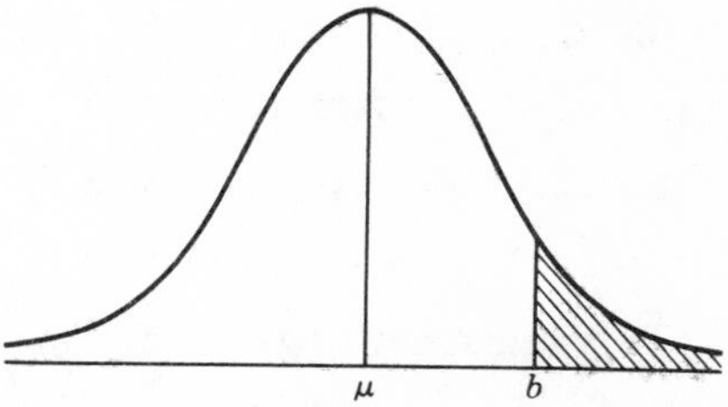

Fig. 3.9b. Probab lity of the event $x > b$.

The probability of the event $a < x \leqq b$ is given by

$$P(a < x \leqq b) = P(x > a) - P(x > b).$$

Both of the probabilities on the right-hand side of the equality can be found by entering Appendix Table 1. For example, let $\mu = 2$, $\sigma^2 = 9$, $b = 8$, and $a = 5$.

$$P(5 < x \leqq 8) = \int_5^8 \frac{1}{\sqrt{2\pi}\,(3)} e^{-(x-2)^2/2(9)}\,dx = P(x > 5) - P(x > 8)$$

$$= \int_{(5-2)/3=1}^{\infty} \frac{1}{\sqrt{2\pi}} e^{-z^2/2}\,dz - \int_{(8-2)/3=2}^{\infty} \frac{1}{\sqrt{2\pi}} e^{-z^2/2}\,dz$$

$$= 0.1587 - 0.0228 = 0.1359.$$

3.4 Combinations of Normally Distributed Random Variables

In Chapter 2, some basic concepts of moments of random variables which are functions of random variables were presented. This discussion dealt with simple functions of a single random variable, e.g., ax, where a is a constant. It is often important to be able to talk about the distribution of functions of more than one variable. For example, the sample mean has great importance in statistics. In fact, the distribution of linear combinations of random variables, of which the sample mean is a special case, plays an important role. It is the purpose of this section to present the distribution of linear combinations of normally distributed random variables. The fundamental results are as follows: An experiment is performed which consists of n trials. Let $x_1, x_2, \cdots, x_n$ denote the measurable quantity associated with the first trial, second trial, $\cdots$, nth trial, respectively. $x_1, x_2, \cdots, x_n$ are then random variables. If

(1) $x_1, x_2, \cdots, x_n$ are independent random variables having means $\mu_1, \mu_2, \cdots, \mu_n$ and variances $\sigma_1^2, \sigma_2^2, \cdots, \sigma_n^2$ respectively,[1]
(2) $a_1, a_2, \cdots, a_n$ are constants, and
(3) $y = a_1x_1 + a_2x_2 + \cdots + a_nx_n$,

then y is a random variable having the following three properties:

1. $$\mu_y = E(y) = E(a_1x_1 + a_2x_2 + \cdots + a_nx_n)$$
$$= a_1\mu_1 + a_2\mu_2 + \cdots + a_n\mu_n.$$

This statement says that the expected value of a linear combination of random variables is equal to the linear combination of the

[1] A constant can also be considered as a degenerate random variable having mean equal to the constant and variance equal to zero.

expected values. Furthermore, this result is valid even if the x's are dependent.

2. $\sigma_y^2 = E(y - \mu_y)^2 = \sigma^2_{(a_1x_1 + a_2x_2 + \cdots + a_nx_n)}$

$$= a_1^2\sigma_1^2 + a_2^2\sigma_2^2 + \cdots + a_n^2\sigma_n^2.$$

This statement says that the variance of a linear combination of independent random variables is equal to the sum of the products of variances and squared constants.

3. Furthermore, if $x_1, x_2, \cdots, x_n$ are normally distributed, then y is also normally distributed with mean μ_y and variance σ_y^2.*

The ensuing sections will deal with examples of applications of these results.

3.5 The Standardized Normal Random Variable

In Section 3.3, the standardized normal random variable, i.e., that random variable having mean 0 and variance 1, was obtained by making a transformation of the normal random variable x having mean μ and variance σ^2. As the first example, this result will be obtained again. Consider the random variable

$$y = \frac{x - \mu}{\sigma} = \left(\frac{1}{\sigma}\right)x - (1)\frac{\mu}{\sigma}$$

and let $a_1 = 1/\sigma$ and $a_2 = -1$. (μ/σ will be considered as a degenerate random variable with mean μ/σ and variance 0.) From the results on linear combinations of random variables

$$E(y) = \left(\frac{1}{\sigma}\right)\mu - (1)\frac{\mu}{\sigma} = 0;$$

$$\sigma_y^2 = \left(\frac{1}{\sigma^2}\right)\sigma^2 - 0 = 1.$$

Furthermore, since x and μ/σ are normally distributed (μ/σ being considered a normally distributed random variable with mean equal to μ/σ and variance 0), y is also normally distributed with mean 0 and variance 1. Thus, if the difference, obtained by subtracting the mean of a normally distributed random variable from the random variable itself, is divided by the standard deviation of the variable [i.e., $y = (x - \mu)/\sigma$], the resulting random variable is normally distributed with zero mean and unit standard deviation.

* A constant can also be considered as a degenerate normally distributed random variable with mean equal to the constant and variance equal to zero.

3.6 The Distribution of the Sample Mean

Another example of an application of the results of Section 3.4 is for the case of the sample mean $\bar{x}$. Let

$$y = \bar{x} = \frac{x_1 + x_2 + \cdots + x_n}{n} = \frac{1}{n}x_1 + \frac{1}{n}x_2 + \cdots + \frac{1}{n}x_n$$

where $x_1, x_2, \cdots, x_n$ are a random sample of n observations (x's are independent and identically distributed). Call the common mean μ and the common variance σ^2. The a's are all $1/n$ so that

$$E(y) = \left(\frac{1}{n}\right)\mu + \left(\frac{1}{n}\right)\mu + \cdots + \left(\frac{1}{n}\right)\mu = \mu\,;$$

$$\sigma_y^2 = \left(\frac{1}{n}\right)^2\sigma^2 + \left(\frac{1}{n}\right)^2\sigma^2 + \cdots + \left(\frac{1}{n}\right)^2\sigma^2 = \frac{\sigma^2}{n}.$$

Furthermore, if $x_1, x_2, \cdots, x_n$ are normally distributed with mean μ and variance σ^2, $\bar{x}$ is normally distributed with mean μ and variance σ^2/n.

In earlier sections, it was shown that by subtracting the mean from a normally distributed random variable and then dividing by its standard deviation, the resulting random variable is normally distributed with mean 0 and variance 1. Therefore $\dfrac{(\bar{x} - \mu)}{\sigma/\sqrt{n}} = (\bar{x} - \mu)\sqrt{n}/\sigma$ is normally distributed with mean zero and variance 1, provided the x's are all independently normally distributed with common mean and variance.

As pointed out in Chapter II, the sample mean is a random variable. It has now been shown that the expected value of this random variable is the same as the expected value of the individual random variables. The variance, however, is reduced to σ^2/n. Furthermore, if the original n random variables are normally distributed, the sample mean has a normal distribution. A comparison of the distribution of the sample mean $\bar{x}$ and one of the original normally distributed random variables x is given in Figure 3.10.

There is a rather important concept involved in discussing the distribution of the sample mean. Whereas the results of n random variables (the individual observations) constitute the observed sample mean, this value may be considered as the result of a single random variable, the sample mean. For example, two experiments are to be performed. Let $x_1, x_2, \cdots, x_5$ be 5 independent normally distributed

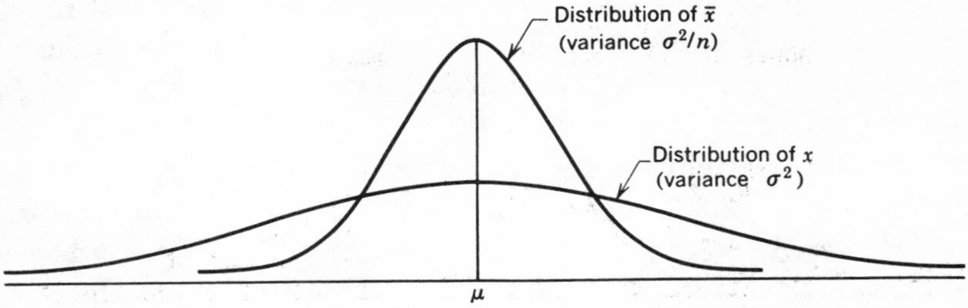

Fig. 3.10. Comparison of the probability distribution of x and the probability distribution of $\bar{x}$.

random variables each with mean 0 and variance 1. The distribution of the sample mean is then also normal with mean 0 but with variance $\frac{1}{5}$. Suppose the outcome of the x's is observed and the value taken on by the random variable $\bar{x}$ is to be computed. The second experiment involves a random variable y which is normally distributed with mean 0 and variance $\frac{1}{5}$. A *single* observation on this random variable is to be taken. An announcement is made that the result of one of the experiments is 0.25, but no mention is made of which experiment led to this result. The problem is to decide whether the experiment involving $\bar{x}$ or the experiment involving y belongs to the 0.25. The answer is, of course, that the experiments were equivalent and that it is impossible to differentiate between the two. The distribution of $\bar{x}$ is identical with the distribution of y. Thus, an observation on the sample mean may be considered as the result of a *single* experiment (a sample of size one) where the random variable has the distribution of the sample mean. Using this probability distribution, the probability of events involving the sample mean may be computed.

3.7 Tolerances

A third example using the results on linear combinations of random variables deals with tolerances. Design specifications on a dimension are usually given as a nominal value plus or minus a tolerance, e.g., 1.530 ± 0.003. If asked exactly what these limits mean, the designer will usually answer that *all* parts should fall within these limits. If he is pressed further, the designer will admit that he really doesn't expect all, but *almost all*, the items to fall within these dimensions. If this statement is questioned, the answer will finally result in stating that no more than a given fraction of the items produced will fall outside these limits. In other words, this dimension is a random variable having a probability distribution with the property that no more than the given fraction will fall outside the lower and upper specification limits.

It is well to distinguish between specification limits and natural tolerance limits. The specification limits are limits that are set somewhat arbitrarily, say by the designer, without regard to what the process can achieve. The natural tolerance limits are the actual capabilities of the process, and can be considered as the limits within which all but a given allowable fraction α of the items produced will fall. If it can be assumed that the dimension is a normally distributed random variable, a good design will usually have a mean value coincident with the nominal value and a standard deviation which will permit only the small allowable fraction α of the items produced to fall outside of the specification limits. In other words, the natural tolerances will coincide with the design specifications. In any event, in order for a process to be acceptable, the natural tolerances must fall *within* the specification limits, which insures that the specification limits will include *at least* the fraction $1 - \alpha$ of the items produced. For example, if the allowable fraction of items falling outside the natural tolerance limits is 27 in 10,000, and the specifications are 1.530 ± 0.003, the process whose dimension is normally distributed with mean 1.530 and standard deviation 0.001 will have natural tolerances which will coincide with the specification limits. This is evident from Figure 3.11.

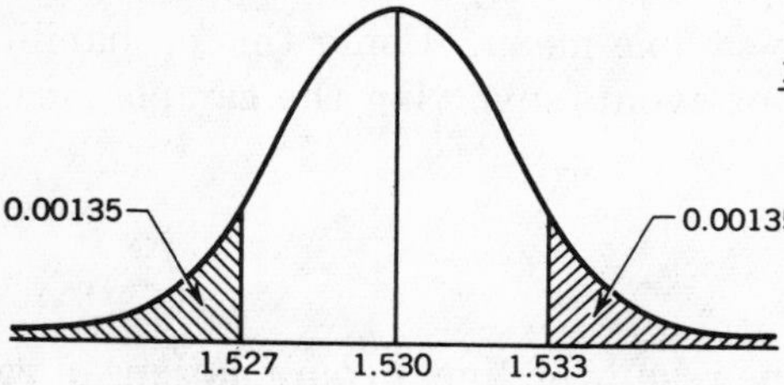

$$\frac{1.533 - 1.530}{\sigma} = K_{0.00135} = 3$$

$$\sigma = 0.001$$

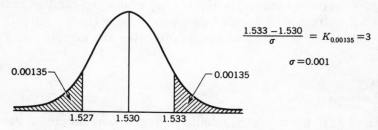

Fig. 3.11. The distribution of a dimension whose natural tolerances coincide with the specifications.

In other words, given symmetrical specifications and an allowable fraction of defectives α the maximum allowable standard deviation (if the dimension of the product is centered at the nominal dimension) enabling the natural tolerances to fall within the specification limits is obtained by finding the normal deviate (percentage point) $K_{\alpha/2}$ corresponding to the fraction above the upper limit, and dividing this number into the difference between the upper specification limit U and the nominal dimension, i.e., $(U - \mu)/K_{\alpha/2}$.

Given the actual standard deviation σ, it is a simple problem to solve the inverse problem, i.e., to find out what fraction of the items will fall outside the specification limits.

It is often the case that a dimension of an assembled product is the sum of the dimensions of several parts. An electrical resistance may be the sum of several electrical resistances. A weight may be the sum of a number of individual weights. The distribution of the individual components may be known, and what is of interest is the distribution of the sum, $y = x_1 + x_2 + \cdots + x_n$, where $x_1, x_2, \cdots, x_n$ are independent components (making up the finished product) having means $\mu_1, \mu_2, \cdots, \mu_n$ and variances $\sigma_1^2, \sigma_2^2, \cdots, \sigma_n^2$, respectively. Referring to the results about linear combinations of independent random variables (Section 3.4), it is evident that all the a's are equal to 1, so that the expected value of y is given by

$$\mu_y = E(y) = \mu_1 + \mu_2 + \cdots + \mu_n$$

and the variance of y is given by

$$\sigma_y^2 = \sigma_1^2 + \sigma_2^2 + \cdots + \sigma_n^2.$$

Furthermore, if all the x's are normally distributed random variables, y is also normally distributed. Hence, if all the μ's and σ^2's are known, the fraction falling outside the specification limit can be obtained.

As an example of simple addition of components into an assembly, consider the problem of assigning individual specification limits to a component shown in Figure 3.12 such that the over-all dimension

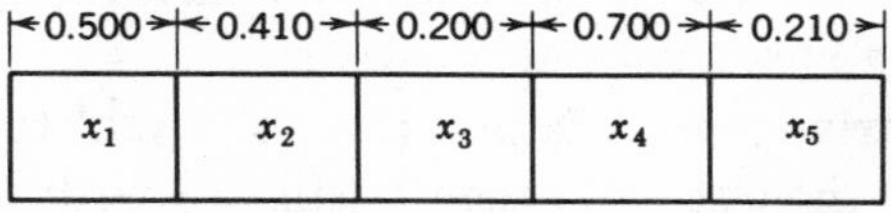

Fig. 3.12. Assembly consisting of five components.

is 2.020 ± 0.030 inches. It will be assumed that the five components are independent normal random variables having means as shown in Figure 3.12 and *common* variance σ^2. It will be assumed that the natural tolerance limits are defined such that the allowable fraction falling outside the over-all tolerance limits as well as the tolerance limits of the individual components is 0.0027.[1] The individual speci-

[1] The implications of this assumption will be discussed later.

fication limits are to be set so that if the natural tolerance limits of the individual components fall within these limits, the natural tolerance limits of the final assembly will fall within the specification limits of the final assembly. It then follows that the fraction falling outside the specification limits of the final assembly will not exceed 0.0027. Let $y = x_1 + x_2 + x_3 + x_4 + x_5$. The expected value of y is given by

$$E(y) = 0.500 + 0.410 + 0.200 + 0.700 + 0.210 = 2.020$$

and the variance of y is given by

$$\sigma_y^2 = 5\sigma^2.$$

Furthermore, if the natural tolerances are to coincide with the specification limits, since y is normally distributed as shown in Figure 3.13, the normal deviate (percentage point) corresponding to an area of

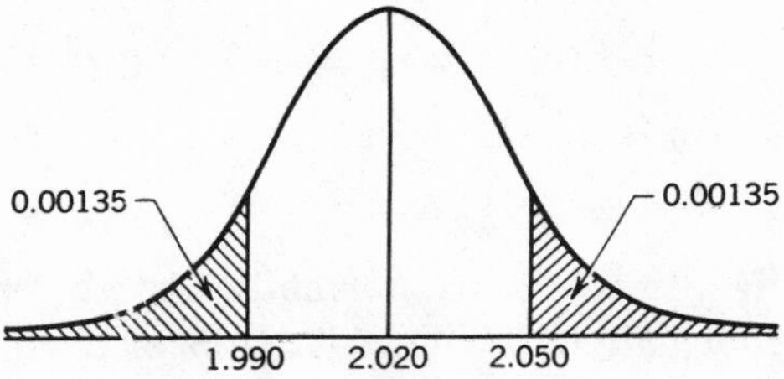

Fig. 3.13. The distribution of y.

0.00135 is given by $K_{0.00135} = 3$ and must be equal to $(2.050 - 2.020)/\sigma_y$. Solving the equation $3 = (2.050 - 2.020)/\sigma_y$ for σ_y, it is evident that σ_y equals 0.01. $\sigma_y^2 = 0.0001$ is the maximum allowable variance for the final assembly which will allow the natural tolerances to fall within the design specifications.

Thus, if the variance of y is less than or equal to 0.0001, the natural tolerance of the over-all process will fall within the given specification limits.

Using the relation $\sigma_y^2 = 5\sigma^2$, it follows that $\sigma^2 = 0.0001/5 = 0.00002$ is the maximum allowable variance for the individual components. The corresponding standard deviation is given by $\sigma = 0.0045$. Therefore, if the standard deviation of the individual components is less than or equal to 0.0045, the natural tolerances of the final assembly will fall within the given specification limits. Since natural tolerances for the individual components correspond to an α of 0.0027, the limits are given by $\mu \pm 3\sigma$. Thus, the widest permissible natural tolerances of the individual components are equal to $\mu \pm 3(0.0045)$ and will contain all but 0.0027 of the items produced when σ equals its maximum value of 0.0045. Hence the specifications for the components which contain the natural tolerances that insure the inclusion of the natural tolerances of the final assembly within the over-all specification limits are as follows:

Component 1	0.500 ± 0.0135
Component 2	0.410 ± 0.0135
Component 3	0.200 ± 0.0135
Component 4	0.700 ± 0.0135
Component 5	0.210 ± 0.0135

The choice of the value for the allowable fraction of defectives defining the natural tolerance limits is irrelevant to the solution of the problem, provided that the *same* value pertains to the individual components as well as the over-all unit. This is most easily seen by choosing a fraction $\alpha > 0$ which pertains to both sets of limits. The natural tolerances of the final assembly should be such that no more than α of the values will fall outside of the specification limits of 2.020 ± 0.030. Similarly, the natural tolerances of the individual components should be such that no more than α of the values will fall outside the determined specification limits.

As before, let

$$y = x_1 + x_2 + x_3 + x_4 + x_5$$

so that the expected value of y is given by

$$E(y) = 0.500 + 0.410 + 0.200 + 0.700 + 0.210 = 2.020$$

and the variance of y is given by

$$\sigma_y^2 = 5\sigma^2.$$

Furthermore, if the natural tolerances are to coincide with the specification limits, since y is normally distributed as shown in

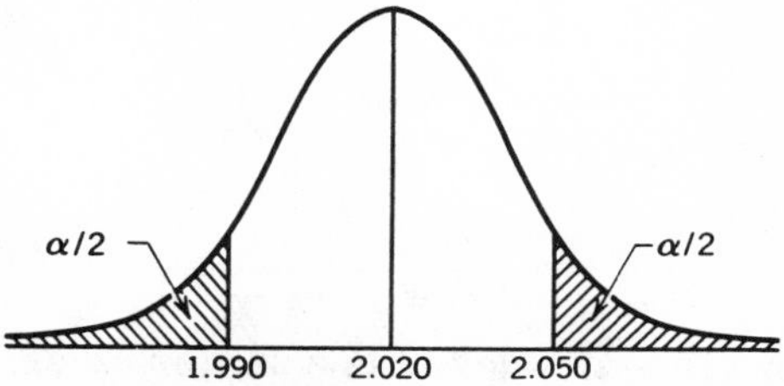

Fig. 3.14. The distribution of the final assembly.

Figure 3.14, the normal deviate (percentage point) corresponding to an area of $\alpha/2$ is given by $K_{\alpha/2}$ and must equal $K_{\alpha/2} = (2.050 - 2.020)/\sigma_y$. Solving this equation for σ_y, we find $\sigma_y = 0.030/K_{\alpha/2}$. Hence, $\sigma_y^2 = (0.030/K_{\alpha/2})^2$ is the maximum allowable variance of the final assembly allowing the natural tolerances to fall within the specification limits. Since $\sigma_y^2 = 5\sigma^2$, it follows that $\sigma^2 = (0.030)^2/5K_{\alpha/2}^2$ is the maximum allowable variance of the individual components allowing the natural tolerances of the final assembly to fall within the given specification limits. The corresponding standard deviation is given by $\sigma = 0.030/\sqrt{5}\,K_{\alpha/2}$.

The widest permissible natural tolerances of the individual components are then given by $(\mu \pm K_{\alpha/2}\sigma) = (\mu \pm 0.030/\sqrt{5})$ and will contain all but α of the items produced when σ equals its maximum value of $0.030/\sqrt{5}\,K_{\alpha/2}$. Hence, the specifications for the components which contain the natural tolerances that insure the inclusion of the natural tolerances of the final assembly within the over-all specification limits are as follows:

Component 1	0.500 ± 0.0135
Component 2	0.410 ± 0.0135
Component 3	0.200 ± 0.0135
Component 4	0.700 ± 0.0135
Component 5	0.210 ± 0.0135

Note that the results are the same as those obtained previously. This observation has an interesting implication about tolerances. Consider the range of the natural tolerances of the final assembly $(X_{max} - X_{min})_{sum}$. Let $(X_{max} - X_{min})_i$ represent the range of the natural tolerance of the ith component. If the same but unknown value of α pertains to the over-all assembly, as well as to the individual components, a relationship between the $(X_{max} - X_{min})_{sum}$ and the $(X_{max} - X_{min})_i$ exists. This is found by recognizing that

$$(X_{max} - X_{min})_{sum} = 2K_{\alpha/2}\,\sigma_{sum}$$

and

$$(X_{max} - X_{min})_i = 2K_{\alpha/2}\,\sigma_i$$

for all i.

Using the equality

$$\sigma_{sum} = \sqrt{\sigma_1^2 + \sigma_2^2 + \cdots + \sigma_n^2}$$

and writing the standard deviations in terms of the appropriate $(X_{max} - X_{min})$, the equation

$$(X_{max} - X_{min})_{sum} = \sqrt{(X_{max} - X_{min})_1^2 + (X_{max} - X_{min})_2^2 + \cdots + (X_{max} - X_{min})_n^2}$$

is obtained. Applying this result, the previous example leads to

$$0.060 = \sqrt{5(X_{max} - X_{min})^2}$$

so that $(X_{max} - X_{min})$ for each component is 0.0270. The values 0.060 and 0.0270 are to be interpreted as the largest natural tolerance ranges falling within the appropriate specification limits. Thus, the range of the specification limits for the individual components must equal 0.0270, which is the result obtained previously by the other methods.

Another application of the results on linear combinations of independent random variables is to the problem of the clearance between components. Consider the following example: The mean external diameter of a shaft is $\mu_s = 1.048$ inches and the standard deviation is $\sigma_s = 0.0013$ inches. The mean inside diameter of the mating bearing is $\mu_b = 1.059$ inches and the standard deviation is $\sigma_b = 0.0017$ inches. Assuming that both the external shaft diameters and the internal bearing diameters are normally distributed and independent, what is the minimum clearance that can be expected to occur by random matching of shafts and bearings in assembly? The minimum clearance will be defined as that value L such that P (clearance $< L$) = 0.00135.[1] In other words, L is defined as the natural tolerance, and is compared with the design specification. Consider the diagram in Figure 3.15.

It is evident that what is of interest is the difference between the

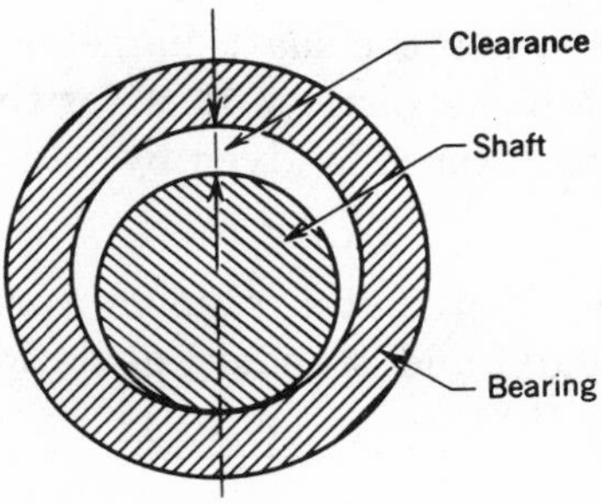

Fig. 3.15. Shaft and mating bearing.

[1] The choice of the value 0.00135 was arbitrary. The same method of analysis would be used if any other probability were assigned with the final result depending upon the particular choice.

two random variables, the diameter of the bearing, b, and the diameter of the shaft, s. This is the clearance. Let the clearance be denoted by y so that $y = (b - s)$. Using the results on linear combinations of random variables with $a_1 = 1$ and $a_2 = -1$, the expected value of the clearance is given by

$$E(y) = \mu_b - \mu_s = 0.011$$

and the variance of the clearance is given by

$$\sigma_y^2 = (1)^2\sigma_b^2 + (-1)^2\sigma_s^2 = \sigma_b^2 + \sigma_s^2 = (0.0017)^2 + (0.0013)^2 = 0.00000458.$$

Furthermore the clearance is normally distributed.

Thus, the distribution of clearance is as shown in Figure 3.16.

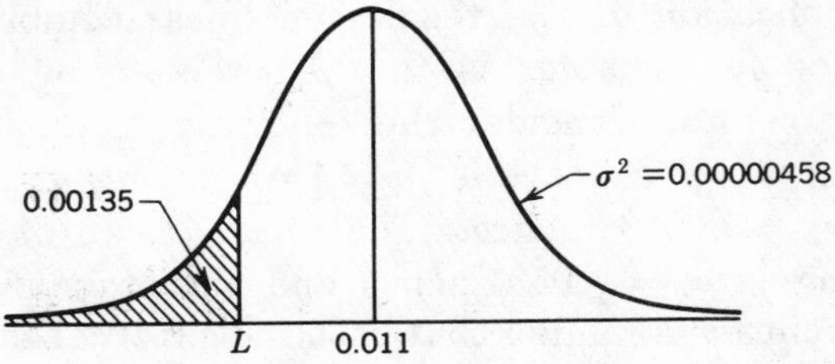

Fig. 3.16. Distribution of clearance.

Consequently, from Appendix Table 1, the normal deviate (percentage point) corresponding to 0.99865 (the area above L) is found to be $K_{0.99865} = -3$ and must be equal to

$$-3 = K_{0.99865} = \frac{L - \mu}{\sigma} = \frac{L - 0.011}{0.00214}.$$

Solving this equation for L it is evident that $L = 0.011 - 0.00642 = 0.00458$.

Interference occurs when the shaft diameter exceeds the bearing diameter. Using the same distributions for the shaft and bearing, the probability of interference is given by

$$P(y < 0)$$

where y is a normally distributed random variable (the clearance) with mean equal to 0.011 and standard deviation 0.00214.

$$P(y < 0) = P\left(z < \frac{0 - 0.011}{0.00214} = -5.1\right)$$

where z is normally distributed with mean zero and standard deviation one.

Since -5.1 is a negative value, the identity

$$P(z < -5) = P(z > 5)$$

is used. Referring to Appendix Table 1, the value 5 exceeds the range of the table so that

$$P(z > 5) = 0$$

for all practical purposes. Therefore, the probability of interference is negligible.

*3.8 Tolerance in "Complex Items"

In the previous section tolerances for components which were the result of direct addition or subtraction were considered. However, similar problems exist for more complex assemblies, where the final assembly is a non-linear function of the components. This situation is handled by approximating this function by a linear function in the region of interest. This is usually accomplished by expanding the function into a Taylor series around the nominal dimensions. For example, in the electrical circuit of Figure 3.17,[1] the voltage E_0 is required to be 60 volts $\pm 3\%$ with components having the following specifications:

$$E_1 = 40 \pm 0.5 \text{ volts};$$
$$N = 2 \text{ to } 1 \pm 1\%;$$
$$K = 3 \pm 2\%.$$

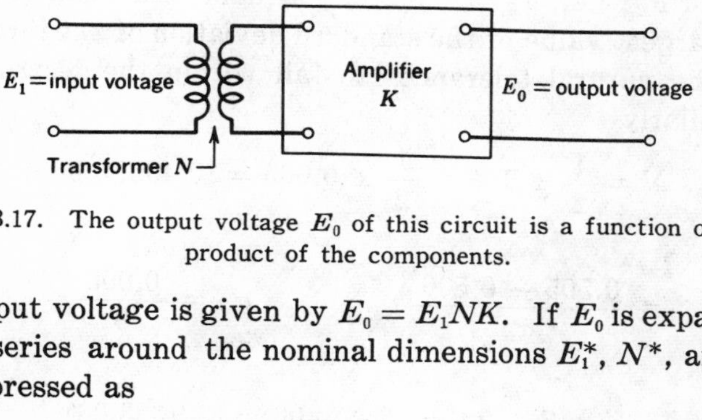

Fig. 3.17. The output voltage E_0 of this circuit is a function of the product of the components.

The output voltage is given by $E_0 = E_1NK$. If E_0 is expanded into a Taylor series around the nominal dimensions E_1^*, N^*, and K^*, E_0 can be expressed as

$$E_0 = E_1^*N^*K^* + (E_1 - E_1^*)\frac{\partial E_0}{\partial E_1}\bigg|_{E_1^*,N^*,K^*} + (N - N^*)\frac{\partial E_0}{\partial N}\bigg|_{E_1^*,N^*,K^*}$$
$$+ (K - K^*)\frac{\partial E_0}{\partial K}\bigg|_{E_1^*,N^*,K^*} + \text{terms of higher order.}$$

Neglecting the terms of higher order,

[1] This example is taken from R. H. Johnson, "How to Evaluate Assembly Tolerance," *Product Engineering*, Jan., 1953.

$$E_0 \cong E_1^* N^* K^* + (E_1 - E_1^*)N^*K^* + (N - N^*)E_1^*K^* + (K - K^*)E_1^*N^*.$$

Furthermore, if these components are normally distributed independent random variables with means at the nominal values, then from the results on linear combinations of random variables, E_0 is approximately normally distributed with mean $E(E_0)$, approximately $E_1^*N^*K^*$; and variance $\sigma^2(E_0)$, approximately

$$(N^*K^*)^2\sigma_{E_1}^2 + (E_1^*K^*)^2\sigma_N^2 + (E_1^*N^*)^2\sigma_K^2.$$

Putting the numerical constants of the example into the expression for the mean, $E(E_0) \cong 40 \times 3 \times \frac{1}{2} = 60$; the variance can be expressed as

$$\sigma_{E_0}^2 \cong \left(\frac{1}{2} \times 3\right)^2 \sigma_{E_1}^2 + (40 \times 3)^2\sigma_N^2 + \left(40 \times \frac{1}{2}\right)^2 \sigma_K^2$$

$$= \frac{9}{4}\sigma_{E_1}^2 + (14{,}400)\sigma_N^2 + 400\sigma_K^2.$$

If the natural tolerances coincide with the design specifications and

$$E_1 = 40 \pm 0.5 = (39.5, 40.5)$$

is interpreted to mean all but 0.0027 of the values to fall within this range, the normal deviate (percentage point) corresponding to 0.00135 is given by $K_{0.00135} = 3$ and must equal $(40.5 - 40)/\sigma_{E_1}$. This expression is solved for σ_{E_1}, i.e., $\sigma_{E_1} = 0.5/3$. σ_{E_1} equal to 0.5/3 can be interpreted as the largest value of the standard deviation of the input voltage allowing the natural tolerances to fall within the design specifications. Similarly

$$N = \frac{1}{2} \pm 1\% = \frac{1}{2} \pm 0.005 = (0.495, 0.505)$$

so that

$$3 = \frac{0.505 - 0.500}{\sigma_N} \qquad \sigma_N = \frac{0.005}{3}.$$

Finally,

$$K = 3 \pm 2\% = 3 \pm 0.06 = (2.94, 3.06)$$

so that

$$3 = \frac{3.06 - 3.00}{\sigma_K} \quad \text{or} \quad \sigma_K = 0.02.$$

σ_N equal to 0.005/3 and σ_K equal to 0.02 can be interpreted as the largest values of the standard deviation of the transformer and amplifier, respectively, allowing the natural tolerances of these components to fall within the design specifications. Hence, the largest value of the variance of the output voltage which still allows the natural tolerances of the individual components to fall within their specifications is given by

$$\sigma^2_{E_0} = \left(\frac{9}{4}\right)\left(\frac{0.25}{9}\right) + (14{,}400)\left(\frac{0.000025}{9}\right) + (400)(0.0004)$$

$$= 0.0625 + 0.040 + 0.16 = 0.2625$$

so that $\sigma_E = 0.512$. If the natural tolerances of the individual components all fall within their specification limits, the natural tolerances for the output voltage will not exceed

$$E_0 = 60 \pm 3(0.512) = 60 \pm 1.536 = 60 \pm 2.56\%,$$

which is within the design specification of $60 \pm 3\%$. Thus the natural tolerances will fall within the design specifications.

A little thought will reveal that the 0.0027 and the corresponding $K_{0.00135} = 3$ really play no role in this problem provided that it is assumed that the allowable fraction of defectives defining the natural tolerance limits is the same for all the components as well as E_0. If $K_{\alpha/2}$ denotes the percentage point corresponding to any allowable fraction of defectives, α, ($\alpha/2$ falling outside of each limit) the maximum standard deviations of the components allowing the natural tolerances to fall within the design specifications are given by

$$\sigma_{E_1} = \frac{0.5}{K_{\alpha/2}}; \qquad \sigma_N = \frac{0.005}{K_{\alpha/2}}; \qquad \sigma_K = \frac{0.06}{K_{\alpha/2}}.$$

The largest value of the variance of the output voltage which still allows the natural tolerances of the individual components to fall within their specifications is given by

$$\sigma^2_{E_0} = \left(\frac{9}{4}\right)\left(\frac{0.25}{(K_{\alpha/2})^2}\right) + (14{,}400)\left(\frac{0.000025}{(K_{\alpha/2})^2}\right) + (400)\left(\frac{0.0036}{(K_{\alpha/2})^2}\right)$$

$$= \frac{0.5625}{K^2_{\alpha/2}} + \frac{0.3600}{K^2_{\alpha/2}} + \frac{1.44}{K^2_{\alpha/2}} = \frac{2.3625}{K^2_{\alpha/2}}$$

so that

$$\sigma_{E_0} = \frac{1.536}{K_{\alpha/2}}.$$

If the natural tolerance of the individual components all fall within their specification limits, the natural tolerances for the output voltage will not exceed

$$E_0 = 60 \pm K_{\alpha/2}\frac{1.536}{K_{\alpha/2}} = 60 \pm 1.536 = 60 \pm 2.56\%,$$

which is exactly the same result as obtained previously.

The function $E_0 = E_1NK$ was expanded into a Taylor series, with the higher order terms neglected. Other functions of n variables $f(x_1, x_2, \cdots, x_n)$ can be expanded into a Taylor series about a particular point, say $x_1^*, x_2^*, \cdots, x_n^*$, i.e.,

$$f(x_1, x_2, \cdots, x_n) = f(x_1^*, x_2^*, \cdots, x_n^*) + (x_1 - x_1^*)\left.\frac{\partial f}{\partial x_1}\right|_{x_1^*, x_2^*, \cdots, x_n^*} + \cdots$$
$$+ (x_n - x_n^*)\left.\frac{\partial f}{\partial x_n}\right|_{x_1^*, x_2^*, \cdots, x_n^*}$$
$$+ \text{ terms of higher order.}$$

Neglecting terms of higher order and assuming that the random variables are independent with means at $x_1^*, x_2^*, \cdots, x_n^*$, respectively, the expected value of the function is approximately given by

$$E[f(x_1, x_2, \cdots, x_n)] \cong f(x_1^*, x_2^*, \cdots, x_n^*)$$

and the variance of the function is approximately given by

$$\sigma^2_{f(x_1, x_2, \cdots, x_n)} \cong \sigma^2_{x_1}\left[\left.\frac{\partial f}{\partial x_1}\right|_{x_1^*, x_2^*, \cdots, x_n^*}\right]^2 + \cdots + \sigma^2_{x_n}\left[\left.\frac{\partial f}{\partial x_n}\right|_{x_1^*, x_2^*, \cdots, x_n^*}\right]^2.$$

Furthermore, if the x's are normally distributed, $f(x_1, x_2, \cdots, x_n)$ is approximately normally distributed.

As a final example,[1] suppose that a phase-shifting synchro is added to the output of the electrical circuit in the previous example as shown in Figure 3.18. The new output voltage is given by

$$E_0 = E_2 \cos\theta + E_1NK \sin\theta$$

where θ refers to the angular relationship of the synchro inputs.

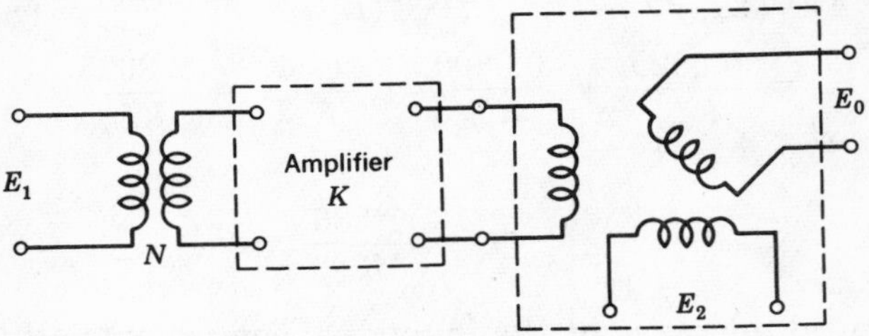

Fig. 3.18. Adding a phase-shifting synchro to the circuit of Figure 3.17 causes the output voltage to become a more complicated function of the components.

[1] This example is taken from R. H. Johnson "How to Evaluate Assembly Tolerances," *Product Engineering*, Jan. 1953.

$$\frac{\partial E_0}{\partial E_1} = NK \sin \theta$$

$$\frac{\partial E_0}{\partial E_2} = \cos \theta$$

$$\frac{\partial E_0}{\partial N} = E_1 K \sin \theta$$

$$\frac{\partial E_0}{\partial K} = E_1 N \sin \theta$$

$$\frac{\partial E_0}{\partial \theta} = -E_2 \sin \theta + E_1 NK \cos \theta.$$

Assigning typical values and specifications to the components

$$E_1 = 40 \pm 0.5 \text{ volts}$$

$$N = 2 \text{ to } 1 \pm 1\%$$

$$K = 3 \pm 2\%$$

$$\theta = 60 \pm \tfrac{1}{4} \text{ degrees } (0.00436 \text{ radian})$$

$$E_2 = 40 \pm 0.4 \text{ volts}$$

and assuming that the components are independently distributed normal random variables with means at the nominal values (points about which the function is expanded) we have that E_0 is approximately normally distributed with mean

$$E(E_0) \cong E_2^* \cos \theta^* + E_1^* N^* K^* \sin \theta^*$$

$$= 40 \times \frac{1}{2} + 40 \times \frac{1}{2} \times 3 \times \frac{\sqrt{3}}{2} = 72 \text{ volts}$$

and variance

$$\sigma_{E_0}^2 \cong \sigma_{E_1}^2 (N^* K^* \sin \theta^*)^2 + \sigma_{E_2}^2 (\cos \theta^*)^2 + \sigma_N^2 (E_1^* K^* \sin \theta^*)^2$$
$$+ \sigma_K^2 (E_1^* N^* \sin \theta^*)^2 + (E_1^* N^* K^* \cos \theta^* - E_2 \sin \theta^*)^2 \sigma_\theta^2$$

$$= \sigma_{E_1}^2 \left(\frac{1}{2} \times 3 \frac{\sqrt{3}}{2} \right)^2 + \sigma_{E_2}^2 \left(\frac{1}{2} \right)^2 + \sigma_N^2 \left(40 \times 3 \frac{\sqrt{3}}{2} \right)^2$$
$$+ \sigma_K^2 \left(40 \times \frac{1}{2} \times \frac{\sqrt{3}}{2} \right)^2 + \sigma_\theta^2 \left(40 \times \frac{1}{2} \times 3 \times \frac{1}{2} - 40 \frac{\sqrt{3}}{2} \right)^2$$

$$= \left(\frac{27}{16} \right) \sigma_{E_1}^2 + \left(\frac{1}{4} \right) \sigma_{E_2}^2 + (10{,}800) \sigma_N^2 + (300) \sigma_K^2 + (21.54) \sigma_\theta^2.$$

If $K_{\alpha/2}$ denotes the percentage point corresponding to any allowable fraction of defectives, α, ($\alpha/2$ falling outside of each limit) the maximum standard deviations of the components allowing the natural tolerances to fall within the design specifications are given by

$$\sigma_{E_1} = \frac{0.5}{K_{\alpha/2}}; \quad \sigma_N = \frac{0.005}{K_{\alpha/2}}; \quad \sigma_K = \frac{0.06}{K_{\alpha/2}};$$

$$\sigma_\theta = \frac{0.00436}{K_{\alpha/2}}; \quad \sigma_{E_2} = \frac{0.4}{K_{\alpha/2}}.$$

The largest value of the variance of the output voltage which still allows the natural tolerances of the individual components to fall within their specifications is given by

$$\sigma^2_{E_0} = \frac{27}{16}\frac{0.25}{K^2_{\alpha/2}} + \frac{1}{4}\frac{0.16}{K^2_{\alpha/2}} + \frac{(10{,}800)\,(0.000025)}{K^2_{\alpha/2}} + (300)\frac{(0.0036)}{K^2_{\alpha/2}} + \frac{(21.54)\,(0.00001901)}{K^2_{\alpha/2}}$$

$$= \frac{1.81}{K^2_{\alpha/2}}$$

so that

$$\sigma_{E_0} = \frac{1.35}{K_{\alpha/2}}.$$

Hence, the design specifications on E_0 should not be less than

$$72 \pm 1.35 \text{ volts.}$$

3.9 The Central Limit Theorem

In many of the previous examples, it was assumed, somewhat arbitrarily, that the random variables were normally distributed. Although in actual practice many random variables are normally distributed, many do not resemble a normal distribution at all, and such assumptions can lead to difficulties. Fortunately, however, there are certain natural factors which, in some cases, tend to make the random variable have a normal distribution. This can be best explained by stating the central limit theorem.

Central Limit Theorem. Let $y = x_1 + x_2 + \cdots + x_n$ where $x_1, x_2, \cdots, x_n$ are identically distributed independent random variables each having mean μ and finite variance σ^2. Then the distribution of $z = (y - n\mu)/\sqrt{n}\,\sigma$ approaches the normal distribution with mean 0 and variance 1 as n becomes infinite.

The theorem is also valid when the random variables are not identically distributed, provided that their variances are not too different. Thus, this theorem implies that the sum of a large number of random variables will be approximately normally distributed regardless of the distribution of the individual random variables. Note furthermore that

$$\frac{y - n\mu}{\sqrt{n}\sigma} = \left(\frac{\bar{x} - \mu}{\sigma}\right)\sqrt{n}$$

so that this theorem also implies that the mean of n identically distributed independent random variables will be approximately normally distributed regardless of the distribution of the individual variables.

The magnitude of n that is required before the results of the theorem begin to hold depends to a great extent on the shape of the distribution of the original variables. In Figure 3.19, it is demon-

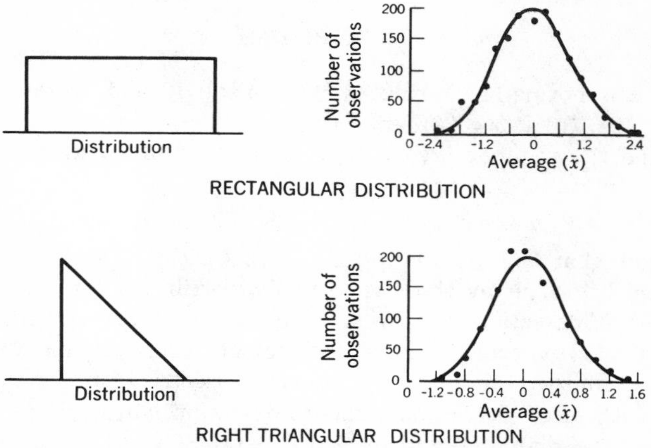

Fig. 3.19. Distribution of the sample average from rectangular and triangular distributions

Reproduced by permission from W. A. Shewhart, "Economic Control of Quality of Manufactured Product," copyright 1931, D.Van Nostrand Co., Inc. Princeton N.J.

strated that even when the underlying distribution is a rectangular or a triangular distribution, the distribution of $\bar{x}$ values from samples of four is approximately normal.

The central limit theorem then, is the key to the importance of the normal distribution in statistics.

In the theory of errors, which deals with the distribution of errors of measurements, errors can usually be assumed to have a normal distribution, since they are usually composed of the sum of many small independent components.

Most of the procedures for estimating parameters or testing hypotheses are dependent upon the sample mean. If the individual random variables going into the sample are normally distributed, then the sample mean is certainly normally distributed, and the distribution theory for normally distributed random variables is relevant. If the individual random variables going into the sample are not normally distributed, but their number is sufficiently large, then the central limit theorem indicates that the sample mean tends toward normality, and the distribution theory for normally distributed random variables is relevant. If the individual random variables going into the sample have an unknown distribution, and their number is not sufficiently large, the distribution of the sample mean *cannot* be assumed to be normal. In such cases, non-parametric techniques should be employed. Such techniques will be presented in Chapter VII.

PROBLEMS

1. If x is normally distributed with mean 10 and variance 4, find (a) $P(8<x<12)$, (b) $P(x<20)$, (c) $P(x>5)$.

2. Find the number k such that for a normally distributed variate x having mean μ and variance σ^2,

$$P(\mu-k\sigma<x<\mu+k\sigma)=0.90, 0.95, 0.99, 0.995.$$

Find k such that $P(x<\mu+k\sigma)=0.95, 0.975, 0.99$.

3. The life of army shoes is normally distributed with mean 12 months and standard deviation 2 months. If 10,000 pairs are issued, how many pairs would be expected to need replacement after $15\frac{1}{2}$ months?

4. The finished diameter on armored electric cable is normally distributed with mean 0.775 and standard deviation 0.010. What is the probability that the diameter will exceed 0.790?

5. The width of a slot on a duralumin forging is normally distributed with mean 0.9000 inch and standard deviation 0.0030 inch. The specification limits were given as 0.9000 ± 0.0050 inch. What percentage of forgings will be defective?

6. The measurement of pitch diameter of thread of a fitting is normally distributed with mean 0.4008 inch and standard deviation 0.0004 inch. The specifications are given as 0.4000 ± 0.0010 inch. What is the probability of a defective occurring?

7. In Problem 5, what is the maximum allowable standard deviation which will permit no more than one in a thousand defectives to be produced when the slot widths are normally distributed with mean 0.9000 inch?

8. In Problem 6, what is the maximum allowable standard deviation which will permit no more than one in a hundred defectives to be produced when the pitch diameters are normally distributed and centered at the nominal dimension?

9. In Problem 4, what must be the mean value of finished diameter if the probability of exceeding 0.790 is to be 0.01?

10. In Problem 6, if the standard deviation of the pitch diameter is 0.0004 inch, what must be the centering to insure that no more than 0.5% of the items produced will exceed the upper specification limit?

11. If the diameters of certain shafts must be less than 1.500 inches to be usable, and if the shafts are produced by a process that gives mean diameter 1.490 inches with standard deviation 0.005 inch, what percentage of shafts must be scrapped? If in addition a shaft must be larger than 1.480 inches, what percentage will be scrapped?

12. In the situation of Problem 4 determine the mean and the standard deviation of sample averages of nine diameters and find what percentage of averages should fall below 0.772. What percentage above 0.790.

13. In the situation of Problem 5 samples of size 5 forgings are observed hourly and their means computed. What percentage of these sample averages will lie outside the specification limits? Determine limits such that the percentage falling outside these limits is 0.27%.

14. In the situation of Problem 6 samples of size 4 pitch diameters are observed. What is the probability that a sample average will fall within the specification limits? What is the probability that an average will exceed 0.4002?

15. Suppose $\bar{x}_1$ and $\bar{x}_2$ are means of two independent random samples of size n. The observations going into a mean are normally distributed with common mean, μ, and with common variance equal to 2. Determine n so that the probability will be 0.99 that $\bar{x}_1$ and $\bar{x}_2$ differ by less than 2, i.e., find n such that

$$P(-2 \leq \bar{x}_1 - \bar{x}_2 \leq 2) = 0.99.$$

16. The standard deviation of the diameters of a shaft and bearing have identical values of 0.001 inch. The mean distance between centers of the parts is 0.003. What is the probability of interference assuming that each dimension of the mating parts is normally distributed?

17. Two resistors are assembled in series. Each is nominally rated at 10 ohms. The resistors are known to be normally distributed about the nominal rating each having a standard deviation equal to 0.5 ohm. What is the probability that an assembly will have a combined dimension in excess of 21 ohms? What must the standard deviations be so that the probability of a combined dimension exceeding 21 ohms will be 0.01?

18. A delivery truck carries loaded cartons of items. If the weight of each carton is approximately normally distributed with mean 50 pounds and standard deviation 5 pounds, how many cartons can the truck carry so that the probability of the total load exceeding 1 ton will be only 0.05?

19. The measurable characteristic of an electrical device is normally distributed with mean 10 units and standard deviation 3 units. The measuring device is subject to error. The probability distribution of this device is normal with mean zero and standard deviation $\frac{1}{2}$ unit. What is the probability of a measurement of a device exceeding 12 units?

20. Two mating parts A and B are approximately normally distributed as follows: $\mu_A = 3.000$ inches, $\mu_B = 3.005$ inches, $\sigma_A = 0.001$ inch, and $\sigma_B = 0.0015$ inch. What is the probability of interference?

21. In Problem 20, what should μ_A be so that the probability of interference is 0.01?

22. Two mating parts A and B are approximately normally distribute as follows: $\mu_A=3.000$ inches and $\mu_B=3.005$ inches. If $\sigma_A=\sigma_B$, determin their common value so that the probability of a clearance less than 0.00C does not exceed 1%.

23. Two 8-ohm resistors are placed in series in such a manner th the final assembly should be 16 ± 0.2 ohms. (a) Assuming each resist is normally and identically distributed, determine specification limits f the components. (b) If the natural tolerance limits, which are defined be those limits within which all but 5% of the items fall, coincide wi the specification limits, what is the probability that the average resistan of five final assemblies will exceed 16.05 ohms?

24. At one stage in the manufacture of an article, a cylindrical r with circular cross section has to fit into a circular socket. Quality contr measurements show that the distributions of rod diameter and sock diameter are normal with constants:

Rod diameter: mean 5.01 centimeters, standard deviation 0.03 centimeter.
Socket diameter: mean 5.11 centimeters, standard deviation 0.04 centimeter.

If components are selected at random for assembly, what proportion rods will not fit?

25. (a) Suppose the life in hours of a certain electronic tube is norma distributed with mean $\mu=160$ hours. The specification limits call for t product to lie between 120 hours and 200 hours with probability 0.8 What is the maximum allowable standard deviation that the process c have and still maintain its quality? (b) If the process is at the value σ found in (a), what is the probability of sample averages of 2 falli outside the specification limits?

26. An emf of E volts drives a current of I amperes through a resistan of R ohms. According to Ohm's law, $E=IR$. Suppose E and R are norma distributed random variables whose natural (3 standard deviation) toleran are given by

$$E=60\pm 3 \text{ volts}, \qquad R=6\pm 0.3 \text{ ohms}.$$

Find the natural tolerances of I under the "customary assumption (List these assumptions.) Start with the expression $I=E/R$.

27. Two circular holes of radii r_1 and r_2 are cut from a rectangu piece of metal of thickness T, width x, and length y. The weight of me remaining in the bar after the holes are cut is

$$W=cT(xy-\pi r_1^2-\pi r_2^2).$$

Suppose all dimensions (T, x, y, r_1, and r_2) are random variables independen and normally distributed and such that 99.7% of each distribution within known specification limits. Find a reasonable approximation the standard deviation of W.

28. It is known that

$$\text{distance} = \text{velocity} \times \text{time}$$

Using the Taylor Expansion around the means of velocity and time, f the expression for the approximate variance of distance in terms of the variance of velocity and the variance of time.

29. The value of the constant of gravity is to be determined from $g=2s/t^2$ by observing a body fall. The experimenter has a choice of setting s at 4 feet with t about $\frac{1}{2}$ second or of setting s at 16 feet with t about 1 second. Assume that the standard deviation of s is 0.002 foot and that the standard deviation of recording time is 0.05 second. If the expected value of s is equal to 4 and the expected value of $t=\frac{1}{2}$, calculate the variance of g. If the expected value of s is equal to 16 and the expected value of $t=1$, calculate the variance of g. Which value of s is better in the sense of smaller variance of determining g?

CHAPTER IV

Other Probability Distributions

4.1 Introduction

Chapter III was concerned with the distribution of a random variable whose density function is

$$\frac{1}{\sqrt{2\pi}\,\sigma} e^{-(x-\mu)^2/2\sigma^2};$$

i.e., the normally distributed random variable with mean μ and variance σ^2. The results on linear combinations of random variables indicated that if $x_1, x_2, \cdots, x_n$ is a random sample of n independent normally distributed random variables, each with mean μ and variance σ^2, the random variable, the sample mean $\left(\bar{x} = \sum_{i=1}^{n} x_i/n\right)$, is normally distributed with mean μ and variance σ^2/n. In other words, the distribution of the sample mean is one which depends on the distribution of the individual random variables comprising the sample mean. The random variable, the sample mean, has a probability distribution which turns out to be normally distributed when the assumptions above are fulfilled. The usefulness of the distribution of the sample mean is evidenced by the following example. The mean, μ, of a normal probability distribution is unknown and is to be estimated on the basis of the results of a random sample. Intuitively, the sample mean $\bar{x}$ is judged as a "good" estimate of the quantity μ. How good an estimate $\bar{x}$ is depends upon the probability distribution of $\bar{x}$. A measure of the "closeness" of $\bar{x}$ to μ is given by

$$\mathrm{P}\,(a \leqq \bar{x} - \mu \leqq b)$$

for any a and b; i.e., it is given by the probability distribution of $\bar{x}$. If the standard deviation σ is known, these probabilities can be evaluated by using the identity

$$\mathrm{P}\,(a \leqq \bar{x} - \mu \leqq b) = \mathrm{P}\left(\frac{a\sqrt{n}}{\sigma} \leqq \frac{\sqrt{n}(\bar{x} - \mu)}{\sigma} \leqq \frac{b\sqrt{n}}{\sigma}\right)$$
$$= \mathrm{P}\left(\frac{a\sqrt{n}}{\sigma} \leqq z \leqq \frac{b\sqrt{n}}{\sigma}\right)$$

where z is a normally distributed random variable with zero mean and unit variance. This last expression can be evaluated for any value of a, b, σ, and n by referring to Appendix Table 1.

Distributions of other random variables which are functions of normally distributed random variables are not necessarily normal. For example, the random variable

$$s^2 = \frac{\sum_{i=1}^{n} (x_i - \bar{x})^2}{n - 1}$$

does not have a normal distribution even though the x's are normal. Most of the remaining sections of the chapter will be devoted to determining the distribution of random variables which are generated from (i.e., functions of) other random variables.

4.2 The Chi-Square Distribution

4.2.1 The Chi-Square Random Variable

This section will be concerned with the distribution of a random variable which consists of sums of squares of other random variables. In particular, an experiment is performed which consists of ν trials. Let $x_1, x_2, \cdots, x_\nu$ denote the measurable quantity associated with the first trial, second trial, $\cdots$, νth trial, respectively. $x_1, x_2, \cdots, x_\nu$ are then random variables. Further, assume that these random variables are independent and normally distributed each with zero mean and unit variance. Let χ^2 equal the sum of squares of these ν random variables, i.e., $\chi^2 = x_1^2 + x_2^2 + \cdots + x_\nu^2$. χ^2 also must be a random variable because it is a function of random variables (see Section 2.3). The set R of all possible values of this random variable is clearly the interval between zero and plus infinity since χ^2 is a sum of squares. All that remains is to determine the probability distribution of this random variable. This probability distribution is not arbitrary since the distribution of the x's was specified (as normal with mean zero and unit variance). The density function of the chi-square random variable is written as

$$f(\chi^2) = \frac{1}{2^{\nu/2}\Gamma\left(\frac{\nu}{2}\right)}(\chi^2)^{(\nu-2)/2} e^{-\chi^2/2} \qquad \text{for } \chi^2 \geqq 0$$

$$= 0 \qquad \text{otherwise}$$

where Γ represents the gamma function.[1]

This probability distribution is known as a chi-square distribution with ν degrees of freedom. The quantity ν is the number of independent random variables whose squares are being added and is designated as the degrees of freedom. ν may also be thought of as a parameter associated with the distribution.

Random variables having an approximate chi-square distribution may arise naturally in practice or may be generated in an exact fashion in the manner previously described. An example of the former is the distribution of the lives of vacuum tubes which often approximates the chi-square distribution or a distribution related to the chi-square. An example of the latter is as follows. A hole is to be drilled in a rectangular table top. The desired location of the hole is specified by giving its distance from two perpendicular edges of the top. The actual location of the hole is obtained by measuring the desired distances from the proper edges, using a steel tape. Measurement errors account for any difference between the desired location and the actual location of the hole. The distribution of measurement errors, then, is of practical importance. If the differences between the actual distance from an edge and its desired position are independent normally distributed random variables each having zero mean and unit standard deviation, the square of the distance of the actual measured point from its desired position has a chi-square distribution with 2 degrees of freedom.

If $x_1, x_2, \cdots, x_\nu$ are independent normally distributed random variables with means $\mu_1, \mu_2, \cdots, \mu_\nu$ and variances $\sigma_1^2, \sigma_2^2, \cdots, \sigma_\nu^2$

[1] By definition,

$$\Gamma\left(\frac{\nu}{2}\right) = \left(\frac{\nu}{2} - 1\right)!$$

$$= \begin{cases} \left(\frac{\nu}{2} - 1\right)\left(\frac{\nu}{2} - 2\right)\cdots 3 \cdot 2 \cdot 1 & \text{for } \nu \text{ even and } \nu > 2 \\ \left(\frac{\nu}{2} - 1\right)\left(\frac{\nu}{2} - 2\right)\cdots \frac{3}{2} \cdot \frac{1}{2}\sqrt{\pi} & \text{for } \nu \text{ odd and } \nu > 2 \end{cases}$$

and, further,

$$\Gamma(1) = 1 \quad \text{and} \quad \Gamma\left(\frac{1}{2}\right) = \sqrt{\pi}.$$

respectively, the sum of squares of the x's does *not* have a chi-square distribution. This resulting random variable fails to fulfill the condition which requires all the x's to have zero mean and unit variance. However, the quantities

$$z_i = \frac{x_i - \mu_i}{\sigma_i} \qquad i = 1, 2, \cdots, \nu$$

do have a normal distribution each with zero mean and unit variance. Hence,

$$\chi^2 = \sum_{i=1}^{\nu} z_i^2 = \sum_{i=1}^{\nu} \left(\frac{x_i - \mu_i}{\sigma_i} \right)^2$$

has a chi-square distribution with ν degrees of freedom.

An example of such a random variable is as follows. A bomber is sent out on a mission. If the lateral and forward distances of the explosion from the target are independent normally distributed random variables each with zero mean (this implies that positive and negative miss distances in each direction are equally likely) and equal variances, the squared distance between the explosion and the target, divided by the variance, has a chi-square distribution with 2 degrees of freedom. This bombing problem is easily extended to three dimensions.

Summarizing then, there are many physical phenomena that have a probability distribution which is approximated by the chi-square distribution without any apparent justification. These situations are usually discovered by experience and by analyzing frequency histograms. Other random variables have chi-square distributions because they are known to satisfy the conditions of the definition, i.e.,

> *if* $x_1, x_2, \cdots, x_\nu$ *are a sample of* ν *independent normally distributed random variables each with zero mean and unit variance, the random variable*
>
> $$\chi^2 = x_1^2 + x_2^2 + \cdots + x_\nu^2$$
>
> *has a chi-square distribution with* ν *degrees of freedom.*

In the former case, the degrees of freedom may be considered as a parameter (constant) of the probability distribution whereas in the latter case the degrees of freedom is the number of independent random variables whose squares are being added.

A sketch of the density function of the chi-square random variable is given in Figure 4.1. It should be noted that this distribution is skewed rather than symmetric as is the normal. Furthermore, the density function has mass only between 0 and ∞. This is evident

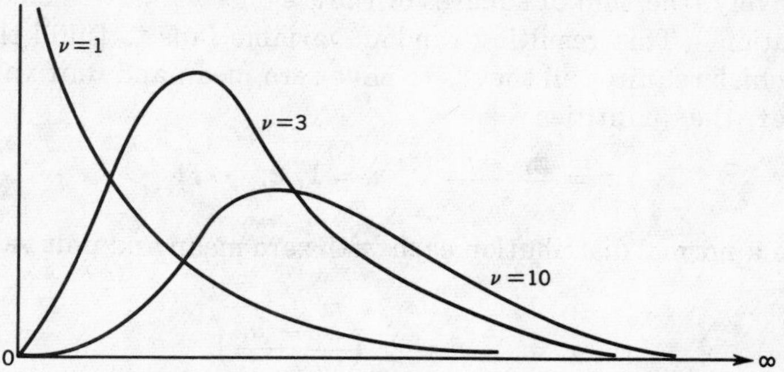

Fig. 4.1 Density function of the χ^2 random variable.

because the random variable is a sum of squares which must necessarily be non-negative. The expected value of the chi-square random variable is given by

$$E(\chi^2) = \nu$$

and the variance of the chi-square random variable is given by

$$\sigma^2_{\chi^2} = 2\nu.$$

In other words these moments are expressed in terms of the degrees of freedom only, the mean equaling the degrees of freedom and the variance equaling twice the degrees of freedom.

The probability that χ^2 is greater than a constant $\chi^2_{\alpha;\nu}$ is represented by

$$P(\chi^2 \geqq \chi^2_{\alpha;\nu}) = \int_{\chi^2_{\alpha;\nu}}^{\infty} f(\chi^2)\, d\chi^2$$

$$= \int_{\chi^2_{\alpha;\nu}}^{\infty} \frac{1}{2^{\nu/2}\Gamma(\nu/2)} (\chi^2)^{(\nu-2)/2} e^{-\chi^2/2}\, d\chi^2 = \alpha.$$

This function has been tabulated, and values of the percentage point $\chi^2_{\alpha;\nu}$ can be found in Appendix Table 2. For example, if $\nu = 5$,

$$P(\chi^2 \geqq \chi^2_{0.10;5}) = P(\chi^2 \geqq 9.236) = 0.10.$$

4.2.2 The Addition Theorem for the Chi-Square Distribution

The chi-square distribution is one of the few probability distributions that has the reproductive property; i.e., if two random variables, each having a chi-square distribution, are independent and are added together, the resulting random variable also has a chi-square distribution. In this property, it is similar to the normal distribution which also has this reproductive feature. This result is stated precisely in the following theorem on the addition of chi-square random variables.

If χ_1^2 and χ_2^2 are independent chi-square random variables with ν_1 and ν_2 degrees of freedom respectively, the sum of these random variables,

$$\chi^2 = \chi_1^2 + \chi_2^2$$

also has a chi-square distribution with

$$\nu = \nu_1 + \nu_2 \quad \textit{degrees of freedom.}$$

This theorem is an immediate consequence of the definitions. The distribution of χ_1^2 is equivalent to the distribution of the sum of squares of ν_1 independent normally distributed random variables each with zero mean and unit variance. Hence, the random variable χ_1^2 can be written as

$$\chi_1^2 = x_1^2 + x_2^2 + \cdots + x_{\nu_1}^2$$

where the x's are independent normally distributed random variables each with zero mean and unit variance.

Similarly χ_2^2 can be written as

$$\chi_2^2 = y_1^2 + y_2^2 + \cdots + y_{\nu_2}^2$$

where the y's are independent normally distributed random variables each with zero mean and unit variance. Furthermore, the x's and the y's are independent since χ_1^2 and χ_2^2 are assumed independent.

The random variable $\chi^2 = \chi_1^2 + \chi_2^2$ can then be written as

$$\chi^2 = x_1^2 + x_2^2 + \cdots + x_{\nu_1}^2 + y_1^2 + y_2^2 + \cdots + y_{\nu_2}^2.$$

It is evident that χ^2 is just the sum of squares of $\nu_1 + \nu_2 = \nu$ independent normally distributed random variables each having zero mean and unit variance. Hence, it follows, from the definition, that the random variable χ^2 must have a chi-square distribution with ν degrees of freedom.

Since the reproductive property of the chi-square distribution holds for the sum of two independent random variables it must hold for the sum of any finite number of independent chi-square random variables.

4.2.3 THE DISTRIBUTION OF THE SAMPLE VARIANCE, s^2

The distribution of the sample mean has been discussed previously. This section will discuss the distribution of the sample variance, s^2, or more correctly, the distribution of a quantity related to this random variable.

Since

$$s^2 = \frac{\sum_{i=1}^{n} (x_i - \bar{x})^2}{n - 1}$$

is called the sample variance, it is natural to expect this random variable to be used as an estimate of the variance, σ^2, of a normal distribution when σ^2 is unknown. The estimation process can be considered as follows: the variance of a normal distribution is unknown, a random sample of n observations is drawn, the outcome of the random variable s^2 is observed, and this value is used as the estimate of σ^2. How good an estimate s^2 is depends upon the probability distribution of s^2. A measure of the closeness of s^2 to σ^2 is given by

$$\mathrm{P}\,(a \leqq s^2/\sigma^2 \leqq b)$$

for any a and b; i.e., it is given by the probability distribution of s^2.[*] It will be shown that the probability distribution of s^2/σ^2 is related to the chi-square distribution.

The particular random variable which has a chi-square distribution is $(n - 1)s^2/\sigma^2$. More precisely,

> *if $x_1, x_2, \cdots, x_n$ are independent normally distributed random variables each having mean μ and variance σ^2,*[†] *i.e., a random sample, the random variable $(n - 1)s^2/\sigma^2$ has a chi-square distribution with $n - 1$ degrees of freedom.*

This may appear to be startling since $(n - 1)s^2/\sigma^2$ is not directly the sum of squares of $(n - 1)$ independent normally distributed random variables each with zero mean and unit variance. However, this result implies that by a suitable change of variables, the random variable

$$\frac{(n - 1)s^2}{\sigma^2} = \frac{\sum_{i=1}^{n} (x_i - \bar{x})^2}{\sigma^2}$$

can be expressed as

$$\sum_{i=1}^{n-1} z_i^2$$

* s^2/σ^2 was used as a measure of closeness instead of $s^2 - \sigma^2$ because the probability distribution of s^2/σ^2 is easily obtained, whereas the probability distribution of $s^2 - \sigma^2$ is rather difficult to obtain.

† This implicitly assumes a structure where an experiment consisting of n trials is performed. x_i is the measurable quantity associated with the ith trial and is a random variable. The assumption about the form of the random variables then follows. This shortened notation will be used throughout the ensuing sections.

where the z's are independent normally distributed random variables each with zero mean and unit variance. This result is not easily proved and will be omitted. A heuristic argument follows: consider the random variable

$$\frac{(n-1)s^2}{\sigma^2} = \frac{(x_1 - \bar{x})^2 + \cdots + (x_n - \bar{x})^2}{\sigma^2} = \frac{\sum_{i=1}^{n} (x_i - \bar{x})^2}{\sigma^2}.$$

Note that if $\bar{x}$ were replaced by μ, forming

$$\frac{\sum_{i=1}^{n} (x_i - \mu)^2}{\sigma^2},$$

this function would have a chi-square distribution with n degrees of freedom since $(x_i - \mu)/\sigma$ are independent normally distributed each with zero mean and unit variance. The function $(n-1)s^2/\sigma^2$ can be broken up as follows:

$$\frac{(n-1)s^2}{\sigma^2} = \frac{\sum_{i=1}^{n} (x_i - \bar{x})^2}{\sigma^2} = \frac{\sum_{i=1}^{n} [(x_i - \mu) - (\bar{x} - \mu)]^2}{\sigma^2}$$

$$= \frac{\sum_{i=1}^{n} (x_i - \mu)^2}{\sigma^2} - \frac{n(\bar{x} - \mu)^2}{\sigma^2}$$

or

$$\frac{\sum_{i=1}^{n} (x_i - \mu)^2}{\sigma^2} = \frac{(n-1)s^2}{\sigma^2} + \left(\frac{\bar{x} - \mu}{\sigma/\sqrt{n}}\right)^2.$$

But

$$\frac{\sum_{i=1}^{n} (x_i - \mu)^2}{\sigma^2}$$

has a chi-square distribution with n degrees of freedom. Also, $\left(\frac{\bar{x} - \mu}{\sigma/\sqrt{n}}\right)^2$ has a chi-square distribution with 1 degree of freedom since $\frac{(\bar{x} - \mu)}{\sigma/\sqrt{n}}$ is a normally distributed random variable with zero mean and unit variance. Hence, it is not surprising that $(n-1)s^2/\sigma^2$ has a chi-square distribution with $n - 1$ degrees of freedom.

As an example of the use of this probability distribution, suppose a random sample of 10 observations is to be drawn where $x_1, x_2, \cdots, x_{10}$ are independent normally distributed random variables each with mean μ and variance σ^2. From Appendix Table 2,

$$\mathrm{P}\left(\frac{9s^2}{\sigma^2} \leqq 16.919\right) = 0.95$$

since $9s^2/\sigma^2$ has a chi-square distribution with 9 degrees of freedon
It will be seen in later chapters that the probability distribution o
$(n-1)s^2/\sigma^2$ can be used to get interval estimates of the varianc
when it is unknown. For example, the probability statement

$$\mathrm{P}\left(\frac{9s^2}{\sigma^2} \leqq 16.919\right) = 0.95$$

can be rewritten as

$$\mathrm{P}\left(\sigma^2 \geqq \frac{9s^2}{16.919}\right) = 0.95,$$

which is really a statement about the variance. Furthermore, thi
probability statement can be used to determine whether the varianc
is equal to a specified value σ_0^2. For example, if the variance is equa
to σ_0^2, then $9s^2/\sigma_0^2$ should usually be less than 16.919.

One further result about the random variable s^2 will be state
without proof.

If $x_1, x_2, \cdots, x_n$ are independent normally distributed random variables with the same mean and variance, the random variables $\bar{x}$ and s^2 are independently distributed.

This implies that the probability distribution of s^2 is unaffected b
announcing the outcome of the random variable $\bar{x}$ even though bot
$\bar{x}$ and s^2 are computed from the same random sample.

4.3 The *t* Distribution

4.3.1 THE *t* RANDOM VARIABLE

An experiment is performed which results in a sequence of trial
or observations. The conditions under which the observations ar
obtained need not necessarily be similar. For example, one observa
tion may consist of a Rockwell hardness reading whereas anothe
observation may consist of the life of a vacuum tube. On the othe
hand, the experiment may consist of determining the life of n vacuun
tubes, each tube manufactured and tested under very controlle
conditions (drawing a random sample). Associated with the experi
ment is the set, S, of all possible outcomes of the experiment
Suppose there exists a random variable (rule) which is normall
distributed with zero mean and unit variance. Denote this randon
variable by x. Suppose further that there exists another randon

variable which is independent of x and has a chi-square distribution with ν degrees of freedom. Denote this random variable by χ^2. Let $t = x\sqrt{\nu}/\sqrt{\chi^2}$ be a function of the random variables x and χ^2. As was pointed out in Section 2.3, random variables are just rules defined over S, so that any function of random variables must itself be a random variable. This is easily seen for the function t. Let R_x and R_{χ^2} be the sets of all values taken on by the random variable x and χ^2, respectively. The rule x assigns to each point in S a point in R_x and the rule χ^2 assigns to the same point in S a point in R_{χ^2}. Thus, for each point in S, there exists a point in each of R_x and R_{χ^2}. By letting $t = x\sqrt{\nu}/\sqrt{\chi^2}$ for each associated point in R_x and R_{χ^2} corresponding to a point in S, the function t assigns to each point in S a point in R, the set of all values of t. Thus, t must be a random variable, as must be any function of random variables. The set of all possible values of the random variable t is given by the interval minus infinity to plus infinity since the values of x lie in this interval and the values of χ^2 are non-negative. All that remains is to determine the probability distribution of t. This probability distribution is not arbitrary since the distributions of x and χ^2 are specified. The density function for the t random variable is written as

$$f(t) = \frac{1}{\sqrt{\pi\nu}} \frac{\Gamma\left(\frac{\nu+1}{2}\right)}{\Gamma\left(\frac{\nu}{2}\right)} \left(1 + \frac{t^2}{\nu}\right)^{-(\nu+1)/2} \qquad -\infty < t < \infty.$$

This probability distribution is known as a t distribution with ν degrees of freedom. The degrees of freedom are those associated with the chi-square component.

Physical phenomena whose probability distributions can be approximated by a t distribution can arise in nature, although this approximation is rarely made. The probability density is difficult to work with and simpler approximations are usually used. The most important applications of the t distributions arise from studying random variables which are generated in the manner previously described; i.e.,

if x is a normally distributed random variable with zero mean and unit variance, and χ^2 is a random variable which is independent of x and has a chi-square distribution with ν degrees of freedom, the random variable

$$t = x\sqrt{\nu}/\sqrt{\chi^2}$$

has a t distribution with ν degrees of freedom.

The usefulness of such random variables in making decisions about the mean of a normal distribution when the variance is unknown will be described in Section 4.3.2.

A sketch of the density function of the t random variable is given in Figure. 4.2.

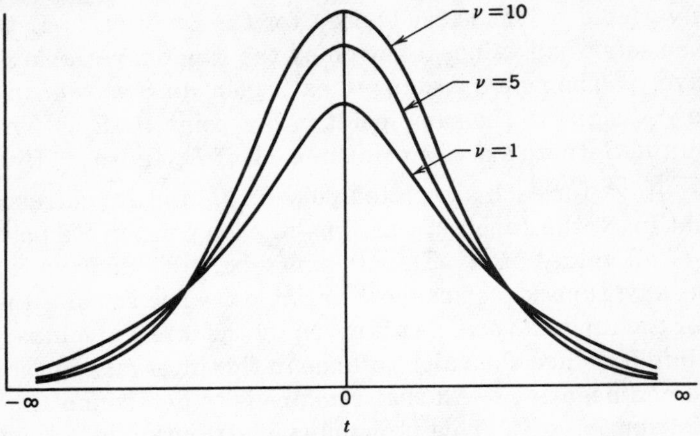

Fig. 4.2 Distribution function of the t random variable.

Like the normal distribution with zero mean, the t distribution is symmetric about zero. In fact, as the degrees of freedom, ν, tend toward infinity, the t distribution tends toward the standardized normal distribution. The expected value of the t random variable is given by $E(t) = 0$ for $\nu > 1$ and the variance of the t random variable is given by $\sigma_t^2 = \nu/(\nu - 2)$ for $\nu > 2$.

The probability that t is greater than a constant $t_{\alpha;\nu}$ is represented by

$$P(t \geqq t_{\alpha;\nu}) = \int_{t_{\alpha;\nu}}^{\infty} f(t)\,dt = \alpha.$$

This function has been tabulated and the values of the percentage point $t_{\alpha;\nu}$ can be found in Appendix Table 3.

For example, if $\nu = 5$,

$$P(t \geqq t_{0.05;5}) = P(t \geqq 2.015) = 0.05;$$

if $\nu = 29$,

$$P(t \geqq t_{0.05;29}) = P(t \geqq 1.699) = 0.05.$$

For $\nu \geqq 30$, $t_{\alpha;\nu}$ is close to K_α, the percentage point of the normal distribution, except for extreme values of α. Since the t distribution is symmetric about zero, values of $t_{\alpha;\nu}$ for $\alpha > 0.50$ can be obtained from the relation $t_{\alpha;\nu} = -t_{1-\alpha;\nu}$.

4.3.2 THE DISTRIBUTION OF $\frac{(\bar{x}-\mu)\sqrt{n}}{s}$.

In Chapter 3, the distribution of the random variable $(\bar{x}-\mu)\sqrt{n}/\sigma$ was shown to be normal with zero mean and unit variance provided the x's which comprise the sample mean are independent and normally distributed each with mean μ and variance σ^2. This random variable is useful for making decisions about μ when σ is known. For example, the probability statement

$$P\left(-1.96 \leqq \frac{(\bar{x}-\mu)\sqrt{n}}{\sigma} \leqq 1.96\right) = 0.95$$

can be rearranged to read

$$P\left(\frac{-1.96\sigma}{\sqrt{n}} \leqq \bar{x}-\mu \leqq \frac{1.96\sigma}{\sqrt{n}}\right) = 0.95.$$

Thus, the difference between the estimate, $\bar{x}$, and the quantity to be estimated, μ, lies between plus and minus $1.96\sigma/\sqrt{n}$ with probability 0.95. The statement has usefulness provided σ is known, usually from past data. However, frequently there is no record of past data so that σ is unknown. In this situation it is natural to replace σ by an estimate, s. Unfortunately, the resulting random variable

$$\frac{(\bar{x}-\mu)\sqrt{n}}{s}$$

is no longer normally distributed with zero mean and unit variance. The purpose of this section is to determine the distribution of this random variable.

If $x_1, x_2, \cdots, x_n$ are independent normally distributed random variables each having mean μ and variance σ^2, the random variable $(\bar{x}-\mu)\sqrt{n}/\sigma$ is normally distributed with zero mean and unit variance, and the random variable $(n-1)s^2/\sigma^2$ has a chi-square distribution with $n-1$ degrees of freedom (see Section 4.2.3). Furthermore, $\bar{x}$ and s^2 are independent random variables (see Section 4.2.3). In accordance with the definition of a t random variable given in Section 4.3.1

$$\frac{\dfrac{(\bar{x}-\mu)\sqrt{n}}{\sigma}\sqrt{n-1}}{\sqrt{\dfrac{(n-1)s^2}{\sigma^2}}} = \frac{(\bar{x}-\mu)\sqrt{n}}{s}$$

has a t distribution with $n-1$ degrees of freedom.

This random variable can be used to make decisions about μ when the standard deviation is unknown. For example, if a sample of size 10 is used, the probability statement

$$P\left(-t_{0.025;\,9} \leqq \frac{(\bar{x}-\mu)\sqrt{10}}{s} \leqq t_{0.025;\,9}\right)$$
$$= P\left(-2.262 \leqq \frac{(\bar{x}-\mu)\sqrt{10}}{s} \leqq 2.262\right) = 0.95$$

can be rearranged to read

$$P\left(\frac{-2.262s}{\sqrt{10}} \leqq \bar{x}-\mu \leqq \frac{2.262s}{\sqrt{10}}\right) = 0.95.$$

The 95% limits on $\bar{x}-\mu$ are now $\pm 2.262s/\sqrt{10}$ instead of $\pm 1.960\sigma/\sqrt{10}$, which were the limits when σ is known.

The distribution of $(\bar{x}-\mu)\sqrt{n}/s$ can also be used to make a decision about whether μ is equal to a specified value μ_0 when σ is unknown. If $\mu = \mu_0$, one would expect the random variable $(\bar{x}-\mu_0)\sqrt{n}/s$ to lie in the interval

$$(-t_{\alpha/2;\,n-1},\ t_{\alpha/2;\,n-1})$$

with probability $1-\alpha$. Thus, for $1-\alpha$ large, a value of this random variable falling outside this interval is an indication that μ is not equal to μ_0.

4.3.3 THE DISTRIBUTION OF THE DIFFERENCE BETWEEN TWO SAMPLE MEANS

Industrial experimentation is often concerned with comparing two treatments. A sample of n_x items is drawn using the standard treatment and a sample of n_y items is drawn using the new treatment. After experimentation, the sample means are judged according to the magnitude of their difference $\bar{x}-\bar{y}$. In order to make reasonable decisions about the effectiveness of these treatments, it is necessary to examine the distribution of $\bar{x}-\bar{y}$.

Let $x_1, x_2, \cdots, x_{n_x}$ be a random sample of n_x independent normally distributed random variables each having mean μ_x and variance σ^2; let $y_1, y_2, \cdots, y_{n_y}$ be a random sample of n_y independent normally distributed random variables each having mean μ_y and variance σ^2 (same as for the x's); and further let all the x's and y's be independent. The distribution of the random variable

$$\bar{x} = \sum_{i=1}^{n_x} x_i/n_x$$

is then normal with mean μ_x and variance σ^2/n_x; the distribution of the random variable

$$\bar{y} = \sum_{i=1}^{n_y} y_i/n_y$$

is also normal with mean μ_y and variance σ^2/n_y, and is independent of $\bar{x}$; the distribution of the random variable

$$\frac{(n_x - 1)s_x^2}{\sigma^2} = \frac{\sum_{i=1}^{n_x} (x_i - \bar{x})^2}{\sigma^2}$$

is chi-square with $n_x - 1$ degrees of freedom and is independent of $\bar{x}$ and $\bar{y}$; and finally, the distribution of the random variable

$$\frac{(n_y - 1)s_y^2}{\sigma^2} = \frac{\sum_{i=1}^{n_y} (y_i - \bar{y})^2}{\sigma^2}$$

is also chi-square with $n_y - 1$ degrees of freedom and is independent of $\bar{x}$, $\bar{y}$, and s_x^2.

Using the results of linear combinations of normally distributed variables, it is clear that $\bar{x} - \bar{y}$ is normally distributed with mean $\mu_x - \mu_y$ and variance $\sigma^2/n_x + \sigma^2/n_y$. Hence, the random variable

$$\frac{\bar{x} - \bar{y} - (\mu_x - \mu_y)}{\sigma\sqrt{\left(\frac{1}{n_x} + \frac{1}{n_y}\right)}}$$

is normally distributed with zero mean and unit variance. If the standard deviation σ is known, the problem is resolved. However, if σ is unknown, as in most practical situations, some further steps are necessary.

Using the reproductive property of the chi-square distribution, the random variable

$$\frac{(n_x - 1)s_x^2}{\sigma^2} + \frac{(n_y - 1)s_y^2}{\sigma^2}$$

has a chi-square distribution with $n_x + n_y - 2$ degrees of freedom, and furthermore is distributed independently of $\bar{x} - \bar{y}$. Hence, in accordance with the definition of a t random variable given in Section 4.3.1, the random variable

$$\frac{\dfrac{\bar{x} - \bar{y} - (\mu_x - \mu_y)}{\sigma\sqrt{\dfrac{1}{n_x} + \dfrac{1}{n_y}}}\sqrt{n_x + n_y - 2}}{\sqrt{[(n_x - 1)s_x^2 + (n_y - 1)s_y^2]/\sigma^2}}$$

$$= \frac{(\bar{x} - \bar{y}) - (\mu_x - \mu_y)}{\sqrt{\dfrac{(n_x - 1)s_x^2 + (n_y - 1)s_y^2}{n_x + n_y - 2}}\sqrt{\dfrac{1}{n_x} + \dfrac{1}{n_y}}}$$

has a t distribution with $n_x + n_y - 2$ degrees of freedom. Note that this random variable does not contain σ.

4.4 The F Distribution

4.4.1 THE F RANDOM VARIABLE

An experiment is performed which consists of a sequence of trials. Associated with the experiment is the set, S, of all possible outcomes of the experiment. Suppose there exists a random variable (rule) which has a chi-square distribution with ν_1 degrees of freedom. Denote this random variable by χ_1^2. Suppose further that there exists another random variable which is independent of χ_1^2 and has a chi-square distribution with ν_2 degrees of freedom. Denote this random variable by χ_2^2. Let

$$F = \frac{\chi_1^2/\nu_1}{\chi_2^2/\nu_2}$$

be a function of the random variables χ_1^2 and χ_2^2; F is also a random variable. The set of all possible values of the random variable F is given by the interval zero to plus infinity since the values of χ_1^2 and χ_2^2 are all non-negative. All that remains is to determine the probability distribution of F. This probability distribution is not arbitrary since the distributions of χ_1^2 and χ_2^2 are specified. The density function of the F random variable is written as

$$f(F) = \frac{\Gamma\left(\frac{\nu_1 + \nu_2}{2}\right)\nu_1^{\nu_1/2}\nu_2^{\nu_2/2}}{\Gamma\left(\frac{\nu_1}{2}\right)\Gamma\left(\frac{\nu_2}{2}\right)} \frac{(F)^{(\nu_1/2)-1}}{(\nu_2 + \nu_1 F)^{(\nu_1+\nu_2)/2}} \quad \text{for } F > 0$$

$$= 0 \quad \text{otherwise.}$$

This probability distribution is known as an F distribution with ν_1 and ν_2 degrees of freedom.

Physical phenomena whose probability distribution can be approximated by an F distribution can arise in nature. In particular, the distribution of the tangent of an angle is often approximated by a distribution related to the F. However, the most important applications of this distribution arise from studying random variables which are generated in the manner previously described; i.e.,

if χ_1^2 and χ_2^2 are independent random variables which have chi-square distributions with ν_1 and ν_2 degrees of freedom, respectively, the random variable

$$F = \frac{\chi_1^2/\nu_1}{\chi_2^2/\nu_2},$$

has an F distribution with ν_1 and ν_2 degrees of freedom.

The usefulness of such random variables in making decisions about the ratio of two sample variances will be described in Section 4.4.2.

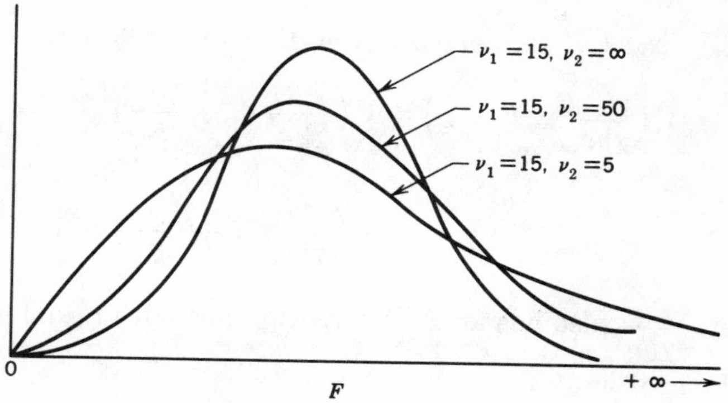

Fig. 4.3 Density of the F random variable.

A sketch of the density function of the F random variable is given in Figure 4.3. Like the chi-square distribution, this distribution is skewed, with probability mass distributed to the right of the origin only. The expected value of the F random variable is given by $E(F) = \nu_2/(\nu_2 - 2)$ for $\nu_2 > 2$ and the variance of the F random variable is given by

$$\sigma_F^2 = \frac{\nu_2^2(2\nu_2 + 2\nu_1 - 4)}{\nu_1(\nu_2 - 2)^2(\nu_2 - 4)} \quad \text{for } \nu_2 > 4.$$

The probability that F is greater than a constant is represented by

$$\text{P}\left(F > F_{\alpha;\,\nu_1,\nu_2}\right) = \int_{F_{\alpha;\,\nu_1,\nu_2}}^{\infty} f(F)\,dF = \alpha.$$

Since the F distribution depends on the two parameters ν_1 and ν_2, a three-way table is needed to tabulate the value of F corresponding to different probabilities and values of ν_1 and ν_2. Appendix Table 4 is such a table, where percentage points are given for values of $\alpha \leqq 50\%$, i.e., the shaded area in Figure 4.4.

If the percentage points for values of $\alpha > 50\%$ are desired, they may be obtained by using the following equality:

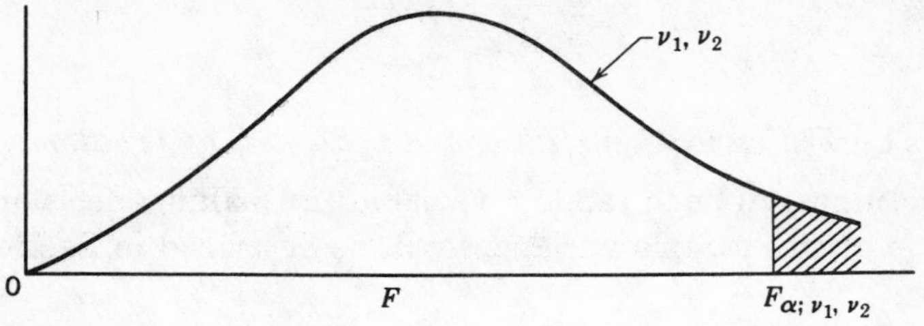

Fig. 4.4 The percentage points for the F distribution.

$$P\left(\frac{\chi_1^2/\nu_1}{\chi_2^2/\nu_2} \geqq F_{\alpha;\nu_1,\nu_2}\right) = P\left(\frac{\chi_2^2/\nu_2}{\chi_1^2/\nu_1} < \frac{1}{F_{\alpha;\nu_1,\nu_2}}\right) = \alpha$$

or

$$P\left(\frac{\chi_2^2/\nu_2}{\chi_1^2/\nu_1} \geqq \frac{1}{F_{\alpha;\nu_1,\nu_2}}\right) = 1 - \alpha.$$

However, $\dfrac{\chi_2^2/\nu_2}{\chi_1^2/\nu_1}$ also has an F distribution but with ν_2 and ν_1 degree of freedom so that

$$P\left(\frac{\chi_2^2/\nu_2}{\chi_1^2/\nu_1} \geqq F_{1-\alpha;\nu_2,\nu_1}\right) = 1 - \alpha.$$

Hence, $F_{1-\alpha;\nu_2,\nu_1} = \dfrac{1}{F_{\alpha;\nu_1,\nu_2}}$

or $$F_{\alpha;\nu_1,\nu_2} = \frac{1}{F_{1-\alpha;\nu_2,\nu_1}}.$$

For example, the percentage point for $\alpha = 0.95$, $\nu_1 = 6$, and $\nu_2 = 8$ $F_{0.95;6,8}$, can be obtained as follows:

$$F_{0.95;6,8} = \frac{1}{F_{0.05;8,6}} = \frac{1}{4.15} = 0.241.$$

4.4.2 The Distribution of the Ratio of Two Sample Variance

A problem arising in industrial experimentation is to compare th variability of two processes. A sample is drawn of n_x items usin process x, and a sample is drawn of n_y items using process y. Th sample variances for both processes are compared, i.e., s_x^2/s_y^2. If thi ratio is close to unity, the variabilities are judged to be nearl equivalent; otherwise not. In order to make good decisions, and t quantify the statement "close to unity," it is necessary to examin the distribution of s_x^2/s_y^2.

Let $x_1, x_2, \cdots, x_{n_x}$ be a random sample of n_x independent normall distributed random variables each having mean μ_x and variance σ_x^2

Let $y_1, y_2, \cdots, y_{n_y}$ be a random sample of n_y independent normally distributed random variables each having mean μ_y and variance σ_y^2; and further let all the x's and y's be independent. The distribution of

$$\frac{(n_x - 1)s_x^2}{\sigma_x^2} = \frac{\sum_{i=1}^{x} (x_i - \bar{x})^2}{\sigma_x^2}$$

is then chi-square with $n_x - 1$ degrees of freedom (see Section 4.2.3). Similarly, the distribution of

$$\frac{(n_y - 1)s_y^2}{\sigma_y^2} = \frac{\sum_{i=1}^{n_y} (y_i - \bar{y})^2}{\sigma_y^2}$$

is also chi-square with $n_y - 1$ degrees of freedom. Furthermore, these two chi-square random variables are independent since the x's and y's are independent. Hence, in accordance with the definition of the F random variable given in Section 4.4.1, the random variable

$$\frac{\dfrac{(n_x - 1)s_x^2}{\sigma_x^2} \Big/ (n_x - 1)}{\dfrac{(n_y - 1)s_y^2}{\sigma_y^2} \Big/ (n_y - 1)} = \frac{s_x^2/\sigma_x^2}{s_y^2/\sigma_y^2}$$

has an F distribution with $n_x - 1$ and $n_y - 1$ degrees of freedom.

*4.5 The Binomial Distribution

*4.5.1 THE BINOMIAL RANDOM VARIABLE

Consider an experiment which consists of n independent trials. At each trial let p be the probability that an event will occur. The probability that r, the number of events occurring in the n trials, is equal to r_1 is given by

$$\mathrm{P}(r = r_1) = \binom{n}{r_1} p^{r_1} q^{n-r_1} \qquad \text{for } r_1 = 0, 1, 2, \cdots, n,$$

where $q = 1 - p$ and $\binom{n}{r_1}$ is the number of combinations of n things taken r_1 at a time. By definition

$$\binom{n}{r_1} = \frac{n!}{r_1!\,(n - r_1)!}$$

and $$n! = n(n-1)(n-2)\cdots 3 \cdot 2 \cdot 1.$$

The expected value of r is given by $E(r) = np$ and the variance of r is given by $\sigma_r^2 = npq$.

A random variable related to the binomial random variable is the fraction of successful events occurring in the n trials, i.e., r/n. The probability distribution of r/n can be obtained from the binomial as follows:

$$P\left(\frac{r}{n} \leqq b\right) = P(r \leqq bn)$$

$$= \sum_{r=0}^{(bn)} \binom{n}{r} p^r q^{n-r}$$

where (bn) is the largest integer less than or equal to bn.

The moments of the random variable r/n are easily obtained using the results of Section 2.10. The expected value of r/n is given by

$$E\left(\frac{r}{n}\right) = \frac{1}{n} E(r) = p$$

and the variance of r/n is given by

$$\sigma_{r/n}^2 = \frac{1}{n^2}\sigma_r^2 = \frac{pq}{n}.$$

The most common application of the binomial theorem in industrial work is in lot-by-lot acceptance inspection where the lot is large compared to the sample size;[1] p is the fraction of defectives in the lot, n is the size of a sample drawn at random from the lot, and r is the observed number of defectives.

The observed number of defectives is some number from 0 to n and hence

$$P(0 \leqq r \leqq n) = \sum_{r=0}^{n} \binom{n}{r} p^r q^{n-r} = 1.$$

As an example, suppose $n = 18$ and $p = 0.10$, the probability of obtaining r defectives, $r = 0, 1, 2, \cdots, 18$, is as follows:

r	*Prob. of r defects*
0	0.150
1	0.300
2	0.284
3	0.168
4	0.070
5	0.022
6	0.005
7	0.001
8	0.000
⋮	⋮
18	0.000

[1] For the binomial distribution to be applicable, the probability of a defective must be constant from trial to trial, and the trials must be independent events. If the lot size is large compared to the sample size, these assumptions are satisfied from a practical point of view.

The distribution function of this binomial distribution is plotted in Figure 4.5.

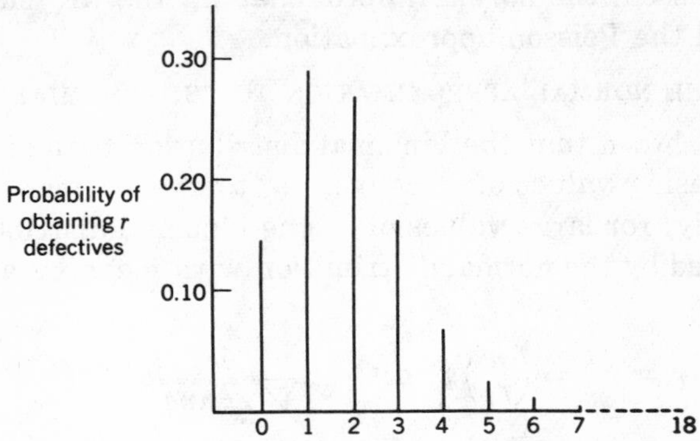

Fig. 4.5 Distribution function for the binomial distribution with $n = 18$ and $p = 0.10$.

*4.5.2 Tables of the Binomial Probability Distribution

The binomial distribution has been tabulated for $n = 1(1)\ 49$ and $p = 0.01\ (0.01)\ 0.50$ in *Tables of the Binomial Probability Distribution*, National Bureau of Standards, Applied Mathematics Series, 6, Washington, 1950. Harry Romig, in a book entitled *50–100 Binomial Tables*, John Wiley & Sons, Inc., New York, 1953, covers the range of n values, $n = 50(5)100$, and the range of p values, $p = 0.01(0.01)0.99$. The most extensive tabulation of the binomial distribution appears in tables of the *Cumulative Binomial Probability Distribution*, Harvard University Press., Cambridge, 1955. This book covers the parameters

$$n = 1(1)50(2)100(10)200(20)500(50)1000$$

and

$$p = 0.01(0.01)0.50,$$

$$\frac{1}{16}, \frac{1}{12}, \frac{1}{8}, \frac{1}{6}, \frac{3}{16}, \frac{5}{16}, \frac{1}{3}, \frac{3}{8}, \frac{5}{12}, \frac{7}{16}.$$

Other values may be obtained by use of the identity

$$P\,(r \leqq r_1) = \sum_{r=0}^{r_1} \binom{n}{r} p^r q^{n-r} = 1 - P\left(F < \frac{(n - r_1)}{(r_1 + 1)} \frac{p}{q}\right)$$

where F is a random variable having the F distribution with $\nu_1 = 2(r_1 + 1)$ and $\nu_2 = 2(n - r_1)$ degrees of freedom, though the usefulness of the identity is restricted by the tables available. For values of the probability distribution outside of this range, the computations may become tedious. Consequently, some form of approxi mation to the distribution is useful. Three types of approximation will be discussed: the normal approximation, the arc sine transfor mation, and the Poisson approximation.

*4.5.3 The normal approximation to the binomial

It can be shown that the binomial distribution for a given value o p and increasing values of n converges to the normal distribution Consequently, for large values of n, the binomial distribution can be approximated by the normal distribution with mean np and variance npq, i.e.,

$$P(r = r_1) = \binom{n}{r_1} p^{r_1} q^{n-r_1} \simeq \frac{1}{\sqrt{2\pi npq}} e^{-(r_1-np)^2/2npq}.$$

In most applications, the cumulative sum is required rather than the single terms, i.e.,

$$P(r_1 \leqq r \leqq r_2) = \sum_{r=r_1}^{r_2} \binom{n}{r} p^r q^{n-r}.$$

This expression can be approximated by

$$P(r_1 \leqq r \leqq r_2) = \sum_{r=r_1}^{r_2} \binom{n}{r} p^r q^{n-r} \simeq \int_{r_1}^{r_2} \frac{1}{\sqrt{2\pi npq}} e^{-(r-np)^2/2npq}\, dr.$$

However, because of the change from a discontinuous to a continuous distribution, a better approximation is given by

$$P(r_1 \leqq r \leqq r_2) = \sum_{r=r_1}^{r_2} \binom{n}{r} p^r q^{n-r} \simeq \int_{r_1-\frac{1}{2}}^{r_2+\frac{1}{2}} \frac{1}{\sqrt{2\pi npq}} e^{-(r-np)^2/2npq}\, dr$$

$$= \int_{(r_1-\frac{1}{2}-np)/\sqrt{npq}}^{(r_2+\frac{1}{2}-np)/\sqrt{npq}} \frac{1}{\sqrt{2\pi}} e^{-z^2/2}\, dz.$$

No exhaustive study has been made of the accuracy of the normal approximation. However, the approximation is known to be poor for $p < 1/(n+1)$ or $p > n/(n+1)$ and outside the interval plus or minus 3 times the standard deviation about the mean, i.e.,

$$r < np - 3\sqrt{npq} \quad \text{or} \quad r > np + 3\sqrt{npq}.$$

The approximation is known to be good for p close to $\frac{1}{2}$, and Hald[1] indicates that the approximation is good when the range of application is limited by the inequality $npq > 9$.

[1] A. Hald, *Statistical Theory with Engineering Applications*, John Wiley & Sons, Inc., New York, 1952.

A similar normal approximation holds for the random variable r/n, the fraction of successes in n trials. r/n is approximately normally distributed with mean p and variance pq/n. Hence, making the correction for continuity

$$P\left(a \leqq \frac{r}{n} \leqq b\right) \simeq \int_{a-\frac{1}{2n}}^{b+\frac{1}{2n}} \frac{\sqrt{n}}{\sqrt{2\pi pq}} e^{-n(x-p)^2/2pq}\, dx$$

$$= \int_{\sqrt{n}\,(a-\frac{1}{2n}-p)/\sqrt{pq}}^{\sqrt{n}\,(b+\frac{1}{2n}-p)/\sqrt{pq}} \frac{1}{\sqrt{2\pi}} e^{-z^2/2}\, dz$$

*4.5.4 THE ARC SINE TRANSFORMATION

It is clear that both the mean and the variance of the binomial distribution depend upon p. In most practical problems, p is never known, and there is some advantage in finding a transformation such that at least the variance of the transformed variable is independent of p. Let

$$h = \frac{r}{n} \quad \text{be the fraction of successes in } n \text{ trials}$$

and

$$x = 2 \text{ arc sin } \sqrt{h}.$$

If h is approximately normally distributed with mean p and variance pq/n, then $x = 2 \text{ arc sin } \sqrt{h}$ is also approximately normally distributed with mean $2 \text{ arc sin } \sqrt{p}$ and variance $1/n$. Hence,

$$P(h_1 \leqq h \leqq h_2) \simeq \int_{2 \text{ arc sin } \sqrt{h_1}}^{2 \text{ arc sin } \sqrt{h_2}} \frac{\sqrt{n}}{\sqrt{2\pi}} e^{-n(h-2 \text{ arc sin } \sqrt{p})^2/2}\, dh.$$

Again, because of the change from a discontinuous to a continuous distribution, a better approximation is given by

$$P(h_1 \leqq h \leqq h_2) \simeq \int_{2 \text{ arc sin } \sqrt{h_1-\frac{1}{2n}}}^{2 \text{ arc sin } \sqrt{h_2+\frac{1}{2n}}} \frac{\sqrt{n}}{\sqrt{2\pi}} e^{-n(h-2 \text{ arc sin } \sqrt{p})^2/2}\, dh$$

$$= \int_{\sqrt{n}\,(2 \text{ arc sin } \sqrt{h_1-\frac{1}{2n}}-2 \text{ arc sin } \sqrt{p})}^{\sqrt{n}\,(2 \text{ arc sin } \sqrt{h_2+\frac{1}{2n}}-2 \text{ arc sin } \sqrt{p})} \frac{1}{\sqrt{2\pi}} e^{-h^2/2}\, dh.$$

There has been no systematic study of this approximation formula, but some scattered results indicate that it is as good as the normal approximation.

The approximate probability distribution of r, the number of successes in n trials is obtained from the relation

$$P(r_1 \leqq r \leqq r_2) = P\left(\frac{r_1}{n} \leqq h \leqq \frac{r_2}{n}\right).$$

*4.5.5 The Poisson Approximation to the Binomial

For fixed p and increasing n, the binomial distribution converges to the normal distribution. If, as n increases, p approaches zero in such a manner that $np = \lambda$ becomes and remains constant, the binomial distribution converges to the Poisson distribution. The probability distribution of the Poisson random variable is given by

$$P(r = r_1) = \frac{e^{-\lambda}\lambda^{r_1}}{(r_1)!} \quad \text{for} \quad r_1 = 0, 1, 2, \cdots$$

The mean and variance of the Poisson random variable are both equal to λ.

The convergence of the binomial distribution to the Poisson distribution is shown as follows: for n sufficiently large, p approaches zero in such a manner that $np = \lambda$ becomes and remains constant. Hence

$$\binom{n}{r_1} p^{r_1} q^{n-r_1} = \frac{n(n-1)\cdots(n-r_1+1)}{r_1!} \frac{\lambda^{r_1}}{n^{r_1}} \left(1 - \frac{\lambda}{n}\right)^{n-r_1}$$
$$= \left\{\left(1 - \frac{1}{n}\right)\left(1 - \frac{2}{n}\right) \cdots \left(1 - \frac{r_1 - 1}{n}\right)\left(1 - \frac{\lambda}{n}\right)^{-r_1}\right\}$$
$$\frac{\lambda^{r_1}}{r_1!}\left(1 - \frac{\lambda}{n}\right)^n.$$

As $n \to \infty$, the terms in braces approach 1, and $\left(1 - \frac{\lambda}{n}\right)^n$ approaches $e^{-\lambda}$. Hence

$$\binom{n}{r_1} p^{r_1} q^{n-r_1} \to \frac{e^{-\lambda}\lambda^{r_1}}{(r_1)!}.$$

This approximation to the binomial is good when p is small and n is large. It is generally considered justifiable to use the Poisson approximation when $p < 0.1$.

PROBLEMS

1. Referring to Section 4.1, find the probability that the sample average shear strength of test spot welds differs from the mean of the distribution by less than 1 pound, i.e., $P(-1.0 \leqq \bar{x} - \mu \leqq 1.0)$, when the individual shear strengths of the spot welds are normally distributed with mean μ and standard deviation 10 pounds, and a sample of nine is observed.

2. In Problem 1, how many observations are required so that

$$P(-1.0 \leq \bar{x} - \mu \leq 1.0) = 0.95\,?$$

3. Referring to Section 4.1, write down the expression for the sample size in terms of σ, $K_{\alpha/2}$, and b so that

$$P(-b \leq \bar{x} - \mu \leq b) = 1 - \alpha\,.$$

4. Verify the 5% point of the chi-square distribution for $\nu = 2$ by direct integration of the density function.

5. Show that $E(\chi^2) = \nu$ and $\sigma^2_{\chi^2} = 2\nu$.

6. If χ^2 has a chi-square distribution with 8 degrees of freedom, find $P(\chi^2 \leq 8)$, $P(\chi^2 \geq 6)$, and $P(4 \leq \chi^2 \leq 9)$.

7. If χ^2 has a chi-square distribution with 20 degrees of freedom, find $P(\chi^2 \leq 20)$, $P(\chi^2 \geq 15)$, and $P(16 \leq \chi^2 \leq 21)$.

8. In the example on drilling a hole in a rectangular table top given in Section 4.2.1, what is the probability that the distance between the measured point and desired location is less than 3 units?

9. In the bombing problem given in Section 4.2.1, what is the probability that the distance between the point of impact and the target exceeds 5 feet if $\sigma^2 = 5$?

10. If χ_1^2 has a chi-square distribution with 8 degrees of freedom and χ_2^2 has an independent chi-square distribution with 3 degrees of freedom, find the probability that $\chi_1^2 + \chi_2^2$ exceeds 12.

11. If χ_1^2 has a chi-square distribution with 4 degrees of freedom and χ_2^2 has an independent chi-square distribution with 5 degrees of freedom, find the probability that $\chi_1^2 + \chi_2^2$ exceeds 11.

12. Find $P(0.462 \leq s^2/\sigma^2 \leq 1.88)$ when s^2 is based upon a random sample of 10 observations on a normally distributed random variable with unknown mean μ and unknown variance σ^2.

13. Referring to Problem 12, how many observations are required to insure that $P(0.462 \leq s^2/\sigma^2 \leq 1.88) \geq 0.95$?

14. The "on" temperature of a thermostatically controlled switch is normally distributed with unknown mean and variance. A random sample is to be drawn and the sample variance computed. How many observations are required to insure that $P(s^2/\sigma^2 \leq 1.83) \geq 0.95$?

15. The sample standard deviation s_1 of the resistance of a certain electrical part is to be computed from a sample of eight parts. The sample standard deviation of another component s_2 which is added to the first making the final assembly, is also computed from a sample of eight parts. It is known that each component is independent and normally distributed with common unknown variance σ^2. An estimate of the variance of the final assembly will be given by $s^2 = s_1^2 + s_2^2$. What is the probability that s^2/σ^2 will be less than 3?

16. In Problem 15, how many observations on each component will be required to insure that $P(s^2/\sigma^2 \leq 3) = 0.95$?

17. Show that $\sigma^2_{r^2} = 2\sigma^4/n$, where $r^2 = \sum_{i=1}^{n} (x_i^2)/n$ and x is normally distributed with zero mean and variance σ^2. *Hint*: Use the relation that $\sigma^2_{\chi^2}=2\nu$.

18. Find $E(s^2)$ and $\sigma^2_{s^2}$, where $s^2 = \sum (x_i - \bar{x})^2/(n-1)$ and x is normally distributed with mean μ and variance σ^2. *Hint*: Use the relations $E(\chi^2)=\nu$ and $\sigma^2_{\chi^2}=2\nu$.

19. Prove that the density function of the t variate approaches the density function of the standardized normal random variable as the number of degrees of freedom ν becomes infinite. Assume that the constant approaches $1/\sqrt{2\pi}$.

20. Verify the 2.5% point of the t distribution for $\nu=1$ by direct integration of the density function.

21. Show that $E(t)=0$ for $\nu>1$ and $\sigma_t^2=\nu/(\nu-2)$ for $\nu>2$.

22. If t has a t-distribution with 8 degrees of freedom, find

$$P(t \geq \tfrac{4}{3}), \qquad P(-\tfrac{8}{3}<t),$$

and

$$P(-3.355 \leq t \leq 3.355).$$

23. If t has a t-distribution with 20 degrees of freedom, find

$$P(t \geq \tfrac{10}{9}), \qquad P(-\tfrac{20}{9} \leq t),$$

and

$$P(-2.845 \leq t \leq 2.845).$$

24. Find $P\{-1 \leq [(\bar{x}-\mu)\sqrt{10}]/s\}$ where $\bar{x}$ and s are based upon 10 observations.

25. Find $P\{-1 \leq [(\bar{x}-\mu)\sqrt{10}]/s\}$ where $\bar{x}$ is based upon 10 observations and s upon 20 different observations.

26. Show that $t^2_{\alpha/2;\,\nu}=F_{\alpha;\,1,\nu}$; i.e., show $P(t^2 \geq t^2_{\alpha/2;\,\nu}) = P(F \geq F_{\alpha;\,1,\nu})$ where $t^2_{\alpha/2;\,\nu}=F_{\alpha;\,1,\nu}$. *Hint*: Use the definitions of the t and F random variables.

27. Show that $E(F)=\nu_2/(\nu_2-2)$ for $\nu_2>2$, and

$$\sigma_F^2=\frac{\nu_2^2(2\nu_2+2\nu_1-4)}{\nu_1(\nu_2-2)^2(\nu_2-4)} \quad \text{for} \quad \nu_2>4.$$

28. Find the value of $F_{0.05;8,9}$ such that $P(F \geq F_{0.05;8,9})=0.05$ where F is a random variable having an F distribution with 8 and 9 degrees of freedom.

29. If F is a random variable having an F distribution with 3 and 6 degrees of freedom, find $P(F \geq 1.5)$, $P(F \geq 2)$, and $P(F \leq 3)$.

30. Find the value of $F_{0.95;9,8}$, $F_{0.05;8,9}$, and $F_{0.90;3,12}$.

31. The variability of two treatments is to be compared. A random sample of 10 is to be obtained from each treatment and the sample variance computed. Find

$$P\left(\frac{s_1^2/\sigma_1^2}{s_2^2/\sigma_2^2} \geq 2.75\right) = P\left(\frac{\sigma_2^2}{\sigma_1^2} \geq 2.75\frac{s_2^2}{s_1^2}\right).$$

32. If $z=\sum_{i=1}^{n}(x_i-\bar{x})^2 \Big/ \sum_{i=1}^{n}(y_i-\bar{y})^2$, where x_i is distributed normally with mean μ_x and variance σ_x^2 and y_i is distributed normally and independent of x with mean μ_y and variance σ_y^2, find $E(z)$.

33. If $x_1, x_2, \cdots, x_n$ are independent normally distributed random variables with mean μ_x and standard deviation σ_x, and $y_1, y_2, \cdots, y_m$ are also independent normally distributed random variables with mean μ_y and standard deviation σ_y (x's and y's are also independent of each other), state whether the following random variables are normally distributed (N), t distributed (t), chi-square distributed (χ^2), F distributed (F), or none of the above (0):

(a) $\dfrac{(\bar{x}-\mu_x)\sqrt{n}}{s_x}$

(b) $\dfrac{(x_1-\mu_x)}{s_x}$

(c) $\dfrac{x_1-\mu_x}{\sigma_x}$

(d) $\dfrac{(n-1)s_x^2}{\sigma_x^2}+\dfrac{\sum_{i=1}^{m}(y_i-\bar{y})^2}{\sigma_y^2}$

(e) s_x^2/s_y^2

(f) $\bar{x}/\bar{y}$

(g) $\sum_{i=1}^{n} x_i^2/\sigma_x^2$

(h) $\dfrac{\bar{x}-\bar{y}}{\sqrt{(\sigma_x^2/n)+(\sigma_y^2/m)}}$

(i) $\dfrac{\bar{x}\sqrt{n}}{s_x}$

34. If the probability of an item being defective is 0.1, what is the probability of obtaining exactly 0, 1, 2, and 3 defectives in a sample of size 3? What is the probability of obtaining less than 2 defectives?

35. If the probability of an item being defective is 0.05, use the Poisson approximation to obtain the probability that a sample of size 50 will contain at least three defectives; exactly zero defectives.

36. Solve Problem 35 using the normal approximation.

37. Solve Problem 35 using the arc sine transformation.

38. Show that the mean of a binomial distribution is np and the variance is npq.

CHAPTER V

Significance Tests

5.1 Introduction

It has already been pointed out that statistics is a science which deals with making decisions in the face of uncertainty. For example, an experimenter may wish to determine whether (1) a new method of sealing vacuum tubes will increase their life, (2) a new alloy will have an increased breaking strength, (3) a new source of raw material has resulted in a change in the variability of output, (4) a special treatment of concrete will have an effect on breaking strength, (5) the average Rockwell hardness of a certain type of material is a fixed value on the B scale, etc. Each of the examples above calls for the making of a decision based upon inferences drawn from a sample. The " uncertainty " is reflected in the fact that the same experiment performed under the same conditions a second time will usually give different sample results. Consequently, if the experiment is performed only once, the decision will be a function of which sample is obtained.

It is important to recognize that statistics is concerned with problems having elements of uncertainty. An experiment is performed, random variables are chosen, and decisions are based upon the values that these random variables take on. Statistics deals with finding procedures, optimum procedures, on which decisions are based. *Procedures are rules of action which are specified before the experiment is performed. All possible values of the chosen random variable are listed together with the decision associated with each outcome.* For example, in the Rockwell hardness problem, suppose that a company research laboratory is to determine whether the average Rockwell hardness of a certain steel is 72 on the B scale. The special steel is to be used in a new product. From long experience in the field the company feels that if the average Rockwell hardness is less than 72, too many units produced will fail, whereas if the average Rockwell hardness exceeds 72, the material is generally difficult to work with.

Several specimens are to be measured. It is recognized that there is a certain amount of variability in all the specimens so that the Rockwell hardness is not a constant. In fact, the Rockwell hardness is a random variable. The company is interested in the expected value of this random variable which, using the customary notion of an expectation, is equivalent to the average Rockwell hardness of *all* the special steel to be used in the manufacture of this new product.[1] A completely specified procedure leading to a decision is to draw a random sample of size 9 and look at the random variable the sample mean, $\bar{X}$. If

$$70.7 \leqq \bar{X} \leqq 73.3,$$

conclude that the average Rockwell hardness is 72. If $\bar{X}$ is not in this interval, conclude that the average Rockwell hardness is not 72. Note that for this procedure the random variable chosen is $\bar{X}$. The sample size chosen is 9. Associated with every value the random variable takes on is a decision, i.e., yes or no. It is very important to recognize that all of this is done before the experiment is performed, before any observations are taken. The value of $n = 9$ and the end points of the interval [70.7, 73.3] are not chosen arbitrarily, but are integrated into the procedure. The choice of these parameters will be discussed in detail later.

There are many other procedures that can be used. To name a few:

Draw a sample of size 9. Observe the random variable, the sample median $\tilde{X}$. If $70 \leqq \tilde{X} \leqq 74$ conclude that the average Rockwell hardness equals 72. Otherwise conclude that it is not 72.

Draw a sample of size 5. Look at the largest observation, $X_{\max}$. If $70 \leqq X_{\max} \leqq 74$, conclude the average Rockwell hardness equals 72. Otherwise conclude that it is not 72.

These are but three of the possible procedures.

The techniques discussed in the ensuing section will provide a basis for the selection of a particular procedure.

[1] If the notion of an expectation is to be equivalent to the average Rockwell hardness of all the special steel to be used in the manufacture of this new product, it is necessary for the probability distribution of Rockwell hardness to remain constant over time. If not, the notion of an expectation is equivalent only to the average Rockwell hardness of the steel on hand at this instant, or until the change occurs.

5.2 The Operating Characteristic Curve

In the previous section, several procedures were presented for determining whether the average Rockwell hardness was 72 on the B scale. These procedures were examples drawn from an infinite number of such procedures, and the question still remains as to which one to choose. It is evident that the rule to choose is that one which leads to the correct conclusion most of the time; i.e., which indicates the average Rockwell hardness is 72 when it really is 72, and which indicates the average Rockwell hardness is not 72 when in reality it is some other number. Since decisions depend upon random variables, which by their nature have probability distributions associated with them, it is rather unusual to be able to make the correct decision *always*.[1]

Generally, there are two types of incorrect decisions that can be made. If the average Rockwell hardness is actually 72, the decision rule of the experimenter can lead to the conclusion that it is not 72. This error is known as the Type I error and the probability of its occurrence is denoted by α. It is also often called the *level of significance*. If the average Rockwell hardness is actually some number other than 72, say 75, the decision rule of the experimenter can lead to the conclusion that it is 72. This error is known as the Type II error and the probability of its occurrence is denoted by β.

To summarize then, the decision about the average Rockwell hardness depends upon the result of a random variable, e.g., if one of the three procedures above are used, the decision is based upon the values the random variables $\overline{X}$, $\tilde{X}$, or X_{max} take on, depending on which rule is chosen. Whichever procedure is chosen, it has the property that it leads to the conclusion that the average Rockwell hardness is 72 when the value taken on by the random variable falls within specified limits. These limits will be referred to as an "acceptance region."[2] Thus, a decision procedure has associated with it a random variable, a sample size, and an acceptance region (which may or may not be

[1] Correct decisions can always be made in the following type of situation. If there were twenty steel specimens and the experimenter were interested in determining whether the average Rockwell hardness of these twenty was equal to 72 (as opposed to the average Rockwell hardness of *all* the steel to be used in the manufacture of the new product), he could measure all the specimens and compute the average. In this case the procedure which concludes that the average Rockwell hardness is 72 when the average of the twenty specimens is 72 (assuming no measurement error) will always lead to the correct conclusion. However, this is a rather pathological example, and not customary in practical situations.

[2] Standard statistical texts refer to the complement of the acceptance region, the region of rejection, as the critical region.

an interval). The probability of the random variable falling outside the acceptance region when the true average Rockwell hardness is 72 is the magnitude of the *error* of Type I (α). The probability of the random variable falling inside the acceptance region when the true average Rockwell hardness is some number other than 72 is the magnitude of the error of Type II (β).

The error of Type II is not a constant then but depends upon the true state of the average Rockwell hardness. If μ denotes the true average Rockwell hardness, then $\beta(\mu)$ is a better notation than the symbol β for the magnitude of the Type II error. In other words, β is a function (in the mathematical sense) of μ. $\beta(\mu)$ can be calculated by finding the probability that the proper random variable falls within the acceptance region when the true average Rockwell hardness is μ. This is equivalent to finding the probability of saying the average Rockwell hardness is 72 when in reality μ is some value other than 72. If a plot is made of the probability of this event occurring as a function of the true average Rockwell hardness, the resulting plot is called an operating characteristic (OC) curve. A sketch of such an OC curve is given in Figure 5.1.

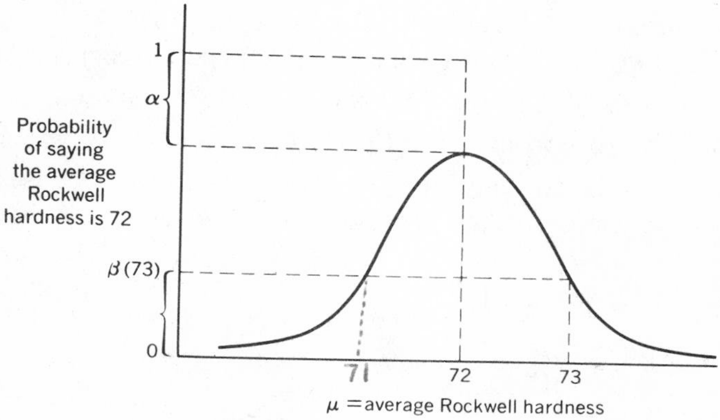

Fig. 5.1. Operating characteristic (OC) curve.

Associated with every decision procedure is an OC curve and procedures are chosen according to these OC curves. Suppose the OC curves of the procedures mentioned above are as indicated in Figure 5.2.

Clearly, the procedure based upon $\bar{X}$ is superior to the procedure based on $\tilde{X}$ except when $\mu = 72$. Thus, the $\bar{X}$ procedure uniformly has a smaller error of Type II than the $\tilde{X}$ procedure, but has a larger error of Type I.

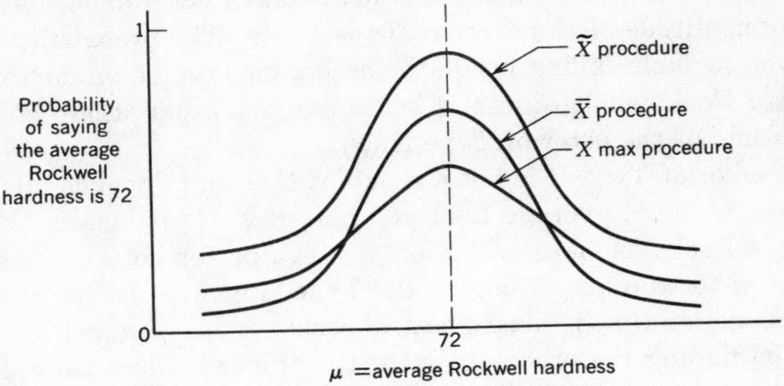

Fig. 5.2. Comparison of several OC curves.

To make these procedures comparable, it is desirable to fix the Type I errors of both procedures and to compare their Type II errors. This can easily be done. It is apparent that the Type I error of either procedure can be made as small as we please by increasing the length of the acceptance interval. For example, in the $\bar{X}$ procedure, if the probability that $\bar{X}$ falls outside the interval [70.7, 73.3] when the true average Rockwell hardness is 72 (the Type I error) is α_1, and the probability that $\bar{X}$ falls outside the interval [60, 90] when the true average Rockwell hardness is 72 is α_2, then α_2 must be less than α_1. A similar example can be given for the $\tilde{X}$ procedure. Consequently, the Type I error of either procedure can be fixed by judiciously choosing the proper interval. If this is done, the OC curves will appear as in Figure 5.3.

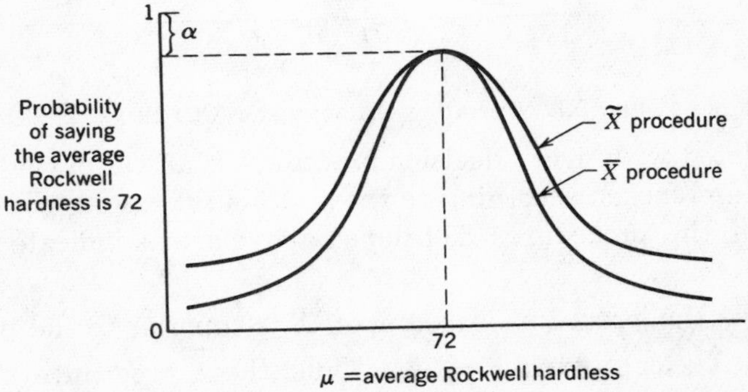

Fig. 5.3. Comparison of the OC curves of the $\bar{X}$ and $\tilde{X}$ procedures for a fixed Type I error.

It is now clear that the $\bar{X}$ procedure is superior to the $\tilde{X}$ procedure. If the true average Rockwell hardness is 72, both procedures will lead to this conclusion with the same probability. However, if the true average Rockwell hardness is *not* 72, the procedure based on $\bar{X}$ will lead uniformly to the correct conclusion with higher probability, i.e., the probability of saying the average Rockwell hardness is 72 when in reality it is not, is smaller for the $\bar{X}$ procedure than for the $\tilde{X}$ procedure; thus, the $\bar{X}$ procedure is better than the $\tilde{X}$ procedure. A similar conclusion is reached if the $\bar{X}$ and X_{max} procedures are compared for a fixed Type I error.

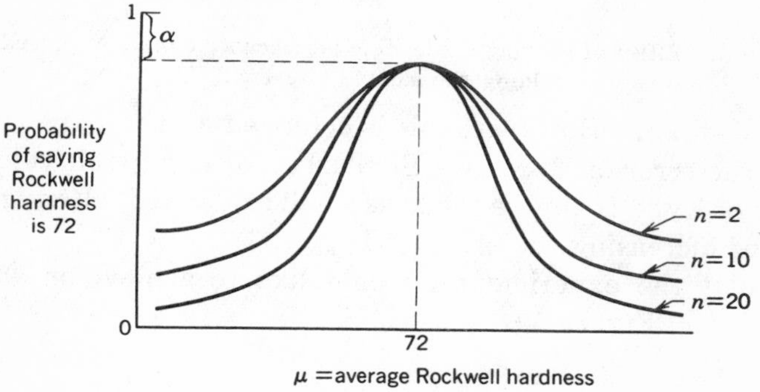

Fig. 5.4. Effect of varying the sample size on the $\bar{X}$ procedure holding the level of significance fixed.

A procedure is said to be *optimum* if, for a fixed sample size, there does not exist any other procedure having the same level of significance whose OC curve lies entirely below the OC curve of the optimum procedure for all values of the abscissa. There may be more than one optimum procedure. All the procedures presented in the ensuing chapters will be optimum.

The discussion above was concerned with a comparison of procedures holding the level of significance and the sample size fixed. The effect of varying the sampling size for a given procedure holding the level of significance fixed can be seen by referring to Figure 5.4. Note that the error of Type II decreases as n increases, as is to be expected. Similarly, the effect of changing the level of significance for a given procedure holding the sample size fixed can be seen by referring to Figure 5.5. This is achieved by changing the end points of the acceptance interval of the stated procedure.

It is evident that for fixed n the error of Type I can be made as

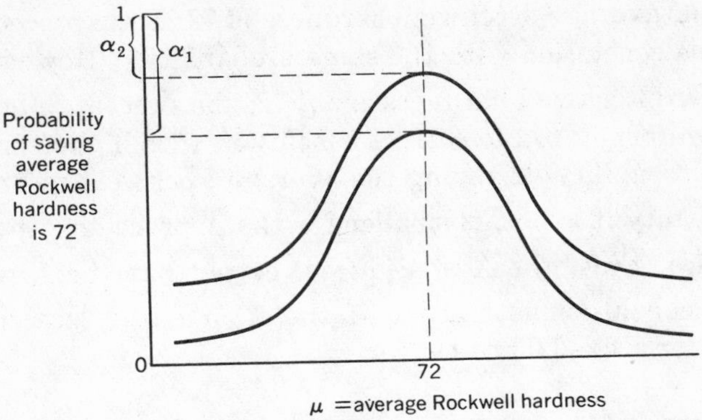

Fig. 5.5. Effect of varying the level of significance on the $\overline{X}$ procedure holding the sample size fixed.

small as desired. However, this is achieved at the expense of increasing the error of Type II. Similarly, for any given value of μ, the error of Type II can be made as small as desired, although at the expense of increasing the error of Type I.

Naturally, the experimenter would like to achieve on OC curve such as that in Figure 5.6.

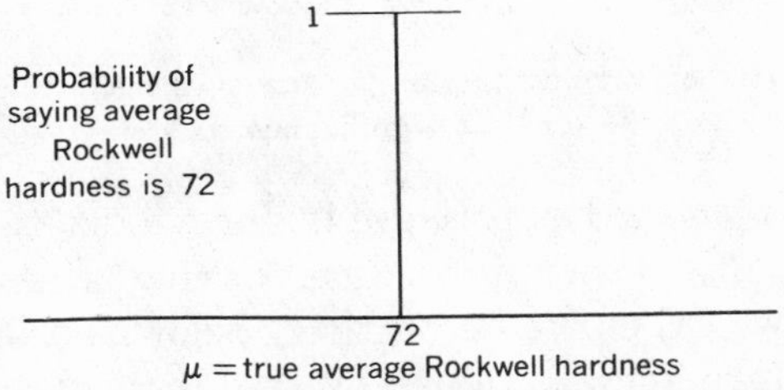

Fig. 5.6. Ideal OC curve.

If the true average Rockwell hardness is 72, the probability of coming to this conclusion is 1. If the true average Rockwell hardness is some number other than 72, the probability of concluding that it is 72 is zero. This OC curve can be achieved only by letting n become very large. This, of course, is entirely too costly, for which reason the experimenter must be satisfied with an approximation to this ideal OC curve. For example, if $\mu = 72.000003$, and the procedure led to the conclusion that $\mu = 72$, generally, this would not result in a serious error. On the other hand, saying that $\mu=72$ when in reality it is 80 may be very serious. Consequently, the experimenter, and only the experimenter, must choose a value for the level of significance which he is willing to tolerate and a value of μ that is important from the practical point of view to detect, together with the error associated with this point. In other words, he must choose two points on the OC curve. These two points completely determine the entire OC curve. Since the OC curve depends upon the sample size n and acceptance constants, this amounts to completely specifying the parameters of the procedure. Very often the experimenter must weigh the cost of taking additional observations against the advantage of decreasing the Type II error. In practice, there are two major limitations to the use of the OC curve. Very often the number of observations is fixed in advance by custom or by the limitations of testing equipment. Indeed, the statistical analysis may be of secondary importance, based on data taken for another purpose. Even in this case, a look at the OC curve is important as it gives an idea of the type of differences one is likely to detect and hence an indication of the sensitivity of the analysis. For example, if it is important to detect an average Rockwell hardness of 74 and a limited sample size is available, the probability of concluding that the average Rockwell hardness is 72 when in reality it is 74 may be 0.70. If this is the case, it is evident that performing the experiment is a waste of time and money unless the sample size can be increased.

In the second place, it often happens that the OC curve depends on parameters in which one is not interested. This could occur in the previous example if the standard deviation were unknown. The OC curve, as will become evident later, depends upon the standard deviation as well as the true mean. However, even if only the general magnitude of the standard deviation is known, the OC curves are very useful in designing experiments. The experimenter must realize that whenever he picks a sample size he is implicitly picking an OC curve and that the more information he has available in making this decision the better. OC curves for most of the standard significance tests are presented in Chapters VI and VII.

5.3 One- and Two-Sided Procedures

At this point it may be well to formalize the discussions of the preceding sections. In the examples presented in Section 5.1, the experimenter is interested in testing a particular hypothesis. For example, he is interested in testing the hypothesis that the true average Rockwell hardness is 72 on the B scale, or he is interested in testing whether a new method of sealing vacuum tubes is the same as the old method, etc. In each of these examples he has an alternative in mind. In the Rockwell hardness example, the alternative may be that the average Rockwell hardness does not equal 72; i.e., it is either less than or greater than 72. This is known as a two-sided alternative and it is associated with a two-sided procedure. This was the alternative considered in the previous sections and is credible. A somewhat less likely alternative is that the experimenter has *a priori* knowledge that the constituents of the new steel can only increase the Rockwell hardness beyond 72 if it has any effect at all. In other words, he discounts the possibility of decreasing the average Rockwell hardness. This alternative is known as a one-sided alternative and the procedure associated with it is known as a one-sided procedure.

In the vacuum tube example an alternative to the hypothesis that the two methods of sealing are equivalent is that the new method either increases or decreases the average life of vacuum tubes. This is a two-sided alternative. Perhaps a more realistic alternative is the one-sided alternative, i.e., the new method can only increase the average life of vacuum tubes. Of course, the realism of the model depends upon the particular situation, and is a problem which the experimenter faces.

A one-sided procedure can also be used in a somewhat different context. Suppose that an average Rockwell hardness below 72 will have no appreciable effect on the performance of the product, and hence, is acceptable. Now, the hypothesis to be tested is that the average Rockwell hardness is 72 or less whereas the alternative is still the one-sided alternative that the average Rockwell hardness is greater than 72.

For one-sided procedures used in either context, the alternatives mentioned dealt with the parameter being greater than a given value. In any given situation, the alternative may be that the parameter is less than the given value.

The two-sided alternatives leave little ambiguity as to the procedure to follow. For example, in the Rockwell hardness problem a logical acceptance region is an interval if the random variable $\overline{X}$ is used. Since $\overline{X}$ is an estimate of the average Rockwell hardness, the hypothesis that the average Rockwell hardness is 72 should be accepted if $\overline{X}$ is close to 72 (above or below). For the one-sided alternative where the possibility of decreasing the average Rockwell hardness is discounted, or in the case where an average Rockwell hardness below 72 is acceptable, the experimenter should reject the hypothesis that the average Rockwell hardness equals 72 if $\overline{X}$ is too large, e.g., $\overline{X} > 73$. A sketch of the OC curve for this procedure is found in Figure 5.7. It is compared with the two-sided procedure with the same level of significance.

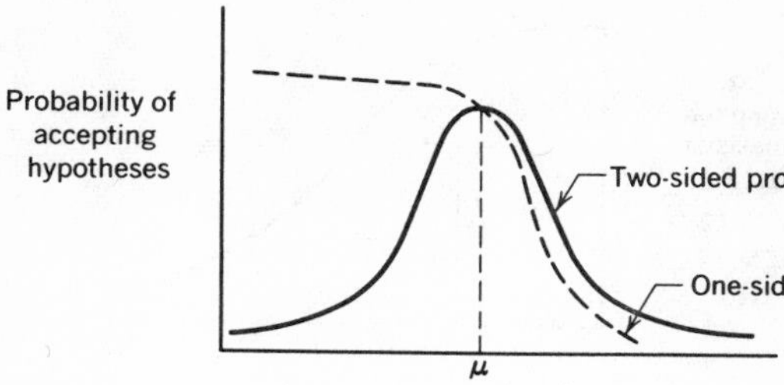

Fig. 5.7. Comparison of the OC curves for one- and two-sided procedures.

It is interesting to note that when the average Rockwell hardness is 72, both procedures are equivalent. When the average Rockwell hardness is greater than 72, the one-sided procedure is better. When the average Rockwell hardness is less than 72, the two-sided procedure has a better chance of concluding that it is not equal to 72. This is to be expected since it has been assumed in using the one-sided procedure that either the average Rockwell hardness *cannot* be less than 72, or in the other situation, that accepting the hypothesis when the average Rockwell hardness is less than 72 is desirable.

In one-sided alternative problems, difficulties often arise as to which procedure to use. For example, suppose the army is presently buying shoes from a manufacturer, and these shoes are known to have an average life of 12 months. A new manufacturer puts in a bid at the same price but states that the average life of his shoes is greater than 12 months. The army is going to run an experiment to ascertain whether or not this claim is valid. They will seek a decision procedure to use with their experimental data.

Using the notation of "testing a hypothesis," two types of hypotheses arise: (1) testing the hypothesis that the average life of the new company's shoes is *less than or equal* to 12 months ($H: \mu \leqq 12$) against the alternative that the average life is *greater than* 12 months ($A: \mu > 12$), and (2) testing the hypothesis that the average life of the new company shoes is *greater than or equal to* 12 months ($H: \mu \geqq 12$) against the alternative that the average life is *less than* 12 months ($A: \mu < 12$).

In the first situation, i.e., $H: \mu \leqq 12$; $A: \mu > 12$, accepting the hypothesis implies a retention of the old manufacturer whereas rejecting the hypothesis implies a switch to the new company. The decision procedure associated with this case should result in an OC curve similar to the plot in Figure 5.8. If the average life of the new

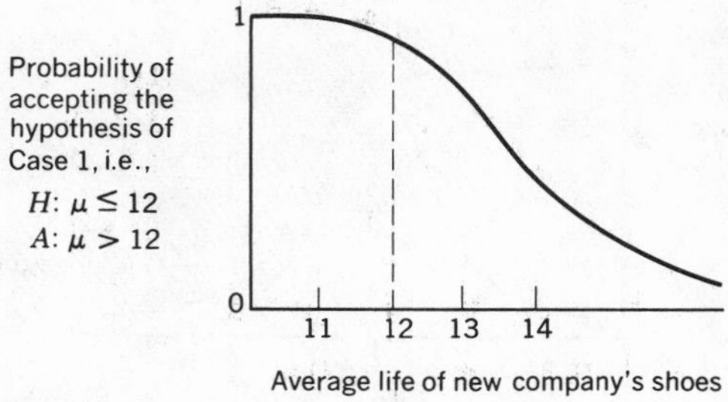

Fig. 5.8. OC curve of the procedure when the army is satisfied with the present manufacturer.

company's shoes is 12 months or less, the probability of accepting the hypothesis, and thereby retaining the old manufacturer, is close to one. If the average life of the shoes is slightly greater than 12 months, the probability of retaining the old manufacturer is still quite high. If the average life of the shoes is substantially greater than 12 months, the probability of accepting the hypothesis is small (rejecting the hypothesis is large) and a switch to the new manufacturer is likely.

The physical situation leading to this hypothesis model is exemplified when the army is very satisfied with the present manufacturer. The old company sends its product on time, they are reliable, their workmanship is good, a switch causes increased red tape and administrative difficulties, etc. In other words, the burden of proof is on the new company to show that its shoes have an average life substantially greater than 12 months. The army is desirous of a switch only if a substantial increase in the average life will result. An example of a procedure which has this property is: Accept the hypothesis that $\mu \leqq 12$ months if $\overline{X} \leqq 15$; otherwise reject it.

In the second situation, i.e., $H: \mu \geqq 12$; $A: \mu < 12$, accepting the hypothesis implies a switch to the new company, whereas rejecting the hypothesis implies a retention of the old manufacturer. The decision procedure associated with this case should result in an OC curve similar to the plot in Figure 5.9.

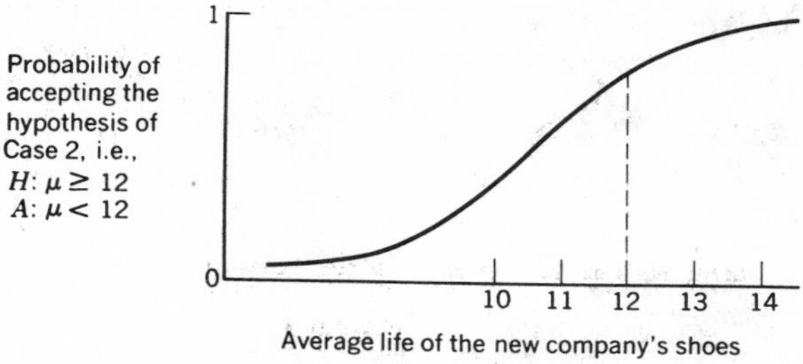

Fig. 5.9. OC curve of the procedure when the army is not satisfied with the present manufacturer.

If the average life of the new company's shoes is 12 months or more, the probability of accepting the hypothesis, and thereby switching to the new manufacturer, is close to one. If the average life of the shoes is slightly less than 12 months, the probability of switching to the new manufacturer is still quite high. If the average life of the shoes is substantially less than 12 months, the probability of accepting the hypothesis is small (rejecting the hypothesis is large) and retaining the old manufacturer is likely.

The physical situation leading to this hypothesis model is exemplified when the army is unhappy with the present manufacturer, and is looking for an opportunity to change. The old company is late with shipments and generally unreliable. In other words, they are willing to change not only if the average life of the new company's shoes exceeds 12 months, but even if their average life is slightly less than 12 months. The burden of proof is on the new manufacturer to show that the average life of his shoes is not substantially less than 12 months. An example of a procedure which has this property is: Accept the hypothesis that $\mu \geqq 12$ if $\overline{X} \geqq 10$; otherwise reject it.

In the two situations just described, the army indicated an inherent preference for one of the companies. A third alternative arises when the army is completely indifferent as to which company it chooses, provided the difference in the average lives of the shoes is small. If the average life of the new company's shoes is only 12 months, the army is indifferent, and is willing to tolerate an equal chance of obtaining other manufacturers. On the other hand, if there is a substantial increase over 12 months the new company should be chosen; if there is a substantial decrease under 12 months the old manufacturer should be retained. The hypotheses of case 1 or case 2 may be used, but with a level of significance of $\frac{1}{2}$, thereby expressing the indifference of the army when the average life of the shoes is only 12 months. The OC curve for the hypothesis of case 1, i.e., $H : \mu \leqq 12$; $A : \mu > 12$, is shown in Figure 5.10 with a level of significance equal to $\frac{1}{2}$.

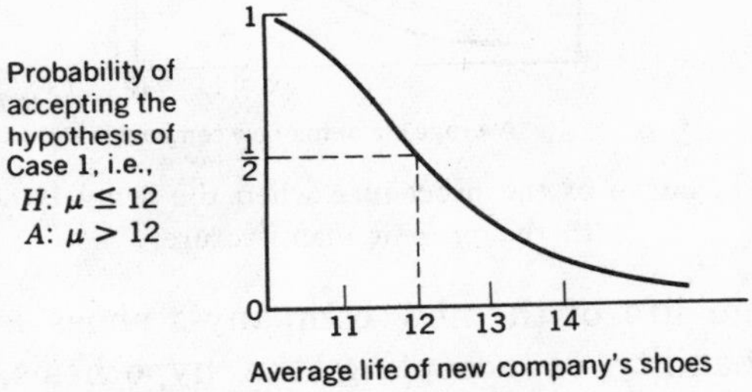

Fig. 5.10. OC curve of the procedure when the army is indifferent.

Thus, if the average life of the new company's shoes is 12 months, the probability of accepting the hypothesis and thereby retaining the old manufacturer is $\frac{1}{2}$. If the average life of the shoes is substantially less than 12 months, the probability of accepting the hypothesis and thereby retaining the old manufacturer is close to one. If the average life of the shoes is substantially greater than 12 months, the probability of accepting the hypothesis and thereby retaining the old manufacturer is quite small, so that a switch will usually result. The procedure which has this property is, as one might expect: Accept the hypothesis that $\mu \leqq 12$ if $\overline{X} \leqq 12$.

If the OC curve of the hypothesis for case 2, i.e., $H : \mu \geqq 12$; $A : \mu < 12$, were plotted, it would appear as in Figure 5.9, but with a level of significance of $\frac{1}{2}$. The procedure which is associated with such an OC curve is: Accept the hypothesis that $\mu \geqq 12$ if $\overline{X} \geqq 12$. Accepting the hypothesis for this case is equivalent to switching to the new company. Thus, if $\overline{X} \geqq 12$, the switch is made. This is equivalent to the procedure associated with Figure 5.10 (case 1). In this situation if $\overline{X} \leqq 12$, the hypothesis is accepted and the old manufacturer retained, or alternatively if $\overline{X} > 12$, the switch is made.

To summarize then, one-sided alternatives can be classified into two groups.

1. There is *a priori* knowledge that certain values of the unknown parameter cannot exist, e.g., the average Rockwell hardness cannot be less than 72.

2. The hypothesis is concerned with statements about values of the unknown parameter being less than or equal to a constant (or alternatively greater than or equal to a constant), e.g., the shoe example.

In either case, it is advantageous to sketch the form of the OC curve and determine a one-sided procedure which will lead to this OC curve.

5.4 Statistical Decision Theory

Analysis of the preceding type of problem can be looked at from a broader point of view using a new concept called decision theory. Instead of considering only two decision problems, i.e., accepting or rejecting the hypothesis, and using the OC curve as a measure of risk, the possible decisions available can be increased, and a more general risk function can be used to evaluate procedures. For example, suppose that in the shoe problem, the army is going to make one of three decisions on the basis of some experimentation. It will give all its business to the present supplier, divide its business equally between the new and old suppliers, or switch entirely to the new supplier. Corresponding to each of these actions and any value of the true average life of the shoes, there is a loss (or gain) to the army which is specified, e.g., if the average life is 11 months and the army diverts its entire order to the new supplier, a particular loss is incurred. This loss is usually expressed in monetary terms. The statistician must choose a procedure which is a rule that specifies

Table 5.1. Loss Function for the Two Decision Problem

	Conclude hypothesis is true	Conclude hypothesis is false
Hypothesis true	0	1
Hypothesis false	1	0

which of the three actions will be selected for every outcome of the random variable associated with the experiment. Thus, for any outcome of the random variable and using a particular decision procedure, one of the three decisions is made, and a loss (or gain) is incurred depending upon the actual average life of the shoes. The expected loss, averaged over all values of the random variable with respect to its probability distribution, is known as the risk. For a fixed decision procedure, the risk when plotted against the average shoe life yields a risk curve. The statistical problem then is to choose optimal procedures from the class of all procedures where optimality is defined relative to the risk curve. This risk curve is not unlike the OC curves studied earlier. In fact, if we return to the two decision problem, i.e., accept or reject the hypothesis, and use the loss function given in Table 5.1, the risk function turns out to be merely the OC curve.

For example, suppose we want to test the hypothesis that the average life of the new supplier's shoes is less than or equal to 12 months against the alternative that it is greater than 12 months. The decision procedure is to accept the hypothesis and not switch to the new supplier if $\overline{X} \leqq 15$ months. If the average life is 13 months, the expected loss using this procedure is given by

$$\begin{aligned} &P(\overline{X} \leqq 15 \text{ given } \mu = 13)(1) + P(\overline{X} > 15 \text{ given } \mu = 13)(0) \\ &\qquad = P(\overline{X} \leqq 15 \text{ given } \mu = 13). \end{aligned}$$

This is just the probability of accepting the hypothesis when the true average life is 13 months, i.e., $\beta(13)$. Similarly, if the average life is 12 months, the expected loss using this procedure is given by

$$\begin{aligned} &P(\overline{X} \leqq 15 \text{ given } \mu = 12)(0) + P(\overline{X} > 15 \text{ given } \mu = 12)(1) \\ &\qquad = P(\overline{X} > 15 \text{ given } \mu = 12). \end{aligned}$$

This is just the probability of rejecting the hypothesis when the hypothesis is true, i.e., α. Hence, it is evident that the risk function using this loss function is just the OC curve.

The general decision theoretic approach requires a great deal from the decision maker. It requires that he be in a position to evaluate numerically, for every possible state of the parameter in the situation under consideration, the consequences of any of the actions that he might take. This is a rather difficult assignment, and it is likely that the experimenter in engineering and the physical sciences is only willing to choose his loss function in terms of an OC curve since the magnitudes of other costs are usually unavailable. As pointed out above, the OC curve is a special case of the more general risk function. This text will confine itself to using the OC curve as the risk function.

CHAPTER VI

Tests of Hypotheses about a Single Parameter

6.1 Test of the Hypothesis that the Mean of a Normal Distribution Has a Specified Value when the Standard Deviation Is Known

6.1.1 Choice of an OC curve

In the previous chapter, the concept of an OC curve was introduced and reference was made to a problem concerning the average Rockwell hardness of a new product made of a special steel. Suppose that information is available indicating that the distribution of Rockwell hardness can be approximated by a normal distribution with *known* standard deviation equal to 2 units on the B scale. This information may be obtained from past data on similar types of steel. The company research laboratory is to design an experiment to ascertain whether the average Rockwell hardness (mean of a normal distribution) is equal to 72. In designing an experiment, the experimenter must consider the operating characteristic curve. This OC curve reflects the risks that the company is willing to tolerate. Instead of designating the entire OC curve, it is sufficient to choose two points on this curve: the probability of concluding the average Rockwell hardness is 72 when it really is 72; and the value of the average Rockwell hardness that it is considered important to detect, together with the risk associated with this point. For example, in determining the first point the experimenter may be seeking a procedure which will conclude that the average Rockwell hardness is 72 when it is actually 72, with probability 0.95. The number 0.95 is not just chosen by chance but rather reflects the Type I error ($\alpha=1-0.95=0.05$) that the experimenter is willing to tolerate. The Type I error can be 0.001, or 0.01, or even 0.50, depending upon the particular situation. If the decision involves determining whether or not to shut down a steel process, the error of Type I should be very small. Closing a steel plant is extremely costly and an error in this decision may be quite important.

At the other end of the spectrum is the case of choosing between two manufacturers who offer similar products at similar prices. If there is no difference in quality of the products, it is not terribly important which manufacturer is chosen, and hence, a large Type I error can be tolerated. In any event, the choice of the magnitude of the Type I error is a decision the experimenter must make, and reflects the risk that he is willing and able to tolerate.

The second point on the OC curve that the experimenter must choose is the value of the average Rockwell hardness that it is considered important to detect, together with the risk associated with this point. Obviously, deciding that an average Rockwell hardness of 72.000007 is different from 72 is not very important, whereas a similar decision about an average Rockwell hardness of 74 may be quite important. It was pointed out in the previous chapter that a correct decision is rarely made all of the time, and consequently, some risk (β) must be associated with concluding that the average Rockwell hardness is 72 when it is actually 74. This risk may be 0.01, 0.1, 0.15, 0.50, etc., depending upon the seriousness of the decision, and is different for different problems. Again, this point on the OC curve, both ordinate and abscissa, is chosen by the experimenter and reflects his risks. This choice is not up to the statistician. Once two points on the OC curve have been found, the proper procedure can be determined.

Formalizing these concepts, the problem of concern is to test the hypothesis that the mean of a normal distribution with known standard deviation is equal to a fixed constant, i.e., $\mu = \mu_0$.[1] Two

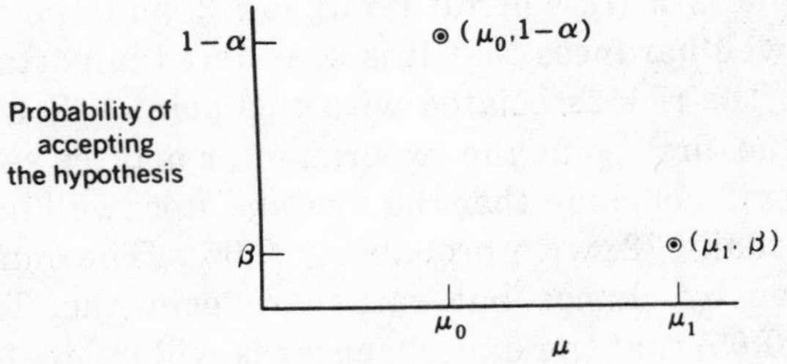

Fig. 6.1. Two points used for determining an OC curve.

[1] If the alternative is one-sided, the hypothesis may also be $\mu \leqq \mu_0$ or $\mu \geqq \mu_0$. The notation $\mu = \mu_0$ will be interpreted to include these cases.

points on the OC curve are chosen. These points are denoted by (μ_0, $1 - \alpha$) and (μ_1, β) and are shown in Figure 6.1.

6.1.2 Tables and Charts for Determining Decision Rules

An experiment consisting of n trials is to be performed. The random variables $x_1, x_2, \cdots, x_n$ are the measurable quantities associated with the $1, 2, \cdots, n$ trials, respectively. It is assumed that the x_i are independent normally distributed random variables, each with unknown mean μ and *known variance* σ^2 (i.e., a random sample). The *optimum* procedure for testing the hypothesis that the mean of the normal distribution has a specified value, i.e., $\mu = \mu_0$, is based upon the random variable U, where

$$U = \frac{(\bar{x} - \mu_0)\sqrt{n}}{\sigma}.$$

A random variable used in this context will be called a *test statistic.* If the value taken on by the test statistic as a result of experimentation falls into the acceptance region, the hypothesis that μ equals μ_0 is accepted; otherwise it is rejected. This section will be devoted to finding the acceptance region and the sample size for the test statistic U, using the tables and charts which are provided. One- and two-sided procedures will be considered.

6.1.2.1 *Tables and charts for two-sided procedures.* Two-sided procedures will be analyzed first. The rules presented assume that the risk function, which is expressed by the OC curve, is symmetric about μ_0. This implies that for any positive constant ε the risk associated with the value $\mu = \mu_0 + \varepsilon$ is the same as the risk associated with $\mu = \mu_0 - \varepsilon$, or an error incurred when $\mu = \mu_0 + \varepsilon$ is as serious as an error incurred when $\mu = \mu_0 - \varepsilon$. This restriction is relaxed later.

The acceptance region for the two-sided procedure is given by the interval $[-K_{\alpha/2}, K_{\alpha/2}]$, where α is the level of significance (Type I error) and $K_{\alpha/2}$ is the $100\,\alpha/2$ percentage point (normal deviate corresponding to $\alpha/2$) of the normal distribution. Thus, the hypothesis that the mean of a normal distribution equals μ_0 is accepted if

$$-K_{\alpha/2} \leqq U \leqq K_{\alpha/2}$$

and rejected if U lies outside this interval. For example, in testing the hypothesis that the average Rockwell hardness of the special steel is 72 on the B scale, the point (72,0.95) on the OC curve is specified. The Type I error is then $\alpha = 0.05$ and $K_{\alpha/2} = K_{0.025} = 1.96$. Hence, the hypothesis that the average Rockwell hardness is 72 is accepted if

$$-1.96 \leqq U \leqq 1.96$$

and rejected if U lies outside this interval.

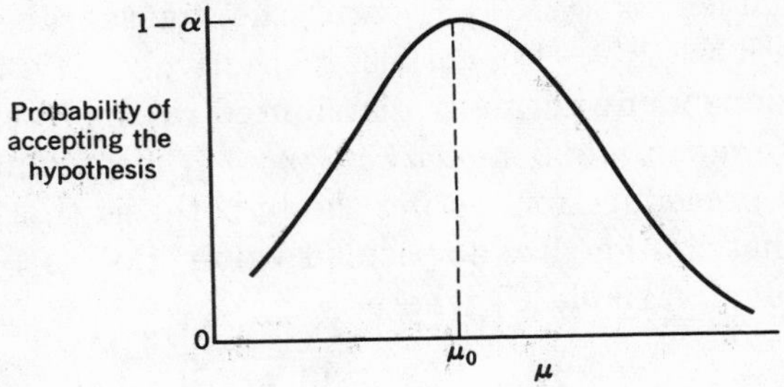

Fig. 6.2. OC curve for testing the hypothesis that the mean of a normal distribution with known standard deviation is μ_0 against a two-sided alternative.

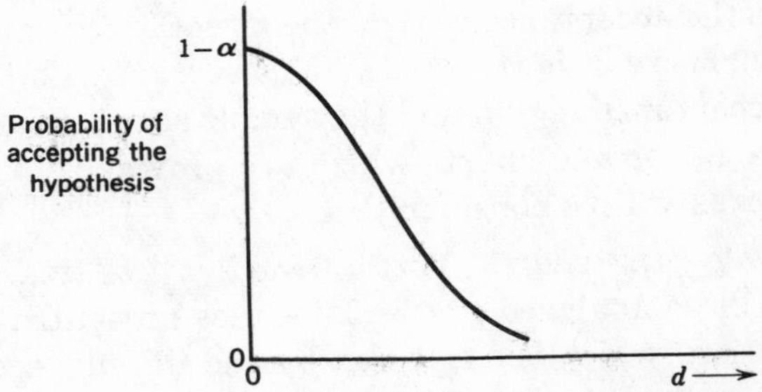

Fig. 6.3. Standardized OC curve for testing the hypothesis that the mean of a normal distribution with known standard deviation is μ_0 against a two-sided alternative.

It is easily verified that this procedure yields the point $(\mu_0, 1 - \alpha)$ on the OC curve. Whenever $\mu = \mu_0$, the test statistic

$$U = \frac{(\bar{x} - \mu_0)\sqrt{n}}{\sigma}$$

is normally distributed with zero mean and unit variance. Hence, if $\mu = \mu_0$

$$P(-K_{\alpha/2} \leqq U \leqq K_{\alpha/2}) = 1 - \alpha.$$

The required sample size is obtained by referring to a set of operating characteristic curves which are provided for this purpose. An OC curve for testing the hypothesis that the mean of a normal distribution is μ_0 against a two-sided alternative is usually a plot of the probability of acceptance against the true mean μ, and is symmetric about μ_0. Such a plot is shown in Figure 6.2 for a fixed sample size n. In order for these curves to be useful in all problems concerning the mean of a normal distribution, regardless of the values of μ_0 and σ, the abscissa scale is changed from μ to d, where

$$d = |\mu - \mu_0|/\sigma.$$[1]

The OC curve of Figure 6.2 is now replotted against d and is shown in Figure 6.3 for the same fixed sample size n. $d = 0$ corresponds to the point $\mu = \mu_0$.

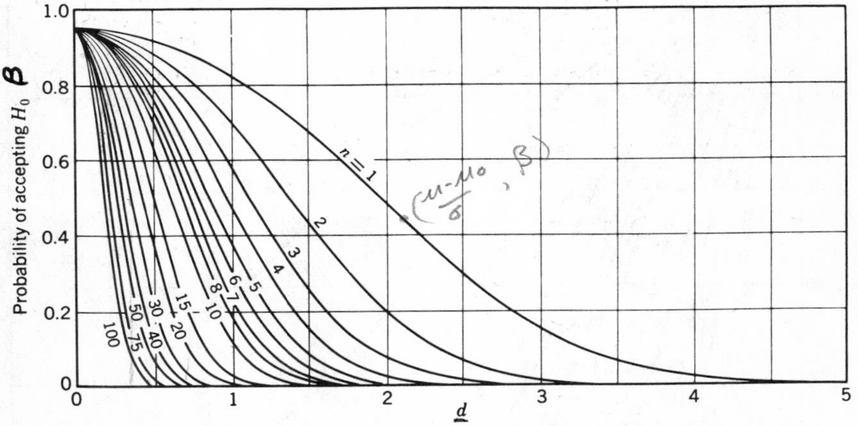

Fig. 6.4. OC curves for different values of n for the two-sided normal test for a level of significance $\alpha=0.05$.

Reproduced by permission from "Operating Characteristics for the Common Statistical Tests of Significance" by Charles L. Ferris, Frank E. Grubbs, Chalmers L. Weaver, Annals of Mathematical Statistics, June, 1946

A set of such OC curves for different values of n appears in Figures 6.4 and 6.5. Figure 6.4 presents the OC curves of the two-sided normal test for a level of significance (Type I error) equal to 0.05. Figure 6.5 presents the OC curves of the two-sided normal test for a level of significance equal to 0.01.

If the level of significance of the procedure is either 0.05 or 0.01, the required sample size n is obtained by entering the appropriate figure (Figure 6.4 if $\alpha = 0.05$ and Figure 6.5 if $\alpha = 0.01$) with the point $\left(\frac{|\mu_1 - \mu_0|}{\sigma}, \beta\right)$ and reading out the value of the sample size associated with the OC curve that passes through this point. Returning to the Rockwell hardness example, if it is stated that the probability of accepting the hypothesis that the average Rockwell hardness is 72

[1] The symbol $|\mu - \mu_0|$ is read as the absolute value of the difference between μ and μ_0.

should not exceed 0.10 when $\mu = 74$,[1] the required sample size is obtained by entering Figure 6.4 with the point $\left(\frac{|74 - 72|}{2}, 0.10\right)$ and reading out a sample size of $n = 11$.

If a non-symmetric risk function is desired, the sample size is obtained as follows: Let the risk associated with $\mu = (\mu_0 + \varepsilon)$ be β^+ and the risk associated with $\mu = (\mu_0 - \varepsilon)$ be β^-. Enter the appropriate figure with the point $d = \varepsilon/\sigma$ and the smaller of the values β^+ or β^-. Read out the sample size associated with the OC curve that passes through this point.

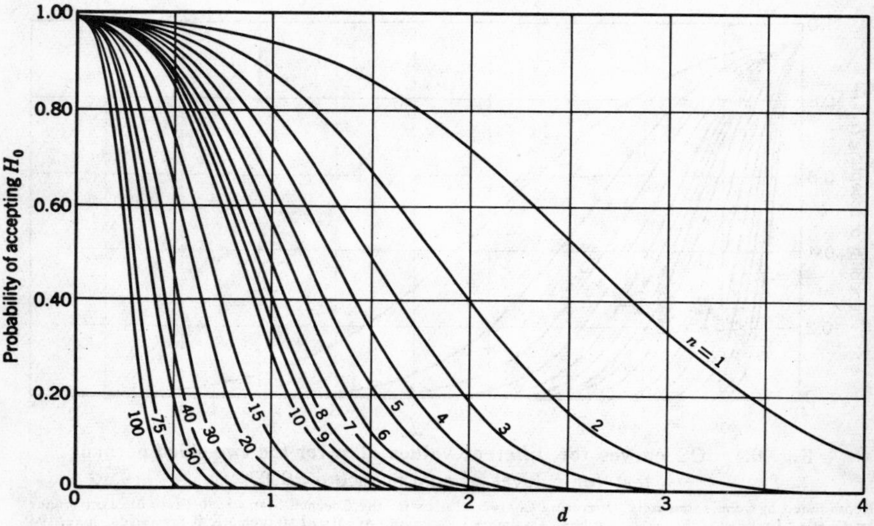

Fig. 6.5. OC curves for different values of n for the two-sided normal test for a level of significance $\alpha = 0.01$.

6.1.2.2 *Summary of two-sided procedures using tables and charts.* The following steps are taken using a two-sided procedure for testing the mean of a normal distribution—σ known.

1. Two points on the OC curve are chosen: the probability of accepting the hypothesis when $\mu = \mu_0$, i.e., $(\mu_0, 1 - \alpha)$ and the probability of accepting the hypothesis when $\mu = \mu_1$, i.e., (μ_1, β) where μ_1 is the value of the mean considered important to detect. On the OC curve with the d scale, these points are translated to $(0, 1 - \alpha)$ and $\left(\frac{|\mu_1 - \mu_0|}{\sigma}, \beta\right)$.

[1] This implies that the Type II error at $\mu = 74$ equals 0.10 since the OC curve is a decreasing function of μ.

2. Find the required sample size n by entering the appropriate figure (Figure 6.4 if $\alpha = 0.05$ and Figure 6.5 if $\alpha = 0.01$) with the point $\left(\frac{|\mu_1 - \mu_0|}{\sigma}, \beta\right)$ and reading out the value of the sample size associated with the OC curve that passes through this point.
3. Determine the acceptance region which is the interval $[-K_{\alpha/2}, K_{\alpha/2}]$.
4. Draw a random sample of n items.
5. Compute the value of the test statistic $U = \frac{(\bar{x} - \mu_0)\sqrt{n}}{\sigma}$.
6. If $-K_{\alpha/2} \leqq U \leqq K_{\alpha/2}$, i.e., $|U| \leqq K_{\alpha/2}$, accept the hypothesis that $\mu = \mu_0$. If U lies outside this interval reject the hypothesis and conclude that the mean is not μ_0.

6.1.2.3 *Tables and charts for one-sided procedures.* If the procedure is one-sided and rejection is desired when the true mean exceeds μ_0 (this being the alternative to the hypothesis that $\mu = \mu_0$ or $\mu \leqq \mu_0$), the acceptance region is given by the interval $[-\infty, K_\alpha]$, where α is the level of significance and K_α is the 100 α percentage point (normal deviate corresponding to α) of the normal distribution. Thus, the hypothesis that $\mu = \mu_0$ (or $\mu \leqq \mu_0$) is accepted if

$$U \leqq K_\alpha$$

and rejected if U lies outside this interval. This acceptance procedure leads to the point $(\mu_0, 1-\alpha)$ on the OC curve. Whenever $\mu = \mu_0$, the test statistic

$$U = \frac{(\bar{x} - \mu_0)\sqrt{n}}{\sigma}$$

is normally distributed with zero mean and unit variance. Hence, if $\mu = \mu_0$,

$$P(U \leqq K_\alpha) = 1 - \alpha.$$

The OC curves necessary for determining the sample size n are given in Figures 6.6 and 6.7 for the one-sided procedure where rejection is desired when the true mean exceeds μ_0. Figure 6.6 presents the OC curves for a level of significance equal to 0.05 and Figure 6.7 presents the OC curves for a level of significance equal to 0.01. As in the two-sided case, in order for these curves to be useful in all such problems concerning the mean of a normal distribution, regardless of the values of μ_0 and σ, the abscissa scale is again changed from μ to d, where d is now defined as

$$d = (\mu - \mu_0)/\sigma.$$

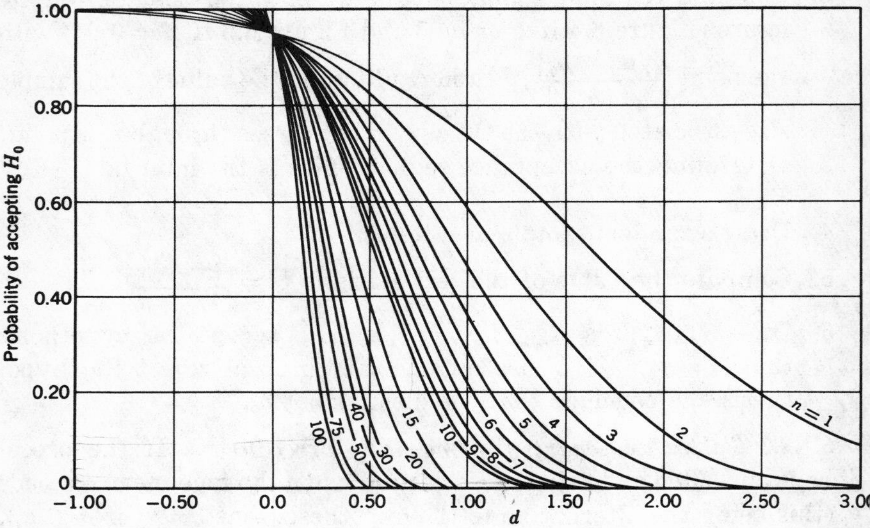

Fig. 6.6. OC curves for different values of n for the one-sided normal test for a level of significance $\alpha = 0.05$.

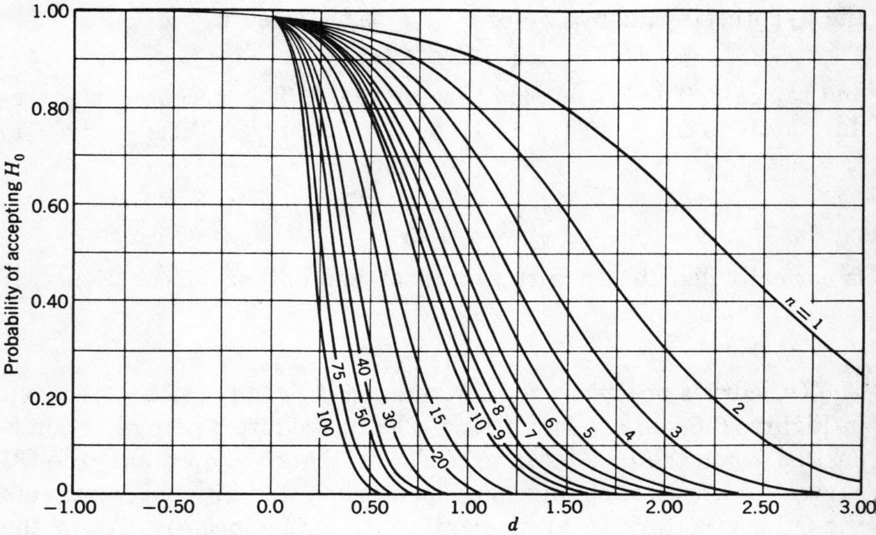

Fig. 6.7. OC curves for different values of n for the one-sided normal test for a level of significance $\alpha = 0.01$.

If the level of significance of the procedure is either 0.05 or 0.01, the required sample size n is obtained by entering the appropriate figure (Figure 6.6 if $\alpha = 0.05$ and Figure 6.7 if $\alpha = 0.01$) with the point

$\left(\frac{\mu_1 - \mu_0}{\sigma}, \beta\right)$ and reading out the value of the sample size associated with the OC curve that passes through this point.

If the procedure is one-sided and rejection is desired when the true mean is less than μ_0 (this being the alternative to the hypothesis that $\mu = \mu_0$ or $\mu \geqq \mu_0$), the acceptance region is given by the interval $[-K_\alpha, \infty]$. Thus, the hypothesis that $\mu = \mu_0$ (or $\mu \geqq \mu_0$) is accepted if

$$U \geqq -K_\alpha$$

and rejected if U lies outside this interval. Clearly, this procedure yields the point $(\mu_0, 1 - \alpha)$ on the OC curve. Whenever $\mu = \mu_0$, the test statistic

$$U = \frac{(\bar{x} - \mu_0)\sqrt{n}}{\sigma}$$

is normally distributed with zero mean and unit variance. Hence, if $\mu = \mu_0$

$$P(U \geqq -K_\alpha) = 1 - \alpha.$$

The OC curves necessary for determining the sample size for the one-sided procedure where rejection is desired when the true mean is less than μ_0 are also given in Figures 6.6 and 6.7. However, the abscissa d is now defined as

$$d = (\mu_0 - \mu)/\sigma.$$

If the level of significance of the procedure is either 0.05 or 0.01, the required sample size n is obtained by entering the appropriate figure (Figure 6.6 if $\alpha = 0.05$ and Figure 6.7 if $\alpha = 0.01$) with the point $\left(\frac{\mu_0 - \mu_1}{\sigma}, \beta\right)$ and reading out the value of the sample size associated with the OC curve that passes through this point.

6.1.2.4 *Summary for one-sided procedures using tables and charts.* The following steps are taken using a one-sided procedure for testing the mean of a normal distribution.

1. Two points on the OC curve are chosen: the probability of accepting the hypothesis when $\mu = \mu_0$, i.e., $(\mu_0, 1 - \alpha)$; and the probability of accepting the hypothesis when $\mu = \mu_1$, i.e., (μ_1, β) where μ_1 is the value of the mean considered important to detect. If rejection is desired when $\mu > \mu_0$, these points are translated to $(0, 1 - \alpha)$ and $\left(\frac{\mu_1 - \mu_0}{\sigma}, \beta\right)$ on the OC curve with

the d scale. If rejection is desired when $\mu < \mu_0$ these points are translated to $(0, 1 - \alpha)$ and $\left(\frac{\mu_0 - \mu_1}{\sigma}, \beta\right)$ on the OC curve with the d scale.

2. Find the required sample size n by entering the appropriate figure (Figure 6.6 if $\alpha = 0.05$ and Figure 6.7 if $\alpha = 0.01$). If rejection is desired when $\mu > \mu_0$, the figure is entered with the point $\left(\frac{\mu_1 - \mu_0}{\sigma}, \beta\right)$ and the value of the sample size associated with the OC curve that passes through this point is read out. If rejection is desired when $\mu < \mu_0$, the figure is entered with the point $\left(\frac{\mu_0 - \mu_1}{\sigma}, \beta\right)$ and the value of the sample size associated with the OC curve that passes through this point is read out.
3. Determine the acceptance region which is the interval $[-\infty, K_\alpha]$ for the alternative $\mu > \mu_0$ and $[-K_\alpha, \infty]$ for the alternative $\mu < \mu_0$.
4. Draw a random sample of n items.
5. Compute the value of the test statistic $U = \frac{(\bar{x} - \mu_0)\sqrt{n}}{\sigma}$.
6. If the alternative is $\mu > \mu_0$, accept the hypothesis that $\mu = \mu_0$ (or $\mu \leqq \mu_0$) whenever $U \leqq K_\alpha$; otherwise reject the hypothesis and conclude that $\mu > \mu_0$.
   If the alternative is $\mu < \mu_0$, accept the hypothesis that $\mu = \mu_0$ (or $\mu \geqq \mu_0$) whenever $U \geqq -K_\alpha$; otherwise reject the hypothesis and conclude that $\mu < \mu_0$.

6.1.2.5 *Tables and charts for operating characteristic curves.* Whenever the sample size is fixed in advance, or after it is chosen, the operating characteristic curves given in Figures 6.4 and 6.5 for two-sided procedures and Figures 6.6 and 6.7 for one-sided procedures can be used to evaluate the risks involved in using these procedures. If the level of significance is either 0.05 or 0.01, the appropriate OC curve is entered with the sample size n and d, and the probability of accepting the hypothesis is read from the curve. The value of d chosen corresponds to the value μ for which the risk is to be evaluated.

The fact that the OC curves presented are for levels of significance equal to 0.01 and 0.05 does not imply that these are the only levels ever used. Often, other values are appropriate so that suitable methods for determining the sample size and acceptance region must be

Table 6.1. Test of the Hypothesis that the Mean of a Normal Distribution Has a Specified Value when the Standard Deviation Is Known

Notation for Hypothesis H: $\mu = \mu_0$	*Test Statistic* $U = (\bar{x} - \mu_0)\sqrt{n}/\sigma$
Criteria for Rejection	*Method for Choosing Sample Size*
$\lvert U \rvert \geqq K_{\alpha/2}$ if we wish to reject when the true mean departs from μ_0 in either direction.	Choose a value of $\mu_1 \neq \mu_0$ for which we wish to reject the hypothesis with given high probability; calculate $d = \lvert \mu_1 - \mu_0 \rvert / \sigma$ and select n from the OC curves of Fig. 6.4 or 6.5.
$U \geqq K_\alpha$ if we wish to reject when the true mean exceeds μ_0.	Choose a value of $\mu_1 > \mu_0$ for which we wish to reject the hypothesis with given high probability; calculate $d = (\mu_1 - \mu_0)/\sigma$ and select n from the OC curves of Fig. 6.6 or 6.7.
$U \leqq -K_\alpha$ if we wish to reject when the true mean is less than μ_0.	Choose a value of $\mu_1 < \mu_0$ for which we wish to reject the hypothesis with given high probability; calculate $d = (\mu_0 - \mu_1)/\sigma$ and select n from the OC curves of Fig. 6.6 or 6.7.

Level of Significance	*For One-sided Tests*	*For Two-sided Tests*
$\alpha = 0.05$	$K_{0.05} = 1.645$	$K_{0.025} = 1.960$
$\alpha = 0.01$	$K_{0.01} = 2.326$	$K_{0.005} = 2.576$

found. Such methods are described in Section 6.1.3. However, it can be said that $\alpha = 0.05$ and 0.01 are most commonly used in industry, although emphasizing that the choice of α is up to the experimenter, rather than the statistician. The level of significance reflects a tolerable risk of making an incorrect decision and can only be chosen by someone involved directly in experimentation.

The OC curves presented are calculated under the assumption that the test statistic U is normally distributed. This condition is fulfilled if the random variables $x_1, x_2, \cdots, x_n$, i.e., the measurable quantities associated with the $1, 2, \cdots, n$ trials respectively, are independent normally distributed random variables. However, the OC curves of the procedures presented are valid for more general forms of underlying probability distributions. Since the test statistic U is related to the sample mean $\bar{x}$, the conclusions of the central limit theorem are applicable. Hence, U is approximately normally distributed, regardless of the form of the distribution of the x's, provided n is sufficiently large.

A summary of the procedures for testing the hypothesis that the mean of a normal distribution has a specified value when the standard deviation is known is given in Table 6.1. For purposes of this summary, criteria for rejecting the hypothesis are given instead of criteria for acceptance.

6.1.3 ANALYTICAL DETERMINATION OF DECISION RULES

6.1.3.1 *Acceptance regions and sample sizes.* Let $x_1, x_2, \cdots, x_n$ be a random sample of n independent normally distributed random variables each with unknown mean μ and known variance σ^2. The optimum procedure for testing the hypothesis that μ equals μ_0 is based upon the test statistic U, where

$$U = \frac{(\bar{x} - \mu_0)\sqrt{n}}{\sigma}.$$

This section will be devoted to finding acceptance regions and expressions for the sample size for both one- and two-sided procedures.

As was indicated in Section 6.1.2 the acceptance region for the two-sided procedure is the interval

$$[-K_{\alpha/2}, K_{\alpha/2}]$$

where α is the level of significance and $K_{\alpha/2}$ is the 100 $\alpha/2$ percentage point of the normal distribution. The expression for the sample size n is given approximately by

$$n \cong \frac{(K_{\alpha/2} + K_\beta)^2\sigma^2}{(\mu_1 - \mu_0)^2}.$$

β is the probability of accepting the hypothesis when $\mu = \mu_1$, i.e., Type II error, and K_β is the 100 β percentage point of the normal distribution. This approximation is good whenever

$$\mathrm{P}\left(z < -K_{\alpha/2} - \sqrt{n}\,\frac{|\mu_1 - \mu_0|}{\sigma}\right)$$

is small compared to β, where z is a normally distributed random variable with zero mean and unit variance.

Returning to the Rockwell hardness example of the previous sections in this chapter, the following quantities are given:

$$\alpha = 0.05, \qquad \mu_0 = 72,$$
$$\beta = 0.10, \qquad \mu_1 = 74,$$
$$\sigma = 2.$$

From Appendix Table 1, $K_{\alpha/2} = K_{0.025} = 1.96$ and $K_\beta = K_{0.10} = 1.28$. Substituting into the expression for n

$$n \cong \frac{(1.96 + 1.28)^2 2^2}{(74 - 72)^2} = 11 \text{ observations.}$$

Checking to see if the approximation is good,

$$P\left(z < -1.96 - \frac{\sqrt{11}\,|74 - 72|}{2}\right) = P(z < -5.28)$$

is negligible and hence $n = 11$ is the solution. This coincides with the value of n found using the OC curves provided.

If the procedure is one-sided and rejection is desired when the true mean exceeds μ_0 (this being the alternative to the hypothesis that $\mu = \mu_0$ or $\mu \leqq \mu_0$) the acceptance region is the interval

$$[-\infty, \quad K_\alpha]$$

where α is the level of significance and K_α is the 100 α percentage point of the normal distribution. If the procedure is one-sided and rejection is desired when the true mean is less than μ_0 (this being the alternative to the hypothesis that $\mu = \mu_0$ or $\mu \geqq \mu_0$) the acceptance region is the interval

$$[-K_\alpha, \quad \infty].$$

For both of these one-sided procedures the expression for the sample size n is given by

$$n = \frac{(K_\alpha + K_\beta)^2 \sigma^2}{(\mu_1 - \mu_0)^2}$$

where β is the probability of accepting the hypothesis when $\mu = \mu_1$, i.e., Type II error, and K_β is the 100 β percentage point of the normal distribution.

It has already been verified that the acceptance regions presented for the one and two-sided procedures yield a probability of $1 - \alpha$ of accepting the hypothesis that $\mu = \mu_0$ when it is true. The derivation of the expression for the sample size for the one-sided procedure where rejection is desired if $\mu > \mu_0$ will follow. The derivations of the sample sizes for the two-sided procedure and for the one-sided procedure where rejection is desired if $\mu < \mu_0$ are similar.

If the procedure is one-sided and rejection is desired when the true mean exceeds μ_0, the acceptance region is given by the interval $[-\infty, K_\alpha]$. Thus, the hypothesis that $\mu = \mu_0$ (or $\mu \leqq \mu_0$) is accepted if

$$U \leqq K_\alpha$$

and the probability of accepting the hypothesis is given by

$$P\left(U = \frac{(\bar{x} - \mu_0)\sqrt{n}}{\sigma} \leqq K_\alpha\right).$$

Two points on the operating characteristic curve are specified, i.e., $(\mu_0, 1 - \alpha)$ and (μ_1, β). The proper choice of the acceptance region returns the point $(\mu_0, 1 - \alpha)$ on the OC curve. The second point requires that the following equality hold, namely

$$P\left(U = \frac{(\bar{x} - \mu_0)\sqrt{n}}{\sigma} \leqq K_\alpha\right) = \beta \quad \text{when } \mu = \mu_1.$$

This is equivalent to stating that the probability of accepting the hypothesis that $\mu = \mu_0$ is β when the mean of the x's is actually equal to μ_1. Whenever the mean of the x's is μ_1, the expected value of $\bar{x}$ is also μ_1 so that the expected value of U is given by

$$E(U) = \frac{(\mu_1 - \mu_0)\sqrt{n}}{\sigma}.$$

Hence, when $\mu = \mu_1$, U is normally distributed with mean equal to $\frac{(\mu_1 - \mu_0)\sqrt{n}}{\sigma}$ and unit variance. It then follows that

$$P(U \leqq K_\alpha) = P\left(z \leqq K_\alpha - \frac{(\mu_1 - \mu_0)\sqrt{n}}{\sigma}\right) = \beta$$

where z is a normally distributed random variable with zero mean and unit variance. This probability statement is illustrated in

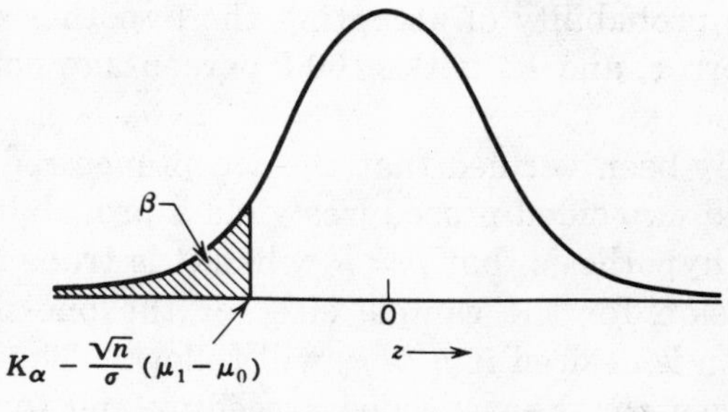

Fig. 6.8. The shaded area indicates $P\left[z \leqq K_\alpha - \frac{(\mu_1 - \mu_0)\sqrt{n}}{\sigma}\right]$.

Figure 6.8. Because of the symmetry of the normal distribution the shaded area of Figure 6.8 is equivalent to

$$P\left(z \geqq \left[\frac{(\mu_1 - \mu_0)\sqrt{n}}{\sigma}\right] - K_\alpha\right)$$

so that

$$P\left(z \geqq \left[\frac{(\mu_1 - \mu_0)\sqrt{n}}{\sigma}\right] - K_\alpha\right) = \beta.$$

Hence, $\left[\frac{(\mu_1 - \mu_0)\sqrt{n}}{\sigma}\right] - K_\alpha = K_\beta$, where K_β is the 100 β percentage point of the normal distribution. Solving this last expression for n leads to

$$n = \frac{(K_\alpha + K_\beta)^2\sigma^2}{(\mu_1 - \mu_0)^2},$$

which is the result mentioned earlier.

6.1.3.2 *The operating characteristic curve.* If the sample size and acceptance region are known, the entire OC curve is easily determined. Let the true mean of the x's be μ so that the expected value of U is

$$E(U) = \frac{(\mu - \mu_0)\sqrt{n}}{\sigma}.$$

U is then normally distributed with mean $\frac{(\mu - \mu_0)\sqrt{n}}{\sigma}$ and unit variance. For the one-sided procedure where rejection is desired when the true mean exceeds μ_0, the expression for the OC curve (probability of accepting the hypothesis) is given by

$$P(U \leqq K_\alpha)$$

which is equivalent to

$$P\left(z \leqq K_\alpha - \frac{(\mu - \mu_0)\sqrt{n}}{\sigma}\right)$$

where z is a normally distributed random variable with zero mean and unit variance. This last expression can be evaluated for any value of μ with the aid of Appendix Table 1. It is interesting to note that for fixed n and K_α this expression for the OC curve is just a function of $d = (\mu - \mu_0)/\sigma$, which is the abscissa of the OC curves plotted in Figure 6.6 and 6.7.

If the procedure is one-sided and rejection is desired when the true mean is less than μ_0, the OC curve is given by the expression

$$P(U \geqq -K_\alpha) = P\left(z \geqq -K_\alpha - \frac{(\mu - \mu_0)\sqrt{n}}{\sigma}\right).$$

Finally, if the procedure is two-sided, the OC curve is given by the expression

$$P(-K_{\alpha/2} \leqq U \leqq K_{\alpha/2})$$
$$= P\left(-K_{\alpha/2} - \frac{(\mu - \mu_0)\sqrt{n}}{\sigma} \leqq z \leqq K_{\alpha/2} - \frac{(\mu - \mu_0)\sqrt{n}}{\sigma}\right).$$

As an example of the use of the curve, we will return to the Rockwell hardness example and the two-sided procedure associated with it. The following data are available or have been obtained:

$$\alpha = 0.05, \qquad \sigma = 2,$$
$$K_{\alpha/2} = K_{0.025} = 1.96, \qquad n = 11,$$
$$\mu_0 = 72.$$

If the actual mean of the normal distribution is $\mu = 73$, the probability of accepting the hypothesis is given by

$$P\left(-1.96 - \frac{(73 - 72)\sqrt{11}}{2} \leqq z \leqq 1.96 - \frac{(73 - 72)\sqrt{11}}{2}\right)$$
$$= P(-3.62 \leqq z \leqq 0.30) = 0.62$$

6.1.4 EXAMPLE

A manufacturer produces a special alloy steel with an average tensile strength of 25,800 psi. A change in the composition of the alloy is said to increase the breaking strength. The standard deviation of the tensile strength is known to be 300 psi and the change in composition is not expected to change this value. If there is no change in the average tensile stength, the manufacturer wants to reach this conclusion with probability 0.99 ($\alpha = 0.01$). If the average tensile strength is increased by as much as 250 psi, the manufacturer is only willing to take a 10% risk in not detecting it ($\beta = 0.10$).

The hypothesis to be tested is that the average tensile strength is unaffected by the change in composition in the material against the alternative that the change in composition increases the average tensile strength. The possibility of the change causing a decrease is discounted. Thus, the acceptance rule is to accept the hypothesis if $U \leqq K_\alpha = K_{0.01} = 2.326$ since the level of significance is given as 0.01. The required sample size is found from the OC curve in Figure 6.7. From the information above $d = (\mu_1 - \mu_0)/\sigma = 250/300 = 0.833$. Entering Figure 6.7, with $d = 0.833$ and a probability of acceptance equal to 0.10, the required sample size is seen to be about 19 observations.

The experiment is performed and the sample mean $\bar{x}$ of the 19 observations is calculated and found to be 26,100 psi. It then follows that

$$U = \frac{(26{,}100 - 25{,}800)\sqrt{19}}{300} = 4.36,$$

which is greater than $K_{0.01} = 2.326$, so the hypothesis is rejected. Thus, it is concluded that the change in composition increases the tensile strength.

It is interesting to comment on this example when the maximum number of observations available is only 7; e.g., suppose funds are provided for no more than 7 specimens. In this case, Figure 6.7 reveals that if the increase in tensile strength is as much as 250 psi, this procedure detects such a difference only 50% of the time. Hence it may be advisable to wait until additional funds become available. If the experiment is performed with only 7 observations, and the procedure results in acceptance of the hypothesis, this does not necessarily imply that an increase in tensile strength has not been effected. Rather, it implies that the increase is not sufficiently large to detect it with a sample of size 7; another way of putting it is that there is insufficient information available to conclude that an increase has occurred.

6.2 Test of the Hypothesis that the Mean of a Normal Distribution Has a Specified Value when the Standard Deviation Is Unknown

6.2.1 Choice of an OC curve

Section 6.1 dealt with the problem of a test for the mean of a normal distribution having a specified value, assuming that the standard deviation is known. Such tests are often quite useful. In repetitive experiments, considerable information about variability may be accumulated and the assumptions of known standard deviation valid. Such a test may be applied quickly; often it is possible to tell at a glance whether an observed mean departs by more than two or three times its standard error from a hypothesized value. However, both the Type I error and the Type II error depend rather sensitively upon the assumed value of σ. In the example of 6.1.4, the standard deviation of tensile strength was assumed to be 300 psi and a test procedure was derived which had an error of the first kind (α) equal to 0.01, and error of the second kind (β) equal to 0.10. If this assumed standard deviation were off by 10% and the true value were, say, 270, then the Type I error (α) would be 0.0049 instead of 0.01 and the Type II error (β) would be 0.074 instead of 0.10.

A common occurrence in experimental work is that all of the information about the variability is contained in the data in hand. In a material testing laboratory a great deal of prior experimental data may be available for hardness experiments of comparable alloys, but in much physical and chemical experimentation, decisions are made on the basis of a single experiment which contains all the relevant information. It is of great importance to provide tests of hypothesis about means which do not depend on a known standard deviation.

In testing the hypothesis that the mean of a normal distribution is equal to μ_0 when the standard deviation is unknown, the t statistic is used. This test-statistic is a function of the sample mean, the sample standard deviation, and the sample size. The abscissae of operating characteristic curves for procedures based upon this t statistic depend upon the quantity $(\mu-\mu_0)/\sigma$. An example of such an OC curve is shown in Figure 6.9 for a two-sided test. In order to determine the

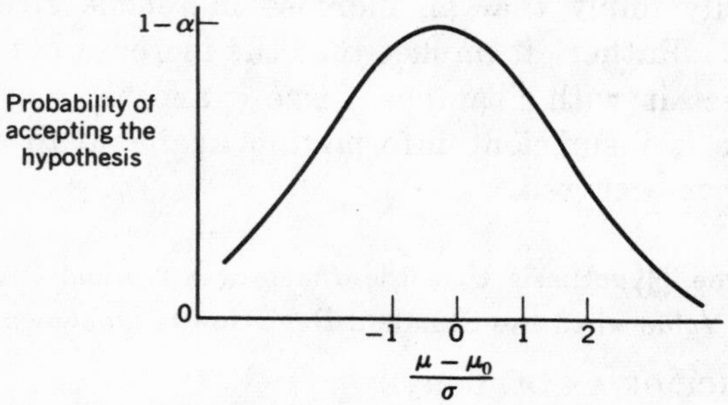

Fig. 6.9. OC curve of the two-sided t test.

acceptance region and sample size, two points on the OC curve must be chosen. These points will be denoted by $(0, 1-\alpha)$ and $\left(\frac{\mu_1-\mu_0}{\sigma}, \beta\right)$. The first point corresponds to choosing the probability of accepting the hypothesis when it is true (when $\mu = \mu_0$) and the second point corresponds to choosing the probability of accepting the hypothesis when $(\mu_1 - \mu_0)/\sigma$ is the value of the ratio of the difference in means to the true stardard deviation considered important to detect. The dependence of this second point on σ may appear to be incongruous at first since the t statistic is used specifically in situations when σ is unknown. However, a little thought will reveal that this is quite reasonable. If the experimental error (σ) is large, small differences in the mean cannot be considered important, whereas if the experimental error is small, these differences tend to become magnified. Hence, it is not the absolute values of the deviations from μ_0 which are important, but rather the relative deviations compared to σ which must be considered. For the purpose of choosing an OC curve a *rough* notion of σ is usually adequate. Before experimentation, such a notion can be obtained from past data or from some *a priori* knowledge. If the OC curve is examined after experimentation, the sample standard deviation can be used as a " rough " estimate. It is emphasized that a rough notion is *not* equivalent to a complete knowledge of the standard deviation.

6.2.2 Tables and Charts for Carrying Out t Tests

An experiment consisting of n trials is to be performed. The random variables $x_1, x_2, \cdots, x_n$ are the measurable quantities associated with the $1, 2, \cdots, n$ trials, respectively. It is assumed that the x_i are independent normally distributed random variables each with unknown mean μ and unknown variance σ^2. The *optimum* procedure for testing the hypothesis that the mean of the normal distribution has a specified value, i.e., $\mu = \mu_0$, is based upon the test statistic t, where

$$t = \frac{(\bar{x} - \mu_0)\sqrt{n}}{s}.$$

If the value taken on by the test statistic as a result of experimentation falls into the acceptance region, the hypothesis that μ equals μ_0 is accepted; otherwise it is rejected. This section will be devoted to finding the acceptance region and the sample size for the test statistic t, using the tables and charts which are provided. One- and two-sided procedures will be considered.

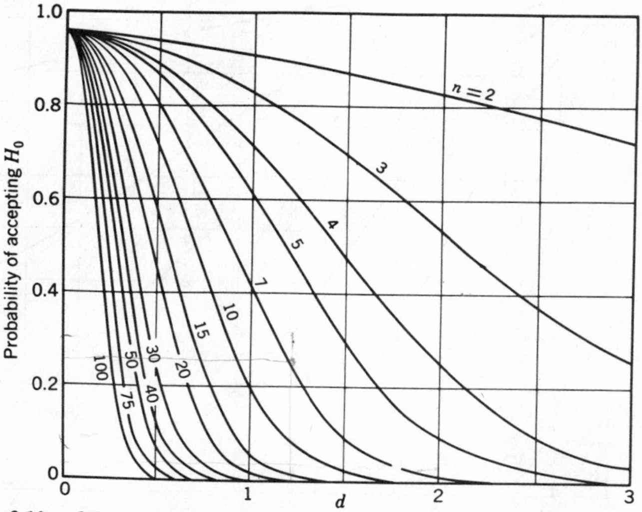

Fig. 6.10. OC curves for different values of n for the two-sided t test for a level of significance $\alpha=0.05$.

6.2.2.1 *Tables and charts for two-sided procedures.* The two-sided procedure will be analyzed first. Symmetry of the OC curve about zero will be assumed. This restriction can be relaxed by using the same methods as were presented in the section on the U test in this connection.

The acceptance region for the two-sided procedure for any sample size, n, is given by the interval

$$[-t_{\alpha/2;\,n-1},\ t_{\alpha/2;\,n-1}]$$

where α is the level of significance (Type I error) and $t_{\alpha/2;\,n-1}$ is the $100\,\alpha/2$ percentage point of the t distribution with $n-1$ degrees of freedom. Thus, the hypothesis that the mean of a normal distribution equals μ_0 is accepted if

$$-t_{\alpha/2;\,n-1} \leqq t \leqq t_{\alpha/2;\,n-1}$$

and rejected if t lies outside this interval. It is easily verified that this procedure yields a probability of $1-\alpha$ of accepting the hypothesis that $\mu = \mu_0$ when it is true. When $\mu = \mu_0$, the test statistic $t = \dfrac{(\bar{x}-\mu_0)\sqrt{n}}{s}$ has a t distribution with $n-1$ degrees of freedom. Hence, if $\mu = \mu_0$

$$P(-t_{\alpha/2;\,n-1} \leqq t \leqq t_{\alpha/2;\,n-1}) = 1-\alpha.$$

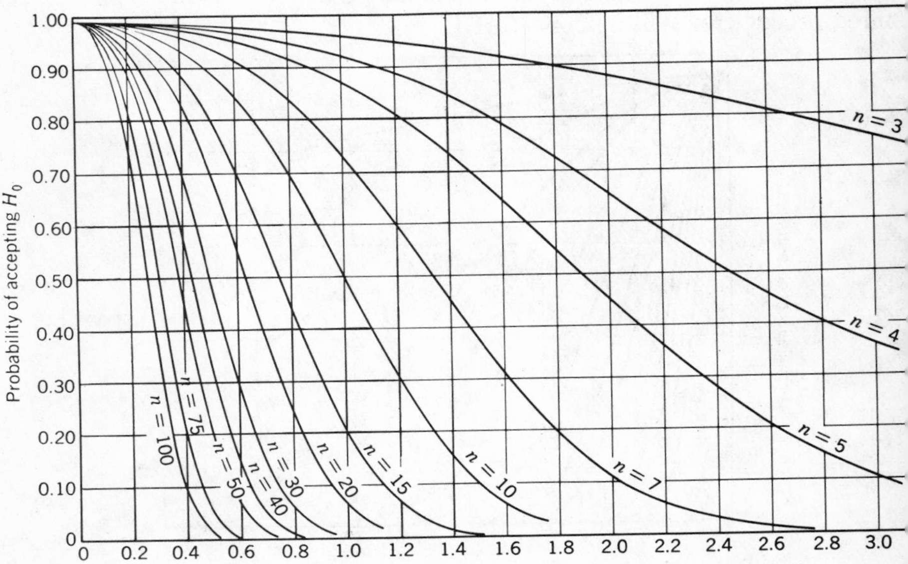

Fig. 6.11. OC curves for different values of n for the two-sided t test for a level of significance $\alpha = 0.01$.

The required sample size is obtained by referring to a set of operating characteristic curves which are provided for this purpose. Figure 6.10 presents the OC curves of the two-sided t test for a level of significance equal to 0.05. Figure 6.11 presents the OC curves of the two-sided t test for a level of significance equal to 0.01. The abscissa scale for these curves is d, where

$$d = \frac{|\mu - \mu_0|}{\sigma}.$$

If the level of significance of the procedure is either 0.05 or 0.01, the required sample size n is obtained by entering the appropriate figure (Figure 6.10 if $\alpha = 0.05$ and Figure 6.11 if $\alpha = 0.01$) with the point $\left(\frac{|\mu_1 - \mu_0|}{\sigma}, \beta\right)$ and reading out the value of the sample size associated with the OC curve that passes through this point.

6.2.2.2 *Summary for two-sided procedures using tables and charts.* The following steps are taken using a two-sided procedure for testing the mean of a normal distribution—σ unknown.

1. Two points on the OC curve are chosen and are translated to the d scales of Figures 6.10 and 6.11. On this scale, these points are denoted by $(0, 1 - \alpha)$ and $\left(\frac{|\mu_1 - \mu_0|}{\sigma}, \beta\right)$.
2. Find the required sample size n by entering the appropriate figure (Figure 6.10 if α=0.05 and Figure 6.11 if α=0.01) with the point $\left(\frac{|\mu_1 - \mu_0|}{\sigma}, \beta\right)$ and then reading out the value of the sample size associated with the OC curve that passes through this point.
3. Determine the acceptance region which is the interval
$$[-t_{\alpha/2; n-1}, t_{\alpha/2; n-1}].$$
4. Draw a random sample of n items.
5. Compute the value of the test statistic.
6. If $-t_{\alpha/2; n-1} \leqq t \leqq t_{\alpha/2; n-1}$, i.e., $|t| \leqq t_{\alpha/2; n-1}$, accept the hypothesis that $\mu = \mu_0$. If t lies outside this interval, reject the hypothesis and conclude that the mean is not μ_0.

6.2.2.3 *Tables and charts for one-sided procedures.* If the procedure is one-sided and rejection is desired when the true mean exceeds μ_0 (this being the alternative to the hypothesis that $\mu = \mu_0$ or $\mu \leqq \mu_0$), the acceptance region for any sample size, n, is given by the interval $[-\infty, t_{\alpha; n-1}]$ where α is the level of significance and $t_{\alpha; n-1}$ is the 100 α

percentage point of the t distribution with $n - 1$ degrees of freedom. Thus, the hypothesis that $\mu = \mu_0$ (or $\mu \leqq \mu_0$) is accepted if

$$t \leqq t_{\alpha; n-1}$$

and rejected if t lies outside this interval. This acceptance procedure has a probability of $1 - \alpha$ of accepting the hypothesis

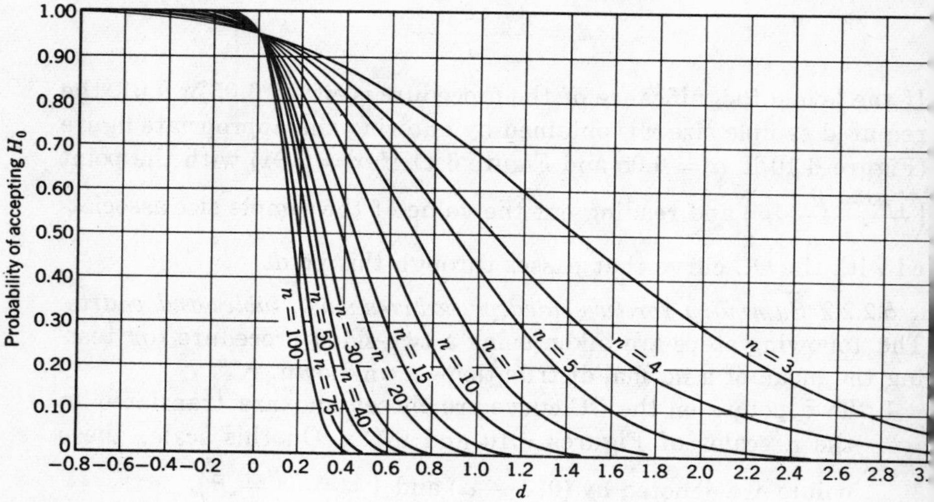

Fig. 6.12. OC curves for different values of n for the one-sided t test for a level of significance $\alpha = 0.05$.

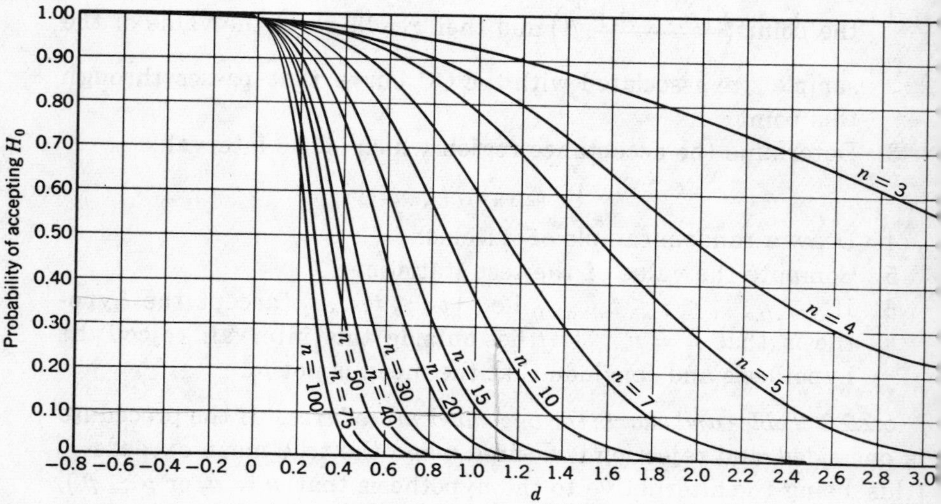

Fig. 6.13. OC curves for different values of n for the one-sided t test for a level of significance $\alpha = 0.01$.

that $\mu = \mu_0$ when it is true. When $\mu = \mu_0$, the test statistic $t = \frac{(\bar{x} - \mu_0)\sqrt{n}}{s}$ has a t distribution with $n - 1$ degrees of freedom. Hence, if $\mu = \mu_0$

$$P\,(t \leqq t_{\alpha;\,n-1}) = 1 - \alpha.$$

The operating characteristic curves for determining the sample size n are given in Figures 6.12 and 6.13 for the one-sided procedure where rejection is desired when the true mean exceeds μ_0. Figure 6.12 presents the OC curves for a level of significance equal to 0.05, and Figure 6.13 presents the OC curves for a level of significance equal to 0.01. The abscissa scale for these curves is d, where

$$d = \frac{(\mu - \mu_0)}{\sigma}.$$

If the level of significance of the procedure is either 0.05 or 0.01, the required sample size n is obtained by entering the appropriate figure (Figure 6.12 if $\alpha = 0.05$ and Figure 6.13 if $\alpha = 0.01$) with the point $\left(\frac{\mu_1 - \mu_0}{\sigma}, \beta\right)$ and reading out the value of the sample size associated with the OC curve that passes through this point.

If the procedure is one-sided and rejection is desired when the true mean is less than μ_0 (this being the alternative to the hypothesis that $\mu = \mu_0$ or $\mu \geqq \mu_0$), the acceptance region for any sample size n is given by the interval $[-t_{\alpha;\,n-1}, \infty]$. Thus, the hypothesis that $\mu = \mu_0$ (or $\mu \geqq \mu_0$) is accepted if

$$t \geqq -\, t_{\alpha;\,n-1}$$

and rejected if t lies outside this interval. This acceptance procedure also has a probability of $1-\alpha$ of accepting the hypothesis that $\mu = \mu_0$ when it is true.

The required sample size is again obtained by referring to Figures 6.12 and 6.13. In order to use these curves for this one-sided procedure, the abscissa scale d is now defined as

$$d = \frac{\mu_0 - \mu}{\sigma}.$$

If the level of significance of the procedure is either 0.05 or 0.01, the required sample size n is obtained by entering the appropriate figure (Figure 6.12 if $\alpha = 0.05$ and Figure 6.13 if $\alpha = 0.01$) with the point $\left(\frac{\mu_0 - \mu_1}{\sigma}, \beta\right)$ and reading out the value of the sample size associated with the OC curve that passes through this point.

6.2.2.4 *Summary for one-sided procedures using tables and charts.* The following steps are taken using a one-sided procedure for testing the mean of a normal distribution—σ unknown.

1. Two points on the OC curve are chosen. If rejection is desired when $\mu > \mu_0$, these points are denoted by $(0, 1 - \alpha)$ and $\left(\frac{\mu_1 - \mu_0}{\sigma}, \beta\right)$. If rejection is desired when $\mu < \mu_0$, these points are translated to the d scale and are then denoted by $(0, 1 - \alpha)$ and $\left(\frac{\mu_0 - \mu_1}{\sigma}, \beta\right)$.
2. Find the required sample size n by entering the appropriate figure (Figure 6.12 if $\alpha = 0.05$ and Figure 6.13 if $\alpha = 0.01$). If rejection is desired when $\mu > \mu_0$, the figure is entered with the point $\left(\frac{\mu_1 - \mu_0}{\sigma}, \beta\right)$ and the value of the sample size associated with the OC curve that passes through this point is read out. If rejection is desired when $\mu < \mu_0$, the figure is entered with the point $\left(\frac{\mu_0 - \mu_1}{\sigma}, \beta\right)$ and the value of the sample size associated with the OC curve that passes through this point is read out.
3. Determine the acceptance region which is the interval $[-\infty, t_{\alpha;n-1}]$ for the alternative $\mu > \mu_0$ and $[-t_{\alpha;n-1}, \infty]$ for the alternative $\mu < \mu_0$.
4. Draw a random sample of n items.
5. Compute the value of the test statistic $t = \frac{(\bar{x} - \mu_0)\sqrt{n}}{s}$.
6. If the alternative is $\mu > \mu_0$, accept the hypothesis that $\mu = \mu_0$ (or $\mu \leqq \mu_0$) whenever $t \leqq t_{\alpha;n-1}$; otherwise reject the hypothesis and conclude that $\mu > \mu_0$. If the alternative is $\mu < \mu_0$, accept the hypothesis that $\mu = \mu_0$ (or $\mu \geqq \mu_0$) whenever $t \geqq -t_{\alpha;n-1}$; otherwise reject the hypothesis and conclude that $\mu < \mu_0$.

6.2.2.5 *Tables and charts for operating characteristic curves.* Whenever the sample size is fixed in advance, or after it is chosen, the operating characteristic curves given in Figures 6.10 and 6.11 for two-sided procedures and Figures 6.12 and 6.13 for one-sided procedures can be used to evaluate the risks involved in using these procedures. If the level of significance is either 0.05 or 0.01, the appropriate OC curve is entered with the sample size n and d, and the probability of accepting the hypothesis is read from the curve. The value of d chosen is related to the value $(\mu - \mu_0)/\sigma$ for which the risk is to be evaluated.

If other levels of significance are desired, Appendix Table 3, which is a table of the percentage points of the t distribution, can be used to determine the acceptance region for a fixed value of n. However, analytical procedures which are beyond the scope of this text must be used to obtain the OC curve.

Although the OC curves are calculated under the assumption that the random variables $x_1, x_2, \cdots, x_n$ are independent normally (or approximately normally) distributed random variables, they are also valid for more general forms of underlying probability distributions, if the sample size is sufficiently large.

A summary of the procedures for testing the hypothesis that the mean of a normal distribution has a specified value when the standard deviation is unknown is given in Table 6.2. For purposes of this summary, criteria for rejecting the hypothesis are given instead of criteria for acceptance.

Table 6.2. Test of the Hypothesis that the Mean of a Normal Distribution Has a Specified Value when the Standard Deviation Is Unknown

Notation for Hypothesis	*Test Statistic*
$H: \mu = \mu_0$	$t = (\bar{x} - \mu_0)\sqrt{n}/s$
Criteria for Rejection	*Method for Choosing Sample Size*
$\|t\| \geqq t_{\alpha/2;n-1}$ if we wish to reject when the true mean departs from μ_0 in either direction.	Determine a value of $d = \|\mu_1 - \mu_0\|/\sigma$ for which we wish to reject the hypothesis with given high probability and enter Fig. 6.10 or 6.11 to find the necessary sample size.
$t \geqq t_{\alpha;n-1}$ if we wish to reject when the true mean exceeds μ_0.	Determine a value of $d = (\mu_1 - \mu_0)/\sigma$ for which we wish to reject the hypothesis with given high probability and enter Fig. 6.12 or 6.13 to find the necessary sample size.
$t \leqq -t_{\alpha;n-1}$ if we wish to reject when the true mean is less than μ_0	Determine a value of $d = (\mu_0 - \mu_1)/\sigma$ for which we wish to reject the hypothesis with given high probability and enter Fig. 6.12 or 6.13 to find the necessary sample size.

Values of $t_{\alpha;\nu}$, the $100\,\alpha$ percentage points of the t distribution for ν degrees of freedom, are given in Appendix Table 3.

6.2.3 EXAMPLE OF t TESTS

(a) In the manufacture of a food product, the label states that the box contains 10 lbs. The boxes are filled by machine and it is of interest to determine whether or not the machine is set properly. Previous experience has indicated that the standard deviation is approximately 0.05 lb., although this is not known precisely. The company feels that the present setting is unsatisfactory if the machine fills the boxes so that the mean weight differs from 10 lbs. by more than 0.1 lb. If this be the case, the probability of detecting a difference of this magnitude must not be less than 0.95 (i.e., $\beta = 0.05$). It is decided to tolerate a level of significance of 5% (i.e., $\alpha = 0.05$). Thus, from Figure 6.10 and $d = 0.1/0.05 = 2$, a sample size of 6 is sufficient to insure rejection with probability 0.95 or greater if the mean weight differs from 10 lbs. by more than 0.1 lb. The data are: 9.99 lbs., 9.99 lbs., 10.00 lbs., 10.11 lbs., 10.09 lbs., 9.95 lbs.

$$\bar{x} = 10.021667.$$[1]

$$s^2 = \frac{\sum (x_i - \bar{x})^2}{n-1} = \frac{\sum x_i^2 - n\bar{x}^2}{n-1} = \frac{602.6229 - 602.6029}{5} = 0.0040;$$

$$s = 0.0632.$$

The hypothesis that the machine setting is 10 lbs. is accepted if $-t_{0.025;\,5} \leqq t \leqq t_{0.025;\,5}$. The value of the test statistic is

$$t = \frac{(\bar{x} - 10)\sqrt{n}}{s} = \frac{(10.0217 - 10)\sqrt{6}}{0.0632} = 0.841.$$

From Appendix Table 3, $t_{\alpha/2;\,n-1} = t_{0.025;\,5} = 2.571$. Thus, the value of the test statistic falls within the acceptance region so that it is unnecessary to change the machine setting.

(b) According to schedule, a given operation is supposed to be performed in 6.4 minutes. A study is performed to determine whether a particular worker conforms to the standard; that is, whether deviations from standard can be regarded as random fluctuations or whether they indicate that the performance achieved deviates systematically, either higher or lower. If the departure is as much as 0.5 minute, a change in the standard should be made with a probability of at least 0.95 (i.e., $\beta = 0.05$). Experience indicates that the standard deviation of such operations should be about 0.4 minute. A level of significance of 1% ($\alpha = 0.01$) will be satisfactory. Thus Figure 6.11 is entered with $d = 0.5/0.4 = 1.25$; a sample size of $n = 15$

[1] It is important to compute $\bar{x}$ to more places than appears necessary for the significance of the final result. The reason becomes clear upon looking at the calculation of s^2 where a subtraction operation is involved.

is found which will guarantee 95% rejection if departures of more than 0.5 from the mean occur. A sample of 15 times of operation of a worker is taken and the data in minutes are found to be: 6.10, 6.65, 7.00, 6.25, 6.35, 6.85, 7.10, 7.35, 7.05, 7.50, 6.90, 6.70, 7.20, 7.15, and 6.95. The sample statistics are found to be as follows:

$$\bar{x} = 6.87333,$$
$$s^2 = 0.16067,$$

and
$$s = 0.4008.$$

The value of the test statistic is

$$t = \frac{(\bar{x} - 6.4)\sqrt{15}}{s} = \frac{(6.87333 - 6.4)\sqrt{15}}{0.4008} = 4.574.$$

This value is compared to $t_{\alpha/2;\,n-1} = t_{0.005;\,14} = 2.977$. Thus, $t = 4.574 > t_{\alpha/2;\,n-1} = 2.977$ and it is concluded that the fluctuation in time cannot be regarded as random and that a change must be made in the standard.

6.3 Test of the Hypothesis that the Standard Deviation of a Normal Distribution Has a Specified Value

6.3.1 Choice of an OC curve

The earlier sections of this chapter dealt with the problem of tests of the mean of a normal distribution. Another common problem is to test the hypothesis that the standard deviation of the normal distribution is equal to σ_0.[1] The test procedure is based upon the chi-square statistic. This test statistic is a function of the sample

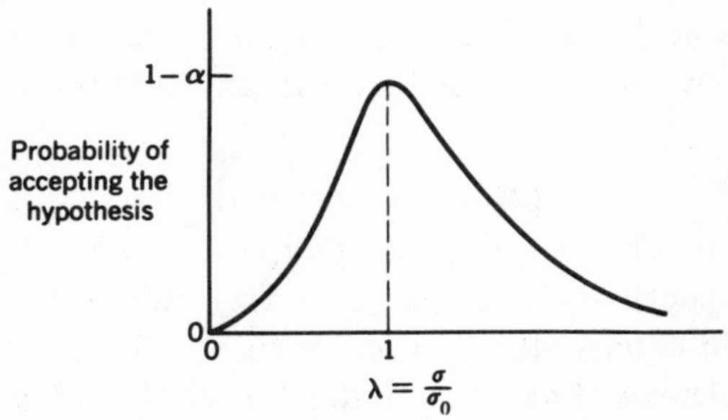

Fig. 6.14. OC curve of the two-sided chi-square test.

variance s^2, or more particularly, of the ratio of s^2 to σ_0^2. The abscissae of operating characteristic curves for procedures based upon this chi-square statistic depend upon the quantity $\lambda = \sigma/\sigma_0$.

[1] This is equivalent to testing the hypothesis that the variance of the normal distribution equals σ_0^2.

An example of such an OC curve is shown in Figure 6.14 for a two-sided test. In order to determine the acceptance region and sample size, two points on the OC curve must be chosen. These points will be denoted by $(1, 1 - \alpha)$ and $\left(\frac{\sigma_1}{\sigma_0}, \beta\right)$. The first point corresponds to choosing the probability of accepting the hypothesis when it is true (when $\sigma = \sigma_0$) and the second point corresponds to choosing the probability of accepting the hypothesis when $\sigma = \sigma_1$ is the value of the true standard deviation considered important to detect.

6.3.2 Charts and tables to design tests of dispersion

An experiment consisting of n trials is to be performed. The random variables $x_1, x_2, \cdots, x_n$ are the measurable quantities associated with the $1, 2, \cdots, n$ trials, respectively. It is assumed that the x_i are independent normally distributed random variables each with unknown mean μ and unknown variance σ^2. The *optimum* procedure for testing the hypothesis that the standard deviation of the normal distribution has a specified value, i.e., $\sigma = \sigma_0$, is based upon the test statistic χ^2, where

$$\chi^2 = \frac{(n-1)s^2}{\sigma_0^2}.$$

If the value taken on by the test statistic as a result of experimentation falls into the acceptance region, the hypothesis that $\sigma = \sigma_0$ is accepted; otherwise it is rejected. This section will be devoted to finding the acceptance region and the sample size for the test statistic χ^2, using the tables and charts which are provided. One- and two-sided procedures will be considered.

6.3.2.1 *Tables and charts for two-sided procedures.* The acceptance region for the two-sided procedure for any sample size n is given by the interval

$$[\chi^2_{1-\alpha/2;\, n-1}, \chi^2_{\alpha/2;\, n-1}]$$

where α is the level of significance (Type I error) and $\chi^2_{1-\alpha/2;\, n-1}$ and $\chi^2_{\alpha/2;\, n-1}$ are respectively the $100(1 - \alpha/2)$ and $100\ \alpha/2$ percentage points of the chi-square distribution with $n - 1$ degrees of freedom. Thus, the hypothesis that the standard deviation of a normal distribution equals σ_0 is accepted if

$$\chi^2_{1-\alpha/2;\, n-1} \leqq \chi^2 \leqq \chi^2_{\alpha/2;\, n-1}$$

and rejected if χ^2 lies outside this interval. It is easily verified that this procedure yields a probability of $1 - \alpha$ of accepting the hypothesis that $\sigma = \sigma_0$ when it is true. When $\sigma = \sigma_0$, the test statistic

$\chi^2 = (n-1)s^2/\sigma_0^2$ has a chi-square distribution with $n-1$ degrees of freedom. Hence, if $\sigma = \sigma_0$,

$$P(\chi^2_{1-\alpha/2;\,n-1} \leqq \chi^2 \leqq \chi^2_{\alpha/2;\,n-1}) = 1 - \alpha.$$

The required sample size is obtained by referring to a set of operating characteristic curves which are provided for this purpose. Figure 6.15 presents the OC curves of the two-sided chi-square test

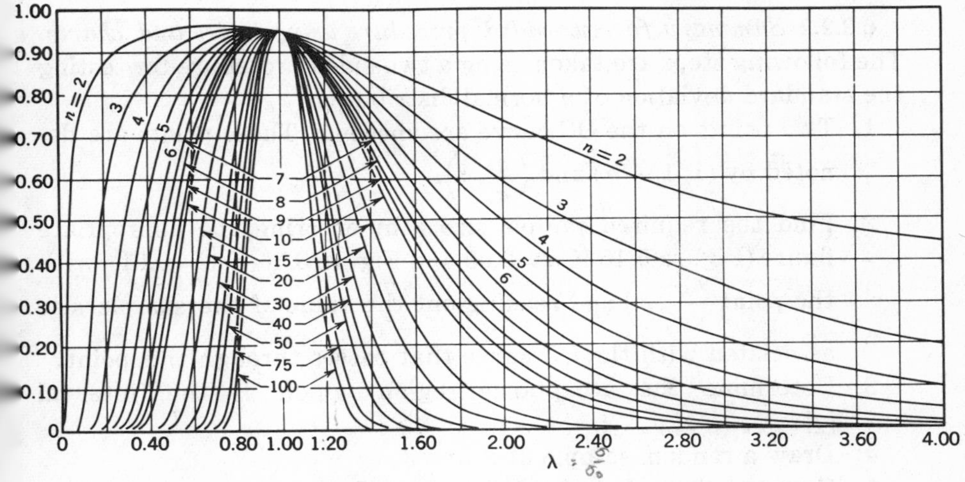

Fig. 6.15. OC curves for different values of n for the two-sided chi-square test for a level of significance $\alpha = 0.05$.

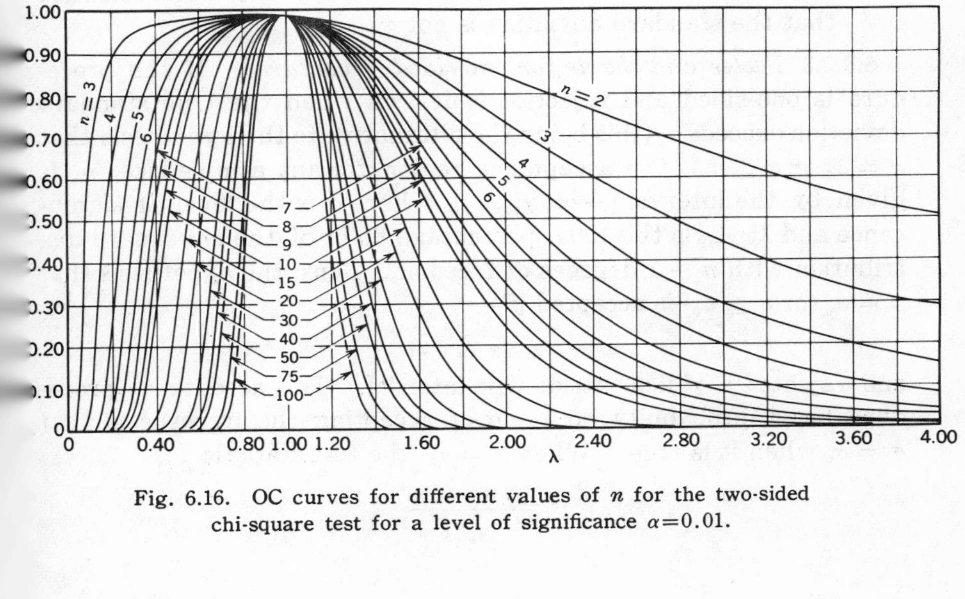

Fig. 6.16. OC curves for different values of n for the two-sided chi-square test for a level of significance $\alpha = 0.01$.

for a level of significance equal to 0.05. Figure 6.16 presents the OC curves of the two-sided chi-square test for a level of significance equal to 0.01. If the level of significance of the procedure is either 0.05 or 0.01, the required sample size n is obtained by entering the appropriate figure (Figure 6.15 if $\alpha=0.05$ and Figure 6.16 if $\alpha=0.01$) with the point $\left(\frac{\sigma_1}{\sigma_0}, \beta\right)$ and reading out the value of the sample size associated with the OC curve that passes through this point.

6.3.2.2 *Summary for two-sided procedure using tables and charts.* The following steps are taken using a two-sided procedure for testing the standard deviation of a normal distribution.

1. Two points on the OC curve are chosen. These points are denoted by $(1, 1 - \alpha)$ and $\left(\frac{\sigma_1}{\sigma_0}, \beta\right)$.
2. Find the required sample size n by entering the appropriate figure (Figure 6.15 if $\alpha=0.05$ and Figure 6.16 if $\alpha=0.01$) with the point $\left(\frac{\sigma_1}{\sigma_0}, \beta\right)$ and reading out the value of the sample size associated with the OC curve that passes through this point.
3. Determine the acceptance region which is the interval $[\chi^2_{1-\alpha/2;\,n-1}, \chi^2_{\alpha/2;\,n-1}]$.
4. Draw a random sample of n items.
5. Compute the value of the test statistic.
6. If $\chi^2_{1-\alpha/2;\,n-1} \leqq \chi^2 \leqq \chi^2_{\alpha/2;\,n-1}$, accept the hypothesis that $\sigma = \sigma_0$. If χ^2 lies outside this interval, reject the hypothesis and conclude that the standard deviation is not σ.

6.3.2.3 *Tables and charts for one-sided procedures.* If the procedure is one-sided and rejection is desired when the true standard deviation exceeds σ_0 (this being the alternative to the hypothesis that $\sigma = \sigma_0$ or $\sigma \leqq \sigma_0$), the acceptance region for any sample size, n, is given by the interval $[-\infty, \chi^2_{\alpha;\,n-1}]$, where α is the level of significance and $\chi^2_{\alpha;\,n-1}$ is the 100 α percentage point of the chi-square distribution with $n - 1$ degrees of freedom. Thus, the hypothesis that $\sigma = \sigma_0$ (or $\sigma \leqq \sigma_0$) is accepted if

$$\chi^2 \leqq \chi^2_{\alpha;\,n-1}$$

and rejected if χ^2 lies outside this interval. This acceptance procedure has a probability of $1 - \alpha$ of accepting the hypothesis that $\sigma = \sigma_0$ when it is true. When $\sigma = \sigma_0$, the test statistic

$$\chi^2 = \frac{(n-1)s^2}{\sigma_0^2}$$

has a chi-square distribution with $n - 1$ degrees of freedom. Hence, if $\sigma = \sigma_0$,

$$P(\chi^2 \leqq \chi^2_{\alpha;\, n-1}) = 1 - \alpha.$$

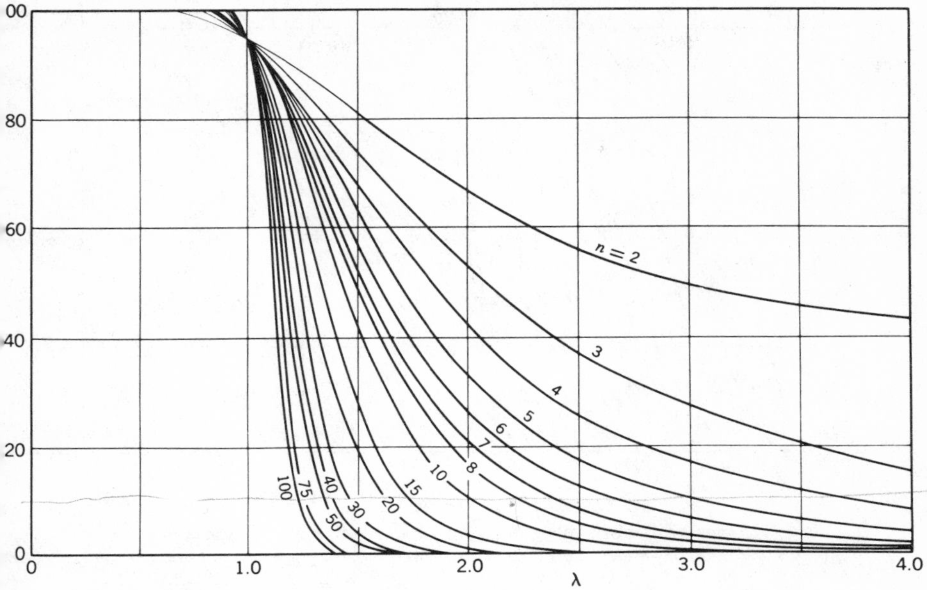

Fig. 6.17. OC curves for different values of n for the one-sided (upper tail) chi-square test for a level of significance $\alpha = 0.05$.

Reproduced by permission from "Operating Characteristics for the Common Statistical Tests of Significance" by Charles L. Ferris, Frank E. Grubbs, Chalmers L. Weaver, Annals of Mathematical Statistics, June, 1946

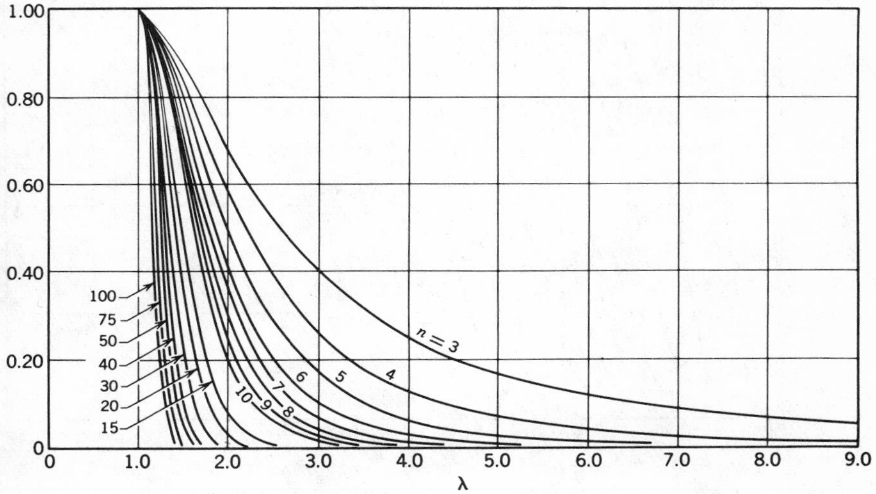

Fig. 6.18. OC curves for different values of n for the one-sided (upper tail) chi-square test for a level of significance $\alpha = 0.01$.

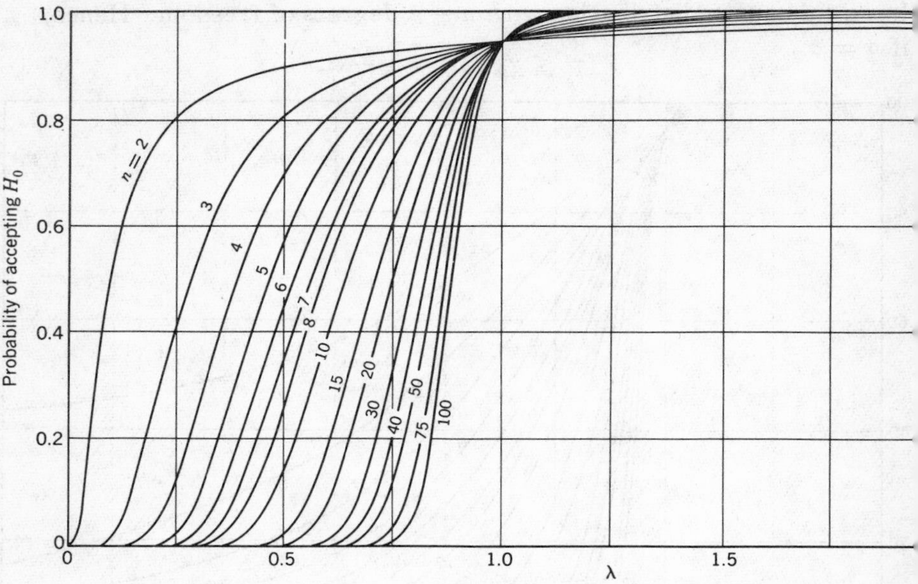

Fig. 6.19. OC curves for different values of n for the one-sided (lower tail) chi-square test for a level of significance $\alpha=0.05$.

Reproduced by permission from "Operating Characteristics for the Common Statistical Tests of Significance" by Charles L. Ferris, Frank E. Grubbs, Chalmers L. Weaver, Annals of Mathematical Statistics, June, 1946

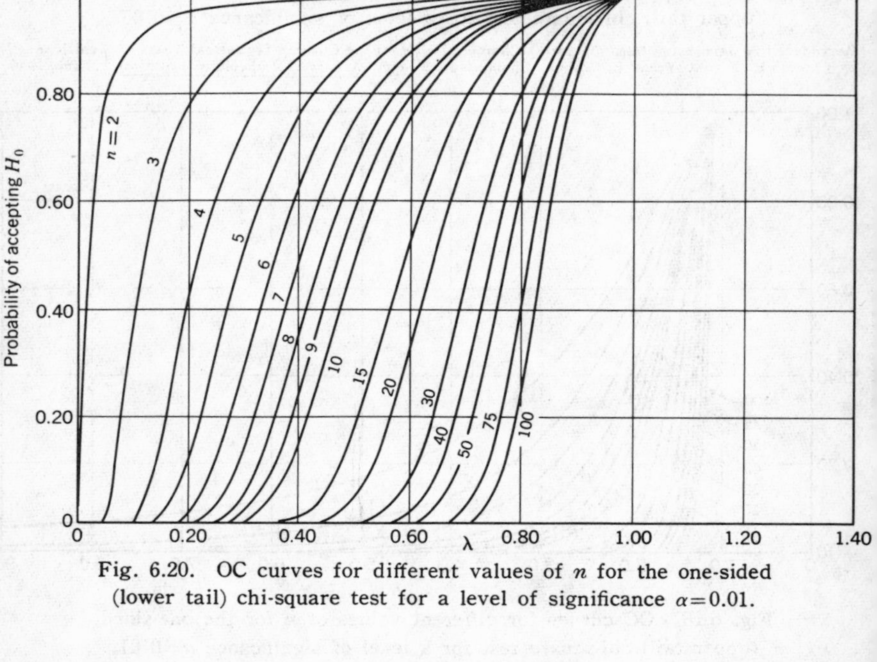

Fig. 6.20. OC curves for different values of n for the one-sided (lower tail) chi-square test for a level of significance $\alpha=0.01$.

The operating characteristic curves for determining the sample size n are given in Figures 6.17 and 6.18 for the one-sided procedure where rejection is desired when the true standard deviation exceeds σ_0. Figure 6.17 presents the OC curves for a level of significance equal to 0.05 and Figure 6.18 presents the OC curves for a level of significance equal to 0.01. If the level of significance of the procedure is either 0.05 or 0.01, the required sample size n is obtained by entering the appropriate figure (Figure 6.17 if $\alpha = 0.05$ and Figure 6.18 if $\alpha = 0.01$) with the point $\left(\frac{\sigma_1}{\sigma_0}, \beta\right)$ and reading out the value of the sample size associated with the OC curve that passes through this point.

If the procedure is one-sided and rejection is desired when the true standard deviation is less than σ_0 (this being the alternative to the hypothesis that $\sigma = \sigma_0$ or $\sigma \geqq \sigma_0$), the acceptance region for any sample size, n, is given by the interval $[\chi^2_{1-\alpha;\,n-1}, \infty]$. Thus, the hypothesis that $\sigma = \sigma_0$ (or $\sigma \geqq \sigma_0$) is accepted if

$$\chi^2 \geqq \chi^2_{1-\alpha;\,n-1}$$

and rejected if χ^2 lies outside this interval. This acceptance procedure also has a probability of $1 - \alpha$ of accepting the hypothesis that $\sigma = \sigma_0$ when it is true.

The operating characteristic curves for determining the sample size n are given in Figures 6.19 and 6.20 for the one-sided procedure where rejection is desired when the true standard deviation is less than σ_0. Figure 6.19 presents the OC curves for a level of significance equal to 0.05 and Figure 6.20 presents the OC curves of this one-sided procedure for a level of significance equal to 0.01. If the level of significance of the procedure is either 0.05 or 0.01, the required sample size n is obtained by entering the appropriate figure (Figure 6.19 if $\alpha = 0.05$ and Figure 6.20 if $\alpha = 0.01$) with the point $\left(\frac{\sigma_1}{\sigma_0}, \beta\right)$ and reading out the value of the sample size associated with the OC curve that passes through this point.

6.3.2.4 *Summary for one-sided procedures using tables and charts.* The following steps are taken using a one-sided procedure for testing the standard deviation of a normal distribution.

1. Two points on the OC curve are chosen. These points are denoted by $(1, 1 - \alpha)$ and $\left(\frac{\sigma_1}{\sigma_0}, \beta\right)$.
2. Find the required sample size n by entering the appropriate

figure. If rejection is desired when $\sigma > \sigma_0$, Figure 6.17 (if $\alpha = 0.05$) or Figure 6.18 (if $\alpha = 0.01$) is entered with the point $\left(\frac{\sigma_1}{\sigma_0}, \beta\right)$ and the value of the sample size associated with the OC curve that passes through this point is read out. If rejection is desired when $\sigma < \sigma_0$, Figure 6.19 (if $\alpha = 0.05$) or Figure 6.20 (if $\alpha = 0.01$) is entered with the point $\left(\frac{\sigma_1}{\sigma_0}, \beta\right)$ and the value of the sample size associated with the OC curve that passes through this point is read out.

3. Determine the acceptance region which is the interval $[-\infty, \chi^2_{\alpha; n-1}]$ for the alternative $\sigma > \sigma_0$ and $[\chi^2_{1-\alpha; n-1}, \infty]$ for the alternative $\sigma < \sigma_0$.
4. Draw a random sample of n items.
5. Compute the value of the test statistic $\chi^2 = \dfrac{(n-1)s^2}{\sigma_0^2}$.
6. If the alternative is $\sigma > \sigma_0$, accept the hypothesis that $\sigma = \sigma_0$ (or $\sigma \leqq \sigma_0$) whenever $\chi^2 \leqq \chi^2_{\alpha; n-1}$; otherwise reject the hypothesis and conclude that $\sigma > \sigma_0$. If the alternative is $\sigma < \sigma_0$, accept the hypothesis that $\sigma = \sigma_0$ (or $\sigma \geqq \sigma_0$) whenever $\chi^2 \geqq \chi^2_{1-\alpha; n-1}$; otherwise reject the hypothesis and conclude that $\sigma < \sigma_0$.

6.3.2.5 *Tables and charts for operating characteristic curves.* Whenever the sample size is fixed in advance, or after it is chosen, the operating characteristic curves given in Figures 6.15 and 6.16 for two-sided procedures and Figures 6.17, 6.18, 6.19, and 6.20 for one-sided procedures can be used to evaluate the risks involved in using these procedures. If the level of significance is either 0.05 or 0.01, the appropriate OC curve is entered with the sample size n and λ, and the probability of accepting the hypothesis is read from the curve. The value of λ chosen corresponds to the value of σ/σ_0 for which the risk is to be evaluated.

If other levels of significance are desired, Appendix Table 2, which is a table of the percentage points of the chi-square distribution, can be used to determine the acceptance region for a fixed value of n. However, analytical procedures which are presented in Section 6.3.3 must be used to obtain the OC curve.

A summary of the prodedures for testing the hypothesis that the standard deviation of a normal distribution has a specified value is given in Table 6.3. For purposes of this summary, criteria for rejecting the hypothesis are given instead of criteria for acceptance.

Table 6.3. Test of a Hypothesis that the Standard Deviation of a Normal Distribution Has a Specified Value

Notation for Hypothesis	*Test Statistic*
$H : \sigma = \sigma_0$	$\chi^2 = (n-1)s^2/\sigma_0^2$
Criteria for Rejection	*Method for Choosing Sample Size*
$\chi^2 \geqq \chi^2_{\alpha/2;\,n-1}$ or $\chi^2 \leqq \chi^2_{1-\alpha/2;\,n-1}$ if we wish to reject when the true standard deviation differs from σ_0 in either direction.	Choose a value of the relative error, $\lambda = \sigma_1/\sigma_0$, for which we wish to reject the hypothesis with given high probability; enter Fig. 6.15 or 6.16 to find the required sample size.
$\chi^2 \geqq \chi^2_{\alpha;\,n-1}$ if we wish to reject when the true standard deviation exceeds σ_0.	Choose a value $\sigma_1 > \sigma_0$ for which we wish to reject the hypothesis with given high probability; calculate $\lambda = \sigma_1/\sigma_0$ and enter Fig. 6.17 or 6.18 to find the required sample size.
$\chi^2 \leqq \chi^2_{1-\alpha;\,n-1}$ if we wish to reject when the true standard deviation is less than σ_0.	Choose a value $\sigma_1 < \sigma_0$ for which we wish to reject the hypothesis with given high probability; calculate $\lambda = \sigma_1/\sigma_0$ and enter Fig. 6.19 or 6.20 to find the required sample size.

Values of $\chi^2_{\alpha;\,\nu}$, the 100 α percentage points of the chi-square distribution for ν degrees of freedom are given in Appendix Table 2.

6.3.3 ANALYTICAL TREATMENT FOR CHI-SQUARE TESTS

In Section 4.2.3 it was pointed out that the quantity

$$\frac{(n-1)s^2}{\sigma^2}$$

has a chi-square distribution with $n-1$ degrees of freedom, provided σ^2 is the variance of the x's which constitute s^2. If we are interested in testing σ_0^2 against alternatives of the sort $\sigma^2 > \sigma_0^2$, the decision rule is to reject when $(n-1)s^2/\sigma_0^2$ is large. If we want a test with a Type I error equal to α, the rule of acceptance is to accept the hypothesis whenever

$$\frac{(n-1)s^2}{\sigma_0^2} \leqq \chi^2_{\alpha;\,n-1}$$

or equivalently

$$s^2 \leqq \frac{\chi^2_{\alpha;\,n-1}\,\sigma_0^2}{n-1}.$$

It is clear that when $\sigma = \sigma_0$,

$$P\left(\frac{(n-1)s^2}{\sigma_0^2} \leqq \chi^2_{\alpha;\,n-1}\right) = 1 - \alpha.$$

The OC curve of the test can be computed easily from a table of the chi-square distribution. Suppose that the true value of the variance is σ^2. Then

$$\frac{(n-1)s^2}{\sigma^2}$$

has a chi-square distribution with $n-1$ degrees of freedom, and the probability of accepting the hypothesis is given by

$$P\left(\frac{(n-1)s^2}{\sigma_0^2} \leqq \chi^2_{\alpha;\,n-1}\right) = P\left(s^2 \leqq \frac{\chi^2_{\alpha;\,n-1}\sigma_0^2}{n-1}\right)$$
$$= P\left(\frac{(n-1)s^2}{\sigma^2} \leqq \chi^2_{\alpha;\,n-1}\frac{\sigma_0^2}{\sigma^2}\right);$$

if we let $\lambda = \sigma/\sigma_0$, we have the expression for the OC curve given by $P(\chi^2 \leqq \chi^2_{\alpha;\,n-1}/\lambda^2)$, where χ^2 denotes a random variable having a chi-square distribution with $n-1$ degrees of freedom. Note that aside from the error of the first kind and number of observation, the OC curve depends only on λ. If $\lambda = 1$, that is, the hypothesis $\sigma = \sigma_0$ is true, then

$$P\left(\chi^2 \leqq \frac{\chi^2_{\alpha;\,n-1}}{\lambda^2}\right) = P(\chi^2 \leqq \chi^2_{\alpha;\,n-1}) = 1 - \alpha.$$

The expression in this form permits easy solution of the sample size problem. Suppose that a particular value of σ_1 is given, which we want to be fairly sure to reject. If $\lambda_1 = \sigma_1/\sigma_0$, the condition is that

$$P\left(\frac{(n-1)s^2}{\sigma_0^2} \leqq \chi^2_{\alpha;\,n-1}\right) = P\left(\chi^2 \leqq \frac{\chi^2_{\alpha;\,n-1}}{\lambda_1^2}\right) = \beta$$

where β is the Type II error associated with λ_1.

This last expression implies that

$$\frac{\chi^2_{\alpha;\,n-1}}{\lambda_1^2} = \chi^2_{1-\beta;\,n-1}$$

or

$$\lambda_1^2 = \frac{\chi^2_{\alpha;\,n-1}}{\chi^2_{1-\beta;\,n-1}}.$$

We can determine the sample size for this problem by going to a table of the percentage points of the chi-square distribution, Appendix Table 2, and finding the degrees of freedom for which the ratio of the 100 αth percentage point to the 100 $(1-\beta)$th percentage point is equal to λ_1^2. For example, if $\alpha = 0.01$, $\beta = 0.05$, and $\lambda_1^2 = 4.2$, Appendix Table 2 indicates that the 15 degrees of freedom have values of $\chi^2_{0.01;15} = 30.578$ and $\chi^2_{0.95;15} = 7.261$ such that $\lambda_1^2 = 30.578/7.261 = 4.2$. Hence the required sample size is 16. Similar analytic treatment can be provided when we are interested in lower-tailed or two-tailed tests.

If rejection is desired when $\sigma < \sigma_0$, the expression for the OC curve (probability of accepting the hypothesis) is given by

$$P\left(\chi^2 \geqq \frac{\chi^2_{1-\alpha;n-1}}{\lambda^2}\right)$$

where χ^2 denotes a random variable having a chi-square distribution with $n - 1$ degrees of freedom and $\lambda = \sigma/\sigma_0$. The expression required for determining the sample size is given by

$$\lambda_1^2 = \frac{\chi^2_{1-\alpha;n-1}}{\chi^2_{\beta;n-1}}$$

If the two-sided chi-square test is used, the expression for the OC curve (probability of accepting the hypothesis) is given by

$$P\left(\frac{\chi^2_{1-\alpha/2;n-1}}{\lambda^2} \leqq \chi^2 \leqq \frac{\chi^2_{\alpha/2;n-1}}{\lambda^2}\right)$$

where χ^2 denotes a random variable having a chi-square distribution with $n - 1$ degrees of freedom and $\lambda = \sigma/\sigma_0$. The expression required for determining the sample size is given approximately by

$$\lambda_1^2 = \frac{\chi^2_{\alpha/2;n-1}}{\chi^2_{1-\beta;n-1}}$$

when $P\left(\chi^2 \leqq \frac{\chi^2_{1-\alpha/2;n-1}}{\lambda_1^2}\right)$ is small compared to β or by

$$\lambda_1^2 = \frac{\chi^2_{1-\alpha/2;n-1}}{\chi^2_{\beta;n-1}}$$

when $P\left(\chi^2 \geqq \frac{\chi^2_{\alpha/2;n-1}}{\lambda_1^2}\right)$ is small compared to β.

6.3.4 Example

The standard deviation of a dimension of a standard product is $\sigma_0 = 0.1225$ inch. This standard product is somewhat obsolete and a new product is under consideration. It will be adopted if the variability of this dimension is not substantially larger than that of the standard. If the standard deviation, σ, of the new product is as large as 0.2450 inch, the probability of adopting this product should not exceed 1% ($\beta = 0.01$). The test is to be made at the 1% level of significance. The hypothesis to be tested, then, is that $\sigma \leqq \sigma_0 = 0.1225$ against the one-sided alternative that $\sigma > \sigma_0$. From Figure 6.18 it appears that a sample of approximately 25 will give a probability of acceptance of 0.01 for $\lambda = 0.2450/0.1225 = 2$. A sample of 25 items is drawn and s^2 is found to be 0.0384. Hence, the value of the test statistic, $\chi^2 = (n - 1)s^2/\sigma_0^2$, is 61.4, which exceeds $\chi^2_{0.01;\,24} = 42.980$, so that the hypothesis is rejected. The new product is not adopted.

Suppose the company allotted a total of only 7 observations for the experiment and kept the 1% level of significance. The hypothesis that $\sigma \leqq 0.1225$ is accepted if $\chi^2 = (n - 1)s^2/\sigma_0^2 \leqq \chi^2_{0.01;\,6} = 16.812$; otherwise it is rejected. Suppose further that s^2 computed from the seven observations was equal to the value of s^2 found for the 25 observations, namely 0.0384. The value of the test statistic is now 15.4, which is less than $\chi^2_{0.01;\,6} = 16.812$ so that the hypothesis is accepted. The new product is adopted. A reference to Figure 6.18 reveals that for $n = 7$ and $\lambda = 2$, the probability of accepting the hypothesis is approximately 35%. In other words, if the standard deviation of the new product is as large as 0.2450, only about 2 out of 3 times will the experimental procedure result in its detection. The hypothesis will be accepted 1 out of 3 times. Thus, even if the hypothesis is accepted, as in this example when $n = 7$, one is unable to feel secure that the standard deviation is not as large as 0.2450.

PROBLEMS

1. The lot average warpwise breaking strength of a certain type of cloth is required to be not less than 180 pounds per square inch. The standard deviation based on past experience is 5 psi. A shipment of a lot of this cloth is received from a supplier and specimens are withdrawn from three pieces. These are tested with the following results:

first piece:	182 psi
second piece:	172 psi
third piece:	177 psi

Using a 5% level of significance, should the lot be accepted? What is the probability of accepting a lot which has a lot average warpwise breaking strength of 170 psi?

2. A manufacture of synthetic rubber claims that the average hardness of his rubber is 64.3 degrees Shore. Past experience indicates that the standard deviation of hardness is 2 degrees Shore. It is felt that this claim may be an overestimate or an underestimate, so an experiment is to be performed. If the true average (μ) is 64.3 degrees, the probability should be 0.95 of reaching this conclusion. Furthermore, if the average strengths differ by as much as ± 1 degree, the procedure should result in the conclusion that μ does not equal 64.3 degrees with probability greater than 0.80. (a) What is the necessary sample size for this experiment? (b) If $\bar{x}=65$ degrees and n is that value found in (a), should we conclude that the hardness is 64.3? (c) Compute exactly the probability of concluding that the average hardness of rubber is 64.3 degrees Shore when the true average hardness is 65 degrees Shore using the sample size found in (a). Do not use the OC curves in the text but compute the exact probability. You can check your answer with the value read off the curve.

3. In a chemical process it is very important that a certain solution to be used as a reactant has a *p*H of 8.30. A method for determining *p*H is available which for solutions of this type is known to give measurements which are approximately normally distributed with a known standard deviation of $\sigma=0.02$. If the *p*H is really 8.30, we wish to design the test so that this conclusion will be reached with probability equal to 0.95. On the other hand, if the *p*H differs from 8.30 by 0.03 (in either direction) we want the probability of picking up such a difference to exceed 0.95. (a) What is the test procedure that should be used? (b) What is the required sample size? (c) If $\bar{x}=8.31$, what is your conclusion? (d) If the actual *p*H is 8.32, what is the probability of concluding that the *p*H is not 8.30 using the above procedure? (Calculate this probability directly but you may check your result using the OC curves in the text.)

4. Primers are used for initiating charges of explosives. The density of the primers affects this initiating power. For a certain purpose it was desirable that the average density of the primers always exceed 1.50 g/cc. A procedure was required to determine whether a batch was acceptable. The standard deviation was known from past experience to be 0.03 and the average density in normal operation was 1.54.

To ensure that good batches were nearly always accepted and bad batches nearly always rejected, it was decided that the following requirements should be satisfied: (1) If the mean density was at the normal value of 1.54, there should be a 95% chance of acceptance. (2) If the mean density was as low as 1.50, there should be a 98% chance of rejection.

(a) Determine the necessary sample size and state the required decision rule that has the risks given above.

(b) If the mean density was 1.52, what is the probability of accepting the lot? Do not use the figure in the text to get the solution to this part of the problem. Calculate it directly.

(c) If $\bar{x}=1.53$, would you accept the lot?

5. A shoe manufacturer claims that he can supply, at the same price now being paid, shoes that will give longer wear than the shoes now being used by the United States Army. Army records show that, for such troops, the average life of shoes is 12 months, with a standard deviation of two months. The Army has had poor relations with the present supplier and will welcome a change.

To test the manufacturer's claim, the Army issues 80 pairs of the new shoes at random among troops. On the average, these 80 pairs wear out in $12\frac{1}{2}$ months. Should the Army accept the manufacturer's claim at the 5% level of significance? Does the OC curve of the test afford reasonable protection?

6. A certain type of rocket has been kept in storage for two years. At the time of its acceptance, it was determined that the average range of these rockets was 2,500 yards with a standard deviation of 150 yards. It is necessary to make a decision regarding the disposition of the lot. If the average range has diminished as much as 100 yards, such a change should be detected with probability 0.80. How many observations are required to run a test at the 5% level of significance? If the sample mean $\bar{x}$ is 2,401 yards, should the rockets be discarded?

7. The average shear strength of test spot welds is 400 psi and the standard deviation is 20 psi. An additive is being considered which is said to increase the average shear strength. An experiment is to be run to ascertain the properties of the new material. If the average shear strength is increased as much as 30 psi, such a change should be detected with probability 0.99. If there is no change, this should be determined with probability 0.95. (a) How many observations are required? (b) If the sample mean is 460 psi, should one conclude that the average shear strength has been increased?

8. In Problem 1, how many observations are necessary so that if the average lot warpwise breaking strength is 170 psi, there will be a probability of at least 0.99 of detecting it? If $\bar{x}$ were 177 psi and the sample size is as found above, would you accept the lot?

9. In Problem 2, suppose that a sample of size 10 is used. If $\mu=65$ degrees, what is the probability of accepting the hypothesis that the hardness is 64.3? If the true average hardness is 64 degrees, what is the probability of accepting the hypothesis? Use the analytical method to get your result.

10. In Problem 5, how many pairs of shoes are required so that if the average life, μ, is actually $11\frac{1}{2}$ months, it will be discovered with probability 0.9, i.e., the hypothesis that $\mu \geqq 12$ months will be rejected? If such a sample leads to $x=11\frac{1}{2}$ months, would you accept the hypothesis?

11. Several complaints relative to the quality of standard No. 5 screened coke have been received by a coke plant foreman. It is suggested that the porosity factor, x, measured in pounds weight differences in dry and soaked coke, may be at fault. The shipments were purchased on the assertion of an average porosity factor of 2 pounds. From the producer's point of view, both large and small average porosity factors are undesirable. It is known from past experience that x is normal with variance $\frac{1}{4}$ pound. Test the hypothesis that the shipments actually did conform with the claim above at the 5% level. Does the choice of sample size afford good protection against accepting the hypothesis when in fact the true average is as great as 2.10 pounds or as small as 1.90 pounds? The data are as follows:

2.16	2.07	2.34	1.98	1.97	1.89
2.19	2.23	2.15	2.47	2.31	1.94
2.31	1.86	2.25	2.14	2.15	2.16
2.30	2.48	2.11	2.15	2.24	2.04
2.25	1.90	2.04	2.09	2.08	2.25

12. In Problem 11, suppose it is desired that the assertion be made, with probability not exceeding 0.10, that the average porosity factor is 2 pounds when the true average is as great as 2.10 pounds. How large a sample size is required if the test is to be run at the 5% level of significance? Solve graphically and analytically.

13. The owner of a large number of taverns buys 30 gallon kegs of beer in large lots (hundreds of kegs at a time). Since the beer is sold in 25 cent glasses, he is interested in whether the average number of gallons in the kegs is adequate. Assume that the process which fills the kegs is such that the number of gallons in a keg is a normally distributed random variable with standard deviation equal to $\frac{2}{3}$ gallon. If the average size is as small as 29 gallons, the owner would like to reject a lot with probability exceeding 0.9. If the average size is as stipulated (30 gallons), he would like to accept a lot with probability 0.95. (a) How many kegs should he be prepared to sample? (b) If $\bar{x}$ is 28.5 gallons, should the lot be accepted or rejected?

14. A manufacturer of light bulbs has developed a new production process which he hopes will increase the mean efficiency (in lumens per watt) of his product from the present mean of 9.5. The results of an experiment conducted on ten bulbs are given below. Should the manufacturer believe that the efficiency has been increased? Use a 1% level of significance. What is the probability of saying that there was no increase if the efficiency were 11? (Use s as the estimate of σ for OC curve purposes.)

9.278	12.045
9.971	13.024
10.250	9.871
11.461	11.578
11.515	10.851

15. A supplier of cloth claims that his product has an average warpwise breaking strength of 90 pounds or better. The company for which you work is anxious to buy this cloth, but decides first to test his claim. If the test leads to acceptance of the claim, it will buy the material. Values of average warpwise strength slightly less than 90 pounds are not considered important to guard against, although values substantially less are critical. From past experience with this kind of material, the standard deviation of the warpwise breaking strength is known to be approximately 12 pounds. (a) The experiment calls for a sample of 18 pieces of cloth. Determine the test procedure. (b) What is the risk under (a) of advising the company to buy the yarn, when its average warpwise breaking strength is actually only 86 pounds? (c) If the cloth has actually an average warp-wise breaking strength of 90 pounds, what is the risk that you might reject the claim? (d) You carry out the test as designed in (a) and find that the average strength for the sample of cloth tested is 85 pounds. Would you or would you not advise the company to buy the yarn? $s=12.5$.

16. Suppose it is desired to test the hypothesis that the melting point of an alloy is 1,200 degrees centigrade. If the melting point differs by more than 20 degrees, it is desired to change the composition. Assume normality, $\alpha=0.01$, $\beta=0.05$, and that σ is approximately 15 degrees. How many determinations should be made? What will the rejection rule be?

17. In the manufacture of cylindrical rods with a circular cross section which fit into a circular socket, the lathe is set so that the mean diameter is 5.01 centimeters. It was necessary to test the accuracy of the machine setting. A sample of size 10 was taken with the results as follows:

5.036	5.031
5.085	5.064
4.991	4.942
4.935	5.051
4.999	5.011

(a) Test the hypothesis that the setting is correct at the 5% level of significance. (b) If the setting is off by 0.01 centimeter, what is the approximate probability of detecting such a shift? (c) How many observations are required to insure, (with probability 0.95), that a setting off by 0.01 centimeter will be detected?

18. It is desired to test whether a special treatment on concrete has an effect (in either direction) on the average strength of this concrete which is known to be 3,000 psi. A small experiment is to be made from a given batch of raw materials using the new treatment. The procedure is to be such that if the new concrete has the same tensile strength as the old, the probability will be 0.95 of reaching this conclusion. Moreover, if the average strength differs by as much as 250 psi, the procedure should result in the conclusion that they differ with probability greater than 0.75. A very rough estimate of the standard deviation is 200 psi. (a) What is the necessary sample size? (b) Suppose that

$$\sum_{i=1}^{n}(x_i-\bar{x})^2=40{,}103\,, \qquad \bar{x}=3{,}393\,,$$

and n is that value found in (a), should we conclude the special treatment has no effect on the strength?

19. In Problem 14, how many observations are required so that if the mean efficiency is increased by as much as 1 lumen per watt, it will be detected with probability 0.90? A rough estimate of σ is $\frac{1}{2}$ lumen per watt. If $\bar{x}=11$, $s=0.503$, and the number of observations are as found above, would the hypothesis be accepted?

20. A certain plastic company supplies plastic sheets for industrial use. A new type plastic has been produced and the company would like to claim that the average stress resistance of this new product is at least

30.0 where stress resistance, x, is measured in pounds per square inch necessary to crack the sheet. There is reason to believe that x is normally distributed. The following random sample was drawn off the production line. Based on this sample would you say, at the 1% level, that the company should make its claim?

30.1	32.7	22.5	27.5
27.7	29.8	28.9	31.4
31.2	24.3	26.4	22.8
29.1	33.4	32.5	21.7

21. A manufacturer of gauges claims that the standard deviation in the use of his gauge is 0.0003. Unknown to an analyst who is using this gauge, you have him measure the same items eight times during the course of his regular testing duties and find that the sample standard deviation of the eight measurements is 0.0006. (a) In view of these results, would you be justified in rejecting the manufacturer's claim? Use a 1% level of significance. (b) What is the probability of rejecting the hypothesis that $\sigma = 0.0003$ when the true standard deviation is 0.0005? Get both the analytical and graphical solution.

22. You wish to test whether the standard deviation of measurements made by a certain thermometer is 0.010 degree against the alternative that it is greater than 0.010 degree. A "tolerance standard deviation" of 0.005 degree is given and you wish to have a 0.90 probability of rejection when this tolerance is exceeded. How many measurements would you make and what would your acceptance criteria be if you used a 5% level of significance? Would you accept the hypothesis if $s = 0.012$?

23. Metal panels of standard thickness were coated with a known thickness of film. The coated panels were then stored and the strength of the resin film was assessed by a machine which screwed a steel ball into the panel, thereby stretching the film on the reverse side until the film ruptured. When rupture occurred, electric contact was completed through the panel and a relay operated to stop the machine. The depth of penetration of the ball was a measure of the film strength. The reproducibility of the original method, as measured by the variance, was known *accurately* to be 16 units from a large number of trials but was not considered satisfactory, and various experiments were conducted with a view to improvement. For these experiments panels were prepared under standard conditions. In one typical experiment an attempt was made to improve reproducibility by specially preparing the surface of the panels prior to coating them with the resin. The level of significance chosen was $\alpha = 0.05$. It was decided that a suitable test would be one which would give a probability of 0.90 of detecting a reduction in variance to half its normal value. (a) What is the test procedure that should be used? (b) What is the required sample size? (c) If $s^2 = 12.2$, what is your conclusion?

24. A particular method of analysis had been in use for a certain substance, and a proposal was made to replace it by another method which was considered to have certain advantages. A question to be decided by experimentation was whether the new method was as precise as the old one. From experience it was known that the standard deviation of the old method was 0.04 unit, and it was decided that the new method would have sufficient advantages in rapidity and economy of materials to justify its adoption, provided that it was not markedly less precise than the old. The test was to be designed so that if the the standard deviation of the new method was as great as 0.10 unit the risk would not be more than 0.01 of failing to discover such a change when the test of significance was made at the $\alpha=0.05$ level. (a) Determine the required number of observations. (b) If the true value of the standard deviation, σ, was 0.07, what is the probability of accepting the hypothesis? Solve analytically. (c) If $s=0.06$, would you accept the hypothesis?

25. A manufacturer of gauges claims that the standard deviation in the use of his gauge is $\sigma_0=0.0003$. Unknown to an analyst who is using this gauge, you have him measure the same items a number of times during the course of his regular testing duties. You do not wish to run a risk of more than 1% of contradicting the manufacturer's claims when they are true. However, if the actual standard deviation σ_1 is such that $\sigma_1/\sigma_0=2$, you wish to conclude that the manufacturer's claims are false with probability greater than 0.8. (a) What is the required sample size? (b) If $(n-1)s^2/(0.0003)^2$ is 25.3, would you accept or reject the manufacturer's claims?

26. A producer claims that the lengths of the parts he manufactures have a standard deviation of 0.05 inch. You measure a sample of ten parts and find that their standard deviation is 0.10 inch. Could you reasonably reject the producer's claim? Discuss the OC curve for your test.

27. Given a sample of size 16 from a normal distribution with $\sum(x-\bar{x})^2 = 0.225$, test the hypothesis that $\sigma=0.1$ with error of the first kind equal to 0.01. If $\sigma=0.2$, what is the probability that you will accept the hypothesis that $\sigma=0.1$?

28. A pharmaceutical house produces a certain drug item whose weight has an inherent standard deviation of 5 milligrams. The company's research team has proposed a new method of producing the drug. However, this entails some costs, and will be adopted only if there is strong evidence the standard deviation of the weight of the items is less than 5 milligrams. In fact, if the new standard deviation is as small as 3 milligrams, the switch should be made with probability exceeding 0.8. Assuming weight to be normally distributed, how many observations are required if $\alpha=0.01$? The company runs the experiment using $n=10$. The data in grams are given below. With these data would the new method be adopted? Was the choice of n good, fair or poor?

5.728	5.731
5.722	5.719
5.727	5.724
5.718	5.726
5.723	5.722

29. The military is developing a new ballistic missile and is interested in determining whether it is precise with respect to the short or long distance (e.g., only the x coordinate of distance). It is assumed that this distance is a normally distributed random variable. Since observation of distances of misses can lead to a correction in aiming, thereby implying that the mean of the distribution can be adjusted, the primary problem deals with the variance of the distribution. It is desired that when aimed properly at least 95% of the misses lie within 20 miles of the target. (a) What is the maximum allowable standard deviation? (b) If the true standard deviation of the missile distance is twice the value found in (a), the missile should be returned to the prime contractor for further development. This is to be determined by experimentation and the Defense Department is willing to assume a risk of no more than 10%. On the other hand, if the missile is satisfactory, it should be ascertained with probability no less than 0.95. What is the required sample size? (c) If $s^2=225$, what should be the disposition of the missile? (d) If σ is $\frac{3}{2}$ the value found in (a), determine analytically the probability of accepting the missile.

30. Derive the expression for the sample size of the two-sided procedure for testing the hypothesis that $\mu=\mu_0$ when the standard deviation is known.

31. Derive the expression for the sample size of the two-sided procedure for testing the hypothesis that $\sigma=\sigma_0$.

CHAPTER VII

Tests of Hypotheses about Two Parameters

7.1 Test of the Hypothesis that the Means of Two Normal Distributions Are Equal when Both Standard Deviations Are Known

7.1.1 Choice of an OC curve

The tests of significance discussed so far in this text are concerned with determining whether a parameter of a probability distribution is equal to a specified value. Another problem which often arises in experimental work is to determine whether two probability distributions have some parameters which are the same, without specifying the common values of these parameters. The simplest example of such tests is testing for the equality of two means.

Suppose that an experiment is to be performed to determine whether surface finish has an effect on the endurance limit of steel; in particular, several unpolished and polished smooth-turned specimens are to be compared. The data will consist of the experimentally determined endurance limits (reverse bending) from both kinds of specimens. The statistical problem is to decide whether the expected values of the observations from each set are equal or are different, i.e., to determine whether the average endurance limit of the polished specimen differs from the average endurance limit of the unpolished specimens. This is equivalent to finding whether surface finish has a real effect.

This section will be concerned with testing the hypothesis that means of two normal distributions, μ_x and μ_y, are equal when both standard deviations, σ_x and σ_y, are known. The statistic used for testing the hypothesis that $\mu_x = \mu_y$ (or $\mu_x - \mu_y = 0$) is a function of the difference between the sample means. The abscissae of operating characteristic curves for procedures based upon this test statistic naturally depend upon the difference in the true means, i.e., $\mu_x - \mu_y$. In order to determine the acceptance region and sample size, two points on the OC curve must be chosen. These points will be denoted by $(0, 1-\alpha)$ and $(\mu_x - \mu_y, \beta)$. The first point corresponds to choosing the probability of accepting the hypothesis when it is true (when $\mu_x = \mu_y$) and the second point corresponds to choosing the probability of accepting the hypothesis when $\mu_x - \mu_y$ is the value of the difference in means considered important to detect.

7.1.2 Tables and charts for determining decision rules

An experiment consisting of a set of n_x trials and a set of n_y trials is to be performed. Let the random variables $x_1, x_2, \cdots, x_{n_x}$ denote the measurable quantities associated with the set of n_x trials. It is assumed that the x_i are independent normally distributed random variables each with unknown mean μ_x and known standard deviation σ_x. Let the random variables $y_1, y_2, \cdots, y_{n_y}$ denote the measurable quantities associated with the set of n_y trials. It is assumed that the y_i are independent normally distributed random variables each with unknown mean μ_y and known standard deviation σ_y. Furthermore, let all the x's and y's be independent. The *optimum* procedure for testing the hypothesis that $\mu_x = \mu_y$ is based upon the random variable U, where

$$U = \frac{\bar{x} - \bar{y}}{\sqrt{\sigma_x^2/n_x + \sigma_y^2/n_y}}.$$

If the value taken on by the test statistic as a result of experimentation falls into the acceptance region, the hypothesis that $\mu_x = \mu_y$ is accepted; otherwise it is rejected. This section will be devoted to finding the acceptance region and sample size for the test statistic U, using the tables and charts which are provided. One- and two-sided procedures will be considered.

7.1.2.1 *Tables and charts for two-sided procedures.* Two-sided procedures will be analyzed first. Symmetry of the OC curve about zero will be assumed. This restriction can be relaxed by using the same methods as were presented in Section 6.1.2.1 in connection with the single parameter problem. The acceptance region for the two-sided procedure is given by the interval $[-K_{\alpha/2}, K_{\alpha/2}]$ where α is the level of significance and $K_{\alpha/2}$ is the 100 $\alpha/2$ percentage point of the normal distribution. Thus, the hypothesis that $\mu_x = \mu_y$ is accepted if

$$-K_{\alpha/2} \leqq U \leqq K_{\alpha/2}$$

and rejected if U lies outside this interval. It is easily verified that this procedure yields the point $(0, 1 - \alpha)$ on the OC curve. Whenever $\mu_x = \mu_y$, the test statistic

$$U = \frac{\bar{x} - \bar{y}}{\sqrt{\sigma_x^2/n_x + \sigma_y^2/n_y}}$$

is normally distributed with zero mean and unit variance. Hence, if the hypothesis is true, i.e., $\mu_x = \mu_y$,

$$\mathrm{P}(-K_{\alpha/2} \leqq U \leqq K_{\alpha/2}) = 1 - \alpha.$$

The required sample size is obtained by referring to the set of operating characteristic curves used in Chapter VI, namely Figures 6.4 and 6.5. In order to use these curves the abscissa scale, d, is now defined as

$$d = \frac{|\mu_x - \mu_y|}{\sqrt{\sigma_x^2 + \sigma_y^2}}$$

and it is necessary to choose $n_x = n_y = n$. If the level of significance of the procedure is either 0.05 or 0.01, the required sample size n is obtained by entering the appropriate figure (Figure 6.4 if $\alpha = 0.05$ and Figure 6.5 if $\alpha = 0.01$) with the point $\left(\frac{|\mu_x - \mu_y|}{\sqrt{\sigma_x^2 + \sigma_y^2}}, \beta\right)$ and reading out the value of the sample size associated with the OC curve that passes through this point.

Occasionally, the cost of obtaining observations from one distribution may be substantially greater than from the other as, for example, in comparing an experimental item with a production item or in comparing a use test with a specification test. In situations such as these, it may be undesirable to choose n_x equal to n_y. A proper choice of n_x and n_y, having the property that the resultant OC curve passes through the two prescribed points, can be found by a method of trial and error. One guesses at values of n_x and n_y. An equivalent value of n is computed from the expression

$$n = \frac{\sigma_x^2 + \sigma_y^2}{\sigma_x^2/n_x + \sigma_y^2/n_y}\,.$$

The OC curve of Figure 6.4 or 6.5 for

$$n = \frac{\sigma_x^2 + \sigma_y^2}{\sigma_x^2/n_x + \sigma_y^2/n_y}$$

is entered with $|\mu_x - \mu_y|/\sqrt{\sigma_x^2 + \sigma_y^2}$ and the probability of accepting the hypothesis is read from the curve. If this probability is β, the chosen values of n_x and n_y are satisfactory. If this probability is not β, changes in n_x and or n_y are made and the procedure is repeated.

7.1.2.2 *Summary for two-sided procedures using tables and charts.* The following steps are taken using a two-sided procedure for testing the hypothesis that $\mu_x = \mu_y$ with both σ_x and σ_y known.

1. Two points on the OC curve are chosen; the probability of accepting the hypothesis when it is true, i.e., $(0, 1 - \alpha)$, and the probability of accepting the hypothesis when $\mu_x - \mu_y$ is the value of the difference in means considered important to detect, i.e., $(\mu_x - \mu_y, \beta)$. On the OC curve with the d scale, these points are translated to $(0, 1 - \alpha)$ and

$$\left(\frac{|\mu_x - \mu_y|}{\sqrt{\sigma_x^2 + \sigma_y^2}}, \beta\right).$$

2. If $n_x = n_y = n$, the required sample size n is found by entering the appropriate figure (Figure 6.4 if $\alpha = 0.05$ and Figure 6.5 if $\alpha = 0.01$) with the point

$$\left(\frac{|\mu_x - \mu_y|}{\sqrt{\sigma_x^2 + \sigma_y^2}}, \beta\right)$$

and reading out the value of the sample size associated with the OC curve that passes through this point. If $n_x \neq n_y$, the appropriate values of n_x and n_y are found by trial and error using an equivalent value of n given by

$$n = \frac{\sigma_x^2 + \sigma_y^2}{\sigma_x^2/n_x + \sigma_y^2/n_y}.$$

3. Determine the acceptance region which is the interval $[-K_{\alpha/2}, K_{\alpha/2}]$.
4. Draw a random sample of n_x and n_y items.
5. Compute the value of the test statistic

$$U = \frac{\bar{x} - \bar{y}}{\sqrt{\sigma_x^2/n_x + \sigma_y^2/n_y}}.$$

6. If $-K_{\alpha/2} \leq U \leq K_{\alpha/2}$, i.e., $|U| \leq K_{\alpha/2}$, accept the hypothesis that $\mu_x = \mu_y$. If U lies outside this interval, reject the hypothesis and conclude that $\mu_x \neq \mu_y$.

7.1.2.3 *Summary for one-sided procedures using tables and charts.* The use of the tables and charts for one-sided procedures for testing the equality of two means is similar to their use for the two-sided procedures. Hence, only a summary will be given.

The following steps are taken using a one-sided procedure for testing the hypothesis that $\mu_x = \mu_y$ with both σ_x and σ_y known.

1. Two points on the OC curve are chosen; the probability of accepting the hypothesis when it is true, i.e., $(0, 1 - \alpha)$, and the probability of accepting the hypothesis when $\mu_x - \mu_y$ is the value of the difference in means considered important to detect, i.e., $(\mu_x - \mu_y, \beta)$. If rejection is desired when $\mu_x > \mu_y$, these points are translated to $(0, 1 - \alpha)$ and

$$\left(\frac{\mu_x - \mu_y}{\sqrt{\sigma_x^2 + \sigma_y^2}}, \beta\right)$$

on the OC curve with the d scale. If rejection is desired when $\mu_x < \mu_y$, these points are translated to $(0, 1 - \alpha)$ and

$$\left(\frac{\mu_y - \mu_x}{\sqrt{\sigma_x^2 + \sigma_y^2}}, \beta\right)$$

on the OC curve with the d scale.

2. If $n_x = n_y = n$, the required sample size is found by entering the appropriate figure (Figure 6.6 if $\alpha = 0.05$ and Figure 6.7 if $\alpha = 0.01$). If rejection is desired when $\mu_x > \mu_y$, the figure is entered with the point

$$\left(\frac{\mu_x - \mu_y}{\sqrt{\sigma_x^2 + \sigma_y^2}}, \beta\right)$$

and the value of the sample size associated with the OC curve that passes through this point is read out. If rejection is desired when $\mu_x < \mu_y$, the figure is entered with the point

$$\left(\frac{\mu_y - \mu_x}{\sqrt{\sigma_x^2 + \sigma_y^2}}, \beta\right)$$

and the value of the sample size associated with the OC curve that passes through this point is read out. If $n_x \neq n_y$ the appropriate values of n_x and n_y are found by trial and error using an equivalent value of n given by

$$n = \frac{\sigma_x^2 + \sigma_y^2}{\sigma_x^2/n_x + \sigma_y^2/n_y}$$

as described in Section 7.1.2.1.

3. Determine the acceptance region which is the interval

$[-\infty, K_\alpha]$ for the alternative $\mu_x > \mu_y$ and $[-K_\alpha, \infty]$ for the alternative $\mu_x < \mu_y$.

4. Draw a random sample of n_x and n_y items.
5. Compute the value of the test statistic

$$U = \frac{\bar{x} - \bar{y}}{\sqrt{\sigma_x^2/n_x + \sigma_y^2/n_y}}.$$

6. If the alternative is $\mu_x > \mu_y$, accept the hypothesis that $\mu_x = \mu_y$ (or $\mu_x \leqq \mu_y$) whenever $U \leqq K_\alpha$; otherwise reject the hypothesis and conclude that $\mu_x > \mu_y$. If the alternative is $\mu_x < \mu_y$, accept the hypothesis that $\mu_x = \mu_y$ (or $\mu_x \geqq \mu_y$) whenever $U \geqq -K_\alpha$; otherwise reject the hypothesis and conclude that $\mu_x < \mu_y$.

7.1.2.4 *Tables and charts for operating characteristic curves.* If $n_x = n_y = n$, and this sample size, n, is fixed in advance, or after it is chosen, the operating characteristic curves given in Figures 6.4 and 6.5 for two-sided procedures and Figures 6.6 and 6.7 for one-sided procedures can be used to evaluate the risks involved in using this procedure. If the level of significance is either 0.05 or 0.01, the appropriate OC curve is entered with the sample size, n, and d, and the probability of accepting the hypothesis is read from the curve. The value of d chosen corresponds to the value of $\mu_x - \mu_y$ for which the risk is to be evaluated.

If $n_x \neq n_y$ and these values are fixed in advance, or after they are chosen, the OC curve can still be used to evaluate the risks as follows: The OC curve corresponding to an equivalent value of

$$n = \frac{\sigma_x^2 + \sigma_y^2}{\sigma_x^2/n_x + \sigma_y^2/n_y}$$

is entered with d calculated for the value of $\mu_x - \mu_y$ for which the risk is to be evaluated; the desired risk (probability of accepting the hypothesis) is read from the curve.

If values of α other than 0.05 or 0.01 are desired, Appendix Table 1 (Table of the Percentage Points of the Normal Distribution) can be used to determine the acceptance region. However, analytical procedures which are given in Section 7.1.3 must be used to obtain the OC curve.

The OC curves presented are calculated under the assumption that the test statistic U is normally distributed. This assumption is satisfied if the conditions stated at the beginning of Section 7.1.2 are fulfilled. It is also satisfied for more general forms of underlying probability distribution if the sample size is sufficiently large because the conclusions of the central limit theorem are applicable (U is a function of sample means).

A summary of the procedures for testing the hypothesis that $\mu_x = \mu_y$ when both σ_x and σ_y are known is given in Table 7.1.

Table 7.1. Test of the Hypothesis that the Means of Two Normal Distributions Are Equal when Both Standard Deviations Are Known

Notation for the Hypothesis $H: \mu_x = \mu_y$	*Test Statistic* $U = (\bar{x} - \bar{y})/\sqrt{\sigma_x^2/n_x + \sigma_y^2/n_y}$
Criteria for Rejection	*Method for Choosing Sample Size, Assuming* $n_x = n_y = n$
$\lvert U \rvert \geqq K_{\alpha/2}$ if we wish to reject when the true difference in means is either positive or negative.	Choose a value of $\mu_x - \mu_y$ for which we wish to reject the hypothesis with given high probability; calculate* $d = \dfrac{\lvert \mu_x - \mu_y \rvert}{\sqrt{\sigma_x^2 + \sigma_y^2}}$ and select n from the OC curves of Fig. 6.4 or 6.5.
$U \geqq K_\alpha$ if we wish to reject when $\mu_x > \mu_y$.	Choose a value of $\mu_x - \mu_y > 0$ for which we wish to reject the hypothesis with given high probability; calculate* $d = \dfrac{\mu_x - \mu_y}{\sqrt{\sigma_x^2 + \sigma_y^2}}$ and select n from the OC curves of Fig. 6.6 or 6.7.
$U \leqq -K_\alpha$ if we wish to reject when $\mu_x < \mu_y$.	Choose a value of $\mu_x - \mu_y < 0$ for which we wish to reject the hypothesis with given high probability; calculate* $d = \dfrac{\mu_y - \mu_x}{\sqrt{\sigma_x^2 + \sigma_y^2}}$ and select n from the OC curves of Fig. 6.6 or 6.7.

Values of K_α, the 100 α percentage points of the normal distribution, are given in Appendix Table 1.

7.1.3 ANALYTICAL DETERMINATION OF DECISION RULES

7.1.3.1 *Acceptance regions and sample sizes.* Let $x_1, x_2, \cdots, x_{n_x}$ be a set of n_x independent random variables each with unknown

* In order to find the required sample size for the OC curves given, it is necessary to choose $n = n_x = n_y$. However, if n_x and n_y are fixed in advance, $n_x \neq n_y$, the resulting protection using the above rule can be obtained by entering the curves using d and $n = (\sigma_x^2 + \sigma_y^2)/[\sigma_x^2/n_x + \sigma_y^2/n_y]$.

mean μ_x and known standard deviation σ_x. Let $y_1, y_2, \cdots, y_{n_y}$ be a set of n_y independent random variables each with unknown mean μ_y and known standard deviation σ_y. Furthermore, let all the x's and y's be independent. The optimum procedure for testing the hypothesis that $\mu_x = \mu_y$ is based upon the test statistic U, where

$$U = \frac{\bar{x} - \bar{y}}{\sqrt{\sigma_x^2/n_x + \sigma_y^2/n_y}}.$$

This section will be devoted to finding acceptance regions and expressions for the sample size for both one- and two-sided procedures.

Under the above-mentioned assumptions (or from the central limit theorem) $\bar{x}$ is normally distributed with mean μ_x and variance σ_x^2/n_x and $\bar{y}$ is normally distributed with mean μ_y and variance σ_y^2/n_y. Moreover, $\bar{x}$ and $\bar{y}$ are independent. It follows from the result on linear combinations of normally distributed random variables that the difference of two independent normal variables is normal with mean equal to the difference of the means and variance equal to the sum of the variances. Therefore, the quantity $\bar{x} - \bar{y}$ is normally distributed with mean $\mu_x - \mu_y$ and variance equal to $\sigma_x^2/n_x + \sigma_y^2/n_y$, and the quantity

$$\frac{\bar{x} - \bar{y} - (\mu_x - \mu_y)}{\sqrt{\sigma_x^2/n_x + \sigma_y^2/n_y}}$$

has a normal distribution with zero mean and unit variance. Thus, if the hypothesis is true, i.e., $\mu_x = \mu_y$,

$$U = \frac{\bar{x} - \bar{y}}{\sqrt{\sigma_x^2/n_x + \sigma_y^2/n_y}}$$

has a normal distribution with zero mean and unit variance, and probabilities associated with various values of U can be obtained directly from tables of the percentage points of the normal distribution (Appendix Table 1). Hence, all of the following acceptance regions given by

1. $[-K_{\alpha/2}, K_{\alpha/2}]$ for the two-sided procedure;
2. $[-\infty, K_\alpha]$ for the one-sided procedure where rejection is desired when $\mu_x > \mu_y$;
3. $[-K_\alpha, \infty]$ for the one-sided procedure where rejection is desired when $\mu_x < \mu_y$

have a probability of accepting the hypothesis when it is true (level of significance) equal to $1 - \alpha$, where K_α is the $100\,\alpha$ percentage point of the normal distribution.

For the two-sided procedure the expression for the sample size

$n = n_x = n_y$ is given approximately by

$$n \cong \frac{(K_{\alpha/2} + K_\beta)^2(\sigma_x^2 + \sigma_y^2)}{(\mu_x - \mu_y)^2}$$

where β is the probability of accepting the hypothesis when the true difference in means equals $\mu_x - \mu_y$, i.e., the Type II error, and K_β is the 100 β percentage point of the normal distribution. This approximation is good whenever

$$P\left(z < -K_{\alpha/2} - \frac{|\mu_x - \mu_y|\sqrt{n}}{\sqrt{\sigma_x^2 + \sigma_y^2}}\right)$$

is small compared to β, where z is a normally distributed random variable with zero mean and unit variance.

For one-sided procedures, the expression for the sample size $n = n_x = n_y$ is given by

$$n = \frac{(K_\alpha + K_\beta)^2(\sigma_x^2 + \sigma_y^2)}{(\mu_x - \mu_y)^2}.$$

The derivation of the expressions for the sample size is similar to that given in Section 6.1.3.1 and, hence, will be omitted.

7.1.3.2 *The operating characteristic curve.* If n_x and n_y and the acceptance region are known, the entire OC curve is easily determined. Let the true difference in means of the x's and y's be given by $\mu_x - \mu_y$. Thus, the expected value of U is

$$E(U) = \frac{\mu_x - \mu_y}{\sqrt{\sigma_x^2/n_x + \sigma_y^2/n_y}}.$$

U is then normally distributed with mean

$$\frac{\mu_x - \mu_y}{\sqrt{\sigma_x^2/n_x + \sigma_y^2/n_y}}$$

and unit variance. For the one-sided procedure where rejection is desired when $\mu_x > \mu_y$, the expression for the OC curve (probability of accepting the hypothesis) is given by

$$P(U \leq K_\alpha)$$

which is equivalent to

$$P\left(z \leq K_\alpha - \frac{(\mu_x - \mu_y)\sqrt{n}}{\sqrt{\sigma_x^2 + \sigma_y^2}}\right) \quad \text{where } n_x = n_y = n.$$

This last expression can be evaluated for any value of $\mu_x - \mu_y$ with the aid of Appendix Table 1. It is interesting to note that for fixed

n and K_α this expression for the OC curve is just a function of

$$d = \frac{\mu_x - \mu_y}{\sqrt{\sigma_x^2 + \sigma_y^2}}$$

and d is the abscissa of the OC curves plotted in Figures 6.6 and 6.7.

For the single parameter problem discussed in Chapter VI, the OC curve for testing the hypothesis that $\mu = \mu_0$ was given by

$$P\left(z \leqq K_\alpha - \frac{(\mu - \mu_0)\sqrt{n}}{\sigma}\right) \quad \text{or} \quad P(z \leqq K_\alpha - d\sqrt{n}),$$

where d was defined as $(\mu - \mu_0)/\sigma$.

For testing the hypothesis that $\mu_x = \mu_y$, the OC curve described above is given by

$$P\left(z \leqq K_\alpha - \frac{(\mu_x - \mu_y)\sqrt{n}}{\sqrt{\sigma_x^2 + \sigma_y^2}}\right) \quad \text{or} \quad P(z \leqq K_\alpha - d\sqrt{n}),$$

where d is presently defined as

$$\frac{\mu_x - \mu_y}{\sqrt{\sigma_x^2 + \sigma_y^2}}.$$

It is now evident why the same OC curves given in Figures 6.6 and 6.7 can be used for both problems.

If the procedure is one-sided and rejection is desired when $\mu_x < \mu_y$, the OC curve is given by the expression

$$P(U \geqq -K_\alpha) = P\left(z \geqq -K_\alpha - \frac{(\mu_x - \mu_y)\sqrt{n}}{\sqrt{\sigma_x^2 + \sigma_y^2}}\right)$$

where $n_x = n_y = n$.

Finally, if the procedure is two-sided, the OC curve is given by the expression

$$P(-K_{\alpha/2} \leqq U \leqq K_{\alpha/2})$$
$$= P\left(-K_{\alpha/2} - \frac{(\mu_x - \mu_y)\sqrt{n}}{\sqrt{\sigma_x^2 + \sigma_y^2}} \leqq z \leqq K_{\alpha/2} - \frac{(\mu_x - \mu_y)\sqrt{n}}{\sqrt{\sigma_x^2 + \sigma_y^2}}\right)$$

where $n_x = n_y = n$.

7.1.4 EXAMPLE

In Section 7.1.1 we discussed an experiment to determine whether surface finish has an effect on the endurance limit of steel. There exists a theory that polishing increases the average endurance limit (reverse bending); the possibility of it decreasing the average endurance limit is discounted. An experiment is performed on 0.4%

carbon steel using both unpolished and polished smooth-turned specimens. The finish on the smooth-turned polished specimens was obtained by polishing with No. 0 and No. 00 emery cloth.

If we denote the average endurance limit of polished specimens by μ_x and the average endurance limit of unpolished specimens by μ_y, the hypothesis to be tested is that $\mu_x = \mu_y$ against the alternative that $\mu_x > \mu_y$. If polishing fails to have an effect, this should be noted with probability 0.99 ($\alpha = 0.01$). Moreover, it is important, from a practical point of view, to detect a change, if it is as much as 7,500 psi, with probability at least equal to 0.9. Furthermore, polishing should not have any effect on the standard deviation of the endurance limit, which is known from the performance of numerous endurance limit experiments to be 4,000 psi. From Figure 6.7 (1% level of significance) with $d = 7{,}500/\sqrt{(4{,}000)^2 + (4{,}000)^2} = 1.33$, we find that about eight observations on each group of specimens are required to have a probability of 0.9 of detecting a difference as large as 7,500 psi. The data are as follows:

Endurance limit for polished 0.4% *carbon steel* (x)	*Endurance limit for unpolished* 0.4% *carbon steel* (y)
86,500	82,600
91,900	82,400
89,400	81,700
84,000	79,500
89,900	79,400
78,700	69,800
87,500	79,900
83,100	83,400
$\bar{x} = 86{,}375$	$\bar{y} = 79{,}838$

$$U = \frac{6537}{\sqrt{(4{,}000)^2/8 + (4{,}000)^2/8}} = 3.27; \qquad K_{0.01} = 2.326$$

The acceptance region is given by the interval $[-\infty, 2.326]$. Since $U = 3.27$ exceeds 2.326, the hypothesis is rejected and we conclude that polished 0.4% carbon steel specimens have a higher mean endurance limit than unpolished ones.

7.2 Test of the Hypothesis that the Means of Two Normal Distributions Are Equal, Assuming that the Standard Deviations Are Unknown but Equal

7.2.1 Choice of an OC curve

This is the same problem as the one discussed in Section 7.1 except that the standard deviations are now assumed to be unknown but equal, i.e., $\sigma_x = \sigma_y = \sigma$, where σ is unknown.[1] In testing the hypothesis that $\mu_x = \mu_y$ in this situation, a t statistic is used. This test statistic is a function of the difference in the sample means, the sample standard deviation, and the sample size. The abscissae of operating characteristic curves for procedures based upon this t statistic depend upon the quantity $(\mu_x - \mu_y)/\sigma$. As in the t test for a single parameter, it is not the absolute value of the difference in means which is important for OC curve purposes, but rather the relative difference, relative to the inherent variability, σ. In order to determine the acceptance region and sample size, two points on the OC curve must be chosen. These points will be denoted by $(0, 1 - \alpha)$ and $\left(\frac{\mu_x - \mu_y}{\sigma}, \beta\right)$. The first point corresponds to choosing the probability of accepting the hypothesis when it is true (when $\mu_x = \mu_y$), and the second point corresponds to choosing the probability of accepting the hypothesis when $(\mu_x - \mu_y)/\sigma$ is the ratio of the difference in means to the standard deviation considered important to detect. For the purpose of choosing an OC curve a *rough* notion of σ is usually adequate. As indicated previously, such a notion can be obtained from past data or some *a priori* knowledge before experimentation; or if the OC curve is examined after experimentation, the sample estimate of the standard deviation can be used as a "rough" estimate.

7.2.2 TABLES AND CHARTS FOR CARRYING OUT TWO-SAMPLE t TESTS

An experiment consisting of a set of n_x trials and a set of n_y trials is to be performed. Let the random variables $x_1, x_2, \cdots, x_{n_x}$, and $y_1, y_2, \cdots, y_{n_y}$ denote the measurable quantities associated with the set of n_x and n_y trials, respectively. It is assumed that the x_i and y_i are independent normally distributed random variables. The mean of each of the x's is μ_x (unknown) and the mean of each of the y's is μ_y (unknown); the standard deviation of the x's is σ_x and the standard deviation of the y's is σ_y, where σ_x and σ_y are both unknown but equal to a common value, σ. The *optimum* procedure for testing the hypothesis that $\mu_x = \mu_y$ is based upon the random variable t, where

$$t = \frac{\bar{x} - \bar{y}}{\sqrt{\dfrac{1}{n_x} + \dfrac{1}{n_y}}\sqrt{\dfrac{\sum_{i=1}^{n_x} (x_i - \bar{x})^2 + \sum_{i=1}^{n_y} (y_i - \bar{y})^2}{n_x + n_y - 2}}}.$$

[1] A procedure for testing the hypothesis that $\sigma_x = \sigma_y$ is given in Section 7.6.

If the value the test statistic takes on as a result of experimentation falls into the acceptance region, the hypothesis that $\mu_x = \mu_y$ is accepted; otherwise it is rejected. This section will be devoted to finding the acceptance region and sample size for the test statistic t, using the tables and charts which are provided. One- and two-sided procedures will be considered.

7.2.2.1 *Tables and charts for two-sided procedures.* Two-sided procedures will be analyzed first. Symmetry of the OC curve about zero will be assumed. This restriction can be relaxed by using the same methods that were presented in Section 6.1.2.1. The acceptance region for the two-sided procedure for any sample sizes, n_x and n_y, is given by the interval $[-t_{\alpha/2;n_x+n_y-2}, t_{\alpha/2;n_x+n_y-2}]$, where α is the level of significance and $t_{\alpha/2;n_x+n_y-2}$ is the 100 $\alpha/2$ percentage point of the t distribution with $n_x + n_y - 2$ degrees of freedom. Thus, the hypothesis that $\mu_x = \mu_y$ is accepted if

$$-t_{\alpha/2;n_x+n_y-2} \leqq t \leqq t_{\alpha/2;n_x+n_y-2}$$

and rejected if t lies outside this interval. It is easily verified that this procedure yields a probability of $1 - \alpha$ of accepting the hypothesis that $\mu_x = \mu_y$ when it is true. It was shown in Section 4.3.3 that when $\mu_x = \mu_y$, the test statistic

$$t = \frac{\bar{x} - \bar{y}}{\sqrt{\dfrac{1}{n_x} + \dfrac{1}{n_y}}\sqrt{\dfrac{\sum_{i=1}^{n_x} (x_i - \bar{x})^2 + \sum_{i=1}^{n_y} (y_i - \bar{y})^2}{n_x + n_y - 2}}}$$

has a t distribution with $n_x + n_y - 2$ degrees of freedom. Hence, if $\mu_x = \mu_y$

$$P(-t_{\alpha/2;n_x+n_y-2} \leqq t \leqq t_{\alpha/2;n_x+n_y-2}) = 1 - \alpha.$$

The required sample size is obtained by referring to the set of operating characteristic curves used in Chapter VI, namely Figures 6.10 and 6.11. In order to use these curves the abscissa scale, d, is now defined as

$$d = \frac{|\mu_x - \mu_y|}{2\sigma}$$

and it is necessary to choose $n_x = n_y = n$. If the level of significance of the procedure is either 0.05 or 0.01, the required sample size n is obtained by entering the appropriate figure (Figure 6.10 if $\alpha = 0.05$

and Figure 6.11 if $\alpha = 0.01$) with the point $\left(\frac{|\mu_x - \mu_y|}{2\sigma}, \beta\right)$ and reading out the value associated with the OC curve that passes through this point. Denote this quantity by n'. The required sample size, n, is given by the expression

$$n = (n' + 1)/2.$$

7.2.2.2 *Summary for two-sided procedures using tables and charts.* The following steps are taken using a two-sided procedure for testing the hypothesis that $\mu_x = \mu_y$ with both σ_x and σ_y unknown but equal to σ.

1. Two points on the OC curve are chosen, namely, $(0, 1 - \alpha)$ and $\left(\frac{\mu_x - \mu_y}{\sigma}, \beta\right)$. These points are translated to the d scale and are denoted by $(0, 1 - \alpha)$ and $\left(\frac{|\mu_x - \mu_y|}{2\sigma}, \beta\right)$ on this scale.
2. If $n_x = n_y = n$, the required sample size n is found by entering the appropriate figure (Figure 6.10 if $\alpha = 0.05$ and Figure 6.11 if $\alpha = 0.01$) with the point $\left(\frac{|\mu_x - \mu_y|}{2\sigma}, \beta\right)$ and reading out the value associated with the OC curve that passes through this point. Denote the quantity by n'. The required sample size, n, is given by the expression $n = (n' + 1)/2$.
3. Determine the acceptance region which is the interval

$$[-t_{\alpha/2;n_x+n_y-2}, t_{\alpha/2;n_x+n_y-2}].$$

4. Draw a random sample of n_x and n_y items.
5. Compute the value of the test statistic

$$t = \frac{\bar{x} - \bar{y}}{\sqrt{\frac{1}{n_x} + \frac{1}{n_y}}\sqrt{\frac{\sum_{i=1}^{n_x}(x_i - \bar{x})^2 + \sum_{i=1}^{n_y}(y_i - \bar{y})^2}{n_x + n_y - 2}}}.$$

6. If $t_{\alpha/2;n_x+n_y-2} \leqq t \leqq t_{\alpha/2;n_x+n_y-2}$, i.e., $|t| \leqq t_{\alpha/2;n_x+n_y-2}$, accept the hypothesis that $\mu_x = \mu_y$. If t lies outside this interval, reject the hypothesis and conclude that $\mu_x \neq \mu_y$.

7.2.2.3 *Summary for one-sided procedures using tables and charts.* The use of the tables and charts for one-sided procedures for testing the equality of two means is similar to their use for the two-sided procedures. Hence, only a summary will be given.

The following steps are taken using a one-sided procedure for testing the hypothesis that $\mu_x = \mu_y$ with both σ_x and σ_y unknown but equal to σ.

1. Two points on the OC curve are chosen, namely $(0, 1 - \alpha)$ and $(\mu_x - \mu_y, \beta)$. If rejection is desired when $\mu_x > \mu_y$, these points are translated to $(0, 1 - \alpha)$ and $\left(\frac{\mu_x - \mu_y}{2\sigma}, \beta\right)$ on the OC curve with the d scale. If rejection is desired when $\mu_x < \mu_y$, these points are translated to $(0, 1 - \alpha)$ and $\left(\frac{\mu_y - \mu_x}{2\sigma}, \beta\right)$ on the OC curve with the d scale.
2. If $n_x = n_y = n$, the required sample size n is found by entering the appropriate figure (Figures 6.12 and 6.13). If rejection is desired when $\mu_x > \mu_y$, the figure is entered with the point $\left(\frac{\mu_x - \mu_y}{2\sigma}, \beta\right)$ and the value associated with the OC curve that passes through this point is read out. If rejection is desired when $\mu_x < \mu_y$, the figure is entered with the point $\left(\frac{\mu_y - \mu_x}{2\sigma}, \beta\right)$ and the value associated with the OC curve that passes through this point is read out. Denote these quantities by n'. The required sample size, n, is obtained from the expression $n = (n' + 1)/2$.
3. Determine the acceptance region which is that interval $[-\infty, t_{\alpha;n_x+n_y-2}]$ for the alternative $\mu_x > \mu_y$ and $[-t_{\alpha;n_x+n_y-2}, \infty]$ for the alternative $\mu_x < \mu_y$.
4. Draw a random sample of n_x and n_y items.
5. Compute the value of the test statistic

$$t = \frac{\bar{x} - \bar{y}}{\sqrt{\dfrac{1}{n_x} + \dfrac{1}{n_y}}\sqrt{\dfrac{\sum_{i=1}^{n_x} (x_i - \bar{x})^2 + \sum_{i=1}^{n_y} (y_i - \bar{y})^2}{n_x + n_y - 2}}}.$$

6. If the alternative is $\mu_x > \mu_y$, accept the hypothesis that $\mu_x = \mu_y$ (or $\mu_x \leqq \mu_y$) whenever $t \leqq t_{\alpha;n_x+n_y-2}$; otherwise reject the hypothesis and conclude that $\mu_x > \mu_y$. If the alternative is $\mu_x < \mu_y$, accept the hypothesis that $\mu_x = \mu_y$ (or $\mu_x \geqq \mu_y$) whenever $t \geqq -t_{\alpha;n_x+n_y-2}$; otherwise reject the hypothesis and conclude that $\mu_x < \mu_y$.

7.2.2.4 *Tables and charts for operating characteristic curves.* If

$n_x = n_y = n$ and this sample size, n, is fixed in advance, or after it is chosen, the operating characteristic curves given in Figures 6.10 and 6.11 for two-sided procedures and Figures 6.12 and 6.13 for one-sided procedures can be used to evaluate the risks involved in using these procedures. If the level of significance is either 0.05 or 0.01, the appropriate OC curve associated with the value $n' = 2n - 1$ is entered with d, and the probability of accepting the hypothesis is read from the curve. The value of d chosen is related to the value $(\mu_x - \mu_y)/\sigma$ for which the risk is to be evaluated.

Table 7.2. Test of the Hypothesis that the Means of Two Normal Distributions Are Equal, Assuming that the Standard Deviations Are Unknown but Equal

Notation for the Hypothesis	*Test Statistic*
$H: \mu_x = \mu_y$	$t = \dfrac{\bar{x} - \bar{y}}{\sqrt{\dfrac{1}{n_x} + \dfrac{1}{n_y}}\sqrt{\dfrac{\sum_{i=1}^{n_x}(x_i - \bar{x})^2 + \sum_{i=1}^{n_y}(y_i - \bar{y})^2}{n_x + n_y - 2}}}$
Criteria for Rejection	*Method for Choosing Sample Size, Assuming $n_x = n_y = n$*
$\lvert t \rvert \geqq t_{\alpha/2;n_x+n_y-2}$ if we wish to reject when μ_x is not equal to μ_y.	Choose a value of $(\mu_x - \mu_y)/\sigma$ for which we wish to reject the hypothesis with given high probability. Calculate $d = \lvert \mu_x - \mu_y \rvert / 2\sigma$ and enter Fig. 6.10 or 6.11 to find n'. The required sample size is $(n'+1)/2$.
$t \geqq t_{\alpha;n_x+n_y-2}$ if we wish to reject when $\mu_x > \mu_y$.	Choose a value of $(\mu_x - \mu_y)/\sigma$ for which we wish to reject the hypothesis with given high probability. Calculate $d = (\mu_x - \mu_y)/2\sigma$ and enter Fig. 6.12 or 6.13 to find n'. The required sample size is $(n'+1)/2$.
$t \leqq -t_{\alpha;n_x+n_y-2}$ if we wish to reject when $\mu_x < \mu_y$.	Choose a value of $(\mu_x - \mu_y)/\sigma$ for which we wish to reject the hypothesis with given high probability. Calculate $d = (\mu_y - \mu_x)/2\sigma$ and enter Fig. 6.12 or 6.13 to find n'. The required sample size is $(n'+1)/2$.

Values of $t_{\alpha;\nu}$, the 100α percentage points of the t distribution with ν degrees of freedom are given in Appendix Table 3.

If other levels of significance are desired, Appendix Table 3, which is a table of the percentage points of the t distribution, can be used to determine the acceptance region for a fixed value of n. However, analytical procedures, which are beyond the scope of this text, must be used to obtain the OC curve.

Although the OC curves are calculated under the assumption that the x's and y's are independent normally (or approximately normally) distributed random variables, they are also valid for more general forms of underlying probability distributions if the sample sizes are sufficiently large.

A summary of the procedures for testing the hypothesis that $\mu_x = \mu_y$ when both σ_x and σ_y are unknown but equal is given in Table 7.2.

7.2.3 Example

A manufacturer of electric irons produces these items in two plants. Both plants have the same suppliers of small parts. A saving can be made by purchasing thermostats for plant B from a local supplier. A single lot was purchased from the local supplier and it was desired to test whether or not these new thermostats were as accurate as the old. The thermostats were to be tested on the irons on the 550°F setting, and the actual temperatures were to be read to the nearest 0.1° with a thermocouple. The level of significance (α) chosen is 5%. With the old supplier, very few complaints were received, and the manufacturer feels that a switch is undesirable if the average temperature changes by more than 10.5°, with the risk of making an incorrect decision not exceeding 0.10. The order of magnitude of the standard deviation is roughly 10° for the old supplier, and there is no reason to suspect that it will be different for the new supplier. For $d = 10.5/(2 \times 10) = 0.525$, and from Figure 6.10, $n' = 45$, corresponding to a probability of 0.9 of detecting a change of 10.5° or more. Hence, $n = 23$. The data are:

New supplier x(degrees F)		*Old supplier y(degrees F)*	
530.3	559.1	559.7	564.6
559.3	555.0	534.7	554.5
549.4	538.6	554.8	553.0
544.0	551.1	545.0	538.4
551.7	565.4	544.6	548.3
566.3	554.9	538.0	552.9
549.9	550.0	550.7	535.1
556.9	554.9	563.1	555.0
536.7	554.7	551.1	544.8
558.8	536.1	553.8	558.4
538.8	569.1	538.8	548.7
543.3			560.3

$$\sum_{i=1}^{n_x} x_i = 12{,}674.3 \qquad \sum_{i=1}^{n_y} y_i = 12{,}648.3$$

$$\bar{x} = 551.056522 \qquad \bar{y} = 549.926086$$

$$\sum_{i=1}^{n_x} x_i^2 = 6{,}986{,}586.07 \qquad \sum_{i=1}^{n_y} y_i^2 = 6{,}957{,}333.23$$

$$\sum_{i=1}^{n_x} (x_i - \bar{x})^2 = \sum_{i=1}^{n_x} x_i^2 - n\bar{x}^2 = 2{,}330.391$$

$$\sum_{i=1}^{n_y} (y_i - \bar{y})^2 = \sum_{i=1}^{n_y} y_i^2 - n\bar{y}^2 = 1{,}703.130$$

$$\frac{\sum_{i=1}^{n_x} (x_i - \bar{x})^2 + \sum_{i=1}^{n_y} (y_i - \bar{y})^2}{n_x + n_y - 2} = \frac{4{,}033.521}{44} = 91.670932$$

The hypothesis that $\mu_x = \mu_y$ is accepted if $|t| \leqq t_{\alpha/2;n_x+n_y-2}$;

$$|t| = \frac{551.056522 - 549.926086}{\sqrt{2/23}\sqrt{91.670932}}$$

$$= 0.400 \leqq t_{\alpha/2;n_x+n_y-2} = t_{0.025;44} = 2.02.$$

Therefore, we accept the hypothesis at the 5% level of significance that there is no difference in the mean temperatures of the two suppliers, and a switch is made.

7.3 Test of the Hypothesis that the Means of Two Normal Distributions Are Equal, Assuming that the Standard Deviations Are Unknown and Not Necessarily Equal

7.3.1 Test Procedure

We are often unable to assume that σ_x equals σ_y (both unknown). Unfortunately, an exact procedure based upon a t statistic is unavailable to cover this situation. However, a procedure exists, based upon a test statistic t', which has the property that when $\mu_x = \mu_y$, t' has an approximate t distribution. The statistic t' is given by

$$t' = \frac{\bar{x} - \bar{y}}{\sqrt{s_x^2/n_x + s_y^2/n_y}}$$

and the associated degrees of freedom are

$$\nu = \frac{(s_x^2/n_x + s_y^2/n_y)^2}{\dfrac{(s_x^2/n_x)^2}{(n_x + 1)} + \dfrac{(s_y^2/n_y)^2}{(n_y + 1)}} - 2.$$

The probability distribution of t' has not been determined when μ_x does not equal μ_y. Hence, only one point on the OC curve can be guaranteed, namely, the probability of accepting the hypothesis that $\mu_x = \mu_y$ when it is true. A summary of this procedure is given in Table 7.3.

Table 7.3. Test of the Hypothesis that the Means of Two Normal Distributions Are Equal when the Standard Deviations Are Unknown and Not Necessarily Equal

Notation for the Hypothesis	*Test Statistic*
$H: \mu_x = \mu_y$	$t' = \dfrac{\bar{x} - \bar{y}}{\sqrt{s_x^2/n_x + s_y^2/n_y}}$
Criteria for Rejection $\mid t' \mid \geqq t_{\alpha/2;\nu}$ if we wish to reject when μ_x is not equal to μ_y.	*Formula for Obtaining the Degrees of Freedom* ν $\nu = \dfrac{(s_x^2/n_x + s_y^2/n_y)^2}{(s_x^2/n_x)^2/(n_x+1) + (s_y^2/n_y)^2/(n_y+1)} - 2$
$t' \geqq t_{\alpha;\nu}$ if we wish to reject when $\mu_x > \mu_y$.	
$t' \leqq - t_{\alpha;\nu}$ if we wish to reject when $\mu_x < \mu_y$.	

Values of $t_{\alpha;\nu}$, the 100α percentage point of the t distribution with ν degrees of freedom, are given in Appendix Table 3.

7.3.2 Example

A manufacturer of automobile crankshafts was troubled with the problem of bend in the final shaft. Bend may be caused by the length and weight of the shaft, by improper setup of machine tools, by the heat treating process, or by some combination of these causes. It is suspected that nitriding the shaft, i.e., the process that hardens the surface of the shaft by a heat and nitrous oxide reaction with the steel, is the main cause for bend. Twenty-five shafts were measured before nitriding at the front main center journal by a dial indicator gauge accurate to the 0.0001 inch. Similarly, another 25 shafts were measured at the same spot after the nitriding operation. It cannot be assumed that the variability in the bend is the same before and after nitriding. We test the hypothesis that the mean value of the bend is the same before and after nitriding, wishing to reject it if the average bend after nitriding is larger. Hence, the acceptance region is $[- t_{\alpha;n_x+n_y-2}, \infty]$. Denoting by x the values before nitriding and by y the values after nitriding, the following results were obtained:

$$\sum_{i=1}^{25} x_i = 203 \times 10^{-4} \qquad \sum_{i=1}^{25} y_i = 453 \times 10^{-4}$$

$$\bar{x} = 0.0008120 \qquad \bar{y} = 0.001812$$

$$\sum_{i=1}^{25} x_i^2 = 2{,}933 \times 10^{-8} \qquad \sum_{i=1}^{25} y_i^2 = 14{,}362 \times 10^{-8}$$

$$s_x^2 = \frac{2{,}933 \times 10^{-8} - 25(0.0008120)^2}{24} \qquad s_y^2 = \frac{14{,}362 \times 10^{-8} - 25(0.001812)^2}{24}$$

$$= 53.5 \times 10^{-8} \qquad = 256 \times 10^{-8}$$

$$\frac{s_x^2}{n_x} = 2.14 \times 10^{-8} \qquad \frac{s_y^2}{n_y} = 10.24 \times 10^{-8}$$

$$\nu = \frac{[(2.14)10^{-8} + (10.24)10^{-8}]^2}{(2.14)^2(10^{-16})/26 + (10.24)^2(10^{-16})/26} - 2 = 34$$

$$t' = \frac{0.0008120 - 0.001812}{\sqrt{(2.14)10^{-8} + (10.24)10^{-8}}} = -2.84$$

$$t_{0.05;34} = 1.69.$$

Since $t' < -1.69$ we reject the hypothesis at the 5% significance level that the means are the same before and after nitriding, and we conclude that the average bend is larger after nitriding.

7.4 Test for Equality of Means when the Observations Are Paired

7.4.1 TEST PROCEDURE

An important special case arises when the observations are taken in pairs, each pair being taken under the same experimental conditions, with the conditions varying from pair to pair. Consider an experiment comparing the corrosion of pipe with two kinds of coating. Many factors besides the coating may effect the corrosion rate: type of soil, length of time of burial, angle of burial, etc. If we were to take a number of specimens of pipe with each kind of coating and bury them in various kinds of soil for various lengths of time, an analysis by the method of the last section might be misleading. If the experiment were not arranged carefully, one type of coating might be associated with a particular type of soil or one type of coating might be left in the ground longer than the others, and the apparent difference in coating might in fact be due to some other factor. There are two ways to guard against such bias, namely, control and randomization. Control is achieved by taking observations in pairs, each pair to consist of specimens of each of the two coatings and thus one pair of specimens might be buried for five years in clay, another eight years in sand, etc. It is practically impossible to control all variables; such things as size of pipe, orientation, etc., may be of some importance but are not major factors. These factors should be randomized, i.e., assigned to the test specimen by a random device such as coin tossing or random numbers. If this is done, the differences between the specimens will be due to the difference in coating with the effect of randomized factors contributing to increasing the experimental error.

The test procedure involves the difference between each pair of observations. Let $(x_1, y_1), (x_2, y_2), \cdots, (x_n, y_n)$ be a set of n paired random variables which denote the measurable quantities associated with the 1, 2, ⋯, n trials, respectively. Form the difference between each pair of observations, i.e., $d_1 = x_1 - y_1$, $d_2 = x_2 - y_2, \cdots$, $d_n = x_n - y_n$. It is assumed that these differences are independent normally distributed random variables each with common unknown mean μ_d and common unknown standard deviation σ_d.* In the corrosion example μ_d is the difference in effect of the two kinds of coating. It is likely to be the same from pair to pair even if the actual means of the x's and y's differ from pair to pair due to changing experimental conditions. σ_d, although unknown, depends upon σ_x^2 and σ_y^2 (and actually equals $\sqrt{\sigma_x^2+\sigma_y^2}$ if x and y are independent) with σ_x not necessarily equal to σ_y. Hence, the standard deviation of observations with one type of coating may differ from the standard deviation of the observations with the other type coating, and both may be unknown, but these values must be constant from pair to pair.

With these assumptions it is evident that testing for the equality of means is equivalent to testing the hypothesis that $\mu_d = 0$ when the standard deviation, σ_d, is unknown. Hence, the results of Section 6.2, testing the hypothesis that the mean of a normal distribution has a specified value when the standard deviation is unknown, are applicable. The test statistic t of Section 6.2 becomes, on replacing the x's by d's,

$$t = \frac{\bar{d}\sqrt{n}}{s_d}$$

* A sufficient condition for this assumption to hold true is that all x_i and y_i be independent normally distributed random variables; for the ith pair, the expected value of x_i is μ_i and the expected value of y_i is $\mu_i + \mu_d$, i.e., the expected values differ by a constant independent of i; the standard deviation of x_i is σ_x and the standard deviation of y_i is σ_y, with σ_x not necessarily equal to σ_y but σ_x and σ_y constant for all i.

where
$$\bar{d} = \frac{\sum_{i=1}^{n} d_i}{n}$$

and
$$s_d = \sqrt{\frac{\sum_{i=1}^{n}(d_i - \bar{d})^2}{n-1}}.$$

Thus, for the case of paired observations the reader is referred to Section 6.2 using the d's as the random variables of interest.

It is important to note that the assumptions necessary for the two-sample t test given in Section 7.2 automatically satisfy the requirements for the use of paired observations when the observations are paired at random; i.e., an x and y are chosen at random and paired. Since all the x_i have the same distribution and all the y_i have the same distribution, the difference in means between any x and y is the same regardless of which x and y are chosen. The effect of this random pairing is seen by noting that the degrees of freedom used for the two-sample t test as suggested in Section 7.2 is $n_x + n_y - 2$ or $2(n-1)$ if $n_x = n_y = n$; whereas the degrees of freedom for the randomly paired observations is just $n-1$. Since the OC curve of the procedure is essentially related to the degrees of freedom, we see that unnecessary random pairing results in a poorer OC curve. However, if the problem described in Section 7.4.1, which required pairing, is solved using the two-sample t test, the results are meaningless. They are meaningless because the denominator of the two-sample t-test statistic,

$$\sqrt{\frac{2}{n}}\sqrt{\frac{\sum (x_i - \bar{x})^2 + \sum (y_i - \bar{y})^2}{2n-2}},$$

is not an estimate of the standard deviation of $\bar{x} - \bar{y}$ when biases exist. If the two-sample t test is applied when paired observations are called for, it is evident that as the bias increases, the sum of squares above becomes larger, and the more difficult it is to judge the distribution means significant even when they do differ. On the other hand,

$$\sqrt{\frac{\sum (d_i - \bar{d})^2}{n(n-1)}}$$

is always an estimate of the standard deviation of $\bar{d}$ even if a bias exists so that the test statistic based on the d's is appropriate.

Hence, if pairing is inherent in the problem, the paired observations are necessary for analysis of the data; if pairing is not required, it should not be used.

7.4.2 EXAMPLE[1]

In Section 7.4.1 we discussed an experiment comparing the corrosion of pipe with two kinds of coating. Pairs of specimens are to be inspected for the amount of corrosion. One specimen of each type of coating is to be included in each pair; each pair is to be buried in the same soil, at the same depth, in a similar position, and for the same length of time. It is important, from a practical point of view, to detect a difference in the average depth of maximum pits for each coating of 0.011 inch with probability greater than or equal to 0.95. The test is to be run at a 5% level of significance. From past experience, the standard deviation of the depth of maximum pit on similar coatings is known to have a very approximate value of 0.008 inch. Hence σ_d is approximately equal to 0.0113 inch, i.e.,

$$\sigma_d = \sqrt{\sigma_x^2 + \sigma_y^2} = \sqrt{(0.008)^2 + (0.008)^2} = 0.0113.$$

From Figure 6.10 with $d = \dfrac{|0.011 - 0|}{0.0113} = 0.97$ we find that about fifteen pairs of observations (fifteen differences) are required to have a probability of 0.95 of detecting a difference as large as 0.011 inch. The results of the sample are as follows:

Depth of Maximum Pits (Expressed in Thousandths of an Inch)

Coating A	Coating B	Difference
73	51	+22
43	41	+ 2
47	43	+ 4
53	41	+12
58	47	+11
47	32	+15
52	24	+28
38	43	− 5
61	53	+ 8
56	52	+ 4
56	57	− 1
34	44	−10
55	57	− 2
65	40	+25
75	68	+ 7

[1] This example is based upon an example taken from H. A. Freeman, *Industrial Statistics*, John Wiley & Sons, Inc., New York, 1942, p.8.

We find that

$$\bar{d} = \frac{\sum_{i=1}^{n} d_i}{n} = \frac{120}{15} = 8$$

and the estimate of the variance,

$$s_d = \sqrt{\frac{\sum_{i=1}^{n} (d_i - \bar{d})^2}{n-1}}$$
$$= \sqrt{121.571}\,.$$

Thus, the test statistic t is equal to

$$t = \frac{8\sqrt{15}}{\sqrt{121.571}} = 2.810.$$

This value is compared with $t_{\alpha/2;n-1}$, for a 5% level of significance. We find that $t_{0.025;14} = 2.145$. Thus,

$$2.810 > t_{0.025;14} = 2.145$$

and we conclude that there is a significant difference in corrosion rates between the two types of coating.

7.5 Non-parametric Tests

The procedures discussed in previous sections of this chapter are, strictly speaking, valid only when the original observations are assumed to come from a normal distribution, and are only approximately valid when the data arise from non-normal distributions which are reasonably well behaved. Unfortunately, it is not possible to give a precise delineation of situations in which they are valid; such judgments must be made by experimenters on the basis of their experience. Recently, more and more procedures have been coming into use which do not depend on the assumption of normality; such tests are called non-parametric tests, and usually assume only continuity of the probability distribution. These procedures are also useful in that they are easily applied.

7.5.1 The sign test

Consider the problem of the corrosion experiment in Section 7.3. If the type of coating had no effect on corrosion, we would expect half of the differences to be positive and half to be negative. If there is a preponderance of plus or minus signs, we would suspect a difference due to coating. The number of plus and minus signs in a sample of n is a random variable having a binomial distribution. r is used to denote the number of times the less frequent sign occurs, with critical values of r for the 5% level and 1% level of significance given in Table 7.4. If the value of r is less than or equal to the critical value given in the table, this is evidence that there is a difference due to coating. Strictly speaking, the hypothesis to be tested is that each difference between the paired observations has a probability distribution (which need not be the same for all differences) with median equal to zero, i.e., $P(x_i - y_i > 0) = \frac{1}{2}$ for all i. If the underlying distributions of the individual observations are symmetric, the sign test can be used to test the hypothesis that $\mu_{x_i} = \mu_{y_i}$ or, equivalently, the mean of the difference, μ_d, equals zero. If it is assumed that the underlying distributions of each pair of observations may differ only in their means, the sign test can also be used to test the hypothesis that $\mu_{x_i} = \mu_{y_i}$, this being equivalent to testing the hypothesis that the probability distributions of each pair are the same. Thus, if r, the number of times the less frequent sign occurs, is less than or equal to the critical value given in the table, the hypothesis that $P(x_i - y_i > 0) = \frac{1}{2}$, or $\mu_{x_i} = \mu_{y_i}$ when applicable, is rejected, and it is concluded that there are treatment effects. Table 7.4 is based upon the binomial distribution with parameter equal to $\frac{1}{2}$. Because of the discreteness of the binomial distribution, the critical values of r given in Table 7.4 do not correspond exactly to the stated levels of significance. The test, in fact, is stricter than the level indicated.

It is interesting to note the results of applying the sign test to the data presented in Section 7.4.2. The number of negative differences (the less frequent sign) is equal to 4; i.e., $r=4$. From Table 7.4, we find that for $n = 15$, the critical value at the 5% level is 3 and we conclude, using the sign test, that there is no difference in the coating. This apparent contradiction to the results of the paired t test for equality of means is due to the fact that the sign test is not as sensitive a test, since it is the sign and not the magnitude of the difference which is taken into account.

Although ties should not occur because the data are assumed to be continuous, the discreteness of measuring instruments allow for such phenomena. In such cases, a tie counts as half a plus and half a minus.

Table 7.4. Critical Values of r for the Sign Test
(Two-tail percentage points for the binomial distribution with parameter equal to 0.5)[1]

n	1%	5%	n	1%	5%
6		0	26	6	7
7		0	27	6	7
8	0	0	28	6	8
9	0	1	29	7	8
10	0	1	30	7	9
11	0	1	31	7	9
12	1	2	32	8	9
13	1	2	33	8	10
14	1	2	34	9	10
15	2	3	35	9	11
16	2	3	36	9	11
17	2	4	37	10	12
18	3	4	38	10	12
19	3	4	39	11	12
20	3	5	40	11	13
21	4	5	41	11	13
22	4	5	42	12	14
23	4	6	43	12	14
24	5	6	44	13	15
25	5	7	45	13	15

For values of $n > 45$, values of r may be obtained by taking the nearest integer less than $(n - 1)/2 - 1.2879\sqrt{n+1}$ for 1% level and $(n - 1)/2 - 0.9800\sqrt{n+1}$ for 5% level.

Table 7.4 can also be used for one-sided procedures. Let r denote either the number of plus signs or the number of minus signs depending on which of the two is expected to appear least frequently according to the alternative to the hypothesis. In other words, if the number of minus signs is expected to be less than the number of plus signs, let r represent the number of minus signs. If r is less than or equal to the critical value in the table, the hypothesis is rejected at a level of significance equal to half the value ($\frac{1}{2}$% or $2\frac{1}{2}$%) given in the table heading.

[1] Reproduced from Dixon and Mood, "The Statistical Sign-Test," *Journal of the American Statistical Association*, 1946, Vol. 41, p. 560.

7.5.2 The Wilcoxon Signed Rank Test

The sign test in the previous section is simple to apply but takes no account of the magnitude of the differences whatever. A better non-parametric test can be based on the signed rank of the differences; i.e., differences are first ranked without regard to sign and then these ranks are given the corresponding sign of the difference. The hypothesis to be tested in this problem is the same as for the sign test, i.e., $P(x_i - y_i > 0) = \frac{1}{2}$ or $\mu_{x_i} = \mu_{y_i}$ for all i, provided the required assumptions that the distributions are symmetric or differ only in their means are fulfilled.

For the corrosion data, this ranking would be as follows:

Difference	*Rank*	*Signed Rank*
22	13	13
2	$2\frac{1}{2}$	$2\frac{1}{2}$
4	$4\frac{1}{2}$	$4\frac{1}{2}$
12	11	11
11	10	10
15	12	12
28	15	15
−5	6	−6
8	8	8
4	$4\frac{1}{2}$	$4\frac{1}{2}$
−1	1	−1
−10	9	−9
−2	$2\frac{1}{2}$	$-2\frac{1}{2}$
25	14	14
7	7	7

The smallest observation is given rank 1 and the ties are assigned average ranks. Affix to each rank the sign of the difference.

If there is no difference between coatings, we would expect the sum of the positive ranks approximately to equal numerically the sum of the negative ranks. Since

$$1 + 2 + 3 + \cdots + n = \frac{n(n+1)}{2},$$

the absolute value of the sum of either the positive or negative signed ranks should be near $n(n+1)/4$ or about 60 for this data. Actually the sum of negative ranks is $18\frac{1}{2}$ and the sum of positive ranks is $101\frac{1}{2}$. To see whether such a difference could arise by chance or whether it is indicative of a significant difference, we refer the absolute value of the smaller sum of ranks (T) to Table 7.5. If T is less than or equal to the critical value given in the table, the hypothesis is rejected. In this case, $T = 18\frac{1}{2}$ is less than 25, which is the critical value corresponding to a 5 level of significance for $n = 15$, so that the hypothesis is rejected at the 5 level of significance, and it is concluded that there is a difference due to coating.

Table 7.5 can also be used for one-sided procedures. Let T denote either the sum of the positive ranks or the absolute value of the sum of the negative ranks, depending on which of the two is expected to be smaller according to the alternative to the hypothesis. In other words, if the sum of the positive ranks is expected to be less than the absolute value of the negative ranks, let T represent the sum of the positive ranks. If T is less than or equal to the critical value in the table, the hypothesis is rejected at a level of significance equal to half the value ($2\frac{1}{2}\%$, 1%, $\frac{1}{2}\%$) given in the table heading.

Table 7.5. Significance Points for the Absolute Value of the Smaller Sum of Signed Ranks (T) Obtained from Paired Observations[1]

n	5%	2%	1%
6	0	—	—
7	2	0	—
8	4	2	0
9	6	3	2
10	8	5	3
11	11	7	5
12	14	10	7
13	17	13	10
14	21	16	13
15	25	20	16
16	30	24	20
17	35	28	23
18	40	33	28
19	46	38	32
20	52	43	38
21	59	49	43
22	66	56	49
23	73	62	55
24	81	69	61
25	89	77	68

For $n > 25$, T is approximately normally distributed with mean $n\left(\frac{n+1}{4}\right)$ and variance $\frac{n}{24}(2n+1)(n+1)$.

[1] Reprinted by permission of Frank Wilcoxon and the American Cyanimid Co.

7.5.3 A TEST FOR TWO INDEPENDENT SAMPLES

The sign test and the signed rank test are applicable to paired samples. When observations are not paired and the two probability distributions are assumed to differ only in their means, a useful procedure for testing the hypothesis that the means are the same is as follows: First arrange all observations from both samples in order of magnitude and rank them. For the data on endurance limit of Section 7.1.4, this would appear as follows:

Sample	*Endurance Limit*	*Rank*
2	69,800	1
1	78,700	2
2	79,400	3
2	79,500	4
2	79,900	5
2	81,700	6
2	82,400	7
2	82,600	8
1	83,100	9
2	83,400	10
1	84,000	11
1	86,500	12
1	87,500	13
1	89,400	14
1	89,900	15
1	91,900	16

If there is no difference between samples, we would expect them to intermingle in a regular way and, if they are both of the same size, we would expect the sum of ranks to be about the same for both. In this case the ranks in sample 2 are 1, 3, 4, 5, 6, 7, 8, and 10, which total 44; the ranks in sample 1 total 92.

To see whether the difference is significant, we enter Table 7.6 with

$$R_1 = \text{sum of ranks of smaller sample}$$

and

$$R_1' = n_1(n_1 + n_2 + 1) - R_1$$

and reject if either is less than or equal to the tabled critical value, $R_1^{+\,+}$, where n_1 is the size of sample 1 and n_2 the size of sample 2. If the samples are unequal in size, sample 1 is to be the smaller; thus $n_1 \leqq n_2$. In this case $R_1 = 44$ and $R_1' = 92$. The 5% point for $n_1 = 8$ and $n_2 = 8$ is 49. Hence the departure is significant at the 5% level (since $R_1 < 49$).

In the case of ties, average ranks are assigned. Table 7.6 can also be used for one-sided procedures. If under the alternative to the hypothesis, R_1 should be large, reject if R_1' is less than or equal to the critical value given in the table. If under the alternative, R_1 should be small, reject if R_1 is less than or equal to the critical value. The level of significance equals half the values ($2\frac{1}{2}\%$ and $\frac{1}{2}\%$) given in the table heading.

Table 7.6. 5% Critical Points of Rank Sums (R_1^{++})*

n_2 \ $n_1 \rightarrow$	2	3	4	5	6	7	8	9	10	11	12	13	14	15
4			10											
5		6	11	17										
6		7	12	18	26									
7		7	13	20	27	36								
8	3	8	14	21	29	38	49							
9	3	8	15	22	31	40	51	63						
10	3	9	15	23	32	42	53	65	78					
11	4	9	16	24	34	44	55	68	81	96				
12	4	10	17	26	35	46	58	71	85	99	115			
13	4	10	18	27	37	48	60	73	88	103	119	137		
14	4	11	19	28	38	50	63	76	91	106	123	141	160	
15	4	11	20	29	40	52	65	79	94	110	127	145	164	185
16	4	12	21	31	42	54	67	82	97	114	131	150	169	
17	5	12	21	32	43	56	70	84	100	117	135	154		
18	5	13	22	33	45	58	72	87	103	121	139			
19	5	13	23	34	46	60	74	90	107	124				
20	5	14	24	35	48	62	77	93	110					
21	6	14	25	37	50	64	79	95						
22	6	15	26	38	51	66	82							
23	6	15	27	39	53	68								
24	6	16	28	40	55									
25	6	16	28	42										
26	7	17	29											
27	7	17												
28	7													

* Reprinted by permission from Colin White, "The Use of Ranks in a Test of Significance for Comparing Two Treatments," *Biometrics*, 1952, Vol. 8, p. 37.

Table 7.6. (Cont.) 1% Critical Points of Rank Sums (R_1^{++})

n_2 \ $n_1 \rightarrow$	2	3	4	5	6	7	8	9	10	11	12	13	14	15
5				15										
6			10	16	23									
7			10	17	24	32								
8			11	17	25	34	43							
9		6	11	18	26	35	45	56						
10		6	12	19	27	37	47	58	71					
11		6	12	20	28	38	49	61	74	87				
12		7	13	21	30	40	51	63	76	90	106			
13		7	14	22	31	41	53	65	79	93	109	125		
14		7	14	22	32	43	54	67	81	96	112	129	147	
15		8	15	23	33	44	56	70	84	99	115	133	151	171
16		8	15	24	34	46	58	72	86	102	119	137	155	
17		8	16	25	36	47	60	74	89	105	122	140		
18		8	16	26	37	49	62	76	92	108	125			
19	3	9	17	27	38	50	64	78	94	111				
20	3	9	18	28	39	52	66	81	97					
21	3	9	18	29	40	53	68	83						
22	3	10	19	29	42	55	70							
23	3	10	19	30	43	57								
24	3	10	20	31	44									
25	3	11	20	32										
26	3	11	21											
27	4	11												
28	4													

7.6 Test of the Hypothesis that the Standard Deviations of Two Normal Distributions Are Equal

7.6.1 CHOICE OF AN OC CURVE

It frequently is of interest to test the hypothesis that the standard deviations of two normal distributions are equal.[1] This test may be preliminary to a two-sample t test or may arise in circumstances in which the mean can be set but the variability cannot, e.g., in machine settings. The test procedure is based upon an F statistic. This test statistic is a function of the ratio of sample variances. The abscissae of operating characteristic curves for procedures based upon this

[1] This is equivalent to testing the hypothesis that the variances of two normal distributions are equal.

F statistic depend upon the quantity $\lambda = \sigma_x/\sigma_y$, the true ratio of the standard deviations.

In order to determine the acceptance region and sample size, two points on the OC curve must be chosen. These points will be denoted by $(1, 1 - \alpha)$ and $\left(\frac{\sigma_x}{\sigma_y}, \beta\right)$. The first point corresponds to choosing the probability of accepting the hypothesis when it is true (when $\sigma_x = \sigma_y$), and the second point corresponds to choosing the probability of accepting the hypothesis when σ_x/σ_y is the ratio of the true standard deviations considered important to detect.

7.6.2 Charts and Tables for Carrying Out F Tests

An experiment consisting of a set of n_x trials and a set of n_y trials is to be performed. Let the random variables $x_1, x_2, \cdots, x_{n_x}$ denote the measurable quantities associated with the set of n_x trials. It is assumed that the x_i are independent normally distributed random variables each with unknown mean μ_x and unknown standard deviation σ_x. Let the random variables $y_1, y_2, \cdots, y_{n_y}$ denote the measurable quantities associated with the set of n_y trials. It is assumed that the y_i are independent normally distributed random variables each with unknown mean μ_y and unknown standard deviation σ_y. Furthermore, let all the x's and y's be independent. The optimum procedure for testing the hypothesis that $\sigma_x = \sigma_y$ is based upon the random variable

$$F = s_x^2/s_y^2$$

where $s_x^2 = \dfrac{\sum_{i=1}^{n_x} (x_i - \bar{x})^2}{n_x - 1}$ and $s_y^2 = \dfrac{\sum_{i=1}^{n_y} (y_i - \bar{y})^2}{n_y - 1}$. If the value taken on by the test statistic as a result of experimentation falls into the acceptance region, the hypothesis that $\sigma_x = \sigma_y$ is accepted; otherwise it is rejected. This section will be devoted to finding the acceptance region and the sample size for the test statistic F using the tables and charts which are provided. One- and two-sided procedures will be considered.

7.6.2.1 *Tables and charts for two-sided procedures.* The acceptance region for the two-sided procedure for any sample sizes, n_x and n_y, is given by the interval $[F_{1-\alpha/2;n_x-1,n_y-1}, F_{\alpha/2;n_x-1,n_y-1}]$, which is equivalent (see Section 4.4.1) to the interval $[1/F_{\alpha/2;n_y-1,n_x-1}, F_{\alpha/2;n_x-1,n_y-1}]$, where α is the level of significance and $F_{\alpha/2;\nu_1,\nu_2}$ is the 100 $\alpha/2$

percentage point of the F distribution with ν_1 and ν_2 degrees of freedom.[1] Thus, the hypothesis that $\sigma_x = \sigma_y$ is accepted if

$$1/F_{\alpha/2;n_y-1,\,n_x-1} \leqq F \leqq F_{\alpha/2;n_x-1,\,n_y-1}$$

and rejected if F lies outside this interval. It is easily verified that this procedure yields a probability of $1 - \alpha$ of accepting the hypothesis that $\sigma_x = \sigma_y$ when it is true. In Section 4.4.2, it was shown that whenever $\sigma_x = \sigma_y$, the test statistic $F = s_x^2/s_y^2$ has an F distribution with $n_x - 1$ and $n_y - 1$ degrees of freedom. Hence, if $\sigma_x = \sigma_y$

$$\mathrm{P}(1/F_{\alpha/2;n_y-1,\,n_x-1} \leqq F \leqq F_{\alpha/2;n_x-1,\,n_y-1}) = 1 - \alpha.$$

The required sample size is obtained by referring to a set of operating characteristic curves which are provided for this purpose. Figure 7.1 presents the OC curves of the two-sided F test for a level of

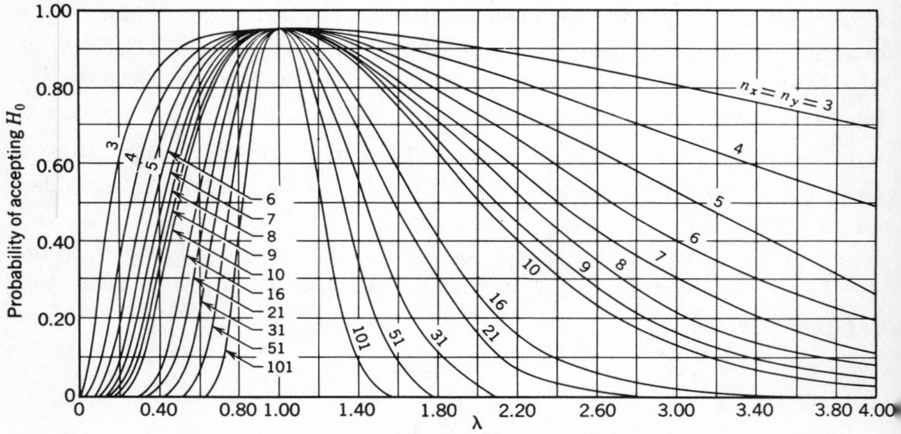

Fig. 7.1. OC curves for different values of n for the two-sided F test for a level of significance $\alpha=0.05$.

significance equal to 0.05. Figure 7.2 presents the OC curves of the two-sided F test for a level of significance equal to 0.01. In order to use these curves it is necessary to choose $n_x = n_y = n$. If the level of significance of the procedure is either 0.05 or 0.01, the required sample size n is obtained by entering the appropriate figure (Figure 7.1 if $\alpha = 0.05$ and Figure 7.2 if $\alpha = 0.01$) with the point $\left(\frac{\sigma_x}{\sigma_y}, \beta\right)$, and reading out the value of the sample size associated with the OC curve that passes through this point.

[1] If the notation x is assigned to the data so that s_x^2 denotes the larger sample variance, the acceptance region is also given by the interval $[-\infty, F_{\alpha/2;\,n_x-1,\,n_y-1}]$.

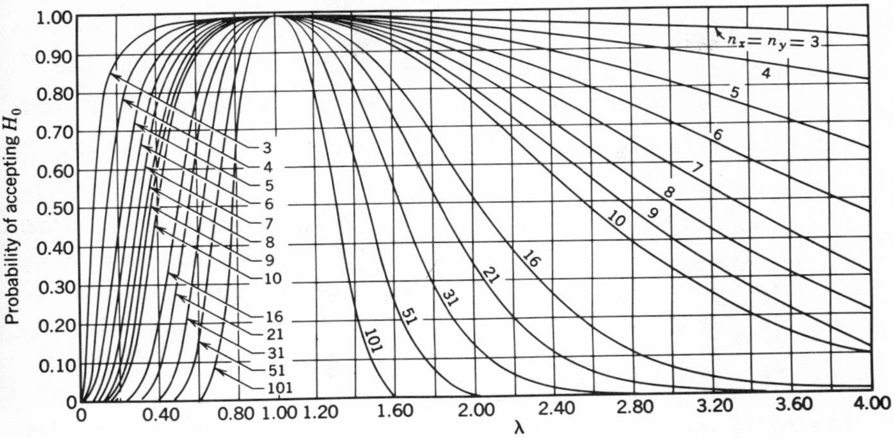

Fig. 7.2. OC curves for different values of n for the two-sided F test for a level of significance $\alpha = 0.01$.

7.6.2.2 *Summary for two-sided procedures using tables and charts.* The following steps are taken using a two-sided procedure for testing the hypothesis that $\sigma_x = \sigma_y$.

1. Two points on the OC curve are chosen. These points are denoted by $(1, 1 - \alpha)$ and $\left(\frac{\sigma_x}{\sigma_y}, \beta\right)$.
2. If $n_x = n_y = n$, the required sample size, n, is found by entering the appropriate figure (Figure 7.1 if $\alpha = 0.05$ and Figure 7.2 if $\alpha = 0.01$) with the point $\left(\frac{\sigma_x}{\sigma_y}, \beta\right)$ and reading out the value of the sample size associated with the OC curve that passes through this point.
3. Determine the acceptance region which is the interval

$$[1/F_{\alpha/2;n_y-1,\,n_x-1},\ F_{\alpha/2;n_x-1,\,n_y-1}].$$

4. Draw a random sample of n_x and n_y items.
5. Compute the value of the test statistic.
6. If $1/F_{\alpha/2;n_y-1,\,n_x-1} \leqq F \leqq F_{\alpha/2;n_x-1,\,n_y-1}$ accept the hypothesis and conclude that $\sigma_x = \sigma_y$.

7.6.2.3 *Tables and charts for one-sided procedures.* The notation x and y is arbitrary in a one-sided test. Let x denote the symbol for the variable with possible larger standard deviation so that the alternative is always $\sigma_x > \sigma_y$. The acceptance region for any sample sizes, n_x and n_y, to test the hypothesis that $\sigma_x = \sigma_y$ (or $\sigma_x \leqq \sigma_y$) is given by the interval $[-\infty, F_{\alpha;n_x-1,n_y-1}]$, where α is the level of significance and $F_{\alpha;n_x-1,n_y-1}$ is the 100α percentage point of the F distribution with $n_x - 1$ and $n_y - 1$ degrees of freedom. Thus, the hypothesis that $\sigma_x = \sigma_y$ (or $\sigma_x \leqq \sigma_y$) is accepted if

$$F \leqq F_{\alpha;n_x-1,n_y-1}$$

and rejected if F lies outside this interval. This acceptance procedure has a probability of $1 - \alpha$ of accepting the hypothesis that $\sigma_x = \sigma_y$ when it is true. Whenever $\sigma_x = \sigma_y$, the test statistic $F = s_x^2/s_y^2$ has an F distribution with $n_x - 1$ and $n_y - 1$ degrees of freedom. Hence, if $\sigma_x = \sigma_y$,

$$P(F \leqq F_{\alpha;n_x-1,n_y-1}) = 1 - \alpha.$$

The operating characteristic curves for determining the sample size are given in Figures 7.3 and 7.4. In order to use these curves

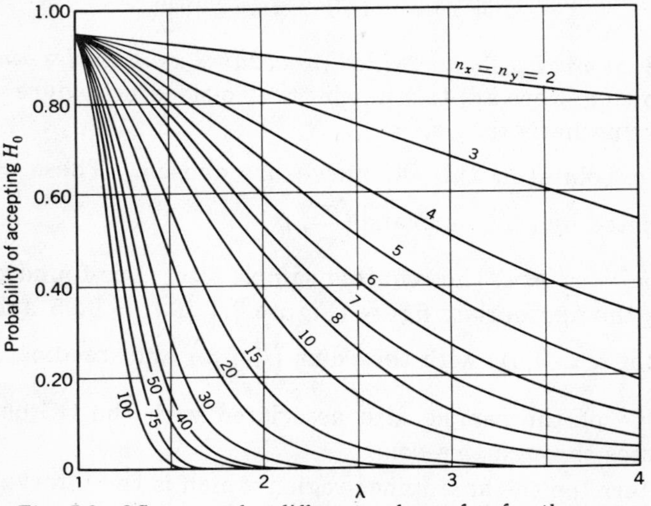

Fig. 7.3. OC curves for different values of n for the one-sided F test for a level of significance α=0.05.

Reproduced by permission from "Operating Characteristics for the Common Statistical Tests of Significance" by Charles L. Ferris, Frank E. Grubbs, Chalmers L. Weaver, Annals of Mathematical Statistics, June, 1946

it is necessary to choose $n_x = n_y = n$. Figure 7.3 presents the OC curves for a level of significance equal to 0.05, and Figure 7.4 presents the OC curves for a level of significance equal to 0.01. If the level of significance of the procedure is either 0.05 or 0.01, the required sample size, n, is obtained by entering the appropriate figure (Figure 7.3 if $\alpha = 0.05$ and Figure 7.4 if $\alpha = 0.01$) with the point $\left(\frac{\sigma_x}{\sigma_y}, \beta\right)$ and reading out the sample size associated with the OC curve that passes through this point.

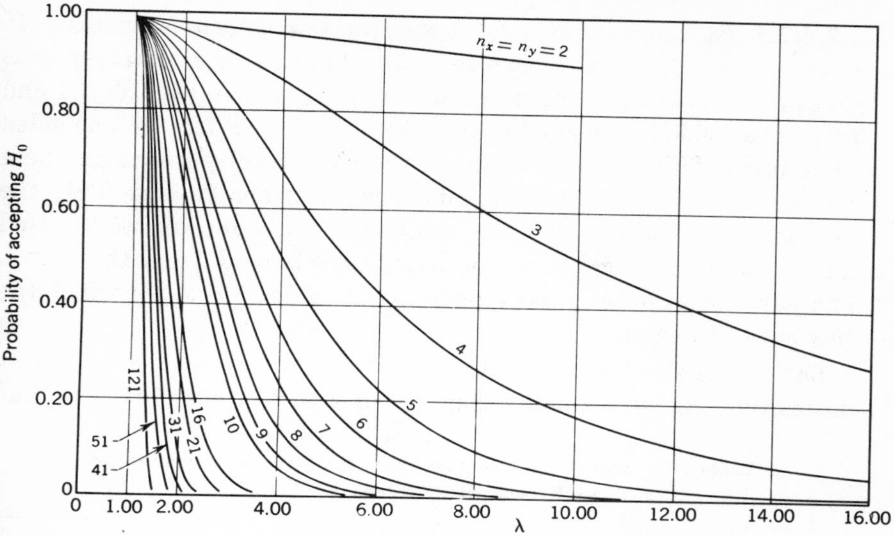

Fig. 7.4. OC curves for different values of n for the one sided F test for a level of significance $\alpha = 0.01$.

7.6.2.4 *Summary for one-sided procedures using tables and charts.* Let x denote the symbol for the variable with possible larger standard deviation so that the alternative is always $\sigma_x > \sigma_y$. The following steps are taken using a one-sided procedure for testing the hypothesis that $\sigma_x = \sigma_y$ (or $\sigma_x \leqq \sigma_y$) with the alternative that $\sigma_x > \sigma_y$.

1. Two points on the OC curve are chosen. These points are denoted by $(1, 1 - \alpha)$ and $\left(\frac{\sigma_x}{\sigma_y}, \beta\right)$.
2. If $n_x = n_y = n$, the required sample size, n, is found by entering the appropriate figure (Figure 7.3 if $\alpha = 0.05$ and Figure 7.4 if $\alpha = 0.01$) with the point $\left(\frac{\sigma_x}{\sigma_y}, \beta\right)$ and reading out the value of the sample size associated with the OC curve that passes through this point.
3. Determine the acceptance region which is the interval
$$[-\infty, F_{\alpha;n_x-1, n_y-1}].$$
4. Draw a random sample of n_x and n_y items.
5. Compute the value of the test statistic.
6. If $F \leqq F_{\alpha;n_x-1, n_y-1}$, accept the hypothesis that $\sigma_x = \sigma_y$ (or $\sigma_x \leqq \sigma_y$). If F lies outside this interval, reject the hypothesis and conclude that $\sigma_x > \sigma_y$.

7.6.2.5 *Tables and charts for operating characteristic curves.* If $n_x = n_y = n$ and this sample size is fixed in advance, or after it is chosen, the operating characteristic curves given in Figures 7.1 and 7.2 for two-sided procedures, and Figures 7.3 and 7.4 for one-sided procedures can be used to evaluate the risks involved in using these procedures. If the level of significance is either 0.05 or 0.01, the OC curve associated with the sample size n is entered with λ, and the probability of accepting the hypothesis is read from the curve. The value of λ chosen corresponds to the value of σ_x/σ_y for which the risk is to be evaluated.

The F statistic for testing the hypothesis that $\sigma_x = \sigma_y$ is relatively insensitive. If we want to detect a λ of 2 with probability 0.10 (and

Table 7.7. Test of the Hypothesis that the Standard Deviations of Two Normal Distributions Are Equal

Notation for the Hypothesis	*Test Statistic*
$H: \sigma_x^2 = \sigma_y^2$ or $\sigma_x = \sigma_y$	$F = \dfrac{\sum_{i=1}^{n_x} (x_i - \bar{x})^2/(n_x - 1)}{\sum_{i=1}^{n_y} (y_i - \bar{y})^2/(n_y - 1)} = \dfrac{s_x^2}{s_y^2}$
Criteria for Rejection	*Method for Choosing Sample Sizes*
$F \leqq 1/F_{\alpha/2;n_y-1,n_x-1}$ or $F \geqq F_{\alpha/2;n_x-1,n_y-1}$ if we wish to reject when σ_x is not equal to σ_y.	The OC curve depends on $\lambda = \sigma_x/\sigma_y$ and is given in Fig. 7.1 and 7.2 for the case $n_x = n_y = n$. Choose a value of λ for which we wish to reject the hypothesis with given high probability. Enter Fig. 7.1 or 7.2 to find the required sample size.
The notation x and y is arbitrary in a one-sided test; let x be the symbol for the variable with possible larger variance. $F \geqq F_{\alpha;n_x-1,n_y-1}$ if we wish to reject when $\sigma_x > \sigma_y$.	The OC curve depends on $\lambda = \sigma_x/\sigma_y$ and is given in Fig. 7.3 and 7.4 for the case $n_x = n_y = n$. OC curves for $n_x \neq n_y$ are given in Fig. 10.10—10.17 for $n_x - 1 = \nu_1 = 1, 2, \cdots, 8$ and various values of $n_y - 1 = \nu_2$. Choose a value of λ for which we wish to reject the hypothesis with given high probability. Enter the appropriate figure to find the required sample size.

Values of $F_{\alpha;\nu_1,\nu_2}$, the 100 α percentage points of the F distribution with ν_1 and ν_2 degrees of freedom are given in Appendix Table 4.

$\alpha = 0.05$), Figure 7.1 reveals that about 26 observations are necessary; a λ of 1.6 requires 50 observations. To detect a 10 or 20% change in σ requires hundreds of observations.

If levels of significance other than 0.05 or 0.01 are desired, Appendix Table 4, which is a table of the percentage points of the F distribution, can be used to determine the acceptance region for fixed values of n_x and n_y. However, analytical methods which are presented in Section 7.6.3 must be used to obtain the OC curve.

A summary of the procedures for testing the hypothesis that $\sigma_x = \sigma_y$ is given in Table 7.7.

*7.6.3 Analytical treatment for tests

In Section 4.4.2 it was pointed out that the quantity

$$\frac{s_x^2/\sigma_x^2}{s_y^2/\sigma_y^2} = \frac{s_x^2}{s_y^2} \cdot \frac{\sigma_y^2}{\sigma_x^2}$$

has an F distribution with $n_x - 1$ and $n_y - 1$ degrees of freedom. If we are interested in testing the hypothesis that $\sigma_x = \sigma_y$ against the alternative that $\sigma_x > \sigma_y$, the test statistic s_x^2/s_y^2 has an F distribution with $n_x - 1$ and $n_y - 1$ degrees of freedom whenever $\sigma_x = \sigma_y$. If we want a test with a Type I error equal to α, the rule of acceptance is to accept the hypothesis if

$$\frac{s_x^2}{s_y^2} \leqq F_{\alpha;n_x-1,n_y-1}.$$

It is clear that whenever $\sigma_x = \sigma_y$

$$\mathrm{P}\left(\frac{s_x^2}{s_y^2} \leqq F_{\alpha;n_x-1,n_y-1}\right) = 1 - \alpha.$$

The OC curve of the test can be computed easily from a table of the F distribution. Suppose that the true value of the ratio of the standard deviations is $\lambda = \sigma_x/\sigma_y$. Then,

$$\frac{s_x^2}{s_y^2} \cdot \frac{\sigma_y^2}{\sigma_x^2} = \frac{s_x^2}{s_y^2} \cdot \frac{1}{\lambda^2}$$

has an F distribution with $n_x - 1$ and $n_y - 1$ degrees of freedom; and the probability of accepting the hypothesis is given by

$$\mathrm{P}\left(\frac{s_x^2}{s_y^2} \leqq F_{\alpha;n_x-1,n_y-1}\right) = \mathrm{P}\left(\frac{s_x^2}{s_y^2} \cdot \frac{1}{\lambda^2} \leqq \frac{F_{\alpha;n_x-1,n_y-1}}{\lambda^2}\right)$$

$$= \mathrm{P}\left(F \leqq \frac{F_{\alpha;n_x-1,n_y-1}}{\lambda^2}\right)$$

where F now denotes any random variable having an F distribution with $n_x - 1$ and $n_y - 1$ degrees of freedom.[1]

$\mathrm{P}\left(F \leqq \dfrac{F_{\alpha;n_x-1,n_y-1}}{\lambda^2}\right)$ is the expression for the OC curve.

The expression in this form permits easy solution of the sample size problem. Suppose that a particular value of λ, say λ_1, is given which we want to be fairly sure to reject. The condition is that

$$\mathrm{P}\left(\frac{s_x^2}{s_y^2} \leqq F_{\alpha;n_x-1,n_y-1}\right) = \mathrm{P}\left(F \leqq \frac{F_{\alpha;n_x-1,n_y-1}}{\lambda_1^2}\right) = \beta$$

where F is a random variable having an F distribution and β is the Type II error associated with λ_1. This last expression implies that

$$\frac{F_{\alpha;n_x-1,n_y-1}}{\lambda_1^2} = F_{1-\beta;n_x-1,n_y-1} = \frac{1}{F_{\beta;n_y-1,n_x-1}}$$

or

$$\lambda_1^2 = (F_{\alpha;n_x-1,n_y-1})(F_{\beta;n_y-1,n_x-1}).$$

If $n_x = n_y = n$, we have

$$\lambda_1^2 = (F_{\alpha;n-1,n-1})(F_{\beta;n-1,n-1}).$$

We can determine the sample size for this problem by going to the table of the F distribution, Appendix Table 4, and finding the degrees of freedom for which the product of the 100 α percentage point and the 100 β percentage point is equal to λ_1^2. The sample size n is one more than the degrees of freedom. For example, if $\alpha = 0.05$, $\beta = 0.10$, and $\lambda_1^2 = 3.0$, Appendix Table 4 indicates that for 30 degrees of freedom, $F_{0.05;30,30} = 1.84$ and $F_{0.10;30,30} = 1.61$ and $\lambda_1^2 = (1.84)(1.61) = 3.0$. Hence the required sample size $n_x = n_y = n = 31$. Similar analytic treatment can be provided for two-sided procedures.

For a two-sided procedure, the expression for the OC curve (probability of accepting the hypothesis) is given by

$$\mathrm{P}\left(\frac{1}{(F_{\alpha/2;n_y-1,n_x-1})\lambda^2} \leqq F \leqq \frac{F_{\alpha/2;n_x-1,n_y-1}}{\lambda^2}\right)$$

where F denotes a random variable having an F distribution and $\lambda = \sigma_x/\sigma_y$. When $n_x = n_y = n$, the expression required for determining the sample size is given approximately by

[1] In the immediately preceding sections F denoted the quantity s_x^2/s_y^2 which has an F distribution only when $\sigma_x = \sigma_y$. In the remainder of this section (Section 7.6.3) F will denote a random variable which has an F distribution with $n_x - 1$ and $n_y - 1$ degrees of freedom, and must not be confused with the quantity s_x^2/s_y^2.

$$\lambda_1^2 \cong (F_{\alpha/2;n-1,n-1})(F_{\beta;n-1,n-1})$$

when $P\left(F \leqq \frac{1}{(F_{\alpha/2;n-1,n-1})\lambda_1^2}\right)$ is small compared to β and by

$$\lambda_1^2 \cong \frac{1}{(F_{\alpha/2;n-1,n-1})(F_{\beta;n-1,n-1})}$$

when $P\left(F \geqq \frac{F_{\alpha/2;n-1,n-1}}{\lambda_1^2}\right)$ is small compared to β.

7.6.4 EXAMPLE

The standard deviation of a particular dimension of a metal component is small enough so that it is satisfactory in subsequent assembly; a new supplier of metal plate is under consideration and will be preferred if the standard deviation of his product is not larger, because the cost of his product is lower than that of the present supplier. Hence, the hypothesis to be tested is that the standard deviation of the new supplier is less than or equal to the standard deviation of the old supplier and the alternative is that the standard deviation of the new supplier exceeds that of the old supplier. A one-sided procedure is called for and the symbol x is reserved for the new supplier. Without consulting an OC curve, it is decided to base the decision on a sample of 100 items from each supplier since data on dimensions are relatively easy to obtain, and it is known that small numbers of observations give relatively little protection against erroneous decisions. The test is to be run at the 5% level of significance. The following are computed from the data:

New Supplier: $s_x^2 = 0.00041$

Current Supplier: $s_y^2 = 0.00057$

$$F = \frac{0.00041}{0.00057} = 0.72.$$

Since the value of $F = 0.72$ is less than 1, it does not exceed $F_{0.05;99,99}$, and the hypothesis of equality of standard deviations is accepted.

It is interesting to examine the implications of this procedure with respect to the OC curve. From Figure 7.3, it is seen that if $\lambda = 1.3$, the probability of accepting the hypothesis is approximately 0.10. Hence, since the experimental data led to acceptance of the hypothesis, it is safe to assume (with a chance of error smaller than 10%) $\lambda = \sigma_x/\sigma_y$ does not exceed 1.3. If ratios of standard deviations less than 1.3 are relatively unimportant and the above-mentioned risks are satisfactory, a sample of 100 observations from each supplier is sufficient.

Table 7.8. Upper 1 Percentage Points of the Ratio of the Largest to the Sum of k Independent Estimates of Variance, Each of Which is Based on n Observations[1]

k \ n	2	3	4	5	6	7	8	9	10	11	17	37	145	∞
2	0.9999	0.9950	0.9794	0.9586	0.9373	0.9172	0.8988	0.8823	0.8674	0.8539	0.7949	0.7067	0.6062	0.5000
3	0.9933	0.9423	0.8831	0.8335	0.7933	0.7606	0.7335	0.7107	0.6912	0.6743	0.6059	0.5153	0.4230	0.3333
4	0.9676	0.8643	0.7814	0.7212	0.6761	0.6410	0.6129	0.5897	0.5702	0.5536	0.4884	0.4057	0.3251	0.2500
5	0.9279	0.7885	0.6957	0.6329	0.5875	0.5531	0.5259	0.5037	0.4854	0.4697	0.4094	0.3351	0.2644	0.2000
6	0.8828	0.7218	0.6258	0.5635	0.5195	0.4866	0.4608	0.4401	0.4229	0.4084	0.3529	0.2858	0.2229	0.1667
7	0.8376	0.6644	0.5685	0.5080	0.4659	0.4347	0.4105	0.3911	0.3751	0.3616	0.3105	0.2494	0.1929	0.1429
8	0.7945	0.6152	0.5209	0.4627	0.4226	0.3932	0.3704	0.3522	0.3373	0.3248	0.2779	0.2214	0.1700	0.1250
9	0.7544	0.5727	0.4810	0.4251	0.3870	0.3592	0.3378	0.3207	0.3067	0.2950	0.2514	0.1992	0.1521	0.1111
10	0.7175	0.5358	0.4469	0.3934	0.3572	0.3308	0.3106	0.2945	0.2813	0.2704	0.2297	0.1811	0.1376	0.1000
12	0.6528	0.4751	0.3919	0.3428	0.3099	0.2861	0.2680	0.2535	0.2419	0.2320	0.1961	0.1535	0.1157	0.0833
15	0.5747	0.4069	0.3317	0.2882	0.2593	0.2386	0.2228	0.2104	0.2002	0.1918	0.1612	0.1251	0.0934	0.0667
20	0.4799	0.3297	0.2654	0.2288	0.2048	0.1877	0.1748	0.1646	0.1567	0.1501	0.1248	0.0960	0.0709	0.0500
24	0.4247	0.2871	0.2295	0.1970	0.1759	0.1608	0.1495	0.1406	0.1338	0.1283	0.1060	0.0810	0.0595	0.0417
30	0.3632	0.2412	0.1913	0.1635	0.1454	0.1327	0.1232	0.1157	0.1100	0.1054	0.0867	0.0658	0.0480	0.0333
40	0.2940	0.1915	0.1508	0.1281	0.1135	0.1033	0.0957	0.0898	0.0853	0.0816	0.0668	0.0503	0.0363	0.0250
60	0.2151	0.1371	0.1069	0.0902	0.0796	0.0722	0.0668	0.0625	0.0594	0.0567	0.0461	0.0344	0.0245	0.0167
120	0.1225	0.0759	0.0585	0.0489	0.0429	0.0387	0.0357	0.0334	0.0316	0.0302	0.0242	0.0178	0.0125	0.0083
∞	0	0	0	0	0	0	0	0	0	0	0	0	0	0

Table 7.8. Upper 5 Percentage Points of the Ratio of the Largest to the Sum of k Independent Estimates of Variance, Each of Which is Based on n Observations[1]

k \ n	2	3	4	5	6	7	8	9	10	11	17	37	145	∞
2	0.9985	0.9750	0.9392	0.9057	0.8772	0.8534	0.8332	0.8159	0.8010	0.7880	0.7341	0.6602	0.5813	0.5000
3	0.9669	0.8709	0.7977	0.7457	0.7071	0.6771	0.6530	0.6333	0.6167	0.6025	0.5466	0.4748	0.4031	0.3333
4	0.9065	0.7679	0.6841	0.6287	0.5895	0.5598	0.5365	0.5175	0.5017	0.4884	0.4366	0.3720	0.3093	0.2500
5	0.8412	0.6838	0.5981	0.5441	0.5065	0.4783	0.4564	0.4387	0.4241	0.4118	0.3645	0.3066	0.2513	0.2000
6	0.7808	0.6161	0.5321	0.4803	0.4447	0.4184	0.3980	0.3817	0.3682	0.3568	0.3135	0.2612	0.2119	0.1667
7	0.7271	0.5612	0.4800	0.4307	0.3974	0.3726	0.3535	0.3384	0.3259	0.3154	0.2756	0.2278	0.1833	0.1429
8	0.6798	0.5157	0.4377	0.3910	0.3595	0.3362	0.3185	0.3043	0.2926	0.2829	0.2462	0.2022	0.1616	0.1250
9	0.6385	0.4775	0.4027	0.3584	0.3286	0.3067	0.2901	0.2768	0.2659	0.2568	0.2226	0.1820	0.1446	0.1111
10	6.6020	0.4450	0.3733	0.3311	0.3029	0.2823	0.2666	0.2541	0.2439	0.2353	0.2032	0.1655	0.1308	0.1000
12	0.5410	0.3924	0.3264	0.2880	0.2624	0.2439	0.2299	0.2187	0.2098	0.2020	0.1737	0.1403	0.1100	0.0833
15	0.4709	0.3346	0.2758	0.2419	0.2195	0.2034	0.1911	0.1815	0.1736	0.1671	0.1429	0.1144	0.0889	0.0667
20	0.3894	0.2705	0.2205	0.1921	0.1735	0.1602	0.1501	0.1422	0.1357	0.1303	0.1108	0.0879	0.0675	0.0500
24	0.3434	0.2354	0.1907	0.1656	0.1493	0.1374	0.1286	0.1216	0.1160	0.1113	0.0942	0.0743	0.0567	0.0417
30	0.2929	0.1980	0.1593	0.1377	0.1237	0.1137	0.1061	0.1002	0.0958	0.0921	0.0771	0.0604	0.0457	0.0333
40	0.2370	0.1576	0.1259	0.1082	0.0968	0.0887	0.0827	0.0780	0.0745	0.0713	0.0595	0.0462	0.0347	0.0250
60	0.1737	0.1131	0.0895	0.0765	0.0682	0.0623	0.0583	0.0552	0.0520	0.0497	0.0411	0.0316	0.0234	0.0167
120	0.0998	0.0632	0.0495	0.0419	0.0371	0.0337	0.0312	0.0292	0.0279	0.0266	0.0218	0.0165	0.0120	0.0083
∞	0	0	0	0	0	0	0	0	0	0	0	0	0	0

[1] Reproduced with permission from C. Eisenhart, M.W. Hastay, W. A. Wallis, *Techniques of Statistical Analysis*, Chapter 15, McGraw-Hill Book Company, Inc., New York, 1947.

7.7 Cochran's Test for the Homogeneity of Variances

It is often necessary to decide whether several variances are equal, i.e., whether $\sigma_1^2 = \sigma_2^2 = \cdots = \sigma_k^2$. The F test can be used for $k=2$, but a different procedure is required for larger values of k. Such a procedure is given by Cochran's test. Let $s_1^2, s_2^2, \cdots, s_k^2$ be independent estimates of $\sigma_1^2, \sigma_2^2, \cdots, \sigma_k^2$, respectively, each based upon n independent normally distributed random variables; and let

$$g = \frac{\text{largest } s^2}{s_1^2 + s_2^2 + \cdots + s_k^2}$$

be the ratio of the largest s^2 to their total. The hypothesis that $\sigma_1^2 = \sigma_2^2 = \cdots = \sigma_k^2$ is accepted if

$$g \leqq g_\alpha$$

where g_α is given in Table 7.8 for levels of significance, α, equal to 0.05 and 0.01. This table is entered with n, the number of observations within each group, and k, the number of variances being considered.

PROBLEMS

1. Two machines, x and y, are used for filling bottles and are supposed to fill with a net volume of 5.65 liters. The filling process is normal, machine x having a standard deviation of 0.014 liter and machine y, 0.016. A question of whether or not the two machines are doing the same job has arisen. The control engineer claims that because of the relative agreement of standard deviations and the results of other measures, the two are both filling on the average with the same amount whether or not this amount is the desired 5.65 liters. A random sample is taken from each machine. Compute the sample means and state whether or not you think the control engineer is right. Then test at the 5% level and compare your decisions. Assuming equal sample size, how many observations would be required to offer protection of $\beta=0.05$ when the true difference is 0.05?

x:	5.63	5.64	5.65	5.65	5.62
	5.61	5.66	5.68	5.62	5.60
	5.68	5.64	5.66	5.61	5.68
	5.61	5.67	5.61	5.65	5.63

y:	5.68	5.65	5.59	5.64	5.66
	5.61	5.62	5.64	5.63	5.61
	5.64	5.66	5.60	5.65	5.63
	5.67	5.64	5.60	5.60	5.65
	5.60	5.65	5.60	5.63	5.60

2. An experiment is performed to test the difference in effectiveness of two methods of cultivating wheat. Ten patches of ground are treated with shallow plowing and fifteen with deep plowing. The sample mean yield per acre of the first group is 40.3 bushels and the sample mean for the second group is 44.7. Assume that the standard deviation of the shallow planting is 0.6 bushel and for the deep is 0.8. Test for equality of treatment at the 5% level. Sketch the OC curve of this test.

3. A new type of ammunition was submitted for test. Twenty rounds of standard ammunition and ten rounds of test ammunition were fired in random sequence. The sample average barrel pressure of the standard was 41,900 psi and the sample average pressure for the test was 44,200 psi. The standard deviation for both the standard and test was known to be 2,050 psi. (a) Test the hypothesis that the average barrel pressures are the same at the 5% level. (b) What difference in the expected values of the barrel pressures will be detected with probability 0.80 using the procedure in (a)?

4. A manufacturer of wooden ladders claims that the side pieces for the ladders have greater shear (parallel to the grain of the wood) if the raw lumber is air dried rather than kiln dried. However, his operating expenses could be appreciably reduced by using kiln dried lumber. Therefore, the manufacturer has decided to conduct a test at the 1% level to determine whether or not the air dried lumber has greater shear than the kiln dried. He is anxious to switch to kiln dried lumber if the experimental results justify this action. However, he feels that he mus continue to use air dried raw material if the shear of the kiln dried material is as much as 160 psi less than that of the air dried, and he desire that the risk of making an incorrect switch in this situation not excee 5%. From past experience with air dried lumber he estimates the standard deviation of its shear to be 30 psi. He believes that the standar deviation for the shear of kiln dried material is the same. (a) Determin the appropriate test for the manufacturer to use. (b) Determine the sampl size required. (c) Assuming the following data results from the test i (a), using the sample size found in (b), what decision would the manufacturer make?

Air dried (psi)	*Kiln dried (psi)*
$\bar{x}=1{,}170$	$\bar{y}=1{,}105$

5. In a certain experimental laboratory a method x_1 for producin gasoline from a crude oil is being investigated. Before completing experimentation, a new method x_2 is proposed. All other things being equal it was decided to abandon x_1 in favor of x_2 only if the average yield o the latter was substantially greater. The yield of both processes is assume to be normally distributed. However, there has been insufficient time t ascertain their true standard deviations, although there appears to be n reason why they cannot be assumed equal. Cost considerations impos size limits on the size of samples that can be obtained. If a 1% risk i all that is allowed, what would be your recommendation based on th following random samples? The numbers represent percent yield of crud oil.

x_1:	23.2,	26.6,	24.4,	23.5,	22.6,	25.7,	25.5
x_2:	25.7,	27.7,	26.2,	27.9,	25.0,	21.4,	26.1

Using the test procedure above, what values of $(\mu_{x_2}-\mu_{x_1})/\sigma$ will be detecte with probability 0.90?

6. In a study of reflex reactions the strength of patellar reflex measured in degrees of arc for 15 men under two conditions yielded the following results:

Tensed	*Relaxed*
31	35
19	14
22	19
26	29
36	34
30	26
29	19
36	37
34	24
33	27
19	14
19	19
26	30
15	7
18	13

(a) Assuming equal variances, test the hypothesis of equality of means a the 1% level. (b) Assuming equal sample sizes, how many observation are required to detect a difference in the time means equal to 8 with prob ability 0.90 ? Assume the common standard deviation is approximately 6.

7. The vice-president for production of a transistor manufacturing company which operates two plants reviews the production of the plants and notes that Plant A appears to produce consistently more per day than Plant B, even though both are designed to produce the same number of transistors in a given period of time. In a test to determine whether or not Plant A, on the average, produces more per day than Plant B, the vice-president specifies a level of significance of 0.01. If the average production of Plant A exceeds that of Plant B by 250 units, the vice-president indicates that the test should detect this difference with probability 0.90. From past experience it is known that the standard deviation of the quantity of daily production in each plant is equal and its approximate magnitude is 110 units.

(a) How many observations from each plant are required to insure the risks above ? (b) If the sample size is as found in (a) and the data are as follows:

Plant A	*Plant B*
x=daily production	y=daily production
$\bar{x}=2{,}800$	$\bar{y}=2{,}680$
$\Sigma(x_i-\bar{x})^2=103{,}600$	$\Sigma(y_i-\bar{y})^2=99{,}800$

would you conclude that Plant A produces more than Plant B ?

8. A new cure has been developed for Portland cement. Tests are run to determine if the new cure has an effect (positive or negative) on the strength. A single batch has been produced and subjected to both standard and experimental cures. The compressive strengths (psi) are given below:

Standard cure	*Experimental cure*
4,125	4,250
4,225	3,950
4,025	3,900
3,900	4,075
3,875	4,550
3,825	4,450
3,975	4,150
3,800	4,550
3,775	3,700
3,850	4,250

(a) Test for effect on strength of change in cure. (b) Assuming that a common value of σ is approximately 220, and using the procedure in (a), find the sample size necessary to detect a change of 360 psi with probability 0.95. (Assume $\alpha=0.05$).

9. It is argued that the resistance of wire A is greater than the resistance of wire B. You make tests on each wire with the following results:

Wire A	*Wire B*
0.140 ohm	0.135 ohm
0.138	0.140
0.143	0.136
0.142	0.142
0.144	0.138
0.137	0.140

Assuming equal variances, what conclusions do you draw? Justify you answer. (*Remark*: It may help to choose a convenient computing origin.

10. You wish to determine whether the shear strength of yarn A i different from that of yarn B. The standard deviation in shear strengtl is known to be somewhere near ten pounds for both types of yarn. Yo are willing to run a maximum risk of 0.05 of saying that the yarns ar different when they are actually the same. On the other hand, you do nc wish to run a risk of more than 0.10 of saying that they are the sam when actually the average strengths differ by as much as 15 pounds.

(a) Determine the sample size and state the test procedure yo would use in your test.

(b) If $\bar{x}_A - \bar{x}_B = 6.3$ and $\sqrt{\frac{2}{n}}\sqrt{\frac{\Sigma(x_A-\bar{x}_A)^2+\Sigma(x_B-\bar{x}_B)^2}{2n-2}}=3.1$, woul you conclude that the average strengths are equal?

11. A laboratory test was devised for estimating the ease of filterin a certain product on the plant scale. The test consisted of measuring th time taken to filter a given volume of material under standard conditions Six samples of plant magma were taken during spells when filtering wa judged to be in State B and State A. The variance, although unknown is assumed to be the same for both cases, and it can be assumed tha the variables are normally distributed.

Filtration on Plant

State A	*State B*
8	9
10	10
12	10
13	4
9	7
14	9

(a) It is required to test at the 5% significance level whether the differ ence is significant. (b) Assuming that the calculated sample variance is good estimate of the true variance, what is the probability of acceptin the hypothesis when the true means differ by 5?

12. Suppose you are in a laboratory supervising a number of experi mental scientists. One young man named Oscar reveals that he has mad a great discovery. By placing element A in steel he can increase its ten sile strength. You are somewhat skeptical so the scientist presents th data to you and explains that he has made a t test (assuming equal vari ances) on the data, which verifies his conclusion. After looking at th results you ask him if the basic assumptions regarding the t test are satis fied. He looks amazed and says he thought that this test can always b used. You mention that this is not true and proceed to point out all th assumptions. In fact, you show him that one assumption is definitely no satisfied, perform the t test, mention the assumptions, and prove that a least one assumption is not valid, given the following data.

Tensile Strength

Steel with A	*Steel without A*
4	5
14	7
6	5
12	7
8	6
4	6
4	6
12	

If the t'-test is used, would the results be changed?

13. The following are 16 independent determinations of the melting point of a compound, eight made by Analyst I and eight made by Analyst II. Data are in degrees Centigrade.

Analyst I	*Analyst II*
164.4	163.5
169.7	162.8
169.2	163.0
169.5	163.2
161.8	160.7
168.7	161.5
169.5	160.9
163.9	162.1

Would you conclude from these data that there was a tendency for one analyst to get higher results than the other? Test at the 5% level. (Assume that the standard deviation of each analyst is the same.) Would your conclusion be different if you did not assume that the standard deviations of the analysts were the same?

14. In Problem 13 how many observations by each analyst would be required to have a probability of 0.90 of detecting an analyst bias of 1.5 degrees Centigrade. (Assume the common value of σ is approximately 3.20 degrees Centigrade.)

15. Test this hypothesis of equality of means in Problem 13, assuming normality but without assuming equal variances.

16. An engineer in the design section of an aircraft manufacturing plant has presented theoretical evidence that painting the exterior of a particular combat airplane reduces its cruising speed. He convinces the chief design engineer that the next nine aircraft off the assembly line should be test flown to determine cruising speed prior to paint, then painted, and finally test flown to ascertain cruising speed after they are painted. The following data are obtained.

Cruising Speed (knots)

Aircraft	*Not Painted*	*Painted*
1	426	416
2	418	400
3	424	420
4	438	431
5	440	432
6	421	404
7	412	398
8	409	405
9	427	422

Design a test at the 5% level and complete the computation necessary to evaluate the design engineer's evidence. Assume normality.

17. In Problem 16, how many observations would be required to make the test procedure indicate rejection with probability 0.99 if the true cruising speed drops by 12.5 knots? Assume that the standard deviation of differences (σ_d) is about 5 knots.

18. A group of men are subjected to a slight change in diet and are weighed before and after a lapse of three months. (Weight in pounds)

Man	*Before*	*After*
1	162	166
2	191	196
3	138	136
4	182	190
5	159	160

(a) Test the hypothesis that there is no gain in weight. (b) How many men would be needed to have a probability of 0.95 of rejecting the hypothesis if the true mean weight changes by 3 pounds? (Assume σ_d is approximately 3 for this calculation.)

19. A new type of mold for concrete has been developed. Suppose that the new mold has certain advantages over the standard mold such as faster hardening, etc., but some people have expressed doubts as to the final strength of the finished product. It is economically feasible to take only three observations on each mold, and the data are presented below.

Compressive Strength (psi)

Batch	*Standard mold*	*New mold*
1	4,680	4,020
2	4,650	3,940
3	4,520	3,980

(a) Determine if there is a loss of compressive strength due to the use of the new mold. (b) Assuming σ_d is approximately equal to 166, how many observations would be necessary to detect a loss of 400 psi compressive strength at least 90% of the time? (c) Discuss the OC curve for your test.

20. Two analysts took repeated readings on the hardness of city water. Determine whether one analyst has a tendency to read higher than the other. Use both normal and non-parametric methods.

Coded Measures of Hardness

Analyst A x	*Analyst B* y
0.46	0.82
0.62	0.61
0.37	0.89
0.40	0.51
0.44	0.33
0.58	0.48
0.48	0.23
0.53	0.25
	0.67
	0.88

21. Ten pairs of duplicate spectrochemical determinations for nickel are presented below. The readings in the first column were taken with one type of measuring instrument and those in the second column were taken with another type.

Sample	*Duplicates*	
1	1.94	2.00
2	1.99	2.09
3	1.98	1.95
4	2.03	2.07
5	2.03	2.08
6	1.96	1.98
7	1.95	2.03
8	1.96	2.03
9	1.92	2.01
10	2.00	2.12

(a) Determine at the 5% level of significance whether the different equipment leads to different results. Use the signed rank test. (b) Determine at the 0.2% (approximately) level of significance whether the different equipment leads to different results. Use the sign test.

22. In research on rocket propellants aimed at reducing the delay time between the application of the firing current and explosion, it was thought that the substitution of a grade T of propellant for the normal grade C should have a favorable effect. The approximate standard deviation of the normal grade C (as well as the grade T) is 0.03. An experiment was planned in which n shots would be made with the C grade and n shots with the T grade. (a) A reduction in the difference of means of 0.06 second would make the necessary change in manufacture worthwhile. It was important, however, not to recommend grade T if no improvement really resulted and α was therefore set at 0.01. The risk β of failing to detect a reduction of 0.06 second was set at 0.05. It is assumed that the observations are normally distributed. How many observations are needed to give the risks above? (b) If $\overline{C}=0.261$ second, $\overline{T}=0.250$ second, $s_C^2=0.0128$ second, and $s_T^2=0.0132$ second, would you adopt grade T? (c) If all the T values were smaller than the C values, except for one observation which was the maximum of all the observations, test the hypothesis of equality of means using a non-parametric test.

23. Two treatments are to be compared at the 5% level of significance. The data are definitely not "normal" and are as follows:

Treatment A	*Treatment B*
16.3	17.0
14.7	18.7
12.3	17.5
13.5	17.9
16.0	18.3
17.1	18.0
17.3	16.9
15.2	

Determine whether the treatments differ.

24. Perform a non-parametric test using the data of Problem 16. Use the 2.5% level.

25. The following are the burning times in seconds of floating smoke pots of two different types:

Type I		*Type II*	
481	572	526	537
506	561	511	582
527	501	556	605
661	487	542	558
501	524	491	578

(a) Would you conclude that one type smoke pot tends to burn longer than the other, assuming the standard deviations are the same? (b) Suppose that the standard deviations could not be assumed to be equal. Would your conclusion be changed? (c) Suppose that the necessary normality assumption could not be made, and analyze the data using non-parametric methods.

26. An experiment was run to see if the amount of metal removed was the same for two different temperatures of the pickling bath. The data are below; each observation represents the thickness of metal lost, expressed in 0.001 inch.

Amount of Metal Removed

Temperature:	*90°*	*120°*
	2.5	2.1
	2.7	2.4
	2.9	2.0
	2.7	1.9
	2.6	2.1
	2.4	2.0

(a) Test the hypothesis of no temperature effect at the 5% level, assuming normality and common variance. (b) Perform a non-parametric test at the 5% level of significance.

27. The following table gives percent loss in tensile strength, following immersion in a corrosive solution, of paired samples of an alloy; one specimen subject to stress and one not.

Test number	*Unstressed*	*Stressed*
1	6.4	9.2
2	4.6	7.9
3	4.6	7.3
4	6.4	8.0
5	3.2	5.7
6	5.2	7.6
7	6.5	5.7
8	4.9	4.1
9	4.3	8.1
10	5.6	6.5
11	3.7	6.9
12	4.6	6.0

What conclusions would you draw from the data about the effect of stressing upon loss in tensile strength? (Use a non-parametric test.)

28. Suppose an experiment is to be performed to determine whether a treatment has an effect (either positive or negative) on the wear resistance of a certain material. The experimenter labels two specimens from the first test-piece $1H$ and $1T$, those from the second test-piece $2H$ and $2T$, and so on for nine pairs of test-pieces. He then tosses a coin nine times.

If the result of the first toss is a head, he sets aside specimen $1H$ for treatment; if a tail, he sets aside specimen $1T$. The result of the second toss decides the fate of specimens $2H$ and $2T$, and so on. The treatment is then applied to the selected nine specimens, and the abrasion resistance of treated and untreated specimens is assessed by the wear test machine. The nine differences, abrasion resistance of treated specimen minus abrasion resistance of untreated specimen, were:

2.6, 3.1, −0.2, 1.7, 0.6, 1.2, 2.2, 1.1, −0.1

(a) Using the sign test, determine if treatment has an effect on the average wear resistance at the 5% (approximately) level of significance. (b) Using the signed rank test, determine whether treatment has an effect on the average wear resistance at the 5% (approximately) level of significance.

29. It is desired to know what type of filter should be used over the screen of a cathode ray oscilloscope in order to have a radar operator easily pick out targets on the presentation. A test to accomplish this has been set up. A noise is first applied to the scope in order to make it difficult to pick out a target. A second signal, representing the target, is put into the scope, and its intensity is increased from zero until detected by the observer. The intensity setting at which the observer first notices the target signal is then recorded. This experiment is repeated again, with a different observer, on the assumption that all people do not see in exactly the same manner. After a set of readings has been taken with one type of filter on the scope, another set of readings, using the same observers, is taken with a different type of filter. The numerical value of each reading listed in the table of data is proportional to the target intensity at the time the operator first detects the target.

Observer	*Filter No. 1*	*Filter No. 2*
1	90	88
2	87	90
3	93	97
4	96	87
5	94	90
6	88	96
7	90	90
8	84	90
9	101	100
10	96	93
11	90	95
12	82	86
13	93	89
14	90	92
15	96	98
16	87	95
17	99	102
18	101	105
19	79	85
20	98	97
21	81	88

(a) Assuming normality, test the hypothesis that the filters are the same ($\alpha=0.05$). (b) Use a non-parametric procedure to test the hypothesis that the filters are the same ($\alpha=0.05$).

30. A small motor is being manufactured for a special purpose. The important characteristic of the motor in this application is its starting torque. Two testers are used to evaluate the starting torque of the motors. It is desired to know whether the two testers are obtaining equivalent results or if, because of slight differences in their testing procedure, one of them is obtaining consistently higher results than the other. The results are as follows:

Motor	*A*	*B*
1	0.41	0.38
2	0.45	0.40
3	0.36	0.32
4	0.49	0.50
5	0.39	0.31
6	0.52	0.54
7	0.38	0.32
8	0.43	0.36

(a) Using the sign test, determine whether the testers differ at the 5% level of significance. (b) Using the signed rank test, determine whether the testers differ at the 5% level of significance.

31. In a study of sex differences in mathematical ability, ten brother-sister pairs were used, resulting in the following scores:

Boys:	92	84	93	91	93	90	86	89	91	88
Girls:	88	85	82	90	81	93	87	92	86	85

Test the hypothesis that there are no sex differences at the 5% level by both normal and non-parametric methods.

32. It is desired to test whether a special treatment of concrete has an effect on the average strength of the concrete. A small experiment is to be made from a given batch of raw materials. Samples are to be divided into two equal groups at random, with group II receiving the special treatment. The procedure is to be such that if there is no difference in the average strengths of the two groups, the probability will be 0.95 of reaching this conclusion. Furthermore, if the average strengths differ by as much as 18 kg/cm², the procedure should result in the conclusion that they differ with probability greater than 0.75. It is felt that the treatment will not have any effect on the variability and consequently the variability for each group can be assumed to be the same. An estimate of the standard deviation is 3 kg/cm². (a) What is the necessary sample size for each group? (b) Suppose that

$$\sum_{i=1}^{n}(x_{1i}-\bar{x}_1)^2=402 \qquad \bar{x}_1=295$$

$$\sum_{i=1}^{n}(x_{2i}-\bar{x}_2)^2=54 \qquad \bar{x}_2=315$$

and n is that value found in (a). Should we conclude that the special treatment has no effect on the strength? (c) Do the data above indicate that $\sigma_1=\sigma_2$?

33. In Problem 4 it was decided to test for equality of variances. Verify that this hypothesis is tenable at the 5% level if $s_x^2=960$ and $s_y^2=1330$.

34. In Problem 22 verify the assumption of equality of variances. What must the ratio of σ_C/σ_T be before there is a probability of 0.9 of detecting a difference in the standard deviations?

35. You wish to determine whether operator B is turning out a product that is more variable than the product of operator A. You are willing to run a maximum risk of 0.05 of saying that B's product is more variable than A's when actually it is not. On the other hand, you do not wish to run a risk of more than 0.20 of saying that the variability of B's product is equal to or less than that of A's when it is actually one and one-half times greater than A's. Determine the sample sizes and acceptance limit you would use in your test.

36. The amount of surface wax on each side of waxed paper bags is believed to be normally distributed. However, there is reason to believe that there is greater variation in the amount on the inner side of the paper than on the outside. A sample of 75 observations of the amount of wax on each side of these bags is obtained and the following data recorded:

Wax in Pounds per Unit Area of Sample

Outside surface	*Inside surface*
$\bar{x}=0.948$	$\bar{y}=0.652$
$\sum x_i^2=91$	$\sum y_i^2=82$

Conduct a test to determine whether or not the variability of the amount of wax on the outer surface is greater than the amount on the inner surface ($\alpha=0.05$).

37. In testing the equality of two variances, determine the sample sizes and acceptance limit to use when you want a maximum risk of 0.05 of saying one variance is greater than the other when they are equal and 0.10 of saying they are the same when one is one and one-half times the other.

38. In testing the equality of two variances, determine the sample sizes and acceptance limit to use when you want a maximum risk of 0.05 of saying one variance is greater than the other when they are equal and 0.05 of saying they are the same when one is 0.40 of the other.

39. The following results are calculated for two samples, each of whose observations are normally distributed:

$$x: \quad n_x=8, \quad \sum x=12, \quad \sum x^2=46$$

$$y: \quad n_y=11, \quad \sum y=22, \quad \sum y^2=80$$

Test the hypothesis $\sigma_x=\sigma_y$ against the alternative $\sigma_x>\sigma_y$, at the 5% level of significance.

40. Let x and y be normally distributed random variables. Suppose a sample of 16 observations is taken from each, and we want to test the hypothesis that $\sigma_x=\sigma_y$ at the 1% level. Give the rejection rule and sketch the OC curve of the test for the three cases: (a) We want to reject only when $\sigma_x>\sigma_y$. (b) We want to reject only when $\sigma_x<\sigma_y$. (c) We want to reject whenever $\sigma_x\neq\sigma_y$.

41. Suppose that 12 trials are made for each of three treatments with the following sample variances:

6.4 1.2 1.8.

Test the hypothesis that the variances are equal at the 1% level.

42. The variability of six machines is to be tested. Five observations are taken on each machine, and the sample variances are calculated. The results are as follows:

12.2, 13.8, 14.6, 11.8, 16.5, 16.0.

Do all the machines have the same variances (use $\alpha=0.05$)?

43. Derive the expression for the sample size of the one-sided procedure for testing the hypothesis that $\mu_x=\mu_y$ when both standard deviations are known.

44. Derive the expression for the sample size of the two-sided procedure for testing the hypothesis that $\mu_x=\mu_y$ when both standard deviations are known.

45. Derive the expression for the sample size of the two-sided procedure for testing the hypothesis that $\sigma_x=\sigma_y$.

CHAPTER VIII

Estimation

8.1 Introduction

Industrial experimentation is often performed to estimate some unknown values. These values are usually parameters (constants) of a probability distribution or functions of these parameters. For example, consider the problem presented in Section 6.2.3. A packaging machine is preset to fill containers with a fixed amount of food. Because there is an inherent variation in the amount which is placed in each container, the manufacturer is often interested in estimating the average weight. High average weights are costly, whereas low average weights are often illegal. Parameters other than the mean may also be of interest. In the filling problem, a determination of the variability in the machine setting is important information. This chapter will be devoted to the problem of estimation and will consider point estimates, confidence interval estimates, and tolerance limit estimates.

8.2 Point Estimation

A point estimate is a single value which is used to estimate the parameter in question. A point estimate rarely coincides with the unknown quantity to be estimated; it usually suffices for the estimate to be " close " to the unknown quantity. Any random variable can be considered to be an estimate of a parameter.[1] A random variable used in this context is often called a *statistic*. For example, if the average value, μ is to be estimated based on a sample of size n, each of the statistics, the sample mean, the sample median, the

[1] The term *estimator* is often applied to the random variable and the word estimate is reserved for the value taken on by the random variable after the experimental data have been inserted. However, we will use the term estimate both for the random variable and its numerical value, distinguishing between the two uses when ambiguity arises.

smallest observation, the largest observation, the second observation, one half the sample mean plus the sample median, etc., can be considered as estimates of μ. How, then, is one to choose between these estimates? The obvious answer is to pick the estimate which is generally "closest" to the true value μ.

The remainder of this section will be devoted to finding "good" estimates of the parameter. In general, the estimation problem may be stated as follows: there exists a random variable x with a density function $f(x, \theta)$, where θ is a parameter to be estimated. On the basis of a random sample of n observations, a function $\hat{\theta}$ of these observations (also a random variable) is desired so that the distribution of this function will be concentrated as closely as possible near the true value of the parameter θ. The statistic $\hat{\theta}$ is then a "good" estimate of θ. There are many estimates of a parameter θ. Intuitively, a good estimate gives results which are close to the value of the quantity being estimated. In the filling machine example, it is desired to estimate the average fill, μ; let $\hat{\theta}_1$ be the sample mean, $\hat{\theta}_2$ be the second observation and $\hat{\theta}_3$ be the largest observation. If the densities of these statistics are as illustrated in Figure 8.1, it is evident that $\hat{\theta}_1$ is the best estimate. In any given situation, the value of the estimate using $\hat{\theta}_2$ (or $\hat{\theta}_3$) may be closer to μ than the value of the estimate using $\hat{\theta}_1$. However, in a long series of trials, the values of $\hat{\theta}_1$ will "generally be nearer" than the values of the other statistics.

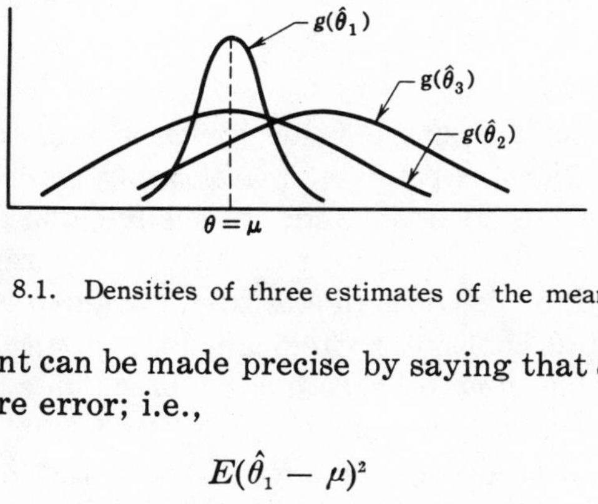

Fig. 8.1. Densities of three estimates of the mean μ.

This statement can be made precise by saying that $\hat{\theta}_1$ has the smallest mean square error; i.e.,

$$E(\hat{\theta}_1 - \mu)^2$$

is the minimum of $E(\hat{\theta}_1 - \mu)^2$, $E(\hat{\theta}_2 - \mu)^2$, and $E(\hat{\theta}_3 - \mu)^2$, where

$$E(\hat{\theta}_i - \mu)^2 = \int_{-\infty}^{\infty} (\hat{\theta}_i - \mu)^2 g_i(\hat{\theta}_i)\, d\hat{\theta}_i$$

is the mean square error of $\hat{\theta}_i$, and $g_i(\hat{\theta}_i)$ is the density function of $\hat{\theta}_i$, $i = 1, 2, 3$. The mean square error can be interpreted as follows: suppose a game is to be played between three players. There exists a probability distribution with mean μ, and this value is known only to a referee. A sample of size n is drawn and player 1 uses $\hat{\theta}_1$ (the sample mean) as his estimate of μ; player 2 uses $\hat{\theta}_2$ (the second observation) as his estimate of μ; and player 3 uses $\hat{\theta}_3$ (the largest observation) as his estimate of μ. Each player pays the other two the square of the difference between his estimate and μ (computed by the referee), multiplied by one dollar. If this game is played over and over again, the mean square error corresponds to the amount of money each player pays to his two opponents. Therefore, the player with the smallest mean square error wins the most money; in this case the largest winner is player 1.

The comparison between two estimates can be made quantitative by comparing their relative efficiency. If a statistic $\hat{\theta}_1$ has mean square error $E(\hat{\theta}_1 - \theta)^2$ and if a second statistic $\hat{\theta}_2$ has mean square error $E(\hat{\theta}_2 - \theta)^2$, the efficiency of $\hat{\theta}_2$ relative to $\hat{\theta}_1$ is defined to be $E(\hat{\theta}_1 - \theta)^2/E(\hat{\theta}_2 - \theta)^2$. If the efficiency of $\hat{\theta}_2$ relative to $\hat{\theta}_1$ is less than one, then $\hat{\theta}_1$ may reasonably be regarded as a better estimate of θ than $\hat{\theta}_2$. It should be noted that if $E(\hat{\theta}) = \theta$, the mean square error of $\hat{\theta}$, $E(\hat{\theta} - \theta)^2$, coincides with the variance of $\hat{\theta}$. Thus, if the weighings in the filling example each have variance σ^2, the mean square error of $\hat{\theta}_1$, the sample mean, coincides with the variance of the sample mean and is equal to σ^2/n. The mean square error of $\hat{\theta}_2$, the second observation, coincides with the variance of the second observation and is equal to σ^2 (which is the variance of any observation). Therefore, the efficiency of $\hat{\theta}_2$ relative to $\hat{\theta}_1$ is $1/n$, and $\hat{\theta}_1$ is a better estimate of θ than $\hat{\theta}_2$. If the sample size is 10 and if the weighings are normally distributed, it can be shown that $E(\hat{\theta}_3) = 1.539\,\sigma + \mu > \mu$ (where $\hat{\theta}_3$ is the largest observation) and that $E(\hat{\theta}_3 - \mu)^2 = 2.712\sigma^2$. (The variance of $\hat{\theta}_3$ and the mean square error of $\hat{\theta}_3$ are not equal.) Thus, the efficiency of $\hat{\theta}_3$ relative to $\hat{\theta}_1$ is

$$\frac{\sigma^2/10}{2.712\sigma^2} < 1$$

so that $\hat{\theta}_1$ can be regarded as a better estimate of θ than $\hat{\theta}_3$.

If $\hat{\theta}$ has the property that $E(\hat{\theta}) = \theta$, $\hat{\theta}$ is said to be an unbiased estimate of θ. The sample mean and the second observation are then unbiased estimates of μ, whereas the largest observation is a biased estimate of μ. In the example presented the best estimate of μ is an unbiased estimate but this should not be taken as a general principle. If $\hat{\theta}_1$ and $\hat{\theta}_2$ have densities as shown in Figure 8.2, it is evident that the biased estimate, $\hat{\theta}_2$, is a better estimate of μ than the unbiased estimate, $\hat{\theta}_1$. Clearly, $\hat{\theta}_2$ has smaller mean square error than $\hat{\theta}_1$.

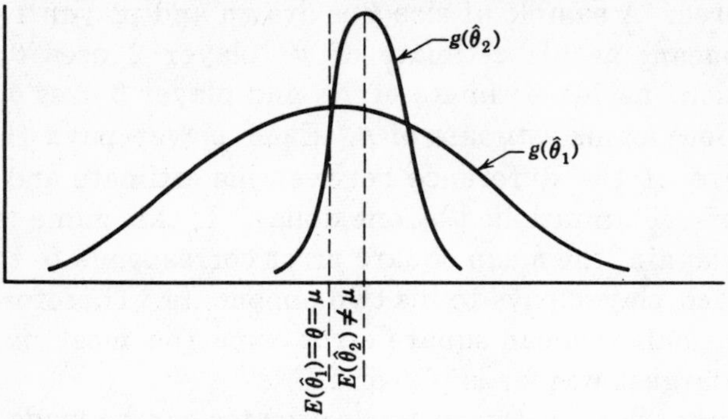

Fig. 8.2. Distribution of two estimates of μ.

8.3 Optimal Estimates

An estimate $\hat{\theta}$ of θ will be called optimal if there does not exist another estimate, $\hat{\theta}'$, whose mean square error is always smaller than the mean square error of $\hat{\theta}$ for all values of θ. The sample mean, $\bar{x}$, is the optimal estimate of the mean, μ, of the normal distribution, and furthermore is unbiased. If the probability distribution is normal with variance σ^2, the estimate

$$s^2 = \frac{\sum_{i=1}^{n}(x_i - \bar{x})^2}{n-1}$$

is unbiased but is not the optimal estimate of σ^2. In fact, the biased estimate

$$s'^2 = \frac{\sum_{i=1}^{n}(x_i - \bar{x})^2}{n+1}$$

has uniformly smaller mean square error and is the optimal estimate of σ^2. However, in practice, s^2 is usually chosen as the point estimate of σ^2, primarily because of the unbiasedness property.

Several general methods which are beyond the scope of this text are available for obtaining point estimates, two of the most useful being the method of moments and the method of maximum likelihood. The method of moments obtains estimates of the desired parameters by equating a sufficient number of sample moments to the moments of the probability distribution. The method of maximum likelihood estimates the parameter as that function of the sample which maximizes the likelihood of obtaining the given sample.

8.4 Confidence Interval Estimation

A point estimate is often inadequate as an estimate of a parameter since it rarely coincides with the parameter. An alternative type of estimate is an interval estimate of the form $[L, U]$, where L is the lower bound and U is the upper bound. The interval estimates considered will be functions of the sample, i.e., random variables, having the property that, prior to experimentation, probability statements about the intervals including the parameter can be made. Therefore, if the parameter to be estimated is denoted by θ, an interval estimate of θ is $[L, U]$ such that there is a given probability, $1-\alpha$, that $L \leq \theta \leq U$. L and U will be called $100\,(1-\alpha)\%$ confidence limits for the given parameter and the interval between them a $100\,(1-\alpha)\%$ confidence interval. The meaning of these statements can be best described by an example such as the problem of determining the average setting μ of a machine for filling packages. Suppose that a 95% confidence interval, $[L, U]$ is to be computed (using the techniques described in the following sections). This implies that, in the long run, 95% of the limits so computed can be expected to include the true average machine setting μ. If in each instance it was asserted that μ lies within the limits computed, correct statements would be expected 95 times in 100 and erroneous statements expected 5 times in 100; i.e., the statement "the true average machine setting lies within the range so computed" has a probability of 0.95 of being correct. But there would be no operational meaning in the following

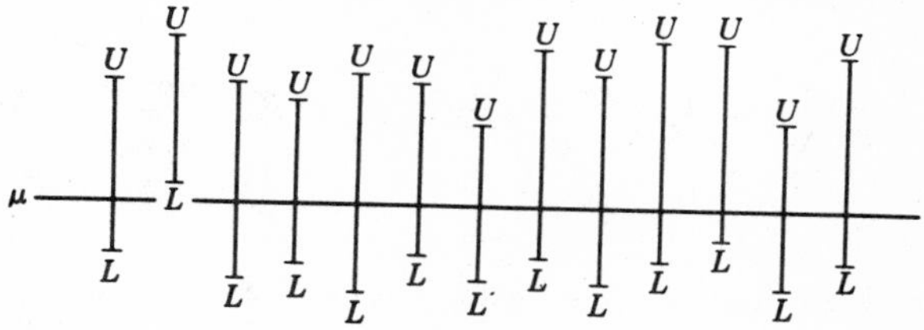

Fig. 8.3. Confidence interval estimates of the true average setting of a weighing machine.

statement made in any one instance: "the probability is 0.95 that the true average setting falls within the limits computed in this case" since μ either does or does not fall within the limits. It should also be emphasized that even in repeated sampling the confidence interval will vary in position and width (if the variability of machine weighings is unknown) from sample to sample. This phenomenon is illustrated in Figure 8.3, where confidence intervals computed from 13 separate trials are shown.

If 95% confidence intervals are obtained, it is expected that 95 times in 100 these intervals will include μ; i.e., if it is stated that μ lies in the interval $[L, U]$, such statements will be correct, on the average, 95 times in 100. In practical situations, the particular confidence interval is obtained only *once*, and the statement is made that it lies in the interval.

8.5 Confidence Interval for the Mean of a Normal Distribution when the Standard Deviation Is Known

Let x be a normally distributed random variable with mean μ and standard deviation σ (known). Consider a random sample of size n. Let $\bar{x}$ be the sample mean. The 100 $(1 - \alpha)$% confidence interval for μ is given by $\bar{x} \pm K_{\alpha/2}\sigma/\sqrt{n}$, where $K_{\alpha/2}$ is the 100 $\alpha/2$ percentage point of the normal distribution. This result is obtained by considering the distribution of $(\bar{x} - \mu)\sqrt{n}/\sigma$, which is normally distributed with mean 0 and standard deviation 1. Hence

$$P\left(-K_{\alpha/2} \leqq \frac{(\bar{x} - \mu)\sqrt{n}}{\sigma} \leqq K_{\alpha/2}\right) = 1 - \alpha.$$

This probability statement is equivalent to

$$P\left(-K_{\alpha/2}\frac{\sigma}{\sqrt{n}} \leqq \bar{x} - \mu \leqq K_{\alpha/2}\frac{\sigma}{\sqrt{n}}\right) = 1 - \alpha$$

or

$$P\left(\bar{x} - K_{\alpha/2}\frac{\sigma}{\sqrt{n}} \leqq \mu \leqq \bar{x} + K_{\alpha/2}\frac{\sigma}{\sqrt{n}}\right) = 1 - \alpha.$$

Thus

$$L = \bar{x} - K_{\alpha/2}\frac{\sigma}{\sqrt{n}} \qquad \text{and} \qquad U = \bar{x} + K_{\alpha/2}\frac{\sigma}{\sqrt{n}}$$

and the 100 $(1 - \alpha)$ % confidence interval is given by

$$\bar{x} \pm K_{\alpha/2}\frac{\sigma}{\sqrt{n}}.$$

Upper one-sided confidence intervals are obtained by letting $L = -\infty$ and replacing $K_{\alpha/2}$ by K_{α}. A 100 $(1 - \alpha)$% upper confidence interval is given by $\bar{x} + K_\alpha \sigma/\sqrt{n}$. This result follows from the probability statement

$$P\left(\frac{(\bar{x} - \mu)\sqrt{n}}{\sigma} \geqq -K_\alpha\right) = 1 - \alpha.$$

Similarly, a $100\,(1 - \alpha)\%$ lower confidence interval is given by

$$\bar{x} - K_\alpha \frac{\sigma}{\sqrt{n}}.$$

It should be noted that there exists a relationship between confidence intervals and significance tests. Suppose we are interested in testing the hypothesis that $\mu = \mu_0$ against the alternative that $\mu \neq \mu_0$. The procedure given in Chapter VI is equivalent to the following: accept the hypothesis that $\mu = \mu_0$ if the confidence interval $\bar{x} \pm K_{\alpha/2}\sigma/\sqrt{n}$ includes μ_0; otherwise, reject it.

8.5.1 EXAMPLE

In the example of Section 6.1.4 the introduction of a new constituent was supposed to increase the average tensile strength of a material. An experiment was run using a sample of size $n = 19$; $\bar{x}$ was found to be 26,100 psi. Tensile strength was assumed to be normally distributed with mean μ and standard deviation $\sigma = 300$ psi. A 95% lower one-sided confidence interval for μ is given by

$$L = \bar{x} - 1.645 \frac{\sigma}{\sqrt{n}}$$

where 1.645 is the 5% point of the normal distribution. Hence

$$L = 26{,}100 - (1.645)\frac{(300)}{\sqrt{19}} = 25{,}986.8.$$

8.6 Confidence Interval for the Mean of a Normal Distribution when the Standard Deviation Is Unknown

Let x be a normally distributed random variable with mean μ and standard deviation σ (unknown). Consider a sample of size n. Let $\bar{x}$ be the sample mean and s the sample standard deviation. The $100\,(1 - \alpha)\%$ confidence interval for μ is given by

$$\bar{x} \pm t_{\alpha/2;n-1} \frac{s}{\sqrt{n}}$$

where $t_{\alpha/2;n-1}$ is the $100\,\alpha/2$ percentage point of the t distribution with

$n - 1$ degrees of freedom. The result is obtained by considering the distribution of $(\bar{x} - \mu)\sqrt{n}/s$ which has been shown in Chapter IV to have a t distribution with $n - 1$ degrees of freedom.

Hence

$$P\left(-t_{\alpha/2;n-1} \leq \frac{(\bar{x} - \mu)\sqrt{n}}{s} \leq t_{\alpha/2;n-1}\right) = 1 - \alpha.$$

This probability statement is equivalent to

$$P\left(-t_{\alpha/2;n-1}\frac{s}{\sqrt{n}} \leq \bar{x} - \mu \leq t_{\alpha/2;n-1}\frac{s}{\sqrt{n}}\right) = 1 - \alpha$$

or

$$P\left(\bar{x} - t_{\alpha/2;n-1}\frac{s}{\sqrt{n}} \leq \mu \leq \bar{x} + t_{\alpha/2;n-1}\frac{s}{\sqrt{n}}\right) = 1 - \alpha.$$

Thus

$$L = \bar{x} - t_{\alpha/2;n-1}\frac{s}{\sqrt{n}} \qquad \text{and} \qquad U = \bar{x} + t_{\alpha/2;n-1}\frac{s}{\sqrt{n}}$$

and the 100 $(1 - \alpha)$ % confidence interval is given by

$$\bar{x} \pm t_{\alpha/2;n-1}\frac{s}{\sqrt{n}}.$$

A $100(1 - \alpha)\%$ upper one-sided confidence interval is given by $\bar{x} + t_{\alpha;n-1}s/\sqrt{n}$, whereas a $100(1 - \alpha)\%$ lower one-sided confidence interval is given by $\bar{x} - t_{\alpha;n-1}s/\sqrt{n}$.

8.6.1 EXAMPLE

In the example of Section 6.2.3, a machine weighing problem was encountered. The experimental results led to $\bar{x} = 10.022$ and $s = 0.0632$. Assume that the weighings are normally distributed. Then, with these values of $\bar{x}$ and s, a 95% confidence interval for μ is given by

$$\bar{x} \pm 2.571\frac{s}{\sqrt{n}}$$

where 2.571 is the $2\frac{1}{2}$ % point of the t distribution with 5 degrees of freedom. Hence, the confidence interval becomes

$$\left[10.022 - (2.571)\frac{(0.0632)}{\sqrt{6}}, \qquad 10.022 + (2.571)\frac{(0.0632)}{\sqrt{6}}\right]$$

$$= [9.956, 10.088].$$

8.7 Confidence Interval for the Standard Deviation of a Normal Distribution

Let x be a normally distributed random variable with mean μ (unknown) and standard deviation σ (unknown). Consider a sample of size n. Let $\bar{x}$ be the sample mean and s, where $s=\sqrt{\sum_{i=1}^{n}(x_i-\bar{x})^2/(n-1)}$, be the sample standard deviation. The $100\,(1-\alpha)\%$ confidence interval for σ is given by

$$\left[s\sqrt{\frac{n-1}{\chi^2_{\alpha/2;n-1}}},\qquad s\sqrt{\frac{n-1}{\chi^2_{1-\alpha/2;n-1}}}\right]$$

where $\chi^2_{\alpha/2;n-1}$ and $\chi^2_{1-\alpha/2;n-1}$ are respectively the $100\,\alpha/2$ and $100\,(1-\alpha/2)$ percentage points of the chi-square distribution with $n-1$ degrees of freedom.

This result is obtained by considering the distribution of $(n-1)s^2/\sigma^2$ which has been shown in Chapter IV to have a chi-square distribution with $n-1$ degrees of freedom. Hence

$$\mathrm{P}\left(\chi^2_{1-\alpha/2;n-1} \leqq \frac{(n-1)s^2}{\sigma^2} \leqq \chi^2_{\alpha/2;n-1}\right) = 1-\alpha.$$

This probability statement is equivalent to

$$\mathrm{P}\left(\frac{1}{\chi^2_{\alpha/2;n-1}} \leqq \frac{\sigma^2}{s^2(n-1)} \leqq \frac{1}{\chi^2_{1-\alpha/2;n-1}}\right) = 1-\alpha$$

or

$$\mathrm{P}\left(s\sqrt{\frac{n-1}{\chi^2_{\alpha/2;n-1}}} \leqq \sigma \leqq s\sqrt{\frac{n-1}{\chi^2_{1-\alpha/2;n-1}}}\right) = 1-\alpha.$$

A $100\,(1-\alpha)\%$ upper one-sided confidence interval is given by

$$U = s\sqrt{\frac{n-1}{\chi^2_{1-\alpha;n-1}}}$$

and a $100\,(1-\alpha)\%$ lower one-sided confidence interval is given by

$$L = s\sqrt{\frac{n-1}{\chi^2_{\alpha;n-1}}}.$$

8.7.1 EXAMPLE

In the example of Section 6.3.3, a problem on the variability of a new product was encountered. A sample of 25 was taken and s^2 was found to be 0.0384. Assuming a normal distribution of the measurements, a 99% confidence interval for σ is given by

$$\left[s\sqrt{\frac{n-1}{\chi^2_{\alpha/2;n-1}}},\qquad s\sqrt{\frac{n-1}{\chi^2_{1-\alpha/2;n-1}}}\right]$$

$$= \left[0.1960\sqrt{\frac{24}{45.558}},\qquad 0.1960\sqrt{\frac{24}{9.886}}\right]$$

$$= [0.1423, 0.3060].$$

8.8 Confidence Interval for the Difference between the Means of Two Normal Distributions when the Standard Deviations Are Both Known

Let x be a normally distributed random variable with mean μ_x and standard deviation σ_x (known) and y be an independent (of x) normally distributed random variable with mean μ_y and standard deviation σ_y (known). Let $\bar{x}$ be the sample mean of n_x observations on x and $\bar{y}$ be the sample mean of n_y observations on y. The 100 $(1 - \alpha)$% confidence interval for $\mu_x - \mu_y$ is given by

$$\bar{x} - \bar{y} \pm K_{\alpha/2}\sqrt{\frac{\sigma_x^2}{n_x} + \frac{\sigma_y^2}{n_y}}$$

where $K_{\alpha/2}$ is the 100 $\alpha/2$ percentage point of the normal distribution. This result is obtained by considering the distribution of

$$\frac{\bar{x} - \bar{y} - (\mu_x - \mu_y)}{\sqrt{\dfrac{\sigma_x^2}{n_x} + \dfrac{\sigma_y^2}{n_y}}}$$

which has previously been shown to have a normal distribution with mean 0 and standard deviation 1. Hence

$$\mathrm{P}\left(-K_{\alpha/2} \leqq \frac{\bar{x} - \bar{y} - (\mu_x - \mu_y)}{\sqrt{\dfrac{\sigma_x^2}{n_x} + \dfrac{\sigma_y^2}{n_y}}} \leqq K_{\alpha/2}\right) = 1 - \alpha.$$

This is clearly equivalent to

$$\mathrm{P}\left(\bar{x} - \bar{y} - K_{\alpha/2}\sqrt{\frac{\sigma_x^2}{n_x} + \frac{\sigma_y^2}{n_y}} \leqq \mu_x - \mu_y \leqq \bar{x} - \bar{y} + K_{\alpha/2}\sqrt{\frac{\sigma_x^2}{n_x} + \frac{\sigma_y^2}{n_y}}\right) = 1 - \alpha$$

so that the 100 $(1 - \alpha)$% confidence interval is given by

$$\bar{x} - \bar{y} \pm K_{\alpha/2}\sqrt{\frac{\sigma_x^2}{n_x} + \frac{\sigma_y^2}{n_y}}.$$

A 100 $(1 - \alpha)$% upper one-sided confidence interval is given by

$$\bar{x} - \bar{y} + K_{\alpha}\sqrt{\frac{\sigma_x^2}{n_x} + \frac{\sigma_y^2}{n_y}}$$

whereas a 100 $(1-\alpha)$% lower one-sided confidence interval is given by

$$\bar{x} - \bar{y} - K_\alpha \sqrt{\frac{\sigma_x^2}{n_x} + \frac{\sigma_y^2}{n_y}}\,.$$

8.8.1 Example

In the example of Section 7.1.4 the effect of polishing on the endurance limit of steel was studied. It was felt that polishing could only have a positive effect on the average endurance limit. Eight specimens of polished steel were measured and their average, $\bar{x}$, was computed to be 86,375 psi. Similarly, 8 specimens of unpolished steel were measured and their average, $\bar{y}$, was computed to be 79,838 psi. All the measurements are assumed to be normally distributed with standard deviation 4,000 psi (both the polished and unpolished specimens). A 90% lower one-sided confidence limit for the difference in means is given by

$$\begin{aligned} L &= \bar{x} - \bar{y} - 1.28 \sqrt{\frac{\sigma_x^2}{n_x} + \frac{\sigma_y^2}{n_y}} \\ &= (86{,}375 - 79{,}838) - 1.28 \sqrt{\frac{(4{,}000)^2}{8} + \frac{(4{,}000)^2}{8}} \\ &= 6{,}537 - 1.28\,(2{,}000) \\ &= 3977. \end{aligned}$$

8.9 Confidence Interval for the Difference between the Means of Two Normal Distributions where the Standard Deviations Are Both Unknown but Equal

Let x be a normally distributed random variable with mean μ_x and standard deviation σ (unknown) and y be an independent (of x) normally distributed random variable with mean μ_y and standard deviation σ (unknown). Let $\bar{x}$ be the sample mean of n_x observations on x, and $\bar{y}$ be the sample mean of n_y observations on y. The 100 $(1-\alpha)$% confidence interval for $\mu_x - \mu_y$ is given by

$$\bar{x} - \bar{y} \pm t_{\alpha/2;n_x+n_y-2} \sqrt{\frac{1}{n_x} + \frac{1}{n_y}} \sqrt{\frac{\sum (x_i - \bar{x})^2 + \sum (y_i - \bar{y})^2}{n_x + n_y - 2}}$$

where $t_{\alpha/2;n_x+n_y-2}$ is the 100 $\alpha/2$ percentage point of the t distribution with $n_x + n_y - 2$ degrees of freedom.

This result is obtained by considering the distribution of

$$\frac{\bar{x} - \bar{y} - (\mu_x - \mu_y)}{\sqrt{\dfrac{1}{n_x} + \dfrac{1}{n_y}} \sqrt{\dfrac{\sum (x_i - \bar{x})^2 + \sum (y_i - \bar{y})^2}{n_x + n_y - 2}}}$$

which has been shown in Chapter IV to have a t distribution with $n_x + n_y - 2$ degrees of freedom. Hence,

$$P\left(-t_{\alpha/2;n_x+n_y-2} \leqq \frac{\bar{x} - \bar{y} - (\mu_x - \mu_y)}{\sqrt{\frac{1}{n_x} + \frac{1}{n_y}}\sqrt{\frac{\sum (x_i - \bar{x})^2 + \sum (y_i - \bar{y})^2}{n_x + n_y - 2}}} \leqq t_{\alpha/2;n_x+n_y-2}\right) = 1 - \alpha.$$

This is clearly equivalent to

$$P\left(\bar{x} - \bar{y} - t_{\alpha/2;n_x+n_y-2}\sqrt{\frac{1}{n_x} + \frac{1}{n_y}}\sqrt{\frac{\sum (x_i - \bar{x})^2 + \sum (y_i - \bar{y})^2}{n_x + n_y - 2}} \leqq \mu_x - \mu_y \leqq \bar{x} - \bar{y} + t_{\alpha/2;n_x+n_y-2}\sqrt{\frac{1}{n_x} + \frac{1}{n_y}}\sqrt{\frac{\sum (x_i - \bar{x})^2 + \sum (y_i - \bar{y})^2}{n_x + n_y - 2}}\right) = 1 - \alpha,$$

so that the $100(1 - \alpha)\%$ confidence interval is given by

$$\bar{x} - \bar{y} \pm t_{\alpha/2;n_x+n_y-2}\sqrt{\frac{1}{n_x} + \frac{1}{n_y}}\sqrt{\frac{\sum (x_i - \bar{x})^2 + \sum (y_i - \bar{y})^2}{n_x + n_y - 2}}.$$

A $100(1 - \alpha)\%$ upper one-sided confidence interval is given by

$$\bar{x} - \bar{y} + t_{\alpha;n_x+n_y-2}\sqrt{\frac{1}{n_x} + \frac{1}{n_y}}\sqrt{\frac{\sum (x_i - \bar{x})^2 + \sum (y_i - \bar{y})^2}{n_x + n_y - 2}};$$

a $100(1 - \alpha)\%$ lower one-sided confidence interval is given by

$$\bar{x} - \bar{y} - t_{\alpha;n_x+n_y-2}\sqrt{\frac{1}{n_x} + \frac{1}{n_y}}\sqrt{\frac{\sum (x_i - \bar{x})^2 + \sum (y_i - \bar{y})^2}{n_x + n_y - 2}}.$$

8.9.1 Example

In the example of Section 7.2.3, a comparison of two manufacturers' thermostats was made. Twenty-three thermostats from the new supplier were measured and their average, $\bar{x}$, computed to be 551.491. Similarly 23 thermostats from the old supplier were measured and their average, $\bar{y}$, computed to be 549.926. It is assumed that the thermostat readings of each manufacturer are independent normally distributed random variables with common unknown standard deviation. The quantity

$$\frac{\sum (x_i - \bar{x})^2 + \sum (y_i - \bar{y})^2}{n_x + n_y - 2}$$

was found to be 87.683. A 90% confidence interval for the difference in means is given by

$$551.491 - 549.926 \pm t_{0.05;44} \sqrt{\frac{1}{23} + \frac{1}{23}} \sqrt{87.683}$$

$$= 1.565 \pm 1.680\ (0.2949)\ (9.3639)$$
$$= [-3.074, \quad 6.204].$$

8.10 Confidence Interval for the Ratio of Standard Deviations of Two Normal Distributions

Let x be a normally distributed random variable with mean μ_x and standard deviation σ_x, and y be an independent (of x) normally distributed random variable with mean μ_y and standard deviation σ_y. Let s_x^2 be the sample variance based on the n_x observations on x, and let s_y^2 be the sample variance based on the n_y observations on y. The $100(1-\alpha)\%$ confidence interval for σ_x/σ_y is given by

$$\left[\frac{s_x}{s_y}\sqrt{\frac{1}{F_{\alpha/2;n_x-1,n_y-1}}}, \qquad \frac{s_x}{s_y}\sqrt{F_{\alpha/2;n_y-1,n_x-1}}\right]$$

where $F_{\alpha/2;n_x-1,n_y-1}$ is the $100\,\alpha/2$ percentage point of the F distribution with $n_x - 1$ and $n_y - 1$ degrees of freedom and $F_{\alpha/2;n_y-1,n_x-1}$ is the $100\,\alpha/2$ percentage point of the F distribution with n_y-1 and n_x-1 degrees of freedom.

This result is obtained by considering the distribution of

$$\frac{s_y^2/\,s_x^2}{\sigma_y^2/\sigma_x^2}$$

which has been shown in Chapter IV to have an F distribution with $n_y - 1$ and $n_x - 1$ degrees of freedom. Hence

$$P\left(F_{1-\alpha/2;n_y-1,n_x-1} \leqq \frac{s_y^2/s_x^2}{\sigma_y^2/\sigma_x^2} \leqq F_{\alpha/2;n_y-1,n_x-1}\right) = 1 - \alpha.$$

This is clearly equivalent to

$$P\left(\frac{s_x}{s_y}\sqrt{F_{1-\alpha/2;n_y-1,n_x-1}} \leqq \frac{\sigma_x}{\sigma_y} \leqq \frac{s_x}{s_y}\sqrt{F_{\alpha/2;n_y-1,n_x-1}}\right) = 1 - \alpha.$$

It has also been shown that

$$F_{1-\alpha/2;n_y-1,n_x-1} = \frac{1}{F_{\alpha/2;n_x-1,n_y-1}}.$$

Hence the 100 (1 − α)% confidence interval is given by

$$\left[\frac{s_x}{s_y}\sqrt{\frac{1}{F_{\alpha/2;n_x-1,n_y-1}}},\qquad \frac{s_x}{s_y}\sqrt{F_{\alpha/2;n_y-1,n_x-1}}\right].$$

A 100 (1 − α)% upper one-sided confidence interval is given by

$$\frac{s_x}{s_y}\sqrt{F_{\alpha;n_y-1,n_x-1}}$$

whereas a 100 (1−α)% lower one-sided confidence interval is given b

$$\frac{s_x}{s_y}\sqrt{\frac{1}{F_{\alpha;n_x-1,n_y-1}}}.$$

8.10.1 Example

In the example of Section 7.6.4, a comparison of the variabilitie of two suppliers was discussed. The new supplier had a sample var ance of $s_x^2 = 0.00041$ based on 100 observations, and the current sup plier a sample variance of $s_y^2 = 0.00057$ also based on 100 observations A 90% lower one-sided confidence limit for σ_x/σ_y is given by

$$\frac{s_x}{s_y}\sqrt{\frac{1}{F_{0.10;99,99}}}$$

$$= \frac{\sqrt{0.00041}}{\sqrt{0.00057}}\sqrt{\frac{1}{1.29}}$$

$$= 0.747.$$

8.11 A Table of Point Estimates and Interval Estimates

Procedure for calculating point estimates and confidence interva estimates are summarized in Table 8.1.

8.12 Statistical Tolerance Limits

The quality of manufactured product is often specified by settin a range, the bounds of which are called tolerance limits. These limit have the property that a certain percentage of the product may be expected to fall within these limits. In Chapter III a study of toler ance limits was made under the assumption that the quality is nor mally distributed with *known* mean, μ, and *known* standard devia tion, σ. Tolerance limits were defined as those limits within whic 100 (1 − α)% of the product falls. These limits were formed by add ing to and subtracting from μ the quantity $K_{\alpha/2}\sigma$. Unfortunately, ir

Parameters	Notation	Qualifying Conditions	Formula	Tables To Find Percentage Points	Point Estimates
Mean of a normal distribution	μ	Known σ	$\bar{x} - K_{\alpha/2} \frac{\sigma}{\sqrt{n}} \leqq \mu \leqq \bar{x} + K_{\alpha/2} \frac{\sigma}{\sqrt{n}}$	Normal table, Table 1	$\bar{x}$
Standard deviation of a normal distribution	σ		$s \sqrt{\frac{n-1}{\chi^2_{\alpha/2;n-1}}} \leqq \sigma \leqq s \sqrt{\frac{n-1}{\chi^2_{1-\alpha/2;n-1}}}$	Chi-square table, Table 2	s
Mean of a normal distribution	μ	Unknown σ	$\bar{x} - t_{\alpha/2;n-1} \frac{s}{\sqrt{n}} \leqq \mu \leqq \bar{x} + t_{\alpha/2;n-1} \frac{s}{\sqrt{n}}$	t table, Table 3	$\bar{x}$
Difference between the means of two normal distributions	$\mu_x - \mu_y$	$x_1, \cdots, x_{n_x}$ are observations in first sample; $y_1, \cdots, y_{n_y}$ are observations in second sample; σ_x and σ_y are known	$\bar{x} - \bar{y} - K_{\alpha/2} \sqrt{\frac{\sigma_x^2}{n_x} + \frac{\sigma_y^2}{n_y}} \leqq \mu_x - \mu_y \leqq \bar{x} - \bar{y} + K_{\alpha/2} \sqrt{\frac{\sigma_x^2}{n_x} + \frac{\sigma_y^2}{n_y}}$		$\bar{x} - \bar{y}$
Difference between the means of two normal distributions	$\mu_x - \mu_y$	$\sigma_x = \sigma_y = \sigma$ Unknown σ	$(\bar{x} - \bar{y}) - t_{\alpha/2;n_x+n_y-2} \sqrt{\frac{1}{n_x} + \frac{1}{n_y}} \sqrt{\frac{\sum (x_i - \bar{x})^2 + \sum (y_i - \bar{y})^2}{n_x + n_y - 2}}$ $\leqq \mu_x - \mu_y \leqq (\bar{x} - \bar{y}) + t_{\alpha/2;n_x+n_y-2} \sqrt{\frac{1}{n_x} + \frac{1}{n_y}} \sqrt{\frac{\sum (x_i - \bar{x})^2 + \sum (y_i - \bar{y})^2}{n_x + n_y - 2}}$		$\bar{x} - \bar{y}$
Ratio of standard deviations of two normal distributions	$\frac{\sigma_x}{\sigma_y}$	s_x is the first sample standard deviation; s_y is the second sample standard deviation	$\frac{s_x}{s_y} \sqrt{\frac{1}{F_{\alpha/2;n_x-1,n_y-1}}} \leqq \frac{\sigma_x}{\sigma_y} \leqq \frac{s_x}{s_y} \sqrt{F_{\alpha/2;n_y-1,n_x-1}}$	F table, Table 4	$\frac{s_x}{s_y}$

Table 8.2. Tolerance Factors For Normal Distributions[1]

Factors K such that the Probability Is γ that at least a Proportion $1 - \alpha$ of the Distribution Will Be Included between $\bar{x} \pm Ks$, where $\bar{x}$ and s Are Estimates of the Mean and the Standard Deviation Computed from a Sample of n.

n \ α	$\gamma = 0.75$					$\gamma = 0.90$				
	0.25	0.10	0.05	0.01	0.001	0.25	0.10	0.05	0.01	0.001
2	4.498	6.301	7.414	9.531	11.920	11.407	15.978	18.800	24.167	30.227
3	2.501	3.538	4.187	5.431	6.844	4.132	5.847	6.919	8.974	11.309
4	2.035	2.892	3.431	4.471	5.657	2.932	4.166	4.943	6.440	8.149
5	1.825	2.599	3.088	4.033	5.117	2.454	3.494	4.152	5.423	6.879
6	1.704	2.429	2.889	3.779	4.802	2.196	3.131	3.723	4.870	6.188
7	1.624	2.318	2.757	3.611	4.593	2.034	2.902	3.452	4.521	5.750
8	1.568	2.238	2.663	3.491	4.444	1.921	2.743	3.264	4.278	5.446
9	1.525	2.178	2.593	3.400	4.330	1.839	2.626	3.125	4.098	5.220
10	1.492	2.131	2.537	3.328	4.241	1.775	2.535	3.018	3.959	5.046
11	1.465	2.093	2.493	3.271	4.169	1.724	2.463	2.933	3.849	4.906
12	1.443	2.062	2.456	3.223	4.110	1.683	2.404	2.863	3.758	4.792
13	1.425	2.036	2.424	3.183	4.059	1.648	2.355	2.805	3.682	4.697
14	1.409	2.013	2.398	3.148	4.016	1.619	2.314	2.756	3.618	4.615
15	1.395	1.994	2.375	3.118	3.979	1.594	2.278	2.713	3.562	4.545
16	1.383	1.977	2.355	3.092	3.946	1.572	2.246	2.676	3.514	4.484
17	1.372	1.962	2.337	3.069	3.917	1.552	2.219	2.643	3.471	4.430
18	1.363	1.948	2.321	3.048	3.891	1.535	2.194	2.614	3.433	4.382
19	1.355	1.936	2.307	3.030	3.867	1.520	2.172	2.588	3.399	4.339
20	1.347	1.925	2.294	3.013	3.846	1.506	2.152	2.564	3.368	4.300
21	1.340	1.915	2.282	2.998	3.827	1.493	2.135	2.543	3.340	4.264
22	1.334	1.906	2.271	2.984	3.809	1.482	2.118	2.524	3.315	4.232
23	1.328	1.898	2.261	2.971	3.793	1.471	2.103	2.506	3.292	4.203
24	1.322	1.891	2.252	2.959	3.778	1.462	2.089	2.489	3.270	4.176
25	1.317	1.883	2.244	2.948	3.764	1.453	2.077	2.474	3.251	4.151
26	1.313	1.877	2.236	2.938	3.751	1.444	2.065	2.460	3.232	4.127
27	1.309	1.871	2.229	2.929	3.740	1.437	2.054	2.447	3.215	4.106
28	1.305	1.865	2.222	2.920	3.728	1.430	2.044	2.435	3.199	4.085
29	1.301	1.860	2.216	2.911	3.718	1.423	2.034	2.424	3.184	4.066
30	1.297	1.855	2.210	2.904	3.708	1.417	2.025	2.413	3.170	4.049
31	1.294	1.850	2.204	2.896	3.699	1.411	2.017	2.403	3.157	4.032
32	1.291	1.846	2.199	2.890	3.690	1.405	2.009	2.393	3.145	4.016
33	1.288	1.842	2.194	2.883	3.682	1.400	2.001	2.385	3.133	4.001
34	1.285	1.838	2.189	2.877	3.674	1.395	1.994	2.376	3.122	3.987
35	1.283	1.834	2.185	2.871	3.667	1.390	1.988	2.368	3.112	3.974
36	1.280	1.830	2.181	2.866	3.660	1.386	1.981	2.361	3.102	3.961
37	1.278	1.827	2.177	2.860	3.653	1.381	1.975	2.353	3.092	3.949
38	1.275	1.824	2.173	2.855	3.647	1.377	1.969	2.346	3.083	3.938
39	1.273	1.821	2.169	2.850	3.641	1.374	1.964	2.340	3.075	3.927
40	1.271	1.818	2.166	2.846	3.635	1.370	1.959	2.334	3.066	3.917
41	1.269	1.815	2.162	2.841	3.629	1.366	1.954	2.328	3.059	3.907
42	1.267	1.812	2.159	2.837	3.624	1.363	1.949	2.322	3.051	3.897
43	1.266	1.810	2.156	2.833	3.619	1.360	1.944	2.316	3.044	3.888
44	1.264	1.807	2.153	2.829	3.614	1.357	1.940	2.311	3.037	3.879
45	1.262	1.805	2.150	2.826	3.609	1.354	1.935	2.306	3.030	3.871
46	1.261	1.802	2.148	2.822	3.605	1.351	1.931	2.301	3.024	3.863
47	1.259	1.800	2.145	2.819	3.600	1.348	1.927	2.297	3.018	3.855
48	1.258	1.798	2.143	2.815	3.596	1.345	1.924	2.292	3.012	3.847
49	1.256	1.796	2.140	2.812	3.592	1.343	1.920	2.288	3.006	3.840
50	1.255	1.794	2.138	2.809	3.588	1.340	1.916	2.284	3.001	3.833

[1] Reproduced with permission from C. Eisenhart, M. W. Hastay, W. A. Wallis, *Techniques of Statistical Analysis*, Chapter 2, McGraw-Hill Book Company, Inc., New York, 1947.

Table 8.2. Tolerance Factors for Normal Distributions (Cont.)

n \ α	γ = 0.95					γ = 0.99				
	0.25	0.10	0.05	0.01	0.001	0.25	0.10	0.05	0.01	0.001
2	22.858	32.019	37.674	48.430	60.573	114.363	160.193	188.491	242.300	303.054
3	5.922	8.380	9.916	12.861	16.208	13.378	18.930	22.401	29.055	36.616
4	3.779	5.369	6.370	8.299	10.502	6.614	9.398	11.150	14.527	18.383
5	3.002	4.275	5.079	6.634	8.415	4.643	6.612	7.855	10.260	13.015
6	2.604	3.712	4.414	5.775	7.337	3.743	5.337	6.345	8.301	10.548
7	2.361	3.369	4.007	5.248	6.676	3.233	4.613	5.488	7.187	9.142
8	2.197	3.136	3.732	4.891	6.226	2.905	4.147	4.936	6.468	8.234
9	2.078	2.967	3.532	4.631	5.899	2.677	3.822	4.550	5.966	7.600
10	1.987	2.839	3.379	4.433	5.649	2.508	3.582	4.265	5.594	7.129
11	1.916	2.737	3.259	4.277	5.452	2.378	3.397	4.045	5.308	6.766
12	1.858	2.655	3.162	4.150	5.291	2.274	3.250	3.870	5.079	6.477
13	1.810	2.587	3.081	4.044	5.158	2.190	3.130	3.727	4.893	6.240
14	1.770	2.529	3.012	3.955	5.045	2.120	3.029	3.608	4.737	6.043
15	1.735	2.480	2.954	3.878	4.949	2.060	2.945	3.507	4.605	5.876
16	1.705	2.437	2.903	3.812	4.865	2.009	2.872	3.421	4.492	5.732
17	1.679	2.400	2.858	3.754	4.791	1.965	2.808	3.345	4.393	5.607
18	1.655	2.366	2.819	3.702	4.725	1.926	2.753	3.279	4.307	5.497
19	1.635	2.337	2.784	3.656	4.667	1.891	2.703	3.221	4.230	5.399
20	1.616	2.310	2.752	3.615	4.614	1.860	2.659	3.168	4.161	5.312
21	1.599	2.286	2.723	3.577	4.567	1.833	2.620	3.121	4.100	5.234
22	1.584	2.264	2.697	3.543	4.523	1.808	2.584	3.078	4.044	5.163
23	1.570	2.244	2.673	3.512	4.484	1.785	2.551	3.040	3.993	5.098
24	1.557	2.225	2.651	3.483	4.447	1.764	2.522	3.004	3.947	5.039
25	1.545	2.208	2.631	3.457	4.413	1.745	2.494	2.972	3.904	4.985
26	1.534	2.193	2.612	3.432	4.382	1.727	2.469	2.941	3.865	4.935
27	1.523	2.178	2.595	3.409	4.353	1.711	2.446	2.914	3.828	4.888
28	1.514	2.164	2.579	3.388	4.326	1.695	2.424	2.888	3.794	4.845
29	1.505	2.152	2.554	3.368	4.301	1.681	2.404	2.864	3.763	4.805
30	1.497	2.140	2.549	3.350	4.278	1.668	2.385	2.841	3.733	4.768
31	1.489	2.129	2.536	3.332	4.256	1.656	2.367	2.820	3.706	4.732
32	1.481	2.118	2.524	3.316	4.235	1.644	2.351	2.801	3.680	4.699
33	1.475	2.108	2.512	3.300	4.215	1.633	2.335	2.782	3.655	4.668
34	1.468	2.099	2.501	3.286	4.197	1.623	2.320	2.764	3.632	4.639
35	1.462	2.090	2.490	3.272	4.179	1.613	2.306	2.748	3.611	4.611
36	1.455	2.081	2.479	3.258	4.161	1.604	2.293	2.732	3.590	4.585
37	1.450	2.073	2.470	3.246	4.146	1.595	2.281	2.717	3.571	4.560
38	1.446	2.068	2.464	3.237	4.134	1.587	2.269	2.703	3.552	4.537
39	1.441	2.060	2.455	3.226	4.120	1.579	2.257	2.690	3.534	4.514
40	1.435	2.052	2.445	3.213	4.104	1.571	2.247	2.677	3.518	4.493
41	1.430	2.045	2.437	3.202	4.090	1.564	2.236	2.665	3.502	4.472
42	1.426	2.039	2.429	3.192	4.077	1.557	2.227	2.653	3.486	4.453
43	1.422	2.033	2.422	3.183	4.065	1.551	2.217	2.642	3.472	4.434
44	1.418	2.027	2.415	3.173	4.053	1.545	2.208	2.631	3.458	4.416
45	1.414	2.021	2.408	3.165	4.042	1.539	2.200	2.621	3.444	4.399
46	1.410	2.016	2.402	3.156	4.031	1.533	2.192	2.611	3.431	4.383
47	1.406	2.011	2.396	3.148	4.021	1.527	2.184	2.602	3.419	4.367
48	1.403	2.006	2.390	3.140	4.011	1.522	2.176	2.593	3.407	4.352
49	1.399	2.001	2.384	3.133	4.002	1.517	2.169	2.584	3.396	4.337
50	1.396	1.969	2.379	3.126	3.993	1.512	2.162	2.576	3.385	4.323

practice, the values μ and σ are rarely known and estimates of these values, $\bar{x}$ and s, respectively, are used. However, where before it could be stated that 95% of a manufactured product lies between $\mu \pm 1.96\sigma$, this statement cannot be extended to the interval $\bar{x} \pm 1.96s$. The quantities $\bar{x}$ and s are random variables, and hence the limits depend upon the particular outcome of the sample. Different samples will lead to different limits. How close these limits are to $\mu \pm 1.96\sigma$ depends upon how good the estimates are.

It is evident then that the fraction of the items included between $\bar{x} \pm K_{\alpha/2}s$ will not always be $1 - \alpha$. However, it is possible to determine a constant K such that in a large series of samples from a normal distribution, a fixed proportion γ of the intervals $\bar{x} \pm Ks$ will include $100(1 - \alpha)\%$ or more of the distribution. Thus, statistical tolerance limits for a normal distribution are given by $[L = \bar{x} - Ks, U = \bar{x} + Ks]$ and have the property that the probability is equal to a preassigned value γ that the interval includes at least a specified proportion $1 - \alpha$ of the statistical distribution. Note that in most practical situations γ is usually a large number close to 1. Statistical tolerance limits should not be confused with confidence intervals for a parameter of the distribution. Confidence limits for the mean of a normal distribution are such that in a given fraction, say 0.95, of the samples from which they are computed, the interval bounded by the limits will include the true mean of the distribution. For confidence interval estimation, 0.95 is also called the confidence coefficient.

Values of K for the parameters $n = 2\,(1)\,50$; $\gamma = 0.75, 0.90, 0.95, 0.99$; and $\alpha = 0.25, 0.10, 0.05, 0.01, 0.001$ are given in Table 8.2.

8.12.1 Example

In the manufacture of electron tubes to be used in stable amplifiers, it is desired to know, with probability 0.99, limits within which 90% of the future tube transconductances lie. The required tests are made on 9 tubes and the transconductances, observed in micromhos, are as follows: 4,330; 4,287; 4,450; 4,295; 4,340; 4,407; 4,295; 4,388; and 4,356. From Table 8.2, the value of K corresponding to $n = 9$, $\gamma = 0.99$ and $\alpha = 0.10$ is 3.822. For this example, $\bar{x}$ is found to be 4,349.8 and s is 56.18. Thus the tolerance limits are given by $4{,}349.8 \pm (3.822)(56.18) = [4{,}135.1,\quad 4{,}564.5]$.

13 One-Sided Statistical Tolerance Limits

Instead of specifying two limits, it is sometimes appropriate to ecify a single limit, $\bar{x} - Ks$, such that a fixed proportion of the dis- ibution will be greater than this limit, or to specify a single nit, $\bar{x} + Ks$, such that a fixed proportion of the distribution will be naller than this limit. Such single limits are called one-sided toler- nce limits.

Values of K for the parameters $n = 3\ (1)\ 25$ and $30\ (5)\ 50$; $\gamma = 0.75$, 90, 0.95, 0.99; and $\alpha = 0.25, 0.10, 0.05, 0.01, 0.001$ are given in Table 3.

8.13.1 Example

A manufacturer of light bulbs would like to specify a single lower nit above which he can be assured, with probability of 0.95, that % of his production will be. A sample of 30 bulbs is taken and the mple mean and sample standard deviation are found to be 987.2 id 5.963 respectively. A value of $K = 3.064$ corresponding to $n =$, $\gamma = 0.95$ and $\alpha = 0.01$ is obtained from Table 8.3 The required wer tolerance limit is given by $\bar{x} - Ks = 987.2 - (3.064)(5.963) =$ 68.9.

14 Distribution-Free Tolerance Limits

There exist tolerance limits which are independent of the form of e underlying distribution. However, tolerance limits based on the ormal distribution are substantially shorter for a fixed sample size an those based on no assumption about the underlying distribution. owever, if the assumption of normality is violated, the limits ob- ined in the manner described above may be irrelevant. Distribu- on-free tolerance limits have limited industrial use since they sually require an enormous sample size for reasonable probability atements. The distribution-free tolerance limits given in this ection are based upon the smallest observation and the largest obser- ation in a sample of size n.

For two-sided tolerance limits, the number of observations required that the probability is γ that at least $100\,(1 - \alpha)\%$ of the distribu- ion will lie between the smallest and largest observations of the ample is given approximately by

$$n \cong \left(\frac{2 - \alpha}{\alpha}\right)\left(\frac{\chi^2_{1-\gamma;4}}{4}\right) + \frac{1}{2}$$

Table 8.3. Tolerance Factors for Normal Distributions

Factors K such that the Probability Is γ that at least a Proportion $1-\alpha$ of the Distribution Will Be Less than $\bar{x}+Ks$ (or Greater than $\bar{x}-Ks$), where $\bar{x}$ and s Are Estimates of the Mean and the Standard Deviation Computed from a Sample of Size n.

n \ α	$\gamma = 0.75$					$\gamma = 0.90$				
	0.25	0.10	0.05	0.01	0.001	0.25	0.10	0.05	0.01	0.001
3	1.464	2.501	3.152	4.396	5.805	2.602	4.258	5.310	7.340	9.651
4	1.256	2.134	2.680	3.726	4.910	1.972	3.187	3.957	5.437	7.128
5	1.152	1.961	2.463	3.421	4.507	1.698	2.742	3.400	4.666	6.112
6	1.087	1.860	2.336	3.243	4.273	1.540	2.494	3.091	4.242	5.556
7	1.043	1.791	2.250	3.126	4.118	1.435	2.333	2.894	3.972	5.201
8	1.010	1.740	2.190	3.042	4.008	1.360	2.219	2.755	3.783	4.955
9	0.984	1.702	2.141	2.977	3.924	1.302	2.133	2.649	3.641	4.772
10	0.964	1.671	2.103	2.927	3.858	1.257	2.065	2.568	3.532	4.629
11	0.947	1.646	2.073	2.885	3.804	1.219	2.012	2.503	3.444	4.515
12	0.933	1.624	2.048	2.851	3.760	1.188	1.966	2.448	3.371	4.420
13	0.919	1.606	2.026	2.822	3.722	1.162	1.928	2.403	3.310	4.341
14	0.909	1.591	2.007	2.796	3.690	1.139	1.895	2.363	3.257	4.274
15	0.899	1.577	1.991	2.776	3.661	1.119	1.866	2.329	3.212	4.215
16	0.891	1.566	1.977	2.756	3.637	1.101	1.842	2.299	3.172	4.164
17	0.883	1.554	1.964	2.739	3.615	1.085	1.820	2.272	3.136	4.118
18	0.876	1.544	1.951	2.723	3.595	1.071	1.800	2.249	3.106	4.078
19	0.870	1.536	1.942	2.710	3.577	1.058	1.781	2.228	3.078	4.041
20	0.865	1.528	1.933	2.697	3.561	1.046	1.765	2.208	3.052	4.009
21	0.859	1.520	1.923	2.686	3.545	1.035	1.750	2.190	3.028	3.979
22	0.854	1.514	1.916	2.675	3.532	1.025	1.736	2.174	3.007	3.952
23	0.849	1.508	1.907	2.665	3.520	1.016	1.724	2.159	2.987	3.927
24	0.845	1.502	1.901	2.656	3.509	1.007	1.712	2.145	2.969	3.904
25	0.842	1.496	1.895	2.647	3.497	0.999	1.702	2.132	2.952	3.882
30	0.825	1.475	1.869	2.613	3.454	0.966	1.657	2.080	2.884	3.794
35	0.812	1.458	1.849	2.588	3.421	0.942	1.623	2.041	2.833	3.730
40	0.803	1.445	1.834	2.568	3.395	0.923	1.598	2.010	2.793	3.679
45	0.795	1.435	1.821	2.552	3.375	0.908	1.577	1.986	2.762	3.638
50	0.788	1.426	1.8 11	2.538	3.358	0.894	1.560	1.965	2.735	3.604

Table 8.3. Tolerance Factors for Normal Distributions (Cont.)

n \ α	$\gamma = 0.95$					$\gamma = 0.99$				
	0.25	0.10	0.05	0.01	0.001	0.25	0.10	0.05	0.01	0.001
3	3.804	6.158	7.655	10.552	13.857					
4	2.619	4.163	5.145	7.042	9.215					
5	2.149	3.407	4.202	5.741	7.501					
6	1.895	3.006	3.707	5.062	6.612	2.849	4.408	5.409	7.334	9.540
7	1.732	2.755	3.399	4.641	6.061	2.490	3.856	4.730	6.411	8.348
8	1.617	2.582	3.188	4.353	5.686	2.252	3.496	4.287	5.811	7.566
9	1.532	2.454	3.031	4.143	5.414	2.085	3.242	3.971	5.389	7.014
10	1.465	2.355	2.911	3.981	5.203	1.954	3.048	3.739	5.075	6.603
11	1.411	2.275	2.815	3.852	5.036	1.854	2.897	3.557	4.828	6.284
12	1.366	2.210	2.736	3.747	4.900	1.771	2.773	3.410	4.633	6.032
13	1.329	2.155	2.670	3.659	4.787	1.702	2.677	3.290	4.472	5.826
14	1.296	2.108	2.614	3.585	4.690	1.645	2.592	3.189	4.336	5.651
15	1.268	2.068	2.566	3.520	4.607	1.596	2.521	3.102	4.224	5.507
16	1.242	2.032	2.523	3.463	4.534	1.553	2.458	3.028	4.124	5.374
17	1.220	2.001	2.486	3.415	4.471	1.514	2.405	2.962	4.038	5.268
18	1.200	1.974	2.453	3.370	4.415	1.481	2.357	2.906	3.961	5.167
19	1.183	1.949	2.423	3.331	4.364	1.450	2.315	2.855	3.893	5.078
20	1.167	1.926	2.396	3.295	4.319	1.424	2.275	2.807	3.832	5.003
21	1.152	1.905	2.371	3.262	4.276	1.397	2.241	2.768	3.776	4.932
22	1.138	1.887	2.350	3.233	4.238	1.376	2.208	2.729	3.727	4.866
23	1.126	1.869	2.329	3.206	4.204	1.355	2.179	2.693	3.680	4.806
24	1.114	1.853	2.309	3.181	4.171	1.336	2.154	2.663	3.638	4.755
25	1.103	1.838	2.292	3.158	4.143	1.319	2.129	2.632	3.601	4.706
30	1.059	1.778	2.220	3.064	4.022	1.249	2.029	2.516	3.446	4.508
35	1.025	1.732	2.166	2.994	3.934	1.195	1.957	2.431	3.334	4.364
40	0.999	1.697	2.126	2.941	3.866	1.154	1.902	2.365	3.250	4.255
45	0.978	1.669	2.092	2.897	3.811	1.122	1.857	2.313	3.181	4.168
50	0.961	1.646	2.065	2.863	3.766	1.096	1.821	2.296	3.124	4.096

where $\chi^2_{1-\gamma;4}$ is the $100(1-\gamma)$ percentage point of the chi-square di tribution with 4 degrees of freedom. In order to find distribution-fr tolerance limits which contain at least 90% of the distribution wit probability 0.99, i.e., $\alpha = 0.10$, $\gamma = 0.99$, and $\chi^2_{0.01;4} = 13.277$, a samp of size

$$n \cong \left(\frac{1.9}{0.1}\right)\left(\frac{13.277}{4}\right) + \frac{1}{2} = 64$$

is required. Thus, if a sample of size 64 is taken, the probability 0.99 that at least 90% of the distribution will lie between the smalle and largest observations of the sample. This is to be compared wit the example of Section 8.12.1 which required a sample of size 9 to b assured with probability 0.99 that at least 90% of the distributic will lie between $\bar{x} \pm 3.822s$ if the distribution is normal.

For one-sided tolerance limits, the number of observations requir so that the probability is γ that at least $100(1-\alpha)\%$ of the distrib tion will exceed the smallest observation (or be less than the large observation) of the sample is given by

$$n = \frac{\log(1-\gamma)}{\log(1-\alpha)}. \ *$$

In order to find the lower distribution-free tolerance limit which co tains at least 99% of the distribution with probability 0.95, i.e $\alpha = 0.01$ and $\gamma = 0.95$, a sample of size

$$n = \frac{\log(0.05)}{\log(0.99)} = 300$$

is required. Thus, if a sample of size 300 is taken, the probability 0.95 that at least 99% of the distribution will exceed the smalle observation of the sample. This is to be compared with the examp of Section 8.13.1 which required a sample of size 30 to be assure with probability 0.95 that at least 99% of the distribution will excee $\bar{x} - 3.064s$ if the distribution is normal.

* The logarithm to any base may be used.

PROBLEMS

1. Suppose that t_1 and t_2 are estimates of θ with $E(t_1)=\theta$, $E(t_2)=\theta/2$, $t_1)=5$, and $\sigma^2(t_2)=2$. Which is a better estimate of θ ? Why ?

2. In sampling from a binomial distribution show that the proportion events observed in a sample is an unbiased estimate of the true proba-lity that an event will occur. What is the variance of the observed pro-rtion ?

3. Suppose that t_1 and t_2 are estimates of θ. $E(t_1)=\theta$ and $E(t_2)>\theta$. ıe variance of $t_1=4$, the variance of $t_2=1$, and $E(t_2-\theta)^2=7$. Which is "better" estimate of θ ? Why ?

4. Prove that $s'^2=\sum_{i=1}^{n}(x_i-\bar{x})^2/n$ is a biased estimate of σ^2.

5. A lower one-sided confidence statement is to be made about the erage life of shoes. How many observations are necessary in order at the lower limit will not differ from the true average life by more an one-half month with probability 0.95 ? Assume that shoe life is nor-ally distributed with standard deviation equal to two months.

6. The distribution of observations from carload lots of a given chemi-l can be shown to be normally distributed about the true average density a given lot with standard deviation 0.005 g/cc. In order to obtain an timate based on the mean of n samples which is within 0.002 g/cc of e true average density for the lot in 90% of the cases, what is the re-ıired sample size ?

7. The warpwise breaking strength of a certain type of cloth was easured on four specimens from a lot with the following results (in psi): 2, 172, 177, 173. The standard deviation based on past experience is 5 i. Find a 99% confidence interval for the lot average warpwise break-g strength.

8. The density of 27 explosive primers was determined and the sample erage density was 1.53; the standard deviation is known to be 0.04. nd a 95% lower confidence interval for the density.

9. The sample average range of a random sample of 100 stored rockets as 2,208 yards. The standard deviation is known to be 50 yards. Find 99% confidence interval for the true average range.

10. In Problem 9, how many observations would be required to make e length of the confidence interval 20 yards ? 5 yards ? 1 yard ?

11. The following data represents porosity measurements on a sample om a shipment of coke. Find a 90% confidence interval for the true ean. Assume normality and $\sigma=\frac{1}{4}$.

2.16	2.07	2.34	1.97	1.97	1.90
2.19	2.23	2.15	2.47	2.31	1.94
2.31	1.86	2.25	2.14	2.15	2.16
2.30	2.48	2.11	2.15	2.24	2.04
2.21	1.91	2.01	2.09	2.07	2.25

12. Work Problem 11 without assuming $\sigma=\frac{1}{4}$.

13. Work Problem 9 assuming that 50 yards is an estimate of σ base on the sample. How much does your answer change?

14. Work Problem 7 without assuming $\sigma=5$.

15. Work Problem 8 assuming that $s=0.04$ based on the sample.

16. Find a 95% confidence interval for the diameter of the rivet base on the data below.

6.68	6.66	6.62	6.72
6.76	6.67	6.70	6.72
6.78	6.66	6.76	6.72
6.76	6.70	6.76	6.76
6.74	6.74	6.81	6.66
6.64	6.79	6.72	6.82
6.81	6.77	6.60	6.72
6.74	6.70	6.64	6.78
6.70	6.70	6.75	6.79

17. If $s=0.04$ based on a sample of 27 primers, find a 95% confidenc interval for σ.

18. Find a 99% confidence interval for σ based on the data of Problem 1(

19. The finished diameter on armored electric cable is normally dis tributed; a sample of 20 yields a mean of 0.780 and a sample standar deviation 0.010. Find 95% confidence intervals for both μ and σ.

20. For a sample of 15, $\sum(x-\bar{x})^2=16.8$. Find upper, lower, and tw sided 95% confidence intervals for σ.

21. Given $s=4$, find a 99% confidence interval for σ if s is based on 1 degrees of freedom; also 25.

22. An experiment is performed to test the difference in effectivenes of two methods of cultivating wheat. Ten patches of ground are treate with shallow plowing and fifteen with deep plowing. The mean yield pe acre of the first group is 40.3 bushels and the mean for the second grou is 44.7. Assume that the standard deviation of the shallow planting is 0. bushel and for the deep it is 0.8. Find a 95% confidence interval for th difference of yield.

23. Twenty rounds of standard ammunition and ten rounds of test am munition were fired in random sequence. The average barrel pressure c the standard was 41,900 psi and the average pressure for the test wa 44,200 psi. The standard deviations for both the standard and test wer known to be 2,050 psi. Find upper, lower, and two-sided 95% confidenc intervals for the change in pressure. Under what practical circumstanc would each be appropriate?

24. In the manufacture of wooden ladders, the side pieces for the lad ders have greater shear (parallel to the grain of the wood) if the ra lumber is air dried rather than kiln dried. The standard deviation of bot can be taken to be 30 psi. A sample of 25 observations of each type wa taken resulting in the following means: air-dried (psi) 1,170 and kiln dried (psi) 1,105. Find a 99% confidence interval for the increase.

25. In research on rocket propellants aimed at reducing the delay tim between the application of the firing current and explosion, tests wer run on a grade T of propellant and a grade C. The standard deviation o both grades can be assumed to be 0.03. If $\overline{C}=0.261$ second and $\overline{T}=0.250$ second for 14 observations on each, find a 90% confidence interval for the change in times.

26. Two analysts took repeated readings on the hardness of city water. Find a 90% confidence interval for the analyst difference assuming unknown but equal variance.

Coded Measures of Hardness

Analyst A x	*Analyst B* y
0.46	0.82
0.62	0.61
0.37	0.89
0.40	0.51
0.44	0.33
0.58	0.48
0.48	0.22
0.53	0.25
	0.67
	0.88

27. The following are 16 independent determinations of the melting point of a compound, eight made by analyst I and eight made by analyst II. Data are in degrees Centigrade.

Analyst I	*Analyst II*
164.4	163.5
169.7	162.8
169.2	163.0
169.5	163.2
161.8	160.7
168.7	161.5
169.5	160.9
163.9	162.1

Find a 99% confidence interval for the mean difference between analysts assuming unknown but equal variances.

28. The following are the burning times in seconds of floating smoke pots of two different types:

Type I		*Type II*	
481	572	526	537
506	561	511	582
527	501	556	605
661	487	542	558
501	524	491	578

Find a 99% confidence interval for the mean difference in burning times assuming unknown but equal variances.

29. Work Problem 25, if we assume unknown but equal variances and obtain $s_C^2=0.00125$ and $s_T^2=0.00132$ from the sample.

30. Let x and y be normally distributed with means μ_x and μ_y; variances σ_x^2 and σ_y^2. Let $\bar{x}$, $\bar{y}$, s_x^2 and s_y^2 be the usual estimates based on n_x and n_y observations. How would you find a confidence interval for $\mu_x-\mu_y$ without assuming $\sigma_x=\sigma_y$?

31. In a study of reflex reactions, the strength of patellar reflex measured in degrees of arc for 15 men under two conditions yielded the following results:

Tensed	*Relaxed*
31	35
19	14
22	19
26	29
36	34
30	26
29	19
36	37
34	24
33	27
19	14
19	19
26	30
15	7
18	13

Find a 95% confidence interval for the change in reflex. *Hint*: use paired t test.

32. The following table gives percent loss in tensile strength, following immersion in a corrosive solution, of paired samples of an alloy. One is subjected to stress and one is not.

Test number	*Unstressed*	*Stressed*
1	6.4	9.2
2	4.6	7.9
3	4.6	7.3
4	6.4	8.0
5	3.2	5.7
6	5.2	7.6
7	6.5	5.7
8	4.9	4.1
9	4.3	8.1
10	5.6	6.5
11	3.7	6.9
12	4.6	6.0

Find a 99% confidence interval for the expected difference in tensile strength.

33. In Problem 29, find a 90% confidence interval for σ_T/σ_C.

34. The following results are calculated for two samples from normal distributions:

$$X: \quad n_x = 8, \quad \Sigma x = 12, \quad \Sigma x^2 = 46$$
$$Y: \quad n_y = 11, \quad \Sigma y = 22, \quad \Sigma y^2 = 80$$

Find 95% confidence interval for σ_x/σ_y.

35. Suppose that 15 trials are made on each of two treatments with the sample standard deviation ratio $s_x/s_y = 3.1$. Find upper, lower, and two-sided confidence intervals for σ_x/σ_y.

36. Let $s_x^2 = 20$ and $s_y^2 = 2$. Find a 99% confidence interval for σ_x/σ_y assuming that the sample variances are based on 25 degrees of freedom. Also 100 and 500.

37. In the manufacture of cylindrical rods with a circular cross section which fit into a circular socket, it is desired to find a tolerance interval for the diameter. A sample of size 10 was taken with the results as follows:

5.036	5.031
5.085	5.064
4.991	4.942
4.935	5.051
4.999	5.011

Find a 99% tolerance interval for diameters with $\gamma=0.95$.

38. The following observations were taken on the net volume in liters of bottles.

5.68	5.65	5.59	5.64	5.66
5.61	5.62	5.64	5.63	5.61
5.64	5.66	5.60	5.65	5.63
5.67	5.64	5.60	5.60	5.65
5.60	5.65	5.60	5.63	5.60

Find a volume interval which will include 90% of the bottles with $\gamma=0.95$.

39. A sample of 40 observations was taken on time of ignition of a rocket propellant with the following results: $\overline{T}=0.250$ second and $s_T^2=0.0132$. Find an interval which will contain 95% of the times with $\gamma=0.99$.

40. In measuring the percent shrinkage on drying, 40 plastic clay test specimens produced the following results:

19.3	20.5	17.9	17.3
15.8	16.9	17.1	19.5
20.7	18.5	22.5	19.1
18.4	18.7	18.8	17.5
14.9	12.3	19.4	16.8
17.3	19.5	17.4	16.3
21.3	23.4	18.5	19.0
16.1	18.8	17.5	18.2
18.6	18.3	16.5	17.4
20.5	16.9	17.5	18.2

Find an interval which will contain 75% of the product with probability 0.95.

41. In the manufacture of forgings used as terminal blocks at the end of an airplane wing span, the critical dimension was the lower limit of the slot width. In particular, it was desired that a single lower limit be specified above which it can be assured with probability 0.95 that at least 99% of the slot widths will lie. A sample of 30 forgings was taken and the sample mean and sample standard deviations were found to be 0.8750 and 0.0015 inch respectively. Find the desired tolerance limit. How many observations would be required to find a non-parametric tolerance interval with these properties?

42. With the data in Problem 38, find a lower tolerance limit which 90% of the volume will exceed with probability 0.95. How many observations would be required for a non-parametric tolerance interval with these properties?

43. With the data of Problem 40, find an upper tolerance limit such that at least 95% of the product will lie below it with probability 0.99.

44. Suppose the largest observation in a sample of 250 observations is 21.7 and the smallest is 18.2. If we are interested in an interval which covers 99% of the product, what probability can we associate with the interval 18.2 to 21.7?

CHAPTER IX

Fitting Straight Lines

9.1 Introduction

Engineering problems often require the presentation of data showing the observed relationship between two variables. This chapter will be devoted to a consideration of some of the problems that arise in expressing such a relationship.

The analysis of pairs of data often requires that they be plotted. If the relationship between the measurements is very nearly exact, a plot by eye may be sufficient. For example, if experimental results showing the relationship between proportional limit and tensile

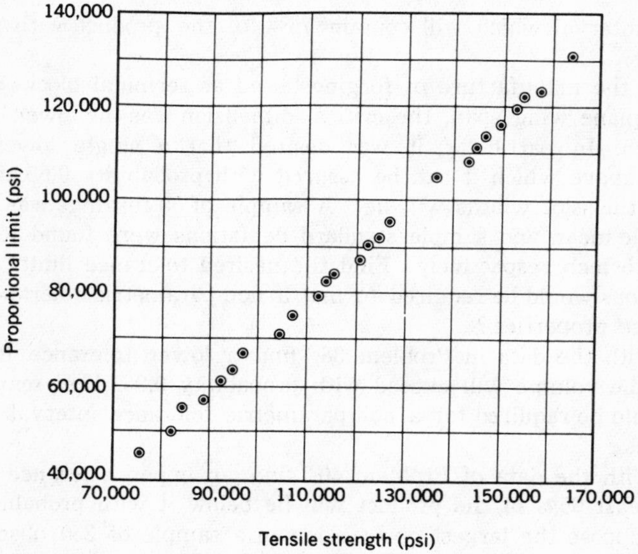

Fig. 9.1. Relationship between proportional limit and tensile strength.

strength of dental alloys are as shown in Figure 9.1, a statistical analysis is probably unnecessary.

Similarly, if experimental data calibrating a new method of determining calcium in the presence of large amounts of magnesium are as shown in Figure 9.2, the analysis is very straightforward.

Different people observing these data would probably draw the same graph, and furthermore, such plots would be representative of the true relationship between the variables. On the other hand, there are instances where the relationship between two variables is

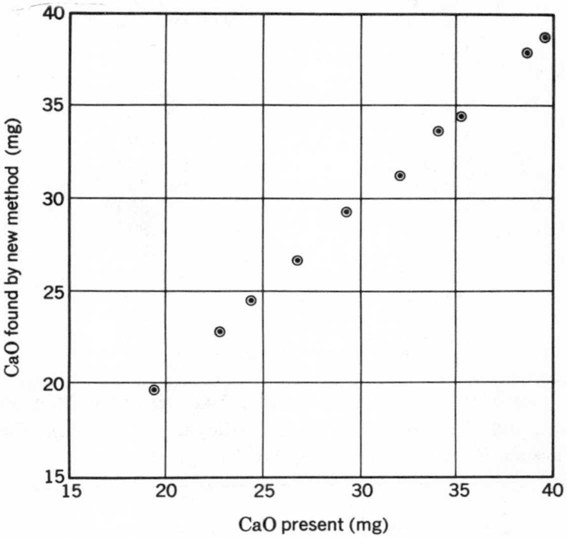

Fig. 9.2. Gravimetric determination of calcium in the presence of magnesium.

less certain, and plotting by eye results in graphs which depend upon the judgment of the people who are involved. Different people observing the *same* data will produce significantly different plots. For example, in a recent paper[1] a statistical analysis was actually carried out on certain mechanical properties of cast and wrought gold dental alloys. Several relationships were found between the mechanical properties; in particular, twenty-five pairs of observations were obtained on proportional limit and tensile strength. The data are given in Table 9.1, and are shown plotted in Figure 9.3.

As a further example, suppose the data obtained for calibration of a new method of determining calcium in the presence of large

[1] S. H. Bush, "A Statistical Analysis of the Mechanical Properties of Cast and Wrought Gold Dental Alloys," *American Society for Testing Materials Bulletin*, Oct. 1952, No. 185, pp. 46–50.

Table 9.1. Mechanical Properties of Dental Gold Alloys*

Tensile Strength, psi	*Proportional Limit, psi*
114,800	72,400
163,800	129,300
167,800	129,600
129,200	92,500
142,200	107,800
128,500	94,800
115,200	89,300
135,700	107,700
86,700	55,000
115,800	87,000
108,200	66,900
90,700	51,700
75,200	49,500
111,500	69,200
130,700	101,400
152,800	128,200
135,700	100,700
140,500	97,800
112,700	84,900
107,300	67,000
130,800	95,300
81,200	49,700
126,000	79,800
100,800	65,700
87,500	58,000

* Data are from S. H. Bush, " A Statistical Analysis of the Mechanical Properties of Cast and Wrought Gold Dental Alloys," *American Society for Testing Materials Bulletin,* Oct. 1952, No. 185, pp. 46–50.

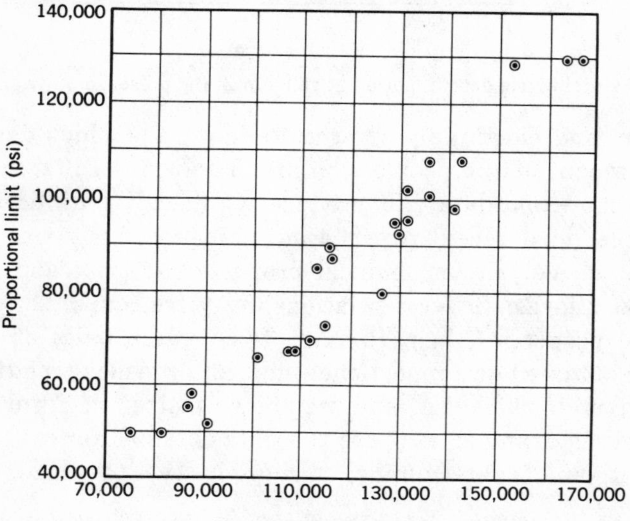

Fig. 9.3. Relationship between proportional limit and tensile strength. (Data are from S. H. Bush, *ASTM Bulletin,* Oct. 1952.)

amounts of magnesium are as shown in Table 9.2 and as plotted in Figure 9.4. Ten different samples containing known amounts of calcium oxide were analyzed by the new method.

Table 9.2. Gravimetric Determination of Calcium in the Presence of Magnesium

CaO Present, mg	CaO Found by New Method, mg
20.0	19.8
22.5	22.8
25.0	24.5
28.5	27.3
31.0	31.0
33.5	35.0
35.5	35.1
37.0	37.1
38.0	38.5
40.0	39.0

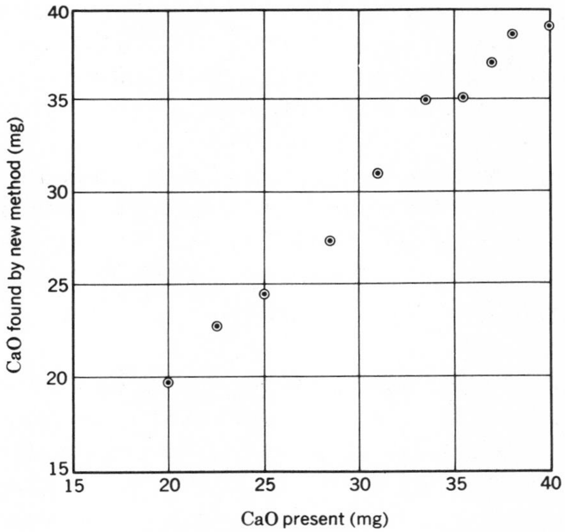

Fig. 9.4. Gravimetric determination of calcium in the presence of magnesium.

In these last two examples, it is evident that plotting by eye is inadequate and that a more systematic method of determining the relationship between two variables is needed. Such a method should not only result in determining a relationship but should also describe how the observed data can be used to predict results about future observations. The following sections will deal with these problems.

9.2 Types of Linear Relationships

There are two different ways in which pairs of measurements can occur: namely, when there is an underlying physical relationship and when there is a degree of association.

In the first case, a functional relationship between y and x is assumed. Observations are made on a random variable y, whereas x is some known constant associated with this random variable. An example of such an experiment is the effect of time of aging on the strength of cement. The time corresponds to the x variate, the values of which are predetermined in the experiment. For a given value of time, the yield (corresponding to the y variate) is the random variable. Most problems which involve time usually fall into the framework above. Time can frequently be measured to a sufficient degree of accuracy so that it can be assumed to be a known constant.

Other examples of a functional relationship are problems which involve calibration against a known standard. A series of observations may be taken by a laboratory on material whose contents are known accurately by design. The random variable y can be considered as the laboratory measurement, whereas the true composition may be regarded as the x variate. The example of calibrating a new method of determining CaO falls into this category. In each of these examples, and in the general situation, interest is centered on determining the average value of the random variable y as a function of the fixed value x. Naturally, x is never known exactly, but it is sufficient to have the error in x small compared to the variability of y.

The degree of association case deals with observations x and y, each of which represents measurements on random variables associated with different characteristics of the same item. Interest is centered on determining the relationship between the two variables for any of a number of reasons. Measurements on one variable, say x, may be relatively inexpensive compared with measurements of y, thereby resulting in a monetary saving if y can be predicted from a knowledge of x. For example, determining abrasion loss is difficult, whereas measuring hardness by means of a Rockwell hardness machine is relatively simple. There exists a degree of association between abrasion loss and hardness. The example of proportional limit and tensile strength falls into this category.

The relationship between two laboratories with respect to measurement on a certain material whose composition is not known accurately is another example where "degree of association" is important. Each laboratory measures the sample for the quantities of the unknown characteristics and the relationship between each laboratory's results is ascertained. Measurements of both laboratories may be regarded as random variables.

In each of these examples, and in this general situation, interest is centered on determining the average value of the random variable y as a function of a given value that the random variable x takes on. Although both of the cases presented have this property, the underlying assumptions are completely different. However, it turns out that the methods of analysis are identical. Hence, once the experimenter recognizes the distinction in models, the formal mechanics can be carried out without regard to the situation that exists.

9.3 Least Squares Estimates of the Slope and Intercept

9.3.1 Formulation of the Problem and Results

Both of the models presented in the previous section have the property that the average value of the random variable y can be expressed as a linear function of a known variate x, i.e.,

$$E(y) = A + Bx.$$

The values of A and B are usually unknown, and are to be estimated using experimental data. The estimates of A and B will be denoted by a and b, respectively. Thus, the estimated relationship will be of the form

$$\tilde{y} = a + bx.$$

Returning to the properties of gold dental alloys, the underlying (unknown) relationship between proportional limit and tensile strength may be expressed as:

$$\text{Expected value of proportional limit} = A + B \times (\text{tensile strength}).$$

A and B are estimated from the experimental data, and the estimated relationship is:

$$\text{Estimated proportional limit} = a + b \times (\text{tensile strength}).$$

In the calibration of CaO, the underlying (unknown) relationship between the amount of CaO present and the amount determined by the new method may be expressed as:

$$\text{Expected value of CaO determined by the new method} = A + B \times (\text{CaO present}).$$

The data will yield an estimated relation of the form:

Estimated value of CaO determined by the new method
$$= a + b \times (\text{CaO present}).$$

The proportional limit versus tensile strength example illustrates the degree of association model, whereas the new CaO method versus the known quantity illustrates the underlying physical relation model.

The usual method of estimating the intercept A and the slope B is the method of least squares. The values of a and b are determined so that the sum of the squares of the deviations of y about the fitted line $\tilde{y} = a + bx$ is a minumum; i. e., so that $(y_1 - \tilde{y}_1)^2 + (y_2 - \tilde{y}_2)^2 + \cdots + (y_n - \tilde{y}_n)^2$ is a minimum. If $(x_1, y_1), (x_2, y_2), \cdots, (x_n, y_n)$ are the n pairs of observations obtained experimentally,

$$b = \frac{(x_1 - \bar{x})(y_1 - \bar{y}) + (x_2 - \bar{x})(y_2 - \bar{y}) + \cdots + (x_n - \bar{x})(y_n - \bar{y})}{(x_1 - \bar{x})^2 + (x_2 - \bar{x})^2 + \cdots + (x_n - \bar{x})^2}$$

$$= \frac{\sum_{i=1}^{n} (x_i - \bar{x})(y_i - \bar{y})}{\sum_{i=1}^{n} (x_i - \bar{x})^2}$$

and $a = \bar{y} - b\bar{x}$ are the least squares estimates of B and A, respectively,

where $$\bar{y} = \frac{y_1 + y_2 + \cdots + y_n}{n} = \frac{\sum_{i=1}^{n} y_i}{n}$$

and $$\bar{x} = \frac{x_1 + x_2 + \cdots + x_n}{n} = \frac{\sum_{i=1}^{n} x_i}{n}.$$

A computational procedure for calculating a and b is given in Section 9.12.

a and b can also be expressed as linear combinations of the y's (the coefficients depending only on the x's which are assumed to be fixed constants). It is then simple to show that a and b are unbiased estimates of A and B, respectively, i.e., $E(a) = A$ and $E(b) = B$. Moreover, if the y's are independent (in a probability sense) and the variance of y is constant, say σ^2, for all x (for a discussion of this point, see Section 9.11), the variance of the estimated line $\tilde{y} = a + bx$ will be less than the variance of any other line whose parameters are expressed as linear combinations of the y's and which are unbiased estimates of A and B. The least squares estimate is therefore also called the best linear unbiased estimate.

*9.3.2 THEORY

The least squares estimates are easily obtained using elementary calculus. Let

$$Q = \sum_{i=1}^{n} (y_i - A - Bx_i)^2.$$

The principle of least squares requires the estimates of A and B to minimize Q. Thus, a and b are the values of A and B which make

$$\frac{\partial Q}{\partial A} = 0$$

and

$$\frac{\partial Q}{\partial B} = 0.$$

These partial derivatives are

$$\frac{\partial Q}{\partial A} = -2\sum_{i=1}^{n} (y_i - A - Bx_i);$$

$$\frac{\partial Q}{\partial B} = -2\sum_{i=1}^{n} x_i (y_i - A - Bx_i).$$

By setting the partial derivatives equal to zero, performing the indicated summations, and replacing the values of A and B by their estimates a and b, the equations

$$n\bar{y} - na - bn\bar{x} = 0$$

and

$$\sum x_i y_i - na\bar{x} - b\sum x_i^2 = 0$$

are obtained.

The first equation leads to

$$a = \bar{y} - b\bar{x},$$

and by substituting this value into the second equation, the desired result is obtained, i.e.,

$$b = \frac{\sum_{i=1}^{n} x_i y_i - n\bar{x}\bar{y}}{\sum_{i=1}^{n} x_i^2 - n\bar{x}^2} = \frac{\sum_{i=1}^{n} (x_i - \bar{x})(y_i - \bar{y})}{\sum_{i=1}^{n} (x_i - \bar{x})^2}.$$

The expressions for b and a as linear combinations of the y's are easily obtained. b can be written as

$$b = \frac{\sum_{i=1}^{n}(x_i - \bar{x})(y_i - \bar{y})}{\sum_{i=1}^{n}(x_i - \bar{x})^2} = \frac{\sum_{i=1}^{n}(x_i - \bar{x})y_i}{\sum_{i=1}^{n}(x_i - \bar{x})^2} - \frac{\bar{y}\sum_{i=1}^{n}(x_i - \bar{x})}{\sum_{i=1}^{n}(x_i - \bar{x})^2}$$

$$= \frac{\sum_{i=1}^{n}(x_i - \bar{x})y_i}{\sum_{i=1}^{n}(x_i - \bar{x})^2} \quad \text{since} \quad \bar{y}\sum_{i=1}^{n}(x_i - \bar{x}) = 0.$$

Let $\dfrac{x_i - \bar{x}}{\sum_{i=1}^{n}(x_i - \bar{x})^2} = c_i$; the c_i are known constants because the x_i are assumed to be known constants and not random variables. Hence,

$$b = \sum_{i=1}^{n} c_i y_i$$

is a linear combination of the y's. Similarly, a can be written as

$$a = \bar{y} - b\bar{x} = \frac{\sum_{i=1}^{n} y_i}{n} - \left[\sum_{i=1}^{n} c_i y_i\right]\bar{x} = \sum_{i=1}^{n}\left(\frac{1}{n} - c_i\bar{x}\right)y_i.$$

Let $(1/n) - c_i\bar{x} = d_i$; the d_i are then fixed constants also. Hence,

$$a = \sum_{i=1}^{n} d_i y_i$$

is a linear combination of the y's.

The property of unbiasedness of b and a can be verified as follows:

$$b = \sum_{i=1}^{n} c_i y_i$$

so that

$$E(b) = \sum_{i=1}^{n} c_i E(y_i) = \sum_{i=1}^{n} c_i(A + Bx_i) = A\sum_{i=1}^{n} c_i + B\sum_{i=1}^{n} c_i x_i$$

$$= 0 + \frac{B\sum_{i=1}^{n}(x_i - \bar{x})\,x_i}{\sum_{i=1}^{n}(x_i - \bar{x})^2} = \frac{B\sum_{i=1}^{n}(x_i - \bar{x})(x_i - \bar{x})}{\sum_{i=1}^{n}(x_i - \bar{x})^2} = B;$$

similarly,

$$a = \bar{y} - b\bar{x}$$

and $\quad E(a) = E(\bar{y} - b\bar{x}) = E(\bar{y}) - E(b\bar{x}) = A + B\bar{x} - B\bar{x} = A.$

9.4 Confidence Interval Estimates of the Slope and Intercept

9.4.1 Formulation of the Problem and Results

The method of least squares for obtaining estimates of the slope and intercept does not depend upon any assumptions about the distribution of y. Furthermore, it was pointed out in Section 9.3.1, that, under very weak conditions, these estimates are the "best linear unbiased point estimates" of the slope and intercept. The least squares estimates have other desirable properties provided the following conditions are satisfied:

(a) the expected value of y is equal to $A + Bx$;
(b) the variance of y is constant for all x; this constant, usually unknown, is denoted by σ^2;
(c) the distribution of y is normal;
(d) the sample is a random sample.

If the conditions above are fulfilled (as will be assumed in most of the remaining sections of this chapter) the estimate, b, of the slope and the estimate, a, of the intercept are normally distributed with means B and A, respectively. The variance of b is given by

$$\sigma_b^2 = \frac{\sigma^2}{\sum_{i=1}^{n} (x_i - \bar{x})^2}$$

and the variance of a is given by

$$\sigma_a^2 = \sigma^2 \left[\frac{1}{n} + \frac{\bar{x}^2}{\sum_{i=1}^{n} (x_i - \bar{x})^2} \right].$$

Therefore, in addition to having b and a as point estimates of B and A, respectively, confidence interval estimates can also be found. If σ^2 is known, b and its variance, σ_b^2, can be used to form a confidence interval for B, and a and its variance, σ_a^2, can be used to form a confidence interval for A; just as the sample mean $\bar{x}$ and its variance $\sigma_{\bar{x}}^2$ are used to form a confidence interval for the mean μ of a normal distribution when the variance σ^2 is known. However, since σ^2 is usually unknown, confidence limits must be found which do not depend upon σ^2. Such a confidence interval for B with confidence coefficient $1 - \alpha$ is given by

$$b \pm t_{\alpha/2;\,n-2} \frac{s_{y\,|\,x}}{\sqrt{\sum_{i=1}^{n} (x_i - \bar{x})^2}}$$

and for A is given by

$$a \pm t_{\alpha/2;\,n-2}\, s_{y\,|\,x} \sqrt{\frac{1}{n} + \frac{\bar{x}^2}{\sum_{i=1}^{n} (x_i - \bar{x})^2}}$$

where $t_{\alpha/2;\,n-2}$ is the 100 $\alpha/2$ percentage point of Student's t distribution with $n - 2$ degrees of freedom (see Appendix Table 3). $s_{y|x}$ is an estimate of the variability about the line and is given by

$$s_{y|x} = \sqrt{\frac{\sum_{i=1}^{n} (y_i - \tilde{y}_i)^2}{n - 2}}.$$

A computational form for $s_{y|x}$ is given in Section 9.12. It is evident that $\sum_{i=1}^{n} (y_i - \tilde{y}_i)^2$ is just the sum of squares of the deviations about the fitted line.

The confidence interval for B is a minimum whenever $\sum_{i=1}^{n} (x_i - \bar{x})^2$ is a maximum. Since the x's are assumed to be fixed, the choice of values of x's is often available. The value of $\sum_{i=1}^{n} (x_i - \bar{x})^2$ is maximized when the x observations are divided equally at the two extreme points of the range of the x's of interest. However, this should be done only when there is strong *a priori* knowledge that the relationship is linear since such a division of the points precludes picking up any non-linear effects. A method of testing for non-linearity is given in Section 9.10.

*9.4.2 Theory

The distributions of b and a are easily obtained. It has already been shown that the estimates of the slope and intercept are linear combinations of independent normally distributed random variables, i.e.,

$$b = \sum_{i=1}^{n} c_i y_i \quad \text{and} \quad a = \sum_{i=1}^{n} d_i y_i$$

where c_i and d_i are constants and the y's are independent normally distributed random variables. It then follows from the results on linear combinations of independent normally distributed random variables that b and a are also normally distributed. The expected value of b is B and the expected value of a is A. The variance of b (a linear combination of independent y's each of which has variance σ^2) is

$$\sigma_b^2 = \sum_{i=1}^{n} c_i^2 \sigma^2 = \frac{\sigma^2}{\sum_{i=1}^{n} (x_i - \bar{x})^2},$$

and the variance of a is

$$\sigma_a^2 = \sum_{i=1}^{n} d_i^2\sigma^2 = \sigma^2 \sum_{i=1}^{n} \left(\frac{1}{n} - c_i\bar{x}\right)^2 = \sigma^2 \sum_{i=1}^{n} \left[\frac{1}{n^2} - \frac{2c_i\bar{x}}{n} + c_i^2\bar{x}^2\right]$$

$$= \sigma^2\left[\frac{1}{n} - 0 + \frac{\bar{x}^2}{\sum_{i=1}^{n}(x_i - \bar{x})^2}\right] = \sigma^2\left[\frac{1}{n} + \frac{\bar{x}^2}{\sum_{i=1}^{n}(x_i - \bar{x})^2}\right].$$

It can be shown that (beyond the scope of this text)

$$\frac{(n-2)s_{y|x}^2}{\sigma^2}$$

has a chi-square distribution with $n - 2$ degrees of freedom, and is distributed independently of a and b. Hence, the expression for the confidence limits is easily derived from the basic definitions of the t distribution; i.e., the ratio of the two independent random variables, the numerator normal with mean zero and standard deviation one and the denominator the square root of a chi-square random variable divided by its degrees of freedom. Thus

$$\left[\frac{b - B}{\sigma \Big/ \sqrt{\sum_{i=1}^{n}(x_i - \bar{x})^2}}\right] \Bigg/ \sqrt{\frac{(n-2)s_{y|x}^2}{\sigma^2(n-2)}} = \frac{b - B}{s_{y|x} \Big/ \sqrt{\sum_{i=1}^{n}(x_i - \bar{x})^2}}$$

has a t distribution with $n - 2$ degrees of freedom.

Hence

$$P\left(-t_{\alpha/2;\,n-2} \leqq \frac{b - B}{s_{y|x} \Big/ \sqrt{\sum_{i=1}^{n}(x_i - \bar{x})^2}} \leqq t_{\alpha/2;\,n-2}\right) = 1 - \alpha$$

and

$$P\left(b - t_{\alpha/2;\,n-2}\frac{s_{y|x}}{\sqrt{\sum_{i=1}^{n}(x_i - \bar{x})^2}} \leqq B \leqq b + t_{\alpha/2;\,n-2}\frac{s_{y|x}}{\sqrt{\sum_{i=1}^{n}(x_i - \bar{x})^2}}\right) = 1 - \alpha$$

or

$$b \pm t_{\alpha/2;\,n-2}\frac{s_{y|x}}{\sqrt{\sum_{i=1}^{n}(x_i - \bar{x})^2}}$$

is the confidence interval for B with confidence coefficient $1 - \alpha$.

The confidence interval for A is found in a similar manner.

9.5 Point Estimates and Confidence Interval Estimates of the Average Value of y for a Given x

9.5.1 FORMULATION OF THE PROBLEM AND RESULTS

Very often estimates of linear relationships are desired in order to get point estimates or interval estimates of the average value of y

corresponding to a given x. For example, in the calcium oxide calibration, the experimenters, in presenting their results to the field, may be interested in estimating by a confidence interval, or a point estimated, the expected value of CaO present as determined by their new method for a fixed amount of CaO present. Another example deals with the relationship between average tensile strength of cement as a function of curing time, t; i.e., tensile strength is related to time by the function $Ce^{-B/t}$, or, in linear form, the expected value of $\ln(\text{strength}) = \ln(C) - B/t$, where $\ln(C)$ corresponds to A, $1/t$ corresponds to x, and B and C are constants. A cement manufacturer is interested in either a point estimate or an interval estimate of the expected tensile strength of his cement after a particular period of time. A point estimate for the expected value of y for a given value of x, say x^*, is given by

$$\tilde{y}^* = a + bx^*.$$

A confidence interval for the average value of y for a given value of x, say x^*, with confidence coefficient $1 - \alpha$ is given by

$$a + bx^* \pm t_{\alpha/2;\,n-2}s_{y\mid x}\sqrt{\frac{1}{n} + \frac{(x^* - \bar{x})^2}{\sum_{i=1}^{n}(x_i - \bar{x})^2}}$$

where values of $t_{\alpha/2;\,n-2}$, the $100\,\alpha/2$ percentage point of the t distribution with $n - 2$ degrees of freedom, can be found in Appendix Table 3. This interval becomes increasingly large as x^* becomes more distant from the value of x used in the experiment. This is intuitively sound since estimates are expected to become poorer when one extrapolates away from the range used in the original experiment. The interval is narrowest whenever $x^* = \bar{x}$.

*9.5.2 THEORY

The expression for the confidence interval of the average value of y for a given x, say x^*, is obtained from the distribution of the quantity $\tilde{y}^*$.

$$\tilde{y}^* = a + bx^* = \sum_{i=1}^{n} d_i y_i + \left(\sum_{i=1}^{n} c_i y_i\right)x^*$$

$$= \sum_{i=1}^{n} (d_i + c_i x^*)y_i.$$

Thus, $\tilde{y}^*$ is a linear combination of independent normally distributed random variables and, therefore, is normally distributed with mean

$$E(\tilde{y}^*) = E(a + bx^*) = A + Bx^*.$$

The variance of $\tilde{y}^*$ (a linear combination of independent y's each of which has variance σ^2) is

$$\begin{aligned}\sigma^2_{\tilde{y}^*} &= \sigma^2 \sum_{i=1}^{n} (d_i + c_i x^*)^2 \\ &= \sigma^2 \sum_{i=1}^{n} [d_i^2 + 2c_i d_i x^* + (c_i x^*)^2] \\ &= \sigma^2 \left[\frac{1}{n} + \frac{\bar{x}^2}{\sum_{i=1}^{n} (x_i - \bar{x})^2} - \frac{2x^*\bar{x}}{\sum_{i=1}^{n} (x_i - \bar{x})^2} + \frac{x^{*2}}{\sum_{i=1}^{n} (x_i - \bar{x})^2} \right] \\ &= \sigma^2 \left[\frac{1}{n} + \frac{(x^* - \bar{x})^2}{\sum_{i=1}^{n} (x_i - \bar{x})^2} \right].\end{aligned}$$

Hence, $\dfrac{\tilde{y}^* - A - Bx^*}{\sigma_{\tilde{y}^*}}$ is normally distributed with mean 0 and variance 1; $\dfrac{(n-2)s^2_{y|x}}{\sigma^2}$ is independently distributed as chi-square with $n - 2$ degrees of freedom; and

$$\frac{\tilde{y}^* - A - Bx^*}{s_{y|x} \sqrt{\dfrac{1}{n} + \dfrac{(x^* - \bar{x})^2}{\sum_{i=1}^{n} (x_i - \bar{x})^2}}}$$

has a t distribution with $n - 2$ degrees of freedom. It follows that

$$P\left(-t_{\alpha/2; n-2} \leq \frac{\tilde{y}^* - A - Bx^*}{s_{y|x} \sqrt{\dfrac{1}{n} + \dfrac{(x^* - \bar{x})^2}{\sum_{i=1}^{n} (x_i - \bar{x})^2}}} \leq t_{\alpha/2; n-2} \right) = 1 - \alpha$$

and the confidence interval for $A + Bx^*$ with confidence coefficient $1 - \alpha$ is given by

$$\tilde{y}^* \pm t_{\alpha/2; n-2} s_{y|x} \sqrt{\frac{1}{n} + \frac{(x^* - \bar{x})^2}{\sum_{i=1}^{n} (x_i - \bar{x})^2}}.$$

9.6 Point Estimates and Interval Estimates of the Independent Variable x Associated with an Observation on the Dependent Variable y

A point estimate or an interval estimate of the independent variable associated with an observation on the dependent variable y is often appropriate when there is an underlying physical relation. For example, in the determination of calcium oxide, the experimenter

Table 9.3. Summary of Point Estimates, Confidence Interval Estimates and Prediction Interval Estimates

Parameter	Symbol for Estimate	Computation Formula for Estimate	$100(1-\alpha)$ Percent Confidence Interval
B	b	$\frac{\sum_{i=1}^{n}(x_i-\bar{x})(y_i-\bar{y})}{\sum_{i=1}^{n}(x_i-\bar{x})^2}$	$b \pm t_{\alpha/2;\,n-2}\frac{s_{y\mid x}}{\sqrt{\sum_{i=1}^{n}(x_i-\bar{x})^2}}$
A	a	$\bar{y}-b\bar{x}$	$a \pm t_{\alpha/2;\,n-2}s_{y\mid x}\sqrt{\frac{1}{n}+\frac{\bar{x}^2}{\sum_{i=1}^{n}(x_i-\bar{x})^2}}$
Average value of y for a given value of $x = x^*$	y^*	$a+bx^*$	$a+bx^* \pm t_{\alpha/2;\,n-2}s_{y\mid x}\sqrt{\frac{1}{n}+\frac{(x^*-\bar{x})^2}{\sum_{i=1}^{n}(x_i-\bar{x})^2}}$
Value of x corresponding to an observed value of y' for the case where there is an underlying physical relationship	x'	$\frac{y'-a}{b}$	$\frac{y'-a}{b} \pm \frac{t_{\alpha/2;\,n-2}s_{y\mid x}}{b}\sqrt{1+\frac{1}{n}+\frac{\left(\frac{y'-a}{b}-\bar{x}\right)^2}{\sum_{i=1}^{n}(x_i-\bar{x})^2}}$ §

Parameter	Symbol for Estimate	Computation Formula for Estimate	$100(1-\alpha)$ Percent Prediction Interval
Future observation on y, y^0, corresponding to $x = x^0$	y^0	$a+bx^0$	$a+bx^0 \pm t_{\alpha/2;\,n-2}s_{y\mid x}\sqrt{1+\frac{1}{n}+\frac{(x^0-\bar{x})^2}{\sum_{i=1}^{n}(x_i-\bar{x})^2}}$

§ Approximate Form

ıay wish to estimate the true amount of CaO present, given a deterıination of this amount using the new method. A point estimate of for a given observation on y, say y', is given by

$$x' = \frac{y' - a}{b}.$$

A confidence interval estimate of x for a given observation in y, ay y', with confidence coefficient $1 - \alpha$ is given by

$$\bar{x} + \frac{b(y' - \bar{y})}{c} \pm t_{\alpha/2;\, n-2} \frac{s_{y\,|\,x}}{c} \sqrt{\left(1 + \frac{1}{n}\right)c + \frac{(y' - \bar{y})^2}{\sum_{i=1}^{n} (x_i - \bar{x})^2}}$$

'here $c = b^2 - t^2_{\alpha/2;\, n-2} \dfrac{s^2_{y\,|\,x}}{\sum_{i=1}^{n} (x_i - \bar{x})^2}$; values of $t_{\alpha/2;\,n-2}$ can be found in ppendix Table 3. An approximate confidence interval is given by he expression

$$\frac{y' - a}{b} \pm \frac{t_{\alpha/2;\, n-2} s_{y\,|\,x}}{b} \sqrt{1 + \frac{1}{n} + \frac{\left(\dfrac{y' - a}{b} - \bar{x}\right)^2}{\sum_{i=1}^{n} (x_i - \bar{x})^2}}.$$

'his approximation is good when $\dfrac{t^2_{\alpha/2;\, n-2} s^2_{y\,|\,x}}{b \sum_{i=1}^{n} (x_i - \bar{x})^2}$ is small. The proofs or the results above are somewhat complicated, and hence will be mitted.

A summary of point estimates and confidence interval estimates is iven in Table 9.3.

.7 Prediction Interval for a Future Observation on the Dependent Variable

9.7.1 FORMULATION OF THE PROBLEM AND RESULTS

Section 9.5 dealt with determining a confidence interval for the nean value of y corresponding to a given x, say x^*. The average 'alue of y is a fixed, but unknown, constant and a confidence interval s appropriate. It is often the case, however, that a confidence tatement about the mean value is unimportant, whereas a probaility statement about a future observation is relevant. For example, n the discussion about the relationship between tensile strength of ement and curing time, it was pointed out that the cement manuacturer is interested in the average tensile strength of his cement fter a particular period of time, i.e., a confidence statement, wheres a builder is interested in the tensile strength of his particular batch of cement to determine whether it will carry the required load. The builder will find a confidence statement inadequate. He requires assurance that after a specified period of time, say 28 days, the probability is $1 - \alpha$ that the tensile strength of his batch of cement will lie in a specified interval. In the CaO determination, the experimenter may be interested in making a probability statement about an interval containing a future determination by the new method for a fixed concentration of CaO rather than making a confidence statement about the average value of CaO determined by the new method for this given concentration.

Finally, in the determination of the mechanical properties of gold dental alloys, the particular dentist using the alloy may be interested in the average value of proportional limit for an alloy having a fixed tensile strength, whereas the patient is interested in the proportional limit of the gold in his mouth for a given tensile strength; i.e., the patient is interested in a prediction interval in which the proportional limit of the gold alloy in his mouth will lie.

The following statement can be made: The probability is $1 - \alpha$ that a future observation y^0 corresponding to x^0 will lie in the interval

$$a + bx^0 \pm t_{\alpha/2;\,n-2}s_{y\,|\,x}\sqrt{1 + \frac{1}{n} + \frac{(x^0 - \bar{x})^2}{\sum_{i=1}^{n}(x_i - \bar{x})^2}}$$

where values of $t_{\alpha/2;\,n-2}$ can be found in Appendix Table 3. Again this interval is narrowest when $x^0 = \bar{x}$, and increases as x^0 becomes more distant from $\bar{x}$.

*9.7.2 Theory

The expression for the prediction interval is obtained from the distribution of the quantity

$$y^0 - \tilde{y}^0.$$

y^0, being a (future) observation, is normally distributed with mean $A + Bx^0$ and variance σ^2. $\tilde{y}^0$, the point on the fitted line corresponding to $x = x^0$, has been shown to be normally distributed with mean $A + Bx^0$ and variance

$$\sigma^2\left[\frac{1}{n} + \frac{(x^0 - \bar{x})^2}{\sum_{i=1}^{n}(x_i - \bar{x})^2}\right] \quad \text{(see Section 9.5.2).}$$

Furthermore, y^0 and $\tilde{y}^0$ are independent in a probability sense since $\tilde{y}^0$ is derived from the first n observations and y^0 corresponds to a future observation. Hence, $y^0 - \tilde{y}^0$ is normally distributed with mean zero and variance

$$\sigma^2\left[1 + \frac{1}{n} + \frac{(x^0 - \bar{x})^2}{\sum_{i=1}^{n}(x_i - \bar{x})^2}\right].$$

Since $(n-2)s^2_{y|x}/\sigma^2$ has a chi-square distribution with $n-2$ degrees of freedom and is distributed independently of $y^0 - \tilde{y}^0$,

$$\frac{y^0 - \tilde{y}^0}{s_{y|x}\sqrt{1 + \frac{1}{n} + \frac{(x^0 - \bar{x})^2}{\sum_{i=1}^{n}(x_i - \bar{x})^2}}}$$

has a t distribution with $n - 2$ degrees of freedom. It then follows that

$$P\left(-t_{\alpha/2;\,n-2} \leq \frac{y^0 - \tilde{y}^0}{s_{y|x}\sqrt{1 + \frac{1}{n} + \frac{(x^0 - \bar{x})^2}{\sum_{i=1}^{n}(x_i - \bar{x})^2}}} \leq t_{\alpha/2;\,n-2}\right) = 1 - \alpha$$

or

$$P\left(a + bx^0 - t_{\alpha/2;\,n-2}s_{y|x}\sqrt{1 + \frac{1}{n} + \frac{(x^0 - \bar{x})^2}{\sum_{i=1}^{n}(x_i - \bar{x})^2}} \leq y^0 \leq a + bx^0 \right.$$

$$\left. + t_{\alpha/2;\,n-2}s_{y|x}\sqrt{1 + \frac{1}{n} + \frac{(x^0 - \bar{x})^2}{\sum_{i=1}^{n}(x_i - \bar{x})^2}}\right) = 1 - \alpha.$$

9.8 Tests of Hypotheses about the Slope and Intercept

It is sometimes useful to determine whether the slope or intercept is equal to some hypothesized value. For example, in the CaO determination, the experimenters may wish to determine whether their new method of determination was "perfect," i.e., $A = 0$, or $B = 1$, or both $A = 0$ and $B = 1$ simultaneously. The fact that experimental results do or do not lead to $a = 0$ and/or $b = 1$ may be attributable to experimental error. Hence, some objective procedures are needed.

The hypothesis that $A = A_0$ is rejected whenever

$$\left|\frac{a - A_0}{s_{y|x}\sqrt{\frac{1}{n} + \frac{\bar{x}^2}{\sum_{i=1}^{n}(x_i - \bar{x})^2}}}\right| \geq t_{\alpha/2;\,n-2}.$$

The hypothesis that $B = B_0$ is rejected whenever

Table 9.4. Significance Tests for Coefficients of Straight Lines

Hypothesis	Test Statistic	Criteria for Rejection	OC Curve	Notation
$A = A_0$	$t = \dfrac{a - A_0}{s_{y\mid x}\sqrt{\dfrac{1}{n} + \dfrac{\bar{x}^2}{\sum (x_i - \bar{x})^2}}}$	$\mid t \mid \geqq t_{\alpha/2;\, n-2}$	Compute $d = \dfrac{\mid A_0 - A_1 \mid}{\sigma\sqrt{\left(\dfrac{1}{n} + \dfrac{\bar{x}^2}{\sum (x_i - \bar{x})^2}\right)(n-1)}}$ and refer to Fig. 6.10 or 6.11, using the curve for $n - 1$.	Here σ is the standard deviation of y about the mean $A + Bx$, $s_{y\mid x}^2$ is the estimate of σ^2, and is given by $s_{y\mid x}^2 = \dfrac{1}{n-2}\sum [y_i - (a + bx_i)]^2$ $= \dfrac{1}{n-2}\left\{\sum (y_i - \bar{y})^2 - \dfrac{[\sum (x_i - \bar{x})(y_i - \bar{y})]^2}{\sum (x_i - \bar{x})^2}\right\}$; A_1 is the value of the intercept for which the probability of acceptance is desired.
$B = B_0$	$t = \dfrac{b - B_0}{s_{y\mid x}\sqrt{\dfrac{1}{\sum (x_i - \bar{x})^2}}}$	$\mid t \mid \geqq t_{\alpha/2;\, n-2}$	Compute $d = \dfrac{\mid B_0 - B_1 \mid}{\sigma\sqrt{\dfrac{n-1}{\sum (x_i - \bar{x})^2}}}$ and refer to Fig. 6.10 or 6.11, using the curve for $n - 1$.	Here B_1 is the value of the slope for which the probability of acceptance is desired.

$$\left| \frac{b - B_0}{s_{y|x}\sqrt{\dfrac{1}{\sum_{i=1}^{n}(x_i - \bar{x})^2}}} \right| \geq t_{\alpha/2;\, n-2}.$$

alues of $t_{\alpha/2;\, n-2}$, the 100 $\alpha/2$ percentage point of the t distribution ith $n - 2$ degrees of freedom, can be found in Appendix Table 3.

The hypothesis that $A = A_0$ and $B = B_0$ simultaneously is rejected henever

$$= [n(a - A_0)^2 + 2n\bar{x}(a - A_0)(b - B_0) + \sum_{i=1}^{n} x_i^2(b - B_0)^2]/2s_{y|x}^2 \geq F_{\alpha;\, 2,\, n-2}$$

here $F_{\alpha;\, 2, n-2}$ is the 100 α percentage point of the F distribution with and $n - 2$ degrees of freedom and can be found in Appendix Table . Whenever one of the hypotheses above is true, the probability (1 − α) that the decision rule will not lead to rejection; i.e., the robability of rejection when the hypothesis is true (the Type I error) α.*

The results about testing the hypotheses that $A = A_0$ or $B = B_0$ ollow from Section 9.4.2 where it was shown that

$$\frac{a - A_0}{s_{y|x}\sqrt{\dfrac{1}{n} + \dfrac{\bar{x}^2}{\sum_{i=1}^{n}(x_i - \bar{x})^2}}}$$

nd

$$\frac{b - B_0}{s_{y|x}\sqrt{\dfrac{1}{\sum_{i=1}^{n}(x_i - \bar{x})^2}}}$$

ave the t distribution with $n - 2$ degrees of freedom.

The OC curves for these procedures are summarized in Table 9.4.

.9 Estimation of the Slope B when A Is Known To Be Zero

Situations frequently arise in which it is possible to assume be- orehand that the intercept A is zero. The expected value of y is iven by

* To get a joint simultaneous confidence interval estimate of A and B with confidence oefficient $1 - \alpha$, the sample values, a, b, and $s_{y|x}^2$, are substituted into the expression or F. The confidence region is defined by $F \leq F_{\alpha;\, 2, n-2}$. To determine whether a iven pair (A_0, B_0) is included in the confidence region, one substitutes these given alues of A_0 and B_0 into F; then, if $F < F_{\alpha;\, 2, n-2}$, the pair (A_0, B_0) is included in the onfidence region.

$$E(y) = Bx.$$

The least squares estimate for B is found by setting the derivative of $\sum_{i=1}^{n} (y_i - Bx_i)^2$ with respect to B equal to zero. The value of B satisfying the equation is given by

$$b = \frac{\sum_{i=1}^{n} x_i y_i}{\sum_{i=1}^{n} x_i^2}.$$

This is similar to the previous result except that the values of x and y are used in the summation rather than the deviations about their mean. For this case, it is very clear that b is normally distributed with mean B and variance $\frac{\sigma^2}{\sum_{i=1}^{n} x_i^2}$ (by noting that the x's are assumed to be fixed constants).

The estimated line is given by

$$\tilde{y} = bx.$$

A confidence interval for B with confidence coefficient $1 - \alpha$ is given by

$$b \pm \frac{t_{\alpha/2; n-1} s_{y|x}}{\sqrt{\sum_{i=1}^{n} x_i^2}}$$

where the value of $t_{\alpha/2; n-1}$, the 100 $\alpha/2$ percentage point of the t distribution with $n - 1$ degrees of freedom, can be found in Appendix Table 3. $s_{y|x}$ is an estimate of the variability about the line and is given by

$$s_{y|x} = \sqrt{\frac{\sum_{i=1}^{n} (y_i - \tilde{y}_i)^2}{n-1}} = \sqrt{\frac{\sum_{i=1}^{n} y_i^2 - \left[\left(\sum_{i=1}^{n} x_i y_i\right)^2 \Big/ \sum_{i=1}^{n} x_i^2\right]}{n-1}}.$$

A point estimate of the expected value of y for a given value of x, say x^*, is given by

$$\tilde{y}^* = bx^*.$$

A confidence interval estimate for the expected value of y for a given value of x, say x^*, with confidence coefficient $1 - \alpha$, is given by

$$bx^* \pm t_{\alpha/2; n-1} s_{y|x} \sqrt{\frac{x^{*2}}{\sum_{i=1}^{n} x_i^2}}$$

where the values of $t_{\alpha/2;\,n-1}$ and $s_{y\,|\,x}$ can be found as described above.

A point estimate of the value of x corresponding to a given observation on y, say y', is given by

$$x' = \frac{y'}{b}.$$

A confidence interval estimate for the value of x corresponding to a given observation on y, say y', with confidence coefficient $1 - \alpha$, is given approximately by

$$\frac{y'}{b} \pm \frac{t_{\alpha/2;\,n-1} s_{y\,|\,x}}{b} \sqrt{1 + \frac{(y'/b)^2}{\sum_{i=1}^{n} x_i^2}}$$

where $t_{\alpha/2;\,n-1}$ and $s_{y\,|\,x}$ are as described above.

A prediction interval, having the property that the probability is $1 - \alpha$ that a future observation y^0 corresponding to x^0 will lie in the interval, is given by

$$bx^0 \pm t_{\alpha/2;\,n-1} s_{y\,|\,x} \sqrt{1 + \frac{(x^0)^2}{\sum_{i=1}^{n} x_i^2}}.$$

Finally, the hypothesis that $B = B_0$ is rejected whenever

$$\left| \frac{b - B_0}{\sqrt{s_{y\,|\,x}^2 \Big/ \sum_{i=1}^{n} x_i^2}} \right| \geq t_{\alpha/2;\,n-1}.$$

The probability of rejection when the hypothesis is true, i.e., the Type I error, is α. The derivations of the mathematical results of the statements above are similar to those presented for the case where the intercept is not assumed to be zero.

9.10 Ascertaining Linearity

In all of the previous sections, it was assumed that the curve to be estimated was linear in x. If the experiment is performed so that there are k values of y for each of the x's, a test for linearity can be made.[1] Let $\bar{y}_i$ be the mean of the k observations on the y_i and $y_{i\nu}$ be the νth of the k observations of y corresponding to x_i. Thus, if there are four y's corresponding to x_3, the values of y_i are y_{31}, y_{32}, y_{33}, y_{34}, and $\bar{y}_3$ is the average of these four observations.

The least squares estimate of the line can be obtained in the usual fashion using the nk individual observations, or the equivalent result may be obtained from the relations

[1] Actually, all that is necessary is that there be more than one value of y for some x.

$$b = \frac{\sum_{i=1}^{n} (x_i - \bar{x})(\bar{y}_i - \bar{\bar{y}})}{\sum_{i=1}^{n} (x_i - \bar{x})^2}$$

and

$$a = \bar{\bar{y}} - b\bar{x}$$

where

$$\bar{\bar{y}} = \frac{\sum_{i=1}^{n} \bar{y}_i}{n} = \frac{\sum_{i=1}^{n} \sum_{\nu=1}^{k} y_{i\nu}}{nk}.$$

The hypothesis of linearity is rejected if

$$F = \frac{k \sum_{i=1}^{n} (\bar{y}_i - \tilde{y}_i)^2/(n-2)}{\sum_{i=1}^{n} \sum_{\nu=1}^{k} (y_{i\nu} - \bar{y}_i)^2/n(k-1)} \geqq F_{\alpha;\, n-2,\, n(k-1)}$$

where values of $F_{\alpha;\, n-2,\, n(k-1)}$, the 100 α percentage point of the F distribution with $n - 2$ and $n(k - 1)$ degrees of freedom are given in Appendix Table 4.

This test is reasonable in that it compares the variability about the fitted line (numerator) with the inherent variability of the y's which is independent of the form of the relationship (denominator). If linearity exists, these variabilities should compare favorably.

Furthermore, if linearity exists, these two variances can be combined to give the estimate

$$s_{y|x}^2 = \frac{k \sum_{i=1}^{n} (\bar{y}_i - \tilde{y}_i)^2 + \sum_{i=1}^{n} \sum_{\nu=1}^{k} (y_{i\nu} - \bar{y}_i)^2}{nk - 2} = \frac{\sum_{i=1}^{n} \sum_{\nu=1}^{k} (y_{i\nu} - \tilde{y}_i)^2}{nk - 2}.$$

Note that this is the same result for $s_{y|x}^2$ that would have been obtained if the least squares relationship had been estimated from the nk individual observations.

The numerator when multiplied by $(n - 2)/\sigma^2$ has a chi-square distribution with $(n - 2)$ degrees of freedom. The denominator is just a weighted average of the *estimates* of σ^2 associated with each x_i (the same value of σ^2 is assumed associated with each x_i) and, when multiplied by $\frac{n(k-1)}{\sigma^2}$, the denominator has a chi-square distribution with $n(k - 1)$ degrees of freedom. Furthermore, the numerator and denominator are independent (in a probability sense). Therefore when these quantities are divided by their respective degrees of freedom and their ratio formed, the resultant expression is just F

ıereby enabling one to conclude that this quantity has an F distri-
ution with $n - 2$ and $n(k - 1)$ degrees of freedom.

The test depends upon having more than one y for some x. If ıere is just one y for each x, a supplementary investigation should e carried out by studying the quantities $(y_i - \tilde{y}_i)/s_{y|x}$. These values re approximately normally distributed with mean 0 and variance 1 or large n provided the relationship is linear with constant varia-ility.

.11 Transformation to a Straight Line

If there is *a priori* knowledge that the relationship is non-linear, r if this has been determined by a test similar to the one in the revious section, it is often possible to transform the data to a linear orm. For example, such a transformation was discussed previously. he relationship between tensile strength of cement and time was iven by

$$y = Ce^{-B/t}.$$

Taking logarithms of both sides, we obtain a linear relation of the orm

$$\text{average value of } \ln(y) = \ln(C) - B/t,$$

here $\ln(C)$ corresponds to A and $1/t$ to x. Generally, then, it is ften possible by simple transformations of the variables to represent he relationship as a straight line in the transformed variates. The ransformations are usually chosen on the basis of a graphical anal-sis of the observations, using all the *a priori* knowledge available bout the theoretical underlying relation.

In some instances, y may have a linear relationship with some unction of x, and no transformation is required. For example, the verage value of y may be written as $B \cos z$ where $\cos z$ corresponds o x and $A = 0$. In any case, the assumptions underlying the use of he previously described techniques must be satisfied for y or the unction of y, whichever is the appropriate variate.

It sometimes occurs that although the underlying relationship is inear, the variability is not constant for all x. Variance stabilizing ransformations are often applied in this case. If the variability in tends to increase linearly with x, plotting the square roots (or igher roots) of the data often stabilizes the variability. If the ariability increases at a higher rate, a logarithm transformation is ometimes appropriate. If the variability tends to increase and then ecrease, an inverse sine transformation often is fruitful. It must

Table 9.5. Worksheet for the Fitting of Straight Lines

General Data:

x represents ________:

$\sum x_i =$ ______; $\quad \bar{x} = \sum x_i/n =$ ______; $\quad (\sum x_i)\bar{x} =$ ______;

$\sum x_i^2 =$ ______;

$\sum x_i^2 - (\sum x_i)\bar{x} = \sum(x_i - \bar{x})^2 =$ ______.

y represents ________:

$\sum y_i =$ ______; $\quad \bar{y} = \sum y_i/n =$ ______; $\quad (\sum y_i)\bar{y} =$ ______;

$\sum y_i^2 =$ ______; $\quad \sum y_i^2 - (\sum y_i)\bar{y} = \sum(y_i - \bar{y})^2 =$ ______.

$n =$ ______; $\quad 1/n =$ ______; $\quad (\sum x_i)(\sum y_i)/n = \bar{x} \sum y_i =$ ______;

$\sum x_i y_i =$ ______;

$\sum x_i y_i - (\sum x_i)(\sum y_i)/n = \sum (x_i - \bar{x})(y_i - \bar{y}) =$ ______;

$[\sum (x_i - \bar{x})(y_i - \bar{y})]^2 =$ ______;

$[\sum(x_i - \bar{x})(y_i - \bar{y})]^2$

Data for Equation of Line:

$$b = \frac{\sum(x_i - \bar{x})(y_i - \bar{y})}{\sum(x_i - \bar{x})^2} = \_\_\_\_\_\_; \quad a = \bar{y} - b\bar{x} = \_\_\_\_\_\_;$$

equation of line $= \tilde{y} = a + bx =$ ______ $+$ ______ x.

Estimation of Standard Deviation:

$$(n-2)\, s_{y|x}^2 = \sum(y_i - \bar{y})^2 - \frac{[\sum(x_i - \bar{x})(y_i - \bar{y})]^2}{\sum(x_i - \bar{x})^2} = \_\_\_\_\_\_;$$

$s_{y|x}^2 =$ ______; $\quad s_{y|x} =$ ______.

Data for Significance Tests and for Confidence Intervals for A and B:

$$\frac{1}{\sum(x_i - \bar{x})^2} = \_\_\_\_\_\_; \quad \sqrt{\frac{1}{\sum(x_i - \bar{x})^2}} = \_\_\_\_\_\_;$$

$$s_{y|x}\sqrt{\frac{1}{\sum(x_i - \bar{x})^2}} = \_\_\_\_\_\_; \quad \bar{x}^2 = \_\_\_\_\_\_;$$

$$\frac{\bar{x}^2}{\sum(x_i - \bar{x})^2} = \_\_\_\_\_\_; \quad \frac{1}{n} + \frac{\bar{x}^2}{\sum(x_i - \bar{x})^2} = \_\_\_\_\_\_;$$

$$\sqrt{\frac{1}{n} + \frac{\bar{x}^2}{\sum(x_i - \bar{x})^2}} = \_\_\_\_\_\_;$$

$$\sqrt{\frac{1}{} \quad \frac{\bar{x}^2}{}}$$

DATA FOR CONFIDENCE INTERVALS FOR AVERAGE VALUE OF y CORRESPONDING TO $x = x^*$:

$x^* = \_\_\_\_$; $(x^* - \bar{x})^2 = \_\_\_\_$;

$$\frac{(x^* - \bar{x})^2}{\sum (x_i - \bar{x})^2} = \_\_\_\_; \quad \frac{1}{n} + \frac{(x^* - \bar{x})^2}{\sum (x_i - \bar{x})^2} = \_\_\_\_;$$

$$\sqrt{\frac{1}{n} + \frac{(x^* - \bar{x})^2}{\sum (x_i - \bar{x})^2}} = \_\_\_\_;$$

$$s_{y|x}\sqrt{\frac{1}{n} + \frac{(x^* - \bar{x})^2}{\sum (x_i - \bar{x})^2}} = \_\_\_\_.$$

DATA FOR PREDICTION INTERVALS FOR A FUTURE VALUE OF y CORRESPONDING TO $x = x^0$:

$$x^0 = \_\_\_\_; \quad (x^0 - \bar{x})^2 = \_\_\_\_; \quad \frac{(x^0 - \bar{x})^2}{\sum (x_i - \bar{x})^2} = \_\_\_\_;$$

$$1 + \frac{1}{n} + \frac{(x^0 - \bar{x})^2}{\sum (x_i - \bar{x})^2} = \_\_\_\_;$$

$$\sqrt{1 + \frac{1}{n} + \frac{(x^0 - \bar{x})^2}{\sum (x_i - \bar{x})^2}} = \_\_\_\_;$$

$$s_{y|x}\sqrt{1 + \frac{1}{n} + \frac{(x^0 - \bar{x})^2}{\sum (x_i - \bar{x})^2}} = \_\_\_\_.$$

DATA FOR CONFIDENCE INTERVALS FOR x CORRESPONDING TO AN OBSERVED VALUE OF $y = y'$ FOR THE CASE WHEN THERE IS AN UNDERLYING PHYSICAL RELATIONSHIP:

$$y' = \_\_\_\_; \quad x' = \frac{y' - a}{b} = \_\_\_\_; \quad \left(\frac{y' - a}{b} - \bar{x}\right)^2 = \_\_\_\_;$$

$$\frac{\left(\frac{y' - a}{b} - \bar{x}\right)^2}{\sum (x_i - \bar{x})^2} = \_\_\_\_;$$

$$1 + \frac{1}{n} + \frac{\left(\frac{y' - a}{b} - \bar{x}\right)^2}{\sum (x_i - \bar{x})^2} = \_\_\_\_;$$

$$\sqrt{1 + \frac{1}{n} + \frac{\left(\frac{y' - a}{b} - \bar{x}\right)^2}{\sum (x_i - \bar{x})^2}} = \_\_\_\_;$$

$$s_{y|x}\sqrt{1 + \frac{1}{n} + \frac{\left(\frac{y' - a}{b} - \bar{x}\right)^2}{\sum (x_i - \bar{x})^2}} = \_\_\_\_.$$

be emphasized that a constant variance is not the only conditio sought, and precautions are still necessary when using the techniqu described previously with the transformed variables. Fortunatel however, the transformation of scale to meet the condition of co stant variability often has the effect of improving the closeness of th distribution to normality; a correlation of variability of y with x i the original scale often supplies excessive skewness which tends t be eliminated after the transformation. But the validity of any a sumption of normality should be watched, for while moderat departures from normality are known not to be serious, any larg departures in the region of more outlying observations are likely t affect the validity of the probability statements made.

9.12 Work Sheets for Fitting Straight Lines

A worksheet for fitting straight lines is given in Table 9.5. Th worksheet has been arranged for ease of computation when using hand calculator. Completion of this table will result in obtaining a the quantities described in the previous sections.

9.13 Illustrative Examples

Example 1. The data on proportional limit and tensile strengt of gold dental alloys given in Table 9.1, Section 9.1, will be analyze The following quantities will be obtained:

1. least squares estimates of A and B;
2. 95% confidence interval estimates of A and B;
3. a point estimate of the average value of the proportional lim (y^*) corresponding to a tensile strength of 129,000 psi (x^*);
4. a 90% confidence interval estimate of the average value of th proportional limit corresponding to a tensile strength of 129,0(psi (x^*);
5. a 95% prediction interval for the proportional limit correspon ing to a tensile strength of 129,000 psi (x^0);
6. a test of the hypothesis that $B = 1.5$ such that the probabilit of accepting the hypothesis when it is true is 0.95.

From the data given on the worksheet for Example 1, we obtain:

1. The least squares estimates of A and B are computed direct on the worksheet; the values of $a = -30{,}793.792$ and $b =$ 0.96982743 are obtained.
2. To compute the confidence interval estimates of A and B, th value of $t_{\alpha/2;\,n-2} = t_{0.025;\,23} = 2.0687$ is obtained from Appendi

Table 3. Substitute into the formulas given in Section 9.4; the confidence interval for A is

$$\begin{aligned} & -30{,}793.792 \pm (2.0687)(6{,}242.8268) \\ &= -30{,}793.792 \pm 12{,}914.536 \\ &= [-43{,}708.328, \quad -17{,}879.256]. \end{aligned}$$

Similarly, the confidence interval estimate for B is

$$\begin{aligned} & 0.96982743 \pm (2.0687)(0.051148740) \\ &= 0.96982743 \pm 0.10581140 \\ &= [0.864, \quad 1.076]. \end{aligned}$$

3. A point estimate of the average value of the proportional limit corresponding to a tensile strength of $x^* = 129{,}000$ is given by

$$\begin{aligned} y^* = a + bx^* &= -30{,}793.792 + (0.96982743)(129{,}000) \\ &= 94{,}313.946. \end{aligned}$$

4. To compute the confidence interval estimate of the average value of the proportional limit associated with a given tensile strength of $x^* = 129{,}000$, the value of $t_{\alpha/2;\,n-2} = t_{0.05;\,23} = 1.7139$ is obtained from Appendix Table 3. One substitutes into the formula given in Section 9.5; the required confidence interval is given by

$$\begin{aligned} & -30{,}793.792 + (0.96982743)(129{,}000) \pm (1.7139)(1{,}321.5532) \\ &= 94{,}313.946 \pm 2{,}265.0100 \\ &= [92{,}048.936, \quad 96{,}578.956]. \end{aligned}$$

5. To compute the prediction interval for the proportional limit corresponding to a tensile strength of $x^0 = 129{,}000$, a value of $t_{\alpha/2;\,n-2} = t_{0.025;\,23} = 2.0687$ is obtained from Appendix Table 3. Then substituting into the formula given in Section 9.7, one obtains for the prediction interval

$$\begin{aligned} & -30{,}793.792 - (0.96982743)(129{,}000) \pm (2.0687)(6{,}300.2909) \\ &= 94{,}313.946 \pm 13{,}033.412 \\ &= [81{,}280.534, \quad 107{,}347.358]. \end{aligned}$$

6. The hypothesis that $B = 1.5$ is rejected at the 95% level if

$$\left| \frac{b - 1.5}{s_{y|x}\sqrt{\dfrac{1}{\sum (x_i - \bar{x})^2}}} \right| = \left| \frac{0.96982743 - 1.5}{0.051148740} \right| = \left| -10.3653 \right|$$

exceeds $t_{\alpha/2;\,n-2} = t_{0.025;\,23} = 2.0687$. Thus

$$|-10.3653| = 10.3653 > 2.0687$$

and therefore the hypothesis that $B = 1.5$ is rejected.

Worksheet for the Fitting of Straight Lines for Example 1

GENERAL DATA:

x represents tensile strength:

$\sum x_i = 2,991,300; \quad \bar{x} = 119,652; \quad (\sum x_i)\,\bar{x} = 357,915,027,600;$

$\sum x_i^2 = 372,419,750,000;$

$\sum x_i^2 - (\sum x_i)\bar{x} = \sum (x_i - \bar{x})^2 = 14,504,722,400.$

y represents proportional limit:

$\sum y_i = 2,131,200; \quad \bar{y} = 85,248; \quad (\sum y_i)\bar{y} = 181,680,537,600;$

$\sum y_i^2 = 196,195,960,000; \; \sum y_i^2 - (\sum y_i)\bar{y} = \sum (y_i - \bar{y})^2 = 14,515,422,400.$

$n = 25; \quad 1/n = 0.04; \quad (\sum x_i)(\sum y_i)/n = \bar{x}\sum y_i = 255,002,342,400;$

$\sum x_i y_i = 269,069,420,000;$

$\sum x_i y_i - (\sum x_i)(\sum y_i)/n = \sum (x_i - \bar{x})(y_i - \bar{y}) = 14,067,077,600;$

$[\sum (x_i - \bar{x})(y_i - \bar{y})]^2 = (1,978,826,722)10^{11};$

$[\sum (x_i - \bar{x})(y_i - \bar{y})]^2$ / [illegible] = [illegible],642,637,670

DATA FOR EQUATION OF LINE:

$b = \dfrac{\sum (x_i - \bar{x})(y_i - \bar{y})}{\sum (x_i - \bar{x})^2} = 0.96982743; \quad a = \bar{y} - b\,\bar{x} = -\,30,793.792;$

equation of line $= \tilde{y} = a + bx = -\,30,793.792 + 0.96982743\,x.$

ESTIMATION OF STANDARD DEVIATION:

$$(n-2)s_{y|x}^2 = \sum (y_i - \bar{y})^2 - \frac{[\sum (x_i - \bar{x})(y_i - \bar{y})]^2}{\sum (x_i - \bar{x})^2} = 872,784,730;$$

$s_{y|x}^2 = 37,947,162.17; \quad s_{y|x} = 6,160.1268.$

DATA FOR SIGNIFICANCE TESTS AND FOR CONFIDENCE INTERVALS FOR A AND B:

$$\frac{1}{\sum (x_i - \bar{x})^2} = (0.68943064)10^{-10}; \quad \sqrt{\frac{1}{\sum (x_i - \bar{x})^2}} = (0.83031960)10^{-5};$$

$$s_{y|x}\sqrt{\frac{1}{\sum (x_i - \bar{x})^2}} = 0.051148740; \quad \bar{x}^2 = 14,316,601,104;$$

$$\frac{\bar{x}^2}{\sum (x_i - \bar{x})^2} = 0.98703034; \quad \frac{1}{n} + \frac{\bar{x}^2}{\sum (x_i - \bar{x})^2} = 1.02703034;$$

$$\sqrt{\frac{1}{n} + \frac{\bar{x}^2}{\sum (x_i - \bar{x})^2}} = 1.01342505;$$

$$s_{y|x}\sqrt{\frac{1}{n} + \frac{\bar{x}^2}{\sum (x_i - \bar{x})^2}} = 6,242.8268.$$

DATA FOR CONFIDENCE INTERVALS FOR AVERAGE VALUE OF y CORRESPONDING TO $x = x^*$:

$$x^* = 129{,}000; \quad (x^* - \bar{x})^2 = 87{,}385{,}104;$$

$$\frac{(x^* - \bar{x})^2}{\sum (x_i - \bar{x})^2} = 0.0060245968; \quad \frac{1}{n} + \frac{(x^* - \bar{x})^2}{\sum (x_i - \bar{x})^2} = 0.046024597;$$

$$\sqrt{\frac{1}{n} + \frac{(x^* - \bar{x})^2}{\sum (x_i - \bar{x})^2}} = 0.21453344;$$

$$s_{y|x}\sqrt{\frac{1}{n} + \frac{(x^* - \bar{x})^2}{\sum (x_i - \bar{x})^2}} = 1{,}321.5532.$$

DATA FOR PREDICTION INTERVALS FOR A FUTURE VALUE OF y CORRESPONDING TO $x = x^0$:

$$x^0 = 129{,}000; \quad (x^0 - \bar{x})^2 = 87{,}385{,}104; \quad \frac{(x^0 - \bar{x})^2}{\sum (x_i - \bar{x})^2} = 0.0060245968;$$

$$1 + \frac{1}{n} + \frac{(x^0 - \bar{x})^2}{\sum (x_i - \bar{x})^2} = 1.04602460;$$

$$\sqrt{1 + \frac{1}{n} + \frac{(x^0 - \bar{x})^2}{\sum (x_i - \bar{x})^2}} = 1.02275344;$$

$$s_{y|x}\sqrt{1 + \frac{1}{n} + \frac{(x^0 - \bar{x})^2}{\sum (x_i - \bar{x})^2}} = 6{,}300.2909.$$

DATA FOR CONFIDENCE INTERVALS FOR x CORRESPONDING TO AN OBSERVED VALUE OF $y = y'$ FOR THE CASE WHEN THERE IS AN UNDERLYING PHYSICAL RELATIONSHIP:

$$y' = \underline{\qquad}; \quad x' = \frac{y' - a}{b} = \underline{\qquad}; \quad \left(\frac{y' - a}{b} - \bar{x}\right)^2 = \underline{\qquad};$$

$$\frac{\left(\frac{y' - a}{b} - \bar{x}\right)^2}{\sum (x_i - \bar{x})^2} = \underline{\qquad};$$

$$1 + \frac{1}{n} + \frac{\left(\frac{y' - a}{b} - \bar{x}\right)^2}{\sum (x_i - \bar{x})^2} = \underline{\qquad};$$

$$\sqrt{1 + \frac{1}{n} + \frac{\left(\frac{y' - a}{b} - \bar{x}\right)^2}{\sum (x_i - \bar{x})^2}} = \underline{\qquad};$$

$$s_{y|x}\sqrt{1 + \frac{1}{n} + \frac{\left(\frac{y' - a}{b} - \bar{x}\right)^2}{\sum (x_i - \bar{x})^2}} = \underline{\qquad}.$$

Example 2. The data on calibrating the new method of determining CaO given in Table 9.2 will be analyzed. The following quantities will be obtained:

1. least squares estimates of A and B;
2. 90% confidence interval estimates of A and B;
3. a point estimate of the average value of the amount of CaO present obtained by the new method (y^*) when there is actually 30 mg present (x^*);
4. a 99% confidence interval estimate of the average value of the amount of CaO present obtained by the new method when there is actually 30 mg present (x^*);
5. a 95% confidence interval estimate of the true amount of CaO present when the new method gives a quantity equal to 29.6 mg (y');
6. a 95% prediction interval for the amount of CaO present determined by the new method when there is actually 30 mg present (x^0);
7. a test of the hypothesis that $B = 1$ such that the probability of accepting the hypothesis when it is true is 0.90;
8. a joint test of the hypothesis that $A = 0$ and $B = 1$ such that the probability of accepting the hypothesis when it is true is 0.90.

From the data given on the worksheet for Example 2, we obtain:

1. The least squares estimates of A and B are computed directly on the worksheet; the values of $a = -0.29277822$ and $b = 1.0065202$ are obtained.
2. To compute the confidence interval estimates of A and B, the value $t_{\alpha/2;\,n-2} = t_{0.05;\,8} = 1.8595$ is obtained from Appendix Table 3. Substitute into the formulas given in Section 9.4; the confidence interval for A is

$$
\begin{aligned}
& -0.29277822 \pm 1.8595\,(1.2611638) \\
&= -0.29277822 \pm 2.3451341 \\
&= [-2.6379123, \quad 2.0523559].
\end{aligned}
$$

Similarly, the confidence interval estimate for B is

$$
\begin{aligned}
& 1.0065202 \pm (1.8595)(0.03968357) \\
&= 1.0065202 \pm 0.073791613 \\
&= [0.93272859, \quad 1.0803118].
\end{aligned}
$$

3. A point estimate of the average value of the amount of CaO present obtained by the new method when there are actually $x^* = 30$ mg present is given by

$$\tilde{y}^* = -\,0.29277822 + (1.0065202)(30) = 29.902828.$$

4. To compute the confidence interval estimate of the average value of the amount of CaO present obtained by the new method when there are actually $x^* = 30$ mg present, the value of $t_{\alpha/2;\,n-2} = t_{0.005;\,8} = 3.3554$ is obtained from Appendix Table 3. One substitutes into the formula given in Section 9.5; the required confidence interval is given by

$$\begin{aligned} &-\,0.29277822 + (1.0065202)(30) \pm (3.3554)(0.26323107) \\ &= 29.902828 \pm 0.88324553 \\ &= [29.019582,\quad 30.786073]. \end{aligned}$$

5. To compute the confidence interval estimate of the true amount of CaO present when the new method gives a quantity $y' = 29.6$ mg, the value of $t_{\alpha/2;\,n-2} = t_{0.025;\,8} = 2.3060$ is obtained from Appendix Table 3. One substitutes into the formula given in Section 9.6; the required confidence interval is

$$\begin{aligned} &\frac{29.6 - (-0.29277822)}{1.0065202} \pm \frac{2.3060}{1.0065202}(0.86274368) \\ &= 29.699134 \pm 1.9765991 \\ &= [27.722535,\quad 31.675733]. \end{aligned}$$

6. To compute the prediction interval for the amount of CaO present determined by the new method corresponding to an actual value $x^0 = 30$ mg present, a value of $t_{\alpha/2;\,n-2} = t_{0.025;\,8} = 2.3060$ is obtained from Appendix Table 3. Then substituting into the formula given in Section 9.7, one obtains for the prediction interval

$$\begin{aligned} &-\,0.29277822 + (1.0065202)(30) \pm (2.3060)(0.86205668) \\ &= 29.902828 \pm 1.9879027 \\ &= [27.914925,\quad 31.890731]. \end{aligned}$$

7. To test the hypothesis that $B = 1$, substitute into the expression given in Section 9.8;

$$\left| \frac{b - 1}{s_{y|x}\dfrac{1}{\sqrt{\sum (x_i - \bar{x})^2}}} \right| = \left| \frac{0.0065202}{0.039683578} \right| = |\,0.164305\,| = 0.164305.$$

This number is compared with the value of $t_{\alpha/2;\,n-2} = t_{0.05;\,8} = 1.8595$ which is obtained from Appendix Table 3. Thus

Worksheet for the Fitting of Straight Lines for Example 2

GENERAL DATA:

x represents CaO present:

$\sum x_i = 311; \quad \bar{x} = \sum x_i/n = 31.1; \quad (\sum x_i)\bar{x} = 9{,}672.1;$

$\sum x_i^2 = 10{,}100;$

$\sum x_i^2 - (\sum x_i)\bar{x} = \sum (x_i - \bar{x})^2 = 427.9.$

y represents CaO found by new method:

$\sum y_i = 310.1; \quad \bar{y} = \sum y_i/n = 31.01; \quad (\sum y_i)\,\bar{y} = 9{,}616.201;$

$\sum y_i^2 = 10{,}055.09; \quad \sum y_i^2 - (\sum y_i)\bar{y} = \sum (y_i - \bar{y})^2 = 438.889.$

$n = 10; \quad 1/n = 0.1; \quad (\sum x_i)(\sum y_i)/n = \bar{x} \sum y_i = 9{,}644.11;$

$\sum x_i y_i = 10{,}074.8;$

$\sum x_i y_i - (\sum x_i)(\sum y_i)/n = \sum (x_i - \bar{x})(y_i - \bar{y}) = 430.69;$

$[\sum(x_i - \bar{x})(y_i - \bar{y})]^2 = 185{,}493.88;$

$[\sum(x_i - \bar{x})(y_i - \bar{y})]^2$

DATA FOR EQUATION OF LINE:

$$b = \frac{\sum(x_i - \bar{x})(y_i - \bar{y})}{\sum(x_i - \bar{x})^2} = 1.0065202; \quad a = \bar{y} - l\bar{x} = -\,0.29277822;$$

equation of line $= \tilde{y} = a + bx = -\,0.29277822 + 1.0065202\,x.$

ESTIMATION OF STANDARD DEVIATION:

$$(n-2)\,s_{y|x}^2 = \sum (y_i - \bar{y})^2 - \frac{[\sum(x_i - \bar{x})(y_i - \bar{y})]^2}{\sum (x_i - \bar{x})^2} = 5.3908086;$$

$s_{y|x}^2 = 0.67385108; \quad s_{y|x} = 0.82088433.$

DATA FOR SIGNIFICANCE TESTS AND FOR CONFIDENCE INTERVALS FOR A AND B:

$$\frac{1}{\sum (x_i - \bar{x})^2} = 0.0023369946; \quad \sqrt{\frac{1}{\sum (x_i - \bar{x})^2}} = 0.048342472;$$

$$s_{y|x}\sqrt{\frac{1}{\sum (x_i - \bar{x})^2}} = 0.039683578; \quad \bar{x}^2 = 967.21;$$

$$\frac{\bar{x}^2}{\sum (x_i - \bar{x})^2} = 2.2603646; \quad \frac{1}{n} + \frac{\bar{x}^2}{\sum (x_i - \bar{x})^2} = 2.3603646;$$

$$\sqrt{\frac{1}{n} + \frac{\bar{x}^2}{\sum (x_i - \bar{x})^2}} = 1.5363478;$$

$\sqrt{1 \quad \bar{x}^2}$

DATA FOR CONFIDENCE INTERVALS FOR AVERAGE VALUE OF y CORRESPONDING TO $x = x^*$:

$x^* = 30; \quad (x^* - \bar{x})^2 = 1.21;$

$$\frac{(x^* - \bar{x})^2}{\sum (x_i - \bar{x})^2} = 0.0028277635; \quad \frac{1}{n} + \frac{(x^* - \bar{x})^2}{\sum (x_i - \bar{x})^2} = 0.10282776;$$

$$\sqrt{\frac{1}{n} + \frac{(x^* - \bar{x})^2}{\sum (x_i - \bar{x})^2}} = 0.32066768;$$

$$s_{y|x}\sqrt{\frac{1}{n} + \frac{(x^* - \bar{x})^2}{\sum (x_i - \bar{x})^2}} = 0.26323107.$$

DATA FOR PREDICTION INTERVALS FOR A FUTURE VALUE OF y CORRESPONDING TO $x = x^0$:

$$x^0 = 30; \quad (x^0 - \bar{x})^2 = 1.21; \quad \frac{(x^0 - \bar{x})^2}{\sum (x_i - \bar{x})^2} = 0.0028277635;$$

$$1 + \frac{1}{n} + \frac{(x^2 - \bar{x})^2}{\sum (x_i - \bar{x})^2} = 1.1028278;$$

$$\sqrt{1 + \frac{1}{n} + \frac{(x^0 - \bar{x})^2}{\sum (x_i - \bar{x})^2}} = 1.0501561;$$

$$s_{y|x}\sqrt{1 + \frac{1}{n} + \frac{(x^0 - \bar{x})^2}{\sum (x_i - \bar{x})^2}} = 0.86205668.$$

DATA FOR CONFIDENCE INTERVALS FOR x CORRESPONDING TO AN OBSERVED VALUE OF $y = y'$ FOR THE CASE WHEN THERE IS AN UNDERLYING PHYSICAL RELATIONSHIP:

$$y' = 29.6; \quad x' = \frac{y' - a}{b} = 29.699134; \quad \left(\frac{y' - a}{b} - \bar{x}\right)^2 = 1.9624255;$$

$$\frac{\left(\frac{y' - a}{b} - \bar{x}\right)^2}{\sum (x_i - \bar{x})^2} = 0.0045861778;$$

$$1 + \frac{1}{n} + \frac{\left(\frac{y' - a}{b} - \bar{x}\right)^2}{\sum (x_i - \bar{x})^2} = 1.1045862;$$

$$\sqrt{1 + \frac{1}{n} + \frac{\left(\frac{y' - a}{b} - \bar{x}\right)^2}{\sum (x_i - \bar{x})^2}} = 1.0509930;$$

$$s_{y|x}\sqrt{1 + \frac{1}{n} + \frac{\left(\frac{y' - a}{b} - \bar{x}\right)^2}{\sum (x_i - \bar{x})^2}} = 0.86274368.$$

$$0.164305 < 1.8595$$

and the hypothesis that $B = 1$ is not rejected at the 90% level.

8. To test the joint hypothesis, $A = 0$ and $B = 1$, substitute into the expression for F given in Section 9.8. For this example, one obtains

$$F = [10(-0.29277822)^2 + 20(31.1)(-0.29277822)(0.0065202) + 10{,}100(0.0065202)^2]/1.3477022$$
$$= 0.0991913/1.3477022$$
$$= 0.0736003.$$

This value of F is compared with a value of $F_{\alpha;\,2,\,n-2} = F_{0.10;\,2,8} = 3.11$ obtained from Appendix Table 4. Thus,

$$F = 0.0736003 < 3.11$$

and the joint hypothesis, $A = 0$ and $B = 1$, is not rejected.

Example 3. As a final example, assume that the data in Table 9.2 are augmented so that, corresponding to a given quantity of CaO present (x_i), there are two values of CaO determined by the new method (y_{i1} and y_{i2}). The data are given below. Using the expressions for a and b given in Section 9.10, i.e.,

$$a = \bar{\bar{y}} - b\bar{x}$$

and

$$b = \frac{\sum_{i=1}^{n} (x_i - \bar{x})(\bar{y}_i - \bar{\bar{y}})}{\sum_{i=1}^{n} (\bar{x}_i - \bar{x})},$$

the least squares line is given by

$$\tilde{y} = -0.67925285 + 1.0146062x.$$

A test for linearity is called for.

x_i	y_{i1}	y_{i2}	$\bar{y}_i$	$\tilde{y}_i$	$(\bar{y}_i - \tilde{y}_i)^2$	$y(_{i1} - \bar{y}_i)^2$	$(y_{i2} - \bar{y}_i)^2$	x_iy_i
20.0	19.8	19.6	19.70	19.61	0.0081	0.0100	0.0100	394.000
22.5	22.8	22.1	22.45	22.15	0.0900	0.1225	0.1225	505.125
25.0	24.5	24.3	24.40	24.69	0.0841	0.0100	0.0100	610.000
28.5	27.3	28.4	27.85	28.24	0.1521	0.3025	0.3025	793.725
31.0	31.0	30.0	30.50	30.77	0.0729	0.2500	0.2500	945.500
33.5	35.0	33.0	34.00	33.31	0.4761	1.0000	1.0000	1139.000
35.5	35.1	35.0	35.05	35.34	0.0841	0.0025	0.0025	1244.275
37.0	37.1	36.8	36.95	36.86	0.0081	0.0225	0.0225	1367.150
38.0	38.5	38.0	38.25	37.88	0.1369	0.0625	0.0625	1453.500
40.0	39.0	40.2	39.60	39.90	0.0900	0.3600	0.3600	1584.000
					1.2024	2.1425	2.1425	10036.275

It is desired to test at the 95% level. Appendix Table 4 is entered and the value of $F_{\alpha; n-2, n(k-1)} = F_{0.05; 8,10} = 3.071$ is obtained. Then substituting into the formula given in Section 9.10, one obtains

$$\frac{2(1.2024)/8}{4.2850/10} = 0.7015 < 3.071.$$

Thus the hypothesis of linearity is accepted.

9.14 Correlation

In engineering problems, interest is sometimes centered in determining the distribution of two related variables and the degree of association between them rather than in estimating one variable from another. In earlier chapters it was indicated that the distribution of a single variate could be characterized by its moments, or approximated by estimates of its moments. In the case of two variables, the joint distribution of these variables is represented not only by the moments of the individual variables, but by some measure of their joint behavior. If the two variables have a bivariate normal distribution, then the measure of their joint behavior is called the correlation coefficient and will be denoted by ρ. Two variables, x and y, are said to have a bivariate normal distribution if the probability of the event ($x \leqq M$ and $y \leqq N$) is written as

$$\mathrm{P}(x \leqq M \text{ and } y \leqq N) = \int_{-\infty}^{N}\int_{-\infty}^{M} \frac{1}{2\pi\sigma_x\sigma_y\sqrt{1-\rho^2}}$$

$$\times\, e^{-\frac{1}{2(1-\rho^2)}\left\{\frac{(x-\mu_x)^2}{\sigma_x^2} - 2\rho\frac{(x-\mu_x)(y-\mu_y)}{\sigma_x\sigma_y} + \frac{(y-\mu_y)^2}{\sigma_y^2}\right\}}\, dx\, dy$$

where μ_x and σ_x^2 are the mean and variance of x, μ_y and σ_y^2 are the mean and variance of y, and ρ is the correlation coefficient. The correlation can be positive or negative. When it is positive, one variable tends to increase as the other increases; when it is negative, one variable tends to decrease as the other increases. ρ lies between the limits -1 and $+1$. A high absolute value of ρ indicates a high degree of association whereas a small absolute value indicates a small degree of association. When the absolute value of ρ is 1, the relationship is perfect. When $\rho = 0$, the variables are independent.

An estimate of ρ is given by the sample correlation coefficient r, which is defined as

$$r = \frac{\sum_{i=1}^{n}(x_i - \bar{x})(y_i - \bar{y})}{\sqrt{\sum_{i=1}^{n}(x_i - \bar{x})^2 \sum_{i=1}^{n}(y_i - \bar{y})^2}}.$$

Data which will lead to a sample correlation coefficient having low positive correlation are indicated in Figure 9.5(a) whereas data which will lead to a sample correlation coefficient having high positive correlation are indicated in Figure 9.5(b).

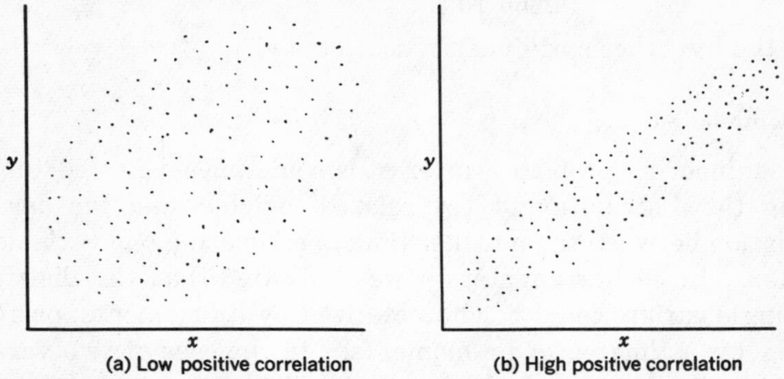

Fig. 9.5. Data leading to low and high positive correlation coefficients.

A test of the hypothesis that $\rho = 0$ is given by rejecting when

$$|t| = \left|\frac{r}{\sqrt{1-r^2}}\right|\sqrt{n-2} \geqq t_{\alpha/2;\,n-2}$$

where values of $t_{\alpha/2;\,n-2}$ are given in Appendix Table 3. It must be emphasized that the discussion above pertains to the correlation coefficient from a bivariate normal distribution. The correlation coefficient is of less value when the bivariate distribution is not normal, but it can still be used as an approximate measure of association.

The sample correlation coefficient can be derived from the slope of the fitted least squares line, e.g.,

$$r = b\sqrt{\frac{\sum (x_i - \bar{x})^2}{\sum (y_i - \bar{y})^2}}.$$

Consequently, it is clear that the sample correlation coefficient does not contain any additional information. In fact, a significance test for $\rho = 0$ is equivalent to testing whether $B = 0$, i.e., whether or not a relationship exists. These relationships indicate that correlation and fitting lines are mathematically equivalent, although these techniques are used for different types of problems. In engineering applications, the correlation coefficient does not play a very important role.

PROBLEMS

1. A chemical reagent (x) is used to obtain a precipitate of a particular substance (y) in a given solution. The data are as follows:

y	x	y	x
8.4	7.2	8.4	6.0
5.4	4.8	9.5	6.7
6.3	5.2	10.4	7.0
6.8	4.9	12.7	8.0
8.0	5.4	10.3	7.3
11.1	6.4	7.0	4.6
12.3	6.8	5.1	4.2
13.3	8.0		

(a) Estimate the line by the method of least squares. (b) If $x=6.9$, estimate the expected amount of precipitate. (c) Determine a 95% confidence limit for the expected amount of precipitate corresponding to $x=6.9$. (d) Find a prediction interval such that the probability is 95% that the value of precipitate corresponding to 6.9 units of reagent will lie in it. (e) Test the hypothesis, at the 5% level of significance, that $B=2$. (f) If the actual slope is 2.5, what is the probability of accepting the hypothesis in (e)?

2. The following results on the tensile strength of specimens of cold drawn copper have been recorded in a laboratory.

y *Tensile strength (psi)*	x *Brinell hardness number*
38,871	104.2
40,407	106.1
39,887	105.6
40,821	106.3
33,701	101.7
39,481	104.4
33,003	102.0
36,999	103.8
37,632	104.0
33,213	101.5
33,911	101.9
29,861	100.6
39,451	104.9
40,647	106.2
35,131	103.1

(a) Estimate the relationship by the method of least squares. (b) Test the hypothesis that $A=-100{,}000$ at the 5% level of significance. (c) Test the hypothesis that $B=1{,}900$ at the 5% level of significance. (d) Get 95% confidence intervals for B; for A; the mean value of tensile strength corresponding to a Brinell hardness of 105.0; and $\sigma_{y|x}$. (e) Find an interval such that the probability is 0.95 that the value of tensile strength for a Brinell hardness of 105.0 will lie in this interval.

3. It has been noted that the number of missing rivets (x) found at aircraft final inspection is associated with the number of errors of alignment (y) also observed at final inspection. A study was undertaken to help estimate the average number of alignment errors from the number of missing rivets. The data are as follows:

x	y	x	y
13	7	9	4
15	7	17	9
17	8	15	8
28	15	13	7
13	8	19	10
14	7	16	9
7	3	25	14
33	18	13	8
20	10	25	13
25	12	15	8

(a) Estimate the relationship by the method of least squares. (b) Estimate the expected number of alignment errors when an aircraft has 12 missing rivets by a point estimate and a 90% confidence interval estimate. (c) Determine a 90% prediction interval for the number of alignment errors when an aircraft has 12 missing rivets.

4. Determining shear strength of spot welds is difficult, whereas measuring weld diameter of spot welds is relatively simple. It would be advantageous if shear strength could be predicted from a measurement of weld diameter. The data are as follows:

	(y) *Shear strength (psi)*	(x) *Weld diameter (0.0001 in.)*
	370	400
	780	800
	1,210	1,250
	1,560	1,600
	1,980	2,000
	2,450	2,500
	3,070	3,100
	3,550	3,600
	3,940	4,000
	3,950	4,000
Totals:	22,860	23,250

The following can be computed from the data:

$$\sum (x_i-\bar{x})^2=1{,}568.625$$

$$\sum (y_i-\bar{y})^2=1{,}546.144$$

$$\sum (x_i-\bar{x})(y_i-\bar{y})=1{,}557.36$$

$$s^2_{y|x}=0.0109$$

(a) Determine the least squares line. (b) Test the hypothesis that the slope (B) equals 1 using a level of significance equal to 0.05. (c) If the true slope equals 1.01, what is the probability of accepting the hypothesis in Problem (b)? (d) Estimate the expected value of shear strength when the weld diameter is 0.2500. (e) Find a prediction interval such that the probability is 95% that the value of shear strength corresponding to a weld diameter of 0.2250 inch will be in it.

5. Consider the data below:

x	1	2	3	4
y	8 10	20 23	31 28	29 28
$\bar{y}$	9	21.5	29.5	28.5

$$\sum_{i=1}^{4} (\bar{y}_i - \bar{\bar{y}})^2 = 267.7\,, \qquad \sum_{i=1}^{4}\sum_{\nu=1}^{2} (y_{i\nu} - \bar{y}_i)^2 = 11.5,$$

$$\sum_{i=1}^{4} (x_i - \bar{x})^2 = 5\,, \qquad \sum_{i=1}^{4} (\bar{y}_i - \tilde{y}_i)^2 = 46.6,$$

$$\sum_{i=1}^{4} (x_i - \bar{x})(\bar{y}_i - \bar{\bar{y}}) = 33.25\,, \qquad \bar{x} = 2.5\,,$$

$$\bar{\bar{y}} = \frac{\sum_{i=1}^{4} \bar{y}_i}{4} = 22.25,$$

x represents the temperature in degrees Centigrade (coded) of a certain part of a process, and the two values of y repeat determinations on the yield at that temperature (coded). We are interested in the effect of temperatures upon the yield. (a) Determine the least squares line. (b) Check for linearity at the 1% level of significance. (c) Test the hypothesis that $B=6$ using a level of significance of 0.05. (d) If the true $B=7$, what is the probability of accepting the hypothesis in (c)? (e) Find a prediction interval such that the probability is 95% that the average value of two determinations corresponding to a temperature of 2.75 degrees Centigrade will lie in it.

6. A screw manufacturer is interested in giving out data to his customers on the relation between nominal and actual lengths. The following results (in inches) were observed:

Nominal x	*Actual* y		
$\frac{1}{4}$	0.262	0.262	0.245
$\frac{1}{2}$	0.496	0.512	0.490
$\frac{3}{4}$	0.743	0.744	0.751
1	0.976	1.010	1.004
$1\frac{1}{4}$	1.265	1.254	1.252
$1\frac{1}{2}$	1.498	1.518	1.504
$1\frac{3}{4}$	1.738	1.759	1.750
2	2.005	1.992	1.992

If the manufacturing process were perfect, it would be expected that $A=0$ and $B=1$. It should be noted that without using regression, one could make a confidence statement about the average of length of "one-inch" screws on the basis of the three numbers in the table; but a much more precise statement can be made by using all the data, together with the fact that linear regression is applicable. (a) Estimate the above relation. (b) For nominal one-inch screws, find a confidence interval for the average actual value of screw length. (c) For a one-inch screw find an interval such that the actual value will lie in this interval with probability 0.95. (d) Check for linearity at the 1% level of significance.

7. The corrosion of a certain metallic substance has been studied in dry oxygen at 500 degrees Centigrade. In this experiment the gain in weight after various periods of exposure was used as a measure of the amount of oxygen which had reacted with the sample. In the table below are given the data.

x (*hours*)	y=*weight gain* (*coded*)
1	1
2	2
3	6

(a) Determine the least squares line. (b) If $x=4$, estimate the expected value of weight gain by a 95% confidence interval. (c) If $x=4$, estimate the weight gain by a 95% prediction interval.

8. The determination of abrasion loss is difficult whereas measuring hardness by means of a Rockwell hardness machine is relatively simple. It would therefore be advantageous if abrasion loss could be predicted from a measurement of hardness. A test was run and data were obtained from 30 observations. Let hardness be denoted by x and abrasion loss by y. The data are summarized as follows:

$$\bar{x}=70.27\,, \qquad \bar{y}=175.4$$

$$\sum(x_i-\bar{x})^2=4{,}300\,, \qquad \sum(y_i-\bar{y})^2=225{,}011$$

$$\sum(x_i-\bar{x})(y_i-\bar{y})=-22{,}946$$

$$s^2_{y|x}=3{,}663\,.$$

The least squares line is

$$\tilde{y}=550.4-5.336x\,.$$

(a) Test the hypothesis that $B=-5$ using a level of significance of 0.05. (b) If the slope equals -6, what is the probability of accepting the hypothesis in (a)? (c) Estimate the expected value of abrasion loss when the hardness is 71 by a 95% confidence interval. (d) Find a prediction interval such that the probability is 95% that the value of abrasion loss corresponding to a hardness of 71 will lie in it.

9. The production of carbon steel (x) and alloy steel (y) in tons for eight quarters during 1953–1954 for a certain steel plant is given below:

x:	1,536	2,004	3,569	4,118	1,621	2,241	3,321	4,432
y:	211	561	341	915	248	589	602	834

(a) Determine the relation between y and x by the method of least squares. (b) Predict with probability 0.90 the number of tons of alloy that would be produced given that 3,000 tons of carbon steel were produced. (c) Estimate by a 90% confidence interval the expected value of the number of tons of alloy steel given that 3,000 tons of carbon steel were produced.

10. A gauge is to be calibrated using several dead weights. Denote by x the standard and by y the actual gauge reading. A sample of ten measurements is taken and the following are computed from the data:

$$\bar{x}=230.5, \qquad \bar{y}=226.1$$

$$\sum(x_i-\bar{x})^2=1{,}560.63, \qquad \sum(y_i-\bar{y})^2=1{,}539.15$$

$$\sum(x_i-\bar{x})(y_i-\bar{y})=1{,}532.41.$$

(a) A completely accurate gauge would be expected to yield a line with slope 1. Test the hypothesis that the slope is 1 at the 5% level of significance. (b) If the true slope is 1.5, what is the probability of accepting the hypothesis in (a)? (c) Estimate by a 95% confidence interval the expected gauge reading when the amount of pressure is 25. (d) Find a prediction interval such that the probability is 95% that the gauge reading will lie in it when the amount of pressure is 25. (e) Estimate the expected amount of pressure by a 95% confidence interval when the gauge reads 25 psi.

11. A test was performed to determine the relationship between the chemical content of a particular constituent (y) in solution and the crystallization temperature (x). The data are as follows:

Degrees Centigrade x	*Grams per liter* y
−1.7	1.1
−0.4	2.3
0.2	3.2
1.1	4.3
2.3	5.4
3.1	6.6
4.2	7.8
5.3	8.8

(a) Estimate the relationship by the method of least squares. (b) Test the hypothesis that $B=0$ at the 1% level of significance. (c) Determine a 95% confidence interval for A and for B. (d) Estimate by a point estimate and by a 99% confidence interval the expected value of y, given $x=5$. (e) Estimate by a 95% prediction interval the amount of phosphorus that will be found in a future trial if the crystallization temperature is 6 degrees Centigrade.

12. The elongation of a piece of boiler plate (y) is related to the amount of force applied in tons per square inch (x). The data are:

x:	1.33	2.68	3.57	4.46	5.35	6.24	7.14	8.93	9.82	10.70
y:	27	50	67	83	101	117	134	150	188	206

(a) Estimate the least squares line. (b) Test the hypothesis that $B=20$ at the 5% level. (c) If the true slope is 25, what is the probability of accepting the hypothesis in (b)? (d) If the elastic limit is reached when $x=16$, predict the expected elongation for this limit by a 95% confidence limit. (e) If the elongation is 175, estimate the amount of force applied by a 95% confidence limit.

13. Under a certain set of conditions, five temperatures were read on a pyrometer and compared with true temperatures. The data are:

Pyrometer readings, °C:	950	975	1,000	1,025	1,050
True temperatures, °C:	1,082	1,118	1,156	1,193	1,231

(a) Estimate the relationship by the method of least squares. (b) Estimate the true temperature for a pyrometer reading of 1,200 degrees Centigrade by a 99% confidence limit. (c) Test the hypothesis that $B=1$ at the 5% level. (d) If the true slope is 1.2, what is the probability of accepting the hypothesis in (c)?

14. A piece of metal $\frac{1}{2}$ inch in diameter and 4 inches in length is subjected to a test to determine how much strain is set up by specified values of twisting movement. The data are:

x-Twisting movement (lb/in.):	100	300	500	700	1,000	1,200	1,360
y-Strain (torsion units):	111	329	551	769	1,103	1,322	1,497

(a) Estimate the relationship by the method of least squares. (b) Determine a 95% confidence interval for the expected value of strain if the elastic limit of 1,400 lb/in. has been reached. (c) Test the hypothesis that the increase in y is 1.25 times the increase in x (i.e., $B=1.25$) at the 5% level of significance. (d) Estimate by a 99% prediction interval the strain if the elastic limit of 1,400 lb/in. has been reached.

15. The gauging laboratory of a manufacturing firm obtained the following data by measuring a group of standard gauge blocks with a micrometer used by the inspection department.

Standard x (inches)	*Micrometer reading y (inches)*
0.5000	0.501
0.5500	0.552
0.6000	0.597
0.6500	0.646
0.7000	0.703
0.7500	0.754
0.8000	0.798
0.8500	0.853
0.9000	0.902
1.0000	1.006

(a) Compute estimates of A and B (where $E(y)=A+Bx$). (b) Plot the data above. Plot the line derived in (a). (c) Determine the confidence interval for B using a 99% confidence coefficient. (d) Test the hypothesis that $B=1$ ($\alpha=0.05$). (e) Determine the confidence interval for the expected value of y, given that $x=x^*=0.8000$. Use $\alpha=0.01$. (f) Estimate the true value x by a 95% confidence interval when the micrometer reading is 0.750.

16. In the construction of mutual characteristic curves for the Type 6J5 triode, the plate voltage is held constant at 100 volts. The grid voltage is varied and the plate current observed. (The grid voltage can be very accurately determined.) The following data are obtained:

Grid voltage (x) (*volts*)	*Plate current (y)* (*milliamperes*)
−4.00	1.32
−3.50	1.73
−3.00	2.71
−2.50	3.42
−2.00	4.59
−1.50	5.41
−1.00	6.33
−0.50	9.29
0.00	12.00

(a) Plot the given raw data. (b) Fit a straight line to the data above. (c) Determine a confidence interval for the slope. Use $\alpha=0.05$. (d) Estimate the plate current by a 95% prediction interval when the grid voltage is −2.21 volts.

17. The following data concern case hardening a machined part by induction hardening at 1,750 degrees Fahrenheit.

Time heated (*seconds*)	*Depth of hardening* (*inches*)
5	0.12
10	0.23
15	0.29
20	0.37
25	0.46
30	0.61

(a) Fit a straight line to the data. (b) Find a confidence interval for the slope. (c) Test the hypothesis that $B=0.2$ for $\alpha=0.01$. (d) What is the probability of accepting the hypothesis in (c) when the true slope is 0.22? (e) Find a 95% prediction interval for the depth of hardening after 20 seconds.

18. The chief production engineer of a manufacturing firm desires to ascertain whether or not there is a linear relation between the number of products assembled by the women who staff his assembly line and the length of time they work without being allowed an interruption. His time study engineer by random sampling obtains the following data:

Time worked without interruption (*hours*)	*Pieces completed*
0.5	56
	61
	54
1.0	123
	101
	110
1.5	176
	160
	151
2.0	200
	215
	201
2.5	270
	243
	251
3.0	340
	370
	353
3.5	401
	426
	418
4.0	432
	461
	428

(a) Using the data above, test for linearity. (b) If linearity is found in (a), estimate the equation of the line. If linearity is not found in (a), how would you determine whether a quadratic relation exists?

19. The following are data given on results of a study on accelerated solidification of ingots.

Depth of solidification	*Time in* (*minutes*)$^{\frac{1}{2}}$
4.8	5.73
6.2	7.40
7.5	8.59
8.5	9.72
9.7	10.90
11.8	12.29

Let y represent the depth of solidification, and let x represent the square root of time of cooling. Suppose that the model that seems likely is $y=A+Bx$. (a) Derive point estimates of A and B by the method of least squares. (b) Construct a 95% confidence interval for B. How is this related to a test of the hypothesis that $B=0$? (c) Estimate by a point estimate and by a 95% confidence interval the expected depth of solidification after 100 minutes. (d) An ingot is cooled for 87 minutes. Estimate the depth of solidification by a 95% confidence interval.

20. The following data represent the results of a calibration of a Heath Model U5A V.T.W.M. Voltmeter against a standard, which is a Sensitive Research Voltmeter. Three readings on the Heath Model were taken for each standard voltage.

Standard x		*Heath* y	
20	19.5	20.4	20.5
30	29.5	30	30.5
40	38.8	39.8	39.8
50	48.5	49	49.5
60	58.5	59	59
70	68	68.5	68.4
80	78.5	79	78.5
90	87.5	88	88
100	97	97	97

a) Estimate the least squares line. (b) Test for linearity at the 5% level of significance. (c) Test the hypothesis that $B=1$ at a 1% level of significance. (d) If $B=1.1$, what is the probability of accepting the hypothesis n (c) ? (e) If $\bar{y}$ based upon a set of Heath meter readings is 58.7, estimate the true reading by a 95% confidence interval. (f) Test, at the 5% level of significance, the hypothesis that the variance of y for each x is constant.

21. The following data represent the results of a calibration test on a Crosby gauge.

Pressure Gauge Calibration Data

	Standard	*Run 1* (*spin-down*)	*Run 2* (*spin-up*)	*Run 3* (*spin-level*)
1.	10	8	6	7
2.	15	14	12	13
3.	20	18	16	17
4.	30	28	26	28
5.	40	40	37	37
6.	45	45	42	43
7.	55	55	51	53
8.	65	65	61	63
9.	70	70	66	69
10.	80	80	76	78
11.	90	90	87	89
12.	95	96	92	94
13.	105	106	101	104
14.	115	116	112	113
15.	120	122	117	119
16.	130	132	129	131
17.	140	143	138	140
18.	145	147	144	145
19.	150	150	148	149

(a) Estimate the least squares line. (b) Test for linearity at the 5% level of significance. (c) Test the hypothesis that $B=1$ at the 5% level of significance. (d) If the average Crosby gauge reading is 89, derive a 95% confidence interval for the reading on the standard.

22. It has been determined that the relation between stress (S) and the number of cycles to failure (N) for a particular type alloy is given by

$$S=\frac{A}{N^m}$$

where A and m are unknown constants. An experiment is run yielding the following data:

Stress (thousand psi)	*N (million cycles to failure)*
55.0	0.223
50.5	0.925
43.5	6.75
42.5	18.1
42.0	29.1
41.0	50.5
35.7	126
34.5	215
33.0	445
32.0	420

(a) Letting $x=\ln S$ and $y=\ln N$, determine the relationship between y and x. (b) Estimate m by a 95% confidence interval. (c) Estimate A by a 95% confidence interval. (d) For a stress of 40,000 psi, estimate by a 90% prediction interval the number of cycles to failure.

23. The following data represent the results of an experiment of small creep tests of a precision cast high temperature alloy. The experiment was run at 725 degrees Centigrade.

Stress (S), tons/in.²	*Fracture-time (T), hours*
19	43
17	141
15	385
13	2099

The underlying relationship is of the form

$$\ln T=A+BS\,.$$

(a) Estimate the relationship above by the method of least squares. (b) For a stress of 16 tons/in.², estimate the fracture time by a point estimate. (c) For a stress of 16 tons/in.², estimate the fracture time by a 95% prediction interval.

24. The following data represent the effect of time on the loss of hydrogen from samples of steel stored at 20 degrees Centigrade.

Time (T), hours	*Hydrogen content (H), (ppm)*	
1	7.7	8.4
2	7.5	8.1
6	6.1	6.8
17	5.7	5.3
30	4.2	4.5

(a) Estimate the relationship $H=A+B\ln T$ by the method of least squares. (b) Test the relationship in (a) at the 10% level of significance. (c) Estimate by a 95% prediction interval the hydrogen content at 20 hours. (d) Estimate by a 95% confidence interval the expected hydrogen content at 20 hours.

25. The following are data given as results of a study of the effect of primary crystallization temperature on the phosphorus content of a solution:

Phosphorus grains per liter (y)	*Primary crystallization temperature, degrees C* (x)
10.9	25
9.3	20
8.2	15
7.5	12
6.2	9
5.8	6
4.2	3
3.9	0
2.8	−3
2.0	−6

(a) Plot the data and draw a freehand line representing the regression. (b) Assuming that the relationship is linear, find the least squares estimate of the regression line of phosphorus content on temperature. (c) Test the hypothesis that $B=0$ at a 5% level of significance. (d) Set up a 95% confidence interval for A. (e) Estimate by a point estimate and by a 95% confidence interval the expected phosphorus content for a primary crystallization temperature of 7 degrees Centigrade. (f) Estimate by a 95% prediction interval the amount of phosphorus that will be found on a future observation if the primary crystallization temperature is 4 degrees Centigrade.

26. Given the following observations of y vs. x:

x	y	
1	−1.9	$\sum x=18$
2	−1.0	$\sum y=2.9$
3	−0.1	$\sum xy=33.1$
5	2.0	$\sum x^2=88$
7	3.9	$\sum y^2=23.83$

suppose that for each x, values of y are normally distributed with mean $A+Bx$ and variance σ^2, and that the y's corresponding to different x's are independent random variables. Find: (a) The least squares regression line of y on x. (b) A 95% prediction interval for a future observation of y corresponding to $x=6$.

27. Suppose for each x the values of y are normally distributed with mean $A+Bx$ and variance σ^2, and the y's corresponding to different x's are independent. Given the following data:

x	y	x^2	y^2	xy
0	−2.8	0	7.84	0
1	−1.7	1	2.89	−1.7
2	−0.9	4	0.81	−1.8
3	0.2	9	0.04	0.6
4	1.1	16	1.21	4.4
5	2.3	25	5.29	11.5
6	3.1	36	9.61	18.6
7	4.2	49	17.64	29.4
8	5.3	64	28.09	42.4
$\sum x_i=36$	$\sum y_i=10.8$	$\sum x_i^2=204$	$\sum y_i^2=73.42$	$\sum xy=103.4$

(a) Find least squares regression of y on x. (b) Find a 95% confidence interval for average y given $x=3$. (c) Find a 95% prediction interval for a future value of y given $x=10$.

28. Suppose for each x the values of y are normally distributed with mean $A+Bx+Cx^2$ and variance σ^2, and the y's corresponding to different x's are independent. Derive the least squares estimate of A, B, and C.

CHAPTER X

Analysis of Variance

10.1 Introduction

In Chapter VII, testing whether the means of two distributions are equal was considered. The t test was used for making such a comparison. This problem is actually a special case of the more general problem of comparing several means. A possible solution to this general problem is to enumerate every possible pair, and test whether each pair differs, using the t test. The disadvantage of this procedure is clear upon a little reflection. Even if the distribution means are the same, we must expect that if there are enough sample means, some will be extremely large and some extremely small. This will lead to one or more fictitious significant differences with probability much higher than α. This distortion in the significance level introduces another difficulty. Suppose that the 45 possible comparisons among ten means have all been made, and four are significant by a 0.05 level t test; the experimenter must now wonder *which* (if any) of these four significant differences are really indicative of true differences.

The appropriate solution to the problem is the analysis of variance. The analysis of variance, however, has a much wider application than this simple generalization, and is probably the most powerful procedure in the field of experimental statistics. To carry out the analysis, it is necessary to formulate a mathematical model in terms of the unknown parameters and the associated random variables. The following sections of this chapter will deal with the concepts involved in the analysis of variance.

10.2 Model for the One-Way Classification

10.2.1 Fixed effects model

An interlaboratory calibration check on horizontal tension testing machines is to be made. Four laboratories are involved in the test and each laboratory makes five tests on No. 16 wire. It can be assumed that the reproducibility among repeat tests is the same for all laboratories. The results of the testing are given in Table 10.1.

Table 10.1. Interlaboratory Calibration, Horizontal Tension Testing Machines

Laboratory					
A	73	73	73	75	75
B	74	74	74	74	75
C	68	69	69	69	70
D	71	71	72	72	73

Before analyzing these data, it is helpful to formulate a mathematical model. This has the advantage of requiring the experimenter to define the effects which he estimates and the interpretation that he will propose.

As a first model, consider these four laboratories as the only laboratories possessing horizontal tension testing machines which are of interest. This may arise when they are the only laboratories within a certain locale, and hence receive all the business from the neighboring industrial establishments. The results of the calibration, then, will apply only to these four laboratories and any conclusions derived will not be extended to laboratories in general.

A mathematical model describing this situation is obtained by assuming that the tension reading of a specimen consists of the sum of two components: a fixed effect due to the machine of the particular laboratory tested and a random effect which represents the variation of the particular observation from the fixed laboratory effect. The random effects are assumed to be independently normally distributed with mean 0 and variance σ^2. Thus, the third specimen from laboratory B has a tensile strength equal to:

74 = fixed effect due to machine in laboratory B + random effect.

All the specimens from laboratory B have the same fixed effect and differ only because of the random effect.

More generally, these assumptions can be summarized as follows:

$$x_{ij} = \zeta_i + \varepsilon_{ij}$$

where x_{ij} is the outcome of the jth specimen corresponding to the

ith treatment (or laboratory in the example given); $i = 1, 2, \cdots, r$; $j = 1, 2, \cdots, c$;[1] ζ_i is the fixed effect attributed to the ith treatment; ε_{ij} is the random effect assumed to be independently normally distributed with mean 0 and variance σ^2.

A few words about the selection of the specimens of wire to be used in the experiment are in order. The tensile strength of wire depends on many factors besides the particular machine used for testing: the temperature, presence of alloys, and so forth. It is important in running experiments which calibrate machines of different laboratories to make sure that differences which are asserted to be due to machines do not arise from some other factors, for example, from the fact that all the specimens used in laboratory A were treated at a higher temperature than the others. In some cases, other factors which affect the wire may be isolated as factors in the experiment and their effect removed by the techniques of two-way, three-way, and other analysis of variance techniques discussed in the subsequent sections. If this is not done, the specimens to be used at each laboratory should be chosen at random with respect to the other factors which are not specified; this random choice will insure that the conclusions are not vitiated by a coincidence of effects. A table of random numbers is useful in assigning specimens to the specific treatments. If these precautions are not taken and the fluctuation of the data is not random, statistical methods based on this assumption should not be used.

An equivalent way of stating the model above is further to subdivide ζ_i into $\zeta_\cdot + \varphi_i$. Thus, the fixed effect ζ_i can be expressed as a component $\zeta_\cdot$ common to all laboratories and a fixed effect φ_i peculiar only to the ith laboratory. This can easily be done by defining $\zeta_\cdot$ as the average of all the ζ_i; i.e.,

$$\zeta_\cdot = \frac{\sum_{i=1}^{r} \zeta_i}{r}$$

and defining φ_i as the deviation of the ζ_i from the average; i.e.,

$$\varphi_i = \zeta_i - \zeta_\cdot\,.$$

Thus φ_i has the property that

[1] Note that the subscript i pertains to the rows which denote the treatments in the one-way classification, there being a total of r rows. The subscript j pertains to the columns which represent the repeat tests, there being a total of c columns (repeat tests).

$$\sum_{i=1}^{r} \varphi_i = 0 .$$

x_{ij} can then be expressed as

$$x_{ij} = \zeta. + \varphi_i + \varepsilon_{ij}$$

where x_{ij} is normally distributed with mean $\zeta. + \varphi_i$ and variance σ^2.

The purpose of the entire experiment is to draw some conclusions about whether the laboratory machines read the same, and, if not, which differ. This is equivalent to making statements about the equality of the φ_i. If the φ_i are equal, the laboratories are the same. Furthermore, if the φ_i are equal, they all must be equal to zero, since $\sum_{i=1}^{r} \varphi_i = 0$.

This model will be called the *fixed effects* model because the φ_i are fixed effects and the conclusion derived from the experiment will be extended only to the treatments (laboratories in this example) considered.

10.2.2 Random effects model

In the calibration example, suppose that a large number of laboratories are to be calibrated instead of four. However, only four laboratories (not necessarily the four of the previous section) are to be included in the experiment. But inferences about all of the laboratories are to be made from the results of the experiment which will be based on a random sample of four chosen from the many.

The mathematical model may still be written as

$$x_{ij} = \zeta. + \varphi_i + \varepsilon_{ij}$$

but the interpretation of the φ_i is different. The φ_i are random variables each having a probability distribution which is the distribution of laboratory effects (from which a random sample of four is to be drawn). The φ_i are usually assumed to be normally distributed with zero mean and variance σ_φ^2. If the experiment were to be repeated, a new random sample would be drawn and different laboratories might be included. In the fixed effects model, repeating the experiment would entail using the same four laboratories. Thus, the φ_i are random variables prior to performing the experiment. Once a laboratory is chosen, the observations within the laboratory differ only because of the random effects ε_{ij} (which are assumed to be normally distributed with mean 0 and variance σ^2 and to be distributed independently of φ_i). The quantity $\zeta.$ is a constant common to all the observations. In this model, knowledge about the particular φ_i is rather useless. Inferences are to be drawn about *all* the laboratories, i.e., the distribution of φ_i which is assumed normal with mean 0 and variance σ_φ^2. If $\sigma_\varphi^2 = 0$, all the laboratories are equivalent; if σ_φ^2 is very large, there are large discrepancies between laboratories. Hence, it is important to estimate σ_φ^2 and to make inferences about its magnitude.

This model will be called the *random effects* model because the φ_i are random variables, and the conclusions derived from the experiment will be extended to all of the treatments (laboratories in the example) from which the random sample was drawn. In this model, it is evident that the x_{ij} are normally distributed with mean $\zeta.$ and variance $\sigma^2 + \sigma_\varphi^2$.

10.2.3 Further examples of fixed effects and of the random effects models

(a) A study was made recently to determine whether the ash content of coal from a certain source varied according to the four origins within the source. The coal was delivered from each origin in separate trucks. Since each truck had a different origin, the coal was sampled from each truck (prior to loading). This one-way analysis of variance fits the framework of the fixed effects model since the four origins are presumably fixed and inferences are to be made about these origins only. Repeated experiments would require the use of the same origins.

(b) A special study was made on lots of metallic oxide. Samples were drawn at random from each of eighteen lots and the specimens were analyzed for the percent metal content by weight. Assuming these eighteen lots to be random samples from a larger number of lots, the problem fits into the random component model. Repeated experiments would use different lots.

10.2.4 Computational procedure, one-way classification

Before carrying out any further analysis, it is important to complete the analysis of variance table. This is done following the computational procedure described below. The results are summarized in the Analysis of Variance Table 10.2.

Computational Procedure

(1) Calculate totals for each treatment:

$$R_1, R_2, \cdots, R_r .$$

Table 10.2. Analysis of Variance Table, One-Way Classification

Source	Sum of Squares	Degrees of freedom D.F.	Mean Square	Expected Mean Square for the Fixed Effects Model	Expected Mean Square for the Random Effects Model
Between treatments...	$SS_3 = c \sum_{i=1}^{r} (\bar{x}_{i.} - \bar{x}_{..})^2$	$r - 1$	$SS_3^* = SS_3/(r-1)$	$\sigma^2 + \dfrac{c \sum_{i=1}^{r} (\varphi_i)^2}{r-1}$	$\sigma^2 + c\sigma_\varphi^2$
Within treatments....	$SS_2 = \sum_{i=1}^{r} \sum_{j=1}^{c} (x_{ij} - \bar{x}_{i.})^2$	$r(c-1)$	$SS_2^* = SS_2/r(c-1)$	σ^2	σ^2
Total	$SS = \sum_{i=1}^{r} \sum_{j=1}^{c} (x_{ij} - \bar{x}_{..})^2$	$rc - 1$	$SS^* = SS/(rc-1)$	—	—

In this table $\bar{x}_{i.} = \sum_{j=1}^{c} \frac{x_{ij}}{c}$; $\bar{x}_{..} = \sum_{i=1}^{r} \sum_{j=1}^{c} \frac{x_{ij}}{rc}$.

(2) Calculate over-all total:

$$T = R_1 + R_2 + \cdots + R_r\,.$$

(3) Compute crude total sum of squares:

$$\sum_{i=1}^{r}\sum_{j=1}^{c} x_{ij}^2 = x_{11}^2 + x_{12}^2 + \cdots + x_{rc}^2\,.$$

(4) Calculate crude sum of squares between treatments:

$$\frac{\sum_{i=1}^{r} R_i^2}{c} = (R_1^2 + R_2^2 + \cdots + R_r^2)/c\,.$$

(5) Calculate correction factor due to mean $= T^2/rc$. From the quantities above compute:

(6) $$SS_3 = (4) - (5) = \sum_{i=1}^{r} \frac{R_i^2}{c} - \frac{T^2}{rc}\,,$$

(7) $$SS = (3) - (5) = \sum_{i=1}^{r}\sum_{j=1}^{c} x_{ij}^2 - \frac{T^2}{rc}\,,$$

(8) $$SS_2 = (7) - (6) = SS - SS_3\,.$$

By referring to Table 10.2 it can be verified that $SS = SS_2 + SS_3$. In fact, this was used in obtaining SS_2 in step 8 in the computational procedure. It should be pointed out that subtracting a constant from each observation will not change the analysis of variance table.

10.2.5 The Analysis of Variance Procedure

10.2.5.1 *A heuristic justification.* Looking at the analysis of variance table, we see that the terms in the "Mean Square" column, SS_3^* and SS_2^*, are unbiased estimates of σ^2 when the hypothesis is true. Thus, in the absence of row effects (in both models) the ratio SS_3^*/SS_2^* should be close to one. When the hypothesis is false, the denominator still estimates σ^2, whereas the numerator tends to overestimate this value. Hence, the analysis of variance procedure involves rejection of the hypothesis when the ratio is too large compared to one. In order to quantify the term "large" it is necessary to obtain the distribution of the ratio SS_3^*/SS_2^*. This is usually done by proving that the terms in the "Sum of Squares" column, SS_3 and SS_2, when divided by σ^2, each have independent chi-square distributions with given degrees of freedom when the hypothesis is true. It then follows that SS_3^*/SS_2^* has an F distribution. The chi-square distributions of SS_3 and SS_2 are usually obtained from the partition theorem which is stated in Section 10.2.5.2. This theorem gives criteria by which "Sums of Squares" can be judged as chi-square random variables. In particular, requirements are given in terms of degrees of freedom so that a method of obtaining degrees of freedom is necessary. The partition theorem is not a requirement for the understanding of the ensuing sections but rather a justification for the characterization of ratios of mean squares as random variables having an F distribution. Similar procedures, as outlined above, are used for the two-way classification presented in the later sections, and the partition theorem provides a rationale for the F test.

Since the "degrees of freedom" play such an important role in the analysis of variance, it is important that some further comments be made. The number of degrees of freedom for n variables may be defined as the number of variables minus the number of linear relations (constraints) between them. For example, the number of degrees of freedom associated with SS_2 is rc minus the number of linear constraints between these variables. For each treatment (row) the sum of the deviations about the row mean must be zero, i.e.,

$$\sum_{j=1}^{c}(x_{ij} - \bar{x}_{i\cdot}) = 0 \quad \text{for} \quad i = 1, 2, \cdots, r .$$

This is equivalent to saying that we have imposed one linear restriction on the observations for each row. Since there are r rows, there are r such constraints. Hence, the number of degrees of freedom associated with SS_2 is $rc - r$ or $r(c - 1)$.

*10.2.5.2 *The partition theorem.* Degrees of freedom having been defined in the previous section, the following partition theorem is stated without proof:

Let $y_1, y_2, \cdots, y_n$ be independent identically distributed normal random variables each with mean 0 and variance 1. Let the sum of squares of the n variables be partitioned into a sum of k sums of squares, $T_1, T_2, \cdots, T_k$, with $\nu_1, \nu_2, \cdots, \nu_k$ degrees of freedom, respectively, so that

$$\sum y_i^2 = T_1 + T_2 + \cdots + T_k .$$

The necessary and sufficient condition that $T_1, T_2, \cdots, T_k$ are independently distributed as chi-square random variables with $\nu_1, \nu_2, \cdots, \nu_k$ degrees of freedom, respectively, is that

$$\nu_1 + \nu_2 + \cdots + \nu_k = n .$$

The importance of this theorem can be seen by considering the random variables x_{ij} in the one-way analysis of variance. Again, if there are no treatment effects (no laboratory differences), each x_{ij}

is normally distributed with mean μ and variance σ^2 (under either model).

Let $y_{ij} = (x_{ij} - \mu)/\sigma$ so that when there are no treatment effects y_{ij} is normally distributed with mean 0 and variance 1.

Thus, the following equalities are obtained:

$$\sum\sum y_{ij}^2 = \frac{\sum\sum(x_{ij} - \mu)^2}{\sigma^2}$$

$$\sum\sum (y_{ij} - \bar{y}..)^2 = \frac{\sum\sum(x_{ij} - \bar{x}..)^2}{\sigma^2}$$

$$(\bar{y}..)^2 = \frac{(\bar{x}.. - \mu)^2}{\sigma^2}$$

$$\sum(\bar{y}_i. - \bar{y}..)^2 = \frac{\sum(\bar{x}_i. - \bar{x}..)^2}{\sigma^2} .$$

It is also evident that

$$\frac{\sum\sum(x_{ij} - \mu)^2}{\sigma^2} = \frac{\sum\sum(x_{ij} - \bar{x}..)^2}{\sigma^2} + \frac{rc(\bar{x} - \mu)^2}{\sigma^2}$$

$$= \frac{SS}{\sigma^2} + \frac{rc(\bar{x} - \mu)^2}{\sigma^2} .$$

However, since $SS = SS_2 + SS_3$, it follows that

$$\frac{\sum\sum(x_{ij} - \mu)^2}{\sigma^2} = \frac{\sum\sum(x_{ij} - \bar{x}_i.)^2}{\sigma^2} + \frac{c\sum(\bar{x}_i. - \bar{x}..)^2}{\sigma^2} + \frac{rc(\bar{x}.. - \mu)^2}{\sigma^2} .$$

Using the equalities about the y's, the following identity holds:

$$\sum\sum y_{ij}^2 = \sum\sum (y_{ij} - \bar{y}_i.)^2 + c\sum(\bar{y}_i. - \bar{y}..)^2 + rc(\bar{y}..)^2 .$$

It follows from the definition of degrees of freedom that

$\sum\sum y_{ij}^2$	has	rc	degrees of freedom ;
$\sum\sum (y_{ij} - \bar{y}_i.)^2$	has	$r(c-1)$	degrees of freedom ;
$c\sum(\bar{y}_i. - \bar{y}..)^2$	has	$r-1$	degrees of freedom ;
$n(\bar{y}..)^2$	has	1	degree of freedom.

Hence, from the partition theorem and from

$$rc = 1 + r(c-1) + (r-1) ,$$

it follows that SS_2/σ^2 and SS_3/σ^2 are independently distributed as chi-square random variables with $r(c-1)$ and $r-1$ degrees of freedom, respectively.

10.2.6 Analysis of the Fixed Effects Model, One-Way Classification

A major problem arising in the analysis of variance for the fixed effects model is in testing the hypothesis that the r effects are all equal. This hypothesis can be written as

$$H: \quad \varphi_1 = \varphi_2 = \varphi_3 = \cdots = \varphi_r .$$

In fact, if the hypothesis is true, each φ_i must equal zero. It is seen by referring to the analysis of variance table, Table 10.2, that the expected mean squares for both the between-treatments and within-treatments sources, $E(SS_3^*)$ and $E(SS_2^*)$, respectively, equal σ^2 if the hypothesis is true. $E(SS_2^*)$ equals σ^2 even if the hypothesis is false. However, if the hypothesis is false, $E(SS_3^*)$ exceeds σ^2 by the amount $\dfrac{c\sum_{i=1}^{n}\varphi_i^2}{r-1}$. Hence, if the ratio SS_3^*/SS_2^* is computed, a value far from unity would indicate that the hypothesis is false. It can be shown (using the partition theorem) that SS_3/σ^2 and SS_2/σ^2 are independently distributed as chi-square random variables with $r - 1$ and $r(c-1)$ degrees of freedom, respectively, when the hypothesis is true. Hence, the ratio SS_3^*/SS_2^* is distributed as an F random variable with $r-1$ and $r(c-1)$ degrees of freedom when the hypothesis is true. Thus, the hypothesis of equality of the φ's is rejected if

$$F = \frac{SS_3^*}{SS_2^*} \geqq F_{\alpha;\, r-1, r(c-1)}$$

where $F_{\alpha;\, r-1, r(c-1)}$ is the upper α percentage point of the F distribution given in Appendix Table 4. Note that a one-sided test is used since this ratio can only be too large if the hypothesis is false.

If the hypothesis

$$H: \quad \varphi_1 = \varphi_2 = \varphi_3 = \cdots = \varphi_r$$

is rejected, it is possible to make statements about which means differ. For a long time there was no satisfactory solution to this problem. However, Tukey[1] and Scheffé[2] have presented solutions whereby it becomes possible to make contrasts of the form $\varphi_m - \varphi_n$. Tukey's procedure leads to probability statements in the form of confidence intervals for the contrasts $\varphi_m - \varphi_n$; i.e., the probability is $1 - \alpha$ that the values $\varphi_m - \varphi_n$ for all such contrasts ($m = 1, 2, \cdots, r$; $n = 1, 2, \cdots, r$) simultaneously satisfy

[1] J. Tukey, *Allowances for Various Types of Error Rates*, unpublished invited address presented before a joint meeting of the Institute of Mathematical Statistics and the Eastern North American Region of the Biometric Society on March 19, 1952, at Blacksburg, Va.

[2] H. Scheffé, "A Method for Judging All Contrasts in the Analysis of Variance," *Biometrika*, June 1953, Vol. 40.

Table 10.3. Table of Factors k^* (5% Significance Level)[1]

ν \ n[a]	2	3	4	5	6	7	8	9	10	11	12	13	14	15	16	17	18	19	20
1	18.0	26.7	32.8	37.2	40.5	43.1	45.4	47.3	49.1	50.6	51.9	53.2	54.3	55.4	56.3	57.2	58.0	58.8	59.6
2	6.09	8.28	9.80	10.89	11.73	12.43	13.03	13.54	13.99	14.39	14.75	15.08	15.38	15.65	15.91	16.14	16.36	16.57	16.77
3	4.50	5.88	6.83	7.51	8.04	8.47	8.85	9.18	9.46	9.72	9.95	10.16	10.35	10.52	10.69	10.84	10.98	11.12	11.24
4	3.93	5.00	5.76	6.31	6.73	7.06	7.35	7.60	7.83	8.03	8.21	8.37	8.52	8.67	8.80	8.92	9.03	9.14	9.24
5	3.64	4.60	5.22	5.67	6.03	6.33	6.58	6.80	6.99	7.17	7.32	7.47	7.60	7.72	7.83	7.93	8.03	8.12	8.21
6	3.46	4.34	4.90	5.31	5.63	5.89	6.12	6.32	6.49	6.65	6.79	6.92	7.04	7.14	7.24	7.34	7.43	7.51	7.59
7	3.34	4.16	4.68	5.06	5.36	5.61	5.82	6.00	6.16	6.30	6.43	6.55	6.66	6.76	6.85	6.94	7.02	7.09	7.17
8	3.26	4.04	4.53	4.89	5.17	5.40	5.60	5.77	5.92	6.05	6.18	6.29	6.39	6.48	6.57	6.65	6.73	6.80	6.87
9	3.20	3.95	4.42	4.76	5.02	5.24	5.43	5.60	5.74	5.87	5.98	6.09	6.19	6.28	6.36	6.44	6.51	6.58	6.65
10	3.15	3.88	4.33	4.66	4.91	5.12	5.30	5.46	5.60	5.72	5.83	5.93	6.03	6.12	6.20	6.27	6.34	6.41	6.47
11	3.11	3.82	4.26	4.58	4.82	5.03	5.20	5.35	5.49	5.61	5.71	5.81	5.90	5.98	6.06	6.14	6.20	6.27	6.33
12	3.08	3.77	4.20	4.51	4.75	4.95	5.12	5.27	5.40	5.51	5.61	5.71	5.80	5.88	5.95	6.02	6.09	6.15	6.21
13	3.06	3.73	4.15	4.46	4.69	4.88	5.05	5.19	5.32	5.43	5.53	5.63	5.71	5.79	5.86	5.93	6.00	6.06	6.11
14	3.03	3.70	4.11	4.41	4.64	4.83	4.99	5.13	5.25	5.36	5.46	5.56	5.64	5.72	5.79	5.86	5.92	5.98	6.03
15	3.01	3.67	4.08	4.37	4.59	4.78	4.94	5.08	5.20	5.31	5.40	5.49	5.57	5.65	5.72	5.79	5.85	5.91	5.96
16	3.00	3.65	4.05	4.34	4.56	4.74	4.90	5.03	5.15	5.26	5.35	5.44	5.52	5.59	5.66	5.73	5.79	5.84	5.90
17	2.98	3.62	4.02	4.31	4.52	4.70	4.86	4.99	5.11	5.21	5.31	5.39	5.47	5.55	5.61	5.68	5.74	5.79	4.84
18	2.97	3.61	4.00	4.28	4.49	4.67	4.83	4.96	5.07	5.17	5.27	5.35	5.43	5.50	5.57	5.63	5.69	5.74	5.79
19	2.96	3.59	3.98	4.26	4.47	4.64	4.79	4.92	5.04	5.14	5.23	5.32	5.39	5.46	5.53	5.59	5.65	5.70	5.75
20	2.95	3.58	3.96	4.24	4.45	4.62	4.77	4.90	5.01	5.11	5.20	5.28	5.36	5.43	5.50	5.56	5.61	5.66	5.71
24	2.92	3.53	3.90	4.17	4.37	4.54	4.68	4.81	4.92	5.01	5.10	5.18	5.25	5.32	5.38	5.44	5.50	5.55	5.59
30	2.89	3.48	3.84	4.11	4.30	4.46	4.60	4.72	4.83	4.92	5.00	5.08	5.15	5.21	5.27	5.33	5.38	5.43	5.48
40	2.86	3.44	3.79	4.04	4.23	4.39	4.52	4.63	4.74	4.82	4.90	4.98	5.05	5.11	5.17	5.22	5.27	5.32	5.36
60	2.83	3.40	3.74	3.98	4.16	4.31	4.44	4.55	4.65	4.73	4.81	4.88	4.94	5.00	5.06	5.11	5.15	5.20	5.24
120	2.80	3.36	3.69	3.92	4.10	4.24	4.36	4.47	4.56	4.64	4.71	4.78	4.84	4.90	4.95	5.00	5.04	5.09	5.13
∞	2.77	3.32	3.63	3.86	4.03	4.17	4.29	4.39	4.47	4.55	4.62	4.68	4.74	4.80	4.84	4.89	4.93	4.97	5.01

[a] Here n is the number of effects being studied.

[1] This table is taken by permission from Joyce M. May, "Extended and Corrected Tables of the Upper Percentage Points of the Studentized Range", *Biometrika*, Vol. 40, Parts 1 and 2, June 1953, p. 236.

Table 10.4. Table of Factors k^* (1% Significance Level)[1]

ν \ n[a]	2	3	4	5	6	7	8	9	10	11	12	13	14	15	16	17	18	19	20
1	90.0	134	164	186	202	216	227	237	246	253	260	266	272	277	282	286	291	295	298
2	14.0	18.9	22.3	24.7	26.6	28.2	29.5	30.7	31.7	32.6	33.4	34.2	34.8	35.5	36.0	36.5	37.0	37.5	38.0
3	8.26	10.56	12.17	13.34	14.25	15.00	15.65	16.20	16.69	17.13	17.53	17.89	18.23	18.54	18.83	19.09	19.33	19.56	19.79
4	6.51	8.08	9.17	9.97	10.58	11.10	11.55	11.93	12.26	12.56	12.84	13.09	13.32	13.53	13.73	13.92	14.09	14.25	14.40
5	5.70	6.97	7.80	8.42	8.91	9.32	9.67	9.97	10.24	10.48	10.70	10.89	11.08	11.24	11.40	11.55	11.68	11.81	11.93
6	5.24	6.32	7.03	7.56	7.97	8.31	8.61	8.87	9.10	9.30	9.49	9.65	9.81	9.95	10.08	10.21	10.32	10.43	10.54
7	4.95	5.92	6.54	7.01	7.37	7.68	7.94	8.17	8.37	8.55	8.71	8.86	9.00	9.12	9.24	9.35	9.46	9.55	9.65
8	4.74	5.63	6.20	6.63	6.96	7.24	7.47	7.68	7.86	8.03	8.18	8.31	8.44	8.55	8.66	8.76	8.86	8.95	9.03
9	4.60	5.42	5.96	6.35	6.66	6.91	7.13	7.33	7.50	7.65	7.79	7.91	8.02	8.13	8.23	8.33	8.41	8.50	8.58
10	4.48	5.26	5.77	6.14	6.43	6.67	6.88	7.06	7.22	7.36	7.49	7.60	7.71	7.81	7.91	7.99	8.08	8.15	8.23
11	4.39	5.14	5.62	5.98	6.25	6.47	6.67	6.84	6.99	7.13	7.25	7.36	7.46	7.56	7.65	7.73	7.81	7.88	7.95
12	4.32	5.04	5.50	5.84	6.10	6.32	6.51	6.67	6.81	6.94	7.06	7.17	7.26	7.36	7.44	7.52	7.60	7.67	7.73
13	4.26	4.96	5.40	5.73	5.98	6.19	6.37	6.53	6.67	6.79	6.90	7.01	7.10	7.19	7.27	7.35	7.42	7.49	7.55
14	4.21	4.89	5.32	5.64	5.88	6.08	6.26	6.41	6.54	6.66	6.77	6.87	6.96	7.05	7.13	7.20	7.27	7.34	7.40
15	4.17	4.83	5.25	5.56	5.80	5.99	6.16	6.31	6.44	6.55	6.66	6.76	6.85	6.93	7.00	7.07	7.14	7.20	7.26
16	4.13	4.78	5.19	5.49	5.72	5.91	6.08	6.22	6.35	6.46	6.56	6.66	6.74	6.82	6.90	6.97	7.03	7.09	7.15
17	4.10	4.73	5.14	5.43	5.66	5.85	6.01	6.15	6.27	6.38	6.48	6.57	6.66	6.73	6.81	6.87	6.94	7.00	7.05
18	4.07	4.70	5.09	5.38	5.60	5.79	5.95	6.08	6.20	6.31	6.41	6.50	6.58	6.65	6.73	6.79	6.85	6.91	6.97
19	4.05	4.66	5.05	5.34	5.55	5.73	5.89	6.02	6.14	6.25	6.34	6.43	6.51	6.58	6.65	6.72	6.78	6.84	6.89
20	4.02	4.63	5.02	5.30	5.51	5.69	5.84	5.97	6.09	6.19	6.28	6.37	6.45	6.52	6.59	6.66	6.71	6.77	6.82
24	3.96	4.54	4.91	5.17	5.37	5.54	5.69	5.81	5.92	6.02	6.11	6.19	6.26	6.33	6.39	6.45	6.51	6.57	6.61
30	3.89	4.45	4.80	5.05	5.24	5.40	5.53	5.65	5.76	5.85	5.93	6.01	6.08	6.14	6.20	6.26	6.31	6.36	6.41
40	3.82	4.36	4.70	4.93	5.11	5.26	5.39	5.50	5.60	5.69	5.77	5.84	5.90	5.96	6.02	6.07	6.12	6.17	6.21
60	3.76	4.28	4.60	4.82	4.99	5.13	5.25	5.36	5.45	5.53	5.60	5.67	5.73	5.78	5.83	5.88	5.93	5.98	6.01
120	3.70	4.20	4.50	4.71	4.87	5.00	5.12	5.21	5.30	5.38	5.44	5.50	5.56	5.61	5.66	5.71	5.75	5.79	5.83
∞	3.64	4.12	4.40	4.60	4.76	4.88	4.99	5.08	5.16	5.23	5.29	5.35	5.40	5.45	5.49	5.53	5.57	5.61	5.64

[a] Here n is the number of effects being studied.

[1] This table is taken by permission from May, "Extended and Corrected Tables of the Upper Percentage Points of the Studentized Range"; and from Hartley, "Corrigenda to Tables of Percentage Points of the Studentized Range."

$$\bar{x}_m - \bar{x}_n - k \leqq \varphi_m - \varphi_n \leqq \bar{x}_m - \bar{x}_n + k$$

where k is a factor obtained from Table 10.3 or 10.4, depending on whether $\alpha = 0.05$ or 0.01 and from the analysis of variance table. The means are said to differ significantly when the appropriate confidence interval fails to include 0. Furthermore, if the interval includes only positive values, the difference is said to differ significantly from zero, and to be positive. Similarly, if the confidence interval includes only negative values, the difference is said to differ significantly from zero, and to be negative. If the interval includes 0, the means are said not to differ significantly. Summarizing then, the F test is made for the homogeneity of the means. If this hypothesis is rejected, Tukey's procedure is applied to determine which of the means differ.

A possible short-cut method for determining the homogeneity of all the treatment effects is to check whether $\bar{x}_{\max} - \bar{x}_{\min} \pm k$ includes 0. If this contrast includes 0, every other contrast of this type includes 0, and hence all the treatment effects are called equal. If the contrast above fails to include zero, the corresponding treatment effects, and perhaps others, differ significantly. This short-cut method has a level of significance exactly equal to α. However, the data in the analysis of variance table are necessary for determining the factor k, and for other results, so that this table should be completed anyway.

The factor k can be obtained from the data in the analysis of variance table and Table 10.3 or 10.4 depending on the level of significance. Table 10.3 or 10.4 is entered with the indices degrees of freedom and the number of treatments studied (r). The proper degrees of freedom are those corresponding to the degrees of freedom of the within treatments source in the analysis of variance table, i.e., $r(c - 1)$. The factor k^* is read from the table; k is obtained from $k = k^*\sqrt{SS_2^*/c}$.

Scheffé's procedure also leads to confidence statements and if we apply his test when the F test of homogeneity is rejected, the level of significance is preserved. On the other hand, if we apply Tukey's test when the F test of homogeneity is rejected, we increase the significance level *slightly*. However, if the only type comparison of interest has the form $\varphi_m - \varphi_n$, Tukey's procedure will result in shorter confidence intervals for $\varphi_m - \varphi_n$ than would Scheffé's procedure.[1]

[1] If other type of contrasts are also desired, e.g., $(\zeta_1 + \zeta_2 + \zeta_3)/3 - (\zeta_4 + \zeta_5 + \zeta_6)/3$, Scheffé's procedure should be used.

Consequently, we are going to use the Tukey method, and proceed as if the significance level suffered a negligible increase.

10.2.7 THE OC CURVE OF THE ANALYSIS OF VARIANCE FOR THE FIXED EFFECTS MODEL

There has been no discussion of the implications of the analysis of variance procedure. This procedure guarantees that if the treatments do not differ, $100(1-\alpha)$ % of the time this conclusion will be reached. On the other hand, there have been no probability statements attached to the procedure if the treatments do differ. For example, suppose that all but one of the testing machines had the same expected value. Denote the expected value of the odd laboratory by M and the other means by N. Furthermore, suppose it were possible to plot the OC curve of the analysis of variance procedure as in Figure 10.1.

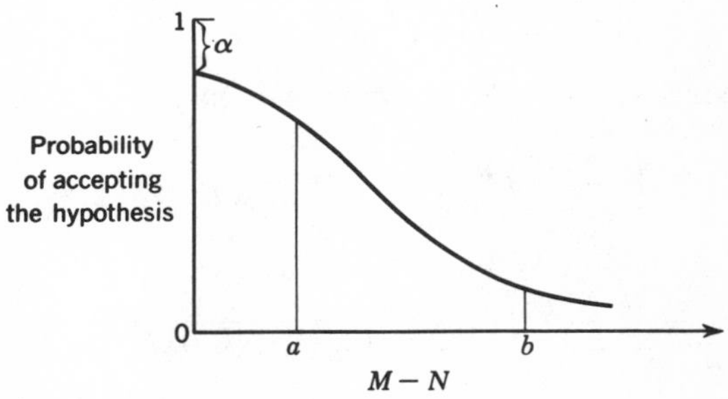

Fig. 10.1. OC curve for laboratory tests.

If $M = N$ (all machines are the same), the probability of accepting the hypothesis is $(1-\alpha)$. If $(M-N) = b$, there is a large probability of rejecting the hypothesis. However, if $(M-N) = a$, the probability of accepting the hypothesis is quite high. If differences of this magnitude are important to detect, a procedure having this OC curve is inadequate. In fact, running an experiment under this setup is rather wasteful of funds if a difference of $(M-N) = a$ is important.

Consequently, some indication of the OC curve for the analysis of variance procedure is important. A problem arises in a suitable choice of an abscissa scale. If laboratories differ, what measure of difference can be used? In the example just described, it was assumed that all but one of the means were the same. Rarely can such an assumption be made in practice. A measure for which an OC curve can be constructed is

$$\Phi = \frac{\sqrt{\sum_{i=1}^{r} \varphi_i^2 \Big/ r}}{\sigma/\sqrt{c}}.$$

Two interpretations for this measure can be given. If the laboratories do not differ, the standard deviation of an observation chosen at random is just σ. On the other hand, if the laboratories do differ, the standard deviation of an observation chosen at random is increased to

$$\sqrt{\sigma^2 + \left(\sum_{i=1}^{r} \varphi_i^2 \Big/ r\right)}.$$

Consequently, if the experimenter chooses a fixed percentage, say P, for the increase in standard deviation of an observation beyond which he wishes to reject the hypothesis of equality of the φ_i, this is equivalent to choosing

$$\frac{\sqrt{\sigma^2 + \left(\sum_{i=1}^{r} \varphi_i^2 \Big/ r\right)}}{\sigma} = 1 + 0.01P$$

(where the increase in σ, P, is expressed in percent) or

$$\frac{\sqrt{\sum_{i=1}^{r} \varphi_i^2 \Big/ r}}{\sigma} = \sqrt{(1 + 0.01P)^2 - 1}$$

so that $$\Phi = \frac{\sqrt{\sum_{i=1}^{r} \varphi_i^2 \Big/ r}}{\sigma/\sqrt{c}} = \left(\sqrt{(1 + 0.01P)^2 - 1}\right)\sqrt{c}\,.$$

If a probability β of rejecting the hypothesis is associated with the chosen value of P, two points on the OC curve are fixed, the other point being the error of Type I associated with $\Phi = 0$.

A second method for determining a critical value of the measure Φ is to find upper and lower bounds for Φ. This information can be obtained if the range (largest minus the smallest) of the φ_i is specified. In the laboratory example this entails choosing the largest difference between any two laboratories which are tolerable, and beyond which the hypothesis of equality of the φ_i should be rejected. Denote the largest difference by W.

$$\Phi\,(\text{min}) = \frac{W}{\sigma}\sqrt{\frac{c}{2r}}$$

$$\Phi\,(\text{max}) = \begin{cases} \dfrac{W}{2\sigma}\sqrt{c} & \text{for } r \text{ even} \\[2ex] \dfrac{W}{2\sigma}\sqrt{\dfrac{c(r^2-1)}{r^2}} & \text{for } r \text{ odd.} \end{cases}$$

Thus a bound on Φ can be obtained if W is given and some estimate of σ is available. Such an estimate can often be obtained from previous data.

OC curves for the analysis of the variance for the fixed effects model are presented in Figures 10.2 to 10.9.* These curves are indexed according to the degrees of freedom for the treatments $\nu_1 = r - 1$ and degrees of freedom for the error term $\nu_2 = r(c - 1)$. These curves can be used directly for finding the probability of accepting the hypothesis given the number of observations used on the experiments, i.e., r and c, or can be used to design an experiment by means of trial and error. In this latter case, if r is fixed (number of treatments) the appropriate value of c can be obtained as follows if two points are specified. A value c is chosen. Φ is then computed and the OC curves are entered with ν_1 and ν_2 for the appropriate value of α. The probability of acceptance corresponding to Φ is read out. If this coincides with the desired value, a solution is reached; if not the next higher value of c is chosen.

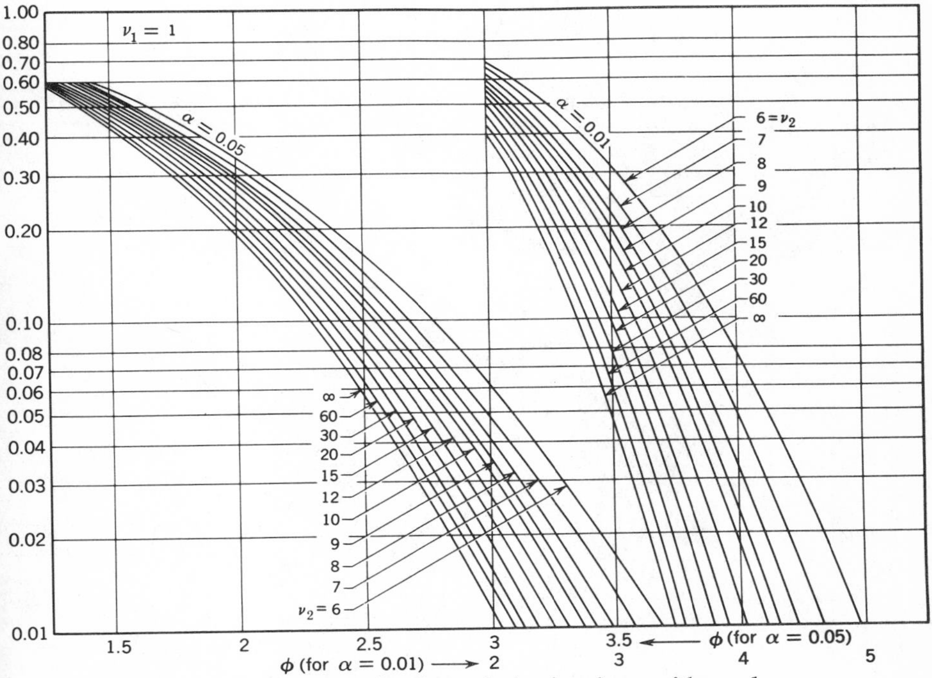

Fig. 10.2. OC curves for the analysis of variance with $\nu = 1$.

* By permission, Figures 10.2–10.9 are from "Power Function for Analysis of Variance Tests Derived from the Non-Central F Distribution," by E.S. Pearson and H.O. Hartley, Biometrika, Vol. 38, Part 1 and 2, June, 1951

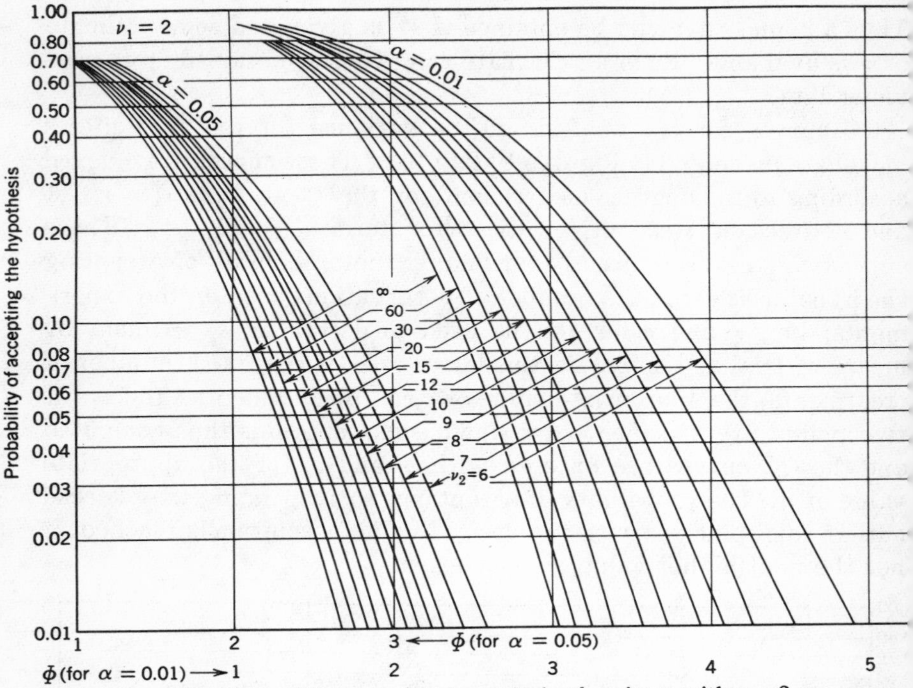

Fig. 10.3. OC curves for the analysis of variance with $\nu = 2$.

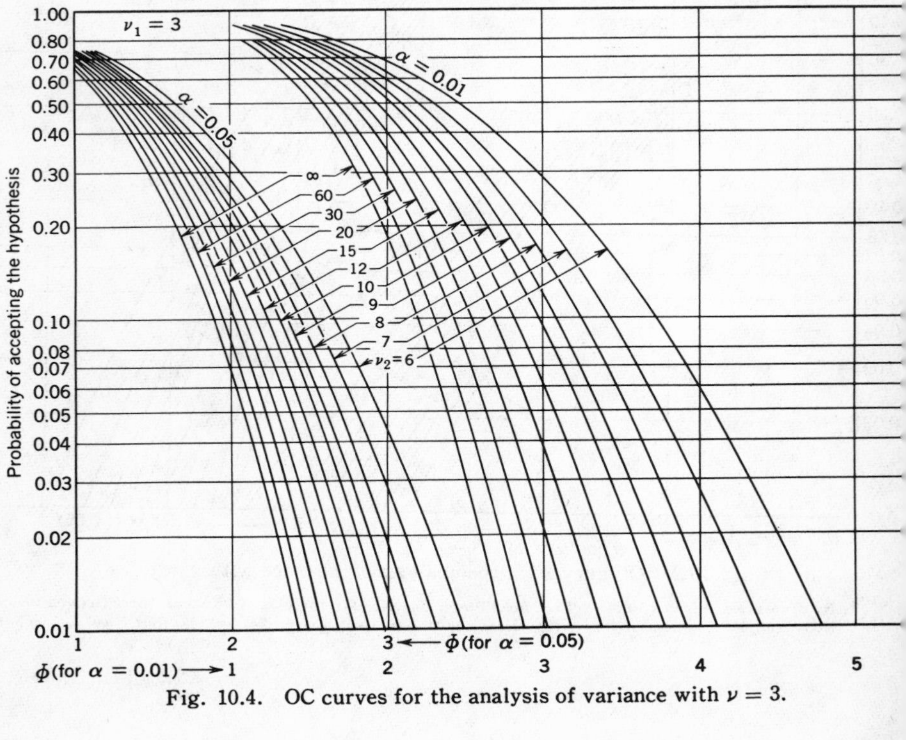

Fig. 10.4. OC curves for the analysis of variance with $\nu = 3$.

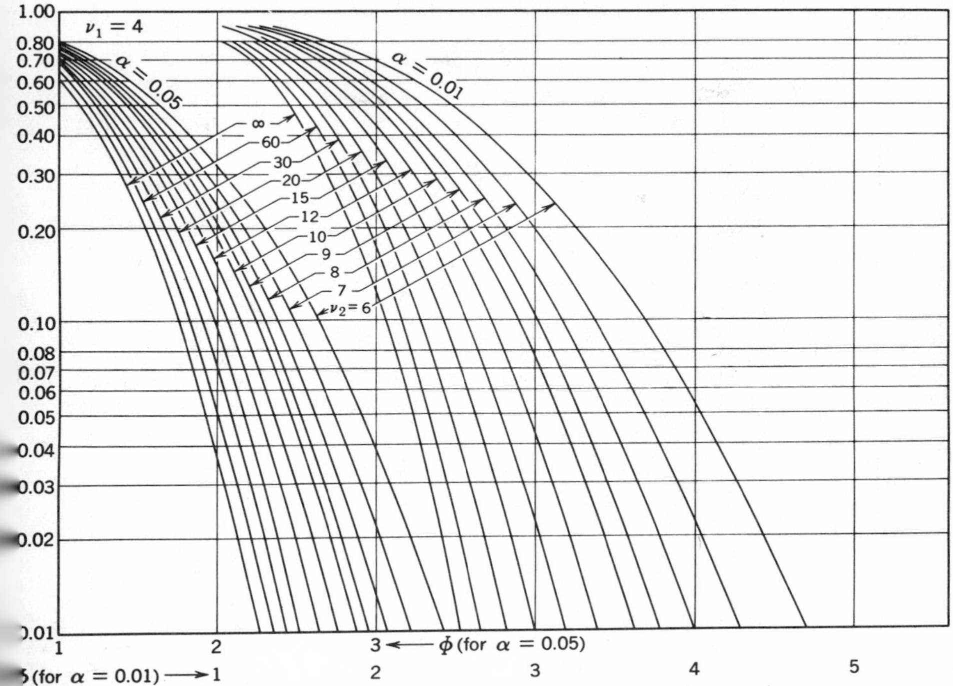

Fig. 10.5. OC curves for the analysis of variance with $\nu = 4$.

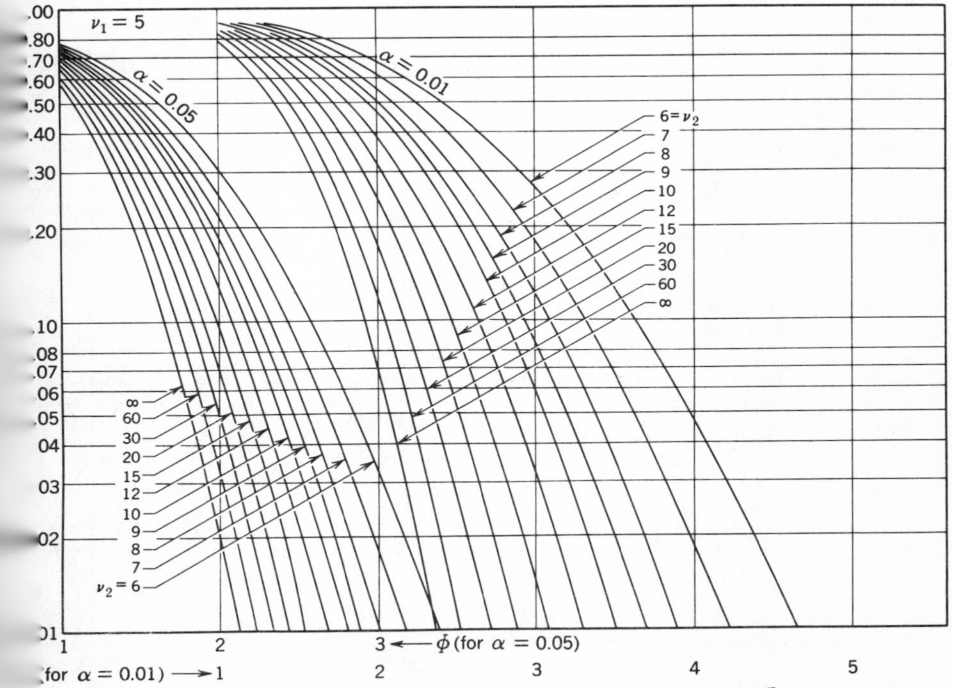

Fig. 10.6. OC curves for the analysis of variance with $\nu = 5$.

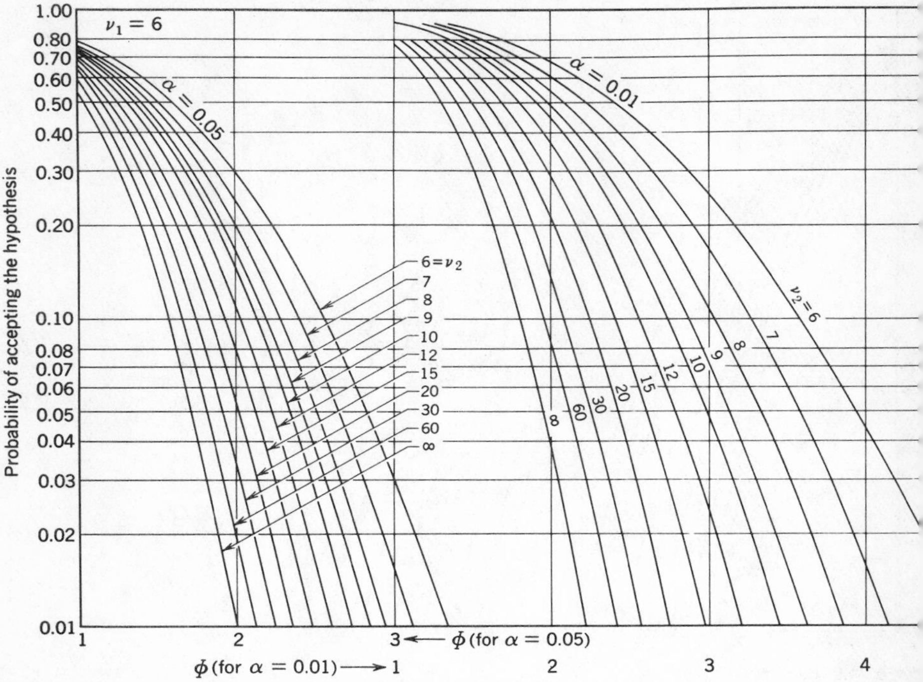

Fig. 10.7. OC curves for the analysis of variance with $\nu = 6$.

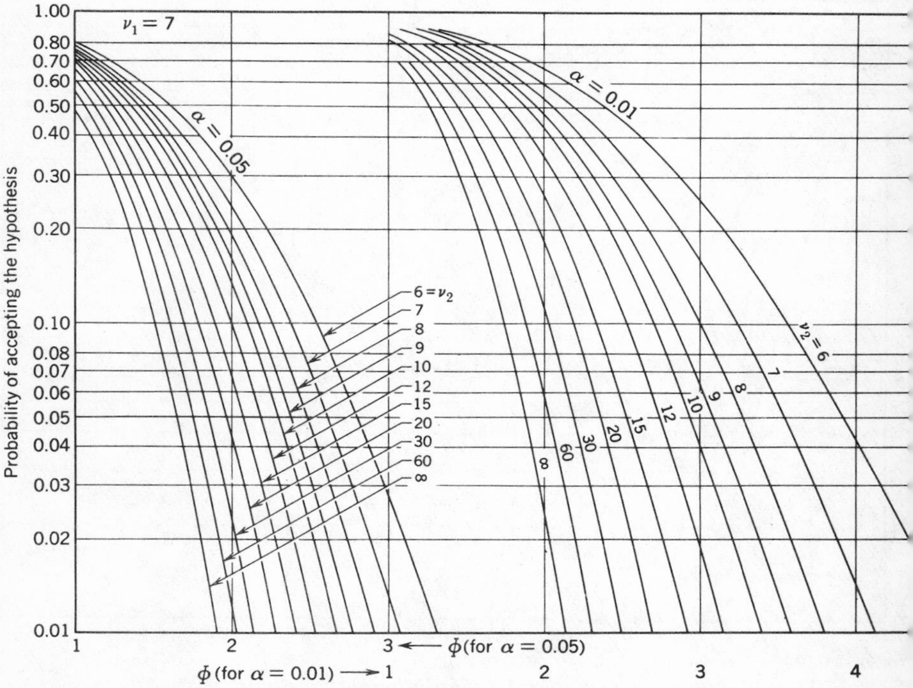

Fig. 10.8. OC curves for the analysis of variance with $\nu = 7$.

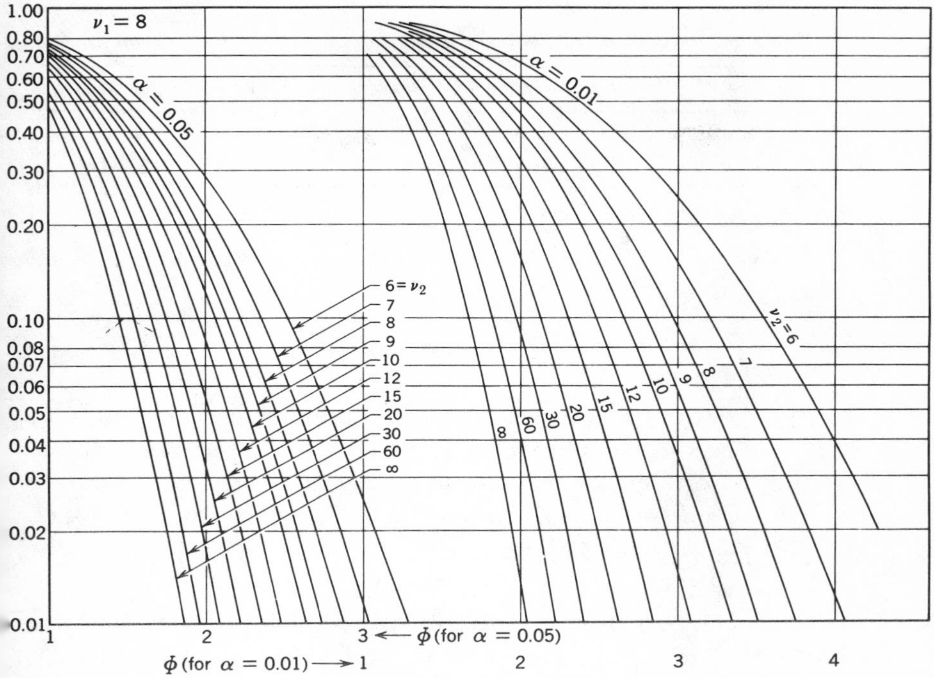

Fig. 10.9. OC curves for the analysis of variance with $\nu = 8$.

10.2.8 Example using the fixed effects model

If the laboratories are treated as fixed effects, the following is an analysis of the data given in Table 10.1.

Computational Procedure

(1) $R_1 = 369, \quad R_2 = 371, \quad R_3 = 345, \quad R_4 = 359.$

(2) $$T = \sum_{i=1}^{4} R_i = 1{,}444$$

(3) $$\sum_{i=1}^{4} \sum_{j=1}^{5} x_{ij}^2 = 104{,}352$$

(4) $$\frac{\sum_{i=1}^{4} R_i^2}{5} = 104{,}341.6$$

(5) $$\frac{T^2}{20} = 104{,}256.8$$

(6) $SS_3 = 84.8$

(7) $SS = 95.2$

(8) $SS_2 = 10.4$

Table 10.5. Analysis of Variance Table for Laboratory Calibration

Source	Sum of Squares	Degrees of Freedom	Mean Square
Between treatments ..	84.8	3	28.27
Within treatments	10.4	16	0.65
Total	95.2	19	5.01

Thus, the value of F for testing whether the laboratories are equal is given by

$$F = \frac{SS_3^*}{SS_2^*} = 43.49.$$

Since $F_{0.05\,;\,3,16} = 3.24$ and $F = 43.49 > 3.24$, the hypothesis that $\varphi_1 = \varphi_2 = \varphi_3 = \varphi_4$ is rejected and we conclude that the laboratories differ. In order to determine the magnitude of these differences, compute k for $r = 4$ and $r(c - 1) = 16$. Here $k^* = 4.05$, where the factor 4.05 is obtained from Table 10.3. Therefore, $k = 4.05\sqrt{0.65/5} = 1.46$ and any two means differing by this much differ significantly at $\alpha = 0.05$. The four laboratory means are: 73.8 for Laboratory A; 74.2 for Laboratory B; 69.0 for Laboratory C; and 71.8 for Laboratory D. Thus the following differences are significant:

Laboratory A from Laboratory C
Laboratory A from Laboratory D
Laboratory B from Laboratory C
Laboratory B from Laboratory D
Laboratory C from Laboratory D.

In obtaining the OC curve for the test, suppose the criterion about the increase in the standard deviation is given; i.e., if the difference in the laboratories is such that a random observation has its standard deviation increased by as much as 20% the resultant value of Φ is

$$[\sqrt{(1.2)^2 - 1}\,]\,\sqrt{5} = \sqrt{(0.44)5} = \sqrt{2.20} = 1.48\,;$$

$\nu_1 = r - 1 = 3$ and $\nu_2 = r(c - 1) = (4)(4) = 16$. Hence, the probability of accepting the hypotheses at the five percent level of significance is found to be approximately 0.43 from Figure 10.4. If, however, prior to the experiment a probability of acceptance of only 0.10 could be tolerated, a sample of 9 observations is required from each laboratory. This is attained by trial and error. If $c = 6$ is chosen first and computations similar to that above are made, the probability of acceptance is found to equal 0.30. Continuing in this manner, the desired probability of acceptance is reached with $c = 9$.

$$\Phi = [\sqrt{(1.2)^2 - 1}]\sqrt{9} = 1.99.$$

Entering Figure 10.4 with $\nu_1 = 3$ and $\nu_2 = 4(8) = 32$, the probability of accepting the hypothesis is approximately 0.10 using the five percent level of significance.

10.2.9 Analysis of the Random Effects Model

In the random effects model, testing the hypothesis about the homogeneity of treatments is equivalent to testing the hypothesis

$$H: \quad \sigma_\varphi^2 = 0.$$

Again referring to the analysis of variance table, Table 10.2, it is evident that both $E(SS_3^*)$ and $E(SS_2^*)$ equal σ^2 if the hypothesis is true. $E(SS_2^*)$ equals σ^2 even if the hypothesis is false, whereas $E(SS_3^*)$ exceeds σ^2 by an amount equal to $c\sigma_\varphi^2$ when the hypothesis is false. Hence, a large value of SS_3^*/SS_2^* compared to unity indicates that the hypothesis is false. Just as in the fixed effects model

$$F = \frac{SS_3^*}{SS_2^*}$$

is distributed as an F random variable with $r - 1$ and $r(c - 1)$ degrees of freedom. Thus, the hypothesis that $\sigma_\varphi^2 = 0$ (implying homogeneity of all treatments) is rejected if

$$\frac{SS_3^*}{SS_2^*} \geqq F_{\alpha;\, r-1, r(c-1)}.$$

In order to determine a measure of how much the treatments differ, the quantity σ_φ^2 can be estimated. The analysis of variance table indicates that SS_2^* is an unbiased estimate of σ^2 and that SS_3^* is an unbiased estimate of $\sigma^2 + c\sigma_\varphi^2$. Assuming that these estimates are equal to the parameters that are being estimated, i.e., $SS_2^* = \sigma^2$ and $SS_3^* = \sigma^2 + c\sigma_\varphi^2$, an estimate of σ_φ^2 can be obtained from these equations. Such an estimate is given by

$$\frac{SS_3^* - SS_2^*}{c}.$$

If this quantity is negative, the estimate is taken to be zero.

10.2.10 The OC Curve for the Random Effects Model

The analysis of variance procedure for the random effects model was presented in Section 10.2.9. The hypothesis that $\sigma_\varphi^2 = 0$ was accepted if the quantity

$$\frac{SS_3^*}{SS_2^*} \leqq F_{\alpha;r-1,r(c-1)}.$$

SS_3^*/SS_2^* has an F distribution whenever $\sigma_\varphi^2 = 0$, and the probability of acceptance is $(1-\alpha)$. In general, the probability of accepting the hypothesis that $\sigma_\varphi^2 = 0$ can be plotted as a function of

$$\lambda = \sqrt{\frac{\sigma^2 + c\sigma_\varphi^2}{\sigma^2}}.$$ [1]

The resulting OC curves are similar to the curves presented in Figures 7.3 and 7.4. Unfortunately these OC curves were presented for equal degrees of freedom and, hence, their use in the analysis of variance is limited. Further OC curves are presented in Figures 10.10 to 10.17, pages 309 to 313, indexed for degrees of freedom $\nu_1 = r - 1$ and $\nu_2 = r(c-1)$.

A physical interpretation of

$$\lambda = \sqrt{\frac{\sigma^2 + c\sigma_\varphi^2}{\sigma^2}} = \sqrt{1 + \frac{c\sigma_\varphi^2}{\sigma^2}}$$

can be obtained as follows: Consider the examples of the laboratories where the same four laboratories are chosen from many. If the laboratories are homogeneous, the standard deviation of an observation chosen at random is just σ. On the other hand, if the laboratories are not homogeneous, the standard deviation of an observation chosen at random is increased to

[1] If the hypothesis is false SS_3^*/SS_2^* does not have the F distribution. However, the quantity

$$\frac{SS_3^*/(\sigma^2 + c\sigma_\varphi^2)}{SS_2^*/\sigma^2}$$

does have an F distribution with $r-1$ and $r(c-1)$ degrees of freedom. Hence, the criterion for acceptance of

$$\frac{SS_3^*}{SS_2^*} \leqq F_{\alpha;(r-1),r(c-1)}$$

is equivalent to

$$\frac{SS_3^*/(\sigma^2 + c\sigma_\varphi^2)}{SS_2^*/\sigma^2} \leqq F_{\alpha;(r-1),r(c-1)} \frac{\sigma^2}{\sigma^2 + c\sigma_\varphi^2}.$$

The probability distribution of

$$\frac{SS_3^*/(\sigma^2 + c\sigma_\varphi^2)}{SS_2^*/\sigma^2}$$

is known and consequently, the probability of acceptance can be plotted as a function of $\lambda = \sqrt{\dfrac{\sigma^2 + c\sigma_\varphi^2}{\sigma^2}}$.

$$\sqrt{\sigma^2 + \sigma_\varphi^2}\,.$$

Consequently, if the experimenter chooses a fixed percentage, say P, for the increase in standard deviation of an observation beyond which he wishes to reject the hypothesis of homogeneity, it is equivalent to choosing

$$\frac{\sqrt{\sigma^2 + \sigma_\varphi^2}}{\sigma} = 1 + 0.01P \quad \text{(where the increase in } \sigma, P, \text{ is expressed in percent)}$$

or

$$\frac{\sigma_\varphi^2}{\sigma^2} = (1 + 0.01P)^2 - 1\,;$$

hence $$\lambda = \sqrt{1 + c[(1 + 0.01P)^2 - 1]}\,.$$

If a probability β of rejecting the hypothesis is associated with the chosen value of P, two points on the OC curve are fixed, the other point being the error of Type I associated with $\lambda = 1$.

The probabilities of acceptance for a given experimental setup can be evaluated in the manner above by choosing a P and then finding the probability of acceptance. On the other hand, if P, β, and r (number of treatments) are chosen in advance, the proper value of c can be found by trial and error in a similar manner as described for the OC curve of the fixed effects model.

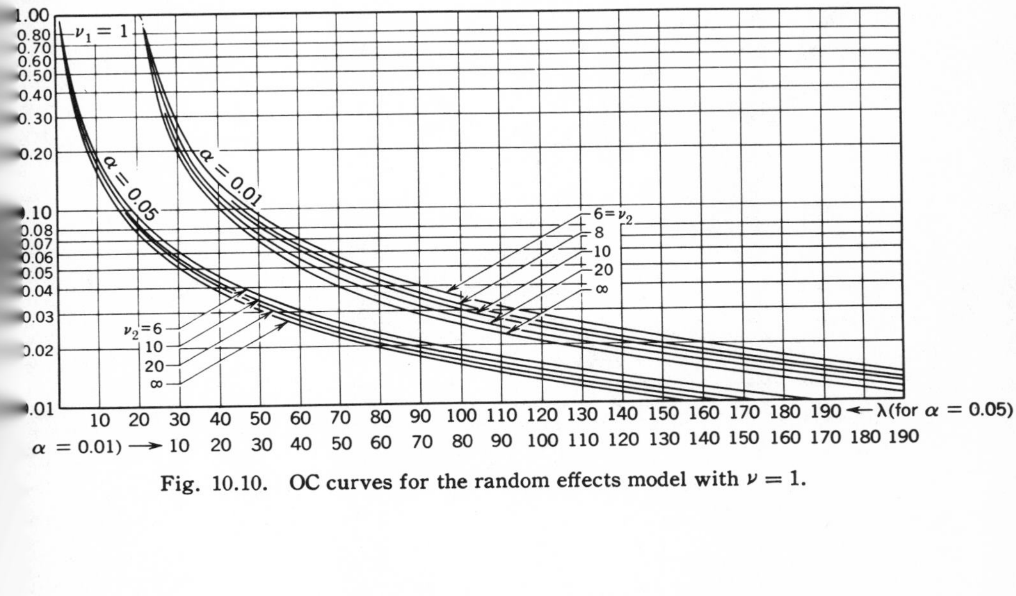

Fig. 10.10. OC curves for the random effects model with $\nu = 1$.

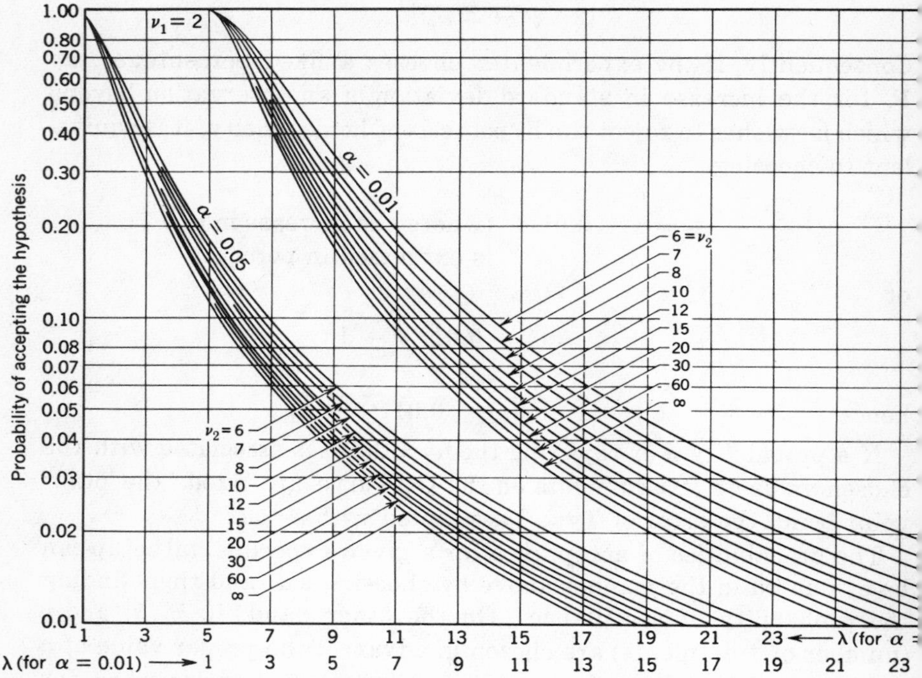

Fig. 10.11. OC curves for the random effects model with $\nu = 2$.

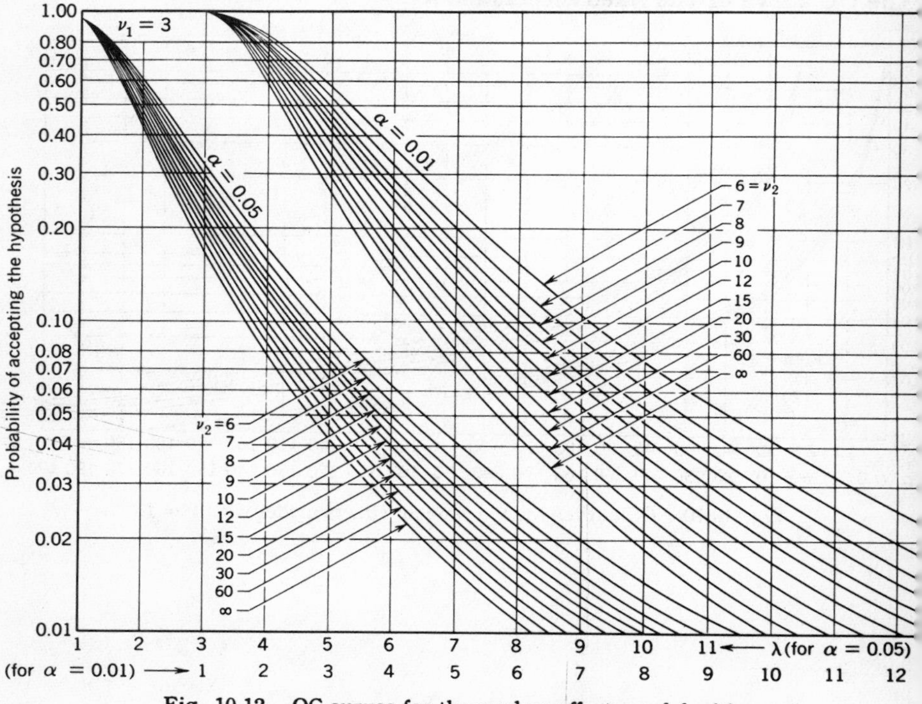

Fig. 10.12. OC curves for the random effects model with $\nu = 3$.

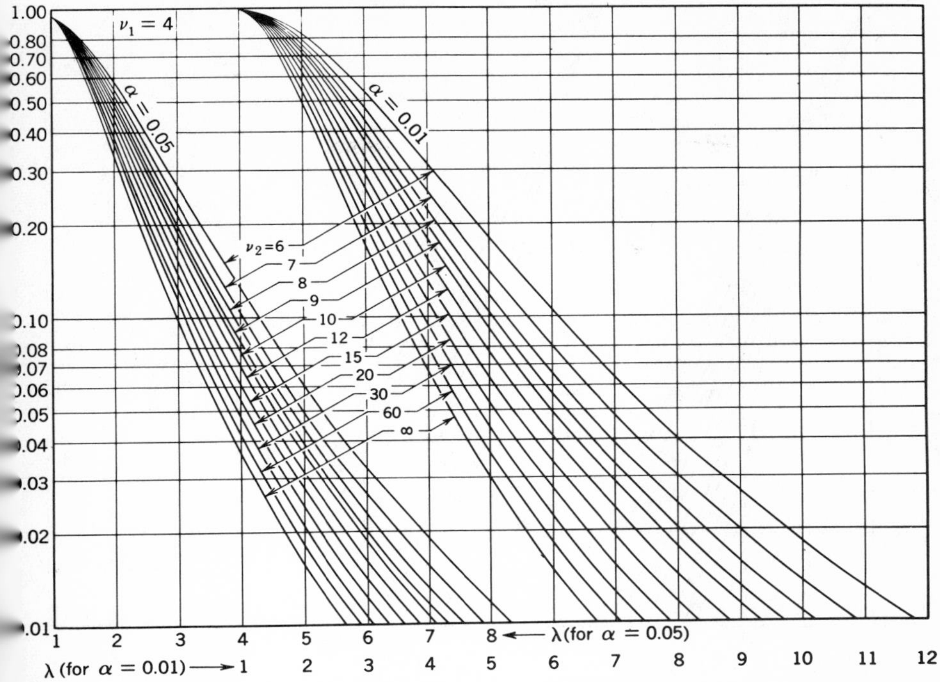

Fig. 10.13. OC curves for the random effects model with $\nu = 4$.

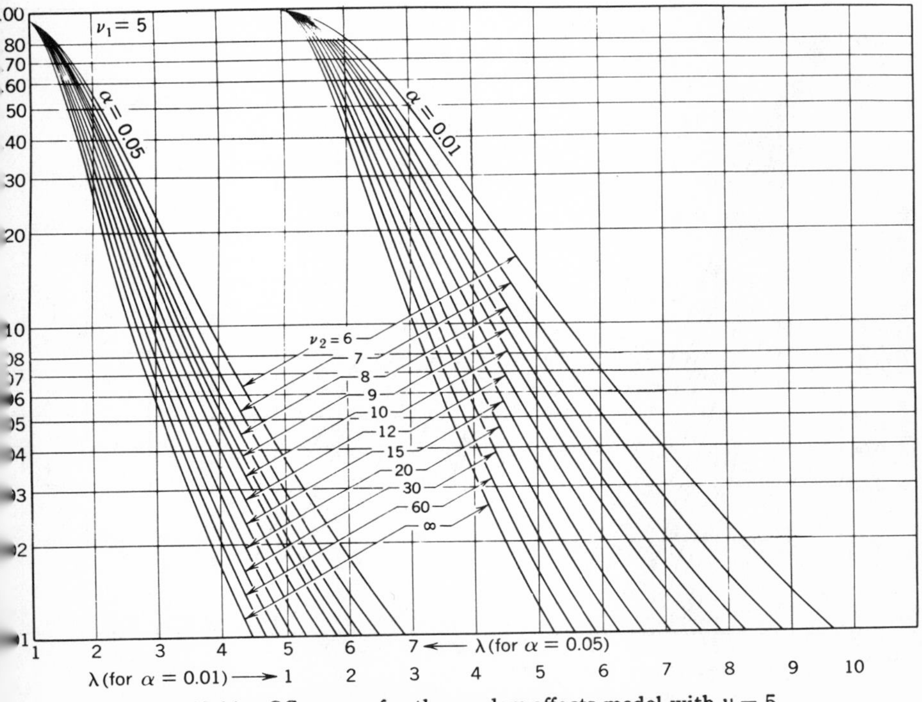

Fig. 10.14. OC curves for the random effects model with $\nu = 5$.

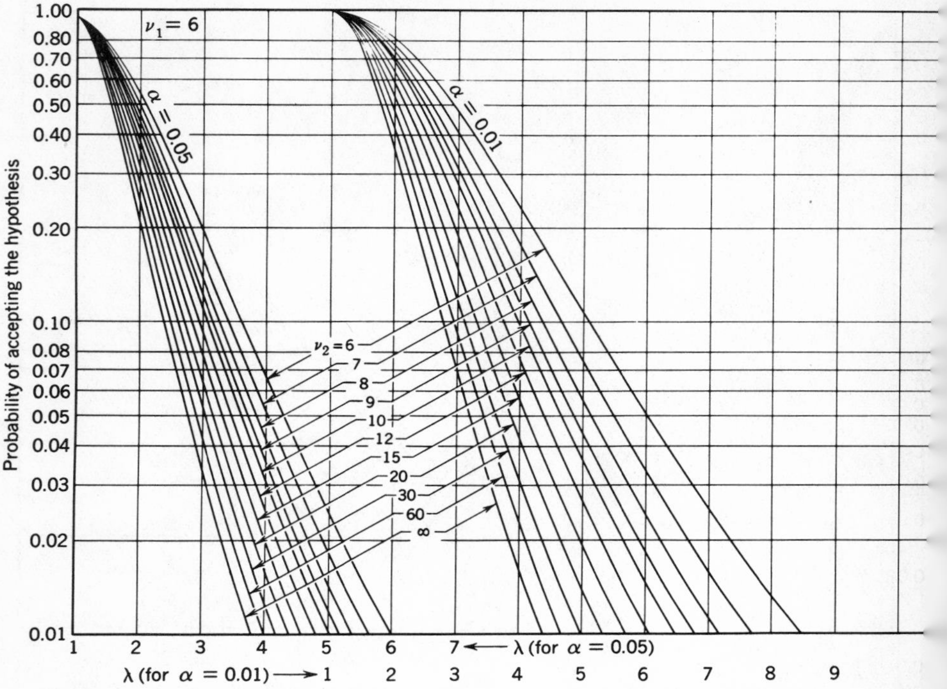

Fig. 10.15. OC curves for the random effects model with $\nu = 6$.

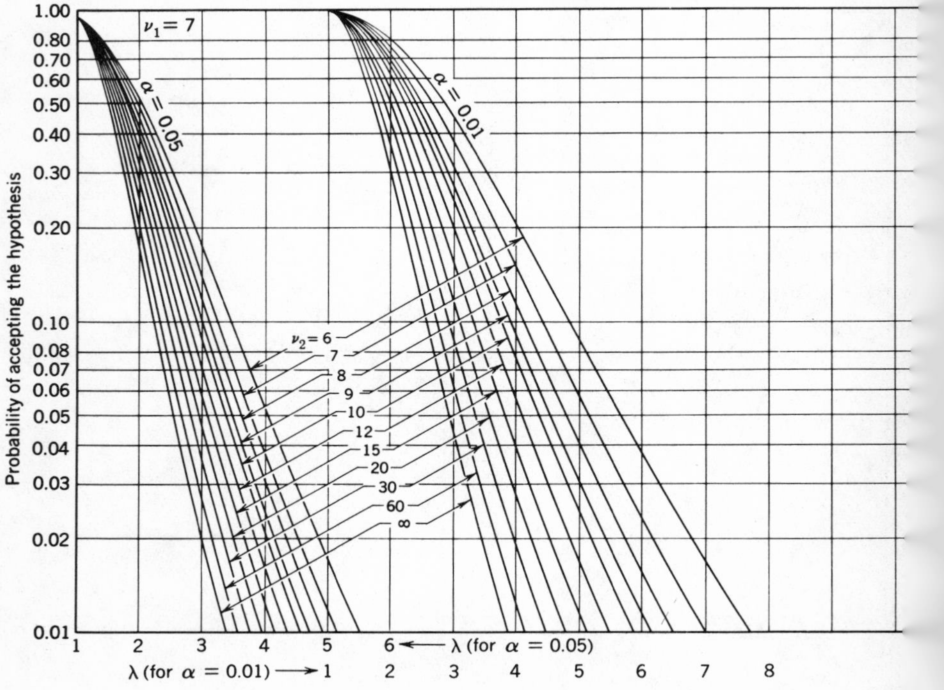

Fig. 10.16. OC curves for the random effects model with $\nu = 7$.

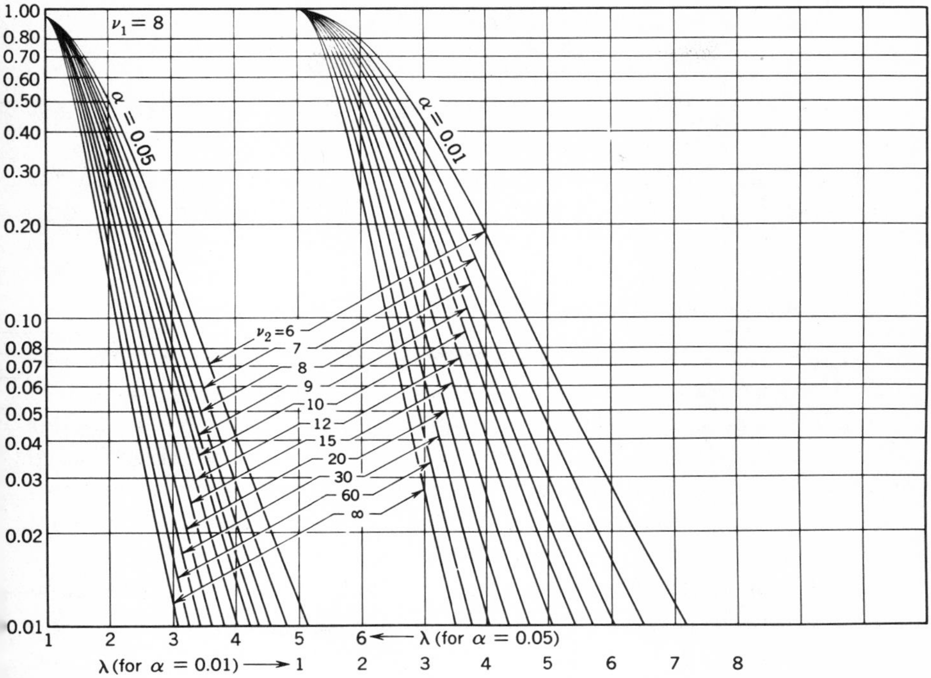

Fig. 10.17. OC curves for the random effects model with $\nu = 8$.

10.2.11 Example using the random effects model

If it is assumed that there were a large number of laboratories to be calibrated and a random sample of four laboratories are chosen, the example of Section 10.2.8 would fit the random effects model. The formal calculations including the computational procedure, analysis of variance table, and F tests are identical with the results found in Section 10.2.8. The fact that $F = SS_3^*/SS_2^*$ exceeded the critical value indicates that all laboratories are not homogeneous. Determining which of the four laboratories included in the experiment differ is irrelevant, and determining which of the laboratories differ among all laboratories is impossible. However, an estimate of the variability of laboratories, σ_φ^2, can be obtained, and is found to be

$$\frac{SS_3^* - SS_2^*}{c} = 5.524.$$

When a 20% increase in the variability of the random variable occurs, the probability of accepting the hypothesis that $\sigma_\varphi^2 = 0$ using the F procedure is 0.59 for a 5 percent level of significance. This value is found as follows:

$$\lambda = \sqrt{1 + 5[(1.2)^2 - 1]}$$
$$= \sqrt{1 + 2.2} = \sqrt{3.2} = 1.79.$$

From Figure 10.12 ($\nu_1 = 3$), the value of 0.59 is read out corresponding to $\lambda = 1.79$.

If only a probability of acceptance of 0.10 can be tolerated for $P = 35\%$, a value of $c = 16$ is required. This value is found by trial and error.

10.2.12 Randomization Tests in the Analysis of Variance

It has been emphasized in the discussion of the F test in the analysis of variance that the random effects ε_{ij} are assumed to be normally and independently distributed. These tests actually have a wider justification if they are considered special cases of what are called permutation tests. Suppose, for example, we are interested in testing the difference between two effects, say percentage of carbon, and suppose that the experimenter takes eight samples (let them be named a, b, c, etc.) and chooses four with a table of random numbers to receive treatment A and the other four treatment B. The results of this experiment are:

A		B	
a	2	f	17
h	15	e	19
b	10	c	18
d	9	g	12

He could test the hypothesis that there is no difference between the treatments by calculating a t test or an F test, or consider it a one-way analysis of variance, calculating an F with 1 and 6 degrees of freedom; the value of F is 6 in this problem. Suppose he is interested in a level of significance of 0.05. Even if he is unwilling to assume normality, he can obtain a test of the hypothesis as follows: Consider all possible pairs of samples that can be formed with the eight numbers *above*; there are $\binom{8}{4} = 70$ different ways for the experiment to turn out (one for each way of selecting four of the eight numbers for treatment A). If there is no difference between treatments, each of these outcomes is equally likely. Suppose we adopt the procedure that we will reject the hypothesis of equality of treatments if we have one of the samples with the four largest differences in the mean; the probability of rejection when the hypothesis is true is $4/70 = 0.057$, since all 70 possibilities are equally likely. If the hypothesis is false, large differences are more likely.

A partial list of the 70 outcomes (starting with most extreme ones) is:

Table 10.6.

A	B	Difference in Means $\bar{x}_A - \bar{x}_B$	F
2 9 10 12	15 17 18 19	−9	14.84
15 17 18 19	2 9 10 12	+9	
2 9 10 15	12 17 18 19	−7.5	6.00
12 17 18 19	2 9 10 15	+7.5	
2 9 10 17	12 15 18 19	−6.5	3.61
12 15 18 19	2 9 10 17	+6.5	
2 9 12 15	10 17 18 19	−6.5	
10 17 18 19	2 9 12 15	+6.5	

Observe that only 4 of the 70 possibilities give values of F as large as that in the particular sample obtained.

The significance level of the test is 0.057. If $F = 6$ is looked up in the table it is seen to be significant at 0.05 level (using the assumption of normality).

This kind of test is called a randomization test. It has the drawback that for medium or large-sized samples the computational burden is enormous. For example, if there are 20 objects divided at random into two groups of 10, then the number of possibilities is $\binom{20}{10} = 184{,}756$, and listing the 5% most extreme cases would be prohibitive.

Both numerical and analytical investigations[1] have shown, however, that for medium and large samples, the distribution of F obtained through enumeration of the possibilities is closely approximated by the tabulated F distribution. So without assuming normality we can test the hypothesis of no treatment effects by making a randomization test, but instead of exactly evaluating the significance level by actual counting we content ourselves with an approximate evaluation, as yielded by the F tables.

10.3 Two-Way Analysis of Variance, One Observation per Combination

Experiments can be conducted in such a way as to study the effects of several variables in the same experiment. For each variable a number of levels may be chosen for study. When observations are made for all possible combinations of levels the experiment is called a factorial experiment. This section is concerned with a factorial experiment where the effects of two variables at several levels (or using several treatments) are to be studied when there is one observation per combination of levels.

[1] The agreement between the two values 0.05 and 0.057 is not merely by coincidence.

10.3.1 Fixed effects model

In this section only fixed levels of the factors will be considered; i.e., the categories or levels of a given factor chosen will be the only ones of interest. Furthermore, for each combination of levels only one observation will be taken. For example, Youden[1] reports the following experiment. A study was made of chemical cells used as a means of setting up a reference temperature. The sealed cells, containing about a pound of chemical, were heated just sufficiently to melt the chemical and were then carefully insulated. The temperature of the triple point was maintained very closely for a day or so. Considerable time was required for the resistance thermometers to attain thermal equilibrium when placed in wells extending into the chemical. Consequently only one thermometer could be employed on a given melting of the cell. Differences in readings found among a number of cells could also arise if the thermometers were not in agreement. When all the thermometers in turn were tried with one cell, the work spread over several days, introducing possible variations in the bridge measurements that did not occur among readings made on the same day. Four chemical cells and four thermometers were chosen for the study. Each thermometer was placed in every chemical cell, the total experiment extending over a period of four days. The possible day effect was randomized out by assigning the day in which a particular thermometer was placed in a given cell by means of a random device.[2] If the purpose of the study was to calibrate these four thermometers and these four chemical cells, and the results interpreted to hold for the chosen thermometers and chemical cells only, these quantities may be considered as fixed effects.

The data are given in Table 10.7. Only one reading is taken for each combination of thermometer and chemical cell. The entries are the readings converted to degrees Centigrade. Only the third and fourth decimal places are given, as the readings agree up to the last two places.

The mathematical model can be written as

$$x_{ij} = \zeta_{ij} + \varepsilon_{ij}; \qquad i = 1, 2, 3, 4; \quad j = 1, 2, 3, 4;$$

where x_{ij} is the temperature reading in the ith chemical cell using the jth thermometer; ζ_{ij} is the mean (of the distribution) tempera-

[1] This example is based on an example appearing in W. J. Youden, *Statistical Methods for Chemists*, John Wiley & Sons, Inc., New York, 1951, pp. 96-98.

[2] Actually, Youden used a design called a Latin Square, which isolated the day effect. The randomization of one day effect is an acceptable alternative.

ture reading in the ith chemical cell using the jth thermometer; ε_{ij} is the random effect which is assumed to be independently normally distributed with mean 0 and variance (unknown) σ^2.

Table 10.7. Pairing of Cells and Thermometers

Chemical Cell	Thermometers I	II	III	IV	Totals	Average
1	36	38	36	30	140	35.0
2	17	18	26	17	78	19.5
3	30	39	41	34	144	36.0
4	30	45	38	33	146	36.5
Totals............	113	140	141	114	508	
Average..........	28.25	35.0	35.25	28.5		

Before proceeding to make inferences about the structure of the ζ_{ij}, it is useful to refer to Table 10.8, which is a table of expected values.

Table 10.8. Table of Expected Values for Temperature Readings

Chemical Cell	Thermometers I	II	III	IV	Average over Chemical Cells	Differences
1	ζ_{11}	ζ_{12}	ζ_{13}	ζ_{14}	$\zeta_{1\cdot}$	$\varphi_1=\zeta_{1\cdot}-\zeta_{\cdot\cdot}$
2	ζ_{21}	ζ_{22}	ζ_{23}	ζ_{24}	$\zeta_{2\cdot}$	$\varphi_2=\zeta_{2\cdot}-\zeta_{\cdot\cdot}$
3	ζ_{31}	ζ_{32}	ζ_{33}	ζ_{34}	$\zeta_{3\cdot}$	$\varphi_3=\zeta_{3\cdot}-\zeta_{\cdot\cdot}$
4	ζ_{41}	ζ_{42}	ζ_{43}	ζ_{44}	$\zeta_{4\cdot}$	$\varphi_4=\zeta_{4\cdot}-\zeta_{\cdot\cdot}$
Average over thermometers:	$\zeta_{\cdot 1}$	$\zeta_{\cdot 2}$	$\zeta_{\cdot 3}$	$\zeta_{\cdot 4}$	$\zeta_{\cdot\cdot}$	
Differences:	$\gamma_1=\zeta_{\cdot 1}-\zeta_{\cdot\cdot}$	$\gamma_2=\zeta_{\cdot 2}-\zeta_{\cdot\cdot}$	$\gamma_3=\zeta_{\cdot 3}-\zeta_{\cdot\cdot}$	$\gamma_4=\zeta_{\cdot 4}-\zeta_{\cdot\cdot}$		

Here

$$\zeta_{i\cdot} = \frac{1}{4}\sum_{j=1}^{4}\zeta_{ij}\,; \quad \zeta_{\cdot j} = \frac{1}{4}\sum_{i=1}^{4}\zeta_{ij}\,; \quad \zeta_{\cdot\cdot} = \frac{1}{16}\sum_{i=1}^{4}\sum_{j=1}^{4}\zeta_{ij}\,.$$

Thus a mean ζ_{ij} can be written

$$\begin{aligned}\zeta_{ij} &= \zeta_{\cdot\cdot} + (\zeta_{i\cdot} - \zeta_{\cdot\cdot}) + (\zeta_{\cdot j} - \zeta_{\cdot\cdot}) + (\zeta_{ij} - \zeta_{i\cdot} - \zeta_{\cdot j} + \zeta_{\cdot\cdot}) \\ &= \zeta_{\cdot\cdot} + \varphi_i + \gamma_j + \eta_{ij}\end{aligned}$$

where $\eta_{ij} = \zeta_{ij} - \zeta_{i\cdot} - \zeta_{\cdot j} + \zeta_{\cdot\cdot}$ and is known as the interaction term. We say that an interaction between two effects, φ_i and γ_j, exists if the joint effect of the two taken simultaneously is different from the sum of their separate effects. Thus, either lye or muriatic acid may be an effective cleanser, but taken together they are ineffective, and would be said to interact. Further, suppose there are five machines together with four workmen and it is desired to test whether machines differ in the number of units produced per day. It is quite possible that the second man may work better on the third machine than any other machine because he has worked for 20 years on such a machine. The resultant increased production cannot be assumed to be a characteristic of this man or of this particular machine, but a combination of only this man with only this machine, and we would say that interaction was present.

Returning to the table of means, we find the following linear relationships:

$$\sum_{i=1}^{4} \varphi_i = 0\,; \quad \sum_{j=1}^{4} \gamma_j = 0\,; \quad \sum_{i=1}^{4} \eta_{ij} = 0\,; \quad \sum_{j=1}^{4} \eta_{ij} = 0.$$

Furthermore, instead of the original 16 means, we have introduced new parameters which will supply the results. By writing $\zeta_{ij} = \zeta_{..} + \varphi_i + \gamma_j + \eta_{ij}$ we can write the mean for the ith chemical cell and jth thermometer as a constant, $\zeta_{..}$, plus an effect due to the ith cell which is constant over all columns, φ_i, plus an effect due to the jth thermometer which is constant over all rows, γ_j, plus the interaction between the ith cell and jth thermometer, η_{ij}. Testing whether (1) the cells are homogeneous, (2) the thermometers are homogeneous, and (3) whether there is any interaction, is equivalent to testing:

(1) $\quad \varphi_1 = \varphi_2 = \varphi_3 = \varphi_4 = 0\,;$

(2) $\quad \gamma_1 = \gamma_2 = \gamma_3 = \gamma_4 = 0\,;$

(3) $\quad \eta_{ij} = 0$

for all i and j.

More generally these assumptions can be summarized as follows:

$$x_{ij} = \zeta_{..} + \varphi_i + \gamma_j + \eta_{ij} + \varepsilon_{ij}$$

where x_{ij} is the outcome of the experiment using the ith row effect and jth column effect, $i = 1, 2, \cdots, r$; $j = 1, 2, \cdots, c$; $\zeta_{..}$ is a general mean; φ_i is the effect of adding the ith row treatment, $\sum_{i=1}^{r} \varphi_i = 0$; γ_j is the effect of adding the jth column treatment, $\sum_{j=1}^{c} \gamma_j = 0$; η_{ij} is the interaction of the ith row treatment with the jth column treatment, $\sum_{i=1}^{r} \eta_{ij} = \sum_{j=1}^{c} \eta_{ij} = 0$; the ε_{ij} are the random effects, which are independently normally distributed each with mean 0 and variance σ^2.

10.3.2 Random effects model

In the chemical cells and thermometer example, a more realistic assumption about the cells and thermometers is that the four used in experimentation were chosen at random from a large number of cells and thermometers. Furthermore, inferences drawn from the experiment are to extend to chemical cells and thermometers of these types in general, rather than be confined to the ones used in the experiment. In other words, the cells and thermometers are actually random variables whose values depend upon the outcome of the selection process. The mathematical model may still be written as

$$x_{ij} = \zeta_{..} + \varphi_i + \gamma_j + \eta_{ij} + \varepsilon_{ij}\,; \quad i = 1, 2, \cdots, r\,;\ j = 1, 2, \cdots, c\,;$$

but the interpretation of the φ_i, the row effect, γ_j, the column effect, and η_{ij}, the interaction effect, are different. The φ_i, γ_j, and η_{ij} are independent random variables all having mean 0 and variance σ^2_φ, σ^2_γ, and σ^2_η, respectively. These random variables are usually assumed to be normally distributed. The outcome of these effects (values associated with the thermometers and cells chosen) depends upon their probability distributions; interest is centered on certain statistical properties characterizing these distributions and not specifically on the particular individuals which happen to be drawn. Consider, for example, the thermometer effects: Is one really interested in (1) whether the four thermometers chosen read differently, or (2) the variation of all thermometers from which the four thermometers may be imagined drawn? If the answer is (1), the thermometer effects must be considered as fixed effects. If the answer is (2), the thermometer effects must be considered as random effects. The interactions, such as chemical celis and thermometers, represent differential effects of one of the factors caused by different levels of the other. In deciding about the interactions among two factors, the interaction effect is considered as a fixed effect if the factors involved in the interaction are fixed effects; otherwise it is treated as a random effect. Of course, in the cell-thermometer example, the interaction may still be assumed zero even if considered as a random effect.

Thus, inferences about row effects, column effects, and interaction effects are equivalent to making inferences about σ^2_φ, σ^2_γ, and σ^2_η, respectively. For example, if $\sigma^2_\varphi = 0$, all of the chemical cells are homogeneous.

10.3.3 Mixed fixed effects and random effects model

The previous sections dealt with the two-way analysis of variance when the factors were fixed effects and random effects, respectively. A third possibility to consider is one effect fixed and one effect random. This model will be denoted as the mixed effect model. In the chemical cells and thermometer problem, the cell effects can be considered as a random sample from a large number of such effects whereas the thermometers can be considered as fixed effects. This implies that the experimental results will lead to conclusions about the distribution of chemical cells. The conclusion about thermometers will extend only to the four thermometers used. It must be emphasized that the decision as to whether a factor is at fixed or random levels is a practical one and not up to the statistician. If the experimenter considers chemical cells as a random factor and wishes to generalize the results of his experiment beyond the few he has chosen, he must take measures to assure that the cells are really chosen at random from the large number of chemical cells available.

The general mathematical model may still be written as

$$x_{ij} = \zeta_{..} + \varphi_i + \gamma_j + \eta_{ij} + \varepsilon_{ij}; \quad i = 1, 2, \cdots, r; \quad j = 1, 2, \cdots, c;$$

where $\zeta_{..}$ is the over-all mean.

φ_i is the row effect and will be taken to be a normally distributed random variable with mean 0 and variance σ_φ^2. This corresponds to chemical cells in the description of the problem given in this section. γ_j is the column effect and will be taken to be the fixed effect of this model. $\sum_{j=1}^{c} \gamma_j = 0$. This corresponds to the thermometers in the description of the problem given in this section. η_{ij} is the interaction effect and will be taken to be a normally distributed random variable with mean 0 and variance σ_η^2. This is a random variable rather than a fixed effect because one of the factors (rows) is a random effect. However, it also seems reasonable to assume that $\sum_{j=1}^{c} \eta_{ij} = 0$ since all the η_{ij} (for fixed i) are in the sample.

Thus, inferences about row effects are equivalent to making inferences about σ_φ^2; inferences about column effects are equivalent to making inferences about γ_j for all j; and inferences about interaction effects are equivalent to making inferences about σ_η^2.

10.3.4 Computational procedure, two-way classification, one observation per combination

Before carrying out any further analyses, it is necessary to complete the analysis of variance table. This is done by following the computational procedure described below. The results are summarized in the analysis of variance table, Table 10.9.

Computational Procedure

(1) Calculate row totals:

$$R_1, R_2, \cdots, R_r.$$

(2) Calculate column totals:

$$C_1, C_2, \cdots, C_c.$$

(3) Calculate over-all total:

$$T = R_1 + R_2 + \cdots + R_r.$$

(4) Calculate crude total sum of squares:

$$\sum_{i=1}^{r} \sum_{j=1}^{c} x_{ij}^2 = x_{11}^2 + x_{12}^2 + \cdots + x_{rc}^2.$$

(5) Calculate crude sum of squares between rows:

$$\sum_{i=1}^{r} \frac{R_i^2}{c} = \frac{(R_1^2 + R_2^2 + \cdots + R_r^2)}{c}.$$

(6) Calculate crude sum of squares between columns:

$$\sum_{j=1}^{c} \frac{C_j^2}{r} = \frac{(C_1^2 + C_2^2 + \cdots + C_c^2)}{r}.$$

(7) Calculate correction factor due to mean $= T^2/rc$.

From the quantities above compute:

(8) $$SS_4 = (6) - (7) = \sum_{j=1}^{c} \frac{C_j^2}{r} - \frac{T^2}{rc},$$

(9) $$SS_3 = (5) - (7) = \sum_{i=1}^{r} \frac{R_i^2}{c} - \frac{T^2}{rc},$$

(10) $$SS = (4) - (7) = \sum_{i=1}^{r} \sum_{j=1}^{c} x_{ij}^2 - \frac{T^2}{rc},$$

(11) $$SS_2 = (10) - (9) - (8) = SS - SS_3 - SS_4.$$

It should be noted that $SS = SS_2 + SS_3 + SS_4$. In fact this was used in obtaining SS_2 in step 11 in the computational procedure.

10.3.5 Analysis of the Fixed Effects Model, Two-Way Classification, One Observation per Combination

In order to determine whether there is a row effect or a column effect it is necessary to test the hypotheses that

$$H: \quad \varphi_1 = \varphi_2 = \cdots = \varphi_r = 0$$

$$H: \quad \gamma_1 = \gamma_2 = \cdots = \gamma_c = 0$$

respectively.

Referring to the analysis of variance table, Table 10.9, the expected mean square for the residual equals σ^2 plus some interaction

Table 10.9. Analysis of Variance Table, Two-Way Classification, One Observation per Combination

Source	Sum of Squares	Degrees of Freedom	Mean Square	Expected Mean Square for the Fixed Effects Model	Expected Mean Square for the Random Effects Model	Expected Mean Square for the Mixed Effects Model
Between columns	$SS_4 = r \sum_{j=1}^{c} (\bar{x}_{.j} - \bar{x}_{..})^2$	$c - 1$	$SS_4^* = SS_4/(c-1)$	$\sigma^2 + \frac{r}{c-1} \sum_{j=1}^{c} \gamma_j^2$	$\sigma^2 + \sigma_\eta^2 + r\sigma_\gamma^2$	$\sigma^2 + \sigma_\eta^2 + \frac{r}{c-1} \sum_{j=1}^{c} \gamma_j^2$
Between rows	$SS_3 = c \sum_{i=1}^{r} (\bar{x}_{i.} - \bar{x}_{..})^2$	$r - 1$	$SS_3^* = SS_3/(r-1)$	$\sigma^2 + \frac{c}{r-1} \sum_{i=1}^{r} \varphi_i^2$	$\sigma^2 + \sigma_\eta^2 + c\sigma_\varphi^2$	$\sigma^2 + c\sigma_\varphi^2$
Residual	$SS_2 = \sum_{i=1}^{r} \sum_{j=1}^{c} (x_{ij} - \bar{x}_{i.} - \bar{x}_{.j} + \bar{x}_{..})^2$	$(r-1)(c-1)$	$SS_2^* = SS_2/(r-1)(c-1)$	$\sigma^2 + \frac{1}{(r-1)(c-1)} \sum_{i=1}^{r} \sum_{j=1}^{c} \eta_{ij}^2$	$\sigma^2 + \sigma_\eta^2$	$\sigma^2 + \sigma_\eta^2$
Total	$SS = \sum_{i=1}^{r} \sum_{j=1}^{c} (x_{ij} - \bar{x}_{..})^2$	$rc - 1$	$SS^* = SS/(rc-1)$	—	—	—

$\bar{x}_{.j} = \sum_{i=1}^{r} x_{ij} \Big/ r; \quad \bar{x}_{i.} = \sum_{j=1}^{c} x_{ij} \Big/ c; \quad \bar{x}_{..} = \sum_{i=1}^{r} \sum_{j=1}^{c} x_{ij} \Big/ rc.$

erms. The expected mean square for the row effects equals σ^2 plus ome row effect terms. Both SS_2^* and SS_3^* estimate σ^2 only if the nteraction terms are zero and there are no row effects, respectively. Ience, in order to get a reasonable test on row effects in the fixed ffects model with one observation per combination, it is necessary o assume that the interaction term is zero, i.e.,

$$x_{ij} = \zeta_{..} + \varphi_i + \gamma_j + \varepsilon_{ij}.$$

similar assumption is necessary to get a reasonable test for column ffects. There is often some *a priori* knowledge available about the hysical situation which allows for this assumption.

In the thermometer and chemical cell example there is no reason o postulate an interaction between chemical cells and thermometers. f there is a ranking among the thermometers, it is certainly not an ffair of the cells, and so we may confidently assume the interactions etween thermometers and chemical cells to be zero, i.e., $\eta_{ij} = 0$ for ll i and j.

Henceforth, the interaction will be assumed to equal zero so that $E(SS_2^*) = \sigma^2$ always. The expected mean squares for both row effects nd the residual sources, $E(SS_3^*)$ and $E(SS_2^*)$ respectively, equal σ^2 f the hypothesis is true. $E(SS_2^*)$ equals σ^2 even if the hypothesis is alse. However, in this event, $E(SS_3^*)$ exceeds σ^2 by the amount $\dfrac{c\sum_{i=1}^{r}\varphi_i^2}{r-1}$. Hence, if the ratio SS_3^*/SS_2^* is computed, a value far from nity would indicate that the hypothesis is false. It can be shown that using the partition theorem) SS_3/σ^2 and SS_2/σ^2 are independently disributed as chi-square random variables with $(r-1)$ and $(r-1)(c-1)$ legrees of freedom, respectively, when the hypothesis is true. Hence the ratio SS_3^*/SS_2^* is distributed as an F random variable with $r-1$ and $(r-1)(c-1)$ degrees of freedom when the hypothesis is rue. Thus, the hypothesis of equality of the φ's is rejected if

$$F = \frac{SS_3^*}{SS_2^*} \geqq F_{\alpha;(r-1),(r-1)(c-1)}.$$

Using an analogous argument, the hypothesis of no column effects, .e., the hypothesis that $\gamma_1 = \gamma_2 = \cdots = \gamma_c = 0$, is rejected if

$$F = \frac{SS_4^*}{SS_2^*} \geqq F_{\alpha;(c-1),(r-1)(c-1)}.$$

If the hypothesis that there are no row effects or no column ffects, or both, is rejected, Tukey's method can again be applied o get confidence intervals on the difference between two row effects or two column effects. The probability is $1 - \alpha$ that the values $(\varphi_m - \varphi_n)$ for all such contrasts (for $m = 1, 2, \cdots, r$; $n = 1, 2, \cdots, r$) simultaneously satisfy

$$\bar{x}_{m\cdot} - \bar{x}_{n\cdot} - k \leqq \varphi_m - \varphi_n \leqq \bar{x}_{m\cdot} - \bar{x}_{n\cdot} + k$$

and the probability is $1 - \alpha$ that the values $(\gamma_f - \gamma_g)$ for all such contrasts (for $f = 1, 2, \cdots, c$; $g = 1, 2, \cdots, c$) simultaneously satisfy

$$\bar{x}_{\cdot f} - \bar{x}_{\cdot g} - k \leqq \gamma_f - \gamma_g \leqq \bar{x}_{\cdot f} - \bar{x}_{\cdot g} + k.$$

Again two mean effects are judged significantly different when 0 is not included in the confidence interval.

The factor k for row effects and for column effects is obtained from the analysis of variance table and from Table 10.3 or Table 10.4 depending on the level of significance. Table 10.3 or 10.4 is entered with the indices, degrees of freedom, and the number of row effects being studied (r) if row effects are of interest. The proper degrees of freedom are those corresponding to the degrees of freedom of the residual source in the analysis of variance table, i.e., $(r - 1)(c - 1)$. The factor k^* is read from the table, and

$$k \text{ (for rows)} = k^*\sqrt{SS_2^*/c}.$$

Similarly, if column effects are of interest, the table is entered with indices degrees of freedom and the number of column effects being studied (c). The factor k^* is read from the table and

$$k \text{ (for columns)} = k^*\sqrt{SS_2^*/r}.$$

Note that the expected mean square column indicates which ratios to use in the F test.

10.3.6 THE OC CURVE OF THE ANALYSIS OF VARIANCE FOR THE FIXED EFFECTS MODEL, TWO-WAY CLASSIFICATION, ONE OBSERVATION PER COMBINATION

If the OC curve for testing the hypothesis about row effects is desired, the appropriate measure for which an OC curve can be constructed is

$$\Phi = \frac{\sqrt{\sum_{i=1}^{r} \varphi_i^2 / r}}{\sigma/\sqrt{c}}.$$

Similarly, if the OC curve for testing the hypothesis about column effects is desired, the appropriate measure for which an OC curve can be constructed is

$$\Phi = \frac{\sqrt{\sum_{j=1}^{c} \gamma_j^2 / c}}{\sigma/\sqrt{r}}.$$

The OC curves as functions of ν_1, ν_2, and Φ are given Figures 10.2 to 10.9. The appropriate values of ν_1 are $(r-1)$ if row effects are being considered and $(c-1)$ if column effects are being considered. ν_2 corresponds to the degrees of freedom associated with the residual and is given by $(r-1)(c-1)$. If the number of row effects and column effects are fixed and only one observation for each combination of levels is allowed, these OC curves can be used to study the probabilities of making the correct decisions as a function of Φ. If more than one observation per combination is allowed, the appropriate number to guarantee a fixed OC curve is given in the results of Section 10.4.

A physical interpretatation of Φ as a measure can be given by finding upper and lower bounds for Φ as a function of the range of the φ_i [$\varphi(\max) - \varphi(\min)$ if row effects are being studied] beyond which rejection of homogeneity should be assured with a given probability. Denote this largest difference by W. Bounds on Φ are given as follows:

$$\Phi(\min) = \frac{W}{\sigma}\sqrt{\frac{c}{2r}}$$

$$\Phi(\max) = \begin{cases} \dfrac{W}{2\sigma}\sqrt{c} & \text{for } r \text{ even} \\ \dfrac{W}{2\sigma}\sqrt{\dfrac{c(r^2-1)}{r^2}} & \text{for } r \text{ odd.} \end{cases}$$

Thus if W is specified and some estimate of σ is available, the probability of accepting the hypothesis of homogeneity of φ_i can be obtained from Figures 10.2 to 10.9.

Similarly, if column effects are being studied,

$$\Phi(\min) = \frac{W}{\sigma}\sqrt{\frac{r}{2c}}$$

$$\Phi(\max) = \begin{cases} \dfrac{W}{2\sigma}\sqrt{r} & \text{for } c \text{ even} \\ \dfrac{W}{2\sigma}\sqrt{\dfrac{r(c^2-1)}{c^2}} & \text{for } c \text{ odd.} \end{cases}$$

10.3.7 Example using the fixed effects model

If the chemical cells and thermometers are treated as fixed effects, the following is an analysis of the data given in Table 10.7. The following computations lead to the analysis of variance table.

(1) $R_1, R_2, R_3, R_4 = 140, 78, 144, 146$

(2) $$C_1, C_2, C_3, C_4 = 113, 140, 141, 114$$

(3) $$T = 140 + 78 + 144 + 146 = 508$$

(4) $$\sum_{i=1}^{4}\sum_{j=1}^{4} x_{ij}^2 = (36)^2 + (38)^2 + \cdots + (33)^2 = 17{,}230$$

(5) $$\sum_{i=1}^{4} \frac{R_i^2}{4} = \frac{(140)^2 + (78)^2 + (144)^2 + (146)^2}{4} = 16{,}934$$

(6) $$\sum_{j=1}^{4} \frac{C_i^2}{4} = \frac{(113)^2 + (140)^2 + (141)^2 + (114)^2}{4} = 16{,}311.5$$

(7) $$T^2/16 = (508)^2/16 = 16{,}129$$

(8) $$SS_4 = 16{,}311.5 - 16{,}129 = 182.5$$

(9) $$SS_3 = 16{,}934 - 16{,}129 = 805$$

(10) $$SS = 17{,}230 - 16{,}129 = 1{,}101$$

(11) $$SS_2 = 1{,}101 - 805 - 182.5 = 113.5$$

Table. 10.10. Analysis of Variance Table for Temperatures of Cells

Source	Sum of Squares	Degrees of Freedom	Mean Square	Test
Between thermometers..	182.5	3	60.83	$F = 4.82$
Between chemical cells..	805	3	268.33	$F = 21.3$
Residual...............	113.5	9	12.61	
Total	1,101.0	15	73.4	

Thus, the values of F for testing the hypotheses that the thermometers and cells are the same are 4.82 and 21.3, respectively. Since both of these values exceed $F_{0.05;3,9} = 3.86$, we conclude that the thermometers differ significantly and the chemical cells also differ significantly. In order to determine the magnitude of these differences, k is computed for thermometers and cells. k^* (for thermometers) $= 4.42$, where the factor 4.42 is obtained from Table 10.3, entering with $n = r = 4$ and $\nu = (r - 1)(c - 1) = 9$. Therefore $k = 4.42\sqrt{12.61/4} = 7.86$ and any two thermometer sample means differing by this much differ significantly at $\alpha = 0.05$. However, none of the thermometer means differ by more than 7.86, so that we are unable to detect the unusual thermometers. Although each confidence interval includes zero, meaning that such pairs *may* be the same, it does not preclude any (or all) of the true differences to be other than zero. Thus, in this situation, which is rather rare, we are unable to pick any unusual thermometers even though the F test indicates that the thermometers differ.

Since there are also four cells, k (for cells) is the same as for thermometers, i.e., k (for cells) $= 7.86$ and any two cell sample means differing by this much differ significantly at $\alpha = 0.05$. From Table 10.7 it is evident that cell number 2 differs from the other three.

If a maximum difference between thermometers of 0.0007 degrees Centigrade ($W = 7$) is specified and σ is approximately equal to 3,

$$\Phi\,(\max) = \frac{7}{6}\sqrt{4} = 2.33,$$

$$\Phi\,(\min) = \frac{7}{3}\sqrt{\frac{4}{8}} = 1.65.$$

Hence the probability of accepting the hypothesis that the thermometers are homogeneous when $W = 7$ lies between 0.11 and 0.41 as is found by entering Figure 10.4 with $\nu_1 = 3$ and $\nu_2 = 9$.

10.3.8 Analysis of the Random Effects Model, Two-Way Classification, One Observation per Combination

In the random effects model, testing the hypothesis about the homogeneity of row effects is equivalent to testing the hypothesis

$$H: \quad \sigma_\varphi^2 = 0$$

and testing the hypothesis about the homogeneity of column effects is equivalent to testing the hypothesis

$$H: \quad \sigma_\gamma^2 = 0.$$

Referring to the analysis of variance table, Table 10.9, the expected mean square for the residual in this model equals $\sigma^2 + \sigma_\eta^2$, i.e., the sum of the error variance and the interaction variance. The expected mean square for the row effects equals $\sigma^2 + \sigma_\eta^2 + c\sigma_\varphi^2$, i.e., the sum of the error variance, interaction variance and a constant times the row effect variance. If the hypothesis of no row effects ($\sigma_\varphi^2 = 0$) is true, the expected mean square for both row effects and the residual sources, $E(SS_3^*)$ and $E(SS_2^*)$ respectively, equals $\sigma^2 + \sigma_\eta^2$. Thus, using the ratio SS_3^*/SS_2^* leads to a reasonable procedure in testing for row effects in this model without making any assumptions about the presence or absence of interaction. This is quite different from the fixed effects model where it was necessary to postulate the absence of interaction. The random effects model, then, leads to a test for row effects whether or not interaction is present, using the ratio SS_3^*/SS_2^* as a statistic. It can be shown that (using the partition theorem) $SS_3/(\sigma^2 + \sigma_\eta^2)$ and $SS_2/(\sigma^2 + \sigma_\eta^2)$ are independently distributed as chi-square random variables with $(r-1)$ and $(r-1)(c-1)$ degrees of freedom, respectively, when the hypothesis is true. Hence, the ratio SS_3^*/SS_2^* is distributed as an F random variable with $r-1$ and $(r-1)(c-1)$ degrees of freedom when the hypothesis is true. Thus, the hypothesis that $\sigma_\varphi^2 = 0$ is rejected if

$$F = \frac{SS_3^*}{SS_2^*} \geqq F_{\alpha;(r-1),(r-1)(c-1)}.$$

Using an analogous argument, the hypothesis of no column effect, i.e., the hypothesis that $\sigma_\gamma^2 = 0$, is rejected if

$$F = \frac{SS_4^*}{SS_2^*} \geqq F_{\alpha;(c-1),(r-1)(c-1)}.$$

In order to determine a measure of how much the row effects differ, the quantity σ_φ^2 can be estimated. The analysis of variance table indicates that SS_2^* is an unbiased estimate of $\sigma^2 + \sigma_\eta^2$ and that SS_3^* is an unbiased estimate of $\sigma^2 + \sigma_\eta^2 + c\sigma_\varphi^2$. Assuming that these estimates are equal to the parameters that are being estimated, i.e., $SS_2^* = \sigma^2 + \sigma_\eta^2$ and $SS_3^* = \sigma^2 + \sigma_\eta^2 + c\sigma_\varphi^2$, an estimate of σ_φ^2 can be obtained from these equations. Such an estimate is given by

$$\frac{SS_3^* - SS_2^*}{c}.$$

If this quantity is negative, the estimate is taken to be zero.

Similarly an estimate of σ_γ^2 is given by

$$\frac{SS_4^* - SS_2^*}{r}.$$

10.3.9 THE OC CURVE FOR THE RANDOM EFFECTS MODEL, TWO-WAY CLASSIFICATION

The hypothesis that $\sigma_\varphi^2 = 0$ is accepted if

$$\frac{SS_3^*}{SS_2^*} \leqq F_{\alpha;(r-1),(r-1)(c-1)}.$$

SS_3^*/SS_2^* has an F distribution whenever $\sigma_\varphi^2 = 0$, and the probability of acceptance is $(1-\alpha)$. In general, the probability of accepting the hypothesis that $\sigma_\varphi^2 = 0$ can be plotted as a function of

$$\lambda = \sqrt{1 + \frac{c\sigma_\varphi^2}{\sigma^2 + \sigma_\eta^2}}.\text{*}$$

* If the hypothesis is false

$$\frac{SS_3^*/(\sigma^2 + \sigma_\eta^2 + c\sigma_\varphi^2)}{SS_2^*/(\sigma^2 + \sigma_\eta^2)}$$

has an F distribution with $(r-1)$ and $(r-1)(c-1)$ degrees of freedom. Since th procedure of accepting the hypothesis when 'he OC curves of Figures 10.10 to 10.17 are again applicable when λ ; defined in this manner. Note that the OC curve depends only on he ratio $\sigma_\varphi^2/(\sigma^2 + \sigma_\eta^2)$.

Similarly if column effects are being studied, the OC curve is a unction of

$$\lambda = \sqrt{1 + \frac{r\sigma_\gamma^2}{\sigma^2 + \sigma_\eta^2}}\,.$$

he OC curves of Figures 10.10 to 10.17 are entered with degrees of reedom $\nu_1 = r - 1$ if row effects are of interest, $\nu_1 = c - 1$ if column ffects are of interest, and $\nu_2 = (r - 1)(c - 1)$.

10.3.10 EXAMPLE USING THE RANDOM EFFECTS MODEL

If the four chemical cells and thermometers are both considered) be random samples from a large number of cells and thermome- ers, the example presented in Section 10.3.7 would fit the random ffects model. The formal calculations including the computational rocedure and the analysis of variance table are identical with the esults of Section 10.3.7. The numerical values of the F test are also he same as are the conclusions drawn from the F test (adapted to e present model). However, in this model it is unnecessary to ssume that there is no interaction between cells and thermometers. n estimate of σ_φ^2 is given by

$$\frac{SS_3^* - SS_2^*}{c} = 63.9$$

nd an estimate of σ_γ^2 is given by

$$\frac{SS_4^* - SS_2^*}{r} = 12.1\,.$$

point on the OC curve when $\dfrac{\sigma_\varphi^2}{(\sigma^2 + \sigma_\eta^2)} = 1.2$ is found as follows: 'he corresponding value of λ is 2.4. Figure 10.12 is entered with

$$\frac{SS_3^*}{SS_2^*} \leqq F_{\alpha;(r-1),(r-1)(c-1)}$$

equivalent to accepting the hypothesis when

$$\frac{SS_3^*/(\sigma^2 + \sigma_\eta^2 + c\sigma_\varphi^2)}{SS_2^*/(\sigma^2 + \sigma_\eta^2)} \leqq F_{\alpha;r-1,(r-1)(c-1)}\left[\frac{(\sigma^2 + \sigma_\eta^2)}{\sigma^2 + \sigma_\eta^2 + c\sigma_\varphi^2}\right].$$

he probability of acceptance can be determined since the probability distribution of e left hand side of the last inequality is known (F distribution). Hence, the OC urves of Figures 10.10 to 10.17 are applicable when

$$\lambda = \sqrt{\frac{\sigma^2 + \sigma_\eta^2 + c\sigma_\varphi^2}{\sigma^2 + \sigma_\eta^2}} = \sqrt{1 + \frac{c\sigma_\varphi^2}{\sigma^2 + \sigma_\eta^2}}\,.$$

$\nu_1 = 3$, $\nu_2 = 9$, and $\lambda = 2.4$. The probability of accepting the hypothesis of chemical cell homogeneity when $\lambda = 2.4$ is then seen to be 0.40.

10.3.11 Analysis of the Mixed Effects Model, Two-Way Classification, One Observation per Combination

In the mixed effects model, the row effects will be assumed as the random effects and the column effects will be assumed as the fixed effects. Testing the hypothesis about the homogeneity of row effects is equivalent to testing the hypothesis

$$H: \quad \sigma_\varphi^2 = 0 \,.$$

Testing the hypothesis about the homogeneity of column effects is equivalent to testing the hypothesis

$$H: \quad \gamma_1 = \gamma_2 = \cdots = \gamma_c = 0 \,.$$

Referring to the analysis of variance table, Table 10.9, the expected mean square for the residual in this model equals $\sigma^2 + \sigma_\eta^2$, i.e., the sum of the error variance and the interaction variance. The expected mean square for the row effects (random effects) equals $\sigma^2 + c\sigma_\varphi^2$, i.e., the sum of the error variance and a constant times the row effect variance. If the hypothesis that σ_φ^2 is zero is true, $E(SS_3^*) = \sigma^2$, whereas $E(SS_2^*) = \sigma^2$ only if there is no interaction present. Hence to test for the row effects (random effects) it is necessary to assume that $\sigma_\eta^2 = 0$. Under this assumption, the ratio SS_3^*/SS_2^* is a reasonable statistic for use in testing for row effects. It can be shown that (using the partition theorem) SS_3/σ^2 and SS_2/σ^2 are independently distributed as chi-square random variables with $(r-1)$ and $(r-1)(c-1)$ degrees of freedom, respectively, when the hypothesis is true. Hence, the ratio SS_3^*/SS_2^* is distributed as an F random variable with $(r-1)$ and $(r-1)(c-1)$ degrees of freedom when the hypothesis is true. Thus, the hypothesis that $\sigma_\varphi^2 = 0$ is rejected if

$$F = \frac{SS_3^*}{SS_2^*} \geqq F_{\alpha;\,(r-1),\,(r-1)(c-1)} \,.$$

In the absence of interaction, an estimate of σ_φ^2 is available, i.e.,

$$\frac{SS_3^* - SS_2^*}{c} \,.$$

In testing for the column (fixed) effects, the problem is somewhat different. The expected mean square, $E(SS_4^*)$, for the column source is given by

$$\sigma^2 + \sigma_\eta^2 + \frac{r}{c-1}\sum_{j=1}^{c} \gamma_j^2 .$$

As before, the expected mean square for the residual sum of squares, $E(SS_2^*)$, is given by $\sigma^2+\sigma_\eta^2$. If there is no column effect, both expected values equal $\sigma^2 + \sigma_\eta^2$. Thus, using the ratio SS_4^*/SS_2^* leads to a reasonable procedure for testing for column (fixed) effects without making any assumption about the presence or absence of interaction. This is quite different from testing for the random effects in this model. It can be shown that (by using the partition theorem) SS_4^*/SS_2^* is distributed as an F random variable with $(c-1)$ and $(r-1)(c-1)$ degrees of freedom when the hypothesis is true. Therefore, the hypothesis that $\gamma_1 = \gamma_2 = \cdots = \gamma_c = 0$ is rejected if

$$F = \frac{SS_4^*}{SS_2^*} \geqq F_{\alpha;\,(c-1),(r-1)(c-1)} .$$

If the hypothesis that there are no column effects is rejected, Tukey's method can again be applied to get confidence intervals for the difference between two column effects. The probability is $1-\alpha$ that the value $(\gamma_f - \gamma_g)$ for *all* such contrasts for $f = 1, 2, \cdots, c$; $g = 1, 2, \cdots, c$ simultaneously satisfy

$$\bar{x}_{\cdot f} - \bar{x}_{\cdot g} - k \leqq \gamma_f - \gamma_g \leqq \bar{x}_{\cdot f} - \bar{x}_{\cdot g} + k .$$

Two column effects are judged significantly different when 0 is not included in the confidence interval.

The factor k for column effects is obtained from the analysis of variance table and from Table 10.3 and Table 10.4 depending on the level of significance. Table 10.3 or Table 10.4 is entered with the indices degrees of freedom (ν) and the number of column effects (c) being studied. The proper degrees of freedom are those corresponding to the degrees of freedom of the residual source in the analysis of variance table, i.e., $(r-1)(c-1)$. The factor k^* is read from the table, and

$$k = k^*\sqrt{SS_2^*/r} .$$

10.3.12 The OC curve of the analysis of variance for the mixed effects model, two-way classification, one observation per cell

In finding the OC curve for the random effects (rows) when the absence of interaction is postulated, the procedures become identical with that given in Section 10.3.9. The OC curve is still a function of λ where

$$\lambda = \sqrt{\frac{\sigma^2 + c\sigma_\varphi^2}{\sigma^2}} = \sqrt{1 + \frac{c\sigma_\varphi^2}{\sigma^2}}.$$

The OC curves of Figures 10.10 to 10.17 are entered with degrees o freedom $\nu_1 = (r - 1)$ and $\nu_2 = (r - 1)(c - 1)$.

The OC curves for testing the hypothesis about the fixed effect (columns) is a function of the measure Φ, where

$$\Phi = \sqrt{\frac{1}{c}\sum_{j=1}^{c} \tau_j^2} \Bigg/ \sqrt{\frac{\sigma^2 + \sigma_\eta^2}{r}}.$$

This is exactly the same procedure as outlined for column effects i Section 10.3.6 with the value σ replaced by $\sqrt{\sigma^2 + \sigma_\eta^2}$.

10.3.13 EXAMPLE USING THE MIXED EFFECTS MODEL

If the chemical cells are treated as a random sample from a larg number of cells and the thermometers are treated as fixed, th chemical cell and thermometer example may be treated as a mixe effects model. The formal calculation including the computation procedure and the analysis of variance table are identical with th results of Section 10.3.7. The numerical values of the F test ar also the same, but in testing for the random effect cells it is neces sary to assume the interaction nonexistent, whereas the test for th thermometers does not depend on any assumptions about the intera tion.

10.4 Two-Way Analysis of Variance, *n* Observations per Combination

This section is very similar to Section 10.3 except that it allow for more than one observation per combination.

10.4.1 DESCRIPTION OF THE VARIOUS MODELS

The following experiment was performed to determine the effec of time of aging on the strength of cement.[1] Three mixes of cemen were prepared and six specimens were made from each mix. Thre specimens from each mix were tested after two days and after seve days. The test specimens were 2-inch cubes that yielded under th indicated load and were in units of 10 pounds. They are shown i Table 10.11.

There are three alternative ways to consider this problem, i.e. fixed effects model, random effects model, or mixed effects model

[1] This example is based on an example appearing in W. J. Youden, *Statistica Methods for Chemists*, John Wiley and Sons, Inc., New York, 1951, pp. 64–65.

Table 10.11. Yield Load For Cement Specimens

	2-Days Test	7-Days Test	$\bar{x}_{i\cdot}$
Mix 1	574 564 550	1,092 1,086 1,065	821.8
Mix 2	524 573 551	1,028 1,073 998	791.2
Mix 3	576 540 592	1,066 1,045 1,055	812.3
$\bar{x}_{\cdot j}$	560.4	1,056.4	

The mixes are actually random components, a sample of three drawn from a large number. The results of the experiment should extend to the distribution of mixes. On the other hand, the times of aging are fixed effects. The conclusions of the experiment will reveal whether the yield loads differ after 2 or 7 days, these periods being fixed. Hence, this model is actually a mixed model, although it could conceivably be analyzed as a fixed effects model (results pertain to these three mixes only) or a random effects model (the two aging periods being chosen at random). There are three observations for each mix and aging combination. This will enable the experimenter to test for the presence of interaction. When only one observation per combination existed, either interaction was assumed to be zero or its presence was intermixed with the error term. Its presence could not be detected analytically.

The mathematical model for this situation is as follows:

$$x_{ijv} = \zeta_{\cdot\cdot} + \varphi_i + \gamma_j + \eta_{ij} + \epsilon_{ijv}: \qquad \begin{aligned} i &= 1, 2, \cdots, r, \\ j &= 1, 2, \cdots, c, \\ v &= 1, 2, \cdots, n, \end{aligned}$$

where $\zeta_{\cdot\cdot}$ is the general mean; ϵ_{ijv} are the experimental errors which are assumed to be independently normally distributed each with mean 0 and variance σ^2.

In the fixed effects model:

φ_i is the effect of adding the ith fixed row treatment,

$$\sum_{i=1}^{r} \varphi_i = 0 .$$

γ_j is the effect of adding the jth fixed column treatment,

$$\sum_{j=1}^{c} \gamma_j = 0 \, .$$

η_{ij} is the fixed interaction of the ith row treatment with the jt column treatment,

$$\sum_{i=1}^{r} \eta_{ij} = \sum_{j=1}^{c} \eta_{ij} = 0 \, .$$

In the random effects model:

φ_i is the effect of adding the ith row treatment and is a rando variable whose probability distribution is the distribution of row trea ments. φ_i is assumed to be normally distributed with mean 0 an variance σ_φ^2.

γ_j is the effect of adding the jth column treatment and is random variable whose probability distribution is the distribution column treatments. γ_j is assumed to be normally distributed wit mean 0 and variance σ_γ^2.

η_{ij} is the interaction of the ith row treatment with the jth colum treatment and is also a random variable. η_{ij} is assumed to be no mally distributed with mean 0 and variance σ_η^2.

In the mixed effects model:

φ_i is the effect of adding the ith row treatment and is a rando variable whose probability distribution is the distribution of ro treatments. φ_i is assumed to be normally distributed with mean and variance σ_φ^2.

γ_j is the effect of adding the jth fixed column treatment,

$$\sum_{j=1}^{c} \gamma_j = 0 \, .$$

η_{ij} is the interaction of the ith random row treatment with the jt fixed column treatment and is a random variable. η_{ij} is assumed to be normally distributed with mean 0 and variance σ_η^2.

10.4.2 Computational Procedure, Two-Way Classification, Observations per Combination

Before carrying out any further analysis it is necessary to com plete the analysis of variance table, Table 10.12. This is done follow ing the computational procedure described below. The results ar summarized in the analysis of variance table.

Computational Procedure

(1) Calculate row totals:

$$R_1, R_2, \cdots, R_r \, .$$

(2) Calculate column totals:

$$C_1, C_2, \cdots, C_c.$$

(3) Calculate within combination totals:

$$w_{11}, w_{12}, \cdots, w_{rc}.$$

(4) Calculate over-all total:

$$T = R_1 + R_2 + \cdots + R_r.$$

(5) Calculate crude sum of squares:

$$\sum_{i=1}^{r}\sum_{j=1}^{c}\sum_{v=1}^{n} x_{ijv}^2 = x_{111}^2 + x_{112}^2 + \cdots + x_{rcn}^2.$$

(6) Calculate crude sum of squares between columns:

$$\sum_{j=1}^{c} \frac{C_j^2}{rn} = \frac{(C_1^2 + C_2^2 + \cdots + C_c^2)}{rn}.$$

(7) Calculate crude sum of squares between rows:

$$\sum_{i=1}^{r} \frac{R_i^2}{cn} = \frac{(R_1^2 + R_2^2 + \cdots + R_r^2)}{cn}.$$

(8) Calculate crude sum of squares between combinations:

$$\sum_{i=1}^{r}\sum_{j=1}^{c} \frac{w_{ij}^2}{n} = \frac{(w_{11}^2 + w_{12}^2 + \cdots + w_{rc}^2)}{n}.$$

(9) Calculate correction factor due to mean $= T^2/rcn$.

From the quantities above compute:

(10) $SS_4 = (6) - (9) = \sum_{j=1}^{c} \frac{C_j^2}{rn} - \frac{T^2}{rcn},$

(11) $SS_3 = (7) - (9) = \sum_{i=1}^{r} \frac{R_i^2}{cn} - \frac{T^2}{rcn},$

(12) $SS_1 = (5) - (8) = \sum_{i=1}^{r}\sum_{j=1}^{c}\sum_{v=1}^{n} x_{ijv}^2 - \sum_{i=1}^{r}\sum_{j=1}^{c} \frac{w_{ij}^2}{n},$

(13) $SS = (5) - (9) = \sum_{i=1}^{r}\sum_{j=1}^{c}\sum_{v=1}^{n} x_{ijv}^2 - \frac{T^2}{rcn},$

(14) $SS_2 = (13) - (12) - (11) - (10) = SS - SS_1 - SS_3 - SS_4.$

10.4.3 Analysis of the Fixed Effects Model, Two-Way Classification, n Observations per Combination

In the situation where there is only one observation per combina-ion (Section 10.3.9), it was pointed out that in order to test for main effects, it was necessary to assume the absence of interaction. In hat model, there was no way to test for the interaction. When there are n $(n > 1)$ observations per combination of level, a test fo interaction does exist. Referring to the analysis of variance table Table 10.12, the expected mean square for the "within combinations" source equals σ^2. The expected mean square for the "interaction" source equals σ^2 plus some interaction effect terms. If there i no interaction present, i.e., $\eta_{ij} = 0$ for all i and j, the expecte mean square for both "interaction effects" and "within combina tions" sources, $E(SS_2^*)$ and $E(SS_1^*)$ respectively, equals σ^2. Thus using the ratio SS_2^*/SS_1^* leads to a reasonable procedure for testin for interaction effects. It can be shown that (using the partitio theorem) SS_2^*/SS_1^* is distributed as an F random variable wit $(r-1)(c-1)$ and $rc(n-1)$ degrees of freedom when the hypothesi of no interaction is true. Thus, the hypothesis that $\eta_{ij} = 0$ for a i and j is rejected if

$$F = \frac{SS_2^*}{SS_1^*} \geqq F_{\alpha;\,(r-1)(c-1),rc(n-1)}.$$

If this hypothesis is rejected, the tests for significance of rov effects and column effects are meaningless under the present formu lation of the problem. For example, suppose that there is an ex periment being performed involving men and machines and there i interaction present between workmen and machines. Furthermor suppose there are also machine effects present. If a particula machine effect is positive, and interaction is present, there is n guarantee that this machine will produce more units consistently Yet, this is what is implied by saying that there are machine effects Similarly, if there are no machine effects, but interaction is presen in the form that a particular man works best on one particula machine, it can be concluded that machines differ when this man i the operator. Consequently, if interaction is present, tests for rov effects and column effects are irrelevant. In this situation, it is possi ble to test whether the set of levels of one treatment differ for given level of the other treatment, e.g. to test whether machine differ when a particular man is the operator rather than test th more general result that machines differ regardless of who is th operator. This is accomplished by performing a one-way analysis o variance over the fixed level (making a one-way analysis of varianc of the data for a particular operator).

If there is more than one observation per cell and it is assume before the experiment is performed that the interaction is zero, th interaction and within-cells sums of squares are combined into

Table 10.12. Analysis of Variance Table, Two-Way Classification, n Observations per Combination

Source	Sum of Squares	Degrees of Freedom	Mean Square	Expected Mean Square for the Fixed Effects Model	Expected Mean Square for the Random Effects Model	Expected Mean Square for the Mixed Effects Model
Between columns	$SS_4 = rn \sum_{j=1}^{c} (\bar{x}_{.j.} - \bar{x}_{...})^2$	$c - 1$	$SS_4^* = SS_4/(c-1)$	$\sigma^2 + \frac{rn}{c-1} \sum_{j=1}^{c} \gamma_j^2$	$\sigma^2 + n\sigma_\eta^2 + rn\sigma_\gamma^2$	$\sigma^2 + n\sigma_\eta^2 + \frac{rn}{c-1} \sum_{j=1}^{c} \gamma_j^2$
Between rows	$SS_3 = cn \sum_{i=1}^{r} (\bar{x}_{i..} - \bar{x}_{...})^2$	$r - 1$	$SS_3^* = SS_3/(r-1)$	$\sigma^2 + \frac{cn}{r-1} \sum_{i=1}^{r} \varphi_i^2$	$\sigma^2 + n\sigma_\eta^2 + cn\sigma_\varphi^2$	$\sigma^2 + nc\sigma_\varphi^2$
Inter-action	$SS_2 = n \sum_{i=1}^{r} \sum_{j=1}^{c} (\bar{x}_{ij.} - \bar{x}_{i..} - \bar{x}_{.j.} + \bar{x}_{...})^2$	$(r-1)(c-1)$	$SS_2^* = SS_2/(r-1)(c-1)$	$\sigma^2 + \frac{n}{(r-1)(c-1)} \sum_{i=1}^{r} \sum_{j=1}^{c} \eta_{ij}^2$	$\sigma^2 + n\sigma_\eta^2$	$\sigma^2 + n\sigma_\eta^2$
Within combi-nation	$SS_1 = \sum_{i=1}^{r} \sum_{j=1}^{c} \sum_{v=1}^{n} (x_{ijv} - \bar{x}_{ij.})^2$	$rc(n-1)$	$SS_1^* = SS_1/rc(n-1)$	σ^2	σ^2	σ^2
Total	$SS = \sum_{i=1}^{r} \sum_{j=1}^{c} \sum_{v=1}^{n} (x_{ijv} - \bar{x}_{...})^2$	$rcn - 1$	$SS^* = SS/(rcn-1)$	—	—	—

In this table $\bar{x}_{.j.} = \sum_{i=1}^{r} \sum_{v=1}^{n} x_{ijv}/rn$; $\bar{x}_{i..} = \sum_{j=1}^{c} \sum_{v=1}^{n} x_{ijv}/cn$; $\bar{x}_{ij.} = \sum_{v=1}^{n} x_{ijv}/n$; $\bar{x}_{...} = \sum_{i=1}^{r} \sum_{j=1}^{c} \sum_{v=1}^{n} x_{ijv}/rcn$.

new row, where $SS'_1 = SS_1 + SS_2$. In all ensuing formulas SS_1 an SS^*_1 should be replaced by SS'_1 and SS'^*_1, respectively.

Table 10.13. Residual with Interaction Assumed Zero

Source	Sum of Squares	Degrees of Freedom	Mean Square	Average Mean Square
Residual	$SS'_1 = SS_1 + SS_2$	$(r-1)(c-1)+rc(n-1) = rcn-r-c+1$	$SS'^*_1 = SS'_1/(rcn-r-c+1)$	σ^2

If the hypothesis that there is no interaction is accepted (or it i assumed to be nonexistent prior to experimentation because o physical considerations), tests for row effects and column effect can be made. Referring to the analysis of variance table, Tabl 10.12, the expected mean square for the "within combination" sourc equals σ^2. The expected mean square for the "row effects" sourc equals σ^2 plus some row effects terms. If there are no row effect present, i.e., $\varphi_i = 0$, $i = 1, 2, \cdots, r$, the expected mean square fo both "row effects" and "within combination" sources, $E(SS^*_3)$ an $E(SS^*_1)$ respectively, equals σ^2. Thus, using the ratio SS^*_3/SS^*_1 lead to a reasonable procedure for testing for row effects. It can be show that (using the partition theorem) SS^*_3/SS^*_1 is distributed as an random variable with $(r - 1)$ and $rc(n - 1)$ degrees of freedom whe the hypothesis of no row effects is true. Thus, the hypothesis tha $\varphi_1 = \varphi_2 = \cdots = \varphi_r = 0$ is rejected if

$$F = \frac{SS^*_3}{SS^*_1} \geqq F_{\alpha;\,(r-1),rc(n-1)}.$$

Using an analogous argument, the hypothesis of no column effects i.e., the hypothesis that $\gamma_1 = \gamma_2 = \cdots = \gamma_c = 0$, is rejected if

$$F = \frac{SS^*_4}{SS^*_1} \geqq F_{\alpha;\,(c-1),rc(n-1)}.$$

If the interactions are not significant and the row effects and/o the column effects are significant, the procedure for comparing ro and/or column effects is exactly the same as that for the two-wa classification, with one observation per cell except for the followin changes.

The factor k for row effects and for column effects is obtaine from the analysis of variance table and from Table 10.3 or Tabl 10.4, depending on the level of significance. Table 10.3 or 10.4 i entered with the indices degrees of freedom and the number of ro effects being studied (r) if row effects are of interest. The prope degrees of freedom are those corresponding to the degrees of freedom of the "within cell" source in the analysis of variance table, i.e., $rc(n-1)$. The factor k^* is read from the table and

$$k\,(\text{for rows}) = k^*\sqrt{SS_1^*/nc}.$$

Similarly, if column effects are of interest, the table is entered with the indices degrees of freedom, and the number of column effects being studied (c). The factor k^* is read from the table and

$$k\,(\text{for columns}) = k^*\sqrt{SS_1^*/nr}.$$

Of course, for finding confidence intervals, the calculated terms that are of interest are $\bar{x}_{i..}$ and $\bar{x}_{.j.}$.

10.4.4 The OC curve of the analysis of variance for the fixed effects model, two-way classification, n observations per cell

The OC curve for this model is obtained in a manner similar to that described in Section 10.3.6. If the OC curve for testing the hypothesis about row effects is desired, the appropriate measure for which an OC curve can be constructed is

$$\Phi = \frac{\sqrt{\sum_{i=1}^{r} \varphi_i^2 / r}}{\sigma/\sqrt{nc}}.$$

Similarly if the OC curve for testing the hypothesis about column effects is desired, the appropriate measure for which an OC curve can be constructed is

$$\Phi = \frac{\sqrt{\sum_{j=1}^{c} \gamma_j^2 / c}}{\sigma/\sqrt{nr}}.$$

Finally if the OC curve for testing the hypothesis about interaction effects is desired, the appropriate measure for which an OC curve can be constructed is

$$\Phi = \frac{1}{\sigma}\sqrt{\frac{n}{(r-1)(c-1)+1}\sum_{i=1}^{r}\sum_{j=1}^{c}\eta_{ij}^2}.$$

The OC curves as functions of ν_1, ν_2 and Φ are given in Figures 10.2 to 10.9. The appropriate values of ν_1 are $(r-1)$ if row effects are being considered, $(c-1)$ if column effects are being considered, and $(r-1)(c-1)$ if interaction effects are being considered; ν_2 corresponds to the degrees of freedom associated with the "within combinations" source and is given by $rc(n-1)$. For fixed r, c, and n, the OC curve of the procedure can be obtained by calculating Φ, ν_1, and ν_2 and entering the appropriate figure. On the other hand, if two points on the OC curve are specified, the appropriate value of n can be obtained by trial and error.

A physical interpretation of Φ as a measure can be given by finding upper and lower bounds for Φ as a function of the range of the $\varphi_i[\varphi(\max) - \varphi(\min)$ if row effects are being studied] beyond which rejection of homogeneity should be assured with a given probability. Denote this largest difference by W. Bounds on Φ are given as follows:

$$\Phi(\min) = \frac{W}{\sigma}\sqrt{\frac{nc}{2r}}$$

$$\Phi(\max) = \begin{cases} \dfrac{W}{2\sigma}\sqrt{nc} & \text{for } r \text{ even} \\ \dfrac{W}{2\sigma}\sqrt{\dfrac{nc(r^2-1)}{r^2}} & \text{for } r \text{ odd.} \end{cases}$$

Similarly, if column effects are being studied

$$\Phi(\min) = \frac{W}{\sigma}\sqrt{\frac{nr}{2c}}$$

$$\Phi(\max) = \begin{cases} \dfrac{W}{2\sigma}\sqrt{nr} & \text{for } c \text{ even} \\ \dfrac{W}{2\sigma}\sqrt{\dfrac{nr(c^2-1)}{c^2}} & \text{for } c \text{ odd.} \end{cases}$$

Finally, if interaction effects are being studied

$$\Phi(\min) = \frac{W}{\sqrt{2}\sigma}\sqrt{\frac{n}{(r-1)(c-1)+1}}$$

$$\Phi(\max) = \begin{cases} \dfrac{W}{2\sigma}\sqrt{\dfrac{nrc}{(r-1)(c-1)+1}} & \text{for } rc \text{ even} \\ \dfrac{W}{2\sigma}\sqrt{\dfrac{n(r^2c^2-1)}{[(r-1)(c-1)+1]rc}} & \text{for } rc \text{ odd.} \end{cases}$$

10.4.5 Example Using the Fixed Effects Model, Two-Way Classification, Three Observations per Combination

Returning to the experiment on the effect of time of aging in the strength of cement described in Section 10.4.1, if time and mixes are considered as fixed effects (making the unrealistic assumption that only the three mixes are of interest), the data can be analyzed according to this model. Interaction cannot be ignored since it is quite possible that the three mixes differ after a long period of time, without differing after a short period. This is equivalent to saying that the effect of an additional period of time is different for the three mixes, i.e., interaction is present.

The following computations lead to the analysis of variance table.

(1) $$R_1, R_2, R_3 = 4{,}931;\ 4{,}747;\ 4{,}874$$

(2) $$C_1, C_2 = 5{,}044;\ 9{,}508$$

(3) $$w_{11}, w_{21}, w_{31}, w_{12}, w_{22}, w_{32} = 1{,}688;\ 1{,}648;\ 1{,}708;\ 3{,}243;\ 3{,}099;\ 3{,}166$$

(4) $$T = 4{,}931 + 4{,}747 + 4{,}874 = 14{,}552$$

(5) $$\sum_{i=1}^{3}\sum_{j=1}^{2}\sum_{v=1}^{3} x_{ijv}^2 = (574)^2 + (564)^2 + \cdots + (1{,}055)^2 = 12{,}882{,}026$$

(6) $$\sum_{j=1}^{2} \frac{C_j^2}{9} = \frac{(5{,}044)^2 + (9{,}508)^2}{9} = 12{,}871{,}556$$

(7) $$\sum_{i=1}^{3} \frac{R_i^2}{6} = \frac{(4{,}931)^2 + (4{,}747)^2 + (4{,}874)^2}{6} = 11{,}767{,}441$$

(8) $$\sum_{i=1}^{3}\sum_{j=1}^{2} \frac{w_{ij}^2}{3} = \frac{(1{,}688)^2 + (1{,}648)^2 + \cdots + (3{,}166)^2}{3} = 12{,}875{,}639$$

(9) $$\frac{T^2}{18} = \frac{211{,}760{,}704}{18} = 11{,}764{,}484$$

(10) $$SS_4 = 12{,}871{,}556 - 11{,}764{,}484 = 1{,}107{,}072$$

(11) $$SS_3 = 11{,}767{,}441 - 11{,}764{,}484 = 2{,}957$$

(12) $$SS_1 = 12{,}882{,}026 - 12{,}875{,}639 = 6{,}387$$

(13) $$SS = 12{,}882{,}026 - 11{,}764{,}484 = 1{,}117{,}542$$

(14) $$SS_2 = 1{,}117{,}542 - 6{,}387 - 2{,}957 - 1{,}107{,}072 = 1{,}126$$

Table 10.14. Analysis of Variance Table for Yield Loads on Cement Specimens

Source	Sum of Squares	Degrees of Freedom	Mean Square	Test
Between times......	1,107,072	1	1,107,072	F=2,079.98
Between mixes	2,957	2	1,478.5	F= 2.78
Interaction	1,126	2	563	F= 1.06
Within combinations	6,387	12	532.25	
Total	1,117,542	17	65,737.8	

The F values in the table for interaction and mixes are well below the critical 5% value for 2 and 12 degrees of freedom, i.e., $F_{0.05;2,12}=3.89$. Hence it is concluded that the data do not give sufficient evidence of the presence of interaction or any mix effect. On the other hand, the F test for times is significant so that it is concluded that there is an effect from the additional five days of aging. Furthermore, a comparison of column means indicates that

$$\bar{x}_{\cdot 2} - \bar{x}_{\cdot 1} - k \leqq \gamma_2 - \gamma_1 \leqq \bar{x}_{\cdot 2} - \bar{x}_{\cdot 1} + k$$

$$(1{,}056.4 - 560.4) - 23.7 = 472.3$$

$$\leqq \text{mean effect for 7 days} - \text{mean effect for 2 days}$$

$$\leqq (1{,}056.4 - 560.4) + 23.7 = 519.7$$

where k is obtained from Table 10.3 using as indices $c = 2$; degrees of freedom $= rc(n - 1) = 12$; $k^* = 3.08$; so that

$$k = 23.7.$$

The following is an example of the use of the OC curves; suppose it is required to find the probability of accepting the hypothesis that there is no time effect when the difference in time effect is as large as 450 psi. Since there are only two time periods, this difference coincides with the range (largest minus smallest). 450 psi corresponds to 45 units so that $W = 45$. An estimate of σ is obtained from the data; σ is approximately equal to $\sqrt{SS_1^*} = 23.2$. In this special case ($c = 2$), $\Phi(\min)$ coincides with $\Phi(\max)$ and is equal to

$$\Phi = \frac{45}{23.2}\sqrt{\frac{(3)(3)}{(2)(2)}} = 2.91.$$

Entering Figure 10.2 with $\nu_1 = 1$, $\nu_2 = 12$, and $\Phi = 2.91$, a probability equal to 0.03 of accepting the hypothesis that there are no time effects is read out.

10.4.6 Analysis of the Random Effects Model, Two-Way Classification, n Observations per Combination

In this model a test for interaction, i.e., $H: \sigma_\eta^2 = 0$, is also available. Referring to the analysis of variance table, Table 10.12, the expected mean square for the "within combinations" source equals σ^2. The expected mean square for the "interaction" source equals σ^2 plus some interaction effect terms. If there is no interaction present, i.e., $\sigma_\eta^2 = 0$, the expected mean square for both "interaction effects" and "within combination" sources, $E(SS_2^*)$ and $E(SS_1^*)$ respectively, equals σ^2. Thus, using the ratio SS_2^*/SS_1^* leads to a reasonable procedure for testing for interaction effects. It can be shown that (using the partition theorem) SS_2^*/SS_1^* is distributed as an F random variable with $(r - 1)(c - 1)$ and $rc(n - 1)$ degrees of freedom when the hypothesis of no interaction is true. Thus, the hypothesis that $\sigma_\eta^2 = 0$ is rejected if

$$F = \frac{SS_2^*}{SS_1^*} \geq F_{\alpha;(r-1)(c-1),rc(n-1)}.$$

In this model if interaction is present, one can still talk about the magnitude of main effects, say σ_φ^2. $\sigma_\eta^2 > 0$ does not imply anything about σ_φ^2. The test for main effects, say row effects, is different from that given in the fixed effects model. Again, referring to the analysis of variance table, Table 10.12, the expected mean square for the "interaction" source equals $\sigma^2 + n\sigma_\eta^2$. The expected mean square for the "row effect" source equals $\sigma^2 + n\sigma_\eta^2 + cn\sigma_\varphi^2$.

If there are no row effects present, i.e., $\sigma_\varphi^2 = 0$, the expected mean square for both "row effects" and "interaction effects" sources, $E(SS_3^*)$ and $E(SS_2^*)$ respectively, equals $\sigma^2 + n\sigma_\eta^2$. Thus, using the ratio SS_3^*/SS_2^* leads to a reasonable procedure for testing for row effects. This is different from the fixed effects model which uses SS_3^*/SS_1^*. It can be shown that (using the partition theorem) SS_3^*/SS_2^* is distributed as an F random variable with $(r-1)$ and $(r-1)(c-1)$ degrees of freedom when the hypothesis of no row effects is true. Thus, the hypothesis that $\sigma_\varphi^2 = 0$ is rejected if

$$F = \frac{SS_3^*}{SS_2^*} \geq F_{\alpha;(r-1),(r-1)(c-1)}.$$

Using an analogous argument, the hypothesis that $\sigma_\gamma^2 = 0$ is rejected if

$$F = \frac{SS_4^*}{SS_2^*} \geq F_{\alpha;(c-1),(r-1)(c-1)}.$$

Using the expected mean square column in the analysis of variance table, estimates of the quantities σ_η^2, σ_φ^2, and σ_γ^2 can be obtained. An estimate of σ_η^2 is given by

$$\frac{SS_2^* - SS_1^*}{n}.$$

An estimate of σ_φ^2 is given by

$$\frac{SS_3^* - SS_2^*}{cn}.$$

An estimate of σ_γ^2 is given by

$$\frac{SS_4^* - SS_2^*}{rn}.$$

If any of these quantities are negative, the estimate is taken to be zero.

10.4.7 The OC Curve of the Random Effects Model, Two-Way Classification, n Observations per Combination

The hypothesis of no interaction is accepted if

$$\frac{SS_2^*}{SS_1^*} \leqq F_{\alpha;(r-1)(c-1),rc(n-1)}.$$

SS_2^*/SS_1^* has an F distribution whenever $\sigma_\eta^2 = 0$, and the probability of acceptance is $(1 - \alpha)$. In general, the probability of accepting the hypothesis that $\sigma_\eta^2 = 0$ can be plotted as a function of $\lambda = \sqrt{1 + n\sigma_\eta^2/\sigma^2}$.* The OC curves of Figures 10.10 to 10.17 are again applicable when λ is defined in this manner. Similarly, testing for row effects involves the statistic SS_3^*/SS_2^* and the OC curve can be expressed as a function of

$$\lambda = \sqrt{1 + \frac{nc\sigma_\varphi^2}{\sigma^2 + n\sigma_\eta^2}}.$$

Finally, testing for column effects involves the statistic SS_4^*/SS_2^*, and the OC curve can be expressed as a function of

$$\lambda = \sqrt{1 + \frac{nr\sigma_\gamma^2}{\sigma^2 + n\sigma_\eta^2}}.$$

The OC curves of Figures 10.10 to 10.17 are entered with degrees of freedom $\nu_1 = (r - 1)(c - 1)$ and $\nu_2 = rc(n - 1)$ if interaction effects are relevant, with $\nu_1 = (r - 1)$ and $\nu_2 = (r - 1)(c - 1)$ if row effects are relevant, and $\nu_1 = (c - 1)$ and $\nu_2 = (r - 1)(c - 1)$ if column effects are desired.

* The quantity

$$\frac{SS_2^*/(\sigma^2 + n\sigma_\eta^2)}{SS_1^*/\sigma^2}$$

always has an F distribution so that the probability of accepting the hypothesis, i.e.,

$$P\left(\frac{SS_2^*}{SS_1^*} \leqq F_{\alpha;(r-1)(c-1),rc(n-1)}\right)$$

can always be expressed in terms of the probability that

$$\frac{SS_2^*/(\sigma^2 + n\sigma_\eta^2)}{SS_1^*/\sigma^2} \leqq F_{\alpha;(r-1)(c-1),rc(n-1)}\left[\frac{\sigma^2}{\sigma^2 + n\sigma_\eta^2}\right].$$

This last expression is just a statement about an F random variable and can be expressed as a function of

$$\lambda = \sqrt{\frac{\sigma^2 + n\sigma_\eta^2}{\sigma^2}} = \sqrt{1 + \frac{n\sigma_\eta^2}{\sigma^2}}.$$

10.4.8 EXAMPLE USING THE RANDOM EFFECTS MODEL

If in the effect of time of aging on the strength of cement example both time and mixes are considered to be random effects, this will provide an example of the model. That mixes are random is a reasonable assumption, but time being random is rather unrealistic. However, for the purpose of the example these assumptions will be made. The formal calculations including the computational procedure and the analysis of variance table are identical with the results of Section 10.4.5. The test for interaction is the same, but the tests for main effects are different, the latter using SS_2^* in the denominator. The results are as follows:

$$\text{mix effect} \quad \frac{SS_3^*}{SS_2^*} = 2.63 < 19.0 = F_{0.05;2,2}$$

$$\text{time effect} \quad \frac{SS_4^*}{SS_2^*} = 1{,}966.38 > 18.5 = F_{0.05;1,2}$$

which leads to the same conclusions as the fixed effects model. An estimate of σ_η^2 is 10.25, an estimate of σ_φ^2 is 152.58, and an estimate of σ_γ^2 is 122,945.44. The fact that the hypothesis that $\sigma_\varphi^2 = 0$ was accepted does not necessarily imply that the hypothesis is true, and hence, an estimate of σ_φ^2 may be desirable. Accepting the hypothesis may just be an indication that there is insufficient evidence to reject it, and must be interpreted according to the OC curve.

10.4.9 ANALYSIS OF THE MIXED EFFECTS MODEL, TWO-WAY CLASSIFICATION, n OBSERVATIONS PER CELL

In the mixed effects model, the row effects will be assumed as the random effects and the column effects will be assumed as the fixed effects. In this model, a test for interaction is also available. η_{ij} is a random variable since one of the factors, row effects, is a random variable. Referring to the analysis of variance table, Table 10.12, the expected mean square for the "within combinations" source equals σ^2. The expected mean square for the "interaction" source equals σ^2 plus some interaction effect terms. If there is no interaction present, i.e., $\sigma_\eta^2 = 0$ the expected mean square for both "interaction effects" and "within combination" sources, $E(SS_2^*)$ and $E(SS_1^*)$ respectively, equals σ^2. Thus, using the ratio SS_2^*/SS_1^* leads to a reasonable procedure for testing for interaction effects. It can be shown that (using the partition theorem) SS_2^*/SS_1^* is distributed as an F random variable with $(r-1)(c-1)$ and $rc(n-1)$ degrees of freedom when the hypothesis of no interaction is true. Thus, the hypothesis that $\sigma_\eta^2 = 0$ is rejected if

$$F = \frac{SS_2^*}{SS_1^*} \geqq F_{\alpha;(r-1)(c-1),rc(n-1)}.$$

Testing for main effects again requires studying the analysis of variance table, Table 10.12. For row effects (random effects), the expected mean square, $E(SS_3^*)$, equals $\sigma^2 + nc\sigma_\varphi^2$. Hence, the logical test is to use the ratio SS_3^*/SS_1^* since $E(SS_1^*) = \sigma^2$. It can be shown that (using the partition theorem) SS_3^*/SS_1^* has an F distribution with $(r - 1)$ and $rc(n - 1)$ degrees of freedom when the hypothesis is true. Thus, the hypothesis that $\sigma_\varphi^2 = 0$ is rejected if

$$F = \frac{SS_3^*}{SS_1^*} \geqq F_{\alpha;(r-1),rc(n-1)}.$$

An estimate of σ_φ^2 is obtained from

$$\frac{SS_3^* - SS_1^*}{nc}.$$

For column effects (fixed effects) the procedure is somewhat different. The expected mean square for columns, $E(SS_4^*)$, equals

$$\sigma^2 + n\sigma_\eta^2 + \frac{rn}{c-1}\sum_{j=1}^{c}\gamma_j^2.$$

Since the expected mean square for the interaction term, $E(SS_2^*)$, equals $\sigma^2 + n\sigma_\eta^2$, the logical test to use is SS_4^*/SS_2^*. It can be shown that (using the partition theorem) SS_4^*/SS_2^* has an F distribution with $(c-1)$ and $(r-1)(c-1)$ degrees of freedom when the hypothesis of no column effect is true. Thus, the hypothesis that $\gamma_1 = \gamma_2 = \cdots = \gamma_c = 0$ is rejected if

$$F = \frac{SS_4^*}{SS_2^*} \geqq F_{\alpha;(c-1),(r-1)(c-1)}.$$

Tukey's method can again be applied to get confidence intervals on the difference between two column effects. The probability is $1 - \alpha$ that the value $(\gamma_f - \gamma_g)$ for *all* such contrasts for $f = 1, 2, \cdots, c$, $g = 1, 2, \cdots, c$ simultaneously satisfies

$$\bar{x}_{\cdot f} - \bar{x}_{\cdot g} - k \leqq \gamma_f - \gamma_g \leqq \bar{x}_{\cdot f} - \bar{x}_{\cdot g} + k.$$

Two mean effects are judged significantly different when 0 is not included in the confidence interval.

The factor k for column effect is obtained from the analysis of variance table and from Table 10.3 or Table 10.4 depending on the level of significance. Table 10.3 or Table 10.4 is entered with the indices, degrees of freedom, and the number of column effects (c)

being studied. The proper degrees of freedom are those corresponding to the degrees of freedom of the "interaction" source in the analysis of variance table, i.e., $(r - 1)(c - 1)$. The factor k^* is read from the table, and

$$k = k^* \sqrt{SS_2^*/nr}.$$

10.4.10 THE OC CURVE OF THE ANALYSIS OF VARIANCE FOR THE MIXED EFFECTS MODEL, TWO-WAY CLASSIFICATION, n OBSERVATIONS PER CELL

The OC curve for the interaction effects is exactly the same as for the interaction in the random effects model, i.e.,

$$\lambda = \sqrt{\frac{\sigma^2 + n\sigma_\eta^2}{\sigma^2}} = \sqrt{1 + \frac{n\sigma_\eta^2}{\sigma^2}}.$$

The OC curves are given in Figures 10.10 to 10.17 and are entered with degrees of freedom $\nu_1 = (r - 1)(c - 1)$ and $\nu_2 = rc(n - 1)$.

The OC curve for the random effects (rows) are found in Figures 10.10 to 10.17 using the index

$$\lambda = \sqrt{\frac{\sigma^2 + nc\sigma_\varphi^2}{\sigma^2}} = \sqrt{1 + \frac{nc\sigma_\varphi^2}{\sigma^2}}.$$

These curves are entered with degrees of freedom $\nu_1 = r - 1$ and $\nu_2 = rc(n - 1)$.

Finally, the OC curves for testing the hypothesis about the fixed effects (columns) is a function of the measure Φ, where

$$\Phi = \frac{\sqrt{\sum_{j=1}^{c} \tau_j^2 \Big/ c}}{\sqrt{(\sigma^2 + n\sigma_\eta^2)/rn}}.$$

This is exactly the same procedure as outlined for column effects in Section 10.4.4 with the value σ replaced by $\sqrt{\sigma^2 + n\sigma_\eta^2}$. The degrees of freedom are $\nu_1 = c - 1$ and $\nu_2 = (r - 1)(c - 1)$.

10.4.11 EXAMPLE USING THE MIXED EFFECTS MODEL

The problem about the effect of aging on the strength of cement presented in Section 10.4.1 is actually a mixed effects model. The mixes are considered to be a random sample from a population of mixes, whereas the two times of aging are fixed effects. The calculations carried out in Section 10.4.3 coincide with the required calculations for this model. However, the numerical values of the F tests are different. The test for interaction is the same, but the test for the random effects involves SS_3^*/SS_1^*, and the test for fixed effects involves SS_4^*/SS_2^*.

Table 10.15. Table of Summary and Tests

One-Way Classification

Source	*Fixed Effects* Test Statistic	*Fixed Effects* Ratio of Expected Mean Squares	*Random Effects* Test Statistic	*Random Effects* Ratio of Expected Mean Squares		
Row effects	$\frac{SS_3^*}{SS^*}$	$\frac{\sigma^2 + \frac{c}{r-1}\sum \varphi_i^2}{\sigma^2}$	$\frac{SS_3^*}{SS_2^*}$	$\frac{\sigma^2 + c\sigma_\varphi^2}{\sigma^2}$		
Column effects	—	—	—	—		
Interaction	—	—	—	—		

Two-Way Classification, One Observation per Combination

Source	*Fixed Effects* Test Statistic	*Fixed Effects* Ratio of Expected Mean Squares	*Random Effects* Test Statistic	*Random Effects* Ratio of Expected Mean Squares	*Mixed Effects* Test Statistic	*Mixed Effects* Ratio of Expected Mean Squares
Row effects	$\frac{SS_3^*}{SS_2^*}$	$\frac{\sigma^2 + \frac{c}{r-1}\sum \varphi_i^2}{\sigma^2}$	$\frac{SS_3^*}{SS_2^*}$	$\frac{\sigma^2 + \sigma_\eta^2 + c\sigma_\varphi^2}{\sigma^2 + \sigma_\eta^2}$	$\frac{SS_3^*}{SS_2^*}$	$\frac{\sigma^2 + c\sigma_\varphi^2}{\sigma^2}$
Column effects	$\frac{SS_4^*}{SS_2^*}$	$\frac{\sigma^2 + \frac{r}{c-1}\sum \gamma_j^2}{\sigma^2}$	$\frac{SS_4^*}{SS_2^*}$	$\frac{\sigma^2 + \sigma_\eta^2 + r\sigma_\gamma^2}{\sigma^2 + \sigma_\eta^2}$	$\frac{SS_4^*}{SS_2^*}$	$\frac{\sigma^2 + \sigma_\eta^2 + \frac{r}{c-1}\sum \gamma_j^2}{\sigma^2 + \sigma_\eta^2}$
Interaction	Interaction assumed to be zero		No test for interaction available but no assumption required about its presence		No test for interaction available. To test for row effects interaction must be assumed zero. No such assumption required for column effects	

Two-Way Classification, n Observations per Combination

Source	*Fixed Effects* Test Statistic	*Fixed Effects* Ratio of Expected Mean Squares	*Random Effects* Test Statistic	*Random Effects* Ratio of Expected Mean Squares	*Mixed Effects* Test Statistic	*Mixed Effects* Ratio of Expected Mean Squares
Row effects	$\frac{SS_3^*}{SS_1^*}$	$\frac{\sigma^2 + \frac{cn}{r-1}\sum \varphi_i^2}{\sigma^2}$	$\frac{SS_3^*}{SS_2^*}$	$\frac{\sigma^2 + n\sigma_\eta^2 + nc\sigma_\varphi^2}{\sigma^2 + n\sigma_\eta^2}$	$\frac{SS_3^*}{SS_1^*}$	$\frac{\sigma^2 + nc\sigma_\varphi^2}{\sigma^2}$
Column Effects	$\frac{SS_4^*}{SS_1^*}$	$\frac{\sigma^2 + \frac{rn}{c-1}\sum \gamma_j^2}{\sigma^2}$	$\frac{SS_4^*}{SS_2^*}$	$\frac{\sigma^2 + n\sigma_\eta^2 + nr\sigma_\gamma^2}{\sigma^2 + n\sigma_\eta^2}$	$\frac{SS_4^*}{SS_2^*}$	$\frac{\sigma^2 + n\sigma_\eta^2 + \frac{rn}{c-1}\sum \gamma_j^2}{\sigma^2 + n\sigma_\eta^2}$
Interaction	$\frac{SS_2^*}{SS_1^*}$	$\frac{\sigma^2 + \frac{n}{(r-1)(c-1)}\sum\sum \eta_{ij}^2}{\sigma^2}$	$\frac{SS_2^*}{SS_1^*}$	$\frac{\sigma^2 + n\sigma_\eta^2}{\sigma^2}$	$\frac{SS_2^*}{SS_1^*}$	$\frac{\sigma^2 + n\sigma_\eta^2}{\sigma^2}$

The results are as follows:

$$\text{mix effect} \quad SS_3^*/SS_1^* = 2.78 < 3.89 = F_{0.05;2,12}$$

$$\text{time effect} \quad SS_4^*/SS_2^* = 1{,}966.38 > 18.5 = F_{0.05;1,2}$$

which leads to the same conclusions as before.

10.5 Summary of Models and Tests

In table 10.15 a summary of the test statistics and the ratios of the expected mean squares is presented.

PROBLEMS

1. It is suspected that four filling machines in a plant are turning out products of non-uniform weight. An experiment is run and the data in ounces are as follows:

Machine					
A	12.25	12.27	12.24	12.25	12.20
B	12.18	12.25	12.26	12.22	12.19
C	12.24	12.23	12.23	12.20	12.16
D	12.20	12.17	12.19	12.18	12.16

(a) Write out the analysis of variance table. (b) Test to see if the machines differ at the 5% level of significance. (c) Test to see if Machine *A* differs from Machine *C*. Note that this is merely testing for the difference between *two* means. (d) If there is a machine effect shown, then compute the "Tukey" contrasts. (e) It is desired that if there is a difference in machines such that a random observation has its standard deviation increased by as much as 25%, the analysis of variance procedure should lead to the rejection of the hypothesis that the machines are the same with probability exceeding 0.90. How many observations on each machine are required?

2. Suppose four different quantities of carbon are to be added to a batch of raw material in the manufacture of steel. Six specimens are taken for each of the four quantities, and it is desired to determine the effect of these percentages on the tensile strength of the resultant steel. The data are given in the following table.

Tensile Strength of Hot Rolled Steel for Different Percentages of Carbon

								$\bar{x}$
Percentage 1:	0.10%	23,050	36,000	31,100	32,650	30,900	31,400	30,850
Percentage 2:	0.20%	41,850	25,650	46,700	34,500	36,650	31,450	36,133
Percentage 3:	0.40%	47,050	43,450	43,000	38,650	41,850	35,450	41,575
Percentage 4:	0.60%	49,650	73,900	66,450	74,550	62,400	63,750	65,117

(a) Does the percentage of carbon have any effect on the tensile strength? Write out the analysis of variance table. (b) If the result of (a) is affirmative, determine which steels differ. (c) If the maximum difference in steels is 10,000 psi, what is the approximate probability of accepting the hypothesis of no differences? Use the " within treatment " mean square as the estimate of σ^2.

3. In evaluating the effect of cloud seeding in Santa Clara County, the following problem was encountered. Past data for one particular station was taken beginning with the year 1922. It was later discovered that the rain gauges had been moved three times since that date. One possible procedure to evaluate the effect of the change in environment is to compare the successive annual precipitation amounts at one station with the corresponding average. Tabulated below are ratios of station precipitation to average precipitation for the near-by stations for each year.

Location 1		*Location 2*		*Location 3*		*Location 4*	
Year	*Ratio*	*Year*	*Ratio*	*Year*	*Ratio*	*Year*	*Ratio*
1922	1.05	1929	1.03	1936	0.98	1943	0.97
1923	1.02	1930	0.96	1937	0.91	1944	0.94
1924	0.93	1931	0.96	1938	0.92	1945	1.02
1925	1.00	1932	0.99	1939	0.86	1946	0.99
1926	1.06	1933	1.04	1940	0.97	1947	0.98
1927	1.01	1934	0.97	1941	1.00	1948	1.05
1928	0.95	1935	1.00	1942	0.90	1949	1.01

Using the analysis of variance, determine if the change in rain gauge location had any effect on the ratio. Write out the analysis of variance table.

4. The tensile strength of a certain rubber seal shows the following variation with different conditions of production. Data are in pounds per square inch.

Condition of Production

A	*B*	*C*	*D*
3,800	4,200	3,800	3,500
4,100	4,200	3,900	3,700
4,000	4,400	3,700	3,600
3,800	4,300	3,800	3,700

(a) Do the production conditions have any effect on the tensile strength of the rubber seal? Write out the analysis of variance table. (b) If the production conditions do differ, use Tukey's method to make confidence statements about which conditions differ. (c) An underlying assumption was that σ^2 was constant for all production conditions. Test this hypothesis. (d) If it is important to detect a maximum difference between production conditions of 375 psi with probability greater than or equal to 0.9, how many observations are required for each condition of production? Use the "within treatment" mean square as the estimate of σ^2.

5. Among the classrooms in the public schools of a given city there are 12 different lighting techniques. Each of these techniques is supposed to provide the same level of illumination. To determine whether or not the illumination they provide is uniform, the following data were compiled from a random sample of four lighting techniques. The classrooms are known to be homogeneous and hence can be discounted as a possible source of variability.

Lighting techniques	*Foot-candles on desk surface*				
1	31	38	38	33	31
2	31	34	27	27	29
3	34	35	39	35	30
4	37	34	27	31	26

(a) Are the lighting techniques homogeneous? Use $\alpha=0.05$. (b) Is it possible to determine which techniques differ, if any, among all 12 lighting techniques? Why? (c) Is it possible to estimate the variability among all lighting techniques? If so, estimate this variability. (d) What is the probability of accepting the hypothesis that the lighting techniques are homogeneous when $\sigma_\varphi^2=40$ and σ^2 is estimated from the "within treatments" mean square?

6. A small laboratory has many thermometers which are used interchangeably to make temperature measurements. To perform an experiment with all the thermometers is too costly, and a random sample of four thermometers is drawn to determine whether thermometers differ. These are placed in a cell which is kept at constant temperature. The data are as follows in degrees Centigrade and are obtained from three readings on each thermometer.

Thermometers

1	2	3	4
0.95	0.33	−2.15	1.05
1.06	−1.46	1.70	1.27
1.96	0.20	0.48	−2.05

(a) Write out the analysis of variance table. (b) Are the thermometers homogeneous? ($\alpha=0.01$) (c) Estimate the variability among the thermometers. (d) How many observations on each thermometer would be required if it is desired to detect a 20% increase in the total variability of a random observation with probability 0.7?

7. An experiment was run to check on the variability of batches of cement briquettes. Five specimens from each of nine batches were analyzed for their breaking strength. The data are as follows:

Breaking Strength (pounds tension) of Nine Batches of Cement Briquettes

1	2	3	4	5	6	7	8	9
553	553	510	520	543	492	542	581	578
550	599	580	559	500	530	550	550	531
568	579	529	539	562	528	580	529	562
541	545	535	510	540	510	545	570	525
537	540	537	540	535	571	520	524	549

(a) Write out the analysis of variance table. (b) Do the batches differ? ($\alpha=0.05$) (c) Estimate the batch variability. (d) What is the probability of accepting the hypothesis that the batches are homogeneous when $\sigma_{\varphi}^2=1{,}000$ and σ^2 is estimated from the "within treatment" mean square?

8. A new cure has been developed for Portland cement. Its production has caused problems since the variability of the compressive strength from batch to batch appears to be excessive. An experiment has been run to study this batch variability. The data are given below.

Batch	*Compressive strength of duplicate measure (psi)*	
1	4,125	4,250
2	4,225	3,950
3	4,025	3,900
4	3,900	4,075
5	3,875	4,550
6	3,825	4,450
7	3,975	4,150
8	3,800	4,550
9	3,775	3,700
10	3,850	4,250

(a) Write out the analysis of variance table. (b) Do the batches differ? (c) Estimate the batch variability. (d) How many observations from each batch would be required if it is desired to detect a 25% increase in the total variability of a random observation with probability 0.8?

9. The following represents the yield of a certain chemical process under varying conditions. The results are excesses over 40.0 pounds, in tenths of a pound.

Temperature	*Concentration of Inert Solvent* A	B	C
Low	4	7	2
Medium....	1	5	0
High.......	5	8	3

Assume there is no interaction between temperature and concentration of inert solvent. (a) Does the concentration of inert solvent have a significant effect on the yield at the 5% level of significance? (b) Does the temperature have a significant effect on yield at the 5% level? (c) If there are differences in yield due to concentration, determine which differ. (d) If there are differences in yield due to temperature, determine which differ. (e) When the maximum difference in yields caused by concentration is $\frac{1}{2}$ pound, what are the bounds on the probability of accepting the hypothesis that there is no concentration effect? Use the residual mean square as the estimate of σ^2.

10. A study has been made on pyrethrum flowers to determine the content of pyrethrin, a chemical used in insecticides. Four methods of extracting the chemical are used and samples are obtained from flowers stored under three conditions, namely, fresh flowers, flowers stored for one year, and flowers stored for one year but treated. It is assumed that there is no interaction present. The data are as follows.

Pyrethrin Content, percent

Storage condition	*Method* *A*	*B*	*C*	*D*
1	1.35	1.13	1.06	0.98
2	1.40	1.23	1.26	1.22
3	1.49	1.46	1.40	1.35

(a) Write out the analysis of variance table. (b) Do the methods of extraction differ? Do the storage conditions effect the content? ($\alpha=0.05$) (c) If there are differences found in (b), determine which differ. (d) When the maximum difference in pyrethrin content due to methods is .25, what are the bounds on the probability of accepting the hypothesis that there is no method effect? Use the residual mean square as the estimate of σ^2.

11. The following is an example of an experiment on the testing of leather soles in a wear tester. The machine consists of four arms attached to a shaft on each of which is placed a shoe. The shaft rotates so that the shoes brush over an abrasive surface, abrading the test specimens. The loss in weight after a given number of cycles is used as a criterion of resistance to abrasion. The position of the shoes is changed after a fixed number of cycles so that each shoe appears on each arm for a fixed number of cycles. The order is randomized. The data are given below where the entries denote the loss in weight (unit of 0.1 milligram) in a run of standard length. Four different types of leather are tested.

Leather specimen	*Arm* 1	2	3	4	*Average*
A	264	260	258	241	255.75
B	208	231	216	185	210.00
C	220	263	219	225	231.75
D	217	226	215	224	220.50
Average:	227.25	245.00	227.00	218.75	229.50

$SS_4=1{,}468.5,\qquad SS_3=4{,}621.5,\qquad SS=7{,}444.0$

(a) Test for specimen effects and position effects at the 5% level of significance. (b) If the specimens differ, determine which are different. (c) Give approximate bounds on the probability of accepting the hypothesis of "no specimen effects" when the largest difference in specimen means is 50. Use the residual mean square as an estimate of σ^2.

12. An experiment was run to test the effect of adding salt to the water used in quenching aluminium casts. These casts are subjected to standard heat treatments and their tensile strength measured. There are five such treatments. The data are given below in units of 1,000 psi.

Heat treatment	*Salt Water quench*	*Water quench*
A	30	31
B	46	44
C	45	47
D	40	43
E	36	35

(a) Write out the analysis of variance table. (b) At the 5% level of significance, determine if there are any heat treatment effects and quenching effects. (c) How does this test on quenching compare with a t-test for quenching effects? (d) If there are any heat treatment effects, determine which differ. (e) If the true difference in tensile strengths due to quenching is 2,000 psi, what is the probability of accepting the hypothesis that there are no quenching effects? Use the residual mean square as an estimate of σ^2. Compare this result with that obtained using the paired t-test.

13. In a large metals processing company which specializes in manufacturing components for sensitive mechanisms, it is noticed that there appears to be considerable variability in the finished items. Each department of this firm conducts its own inspection and has its own standard $\frac{1}{4}$-inch gauge blocks with which to check its measuring devices. It is believed that there may be too much variation in the gauge blocks and/or the various measuring devices. Therefore four gauge blocks and four micrometers are chosen at random, and the following data obtained by the chief quality control engineer.

Gauge blocks	*Micrometer* 1	2	3	4
1	0.0251	0.0251	0.0259	0.0249
2	0.0248	0.0254	0.0254	0.0249
3	0.0247	0.0249	0.0255	0.0245
4	0.0245	0.0248	0.0249	0.0252

(a) Test at the 0.01 level for homogeneity of micrometers and gauge blocks. (b) Estimate the gauge block variability and micrometer variability. (c) What value of micrometer variance will be detected with probability 0.8 using the procedure of (a)? Use the residual mean square as an estimate of σ^2. Assume there is no interaction present.

14. The plant manager has decided to run an experiment to determine whether the material received at different times has the same tensile strength. Five randomly chosen time periods are to be considered and five randomly chosen men work on the material.

Time period	1	2	3	4	5	*Mean*
1	7.6	11.8	17.6	8.8	17.9	12.74
2	21.4	12.9	12.4	15.0	20.6	16.46
3	16.0	9.7	7.4	18.4	16.6	13.62
4	16.0	18.3	23.6	27.4	25.2	22.10
5	23.3	30.5	25.8	24.5	26.6	26.14
Mean:	16.86	16.64	17.36	18.82	21.38	

Analysis of Variance Table

Source	S.S.	D.F.	M.S.
Between time periods ..	660.34	4	165.09
Between men	77.15	4	19.29
Residual	—	—	—
Total:	1,039.39	24	

(a) Test at the 5% level for homogeneity of time periods and homogeneity of men. (b) Estimate the time period variability and the men variability. (c) What value of time variability will be detected with probability 0.75 using the procedure of (a)? Use the residual mean square as an estimate of $\sigma^2+\sigma_\eta^2$.

15. Receivers of a certain type have been causing trouble because of malfunctioning. A series of tests were performed to evaluate the variations of the signal generator output. These receivers had in their design two different types of vacuum tubes, each of which had an influence on the signal generator output. A single receiver was chosen for this study and five vacuum tubes were chosen randomly from each type. The signal generator output in KUV data are as follows.

Vacuum tube, type B	*Vacuum tube, type A* 1	2	3	4	5
1	18	21	25	28	18
2	21	18	14	22	22
3	23	21	21	23	25
4	37	19	21	20	24
5	17	13	20	15	23

(a) Test for tube-type A effects at the 5% level of significance. (b) Test for tube-type B effects at the 5% level of significance. (c) Estimate the component of the variance due to tube-type A. (d) Estimate the component of the variance due to tube-type B. (e) What value of tube variability will be detected with probability 0.8 using the procedure of (a)? Use the residual mean square as an estimate of $\sigma^2+\sigma_\eta^2$.

16. Design engineers of a radio receiver manufacturing firm desire to determine whether or not the output at a given point of a certain type of receiver differs when transistors are used instead of vacuum tubes. Three receivers are selected at random and operated at the given point, first with vacuum tubes. Then they are reassembled, using transistors, and operated. The following table of data is obtained:

Receiver	*Configuration* *Vacuum tubes*	*Transistors*
1	19	17
2	18	19
3	21	23
	Output voltages	

(a) Determine whether or not there is a significant difference between output voltages when vacuum tubes are used as opposed to when transistors are used. (Use $\alpha=0.01$.) (b) If the true difference between vacuum tubes and transistors is 2 volts, what is the probability of concluding there is no difference? Use the residual mean square as an estimate of $\sigma^2+\sigma_\eta^2$.

17. Modification of the carburetor for standard model automobile engines recommended by an engine rebuilder will allegedly increase the mileage (miles per gallon) of the automobile in which this type engine is installed. Six automobiles are chosen at random and are test driven over the same test run, by the same driver, under the same conditions of test (rpm, mixture setting, etc.). The table of data below results.

	Carburetor	
Auto	*Original*	*Modified*
1	17	21
2	19	19
3	17	23
4	15	15
5	16	23
6	18	17

Mileage in mi/gal

(a) Assuming no interaction, determine for $\alpha=0.05$ whether or not there is a significant difference in mileage for the six cars used in the test. (b) Is it possible to test for interaction using the data above? If not, design an experiment for this situation in which a test for interaction between automobiles and the carburetors could be conducted. (DO NOT SOLVE THIS.) (c) Determine at the 5% level whether there are any carburetor effects. (d) Assuming no interaction, what value of the automobile variability will be detected with probability 0.8? Use the residual mean square as an estimate of σ^2.

18. An engineer in the design section of an aircraft manufacturing plant has presented theoretical evidence that painting the exterior of a particular combat airplane reduces its cruising speed. He convinces the chief design engineer that the next nine aircraft off the assembly line should be test flown to determine cruising speed prior to paint, then painted, and finally test flown to ascertain cruising speed after they are painted. The following data are obtained.

	Cruising Speed (knots)	
Aircraft	*Not painted*	*Painted*
1	426	416
2	418	400
3	424	420
4	438	431
5	440	432
6	421	404
7	412	398
8	409	405
9	427	422

(a) Design a test at the 5% level and complete the computations necessary to evaluate the design engineer's evidence. (b) What value of the difference in cruising speeds between the painted and unpainted planes will be detected with probability 0.9? Use the residual sum of squares as an estimate of $\sigma^2+\sigma_\eta^2$.

19. The rate of flow of fuel through three different types of nozzles was the subject of a recent experiment. Five different operators chosen from a group of 25 were used to test each nozzle. The data recorded were as follows:

Cubic Centimeters of Fuel through Three Nozzles for Five Trials

	Nozzle Type		
Operator	*A*	*B*	*C*
1	96.5	96.5	97.1
2	97.4	96.1	96.4
3	96.0	97.9	95.6
4	97.8	96.3	95.7
5	97.2	96.8	97.3

(a) Analyze the data as a two-way analysis of variance and write down the analysis of variance table. (b) Test for effects of operators and nozzle types at the 5% level of significance. (c) Estimate the variability due to operators and determine which nozzles differ, if any. (d) What is the approximate probability of accepting the hypothesis of no nozzle effects when the maximum difference between nozzles is actually 0.4 cc? Use the residual mean square as an estimate of $\sigma^2+\sigma_\eta^2$. (e) Assuming no interaction, what value of operator variability will be detected with probability 0.75? Use the residual mean square as an estimate of σ^2.

20. It is desired to know what type of filter should be used over the screen of a cathode ray oscilloscope in order to have a radar operator easily pick out targets on the presentation. A test to accomplish this has been set up. A noise is first applied to the scope in order to make it difficult to pick out a target. A second signal, representing the target, is put into the scope, and its intensity is increased from zero until detected by the observer. The intensity setting at which the observer first notices the target signal is then recorded. This experiment is repeated again, with a different observer, on the assumption that all people do not see in exactly the same manner. A total of 20 observers is chosen at random. After a set of readings has been taken with one type of filter on the scope, another set of readings, using the same observers, is taken with a different type of filter. The procedure above is then carried out with a third filter. The numerical value of each reading listed in the table of data is proportional to the target intensity at the time the operator first detects the target.

Observer	*Filter No. 1*	*Filter No. 2*	*Filter No. 3*
1	90	88	95
2	87	90	95
3	93	97	89
4	96	87	98
5	94	90	96
6	88	96	81
7	90	90	92
8	84	90	79
9	101	100	105
10	96	93	98
11	90	95	92
12	82	86	85
13	93	89	97
14	90	92	90
15	96	98	87
16	87	95	90
17	99	102	101
18	101	105	100
19	79	85	84
20	98	97	102

(a) Test the hypothesis that the filters are the same with $\alpha=0.01$. (b) What value of the difference in intensity between filters will be detected with probability 0.8 ? Use the residual sum of squares as an estimate of $\sigma^2+\sigma_\eta^2$.

21. Three different procedures for measuring runoff in three types of terrain were employed to determine whether or not these procedures were all of equal value. The tests were carried out over a period of four years, and the results are presented below (coded).

	Procedure A		*Procedure B*		*Procedure C*	
Area D	30	54	−68	−60	−80	18
	−26	81	4	123	30	42
Area E	−50	26	36	−15	−83	−30
	59	−12	−6	122	−42	−55
Area F	38	110	74	−20	−54	−9
	68	60	64	50	−42	63

(a) Write out the analysis of variance table. (b) Test for the significance of interaction. If interaction is absent, test for procedure effects and area effects at $\alpha=0.05$. What would be the physical meaning of interaction in this problem if any exists ? (c) If there is no interaction, determine which procedures differ and which areas differ. Obtain confidence intervals for those effects which differ. (d) How many years of readings are required to detect a maximum difference between procedures equal to 40 units with probability 0.75 ? Obtain this result by using $\Phi(\min)$ and using the "within combinations" mean square as an estimate of σ^2.

22. The design engineer of a home automatic clothes washer is uncertain as to the length of time that he should specify for the washing cycle. However, he believes that it should be no less than 20 minutes and no greater than 30 minutes. With the aid of a home economist he designs and conducts a test in which the following data are obtained.

Soap or detergent	*Washing Time* 20 min	25 min	30 min
	11	12	12
1	10	11	12
	12	13	13
	9	11	13
2	13	14	14
	11	14	16
	12	10	11
3	13	12	12
	11	11	11

Dirt Removed (10^{-2} lb)

Three equally sized and equally soiled loads of clothes were washed for each combination of soap or detergent and washing time in the same machine. (a) Consider the effects as fixed and analyze the above data for interaction at $\alpha=0.01$. If no interaction is present, test for main effects. (b) If there is no interaction present, determine which soaps differ. (c) If interaction is absent, how many observations are required to detect a maximum difference between soaps equal to 0.03 lb with probability 0.8? Obtain this result by using Φ (min) and using the "within combinations" mean square as an estimate of σ^2.

23. An experiment was run to determine the abrasive wear of several laminated bearing materials. The bearings used were either oven dried or moisture saturated, with the shaft used in the experiment made of a high carbon steel. The test was conducted on an Amsler testing machine and it can be assumed that the shaft exhibited negligible wear. The machine ran for a fixed amount of time and the depth of bearing wear in units of 10^{-3} was recorded.

Heat treatment	*Bearing Material* *A*	*B*	*C*	*D*
Oven dried	38	40	38	42
	40	37	41	35
Moisture saturated..	32	38	32	34
	44	42	33	36

(a) Write out the analysis of variance table. (b) Test for interaction. Also test for bearing material effects and heat treatment effects if interaction is absent. If materials differ, determine which differ. (c) If interaction is absent, what is the approximate probability of detecting a maximum difference between bearing materials equal to 0.02? Use the "within combinations" mean square as an estimate of σ^2. If there were four observations per combination taken, what is the approximate probability?

24. An electronic manufacturer was having difficulty with a particular type of tube, where the mutual conductance was running below the bogie value. An experiment was performed to determine the effect of the exhaust variables, plate temperature, and filament lighting on the electrical characteristics of the tubes. Two levels of plate temperature and four levels of filament lighting current were used.

Transconductance

		Filament Lighting Conditions		
	L_1	L_2	L_3	L_4
Plate temperature T_1	3,774	4,710	4,176	4,540
	4,364	4,180	4,140	4,530
	4,374	4,514	4,398	3,964
Plate temperature T_2	4,216	3,828	4,122	4,484
	4,524	4,170	4,280	4,332
	4,136	4,180	4,226	4,390

(a) Write out the analysis of variance table. (b) Test for interaction at $\alpha=0.05$. If none is present test for filament lighting conditions effects and plate temperature effects. If there are filament lighting effects, determine which differ. (c) If interaction is absent, how many observations per combination are required to insure detection of a difference in plate temperatures equal to 300 units with probability 0.75 ? Use the " within combinations " mean square as an estimate of σ^2.

25. An experiment was run to isolate and determine the magnitude of the major sources of variability contributing to differences in the breaking strength of some wool serge material. It has already been ascertained that the material is very homogeneous so that material was not felt to be a contributing factor. The remaining factors which appeared to be causing trouble were operators and machines. Three experienced operators and three Scott Testers were picked at random, and each operator was paired with each machine. The experiment was run on a one yard section of the wool serge and the data are as follows:

Breaking Strength, pounds

	Machines		
Operator	*A*	*B*	*C*
1	110	101	108
	116	102	109
2	112	115	111
	111	106	109
3	114	107	113
	112	109	110

(a) Write out the analysis of variance table. (b) Test for interaction, operator effects and machine effects at a 5% level of significance. (c) Estimate the interaction variability, operator variability and machine variability. (d) How many observations per combination are required to detect a machine variance equal to $2\sigma^2$ with probability 0.8 ? Use the value found in (c) as an estimate of the interaction variance. Use the " within combinations " mean square as an estimate of σ^2.

26. In an electronic components manufacturing firm there are 20 automatic machines which assemble identical transistors. The variability of the measured characteristic is a source of trouble. There are 12 identical test units with which the finished items are checked for proper operation. It is desired to determine whether or not there is homogeneity among the assembly machines and among the test units. Also it is desired to know whether or not there is interaction among the assembly machines and test units. The data below are obtained from three assembly machines and three test units, and are randomly selected.

Assembly machine	*Test Unit* 1	2	3
1	−2.3	−3.7	−3.1
	−3.4	−2.8	−3.2
	−3.5	−3.7	−3.5
2	−3.5	−3.9	−3.3
	−2.6	−3.9	−3.4
	−3.6	−3.4	−3.5
3	−2.4	−3.5	−2.6
	−2.7	−3.2	−2.6
	−2.8	−3.5	−2.5

I_c (the instantaneous transistor collector current) in milliamperes for instantaneous emitter-to-base voltage drop equals 0.4 volt and instantaneous transistor emitter current equals 1.0 milliampere.

(a) Test for main effects and interaction at the 0.05 level. (b) Estimate the test unit variability, the assembly machine variability, and the interaction variability. (c) What value of assembly machine variability can be detected with probability 0.75 using the procedure of (a)? Use the value found in (b) as an estimate of the interaction variability. Use the "within combinations" mean square as an estimate of σ^2. What is your answer if there were five observations per combination?

27. A small motor is being manufactured for a special purpose. The important characteristic of the motor in this application is its starting torque. Several testers are used to evaluate the starting torque of the motors. It is desired to know whether the testers are obtaining equivalent results. An experiment is performed using three testers and eight motors, each chosen at random. Duplicate measurements are obtained. The results are as follows (coded data):

Motor	*Tester* A		B		C	
1	9.60	9.72	11.23	10.90	9.00	10.34
2	10.81	9.90	10.98	10.55	10.81	9.65
3	9.69	10.15	10.10	10.99	9.57	8.57
4	8.43	10.12	11.01	9.68	9.03	9.05
5	8.50	9.87	9.75	8.93	8.58	9.67
6	10.92	9.23	11.14	11.69	10.31	9.76
7	9.98	9.26	9.44	9.89	9.56	11.29
8	10.31	8.95	10.04	12.72	9.80	9.21

(a) Write out the analysis of variance table. (b) Test for interaction and main effects ($\alpha=0.05$). (c) Estimate the motor variance, tester variance, and interaction variance. (d) How many observations are required per combination to detect a tester variance of 0.30 with probability 0.75? Use the value found in (c) as an estimate of the interaction variance. Use the "within combinations" mean square as an estimate of σ^2.

28. A quality assurance engineer for an electronics components manufacturing firm intuitively feels that there is large variability among the 36 ovens used by this firm in testing the life of the various components. To determine whether or not he is correct he chooses a single type component, and obtains the following data for the two temperatures normally used for life testing of this item. The component is operated in an oven until it fails. Three randomly selected ovens are used for the experiment.

	Life of Components (minutes)		
Temperature	*Oven 1*	*Oven 2*	*Oven 3*
550°F	237	208	192
	254	178	186
	246	187	183
600°F	178	146	142
	179	145	125
	183	141	136

(a) From the data given above determine whether or not the engineer is correct in his belief that there is variability among the ovens. Use $\alpha = 0.05$. Test for interaction and temperature effects. (b) Estimate the oven variability. (c) How many observations per combination are required to detect an oven variance equal to $2\sigma^2$ with probability 0.8?

29. There are three time-study engineers employed by a certain company. The chief industrial engineer for whom these men set standard times wants to determine whether or not all three of them essentially follow his prescribed methods in setting standards. Also he wishes to ascertain whether or not the standards set by these three men are influenced by any particular combinations of time-study engineers and machine operators. To obtain this information the chief engineer selects at random four of 25 turret lathe operators, and directs each of the three engineers to set three standard times (each independent of the others) for each operator, all standards being for the same job. The following data are obtained.

	Time Study Engineer		
Operator	1	2	3
1	2.95	2.07	2.18
	2.93	2.99	2.33
	2.42	2.53	2.21
2	2.58	2.59	2.15
	2.26	2.36	2.94
	2.05	2.78	2.55
3	2.47	2.38	2.72
	2.44	2.48	2.85
	2.52	2.82	2.73
4	2.66	2.39	2.67
	2.95	2.61	2.89
	2.27	2.01	2.75

Std. Time, in minutes

(a) Test the hypothesis that the main effect due to time study engineers equals zero. (Use $\alpha = 0.05$.) (b) Are there interaction or operator effects ($\alpha = 0.05$)? (c) If the time study engineers differ, determine which differ. (d) What is the approximate probability of saying that the time study engineers do not differ when they differ by ½ minute? Use the interaction mean square as an estimate of $\sigma^2+n\sigma_\eta^2$.

30. The same experiment as described in Problem 19 is repeated three times, yielding the following results.

Operator	*Nozzles* A	B	C
1	96.6	97.7	97.6
	96.0	96.0	97.5
	94.4	97.7	94.6
2	98.5	96.0	92.2
	97.2	96.0	96.8
	96.5	97.9	94.4
3	97.5	97.1	97.1
	96.0	97.4	95.4
	96.6	97.4	94.7
4	98.3	95.6	97.5
	97.6	96.2	95.2
	96.3	95.3	94.3
5	98.6	97.1	95.4
	97.2	96.0	97.4
	98.3	96.3	98.1

(a) Write out the analysis of variance table and test for the significance of interaction, operator effect, and nozzle effect. What would be the physical meaning of interaction if any existed in this problem? (b) If nozzles differ, determine which nozzles differ. Obtain confidence intervals for those effects which differ. (c) What is the approximate probability of accepting the hypothesis of no nozzle effects when the maximum difference between nozzles is actually 0.4 cc? Use the interaction mean square as an estimate of $\sigma^2+n\sigma_\eta^2$.

31. Five methods have been suggested for mixing concrete. Three batches were mixed with two specimens from each batch used with each method and tested for compression strength (in psi).

Batch	*Method* 1	2	3	4	5
A	4,250 4,260	4,110 4,120	4,390 4,380	4,020 4,040	3,880 3,890
B	4,260 4,250	4,110 4,100	4,380 4,370	4,030 4,020	3,870 3,860
C	4,180 4,160	4,020 4,020	4,220 4,200	3,990 3,980	3,760 3,740

a) Test for interaction at the 5% level. (b) Do the methods have any effect on the strength of the concrete? ($\alpha=0.05$.) (c) If the methods do differ, determine Tukey confidence intervals to make statements about which ones differ.

32. A factory employs a large number of people who assemble electronic tubes. Management is interested in whether there tends to be a variation in the average number of tubes assembled from worker to worker, and is particularly interested in detecting a variation between workers as large as one and one-half times an individual's daily variation. It is decided that 18 observations may be taken according to one of two schemes. (a) Six workers are chosen at random and observed on each of three days.

(b) Three workers are chosen at random and observed on each of si days. Which scheme do you prefer? Why?

33. Two examiners are checked on the time they take to use a certai gauge. Duplicate readings are taken for each examiner. The results ar as follows:

Examiner I	*Examiner II*
a	c
b	d

where $a<b<c<d$. The value of F for the data above is:

$$\frac{2\left\{\left[\frac{a+b}{2}-\frac{a+b+c+d}{4}\right]^2+\left[\frac{c+d}{2}-\frac{a+b+c+d}{4}\right]^2\right\}}{\frac{1}{2}\left\{\left[a-\frac{(a+b)}{2}\right]^2+\left[b-\frac{(a+b)}{2}\right]^2+\left[c-\frac{(c+d)}{2}\right]^2+\left[d-\frac{(c+d)}{2}\right]^2\right\}}$$

$$=\frac{\left[\frac{(a+b)-(c+d)}{2}\right]^2+\left[\frac{(c+d)-(a+b)}{2}\right]^2}{\left(\frac{a-b}{2}\right)^2+\left(\frac{b-a}{2}\right)^2+\left(\frac{c-d}{2}\right)^2+\left(\frac{d-c}{2}\right)^2}$$

$$=\frac{[(a+b)-(c+d)]^2}{(a-b)^2+(c-d)^2}.$$

Using the principle of randomization, enumerate all the possible values o F and accept or reject the hypothesis of equality of means at the $33\frac{1}{3}$% level of significance.

CHAPTER XI

Analysis of Enumeration Data

11.1 Enumeration Data

Most of the text up to this point has dealt with data which came from measurements on a particular variable that was usually on a continuous scale. Examples are such things as dimensions, voltage, and Rockwell hardness. In many experiments it is not possible or useful to make measurements because the data are the number of cases which fall into specified categories. For example, we may count the number of defective items produced by various machines, the number of errors made by several operators, the number of hits and misses made by several fire control devices, or the number of accidents as a function of shift. The quality of manufactured product is commonly characterized by the number or proportion of defective items in a lot or in a production run. Procedures for handling fraction defective data are considered in Chapter XII, Quality Control, and Chapter XIII, Sampling Inspection.

11.2 Chi-Square Tests

Enumeration or attribute data consist of the number of observations in a sample which fall into certain specified categories. The data are usually taken to test hypotheses about the true relative frequency of these categories. To be specific, assume that each sample observation must fall into one and only one of k categories; let $O_1, O_2, \cdots, O_k$ be the observed frequencies for each category. We are interested in testing hypotheses about the true relative frequencies.

Category	*Observed frequency*	*Theoretical frequency*
1	O_1	E_1
2	O_2	E_2
$\vdots$	$\vdots$	$\vdots$
k	O_k	E_k

The test statistic is

$$\chi^2 = \sum_{i=1}^{k} \frac{(O_i - E_i)^2}{E_i}$$

which for large enough samples[1] has an approximate chi-square di tribution. The number of degrees of freedom depends on how th data are used computing the E_i. A small value of χ^2 is associate with good agreement between observed and theoretical values; large value tends to indicate discrepancy. Whether the discrepanc is likely to arise by chance is decided by reference to a table of pe centage points of the chi-square distributions, Appendix Table 2.

11.3 The Hypothesis Completely Specifies the Theoretical Frequency

Assume that each observation can fall into one of k categories ar let $p_1, p_2, \cdots, p_k$ be the true probabilities. p_i is the probability (a random observation falling in the ith category. In this type (problem, the theoretical frequencies are calculated from the formul

$$E_i = Np_i$$

and the chi-square statistic has $k - 1$ degrees of freedom.

For example, the total number of defective units in a day's produ tion was tabulated by shifts. If there is no difference in quality be tween the shifts, the probability of any defect occurring in any shi i is $\frac{1}{3}$. There are 80 observations, so the theoretical frequenc would be $\frac{80}{3} = 26.67$.

	Observed frequency	*Theoretical frequency*
Shift 1	20	26.67
Shift 2	36	26.67
Shift 3	24	26.67
	80	

It is of interest to see whether the variation from shift is due t chance or whether there is a real difference in the occurrence c

[1] The following rules of thumb can be used to assess adequacy of sample size: If χ^2 has from 2 to 15 degrees of freedom, all E_i should be at least 2.5. 2. If χ^2 computed from a 2 × 2 table (which will be discussed in Section 11.4), all E_i should b at least 5; however, if all but one of them is at least 5, the remaining one may be small as 1 with little distortion in significance level.

fectives. That is, we test the hypothesis that the true relative equency of defectives is the same on all shifts, i.e., $p_1=p_2=p_3=\frac{1}{3}$. he test statistic is

$$\chi^2 = \frac{(20-26.67)^2}{26.67} + \frac{(36-26.67)^2}{26.67} + \frac{(24-26.67)^2}{26.67}$$
$$= \frac{44.4889}{26.67} + \frac{87.0489}{26.67} + \frac{7.1289}{26.67}$$
$$= 1.6681 + 3.2639 + 0.26730$$
$$= 5.1993 < \chi^2_{0.05;\,2} = 5.991\ .$$

herefore, we accept the hypothesis and conclude that the data do ot indicate that the frequencies differ.

11.3.1 Dichotomous data

An interesting special case of completely specified frequency curs when each observation can fall into one of two categories. xamples are the toss of a coin, which may be heads or tails, and e quality of an item of product, which may be defective or non- fective. Suppose that we have tossed a coin 100 times with the llowing results:

Heads	40
Tails	60
Number of tosses	100

e want to see whether the coin is fair, whether $p_1 = p_2 = \frac{1}{2}$. The pected number of heads or tails would be 50. By the method of e previous section, we would find

$$\chi^2 = \frac{(10)^2}{50} + \frac{(10)^2}{50} = \frac{200}{50} = 4$$

here χ^2 has one degree of freedom, $\chi^2_{0.05;\,1}=3.841$, so the hypothesis ould be rejected at the 5% level.

Actually, it would not be necessary to apply a chi-square test to ese data as an exact test is available. The number of heads (or e number of tails) is a binomial random variable (see Section 4.5). he probability of obtaining r successes in N trials is given by

$$\binom{N}{r} p^r q^{N-r}$$

here p is the individual trial probability and $q = 1 - p$. An exact lculation of the probability of departing by 10 or more heads from e expected number of 50 for $N = 100$, $p = \frac{1}{2}$ is given by

$$\sum_{r=0}^{40}\binom{100}{r}p^r q^{100-r} + \sum_{r=60}^{100}\binom{100}{r}p^r q^{100-r} = \frac{\sum_{r=0}^{40}\binom{100}{r} + \sum_{r=60}^{100}\binom{100}{r}}{2^{100}}$$

This probability can be evaluated directly. However, for this N, w would be likely to use the normal approximation, noting that

$$\frac{\frac{r}{N} - p}{\sqrt{pq/N}}$$

is approximately normal. The approximate probability of a deviatio as large as 10 heads would be found finally by calculating th standard normal deviate corresponding to 40, which is -2, and th normal deviate corresponding to 60, which is $+2$, and calculating th area in a normal curve outside the interval -2 to $+2$. This prob bility is 0.0456, slightly less than 0.05, so the hypothesis of a fa coin would be rejected at the 0.05 level.

The fact that the result based on the normal approximation to th binomial and the result based on χ^2 are similar is no acciden actually the two methods are equivalent. Chi-square in general the sum of squares of standard normal deviates, and chi-square wi one degree of freedom is just the square of a normal deviate.

In general, consider a sample N classified into one of two ca egories, and let $p_1 = p$ and $p_2 = 1 - p = q$.

Category	*Observed frequency*	*Theoretical frequency*
1	O_1	Np
2	O_2	Nq
Total	N	N

Then O_1 would have a binomial distribution and the quantity

$$\frac{\frac{O_1}{N} - p}{\sqrt{pq/N}} = \frac{O_1 - Np}{\sqrt{Npq}}$$

would be approximately normal. The chi-square statistic would be

$$\chi^2 = \frac{(O_1 - Np)^2}{Np} + \frac{(O_2 - Nq)^2}{Nq}.$$

Since $O_2 = N - O_1$ and $q = 1 - p$,

$$\chi^2 = \frac{(O_1 - Np)^2}{Npq}$$

or just the square of the normal variable above. This special case considered in detail to give an indication of the nature of the chi-square approximation and why it is used.

11.4 Test of Independence in a Two-Way Table

Often frequency data are tabulated according to two criteria, with a view toward testing whether the criteria are associated. Consider the following analysis of the 157 machine breakdowns during a given quarter.

Number of Breakdowns

	Machine A	Machine B	Machine C	Machine D	Total per Shift
Shift 1	10	6	12	13	41
Shift 2	10	12	19	21	62
Shift 3	13	10	13	18	54
Total per machine	33	28	44	52	157

We are interested in whether the same percentage of breakdowns occurs on each machine during each shift or whether there is some difference due perhaps to untrained operators or other factors peculiar to a given shift.

If the number of breakdowns is independent of shifts and machines, the probability of a breakdown occurring in the first shift and in the first machine can be estimated by multiplying the proportion of first shift breakdowns by the proportion of machine A breakdowns.

$$p_{11} = \frac{41}{157} \times \frac{33}{157} = 0.05489 \, .$$

If there are 157 breakdowns, the expected number of breakdowns on this shift and machine is estimated as

$$E_{11} = 157 \times p_{11} = 8.6177 \, .$$

Similarly for the third shift and second machine

$$p_{32} = \frac{54}{157} \times \frac{28}{157} = 0.06134 \, ,$$

$$E_{32} = p_{32} \times 157 = 9.630 \, .$$

This is done for all categories and

$$\chi^2 = \sum_{i=1}^{3} \sum_{j=1}^{4} \frac{(O_{ij} - E_{ij})^2}{E_{ij}}$$

$$= \frac{(10 - 8.6177)^2}{8.6177} + \frac{(6 - 7.3120)^2}{7.3120} + \cdots + \frac{(18 - 17.885)^2}{17.885}$$

$$= 2.02 \, .$$

This is to be compared with the percentage point of χ^2 for $(3 - 1)(4 - 1) = 6$ degrees of freedom, i.e., $\chi^2_{0.05;\,6} = 12.6$. Hence, we accept the hypothesis and conclude that the data do not indicate different percentages of breakdowns on each machine during each shift.

In general, for an r by s table, we have

$$\chi^2 = \sum_{i=1}^{r} \sum_{j=1}^{s} \frac{(O_{ij} - E_{ij})^2}{E_{ij}} \, ,$$

which has an approximate chi-square distribution with $(r - 1)(s - 1)$ degrees of freedom.

11.4.1 Computing form for test of independence in a 2 × 2 table

An important special case of the test of independence arises when both criteria of classification have two categories; the data may be represented in a 2 × 2 table. The data may be shown in the following way:

	First Criteria		
Second Criteria	a	b	$a + b$
	c	d	$c + d$
	$a + c$	$b + d$	

In this case the chi-square statistic, which has one degree of freedom, may be written

$$\chi^2 = \frac{(ad - bc)^2 \, (a + b + c + d)}{(a + b)(a + c)(b + d)(c + d)} \, .$$

As an example, the following data are used to determine whether the same percentage of breakdowns occur on each machine using material from each supplier.

Number of Machine Breakdowns

Source of Raw Materials

Machine	Supplier A	Supplier B	
I	4	9	13
II	15	3	18
	19	12	31

$$\chi^2 = \frac{(12 - 135)^2(31)}{(13)(19)(12)(18)} = 8.8$$

$$\geq \chi^2_{0.05;\,1} = 3.84\,.$$

Therefore we reject the hypothesis that the percentage of breakdowns is the same for each machine using material from each supplier.

11.5 Comparison of Two Percentages

A problem which in many cases is equivalent to the test of independence in a 2×2 table is the problem of testing equality of two percentages. Suppose, following Wallis,[1] we have the following data on two fire control devices.

	Hits	Misses	
Old	3	197	200
New	4	196	200
			400

Suppose we want to test the hypothesis that the percentage of hits is the same, and want to reject when the new is superior, i.e., has a larger percentage of hits. If we let p_E be the observed proportion for the new experimental method, p_S be the observed proportion for the standard, and N be the number of trials with each method, then

$$U = \frac{\sqrt{N}(p_E - p_S)}{\sqrt{(p_E + p_S)\left(1 - \dfrac{p_E + p_S}{2}\right)}}$$

has approximately the normal distribution, and we would reject when

[1] Eisenhart, Hastay, Wallis, *Techniques of Statistical Analysis*, McGraw-Hill Book Company, Inc., New York, 1947.

$$U \geqq K_\alpha .$$

For the data above

$$p_S = 0.015$$
$$p_E = 0.020$$

and

$$U = \frac{\sqrt{200}(0.005)}{\sqrt{(0.020+0.015)\left(1 - \frac{0.035}{2}\right)}} = 0.3795 < K_{0.05} = 1.645$$

and we accept the hypothesis and conclude that the data do not reveal that the new device is better than the old one.

In terms of the notation of the last section,

$$U = \frac{\sqrt{a+b+c+d}\,(ad - bc)}{\sqrt{(a+b)(a+c)(b+d)(c+d)}} .$$

If we are interested in rejecting whenever the proportions differ, we could reject when

$$| U | \geqq K_{\alpha/2}$$

or when $U^2 \geqq K^2_{\alpha/2}$. This is exactly the same as analyzing the data by the method used in Section 11.4.1.

11.6 Confidence Intervals for Proportion

Confidence intervals for a proportion were not included in Chapter VIII, Estimation, since most of the test procedures discussed up to that point depend on the normality of the observations. The topic is of considerable importance in applied statistics. We frequently want to estimate the proportion of people who will vote for a particular candidate, buy a new detergent, prefer movies to television, etc. In manufacturing we need to estimate the proportion of defective items or the proportion of rejects to be expected. If these estimates are made on the basis of a random sample, the number of, say, defects (r) in a sample of N will follow the binomial law (see Section 4.5). That is,

$$\mathrm{P}(r = r_1) = \binom{N}{r_1} p^{r_1} q^{N-r_1} \quad \text{for} \quad r_1 = 0, 1, \cdots, N .$$

The probability distribution depends on the parameter p, which is the true probability that an event will occur or is the proportion of defects in the lot. The point estimate of p is simply the observed proportion $\hat{p} = r/N$.

11.6.1 Exact confidence intervals for p

Let the lower confidence limits for p be denoted by $\underline{p}$ and the upper by $\overline{p}$. It is possible to obtain exact expressions for $\underline{p}$ and $\overline{p}$ in terms of the percentage points of the F distribution given in Appendix Table 4. In particular, 100 $(1 - \alpha)$ percent confidence intervals are given by

$$\overline{p} = \frac{(r + 1)F_{\alpha/2;\, 2(r+1),\, 2(N-r)}}{(N - r) + (r + 1)F_{\alpha/2;\, 2(r+1),\, 2(N-r)}}$$

$$\underline{p} = \frac{r}{r + (N - r + 1)F_{\alpha/2;\, 2(N-r+1),\, 2r}}\,.$$

For example, suppose that we observe 4 defects in a sample of 25. The point estimate $p = 0.16$. To find 95% confidence intervals, we need, for $\alpha = 0.05$, $N = 25$, $r = 4$,

$$F_{0.025;\, 10,\, 42} = 2.37$$

$$F_{0.025;\, 44,\, 8} = 3.82$$

$$\overline{p} = \frac{5(2.37)}{21 + 5(2.37)} = 0.361$$

$$\underline{p} = \frac{4}{4 + 22(3.82)} = 0.045.$$

11.6.2 Normal approximations to confidence intervals

The F table can be used to get confidence limits for most sample sizes which arise in practice. As explained in Section 4.5.3 for large N and values of p not too close to 0 or 1, the normal approximation is adequate. This gives the formulas

$$\overline{p} = \hat{p} + K_{\alpha/2}\sqrt{\frac{pq}{N}}$$

$$\underline{p} = \hat{p} - K_{\alpha/2}\sqrt{\frac{pq}{N}}$$

$$\hat{p} = \frac{r}{N}$$

for 100 $(1 - \alpha)$ percent confidence intervals.

For example, suppose $N = 400$ and $r = 100$; the exact method gives the interval 0.196 to 0.311 and the normal approximation gives 0.194 to 0.306 for the 99% confidence interval.

PROBLEMS

1. The failures in a week's around-the-clock run are distributed over four shift groups as follows:

Shift A	8
Shift B	9
Shift C	7
Shift D	14

Are we justified in saying that Shift D has too many failures compared with the other shifts?

2. In a game of craps, a suspicious player tallies the results of 100 throws by a particular opponent as follows:

Result of throw	*Observed frequency*
11	12
7	15
other	73
	100

Would you say his suspicions are justified? (See Section 2.5 if you don't know the odds.)

3. Two inspectors are testing pumps. It is suspected that one of the men was departing from the standard test procedure with a resulting bias toward reporting an undue number of failures. A record was kept of the same number of tests made by the two men in the same period of time.

Frequency of Pumps Failure

Operator A	168
Operator B	114

Do the operators appear to differ?

4. The past output of a machine indicates that an output of 500 pieces would be graded as follows:

Expected Output

Top grade......	200
High grade	150
Medium grade..	100
Low grade......	50
	500

A new machine is designed and built to do the same job. A sample of 500 pieces from the new machine is graded as follows:

Top grade......	231
High grade	122
Medium grade..	82
Low grade......	65
	500

Can the difference in the pattern of output be ascribed to chance?

5. In a mass production process, a sample of 200 was taken from each day's production and the number of defective items were calculated:

	Number of Defects
Monday	12
Tuesday	16
Wednesday	8
Thursday........	14
Friday	10

Test the hypothesis that the proportion of defects is constant from day to day.

6. An experiment on rockets yields the following data on the characteristics of lateral deflection and ranges.

	Lateral Deflection		
Range	*Left*	*Normal*	*Right*
0–1199......	12	15	14
1200–1799......	9	6	11
1800–2699......	6	17	8

Test the hypothesis that deflection and range are independent.

7. The table below shows the number of defective and acceptable items in samples both before and after the introduction of a modification intended to improve the process of manufacturing. The proportion of defectives has dropped and it is desired to test it, as change is significant at the 0.05 level.

	Defective	*Acceptable*
Before	25	217
After	8	92

8. A study is made on the failures of vacuum tubes. There are four types of failures and two blocks of tubes in the equipment. Test the hypothesis that the type of failure is distributed independently of the location of the tube at the 0.01 level.

	Type of Failure			
	A	*B*	*C*	*D*
Top block	75	10	15	20
Bottom block..	40	30	10	17

9. In a study of failure of tires during a road test of wearing strength the following results were obtained.

	Number of Failures		
	Front	*Rear*	
Left ..	115	65	180
Right..	125	95	220
	240	160	400

Test, at the 5% level, the hypothesis that left-right wear is independent of front-rear wear. (See Acheson J. Duncan, "Chi-Square Tests of Independence and the Comparison of Percentages," *Industrial Quality Control*, June, 1955, p. 9.)

10. The result of testing 164 fuses at two temperature levels is:

Result of Test	Fuses Tested Temp. 1	Temp. 2
Successes	111	92
Failures........	16	9

Is there a significant difference at the 5% level in the proportion of failures at the two levels?

11. The grade distribution of three sections, taught by different instructors, of a course based on this book at Stanford University is:

Instructor	*A*	*B*	*C*	*other*	
L	25	34	36	12	107
B	18	29	27	8	82
T	17	19	22	7	65

Do the instructors differ in the percentages of grades?

12. Fabric is often inspected and graded into three classifications: *A*, *B*, *C*. In a study of four looms, the following results were obtained:

Loom	*Number of Pieces of Fabric in:* *Grade A*	*B*	*C*
1	191	22	13
2	184	26	15
3	157	32	11
4	170	14	8

Is the output homogeneous or does it depend on the loom?

13. A vaccine supposed to reduce absenteeism caused by flu was given to 167 individuals in a plant. 214 people were untreated. The results were as follows:

	Caught flu	*Well*	
Treated....	17	150	167
Untreated..	28	186	214

Was the vaccine worthwhile?

14. Grades in an English course and a statistics course taken simultaneously were as follows for a sample of students:

Statistics grade	*English grade* *A*	*B*	*C*	*Other*
A	20	6	17	12
B	17	16	18	7
C	19	4	15	12
Other	12	8	12	23

Are the grades in statistics and English related?

15. In a sample of 275 items from a process, eight defects were found. Find a 95% confidence interval for the proportion defective in the lot.

16. In a sample of 712 college students, 411 were found to be in favor of selling beer on the campus. Find a 99% confidence interval for the number in favor. Would you be sure that a majority favored it?

17. In a sample of 400 pieces of cloth, 316 were found to be grade *A*. Compare the exact and the normal approximation method for finding 95% confidence intervals for the true proportion of grade *A* cloth.

18. In the test of a vaccine, 17 out of 1,000 people tested were found to have an allergic reaction. Find a 95% confidence interval for the true proportion of people allergic.

19. The proportion of defectives in a random sample of 300 items checked on a go-no-go gauge is 9%. What are the 95% confidence limits for the proportion defective for the total population?

20. In 100 throws of dice the number 11 was found to appear 12 times. Does the 95% confidence interval for the true probability include the probability of obtaining 11, if the dice are unbiased?

CHAPTER XII

Statistical Quality Control: Control Charts

12.1 Introduction

In his book, *Statistical Quality Control*, E. L. Grant states: "Measured quality of manufactured product is always subject to a certain amount of variation as a result of chance. Some stable 'system of chance causes' is inherent in any particular scheme of production and inspection. Variation within this stable pattern is inevitable. The reasons for variation outside this stable pattern may be discovered and corrected····."[1] These variations outside the stable pattern are known as *assignable causes* of quality variation. A process operating in the absence of any assignable causes of erratic fluctuations is said to be in *statistical control*.

To the manufacturer, the primary purpose of a control chart is to provide a basis for *action*. The introduction of a control chart aids in determining the capabilities of the production process. Action is taken when these estimated capabilities are unsatisfactory in relation to the design specifications. Furthermore, once the process capabilities have been determined, and are satisfactory, action is taken only when the control chart indicates that the process has fallen out of control, e.g., assignable causes of variation have entered.

12.2 Obtaining Data from Rational Subgroups

The essential feature of the control chart method is the drawing of inferences about the production process on the basis of samples drawn from the production line. The success of the technique depends upon grouping observations under consideration into subgroups or samples, within which a stable system of chance causes is

[1] E. L. Grant, *Statistical Quality Control*, 2nd ed., McGraw-Hill Book Company, Inc., New York, 1952, p. 3.

operating, and between which the variations may be due to assignable causes whose presence is suspected or considered possible. Order of production is one of the more commonly used bases for obtaining rational subgroups. If items are coming from more than one source, the source may be a basis for rational subgrouping.

The size of the subgroup or sample usually is not less than 4. In industry, 5 seems to be the most common.[1] It is preferable that all samples be of equal size.

12.3 Control Chart for Variables: $\bar{x}$—Charts

The notation in this chapter and in the ensuing chapter will differ from the notation used throughout the book. This is done so that the notation in this chapter on control charts will coincide with that recommended by the American Society for Quality Control. The basic changes are in the symbols for the parameters of the normal distribution. In the previous sections, the symbols μ and σ represented the mean and standard deviation of the normal distribution. In this chapter these symbols will be replaced by $\bar{x}'$ and σ' respectively. The "prime" will always represent a parameter of a probability distribution. $\bar{x}$ will still denote the sample average and will be used as an estimate of the mean of the normal distribution. However, the symbol σ will now be used as an estimate of the standard deviation of the normal distribution, whereas in the previous chapters the symbol actually represented this parameter.

12.3.1 Statistical concepts

In Chapter III it was shown that if observations on an item are normally distributed, with mean $\bar{x}'$ and standard deviation σ', it is possible to find the probability that an observation will lie in a fixed interval by referring to Appendix Table 1. Furthermore, if we denote the variable by x, the probability that x will be within the interval $[\bar{x}' - 3\sigma', \bar{x}' + 3\sigma']$ is 0.9973. In other words, if a plot such as that in Figure 12.1 is made, on the average only 27 observations out of 10,000 will be outside the intervals above, *provided the distribution of the variable is normal with mean $\bar{x}'$ and standard deviation σ'*. If this is the case, it follows that the distribution of averages of subgroups of size n is also normal with mean $\bar{x}'$ and standard deviation $\sigma'_{\bar{x}} = \sigma'/\sqrt{n}$.

[1] When characteristics are classified into two groups, those containing defects and those not containing defects, and no other measurement is recorded, the sample size is usually much larger.

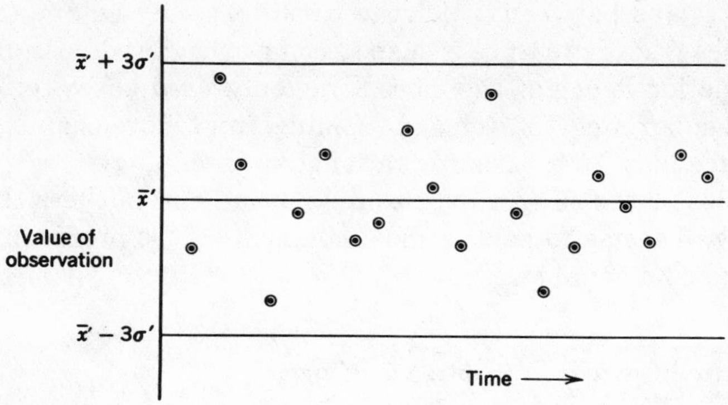

Fig. 12.1. Plot of individual observations.

Denoting the n observations by $x_1, x_2, x_3, \cdots, x_n$, the average of these n observations, $\bar{x}$, is defined as

$$\bar{x} = \frac{x_1 + x_2 + \cdots + x_n}{n}.$$

If, instead of plotting individual values as in Figure 12.1, averages of samples of n are plotted, it is expected that on the average only 27 in 10,000 values of these averages will fall outside the interval

$$\bar{x}' \pm 3\sigma'_{\bar{x}} = \bar{x}' \pm \frac{3\sigma'}{\sqrt{n}}.$$

The plot in Figure 12.2 is referred to as a control chart for $\bar{x}$. Here $\bar{x}' + 3\sigma'/\sqrt{n}$ is referred to as the upper control limit (UCL) while $\bar{x}' - 3\sigma'/\sqrt{n}$ is referred to as the lower control limit (LCL).

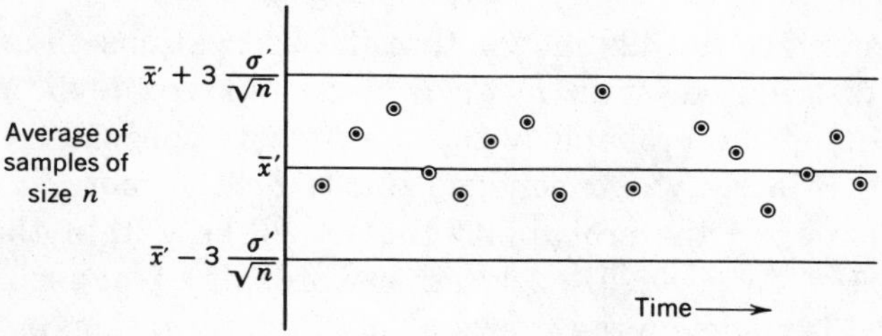

Fig. 12.2. Control chart for $\bar{x}$.

In this discussion it has always been assumed that the underlying distribution is normal. Although in actual practice many distributions of observations are "nearly" normally distributed, many others do not resemble a normal distribution at all. However, if sample averages are considered as the variable plotted on the control chart, the central limit theorem (see Chapter III) enables us to use the properties of the normal distribution even if the form of the underlying distribution is unknown. It is evident, then, that the central limit theorem is one of the keys to the success of the control chart for $\bar{x}$.

12.3.2 ESTIMATES OF $\bar{x}'$

In Figure 12.2, a control chart for $\bar{x}$ is shown, where the control limits are drawn in as functions of $\bar{x}'$ and σ'. In most practical applications $\bar{x}'$ and σ' are not known, and consequently estimates of these parameters must be obtained. It is desirable that these estimates be based on at least 25 subgroups of n observations. Naturally, the larger the number of subgroups, the better the estimates of the parameters of the distribution, provided the process is in control. Suppose the estimates are based upon k subgroups, each of size n. Denoting the averages of the subgroups by $\bar{x}_1, \bar{x}_2, \cdots, \bar{x}_k$, the best estimate of $\bar{x}'$, the mean of the probability distribution, is

$$\bar{\bar{x}} = \frac{\bar{x}_1 + \bar{x}_2 + \cdots + \bar{x}_k}{k}$$

where $\bar{\bar{x}}$ is also the average of all the nk observations. Furthermore, $\bar{\bar{x}}$ is an unbiased estimate of $\bar{x}'$.

12.3.3 ESTIMATE OF σ' BY $\bar{\sigma}$

As was pointed out in the previous section, it is usually necessary to estimate σ', the population standard deviation. Define, for each subgroup,

$$\sigma = \sqrt{\frac{(x_1 - \bar{x})^2 + (x_2 - \bar{x})^2 + \cdots + (x_n - \bar{x})^2}{n}}$$

$$= \sqrt{\frac{x_1^2 + x_2^2 + \cdots + x_n^2 - n\bar{x}^2}{n}}.$$

Further, define $\bar{\sigma} = (\sigma_1 + \sigma_2 + \cdots + \sigma_k)/k$, where $\sigma_1, \sigma_2, \cdots, \sigma_k$ are the values of σ computed from the $1, 2, \cdots, k$ subgroups, respectively. An unbiased estimate of σ' is given by $\bar{\sigma}/c_2$, where values of c_2 for different n can be found in Table 12.1. The quantity c_2 is determined such that $E(\bar{\sigma}/c_2) = \sigma'$ or $E(\bar{\sigma})/\sigma' = c_2$.

To summarize, the estimated central line for the control chart for $\bar{x}$ is $\bar{\bar{x}}$, and the estimated control limits are $\bar{\bar{x}} \pm 3\bar{\sigma}/(\sqrt{n}\, c_2)$. Values of $3/(\sqrt{n}\, c_2) = A_1$ are given in Table 12.1 so that the estimated control limits can be written as $\bar{\bar{x}} \pm A_1\bar{\sigma}$.

Table 12.1. Factors for Computing Control Chart Lines[1]

Number of observations in sample, n	Chart for averages: Factors for control limits			Chart for standard deviations: Factors for central line		Chart for standard deviations: Factors for control limits				Chart for ranges: Factors for central line		Chart for ranges: Factors for control limits				
	A	A_1	A_2	c_2	$1/c_2$	B_1	B_2	B_3	B_4	d_2	$1/d_2$	d_3	D_1	D_2	D_3	D_4
2	2.121	3.760	1.880	0.5642	1.7725	0	1.843	0	3.267	1.128	0.8865	0.853	0	3.686	0	3.267
3	1.732	2.394	1.023	0.7236	1.3820	0	1.858	0	2.568	1.693	0.5907	0.888	0	4.358	0	2.575
4	1.500	1.880	0.729	0.7979	1.2533	0	1.808	0	2.266	2.059	0.4857	0.880	0	4.698	0	2.282
5	1.342	1.596	0.577	0.8407	1.1894	0	1.756	0	2.089	2.326	0.4299	0.864	0	4.918	0	2.115
6	1.225	1.410	0.483	0.8686	1.1512	0.026	1.711	0.030	1.970	2.534	0.3946	0.848	0	5.078	0	2.004
7	1.134	1.277	0.419	0.8882	1.1259	0.105	1.672	0.118	1.882	2.704	0.3698	0.833	0.205	5.203	0.076	1.924
8	1.061	1.175	0.373	0.9027	1.1078	0.167	1.638	0.185	1.815	2.847	0.3512	0.820	0.387	5.307	0.136	1.864
9	1.000	1.094	0.337	0.9139	1.0942	0.219	1.609	0.239	1.761	2.970	0.3367	0.808	0.546	5.394	0.184	1.816
10	0.949	1.028	0.308	0.9227	1.0837	0.262	1.584	0.284	1.716	3.078	0.3249	0.797	0.687	5.469	0.223	1.777
11	0.905	0.973	0.285	0.9300	1.0753	0.299	1.561	0.321	1.679	3.173	0.3152	0.787	0.812	5.534	0.256	1.744
12	0.866	0.925	0.266	0.9359	1.0684	0.331	1.541	0.354	1.646	3.258	0.3069	0.778	0.924	5.592	0.284	1.716
13	0.832	0.884	0.249	0.9410	1.0627	0.359	1.523	0.382	1.618	3.336	0.2998	0.770	1.026	5.646	0.308	1.692
14	0.802	0.848	0.235	0.9453	1.0579	0.384	1.507	0.406	1.594	3.407	0.2935	0.762	1.121	5.693	0.329	1.671
15	0.775	0.816	0.223	0.9490	1.0537	0.406	1.492	0.428	1.572	3.472	0.2880	0.755	1.207	5.737	0.348	1.652
16	0.750	0.788	0.212	0.9523	1.0501	0.427	1.478	0.448	1.552	3.532	0.2831	0.749	1.285	5.779	0.364	1.636
17	0.728	0.762	0.203	0.9551	1.0470	0.445	1.465	0.466	1.534	3.588	0.2787	0.743	1.359	5.817	0.379	1.621
18	0.707	0.738	0.194	0.9576	1.0442	0.461	1.454	0.482	1.518	3.640	0.2747	0.738	1.426	5.854	0.392	1.608
19	0.688	0.717	0.187	0.9599	1.0418	0.477	1.443	0.497	1.503	3.689	0.2711	0.733	1.490	5.888	0.404	1.596
20	0.671	0.697	0.180	0.9619	1.0396	0.491	1.433	0.510	1.490	3.735	0.2677	0.729	1.548	5.922	0.414	1.586
21	0.655	0.679	0.173	0.9638	1.0376	0.504	1.424	0.523	1.477	3.778	0.2647	0.724	1.606	5.950	0.425	1.575
22	0.640	0.662	0.167	0.9655	1.0358	0.516	1.415	0.534	1.466	3.819	0.2618	0.720	1.659	5.979	0.434	1.566
23	0.626	0.647	0.162	0.9670	1.0342	0.527	1.407	0.545	1.455	3.858	0.2592	0.716	1.710	6.006	0.443	1.557
24	0.612	0.632	0.157	0.9684	1.0327	0.538	1.399	0.555	1.445	3.895	0.2567	0.712	1.759	6.031	0.452	1.548
25	0.600	0.619	0.153	0.9696	1.0313	0.548	1.392	0.565	1.435	3.931	0.2544	0.709	1.804	6.058	0.459	1.541
Over 25	$\frac{3}{\sqrt{n}}$	$\frac{3}{\sqrt{n}}$	.....	.....	.....	*	**	*	**	.....	.....	.....	.....	.....	.....	.....

$^*1 - \frac{3}{\sqrt{2n}}$ $\qquad ^{**}1 + \frac{3}{\sqrt{2n}}$

[1] Reproduced by permission from *ASTM Manual on Quality Control of Materials*, American Society for Testing Materials, Philadelphia, Pa., 1951.

Formulas for Central Lines and Control Limits

Statistic	Standards given: Central line	Standards given: Limits	Analysis of past data: Central line	Analysis of past data: Limits
Average, using σ'	$\bar{x}'$	$\bar{x}' \pm A\sigma'$	$\bar{\bar{x}}$	$\bar{\bar{x}} \pm A_1\bar{\sigma}$
Average, using R	—	—	$\bar{\bar{x}}$	$\bar{\bar{x}} \pm A_2\bar{R}$
Standard deviation	$c_2\sigma'$	$B_1\sigma'$, $B_2\sigma'$	$\bar{\sigma}$	$B_3\bar{\sigma}$, $B_4\bar{\sigma}$
Range	$d_2\sigma'$	$D_1\sigma'$, $D_2\sigma'$	$\bar{R}$	$D_3\bar{R}$, $D_4\bar{R}$

12.3.4 Estimate of σ' by $\overline{R}$

Define, as the range (R) of a subgroup of n observations, the difference between the largest and smallest value. Let $\overline{R}$ be the average of the ranges of the k subgroups, i.e.,

$$\overline{R} = \frac{R_1 + R_2 + \cdots + R_k}{k}.$$

Another unbiased estimate of σ' is $\overline{R}/d_2$. Values of d_2 for different values of n can also be found in Table 12.1. The quantity d_2 is determined such that $E(\overline{R}/d_2) = \sigma'$ or $E(\overline{R})/\sigma' = d_2$.

In estimating the control limits for $\bar{x}$ with $\overline{R}$ as an estimate of σ', the limits become $\bar{\bar{x}} \pm 3\overline{R}/(\sqrt{n}d_2)$. Values of $3/(\sqrt{n}d_2) = A_2$ are given in Table 12.1 so that the estimated control limits can be written as $\bar{\bar{x}} \pm A_2\overline{R}$.

It must be pointed out that although $\overline{R}$ is simpler to calculate than $\bar{\sigma}$, the estimate of σ' based on $\bar{\sigma}$ is a better estimate in the sense of having a smaller variance; i.e., it is more efficient.

12.3.5 Starting a control chart for $\bar{x}$

It has been indicated that observations should be classified into rational subgroups, and each subgroup should contain at least 4 observations. The determination of the minimum number of subgroups is a compromise between obtaining the guidance of the control chart as quickly as possible, and the desire for the guidance to be reliable. Usually, at least 25 subgroups should be chosen.

With the data at hand, trial control limits can be calculated. Estimates of $\bar{x}'$ and σ' can be obtained in the manner described previously. The trial control limits are then $\bar{\bar{x}} \pm A_1\bar{\sigma}$ or $\bar{\bar{x}} \pm A_2\overline{R}$, depending on which estimate of σ' is chosen. If σ' and/or $\bar{x}'$ are known, the control limits should be calculated, using the known values; e.g., if both are known, $\bar{x}' \pm 3\sigma'/\sqrt{n}$. Values of $3/\sqrt{n} = A$ are given in Table 12.1, so that the control limits can be written $\bar{x}' \pm A\sigma'$.

Returning to the case where the parameters $\bar{x}'$ and σ' are unknown, and estimates are computed, some modifications may have to be made in the trial control limits. Lack of control is usually indicated by points falling outside these limits. If all the points fall within the limits, we conclude that the process is in control. However, there is still no assurance that assignable causes of variation are absent, and that this is a constant cause system. It merely means that for practical purposes it pays to act as if no assignable causes of variation are present, while realizing that an error of judgment is quite possible.

The trial control limits serve the purpose of determining whether past operations are in control. To continue using these limits as a basis for action on future production may require a revision of the trial limits, especially if a lack of control is exhibited by points falling outside the trial control limits. If the process is not in control, all the points do not come from a stable distribution. However, it is evident that future control limits should be based upon data from a controlled process. As a practical rule, then, those points falling outside the trial control limits are eliminated, and new trial control limits are computed using the remaining points. This procedure may be continued until all points fall within the control limits. There is no theoretical justification for this, other than the fact that those points falling outside the trial control limits are more likely to belong to another probability distribution; i.e., they may be due to some assignable cause.

The control chart when used as a basis for action on future production may be set, using aimed-at values of $\bar{x}'$ and/or σ'. In this case, the control limits are modified by using the aimed-at values as if they were the known values in the computations. If this is done, and in the future the control chart exhibits a lack of control, the trouble may be that the process is not in control using the aimed-at values, but is in control at some other level. For example, suppose that an item is being produced according to a probability distribution which has a mean, $\bar{x}'$, equal to 20. If the aimed-at value is $\bar{x}' = 25$, the control chart based on $\bar{x}' = 25$ will exhibit a lack of control. Suppose the mean can be shifted by a change in machine setting. The interpretation, then, is that there is an assignable cause present, namely, the machine setting, preventing the production process from operating at the aimed-at value. Yet, with the present machine setting, the process is in control at $\bar{x}' = 20$. The term "state of control" should be interpreted in this light. Aimed-at values of $\bar{x}'$ or σ' are often used when they can be achieved by a simple adjustment of the machine.

12.3.6 Relation between Natural Tolerance Limits and Specification Limits

After ascertaining at what level the process is in control by means of the control chart, it remains to determine whether or not the process can meet the specification limits set for the item. In Sections 12.3.3 and 12.3.4 it was shown that σ' can be estimated from $\bar{\sigma}$ or $\bar{R}$; i.e., the estimate of σ' is $\bar{\sigma}/c_2$ or $\bar{R}/d_2$. Furthermore, $\bar{x}'$ can be estimated from $\bar{\bar{x}}$. Moreover, it is known that if $\bar{x}'$ and σ' are really equal to the estimates, almost all the individual items will lie between $\bar{x}' \pm 3\sigma'$, i.e., on the average, 9,973 out of 10,000, provided the distribution is normal. Under this assumption, if the natural tolerances are defined to mean those limits containing all but 27 in 10,000 values, $\bar{x}' \pm 3\sigma'$ will coincide with the natural tolerances. On the other hand, if the natural tolerances are defined to mean those limits containing all but 1 in 1,000, $\bar{x}' \pm 3.29\sigma'$ will coincide with the natural tolerances. Once the meaning of natural tolerances is established and the quantities $\bar{x}'$ and σ' (or their estimates) are determined, the natural tolerance limits can be computed. These natural tolerances are then compared with the specification limits. If the natural tolerances are not included within the specification limits, the process must be readjusted with respect to either $\bar{x}'$ or σ' or both, or the specification limits must be changed. If the natural tolerances are included well within the specification limits, this often signifies unrealistic specification limits, or increased labor and material costs resulting in a product which is "too good." After a period of time, a process in equilibrium usually has the tolerance limits coinciding with the specification limits or operating just within the specification limits.

As an example, suppose the designer indicates specification limits of 20 ± 2, and natural tolerance limits are defined as those limits excluding only 1 in 1,000. If $\bar{x}'$ is 20.40 and $\sigma' = 1$, it follows that on the average 999 out of 1,000 items will fall between $(\bar{x}' - 3.29\sigma', \bar{x}' + 3.29\sigma') = (17.11, 23.69)$. The specification limits allow only (18,22). Consequently, the process above cannot meet the specification limits, and either the process must be changed or the specification limits revised. Of course, in practical situations, $\bar{x}'$ and σ' are never really known and estimates are used. If the estimates are good, these procedures are adequate. However, if the estimates are from small samples, a technique such as that described in Chapter VIII, using statistical tolerance limits, is in order.

12.3.7 INTERPRETATION CONTROL CHARTS FOR $\bar{x}$

As long as the process is in control, almost all the values of $\bar{x}$ will fall within the control limits. In this case, no action need be taken. A point falling outside these limits is a signal to hunt for trouble. In some instances, production may be stopped until a source of trouble has been discovered. Occasionally, approximately 27 in 10,000 times, an error will be made, in that trouble will be sought even though nothing has gone wrong with the process. If something has gone wrong with the process, points will begin to fall outside the control limits. For example, a shift in the mean $\bar{x}'$ (σ' remaining constant) will result in points falling outside the control limits. This is usually indicated by points falling above or below the limits (but rarely both), depending on whether the shift is in the positive or negative direction. On the other hand, an increase in σ' ($\bar{x}'$ remaining constant) will result in points falling above *and* below the control limits. In addition to examining the process, when points fall outside the control limits, the control limits themselves should be re-examined, and perhaps brought up to date.

It has been pointed out that an error may be committed when action is taken after a point falls outside the control limits. There is a small probability that a point will fall outside the control limits even though the process is in control; i.e., if $3\sigma'_{\bar{x}}$ limits are used, the probability is 0.0027. This probability of falling outside the control limits when the process is in control is known as the probability of committing a *Type I error*. Similarly, if the process goes out of control, there is a probability greater than 0 that a point will fall within the control limits. This is known as the probability of a *Type II error*. These two errors are related. A decrease in one results in an increase in the other. An increase in one results in a decrease in the other. The use of $3\sigma'_{\bar{x}}$ limits implies a probability of a Type I error of 0.0027. If the process jumps out of control, and the new level is specified, the probability of a Type II error can be computed. If $2\sigma'_{\bar{x}}$ limits were used instead of $3\sigma'_{\bar{x}}$, the probability of a Type I error is increased to about 0.0455, but the probability of a Type II error is decreased. These are the same concepts that were applied in the chapters on significance tests, i.e., Chapters V, VI and VII. The control chart is a procedure used to test the hypothesis that the process is in control. This procedure has an OC curve with its associated Type I and Type II errors, just as any other procedure.

For example, suppose $\bar{x}'$ is 25, and $\sigma' = 1$, and there are 4 observations in each subgroup. The control limits are then $\bar{x}' \pm A\sigma' = 25 \pm \frac{3}{2}$. The probability of a Type I error is 0.0027. If the mean shifts to 27, the probability of a point falling inside the control limits of $25 \pm \frac{3}{2}$ is 0.1587. On the other hand, if $2\sigma'_{\bar{x}}$ are used as control limits (25 ± 1) the probability of a Type I error is increased to 0.0455, whereas the probability of committing a Type II error is only 0.0228.

It is evident that Type I errors or Type II errors (but not both) can be made as small as is desirable at the expense of increasing the other type error. It is also evident that using $3\sigma'_{\bar{x}}$ limits is conservative from the point of view of considering the Type I error. A point falling outside the control limits is almost sure evidence that the process is no longer in control. On the other hand, one cannot be assured that a small shift has not taken place even if the points fall within the control limits.

12.4 R Charts and σ Charts

12.4.1 Statistical concepts

Although both R and σ do not have normal distributions, both of these functions are random variables. The probability distribution of σ is related to the chi-square distribution, and the distribution of R is approximately related to the chi-square distribution. Furthermore, it can be shown that almost all of the probability distribution is contained within the expected value of these variables plus and/or minus three standard deviations (of these variables). Consequently, for control chart purposes, it is only necessary to calculate both the expected value and standard deviation of these random variables.

The expected value of σ is $c_2\sigma'$. The standard deviation of σ is

$$\sigma_\sigma = [2(n-1) - 2nc_2^2]^{1/2} \frac{\sigma'}{\sqrt{2n}}.$$

The control limits are then $c_2\sigma' \pm 3\sigma_\sigma$ which can be written as

$$\sigma'\left(c_2 \pm \frac{3}{\sqrt{2n}}[2(n-1) - 2nc_2^2]^{1/2}\right).$$

The factors

$$B_2 = c_2 + \frac{3}{\sqrt{2n}}[2(n-1) - 2nc_2^2]^{1/2}$$

and

$$B_1 = c_2 - \frac{3}{\sqrt{2n}}[2(n-1) - 2nc_2^2]^{1/2}$$

can be obtained from Table 12.1. Thus

$$UCL_\sigma = B_2\sigma'; \qquad LCL_\sigma = B_1\sigma'.$$

When σ' is unknown, it can be estimated from $\bar{\sigma}/c_2$, so that the estimated control limits are written as

$$\bar{\sigma}\left\{1 \pm \frac{3}{\sqrt{2n}\,c_2}[2(n-1) - 2nc_2^2]^{1/2}\right\}.$$

The factors

$$B_4 = 1 + \frac{3}{\sqrt{2n}\, c_2}[2(n-1) - 2nc_2^2]^{1/2}$$

and

$$B_3 = 1 - \frac{3}{\sqrt{2n}\, c_2}[2(n-1) - 2nc_2^2]^{1/2}$$

are also given in Table 12.1. The estimated control limits are then

$$UCL_\sigma = B_4\bar{\sigma}; \qquad LCL_\sigma = B_3\bar{\sigma}.$$

Note that the center line is $c_2\sigma'$ if σ' is known, and $\bar{\sigma}$ if σ' is unknown.

The control limits for R are obtained in a similar manner. The expected value of R is $d_2\sigma'$. The standard deviation of R is $\sigma_R = d_3\sigma'$. The control limits are then $d_2\sigma' \pm 3\sigma_R$, which can be written $\sigma'(d_2 \pm 3d_3)$. The factors

$$D_2 = d_2 + 3d_3; \qquad D_1 = d_2 - 3d_3$$

can be obtained from Table 12.1. Thus

$$UCL_R = D_2\sigma'; \qquad LCL_R = D_1\sigma'.$$

When σ' is unknown, it can be estimated from $\bar{R}/d_2$ so that the estimated control limits are written as

$$\bar{R}\left[1 \pm \frac{3d_3}{d_2}\right].$$

The factors

$$D_4 = 1 + \frac{3d_3}{d_2}; \qquad D_3 = 1 - \frac{3d_3}{d_2}$$

can be obtained from Table 12.1. The estimated control limits are then

$$\text{UCL}_R = D_4\bar{R}; \qquad \text{LCL}_R = D_3\bar{R}.$$

The center line is $d_2\sigma'$ if σ' is known, and $\bar{R}$ if σ' is unknown.

12.4.2 Setting up a control chart for R or σ

The range of each subgroup is obtained and $\bar{R}$ calculated from these values. The trial control limits are given by

$$\text{UCL}_R = D_4\bar{R}; \qquad \text{LCL}_R = D_3\bar{R}.$$

If a σ chart is desired, similar calculations are made in accordance with rules described above. If all the points fall inside the trial control limits, no modification is made unless it is desired to reduce the process dispersion. In this case, an aimed-at value of σ' should be used in the calculations. When the $\bar{R}$ (or σ) chart indicates a possible lack of control by points falling outside the trial control limits, it is desirable to estimate the value of σ' that might be attained if the dispersion were brought into control. A method of estimation is to eliminate those points out of control (only those above the UCL_R if points fall both above and below) and recompute the values of the control limits based only on the remaining observations. If more points fall out of control, the procedure is repeated.

The final revised values of $\bar{R}, \bar{\sigma}$, or σ', whichever is appropriate, may also be used to obtain new control limits for $\bar{x}$. This has the effect of tightening the limits on the $\bar{x}$ chart, making them consistent with a σ' that may be estimated from the revised $\bar{R}$ or $\bar{\sigma}$.

Control limits should be revised from time to time as additional data are accumulated.

12.5 Example of $\bar{x}$ and R Chart

The following are the $\bar{x}$ and R values for 20 subgroups of five readings. The specifications for this product characteristic are 0.4037 ± 0.0010. The values given are the last two figures of the dimension reading; i.e., 31.6 should be 0.40316.

Table 12.2. Data on Dimensions

Subgroup	$\bar{x}$	R	Subgroup	$\bar{x}$	R	Subgroup	$\bar{x}$	R
1	34.0	4	8	32.6	13	15	33.8	7
2	31.6	4	9	33.8	19	16	31.6	5
3	30.8	2	10	37.8	6	17	33.0	5
4	33.0	3	11	35.8	4	18	28.2	3
5	35.0	5	12	38.4	4	19	31.8	9
6	32.2	2	13	34.0	14	20	35.6	6
7	33.0	5	14	35.0	4			

$$\sum \bar{x} = 671.0; \qquad \bar{\bar{x}} = 33.6; \qquad \sum R = 124; \qquad \bar{R} = 6.20.$$

The trial control limits are computed from

$$\bar{\bar{x}} \pm A_2\bar{R}; \qquad UCL_{\bar{x}} = 37.2; \qquad UCL_R = (2.115)(6.20)$$
$$\bar{\bar{x}} \pm (0.577)(6.20); \qquad\qquad = 13.1$$
$$\bar{\bar{x}} \pm 3.58. \qquad LCL_{\bar{x}} = 30.0; \qquad LCL_R = 0$$

The control charts for $\bar{x}$ and R with trial limits are shown in Figure 12.3. Since some points fall outside the control limits, the process is assumed to be out of control. Eliminating these points, new control limits are computed:

$$\sum \bar{x} = 566.6; \qquad \bar{\bar{x}} = 33.3$$
$$\sum R = 91; \qquad \bar{R} = 5.06$$

$$UCL_{\bar{x}} = 33.3 + (0.577)(5.06) = 33.3 + 2.92 = 36.2;$$
$$LCL_{\bar{x}} = 33.3 - (0.577)(5.06) = 33.3 - 2.92 = 30.4;$$
$$UCL_R = 2.115(5.06) = 10.7; \quad LCL_R = 2.115(0) = 0.$$

None of the remaining points fall outside the limits. If it is now assumed that the process can be brought into control at this level, we

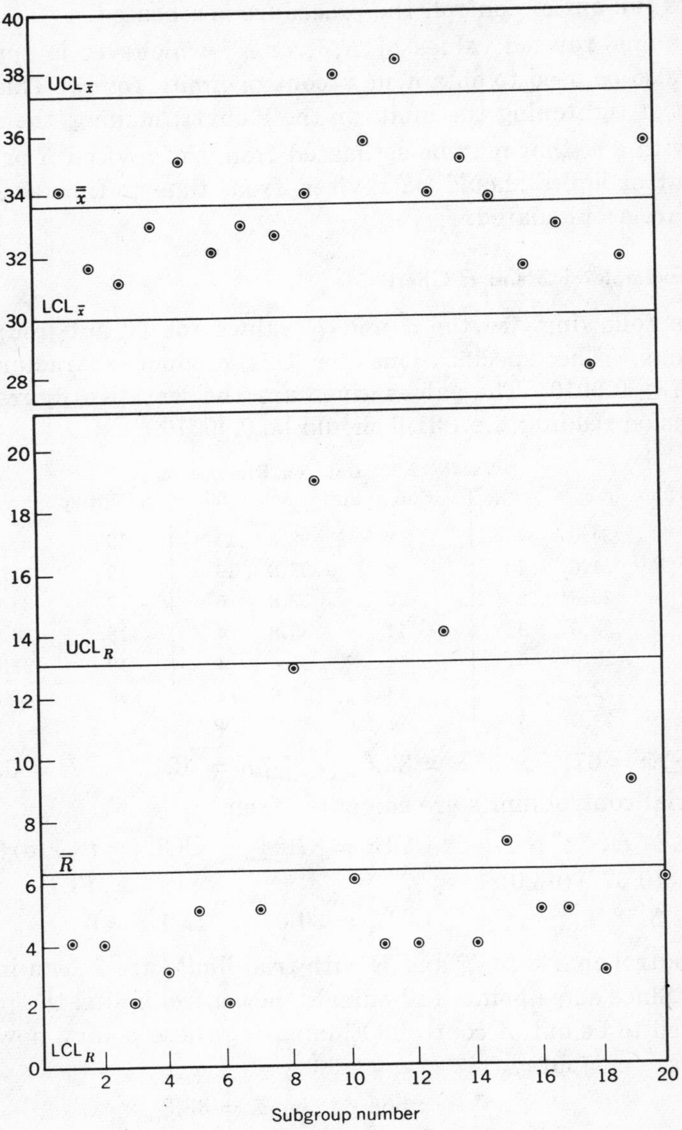

Fig. 12.3. Control chart for $\bar{x}$ and R.

find $\bar{x}' = 33.3$ and $\sigma' = 2.175$. The value of σ' is estimated from $\bar{R}/d_2$;

$$\bar{R}/d_2 = 5.06/2.326 = 2.175$$

and thus

$$(\bar{x}' - 3\sigma', \bar{x}' + 3\sigma') = (\bar{x}' - 6.525, \bar{x}' + 6.525) = (26.8, 39.8).$$

In terms of the actual data, $(\bar{x}' - 3\sigma', \bar{x}' + 3\sigma') = (0.4027, 0.4040)$. Thus (0.4027, 0.4040) are the estimated natural tolerance limits where these limits are defined to include all but 27 in 10,000. The specifications are

$$(0.4037 - 0.0010, 0.4037 + 0.0010) = (0.4027, 0.4047)$$

so that this production process is able to meet these specifications even though the process is not centered at the nominal value of 0.4037, provided the process remains in control at the above level.

12.6 Control Chart for Fraction Defective

12.6.1 Relation between control charts based on variables data and charts based on attributes data

The $\bar{x}$ and R charts are charts for variables, i.e., for quality characteristics that can be measured and expressed in numbers. However, many quality characteristics can be observed only as attributes, i.e., by classifying the item into one of two classes, usually defective or non-defective. Furthermore, with the existing techniques, an $\bar{x}$ and R chart can be used for only one measurable characteristic at a time. For example, if an item consists of 10,000 measurable characteristics, each characteristic is a candidate for an $\bar{x}$ and R chart. However, it would be impossible to have 10,000 such charts, and only the most important and troublesome would be charted. As an alternative to $\bar{x}$ and R charts, and as a substitute when characteristics are measured only by attributes, a control chart based on the fraction defective can be used. This is known as a p chart. A p chart can be applied to quality characteristics that are actually observed as attributes even though they may have been measured as variables. The cost of obtaining attribute data is usually less than that for obtaining variables data. The cost of computing and charting may also be less, since one p chart can apply to any number of characteristics.

12.6.2 Statistical theory

In Chapter IV it was shown that if we have a series of n independent trials and if at each trial the symbol Π represents the probability that the event will occur, the probability that r events occur in the n trials is $\binom{n}{r} \Pi^r(1 - \Pi)^{n-r}$, where $\binom{n}{r}$ is the number of combinations of n things taken r at a time;

$$\binom{n}{r} = \frac{n!}{r!\,(n-r)!}$$

and

$$n! = n(n-1)(n-2)\cdots(3)(2)(1); \qquad 0! = 1.$$

A random variable having this probability distribution is known as a binomial random variable.

To apply the binomial distribution to control charts, denote by p' the probability of an item being defective, so that the probability of obtaining exactly d defectives in a sample of n is $\binom{n}{d}p'^d(1-p')^{n-d}$. The probability of obtaining d_1 defectives or less in a sample of n is

$$\sum_{d=0}^{d_1}\binom{n}{d}p'^d(1-p')^{n-d} = \binom{n}{0}p'^0(1-p')^n + \binom{n}{1}p'^1(1-p')^{n-1} + \cdots + \binom{n}{d_1}p'^{d_1}(1-p')^{n-d_1}.$$

Furthermore, the expected value of the total number of defectives in a sample of n is np', and the standard deviation is $\sqrt{np'(1-p')}$.

If the fraction defective, p, is defined as the ratio of the number of defectives d to the total number of items in the sample n, i.e., d/n, the expected value of the fraction defective is p' and the standard deviation is $\sqrt{p'(1-p')/n}$.

As a working rule, the control limits for the fraction defective are

$$\text{UCL}_p = p' + 3\sqrt{\frac{p'(1-p')}{n}}; \qquad \text{LCL}_p = p' - 3\sqrt{\frac{p'(1-p')}{n}}.$$

It is important to note that the probability of d/n falling within these limits depends on the value of p', even if the process is in control.[1] However, almost all the d/n will fall within the given limits, provided the process is in control. Furthermore, the limits above result in a simple empirical rule and, for large n, result in an accurate approximation. This follows from the fact that d/n is an average. Hence, the central limit theorem can be invoked for large n, thereby implying that d/n is approximately normally distributed.

The exact probability of a point falling within the control limits can be evaluated using the expression

[1] This is quite different from the $\bar{x}$ chart, where the probability of $\bar{x}$ falling between $\bar{x}' \pm 3\sigma'/\sqrt{n}$ is 0.9973 for any values of $\bar{x}'$ and σ', provided the process is in control.

$$P\left(p' - 3\sqrt{\frac{p'(1-p')}{n}} \leqq \frac{d}{n} \leqq p' + 3\sqrt{\frac{p'(1-p')}{n}}\right)$$
$$= P[np' - 3\sqrt{np'(1-p')} \leqq d \leqq np' + 3\sqrt{np'(1-p')}].$$

For a fixed p', the right-hand side of this equation is just the probability that a binomial variable lies between two fixed constants, and can be obtained from the binomial distribution.

If p' is not known, it is usually estimated from past data. The rules are similar to those used in estimating the parameters for the $\bar{x}$ charts. The estimate in this case is d_T/n_T, where d_T is the total number of defectives found in the past data of size n_T.

It often happens in control charts for fraction defective that the size of the subgroup varies. In this case, three possible solutions to the problem are as follows: (1) Compute control limits for every subgroup and show these fluctuating limits on the p chart. (2) Estimate the average subgroup size, and compute one set of limits for this average and draw them on the control chart. This method is approximate and is appropriate only when the subgroup sizes are not too variable. Points near the limits may have to be re-examined in accordance with (1). (3) Draw several sets of control limits on the chart corresponding to different subgroup sizes. This method is also approximate and is actually a cross between (1) and (2). Again, points falling near the limits should be re-examined in accordance with (1).

12.6.3 Starting the control chart

The subgroup size is usually large compared with that used for $\bar{x}$ and R charts. The main reason for this is that if p' is very small, and n is small, the expected number of defectives in a subgroup will be very close to zero.

For each subgroup compute p, where

$$p = \frac{\text{number of defectives in the subgroup}}{\text{number inspected in the subgroup}}.$$

Whenever practicable, no fewer than 25 subgroups should be used to compute trial control limits. These limits are computed from the data by finding the average fraction defective $\bar{p}$, where

$$\bar{p} = \frac{\text{total number of defectives during period}}{\text{total number inspected during period}}$$

and substituting in the expressions

$$\mathrm{UCL}_p = \bar{p} + \frac{3\sqrt{\bar{p}(1-\bar{p})}}{\sqrt{n}}; \qquad \mathrm{LCL}_p = \bar{p} - \frac{3\sqrt{\bar{p}(1-\bar{p})}}{\sqrt{n}}.$$

Inferences about the existence of control or lack of control can be drawn in a manner similar to that described for the $\bar{x}$ and R chart.

12.6.4 Continuing the p Chart

The preliminary plot reveals two facts, namely, whether or not the process is in an apparent state of control and the quality level. If the process appears to be in control, it may be in control at too high a level—i.e., the estimate of p' is higher than can be tolerated. In this case, the production process must be examined, and possible major changes made. On the other hand, the estimate of p' may indicate a good quality level, even though the process may appear to be out of control. In this case, assignable causes should be sought and eliminated. Of course, control limits can be determined on the basis of an aimed-at value of p', but an indicated lack of control must be interpreted with this in mind.

At first glance, points falling below the lower control limit may appear to be desirable. However, this may be attributed to a poor estimate of p' although it may possibly indicate a change for the better in the quality level, and tracking down the assignable cause may enable an improvement to be made in the production process. In either case, points falling below the lower control limit call for a re-examination of the control chart or the production process or both.

Table 12.3. Screw Machine Data

Subgroup	*d*	*p*	*Subgroup*	*d*	*p*	*Subgroup*	*d*	*p*
1	1	0.02	9	1	0.02	18	0	0.00
2	2	0.04	10	0	0.00	19	0	0.00
3	5	0.10	11	0	0.00	20	1	0.02
4	6	0.12	12	1	0.02	21	1	0.02
5	3	0.06	13	0	0.00	22	0	0.00
6	5	0.10	14	1	0.02	23	0	0.00
7	2	0.04	15	0	0.00	24	1	0.02
8	1	0.02	16	2	0.04	25	0	0.00
			17	1	0.02			

$$\bar{p} = \frac{34}{1{,}250} = 0.0272,$$

$$\mathrm{UCL} = 0.0272 + \frac{3\sqrt{(0.0272)(0.9728)}}{7.071} = 0.0963,$$

$$\mathrm{LCL} = 0.$$

12.6.5 Example

A sample of 50 pieces is drawn from the production of the last two hours from a single spindle automatic screw machine and each item is checked by go and no-go gauges for several possible sources of defectives. The number of defective items found in 25 such successive samples are given in Table 12.3.

The control chart for p with trial limits is shown in Figure 12.4. Many points fall outside the control limits, and it appears that an assignable cause of variation was present when the first samples were taken that was not present when the later samples were taken. Recomputing the control limits starting with the seventh sample, we find

$$\bar{p} = \frac{12}{950} = 0.0126,$$

$$\text{UCL} = 0.0126 + \frac{3\sqrt{(0.0126)(0.9874)}}{7.071} = 0.0599,$$

$$\text{LCL} = 0.$$

No points fall outside these limits and hence we use these as the control limits for future production.

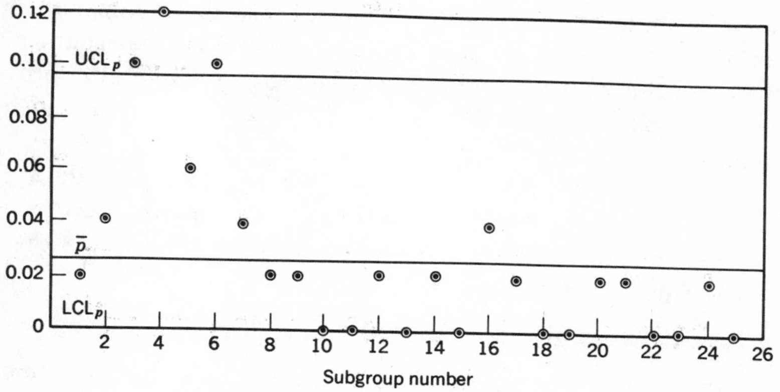

Fig. 12.4. Control chart for p.

12.7 Control Charts for Defects

12.7.1 Difference between a defect and a defective

An item is considered to be defective if it fails to conform to the specifications in any of the characteristics. Each characteristic that does not meet the specifications is a defect. An item is defective if it contains at least one defect.

The c chart is a control chart for defects per unit. The unit considered may be a single item, a group of items, part of an item, etc. The unit is examined and the number of defects found is recorded on the c chart. If for each unit there are numerous opportunities for defects to occur, and if the probability of a defect occurring in a particular spot is small, the statistical theory for the c chart is based on the Poisson distribution. Some examples where c charts are applicable are in counting the number of defective rivets on an airplane wing, the number of imperfections in a piece of cloth, etc.

12.7.2 Statistical Theory

The probability of finding c defects in an item where the number of defects follow the Poisson law is $c'^c e^{-c'}/c!$, where c' is the *average* number of defects per unit. The standard deviation of this random variable is $\sqrt{c'}$.

As a working rule, control limits for the c chart are defined as

$$\text{UCL}_c = c' + 3\sqrt{c'}; \qquad \text{LCL}_c = c' - 3\sqrt{c'}.$$

As in the case of the control chart for the fraction of defectives, these limits do not guarantee that a fixed percentage of the population will lie between them for all values of c' even if the process is in control; i.e., the probability of c falling between these limits depends on the value of c'. However, for practical purposes, it can be assumed that "almost all the values" will fall between the control limits, provided the process is in control. For large n, c is approximately normally distributed.

If c' is not known, it is usually estimated from past data. The rules are similar to those used in estimating the parameters from the $\bar{x}$ charts. The estimate in this case is $c = c_T/N_T$, where c_T is the total number of defects found in N_T units.

12.7.3 Starting and Continuing the c Chart

The discussion on starting and continuing the control chart for fraction defective given in Section 12.6.3 and 12.6.4 is pertinent to the c chart.

12.7.4 Example

The following table gives the number of missing rivets noted at final inspection on 25 airplanes. Read the columns downward and left to right.

Table 12.4. Data on Missing Rivets

10	14	21	15	12
17	7	13	11	18
16	14	10	12	6
20	19	11	8	10
10	16	25	30	8

$$\bar{c} = \frac{353}{25} = 14.12,$$

$$\text{UCL}_c = 14.12 + 3\sqrt{14.12} = 25.40,$$
$$\text{LCL}_c = 14.12 - 3\sqrt{14.12} = 2.84.$$

The control chart for c with trial limits is shown in Figure 12.5. However, the twentieth plane falls outside the control limits, and it

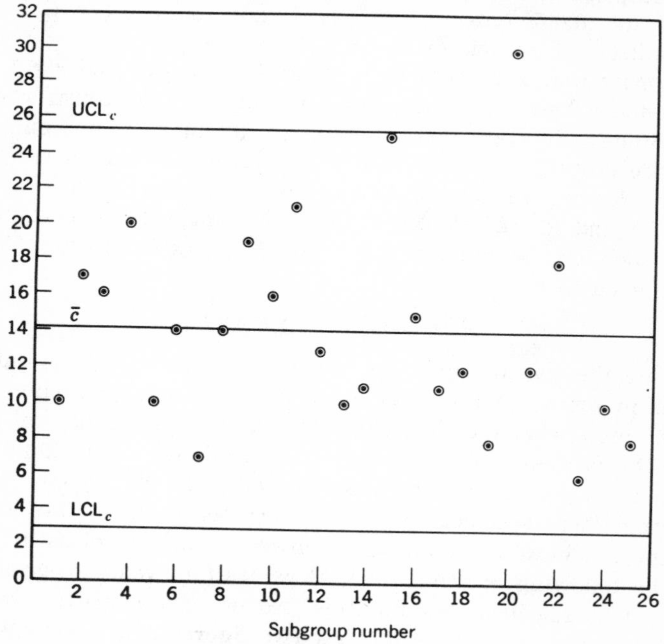

Fig. 12.5. Control chart for c.

would be best to base future calculations on a $\bar{c}$ which did not involve this airplane.

$$\bar{c} = \frac{323}{24} = 13.46,$$

$$\text{UCL}_c = 13.46 + 3\sqrt{13.46} = 24.47,$$
$$\text{LCL}_c = 13.46 - 3\sqrt{13.46} = 2.45.$$

The fifteenth airplane now falls outside the control limits, so that $\bar{c}$ is again recomputed without the results for this plane.

$$\bar{c} = \frac{298}{23} = 12.96,$$

$$\text{UCL}_c = 12.96 + 3\sqrt{12.96} = 23.76,$$

$$\text{LCL}_c = 12.96 - 3\sqrt{12.96} = 2.16.$$

PROBLEMS

1. Suppose probability limits are desired for $\bar{x}$ charts, and probability limits corresponding to a probability of 0.004 of a point falling outside the control limits (0.002 outside each limit) are to be used. Calculate the control limits for a sample of size 5.

2. Suppose that a process is in control at $\bar{x}'=17$ and $\sigma'=2$. A control chart for $\bar{x}$ based on a subgroup size of $n=4$ is constructed using 3-sigma limits. Suppose that a shift in the mean of 2.5 units occurs. What is the probability of the next $\bar{x}$ falling outside the control limits?

3. Samples of five items each are taken from a process at regular intervals. $\bar{x}$ and R are calculated for each sample for a certain quality characteristic, x. The sums of the $\bar{x}$ and R values for the first 25 samples are determined to be:

$$\Sigma\bar{x}=358.50, \qquad \Sigma R=9.80.$$

(a) Compute the control limits for the $\bar{x}$ and R charts. (b) Assuming the process is in control at the level found in (a) (assume estimates are equal to the parameters being estimated), what are the 3-sigma natural tolerance limits of the process? (c) If the specification limits are 14.40 ± 0.45, what conclusions can you draw concerning the ability of the process to produce items within these specifications? (d) What percentage of the lot will fall outside the specification limits if the process is in control as in (b)?

4. The following are the $\bar{x}$ and R values for twenty subgroups of five readings. The specifications for this product characteristic are 0.4037 ± 0.0010. The values given are the last two figures of the dimension reading; i.e., 31.6 should be 0.40316.

Subgroup No.	$\bar{x}$	R	*Subgroup No.*	$\bar{x}$	R
1	34.0	4	11	35.8	4
2	31.6	4	12	35.8	4
3	30.8	2	13	34.0	14
4	33.0	3	14	35.0	4
5	35.0	5	15	33.8	7
6	32.2	2	16	31.6	5
7	33.0	5	17	33.0	5
8	32.6	13	18	33.2	3
9	33.8	19	19	31.8	9
10	35.8	6	20	35.6	6

$\Sigma\bar{x}=671.0 \qquad \bar{\bar{x}}=33.6$

$\Sigma R=124 \qquad \bar{R}=6.2$

$\bar{x}$ chart UCL=37.2

$\bar{x}$ chart LCL=30.0

R chart UCL=13.1

R chart LCL=0

(a) Is this process probably in control? (b) If not, suggest values for UCL and LCL for use with succeeding subgroups. (c) If the process is in control with $\bar{x}'$ and σ' equal to these values used in obtaining the UCL and LCL values in (b), will this process be able to meet the specifications for the dimension? If not, what specification can you expect the process to meet?

5. Control charts for $\bar{x}$ and σ are maintained on the shear strength of spot welds. After 30 subgroups of 4, $\Sigma\bar{x}=12{,}660$ and $\Sigma\sigma=945$. Assume that the process is in control. (a) What are the $\bar{x}$ chart limits? (b) What are the σ chart limits? (c) Estimate the standard deviation for the process. (d) If the minimum specification for this weld is 400 pounds, what percentage of the welds does not meet the minimum specification?

6. Control charts for $\bar{x}$ and R are maintained on the breaking strength in pounds of a certain metal piece. The subgroup size is 4. The values of $\bar{x}'$ and σ' computed after a long period are 7.3 and 1.1, respectively. Compute the values of the 3-sigma limits for the $\bar{x}$ and R charts.

7. Control charts for $\bar{x}$ and R are maintained on resistors (in ohms). The subgroup size is 3. The values of $\bar{x}$ and R are computed for each subgroup. After 20 subgroups, $\Sigma\bar{x}=8{,}620$, and $\Sigma R=910$. (a) Compute the values of the 3-sigma limits for the $\bar{x}$ and R charts. (b) Estimate the value of σ' on the assumption that the process is in statistical control. (c) If the specification limits are 430 ± 30, what conclusions can you draw regarding the ability of the process to produce items within these specifications? (d) If $\bar{x}'$ is increased by 60, what is the probability of a subgroup average falling outside the control limits?

8. (a) Assume that probability limits rather than 3-sigma limits were to be used in Problem 7. Assuming that the value of σ' is as found in Problem 7, what would these limits be on the $\bar{x}$ chart if the probability of a point falling outside each limit is 0.002? (b) How many observations are required in Problem 7 such that $\bar{\bar{x}}$ is within ten units of $\bar{x}'$ with probability greater than 0.99? Assume σ' is as found in Problem 7.

9. Given that the expected value of $\bar{\sigma}=10$, what is the expected value of $\bar{R}$? Note that the term "expected value of $\bar{\sigma}$" means the average value of the variable $\bar{\sigma}$.

10. The range of a subgroup of two items is 4. What is σ for this subgroup? Estimate σ' using both R and σ. Estimate the standard deviation for subgroup averages, i.e., $\sigma'_{\bar{x}}$.

11. A process is in control at $\bar{x}'=17$ and $\sigma'=3$. Subgroups of size 5 are being plotted for an $\bar{x}$ and σ chart. (a) What are the 3-sigma control limits for the $\bar{x}$ chart? (b) The $\bar{x}$ chart (with 3-sigma limits) is to be used as a sampling inspection plan. If a point falls outside the limits, the entire production period between this point and the previous point plotted is rejected. If the process is in control at the given level, what is the probability of rejecting one out of two production periods? (c) Suppose that $\bar{x}'$ was unknown but σ' was known to be equal to 3. How many observations are required so that $\bar{\bar{x}}$ differs from $\bar{x}'$ by no more than one unit with probability greater than 0.99? (d) What is the upper limit for the σ chart so that the probability of a point falling above the upper limit is 0.001? [*Hint*: Note that $n\sigma^2/(\sigma')^2$ has a χ^2 distribution with $(n-1)$ degrees of freedom.]

12. A sample of 50 pieces is drawn from the production of the last two hours from a single spindle automatic screw machine and each item is checked by go and no-go gauges for several possible sources of defectiveness. The number of defective items found in 25 successive samples was: (read each line left to right in order)

1, 2, 5, 6, 3, 5, 2, 1, 1, 0, 0, 1, 0,
1, 0, 2, 1, 0, 0, 1, 1, 0, 0, 1, 0:
total, 34 defectives.

(a) Is this process in control? (b) If the consumer is willing to accept lots of this product that are 2.5% or less defective, will the process in the present condition produce an acceptable product? (c) From looking at the chart of this process, what do you think happened during the run?

13. The following data represent the number of defective transistors produced for a daily production of 2,800 units:

Day	*Defectives*
1	110
2	117
3	112
4	105
5	130
6	120
7	119
8	113

(a) Draw an appropriate control chart for these data. (b) What can you conclude?

14. Assume that a process is in control at $p'=4\%$ and $n=100$. (a) Calculate the 3-sigma control limits for a p chart. (b) What is the probability of a point falling above the upper control limit? (Use the Poisson approximation.) (c) If the process is in control at $p'=4\%$, how many observations, N, are required so that $\bar{p}=T/N$, where T is the total number of defectives found in a sample of size N, is within ± 0.005 of p' with probability greater than 0.95? (Use the normal approximation.) (d) If there is no information about the value of p', how many observations are required so that $\bar{p}$ is within ± 0.005 of p' with probability greater than 0.95? (Use the normal approximation.)

15. The process average has been shown to be 0.03. Your control chart for fraction defective calls for taking daily samples of 800 items. What is the chance that, if the process average should suddenly shift to 0.06 you would catch the shift on the first sample taken after the shift?

16. In "work sampling" the fraction of time that an employee works is often estimated by taking a random sample and finding the ratio of the number of times he is working to the total number of tries, i.e., $p=D/N$. Thus, p is a binomial variable. Assuming the normal approximation, find the number of observations necessary so that the estimated value (p) differs from the true value (p') by no more than ± 0.03 with probability greater than 0.95. (*Hint*: Find the maximum value of the standard deviation of p.)

17. Three-sigma control limits for control charts for percent of defectives are:

$$\text{UCL}_p = 6.20\%,$$
$$\text{LCL}_p = 0,$$

where the process average is 2% defective and the subgroup size is 100. Using the Poisson approximations, find: (a) the probability of a point falling outside these limits if the incoming quality is 2% defective. (b) the probability of a point falling outside these limits if the incoming quality shifts to 5% defective. (c) the probability of at least two points out of ten falling outside these limits if the incoming quality remains at 5% defective.

18. Over the past eight or nine months an assembly line has been producing 400 units of a product per day. Final inspection records show that on an average 20 units a day are defective. A survey of the daily reports shows that on an occasional day as many as 32 or 33 units are defective and that on occasional days as few as one or two are defective. (a) Do these records indicate statistical control of quality? (b) What help might be obtained, if any, from these records in improving quality performance at this line?

19. A control chart on the number of breakdowns in successive lengths of 10,000 feet each of rubber-covered wire leads to a value of $\bar{c}=6.2$. (a) Assuming that the process is in control at this level, what are the 3-sigma control limits? (b) What is the probability of a point falling above the upper control limit?

20. Twenty-five 100-yard pieces of woolen goods were measured and the average number of defects per unit was found to be 3.2. Determine the 3-sigma control limits. What are the probability limits corresponding to a probability of 0.002 of falling outside the upper control limit?

CHAPTER XIII

Sampling Inspection

13.1 The Problem of Sampling Inspection

13.1.1 INTRODUCTION

Sampling inspection is of two kinds, namely, lot-by-lot sampling inspection and continuous sampling inspection. In the former, items are formed into lots, a sample is drawn from the lot, and the lot is either accepted or rejected on the basis of the quality of the sample. This is most appropriate for acceptance inspection. In continuous sampling inspection, current inspection results are used to determine whether sampling inspection or screening inspection is to be used for the next articles to be inspected. Sampling plans are further classified depending on whether the quality characteristics are measured and expressed in numbers, i.e., variables inspection, or whether articles are classified only as defective or non-defective, i.e., attributes inspection.

An alternative to sampling inspection is screening every item. The cost of such a scheme is prohibitive, with perfect quality rarely achieved. Still another advantage of sampling inspection schemes —if properly designed—is that they create more effective pressure for quality improvement, and therefore result in the submission of better quality product for inspection.

Although the factors above were recognized somewhat prior to World War II, and a few sampling inspection schemes were then in use, it was not until the outbreak of hostilities that modern sampling inspection received its impetus. Good sampling plans replaced bad ones, sometimes at the expense of an increase in the total amount of inspection but almost always resulting indirectly in better quality being produced.

Any lot-by-lot sampling plan has as its primary purpose the acceptance of good lots and the rejection of bad lots. It is important to define what is meant by a good lot. Naturally, the consumer would like all of his accepted lots to be free of defectives. On the other hand, the manufacturer will usually consider this to be an unreasonable request since some defectives are bound to appear in the manufacturing process. If the manufacturer screens the lot a few times he may eliminate all the defectives, but at the prohibitive cost of screening. This cost will naturally be reflected in his price to the consumer. Ordinarily, the consumer can tolerate some defectives in his lot, provided the number is not too large. Consequently, the manufacturer and the consumer get together and agree on what constitutes good quality. If lots are submitted at this quality or better, the lot should be accepted; otherwise, rejected. Again this is an imposing task and can be accomplished only at the expense of screening. It is at this point that sampling inspection, with its corresponding advantage of reduced inspection costs, can be instituted. This advantage should not be minimized. Few manufacturers or consumers, whichever has to bear the cost of inspection, can stay in business very long if all lots are screened.

13.1.2 Drawing the Sample

A decision must be made on what shall constitute a lot for acceptance purposes, for each lot must be identified. Each lot should represent, as nearly as possible, the output of one machine or process during one interval of time, so that all parts or products in the lot have been produced under essentially the same conditions. Wherever practicable, parts from different sources or different conditions should not be mixed into one lot. The power of the sampling plans to distinguish between good and bad lots is dependent on the variation from lot to lot, and provision should be made to maintain the identity of, and prevent the mixing of, the different lots.

A sample from each lot supplies the information on which the decision to accept or reject the lot is based. Therefore it is essential that the sample be drawn from each lot in a random manner. A sample is random if every piece in the lot has an equal chance of being selected.[1] This may be accomplished by using a table of

[1] This definition differs from that given in Chapter II, which is inapplicable in situations where the probability distribution always depends upon the results of the previous trials; if the size of the lot is finite, the probability of an item being defective depends upon the outcome of previous trials.

random numbers, a deck of cards, or some such chance method, to insure the equal chance for each piece. Human attempts to randomize a sample without such aids often result in biased samples.

13.2 Lot-by-Lot Sampling Inspection by Attributes

13.2.1 Single sampling plans

13.2.1.1 *Single sampling.* A single sampling procedure can be characterized by the following: one sample of n items is drawn from a lot of N items; the lot is accepted if the number of defectives d in the sample does not exceed c. Here c is referred to as the acceptance number.

Certain risks must be taken if sampling inspection is to be used. A graph of these risks plotted as a function of the incoming lot quality (p') is known as an operating characteristic (OC) curve, and is illustrated in Figure 13.1.

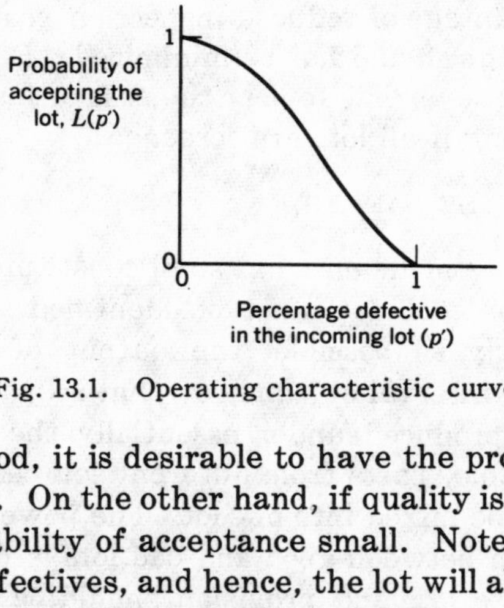

Fig. 13.1. Operating characteristic curve.

If quality is good, it is desirable to have the probability of acceptance, $L(p')$, high. On the other hand, if quality is bad, it is desirable to have the probability of acceptance small. Note that if p' is 0, the lot contains no defectives, and hence, the lot will always be accepted, i.e., $L(p')=1$. If p' is 1 the entire lot is defective so that the sample contains all defectives, thereby insuring that the lot will always be rejected, i.e., $L(p')=0$.

Assume that a quality standard p_g' is established and that all lots better than this standard are considered to be "good" and all lots worse are considered to be "bad". An "ideal" OC curve would then be the form shown in Figure 13.2 with all good lots accepted and all bad ones rejected. Obviously, no sampling plan can have such a curve. The degree of approximation to the *ideal curve* depends on n and c.

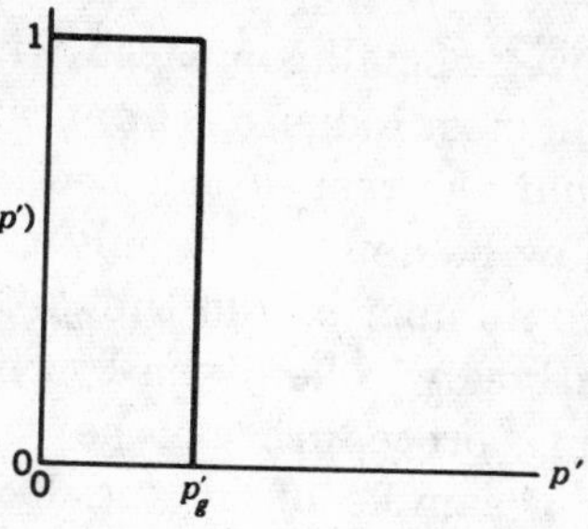

Fig. 13.2. Ideal operating characteristic curve.

If c is held constant, and n is increased, the slope becomes steeper. On the other hand, holding n constant and changing the acceptance number has the effect of shifting the OC curve to the left or right. These concepts are illustrated in Figures 13.3 and 13.4.

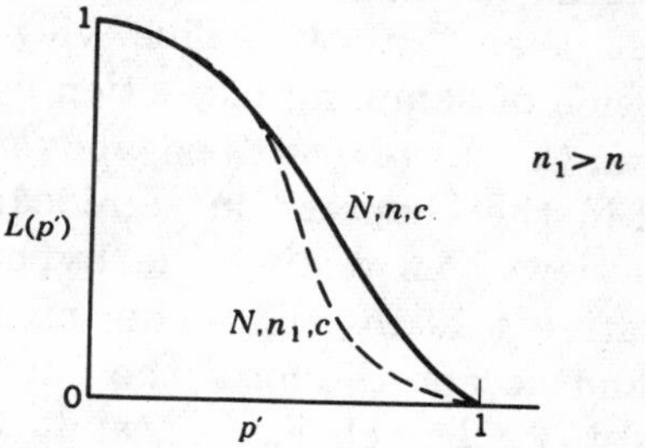

Fig. 13.3. Effect on OC of varying n.

If the lot size N is large compared with the sample size n, the OC curve is essentially independent of the lot size. In other words, if the OC curve of a sampling plan, defined by a sample of size n, a lot of size N, and an acceptance number c, is compared with the OC curve of a sampling plan, defined by a sample of size n, a lot of size N_1 ($N_1 > N$), and an acceptance number c, the difference is negli-

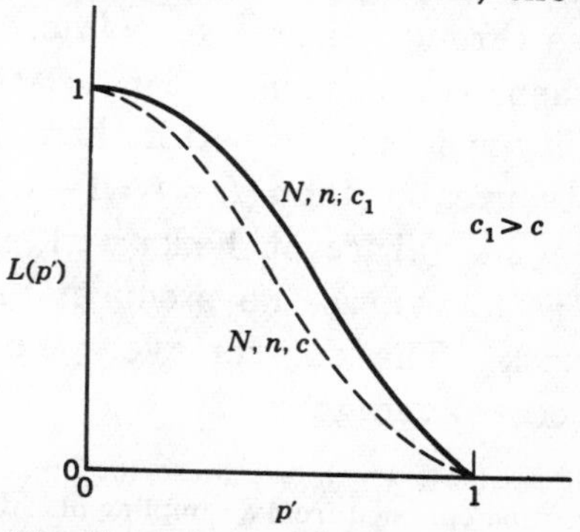

Fig. 13.4. Effect on OC of varying c.

gible if the sample size is small compared with the lot size.[1] Since this is the situation in most industrial applications, we will make this assumption throughout the rest of this section. As a result, a single sampling plan will now be defined by only two numbers, the sample size n and acceptance number c, and the probability distribution of the number of defectives will be given by the binomial distribution.

The single sampling procedure can be seen from another viewpoint. By drawing a sample of n, and looking at the number of defectives d present, the consumer is essentially estimating the fraction defective in the lot; i.e., $p=d/n$ is an estimate of the incoming quality p'. A lot is rejected if this estimate is too high, i.e., if $p>c/n=M$. Therefore, a procedure which says to reject a lot if the number of defectives d is greater than c in a sample of size n is equivalent to rejecting a lot if the estimated fraction defective, $d/n=p$, is greater than $c/n=M$. This viewpoint will be useful later in the discussion of sampling inspection by variables.

The concept of an OC curve described above is identical with the concept presented in the chapters on significance tests. Sampling inspection can be viewed as testing the hypothesis that $p' \leqq p'_o$ against the alternative that $p'>p'_o$. The procedure is to take a sample of size n and accept the lot if the number of defectives d is less than or equal to c. The choice of n and c depends upon the OC curve desired.

13.2.1.2 *Choosing a sampling plan.* The consumer has at his dis posal the choice of one of many OC curves. The one chosen shoul reflect his views as to the cost of making wrong decisions. By varying n and c, he can always find an OC curve which will pas through two preassigned points. In other words, by specifying th points $[p'_1, L(p'_1)]$ and $[p'_2, L(p'_2)]$ an n and a c can be found such tha the OC curve passes through these two points. Before using sam pling inspection, then, the consumer can locate two points p'_1 an p'_2 such that if quality is submitted better than p'_1, he will accept th lot with probability greater than $L(p'_1)=1-\alpha$; if quality is sul mitted worse than p'_2 he will accept the lot with probability less tha $L(p'_2)=\beta$. Here α is known as the producer's risk and β is know as the consumer's risk. The long-run average of submitted qualit is known as the process average.

[1] The notion of taking a sample which is a prescribed percentage of the lot is u satisfactory. The protection obtained from a sampling plan depends upon the absolu size of the sample rather than the relative size.

13.2.1.3 *Calculation of* OC *curves for single sampling plans.* It was pointed out previously that a sampling plan is defined by two numbers: n, the sample size, and c, the acceptance number. Thus the sampling procedure is to accept a lot if the number of defectives d in a sample of n does not exceed c. The OC curve is defined by this procedure. If lot quality is submitted at p', the probability of accepting the lot is given by $L(p')$, where $L(p') = P(0) + P(1) + P(2) + \cdots + P(c)$ and $P(i)$ is the probability of i defectives in the sample of n. Hence,

$$L(p') = \sum_{d=0}^{c} \binom{n}{d} (p')^d (1 - p')^{n-d}$$

$$= (1 - p')^n + \binom{n}{1}(p')(1 - p')^{n-1} + \binom{n}{2}(p')^2(1 - p')^{n-2}$$

$$+ \cdots + \binom{n}{c}(p')^c (1 - p')^{n-c}$$

where

$$\binom{n}{d} = \frac{n!}{d!(n-d)!},$$

and 0! is defined to be equal to 1.

Such calculations as the one above are often cumbersome, and an approximation is desirable. Approximate answers may be obtained very rapidly by the use of the Poisson distribution, which is tabulated in Table 13.1.[1] The Poisson approximation to the binomial is discussed in Chapter IV, Section 4.5.

13.2.1.4 *Example.* What is the probability of accepting a lot whose incoming quality is 4.0% defective, using a sample of size 30 and an acceptance number $c = 1$?

$$L(0.04) = \sum_{d=0}^{1} \binom{30}{d} (0.04)^d (0.96)^{30-d}$$

$$= (0.96)^{30} + \frac{(30)!}{(29)!\,(1)!} 0.04(0.96)^{29} = 0.661.$$

Using the Poisson approximation,

$$np' = 30 \times 0.04 = 1.20; \qquad c = 1; \qquad L(0.04) = 0.663$$

which, in this case, is in good agreement with the correct value.

[1] $L(p')$ is approximately the value read out of Table 13.1 with entries c and np'. The larger the n and smaller the p', the closer the approximate answer is to the true probability.

Table 13.1. Summation of Terms of Poisson's Exponential Binomial Limit[1]
(1,000 × probability of c or less occurrences of event that has an average number of occurrences equal to c' or np')

c' or np' \ c	0	1	2	3	4	5	6	7	8	9
0.02	980	1,000								
0.04	961	999	1,000							
0.06	942	998	1,000							
0.08	923	997	1,000							
0.10	905	995	1,000							
0.15	861	990	999	1,000						
0.20	819	982	999	1,000						
0.25	779	974	998	1,000						
0.30	741	963	996	1,000						
0.35	705	951	994	1,000						
0.40	670	938	992	999	1,000					
0.45	638	925	989	999	1,000					
0.50	607	910	986	998	1,000					
0.55	577	894	982	998	1,000					
0.60	549	878	977	997	1,000					
0.65	522	861	972	996	999	1,000				
0.70	497	844	966	994	999	1,000				
0.75	472	827	959	993	999	1,000				
0.80	449	809	953	991	999	1,000				
0.85	427	791	945	989	998	1,000				
0.90	407	772	937	987	998	1,000				
0.95	387	754	929	984	997	1,000				
1.00	368	736	920	981	996	999	1,000			
1.1	333	699	900	974	995	999	1,000			
1.2	301	663	879	966	992	998	1,000			
1.3	273	627	857	957	989	998	1,000			
1.4	247	592	833	946	986	997	999	1,000		
1.5	223	558	809	934	981	996	999	1,000		
1.6	202	525	783	921	976	994	999	1,000		
1.7	183	493	757	907	970	992	998	1,000		
1.8	165	463	731	891	964	990	997	999	1,000	
1.9	150	434	704	875	956	987	997	999	1,000	
2.0	135	406	677	857	947	983	995	999	1,000	

[1] Reprinted by permission from E. L. Grant, *Statistical Quality Control*, 2nd ed., McGraw-Hill Book Company, Inc., New York, 1952.

Table 13.1. (cont.) Summation of Terms of Poisson's Exponential Binomial Limit

c' or *np'* \ *c*	0	1	2	3	4	5	6	7	8	9
2.2	111	355	623	819	928	975	993	998	1,000	
2.4	091	308	570	779	904	964	988	997	999	1,000
2.6	074	267	518	736	877	951	983	995	999	1,000
2.8	061	231	469	692	848	935	976	992	998	999
3.0	050	199	423	647	815	916	966	988	996	999
3.2	041	171	380	603	781	895	955	983	994	998
3.4	033	147	340	558	744	871	942	977	992	997
3.6	027	126	303	515	706	844	927	969	988	996
3.8	022	107	269	473	668	816	909	960	984	994
4.0	018	092	238	433	629	785	889	949	979	992
4.2	015	078	210	395	590	753	867	936	972	989
4.4	012	066	185	359	551	720	844	921	964	985
4.6	010	056	163	326	513	686	818	905	955	980
4.8	008	048	143	294	476	651	791	887	944	975
5.0	007	040	125	265	440	616	762	867	932	968
5.2	006	034	109	238	406	581	732	845	918	960
5.4	005	029	095	213	373	546	702	822	903	951
5.6	004	024	082	191	342	512	670	797	886	941
5.8	003	021	072	170	313	478	638	771	867	929
6.0	002	017	062	151	285	446	606	744	847	916

	10	11	12	13	14	15	16
2.8	1,000						
3.0	1,000						
3.2	1,000						
3.4	999	1,000					
3.6	999	1,000					
3.8	998	999	1,000				
4.0	997	999	1,000				
4.2	996	999	1,000				
4.4	994	998	999	1,000			
4.6	992	997	999	1,000			
4.8	990	996	999	1,000			
5.0	986	995	998	999	1,000		
5.2	982	993	997	999	1,000		
5.4	977	990	996	999	1,000		
5.6	972	988	995	998	999	1,000	
5.8	965	984	993	997	999	1,000	
6.0	957	980	991	996	999	999	1,000

Table 13.1. (cont.) Summation of Terms of Poisson's Exponential Binomial Limit

c' or np' \ c	0	1	2	3	4	5	6	7	8	9
6.2	002	015	054	134	259	414	574	716	826	902
6.4	002	012	046	119	235	384	542	687	803	886
6.6	001	010	040	105	213	355	511	658	780	869
6.8	001	009	034	093	192	327	480	628	755	850
7.0	001	007	030	082	173	301	450	599	729	830
7.2	001	006	025	072	156	276	420	569	703	810
7.4	001	005	022	063	140	253	392	539	676	788
7.6	001	004	019	055	125	231	365	510	648	765
7.8	000	004	016	048	112	210	338	481	620	741
8.0	000	003	014	042	100	191	313	453	593	717
8.5	000	002	009	030	074	150	256	386	523	653
9.0	000	001	006	021	055	116	207	324	456	587
9.5	000	001	004	015	040	089	165	269	392	522
10.0	000	000	003	010	029	067	130	220	333	458
	10	**11**	**12**	**13**	**14**	**15**	**16**	**17**	**18**	**19**
6.2	949	975	989	995	998	999	1,000			
6.4	939	969	986	994	997	999	1,000			
6.6	927	963	982	992	997	999	999	1,000		
6.8	915	955	978	990	996	998	999	1,000		
7.0	901	947	973	987	994	998	999	1,000		
7.2	887	937	967	984	993	997	999	999	1,000	
7.4	871	926	961	980	991	996	998	999	1,000	
7.6	854	915	954	976	989	995	998	999	1,000	
7.8	835	902	945	971	986	993	997	999	1,000	
8.0	816	888	936	966	983	992	996	998	999	1,000
8.5	763	849	909	949	973	986	993	997	999	999
9.0	706	803	876	926	959	978	989	995	998	999
9.5	645	752	836	898	940	967	982	991	996	998
10.0	583	697	792	864	917	951	973	986	993	997
	20	**21**	**22**							
8.5	1,000									
9.0	1,000									
9.5	999	1,000								
10.0	998	999	1,000							

Table 13.1. (cont.) Summation of Terms of Poisson's Exponential Binomial Limit

c / ′ or np'	0	1	2	3	4	5	6	7	8	9
10.5	000	000	002	007	021	050	102	179	279	397
11.0	000	000	001	005	015	038	079	143	232	341
11.5	000	000	001	003	011	028	060	114	191	289
12.0	000	000	001	002	008	020	046	090	155	242
12.5	000	000	000	002	005	015	035	070	125	201
13.0	000	000	000	001	004	011	026	054	100	166
13.5	000	000	000	001	003	008	019	041	079	135
14.0	000	000	000	000	002	006	014	032	062	109
14.5	000	000	000	000	001	004	010	024	048	088
15.0	000	000	000	000	001	003	008	018	037	070

	10	11	12	13	14	15	16	17	18	19
10.5	521	639	742	825	888	932	960	978	988	994
11.0	460	579	689	781	854	907	944	968	982	991
11.5	402	520	633	733	815	878	924	954	974	986
12.0	347	462	576	682	772	844	899	937	963	979
12.5	297	406	519	628	725	806	869	916	948	969
13.0	252	353	463	573	675	764	835	890	930	957
13.5	211	304	409	518	623	718	798	861	908	942
14.0	176	260	358	464	570	669	756	827	883	923
14.5	145	220	311	413	518	619	711	790	853	901
15.0	118	185	268	363	466	568	664	749	819	875

	20	21	22	23	24	25	26	27	28	29
10.5	997	999	999	1,000						
11.0	995	998	999	1,000						
11.5	992	996	998	999	1,000					
12.0	988	994	997	999	999	1,000				
12.5	983	991	995	998	999	999	1,000			
13.0	975	986	992	996	998	999	1,000			
13.5	965	980	989	994	997	998	999	1,000		
14.0	952	971	983	991	995	997	999	999	1,000	
14.5	936	960	976	986	992	996	998	999	999	1,000
15.0	917	947	967	981	989	994	997	998	999	1,000

Table 13.1. (cont.) Summation of Terms of Poisson's Exponential Binomial Limit

c' or np' \ c	4	5	6	7	8	9	10	11	12	13
16	000	001	004	010	022	043	077	127	193	275
17	000	001	002	005	013	026	049	085	135	201
18	000	000	001	003	007	015	030	055	092	143
19	000	000	001	002	004	009	018	035	061	098
20	000	000	000	001	002	005	011	021	039	066
21	000	000	000	000	001	003	006	013	025	043
22	000	000	000	000	001	002	004	008	015	028
23	000	000	000	000	000	001	002	004	009	017
24	000	000	000	000	000	000	001	003	005	011
25	000	000	000	000	000	000	001	001	003	006
	14	15	16	17	18	19	20	21	22	23
16	368	467	566	659	742	812	868	911	942	963
17	281	371	468	564	655	736	805	861	905	937
18	208	287	375	469	562	651	731	799	855	899
19	150	215	292	378	469	561	647	725	793	849
20	105	157	221	297	381	470	559	644	721	787
21	072	111	163	227	302	384	471	558	640	716
22	048	077	117	169	232	306	387	472	556	637
23	031	052	082	123	175	238	310	389	472	555
24	020	034	056	087	128	180	243	314	392	473
25	012	022	038	060	092	134	185	247	318	394
	24	25	26	27	28	29	30	31	32	33
16	978	987	993	996	998	999	999	1,000		
17	959	975	985	991	995	997	999	999	1,000	
18	932	955	972	983	990	994	997	998	999	1,000
19	893	927	951	969	980	988	993	996	998	999
20	843	888	922	948	966	978	987	992	995	997
21	782	838	883	917	944	963	976	985	991	994
22	712	777	832	877	913	940	959	973	983	989
23	635	708	772	827	873	908	936	956	971	981
24	554	632	704	768	823	868	904	932	953	969
25	473	553	629	700	763	818	863	900	929	950
	34	35	36	37	38	39	40	41	42	43
19	999	1,000								
20	999	999	1,000							
21	997	998	999	999	1,000					
22	994	996	998	999	999	1,000				
23	988	993	996	997	999	999	1,000			
24	979	987	992	995	997	998	999	999	1,000	
25	966	978	985	991	994	997	998	999	999	1,000

13.2.2 Double sampling plans

13.2.2.1 *Double sampling.* A double sampling procedure can be characterized by the following: A sample of n_1 items is drawn from a lot; the lot is accepted if there are no more than c_1 defective items. If there are between $c_1 + 1$ and c_2 defective items, a second sample of size n_2 is drawn; the lot is accepted if there are no more than c_2 defectives in the combined sample of $n_1 + n_2$; the lot is rejected if there are more than c_2 defective items in the combined sample of $n_1 + n_2$.

Thus a lot will be accepted on the first sample if the incoming quality is very good. Similarly, a lot will be rejected on the first sample if the incoming quality is very bad. If the lot is of intermediate quality, a second sample may be required. Double sampling plans have the psychological advantage of giving a second chance to doubtful lots. Furthermore, they can have the additional advantage of requiring fewer total inspections, on the average, than single sampling plans for any given quality protection. On the other side of the ledger, double sampling plans can have the following disadvantages: (1) more complex administrative duties; (2) inspectors will have variable inspection loads depending upon whether one or two samples are taken; (3) the maximum amount of inspection exceeds that for single sampling plans (which is constant).

*13.2.2.2 OC *curves for double sampling plans.* A lot will be accepted if and only if (1) the number of defectives d_1 in the first sample of n_1 does not exceed c_1, i.e., $d_1 \leqq c_1$; (2) there are more than c_1 defectives in the first sample but no more than c_2 defectives and the total number of defectives, $d_1 + d_2$, in the combined sample of $n_1 + n_2$ does not exceed c_2, i.e., if $c_1 < d_1 \leqq c_2$, then $d_1 + d_2 \leqq c_2$.

Thus for any incoming quality p', the probability of accepting the lot is

$$
\begin{aligned}
L(p') &= \mathrm{P}(d_1 \leqq c_1 \text{ given } p') \\
&\qquad + \mathrm{P}(d_1 + d_2 \leqq c_2 \text{ given } c_1 < d_1 \leqq c_2;\, p') \\
&= \mathrm{P}(d_1 \leqq c_1 \mid p') + \mathrm{P}\,(d_1 + d_2 \leqq c_2 \mid c_1 < d_1 \leqq c_2;\, p')
\end{aligned}
$$

where the symbol | reads "given" and $\mathrm{P}(d_1 \leqq c_1 \mid p')$ reads "the probability that d_1 is less than or equal to c_1, given p'." The OC curve is then given by the expression

$$L(p') = \sum_{d_1=0}^{c_1} \binom{n_1}{d_1} (p')^{d_1}(1 - p')^{n_1-d_1}$$

$$+ \sum_{d_1=c_1+1}^{c_2} \sum_{d_2=0}^{c_2-d_1} \binom{n_1}{d_1} (p')^{d_1}(1 - p')^{n_1-d_1} \binom{n_2}{d_2} (p')^{d_2}(1 - p')^{n_2-d_2}$$

$$= \sum_{d_1=0}^{c_1} \binom{n_1}{d_1} (p')^{d_1}(1 - p')^{n_1-d_1}$$

$$+ \sum_{d_1=c_1+1}^{c_2} \left[\binom{n_1}{d_1} (p')^{d_1}(1 - p')^{n_1-d_1} \sum_{d_2=0}^{c_2-d_1} \binom{n_2}{d_2} (p')^{d_2}(1 - p')^{n_2-d_2} \right]$$

*13.2.2.3 *Example.* A large lot of $\frac{3}{4}$-inch screws is submitte for sampling inspection by means of double sampling. If the firs sample of 20 contains no defectives, the lot is to be accepted. If i contains more than two defectives, the lot is to be rejected. Other wise, a second sample of 40 is to be drawn, and the lot is to b accepted if the total number of defectives in both samples does no exceed two. This plan can be characterized by the following table

Sample	Sample Size	Combined Samples		
		Size	Acceptance Number	Rejection Number
First	20	20	0	3
Second	40	60	2	3

Find the probability of accepting the lot if p′ = 5%.

(1) $\sum_{d_1=0}^{0} \binom{n_1}{d_1} (p')^{d_1}(1 - p')^{n_1-d_1} = (1 - p')^{20} = (0.95)^{20} = 0.358$

(2) $\binom{n_1}{1}(p')(1-p')^{n_1-1} \sum_{d_2=0}^{1} \binom{n_2}{d_2} (p')^{d_2}(1 - p')^{n_2-d_2}$

$$= \binom{n_1}{1}(p')(1 - p')^{n_1-1}[(1 - p')^{n_2} + \binom{n_2}{1}(p')(1 - p')^{n_2-1}]$$

$$= \binom{20}{1} 0.05(0.95)^{19}[(0.95)^{40} + \binom{40}{1} 0.05(0.95)^{39}]$$

$$= 0.377[0.129 + 0.270] = 0.150$$

(3) $\binom{n_1}{2}(p')^2(1 - p')^{n_1-2} \sum_{d_2=0}^{0} \binom{n_2}{d_2} (p')^{d_2}(1 - p')^{n_2-d_2}$

$$= \binom{n_1}{2}(p')^2(1 - p')^{n_1-2}(1 - p')^{n_2}$$

$$= \binom{20}{2}(0.05)^2(0.95)^{18}(0.95)^{40} = 0.189 \times 0.129 = 0.02438.$$

Thus, $L(p') = (1) + (2) + (3) = 0.532.$

Using the Poisson approximation we have

(1) 0.368;

(2)* $(0.736 - 0.368)(0.406) = 0.368 \times 0.406 = 0.149$;

(3) $(0.920 - 0.736)(0.135) = 0.184 \times 0.135 = 0.025$.

Thus,

$$L(p') = (1) + (2) + (3) = 0.542.$$

13.2.3 MULTIPLE SAMPLING PLANS

A multiple sampling procedure is represented in Table 13.2.

Table 13.2. Multiple Sampling Plan

Sample	Sample Size	Combined Samples		
		Size	Acceptance Number	Rejection Number
First	n_1	n_1	c_1	r_1
Second	n_2	$n_1 + n_2$	c_2	r_2
Third	n_3	$n_1 + n_2 + n_3$	c_3	r_3
Fourth	n_4	$n_1 + n_2 + n_3 + n_4$	c_4	r_4
Fifth	n_5	$n_1 + n_2 + n_3 + n_4 + n_5$	c_5	r_5
Sixth	n_6	$n_1 + n_2 + n_3 + n_4 + n_5 + n_6$	c_6	r_6
Seventh	n_7	$n_1 + n_2 + n_3 + n_4 + n_5 + n_6 + n_7$	c_7	c_7+1

A first sample of n_1 is drawn. The lot is accepted if there are no more than c_1 defectives; the lot is rejected if there are r_1 or more defectives. Otherwise, a second sample of n_2 is drawn. The lot is accepted if there are no more than c_2 defectives in the combined sample of $n_1 + n_2$. The lot is rejected if there are r_2 or more defectives in the combined sample of $n_1 + n_2$. The procedure is continued in accordance with the table above. If, by the end of the sixth sample, the lot is neither accepted nor rejected, a sample of n_7 is drawn. The lot is accepted if the number of defectives in the combined sample of $n_1 + n_2 + n_3 + n_4 + n_5 + n_6 + n_7$ does not exceed c_7. Otherwise, the lot is rejected. Note that $c_1 < c_2 < \cdots < c_7$ and $c_i < r_i$, for all i. Of course, plans can be devised which permit any number of samples before a decision is reached. A multiple sampling plan will generally involve less inspection, on the average,

* The Poisson table is set up to give $\sum_{d=0}^{c} \binom{n}{d} (p')^d (1 - p')^{n-d}$. To get an individual term, say $\binom{n}{c}(p')^c(1 - p')^{n-c}$, we take

$$\sum_{d=0}^{c} \binom{n}{d} (p')^d (1 - p')^{n-d} - \sum_{d=0}^{c-1} \binom{n}{d} (p')^d (1 - p')^{n-d} = \binom{n}{c} (p')^c (1 - p')^{n-c}.$$

than the corresponding single or double sampling plan guaranteein the same protection. The advantage is quite important, sin inspection costs are directly related to sample sizes. On the oth hand, some of the disadvantages of multiple sampling plans are: (they usually require higher administrative costs than single double sampling plans; (2) the variability of the inspection load intr duces difficulties in scheduling inspection time; (3) higher calib inspection personnel may be necessary to guarantee proper use the plans; (4) adequate storage facilities must be provided for th lot while multiple sampling is being carried on.

13.2.4 Classification of sampling plans

Published tables of sampling plans have been classified into fou categories, namely, by acceptable quality level (AQL), by l tolerance percent defective (LTPD), by point of control, and b average outgoing quality limit (AOQL).

13.2.4.1 *Classification by* AQL. Acceptance procedures based o the acceptable quality level (AQL) generally make use of the proce average to determine the sampling plan to be used. The AQL ma be viewed as the highest percent defective that is acceptable as process average. In normal sampling, a lot at AQL quality wi have a high probability of acceptance. The probability of acce tance, $1 - \alpha$, at the AQL is usually set near the 95% point, and is known as the producer's risk. Thus, a producer has good prote tion against rejection of submitted lots from a process that is at th AQL or better. On the other hand, this type of classification do not specify anything about the protection the consumer has again the acceptance of a lot worse than the AQL.

13.2.4.2 *Classification by* LTPD. The lot tolerance percent defe tive (LTPD) usually refers to that incoming quality above whic there is a small chance that a lot will be accepted. This probabilit is usually taken to be near the 10% point. Thus, a consumer i specting lots submitted from a process that is at the LTPD or wors has a small probability of accepting such lots. This probability known as the consumer's risk. On the other hand, this type c classification does not specify anything about the protection th producer has against the rejection of lots better than the LTPD.

13.2.4.3 *Classification by point of control.* The point of contr is the lot quality for which the probability of acceptance is 0.5 The concept here is one of splitting the risk between producer an consumer. Lots submitted from processes whose quality is bett than the point of control have a probability of acceptance that is higher than 50%. Lots submitted from processes whose quality is worse than the point of control have a probability of acceptance smaller than 50%.

13.2.4.4 *Classification by* AOQL. The average outgoing quality limit (AOQL) does not refer to a point on the OC curve, but rather to the upper limit on outgoing quality that may be expected in the long run when all rejected lots are subjected to 100% inspection, with all defective articles removed and replaced by good articles. The average outgoing quality limit (AOQL) can be computed from the formulas

$$AOQ = p'L(p'); \qquad AOQL = \max_{p'} AOQ$$

where $\max_{p'}$ AOQ is the maximum AOQ over the range $p' = 0$ to $p' = 1$.

Figure 13.5 shows a graphic comparison of these four methods of indexing acceptance plans in relation to a stated quality standard. The four curves shown in this figure are OC curves for single sampling plans that are indexed in their respective tables by the 1.0% defective figure. The most lenient plan (curve IV) is that classified

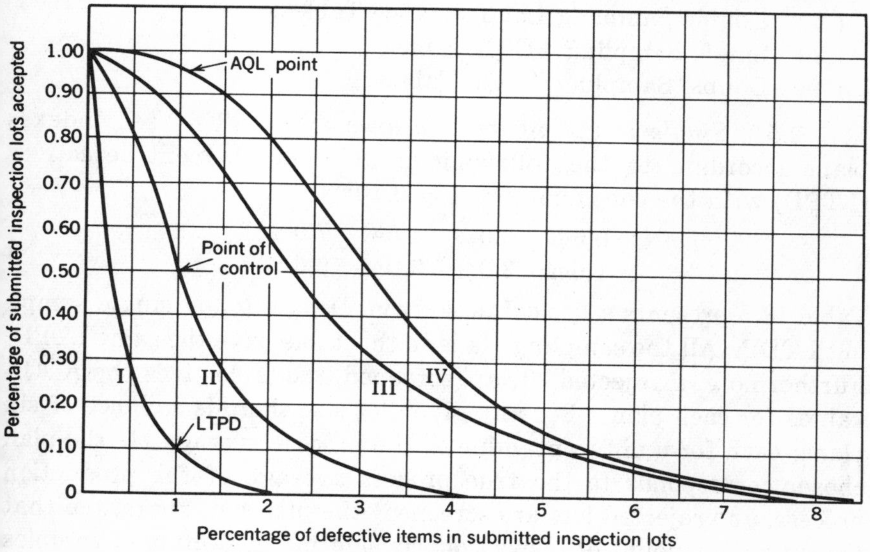

Fig. 13.5. Single sampling classified on
I. 1% LTPD: $n = 225$; $c = 50$.
II. Point of control = 1%: $n = 50$; $c = 1$.
III. 1% AOQL: $n = 75$; $c = 1$.
IV. AQL = 1%: $n = 150$; $c = 4$.

by the AQL; in this plan with an AQL of 1.0%, four defectives ar permitted in a sample of 150. The strictest plan (curve I) is th LTPD plan permitting no defectives in a sample of 225. The poin of control plan (curve II) permits 1 defective in a sample of 150 here the probability of acceptance of a 1.0% defective lot is 0.50 The 1.0% AOQL plan allows 1 defective in a sample of 75; the cor responding OC curve (curve III) is intermediate between the AQI 1.0% and the point of control 1.0% plans.

13.2.5 Dodge-Romig Tables

In 1944, H. F. Dodge and H. G. Romig published a volume o attribute sampling tables called *Sampling Inspection Tables.*[1] Thes tables originated in the Bell Telephone Laboratories and reflect man years of experience with acceptance sampling in the Bell Telephon System. It should be pointed out that these tables were originall prepared for use within the Bell Telephone System. They wer designed primarily to minimize the total amount of inspection, con sidering total sampling inspection and screening inspection of rejecte lots. The Dodge-Romig volume contains four sets of tables, a follows:

(1) Single Sampling Lot Tolerance Tables;
(2) Double Sampling Lot Tolerance Tables;
(3) Single Sampling AOQL Tables;
(4) Double Sampling AOQL Tables.

13.2.5.1 *Single sampling lot tolerance tables.* This set indexe plans according to the following lot tolerance percent defective (LTPD) with the consumer's risk = 0.10:

0.5%	2.0%	4.0%	7.0%
1.0%	3.0%	5.0%	10.0%.

Table 13.3 presents a typical table from Dodge-Romig single sampl ing LTDP. All the sampling plans in this table have the same LTPD Furthermore, if rejected lots are screened, the table gives the AOQI values for each plan. For any given lot size there is a choice of six plans, each for a different value of the process average. If the plar chosen corresponds to the true process average of the productior process, and rejected lots are screened, the plan will guarantee that the average amount of inspection, considering inspection of samples and screening of rejected lots, will be smaller than for any of the

[1] H. F. Dodge and H. G. Romig, *Sampling Inspection Tables*, John Wiley & Sons, Inc., New York, 1944.

Table 13.3. Example of Dodge-Romig Single Sampling Lot Tolerance Tables[1]

(Lot tolerance percent defective = 5.0%; consumer's risk = 0.10)

Process average %	0–0.05			0.06–0.50			0.51–1.00			1.01–1.50			1.51–2.00			2.01–2.50		
Lot size	*n*	*c*	AOQL %	*n*	*c*	AOQL %	*n*	*c*	AOQL %	*n*	*c*	AOQL %	*n*	*c*	AOQL %	*n*	*c*	AOQL %
1–30	All	0	0	All	0	0	All	0	0	All	0	0	All	0	0	All	0	0
31–50	30	0	0.49	30	0	0.49	30	0	0.49	30	0	0.49	30	0	0.49	30	0	0.49
51–100	37	0	0.63	37	0	0.63	37	0	0.63	37	0	0.63	37	0	0.63	37	0	0.63
101–200	40	0	0.74	40	0	0.74	40	0	0.74	40	0	0.74	40	0	0.74	40	0	0.74
201–300	43	0	0.74	43	0	0.74	70	1	0.92	70	1	0.92	95	2	0.99	95	2	0.99
301–400	44	0	0.74	44	0	0.74	70	1	0.99	100	2	1.0	120	3	1.1	145	4	1.1
401–500	45	0	0.75	75	1	0.95	100	2	1.1	100	2	1.1	125	3	1.2	150	4	1.2
501–600	45	0	0.76	75	1	0.98	100	2	1.1	125	3	1.2	150	4	1.3	175	5	1.3
601–800	45	0	0.77	75	1	1.0	100	2	1.2	130	3	1.2	175	5	1.4	200	6	1.4
801–1,000	45	0	0.78	75	1	1.0	105	2	1.2	155	4	1.4	180	5	1.4	225	7	1.5
1,001–2,000	45	0	0.80	75	1	1.0	130	3	1.4	180	5	1.6	230	7	1.7	280	9	1.8
2,001–3,000	75	1	1.1	105	2	1.3	135	3	1.4	210	6	1.7	280	9	1.9	370	13	2.1
3,001–4,000	75	1	1.1	105	2	1.3	160	4	1.5	210	6	1.7	305	10	2.0	420	15	2.2
4,001–5,000	75	1	1.1	105	2	1.3	160	4	1.5	235	7	1.8	330	11	2.0	440	16	2.2
5,001–7,000	75	1	1.1	105	2	1.3	185	5	1.7	260	8	1.9	350	12	2.2	490	18	2.4
7,001–10,000	75	1	1.1	105	2	1.3	185	5	1.7	260	8	1.9	380	13	2.2	535	20	2.5
0,001–20,000	75	1	1.1	135	3	1.4	210	6	1.8	285	9	2.0	425	15	2.3	610	23	2.6
0,001–50,000	75	1	1.1	135	3	1.4	235	7	1.9	305	10	2.1	470	17	2.4	700	27	2.7
0,001–100,000	75	1	1.1	160	4	1.6	235	7	1.9	355	12	2.2	515	19	2.5	770	30	2.8

[1] Reprinted by permission from H. F. Dodge and H. Romig, *Sampling Inspection Tables*, John Wiley and Sons, Inc., New York, 1944.

Table 13.4. Example of Dodge-Romig Double Sampling Lot Tolerance Tables[1]

(Lot tolerance percent defective = 5.0% ; consumer's risk = 0.10)

Process average %	0–0.05						0.06–0.50						2.01–2.50					
	Trial 1		Trial 2			AOQL in %	Trial 1		Trial 2			AOQL in %	Trial 1		Trial 2			AOQL in %
Lot size	n_1	c_1	n_2	n_1+n_2	c_2		n_1	c_1	n_2	n_1+n_2	c_2		n_1	c_1	n_2	n_1+n_2	c_2	
1–30	All	0	..	..	.	0	All	0	..	..	.	0	All	0	..	..	.	0
31–50	30	0	..	..	.	0.49	30	0	..	..	.	0.49	30	0	..	..	.	0.49
51–75	38	0	..	..	.	0.59	38	0	..	..	.	0.59	38	0	..	..	.	0.59
76–100	44	0	21	65	1	0.64	44	0	21	65	1	0.64	44	0	21	65	1	0.64
101–200	49	0	26	75	1	0.84	49	0	26	75	1	0.84	49	0	51	100	2	0.91
201–300	50	0	30	80	1	0.91	50	0	30	80	1	0.91	50	0	100	150	4	1.1
301–400	55	0	30	85	1	0.92	55	0	55	110	2	1.1	85	0	105	190	6	1.3
401–500	55	0	30	85	1	0.93	55	0	55	110	2	1.1	85	1	140	225	7	1.4
501–600	55	0	30	85	1	0.94	55	0	60	115	2	1.1	85	1	165	250	8	1.5
601–800	55	0	35	90	1	0.95	55	0	65	120	2	1.1	120	2	185	305	10	1.6
801–1,000	55	0	35	90	1	0.96	55	0	65	120	2	1.1	120	2	210	330	11	1.7
1,001–2,000	55	0	35	90	1	0.98	55	0	95	150	3	1.3	175	4	260	435	15	2.0
2,001–3,000	55	0	65	120	2	1.2	55	0	95	150	3	1.3	205	5	375	580	21	2.3
3,001–4,000	55	0	65	120	2	1.2	55	0	95	150	3	1.3	230	6	420	650	24	2.4
4,001–5,000	55	0	65	120	2	1.2	55	0	95	150	3	1.4	255	7	445	700	26	2.5
5,001–7,000	55	0	65	120	2	1.2	55	0	95	150	3	1.4	255	7	495	750	28	2.6
7,001–10,000	55	0	65	120	2	1.2	55	0	120	175	4	1.5	280	8	540	820	31	2.7
10,001–20,000	55	0	65	120	2	1.2	55	0	120	175	4	1.5	280	8	660	940	36	2.8
20,001–50,000	55	0	65	120	2	1.2	55	0	150	205	5	1.7	305	9	745	1050	41	2.9
50,001–100,000	55	0	65	120	2	1.2	55	0	150	205	5	1.7	330	10	810	1140	45	3.0

remaining five plans.[1] The process average is usually estimated from past data by taking the ratio of the total number of defectives to the total number inspected. If there are no past data to estimate the process average, the plan should be selected from the right-hand column of the table. This gives good lots a better chance of acceptance.

It must be emphasized that if screening of rejected lots does not take place, these plans still guarantee a stated LTPD.

13.2.5.2 *Double sampling lot tolerance tables.* This set indexes plans according to the same LTPD's as for the single sampling LTPD plans. A typical double sampling LTPD table is shown in Table 13.4. This table has the same features as the single sampling LTPD tables with the exception that it uses double sampling instead of single sampling. The first sample is always smaller than the corresponding single sampling plan, but if a second sample is required, the combined sample size exceeds the sample size of the single sampling plan. However, the average amount of sampling inspection is generally smaller with the double sampling plan. Since the second sample is not always taken (depending on the submitted quality), the process average is usually estimated from the first sample only.

13.2.5.3 *Single sampling* AOQL *tables.* This set indexes plans according to the following average outgoing quality limits (AOQL).

0.1%	1.5%	4.0%
0.25%	2.0%	5.0%
0.5%	2.5%	7.0%
0.75%	3.0%	10.0%
1.0%		

A typical single sampling AOQL table is shown in Table 13.5.

All the sampling plans in this table have the same AOQL, assuming all rejected lots are screened. Furthermore, the table gives the LTPD values for each plan. Like all the Dodge-Romig plans these plans have the property that the average amount of inspection is smallest for that plan which belongs to the class under whose heading the true process average falls.

13.2.5.4 *Double sampling* AOQL *tables.* This set indexes plans according to the same AOQL's as for the single sampling AOQL plans. A typical double sampling AOQL table is shown in Table 13.6. This

[1] The average amount of inspection (AOI) is given by the formula

$$\text{AOI} = nL(p') + N[1 - L(p')].$$

Table 13.5. Example of Dodge-Romig Single Sampling AOQL Tables[1]

(Average outgoing quality limit = 2.0%)

Process average %	0–0.04			0.05–0.40			0.41–0.80			0.81–1.20			1.21–1.60			1.61–2.00		
Lot size	n	c	p_t%	n	c	p_t%	n	c	p_t%	n	c	p_t%	n	c	p_t%	n	c	p_t%
1–15	All	0	....	All	0	....	All	0	....	All	0	....	All	0	....	All	0	....
16–50	14	0	13.6	14	0	13.6	14	0	13.6	14	0	13.6	14	0	13.6	14	0	13.6
51–100	16	0	12.4	16	0	12.4	16	0	12.4	16	0	12.4	16	0	12.4	16	0	12.4
101–200	17	0	12.2	17	0	12.2	17	0	12.2	17	0	12.2	35	1	10.5	35	1	10.5
201–300	17	0	12.3	17	0	12.3	17	0	12.3	37	1	10.2	37	1	10.2	37	1	10.2
301–400	18	0	11.8	18	0	11.8	38	1	10.0	38	1	10.0	38	1	10.0	60	2	8.5
401–500	18	0	11.9	18	0	11.9	39	1	9.8	39	1	9.8	60	2	8.6	60	2	8.6
501–600	18	0	11.9	18	0	11.9	39	1	9.8	39	1	9.8	60	2	8.6	60	2	8.6
601–800	18	0	11.9	40	1	9.6	40	1	9.6	65	2	8.0	65	2	8.0	85	3	7.5
801–1,000	18	0	12.0	40	1	9.6	40	1	9.6	65	2	8.1	65	2	8.1	90	3	7.4
1,001–2,000	18	0	12.0	41	1	9.4	65	2	8.2	65	2	8.2	95	3	7.0	120	4	6.5
2,001–3,000	18	0	12.0	41	1	9.4	65	2	8.2	95	3	7.0	120	4	6.5	180	6	5.8
3,001–4,000	18	0	12.0	42	1	9.3	65	2	8.2	95	3	7.0	155	5	6.0	210	7	5.5
4,001–5,000	18	0	12.0	42	1	9.3	70	2	7.5	125	4	6.4	155	5	6.0	245	8	5.3
5,001–7,000	18	0	12.0	42	1	9.3	95	3	7.0	125	4	6.4	185	6	5.6	280	9	5.1
7,001–10,000	42	1	9.3	70	2	7.5	95	3	7.0	155	5	6.0	220	7	5.4	350	11	4.8
10,001–20,000	42	1	9.3	70	2	7.6	95	3	7.0	190	6	5.6	290	9	4.9	460	14	4.4
20,001–50,000	42	1	9.3	70	2	7.6	125	4	6.4	220	7	5.4	395	12	4.5	720	21	3.9
50,001–100,000	42	1	9.3	95	3	7.6	160	5	5.9	290	9	4.9	505	15	4.2	955	27	3.7

[1] Reprinted by permission from H. F. Dodge and H. Romig, *Sampling Inspection Tables*, John Wiley and Sons, Inc., New York, 1944.

(Average outgoing quality limit = 2.0%)

Process average %	0–0.04						0.05–0.40						1.61–2.00					
	Trial 1		Trial 2				Trial 1		Trial 2				Trial 1		Trial 2			
Lot size	n_1	c_1	n_2	n_1+n_2	c_2	p_t%	n_1	c_1	n_2	n_1+n_2	c_2	p_t%	n_1	c_1	n_2	n_1+n_2	c_2	p_t%
1–15	All	0	..	..	.	...	All	0	..	..	.	...	All	0	..	..	.	...
16–50	14	0	..	..	.	13.6	14	0	..	..	.	13.6	14	0	..	..	.	13.6
51–100	21	0	12	33	1	11.7	21	0	12	33	1	11.7	23	0	23	46	2	10.9
101–200	24	0	13	37	1	11.0	24	0	13	37	1	11.0	27	0	28	55	2	9.6
201–300	26	0	15	41	1	10.4	26	0	15	41	1	10.4	32	0	48	80	3	8.4
301–400	26	0	16	42	1	10.3	26	0	16	42	1	10.3	36	0	69	105	4	7.6
401–500	27	0	16	43	1	10.3	30	0	35	65	2	9.0	60	1	90	150	6	7.0
501–600	27	0	16	43	1	10.3	31	0	34	65	2	8.9	65	1	95	160	6	6.8
601–800	27	0	17	44	1	10.2	31	0	39	70	2	8.8	70	1	120	190	7	6.4
801–1,000	27	0	17	44	1	10.2	32	0	38	70	2	8.7	70	1	145	215	8	6.2
1,001–2,000	33	0	37	70	2	8.5	33	0	37	70	2	8.5	110	2	205	315	11	5.5
2,001–3,000	34	0	41	75	2	8.2	34	0	41	75	2	8.2	160	3	310	470	15	4.7
3,001–4,000	34	0	41	75	2	8.2	38	0	62	100	3	7.3	235	5	415	650	20	4.3
4,001–5,000	34	0	41	75	2	8.2	38	0	62	100	3	7.3	275	6	475	750	23	4.2
5,001–7,000	35	0	40	75	2	8.1	38	0	62	100	3	7.3	280	6	575	855	26	4.1
7,001–10,000	35	0	40	75	2	8.1	38	0	62	100	3	7.3	320	7	645	965	29	4.0
10,001–20,000	35	0	40	75	2	8.1	39	0	66	105	3	7.2	395	9	835	1,230	37	3.9
20,001–50,000	35	0	40	75	2	8.1	43	0	92	135	4	6.6	480	11	1,090	1,570	46	3.7
50,001–100,000	35	0	45	80	2	8.0	43	0	92	135	4	6.6	580	13	1,460	2,040	58	3.5

[1] Reprinted by permission from H. F. Dodge and H. Romig, *Sampling Inspection Tables*, John Wiley and Sons, Inc., New York, 1944.

table has the same features as the single sampling AOQL tables except that it uses double sampling instead of single sampling.

13.2.6 MILITARY STANDARD 105B

13.2.6.1 *History.* There have been various sets of published sam pling tables using the AQL as an index. The first set of tables wa developed in 1943 for Army Ordnance, and with some changes an extensions, became the Army Service Forces tables, used durin the final years of World War II.

An administration manual, *Standard Sampling Inspection Pro cedures,* was issued by the Navy in October, 1945. This manual including tables and OC (operating characteristic) curves for all th acceptance sampling plans given, had been prepared by the Statis tical Research Group of Columbia University during World War II Although the tables and procedures in this manual were similar i many respects to those of the Army Service Forces, sequentia (multiple) sampling plans were included in addition to single an double sampling plans, and there were several other important point of difference. The actual tables from this manual were issued b the Navy Department in April, 1946, as Appendix X to *Genera Specifications for Inspection of Material.* After the unification o the armed services, Appendix X was adopted by the Department o Defense in February, 1949, as JAN-STD-105.

In 1949, a book entitled *Sampling Inspection,* edited by H. A. Free man, M. Friedman, F. Mosteller, and W. A. Wallis[1] was published This book was prepared by the Statistical Research Group of Colum bia University, and with a few minor differences, the tables i *Sampling Inspection* are identical with JAN-STD-105.

JAN-STD-105 was superseded by MIL-STD-105A in 1950. Mino revisions resulted in MIL-STD-105B, dated December, 1958. Thi new version deletes many references to the government and incor porates special procedures for small sample inspection. Discussio of these special procedures will be omitted. It is evident that th new standard, like its predecessor, will doubtless have a grea influence on acceptance procedures used in industry.

MIL-STD-105B is available for private use and is for sale by th Superintendent of Documents, U. S. Government Printing Office Washington 25, D. C. Most of the remaining paragraphs in thi

[1] H. A. Freeman, M. Friedman, F. Mosteller, and W. A. Wallis, *Sampling Inspec tion,* McGraw-Hill Book Company, Inc., New York, 1946.

section on sampling inspection by attributes will be concerned with a discussion of this document.

13.2.6.2 *Classification of defects.* Defects are grouped into three classes, i.e., critical, major, and minor. A critical defect "is one that judgment and experience indicate could result in hazardous or unsafe conditions for individuals using or maintaining the product; or for major end item units of product, such as ships, aircraft, or tanks, a defect that could prevent performance of their tactical functions." A major defect "is a defect, other than critical, that could result in failure, or materially reduce the usability of the unit of product for its intended purpose." A minor defect "is one that does not materially reduce the usability of the unit of product for its intended purpose, or is a departure from established standards having no significant bearing on the effective use or operation of the unit."

13.2.6.3 *Acceptable quality levels.* The acceptable quality level (AQL) "is a nominal value expressed in terms of percent defective or defects per hundred units, whichever is applicable, specified for a given group of defects of a product." The distinction between a defect and a defective is readily seen if we define a defective item as one containing one or more defects. If the percentage defective is relatively small, more than one defect will occur on an item infrequently. Consequently, in this case, the mathematical theory using defects is essentially equivalent to using defectives. The value of the AQL's are

0.015	0.25	2.5	25.0	250.0
0.035	0.40	4.0	40.0	400.0
0.065	0.65	6.5	65.0	650.0
0.10	1.0	10.0	100.0	1,000.0
0.15	1.5	15.0	150.0	

Here AQL values of 10.0 or less are expressed either in percent defective or in defects per hundred units; those over 10.0 are expressed in defects per hundred units only. These points are approximately in geometric progression and multiples of 1, 1.5, 2.5, 4.0 and 6.5. The probabilities of acceptance of submitted lots having these AQL's range from about 0.80 for the smallest sample sizes to about 0.998 for the largest sample sizes. The particular AQL values to be used for a given product shall be specified by the Government. The Government may at its option specify a separate AQL for all the defects of a given class considered collectively or for any particular type of defect considered individually. A single AQL value may be specified for the group of all defects for which a product is inspected, or separate AQL values may be specified for each of two or more subgroups of the defects.

13.2.6.4 *Normal, tightened, and reduced inspection.* The government determines whether to use normal, tightened, or reduced inspection at the start of a contract. During the course of a contract, the government determines whether to use normal, tightened, or reduced inspection as follows:

Normal inspection is used when the estimated process average is not outside the applicable upper and lower limits shown in Table 13.7 (Table II A and II C from MIL-STD-105B). These limits (in percent) are obtained from the relation

$$\text{Upper Limit} = \text{AQL} + 3\sqrt{\frac{\text{AQL}(100 - \text{AQL})}{n}}$$

$$\text{Lower Limit} = \text{AQL} - 3\sqrt{\frac{\text{AQL}(100 - \text{AQL})}{n}}$$

where AQL is read in percent.

Tightened inspection is instituted when the estimated process average exceeds the applicable upper limit shown in Table 13.7 (Table II C of MIL-STD-105B). Normal inspection is reinstated when the estimated process average is equal to or less than the AQL while tightened inspection is in effect. Under tightened inspection, the producer's risk is increased, whereas the consumer's risk is decreased.

In other words, the probability of accepting bad lots (as well as good lots) is decreased. This is accomplished by reducing the acceptance numbers while keeping the sample size fixed.

Reduced inspection is instituted, if the government so desires, provided that all the following conditions are satified.

"Condition A. The preceding 10 lots have been under normal inspection and none have been rejected.

"Condition B. The estimated process average is less than the applicable lower limit shown in [Table 13.7(Table II A of MIL-STD-105B)].

"Condition C. Production is at a steady rate."

Normal inspection is reinstated if any one of the following conditions occurs while reduced inspection is in effect.

"Condition A. A lot is rejected.

"Condition B. The estimated process average is greater than the AQL.

"Condition C. Production becomes irregular or delayed.

"Condition D. The Government deems that normal inspection should be reinstated."

Under reduced inspection, the sample size is decreased, thereby increasing the consumer's risk and decreasing (slightly) the producer's risk.

The use of tightened and reduced inspection is an important feature in the success of MIL-STD-105B. If quality is submitted close to but above the AQL, there is still a relatively large probability that it will be accepted. Although this appears to be harmful on the surface, in reality it is not too damaging. In the first place, rejected lots cost the manufacturer a great deal of money. In fact, too many rejected lots can easily force a manufacturer out of business. For example, few manufacturers can tolerate even a 20% rejection rate for their lots. Second, if lots are constantly being submitted above the AQL, this will be reflected in the process average, which results in tightened inspection. This will cause even more of the manufacturer's lots to be rejected. Consequently, the manufacturer is actually forced to keep his submitted quality better than the agreed-upon level.

On the other hand, if the manufacturer continues to submit quality a great deal better than the AQL, the government is able to use reduced inspection, which results in a saving of inspection costs.

13.2.6.5 *Sampling plans.* Sample sizes are designated by code letters from A to Q. The sample size code letter depends on the inspection level and the lot size. There are three inspection levels, I, II, and III. Unless the government specifies otherwise, inspection level II is used. The sample size code letter applicable to the specified inspection level and for lots of given size is obtained from Table 13.8 (Table III A of MIL-STD-105B).

Since it has been pointed out that the OC curves are essentially independent of lot size (for large lots), it may appear incongruous that one of the entries to the table is lot size. However, a moment's reflection will reveal that to the inspection agency the acceptance of a bad lot is much more serious when the lot size is large than when it is small. Consequently, better protection in the form of steeper OC curves is desired for large lots.

The appropriate master sampling table is selected as follows:

For normal or tightened inspection:

AQL values of 10.0 or less

Single sampling: Table 13.9 (Table IVA of MIL-STD-105B)
Double sampling: Table 13.10 (Table IVB of MIL-STD-105B)
Multiple sampling: Table 13.11 (Table IVC of MIL-STD-105B)

Table 13.7. Limits of the Process Average
(Lower limits for AQL's from 0.015 to 4.0; Table II A of MIL-STD-105B)

Number of sample units included in estimated process average	Acceptable quality levels											
	0.015	0.035	0.065	0.10	0.15	0.25	0.40	0.65	1.0	1.5	2.5	4.0
25–34	★	★	★	★	★	★	★	★	★	★	★	★
35–49	★	★	★	★	★	★	★	★	★	★	★	★
50–74	★	★	★	★	★	★	★	★	★	★	★	★
75–99	★	★	★	★	★	★	★	★	★	★	★	★
100–124	★	★	★	★	★	★	★	★	★	★	★	★
125–149	★	★	★	★	★	★	★	★	★	★	★	★
150–199	★	★	★	★	★	★	★	★	★	★	★	★
200–249	★	★	★	★	★	★	★	★	★	★	★	★
250–299	★	★	★	★	★	★	★	★	★	★	★	0.38
300–349	★	★	★	★	★	★	★	★	★	★	★	0.67
350–399	★	★	★	★	★	★	★	★	★	★	0.05	0.90
400–449	★	★	★	★	★	★	★	★	★	★	0.20	1.09
450–549	★	★	★	★	★	★	★	★	★	★	0.38	1.32
550–649	★	★	★	★	★	★	★	★	★	★	0.56	1.55
650–749	★	★	★	★	★	★	★	★	★	0.11	0.71	1.73
750–899	★	★	★	★	★	★	★	★	★	0.22	0.85	1.91
900–1,099	★	★	★	★	★	★	★	★	0.05	0.34	1.00	2.10
1,100–1,299	★	★	★	★	★	★	★	★	0.13	0.44	1.13	2.27
1,300–1,499	★	★	★	★	★	★	★	0.004	0.20	0.52	1.23	2.40
1,500–1,699	★	★	★	★	★	★	★	0.045	0.25	0.58	1.31	2.50
1,700–1,899	★	★	★	★	★	★	★	0.080	0.29	0.63	1.41	2.59
1,900–2,249	★	★	★	★	★	★	★	0.119	0.34	0.69	1.46	2.68
2,250–2,749	★	★	★	★	★	★	0.021	0.166	0.40	0.77	1.55	2.80
2,750–3,499	★	★	★	★	★	★	0.061	0.217	0.46	0.84	1.65	2.93
3,500–4,999	★	★	★	★	★	0.020	0.109	0.279	0.54	0.94	1.77	3.08
5,000–6,999	★	★	★	★	★	0.056	0.155	0.338	0.61	1.03	1.89	3.23
7,000–8,999	★	★	★	★	0.020	0.188	0.188	0.380	0.66	1.09	1.97	3.33
9,000–10,999	★	★	★	0.005	0.034	0.100	0.210	0.408	0.70	1.13	2.03	3.40
11,000–13,499	★	★	★	0.015	0.046	0.116	0.230	0.434	0.73	1.17	2.08	3.46
13,500–17,499	★	★	0.003	0.024	0.057	0.129	0.248	0.456	0.76	1.20	2.12	3.52
17,500–22,499	★	★	0.011	0.033	0.068	0.144	0.266	0.470	0.79	1.24	2.16	3.58

Table 13.7. (cont.) Limits of the Process Average

(Lower limits for AQL's from 6.5 to 1,000.0; Table II A of MIL-STD-105B)

Number of sample units included in estimated process average	Acceptable quality levels											
	6.5	10.0	15.0	25.0	40.0	65.0	100.0	150.0	250.0	400.0	650.0	1,000.0
25–34	★	★	★	★	5.07	20.47	44.8	82.4	162.7	289.5	509.2	825.3
35–49	★	★	★	1.86	10.72	27.67	53.7	93.3	176.8	307.4	532.0	853.6
50–74	★	★	0.25	5.95	15.90	34.28	61.9	103.3	189.8	323.8	552.9	879.5
75–99	★	★	2.54	8.92	19.66	39.07	67.8	110.6	199.1	335.7	568.0	898.3
100–124	★	1.04	4.02	10.83	22.07	42.14	71.7	115.3	205.2	343.3	577.7	910.4
125–149	★	1.89	5.07	12.18	23.79	44.33	74.4	118.6	209.5	348.7	584.7	918.9
150–199	0.71	2.82	6.20	13.64	25.64	46.69	77.3	122.2	214.1	354.6	592.1	928.2
200–249	1.40	3.67	7.24	14.99	27.34	48.86	80.0	125.5	218.3	360.0	599.0	936.7
250–299	1.88	4.27	7.99	15.95	28.55	50.40	81.9	127.8	221.4	363.8	603.8	942.7
300–349	2.25	4.73	8.55	16.67	29.47	51.57	83.3	129.6	223.7	366.7	607.5	947.3
350–399	2.55	5.10	9.00	17.25	30.17	52.50	84.5	131.0	225.5	369.0	610.5	951.0
400–449	2.79	5.40	9.36	17.72	30.79	53.26	85.4	132.2	227.0	370.9	612.9	954.0
450–549	3.08	5.76	9.80	18.29	31.51	54.18	86.6	133.6	228.8	373.2	615.8	957.6
550–649	3.38	6.13	10.25	18.87	32.25	55.12	87.7	135.0	230.6	375.5	618.8	961.3
650–749	3.61	6.41	10.61	19.33	32.83	55.86	88.7	136.1	232.1	377.3	621.1	964.1
750–899	3.84	6.70	10.95	19.78	33.39	56.58	89.6	137.2	233.5	379.1	623.4	967.0
900–1,099	4.08	7.00	11.32	20.26	34.00	57.35	90.5	138.4	235.0	381.0	625.8	970.0
1,100–1,299	4.29	7.26	11.64	20.67	34.52	58.02	91.3	139.4	236.3	382.7	627.9	972.6
1,300–1,499	4.46	7.46	11.89	20.99	34.93	58.53	92.0	140.2	237.3	384.0	629.6	974.6
1,500–1,699	4.59	7.63	12.09	21.25	35.26	58.95	92.5	140.8	238.1	385.0	630.9	976.3
1,700–1,899	4.70	7.76	12.26	21.46	35.53	59.30	92.9	141.3	238.8	385.9	632.0	†
1,900–2,249	4.82	7.92	12.45	21.71	35.83	59.69	93.4	141.9	239.6	386.8	†	†
2,250–2,749	4.97	8.10	12.68	22.00	36.21	60.16	94.0	142.7	240.5	†	†	†
2,750–3,499	5.13	8.30	12.92	22.32	36.61	60.67	94.6	143.4	†	†	†	†
3,500–4,999	5.33	8.54	13.22	22.70	37.09	61.29	95.4	†	†	†	†	†
5,000–6,999	5.51	8.78	13.50	23.06	37.55	61.88	†	†	†	†	†	†
7,000–8,999	5.64	8.94	13.70	23.32	37.88	†	†	†	†	†	†	†
9,000–10,999	5.73	9.05	13.84	23.50	†	†	†	†	†	†	†	†
11,000–13,499	5.82	9.15	13.96	†	†	†	†	†	†	†	†	†
13,500–17,499	5.89	9.24	†	†	†	†	†	†	†	†	†	†
17,500–22,499	5.96	†	†	†	†	†	†	†	†	†	†	†
22,500 and up	†	†	†	†	†	†	†	†	†	†	†	†

★ Number of sample units included in estimated process average is insufficient for reduced inspection.
† Number of sample units included in estimated process average is too great. Discard older results.

Table 13.7. (cont.) Limits of the Process Average

(Upper limits for AQL's from 0.015 to 4.0; Table IIC of MIL-STD-105B)

Number of sample units included in estimated process average	Acceptable quality levels											
	0.015	0.035	0.065	0.10	0.15	0.25	0.40	0.65	1.0	1.5	2.5	4.0
25–34	‡	‡	‡	‡	‡	‡	‡	5.103	6.52	8.27	11.23	15.05
35–49	‡	‡	‡	‡	‡	‡	3.328	4.383	5.63	7.17	9.82	13.26
50–74	‡	‡	‡	‡	‡	2.155	2.810	3.722	4.81	6.17	8.52	11.62
75–99	‡	‡	‡	‡	1.396	1.858	2.434	3.243	4.22	5.44	7.59	10.43
100–124	‡	‡	‡	0.996	1.248	1.667	2.193	2.935	3.83	4.97	6.98	9.67
125–149	‡	‡	‡	0.911	1.143	1.532	2.021	2.716	3.56	4.64	6.55	9.13
150–199	‡	‡	0.644	0.818	1.030	1.386	1.836	2.481	3.27	4.28	6.09	8.54
200–249	‡	0.410	0.575	0.733	0.926	1.251	1.666	2.264	3.00	3.95	5.67	8.00
250–299	‡	0.374	0.527	0.673	0.851	1.155	1.545	2.110	2.81	3.72	5.36	7.62
300–349	0.219	0.347	0.490	0.627	0.795	1.083	1.453	1.993	2.67	3.54	5.13	7.33
350–399	0.205	0.325	0.460	0.590	0.750	1.025	1.380	1.900	2.55	3.40	4.95	7.10
400–449	0.193	0.307	0.436	0.561	0.714	0.978	1.321	1.824	2.46	3.28	4.80	6.91
450–549	0.179	0.286	0.407	0.525	0.670	0.921	1.249	1.732	2.34	3.14	4.62	6.68
550–649	0.165	0.264	0.377	0.488	0.625	0.863	1.175	1.638	2.23	3.00	4.44	6.45
650–749	0.151	0.247	0.354	0.459	0.589	0.817	1.117	1.564	2.13	2.89	4.29	6.27
750–899	0.143	0.231	0.331	0.430	0.555	0.772	1.061	1.492	2.04	2.78	4.15	6.09
900–1,099	0.131	0.213	0.307	0.400	0.518	0.724	1.000	1.415	1.95	2.66	4.00	5.90
1,100–1,299	0.121	0.197	0.286	0.374	0.485	0.683	0.948	1.348	1.87	2.56	3.87	5.73
1,300–1,499	0.113	0.185	0.270	0.354	0.461	0.651	0.907	1.296	1.80	2.48	3.77	5.60
1,500–1,699	0.107	0.175	0.256	0.337	0.440	0.625	0.874	1.255	1.75	2.42	3.69	5.50
1,700–1,899	0.102	0.167	0.245	0.324	0.424	0.604	0.847	1.220	1.71	2.37	3.59	5.41
1,900–2,249	0.096	0.158	0.233	0.308	0.405	0.579	0.817	1.181	1.66	2.31	3.54	5.32
2,250–2,749	0.089	0.147	0.218	0.290	0.383	0.550	0.779	1.134	1.60	2.23	3.45	5.20
2,750–3,499	0.081	0.136	0.202	0.270	0.358	0.518	0.739	1.083	1.54	2.16	3.35	5.07
3,500–4,999	0.071	0.121	0.182	0.246	0.328	0.480	0.691	1.021	1.46	2.06	3.23	4.92
5,000–6,999	0.062	0.108	0.164	0.222	0.300	0.444	0.645	0.962	1.39	1.97	3.11	4.77
7,000–8,999	0.056	0.098	0.151	0.206	0.280	0.418	0.612	0.920	1.34	1.91	3.03	4.67
9,000–10,999	0.052	0.091	0.142	0.195	0.266	0.400	0.590	0.892	1.30	1.87	2.97	4.60
11,000–13,499	0.048	0.085	0.133	0.185	0.254	0.384	0.570	0.866	1.27	1.83	2.92	4.54
13,500–17,499	0.044	0.080	0.127	0.176	0.243	0.371	0.552	0.844	1.24	1.80	2.88	4.48
17,500–22,499	0.041	0.075	0.119	0.167	0.232	0.356	0.534	0.821	1.21	1.76	2.84	4.42

Table 13.7. (cont.) Limits of the Process Average

(Upper limits for AQL's from 6.5 to 1,000.0; Table IIC of MIL-STD-105B)

Number of sample units included in estimated process average	Acceptable quality levels											
	6.5	10.0	15.0	25.0	40.0	65.0	100.0	150.0	250.0	400.0	650.0	1,000.0
25–34	20.58	27.47	36.39	52.62	74.93	109.53	155.2	217.6	337.3	510.5	790.8	1,174.7
35–49	18.30	24.64	32.93	48.14	69.28	102.33	146.3	206.7	323.2	492.6	768.0	1,146.4
50–74	16.21	22.05	29.75	44.05	64.10	95.72	138.1	196.7	310.2	476.2	747.1	1,120.5
75–99	14.70	20.17	27.46	41.08	60.34	90.93	132.2	189.4	300.9	464.3	732.0	1,101.7
100–124	13.73	18.96	25.98	39.17	57.93	87.86	128.3	184.7	294.8	456.7	722.3	1,089.6
125–149	13.03	18.11	24.93	37.82	56.21	85.67	125.6	181.4	290.5	451.3	715.3	1,081.1
150–199	12.29	17.18	23.80	36.36	54.36	83.31	122.7	177.8	285.9	445.4	707.9	1,071.8
200–249	11.60	16.33	22.76	35.01	52.66	81.14	120.0	174.5	281.7	440.0	701.0	1,063.3
250–299	11.12	15.73	22.01	34.05	51.45	79.60	118.1	172.2	278.6	436.2	696.2	1,057.3
300–349	10.75	15.27	21.45	33.33	50.53	78.43	116.7	170.4	276.3	433.3	692.5	1,052.7
350–399	10.45	14.90	21.00	32.75	49.83	77.50	115.5	169.0	274.5	431.0	689.5	1,049.0
400–449	10.21	14.60	20.64	32.28	49.21	76.74	114.6	167.8	273.0	429.1	687.1	1,046.0
450–549	9.92	14.24	20.20	31.71	48.49	75.82	113.4	166.4	271.2	426.8	684.2	1,042.4
550–649	9.62	13.87	19.75	31.13	47.75	74.88	112.3	165.0	269.4	424.5	681.2	1,038.7
650–749	9.39	13.59	19.39	30.67	47.17	74.14	111.3	163.9	267.9	422.7	678.9	1,035.9
750–899	9.16	13.30	19.05	30.22	46.61	73.42	110.4	162.8	266.5	420.9	676.6	1,033.0
900–1,099	8.92	13.00	18.68	29.74	46.00	72.65	109.5	161.6	265.0	419.0	674.2	1,030.0
1,100–1,299	8.71	12.74	18.36	29.33	45.48	71.98	108.7	160.6	263.7	417.3	672.1	1,027.4
1,300–1,499	8.54	12.54	18.11	29.01	45.07	71.47	108.0	159.8	262.7	416.0	670.4	1,025.4
1,500–1,699	8.41	12.37	17.91	28.75	44.74	71.05	107.5	159.2	261.9	415.0	669.1	1,023.7
1,700–1,899	8.30	12.24	17.74	28.54	44.47	70.70	107.1	158.7	261.2	414.1	668.0	†
1,900–2,249	8.18	12.08	17.55	28.29	44.17	70.31	106.6	158.1	260.4	413.2	†	†
2,250–2,749	8.03	11.90	17.32	28.00	43.79	69.84	106.0	157.3	259.5	†	†	†
2,750–3,499	7.87	11.70	17.08	27.68	43.39	69.33	105.4	156.6	†	†	†	†
3,500–4,999	7.67	11.46	16.78	27.30	42.91	68.71	104.6	†	†	†	†	†
5,000–6,999	7.49	11.22	16.50	26.94	42.45	68.12	†	†	†	†	†	†
7,000–8,999	7.36	11.06	16.30	26.68	42.12	†	†	†	†	†	†	†
9,000–10,999	7.27	10.95	16.16	26.50	†	†	†	†	†	†	†	†
11,000–13,499	7.18	10.85	16.04	†	†	†	†	†	†	†	†	†
13,500–17,499	7.11	10.76	†	†	†	†	†	†	†	†	†	†
17,500–22,499	7.04	†	†	†	†	†	†	†	†	†	†	†
22,500 and up	†	†	†	†	†	†	†	†	†	†	†	†

‡ Normal inspection for these AQL's does not provide sample sizes this small.
† Number of sample units included in estimated process average is too great. Discard older results.

Table 13.8. Sample Size Code Letters
(Table IIIA of MIL-STD-105B)

Lot Size		Inspection Levels		
		I	II	III
2 to	8	A	A	C
9 to	15	A	B	D
16 to	25	B	C	E
26 to	40	B	D	F
41 to	65	C	E	G
66 to	110	D	F	H
111 to	180	E	G	I
181 to	300	F	H	J
301 to	500	G	I	K
501 to	800	H	J	L
801 to	1,300	I	K	L
1,301 to	3,200	J	L	M
3,201 to	8,000	L	M	N
8,001 to	22,000	M	N	O
22,001 to	110,000	N	O	P
110,001 to	550,000	O	P	Q
550,001 and over		P	Q	Q

AQL values greater than 10.0

Single sampling: Table 13.9 (Table IVA of MIL-STD-105B)

For reduced inspection and for all AQL values:

Single sampling: Table 13.12 (Table V of MIL-STD-105B).

The Standard contains the OC curve for each of the plans. Each OC curve corresponds to a single sampling plan, double sampling plan, and multiple sampling plan. These curves are reproduced in Figure 13.6.

13.2.7 Designing your own attribute plan

In many cases, it will not be desirable to select a plan according to the standard procedure of MIL-STD-105B, implying, as it does, a rather arbitrary relation between lot size and sample size. If the characteristics of an OC curve are specified in advance, it is possible to determine the sample size and acceptance number accordingly. This section will deal with procedures for determining sampling plans when two points are specified, and with finding the OC curve of an arbitrary sampling plan.

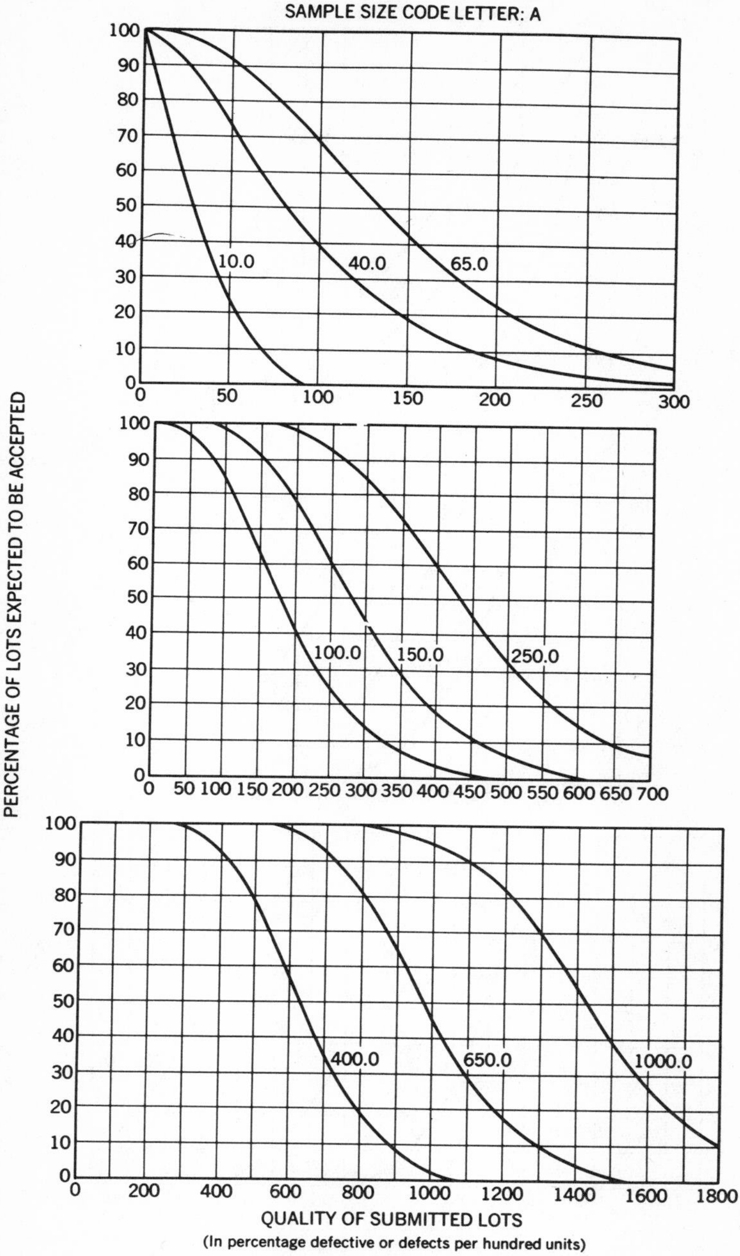

Note: Figures on curves are Acceptable Quality Levels for normal inspection.

Fig. 13.6. OC curves from MIL-STD-105B.

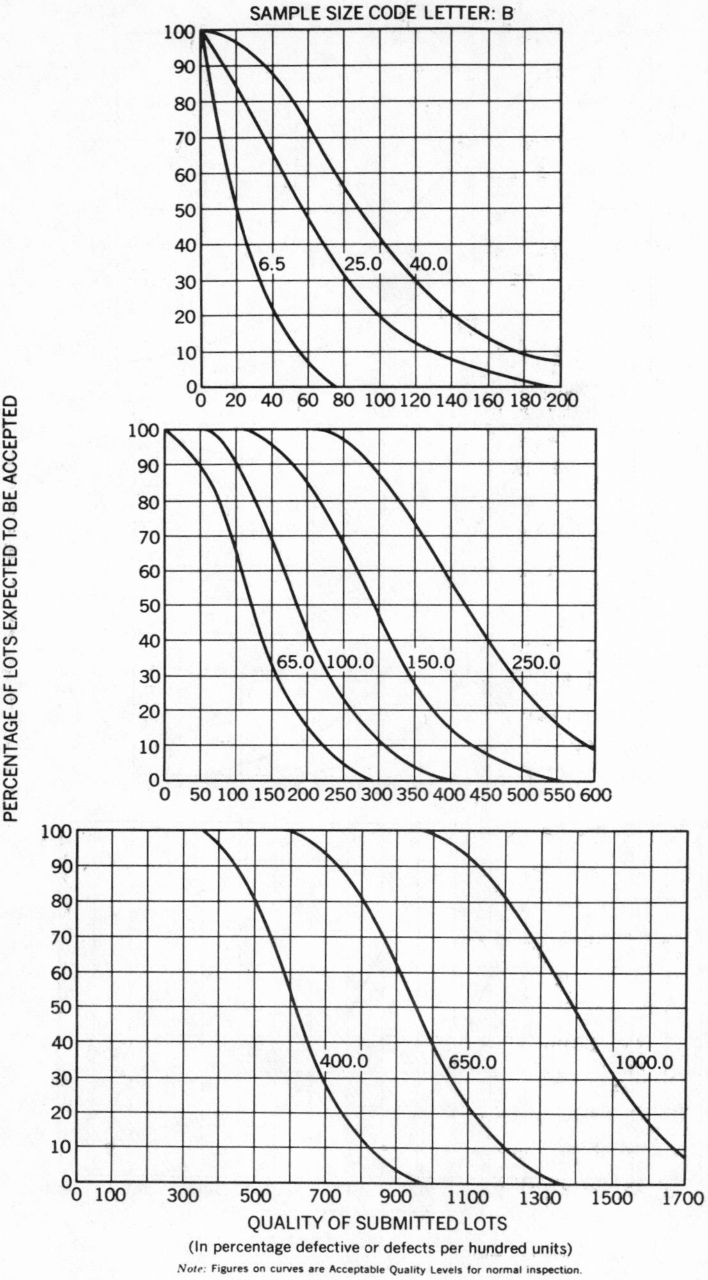

Fig. 13.6. (cont.) OC curves from MIL-STD-105B.

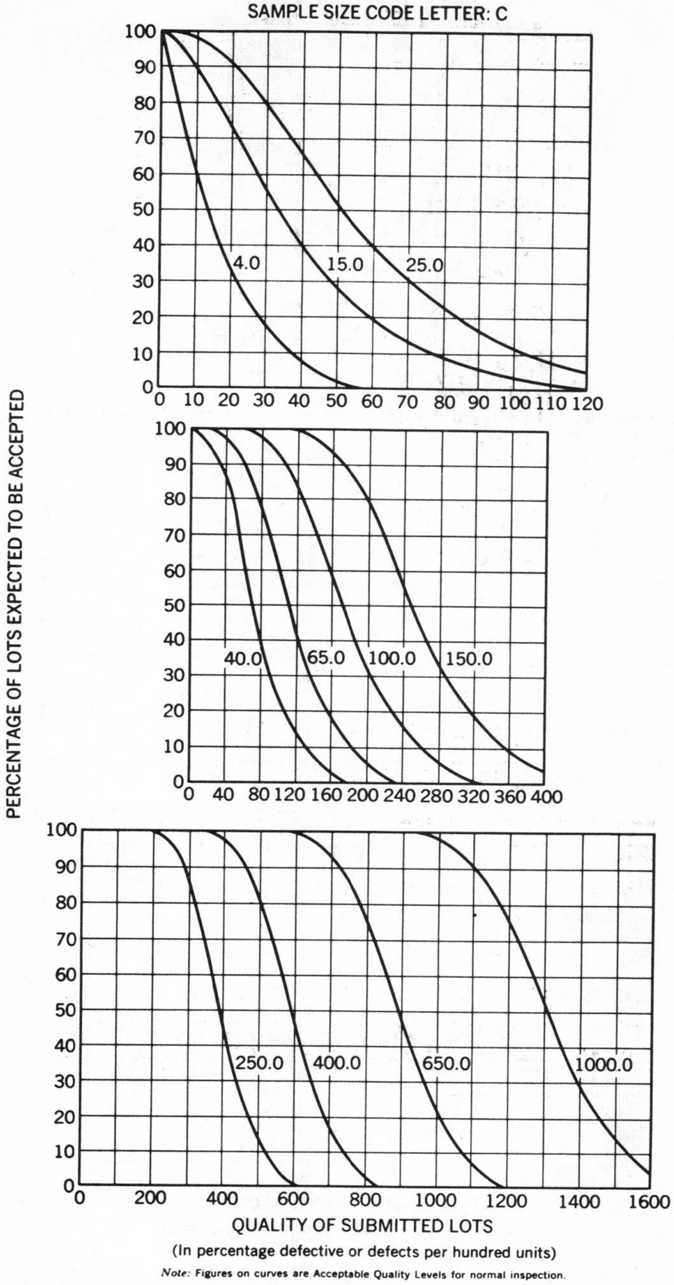

Fig. 13.6. (cont.) OC curves from MIL-STD-105B.

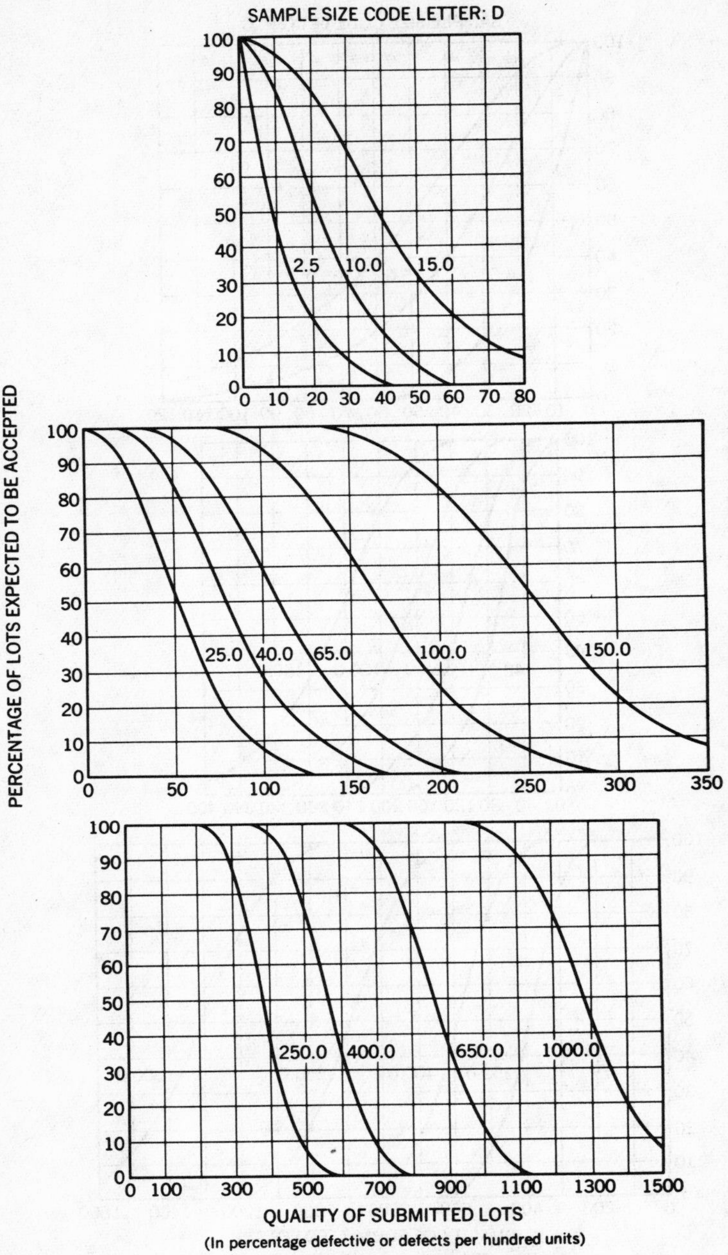

Note: Figures on curves are Acceptable Quality Levels for normal inspection.

Fig. 13.6. (cont.) OC curves from MIL-STD-105B.

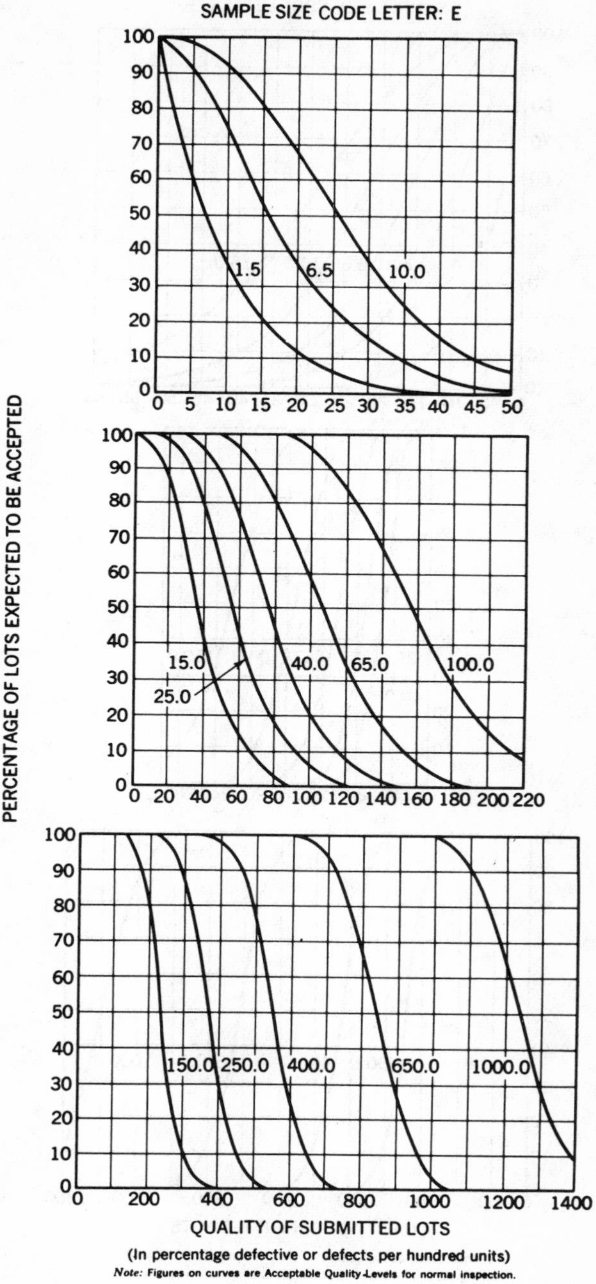

Fig. 13.6. (cont.) OC curves from MIL-STD-105B.

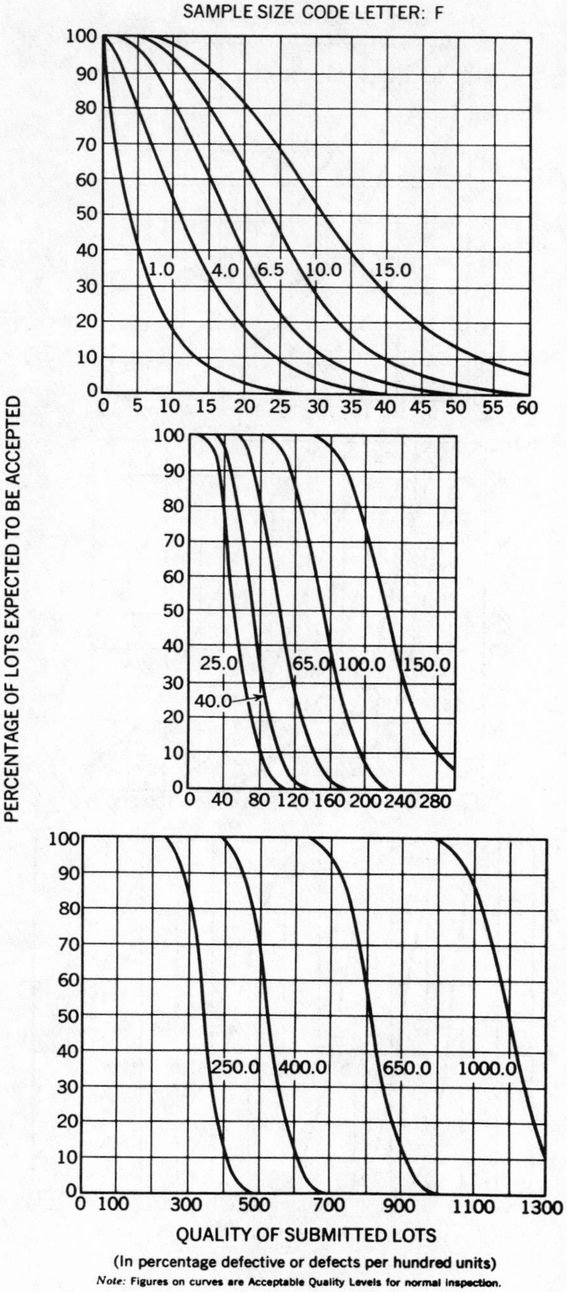

Fig. 13.6. (cont.) OC curves from MIL-STD-105B.

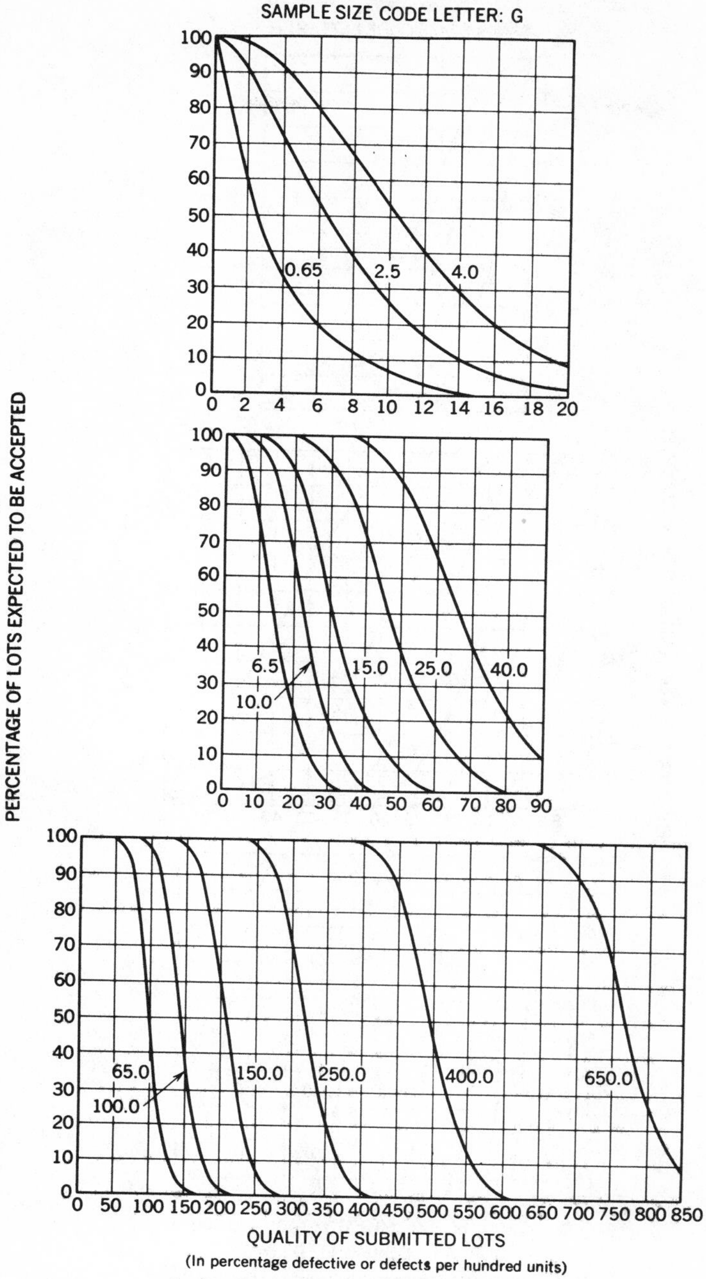

Note: Figures on curves are Acceptable Quality Levels for normal inspection.

Fig. 13.6. (cont.) OC curves from MIL-STD-105B.

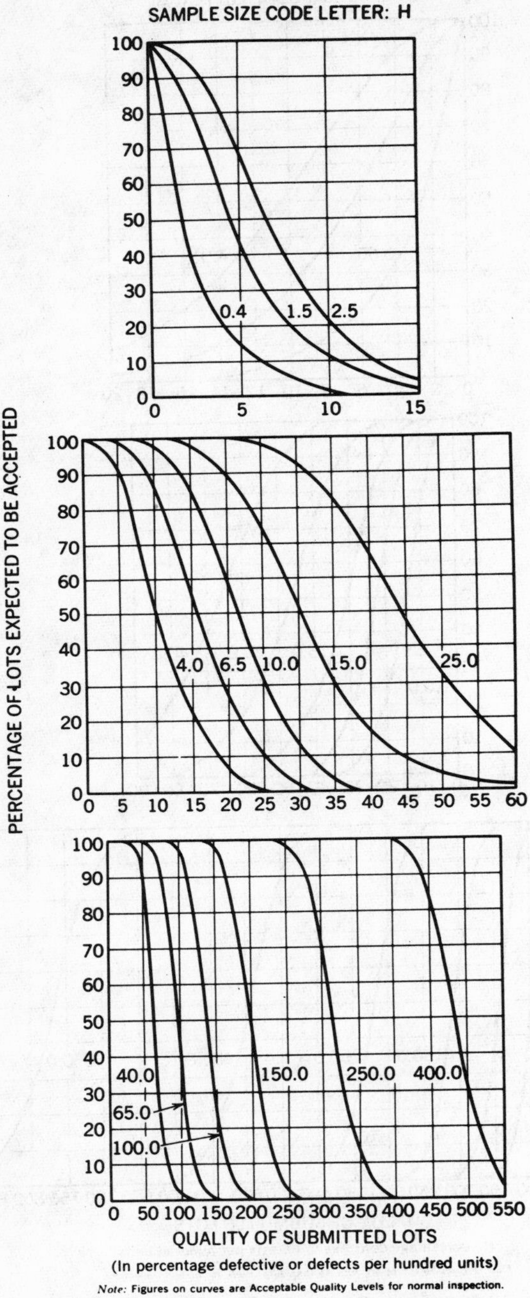

Fig. 13.6. (cont.) OC curves from MIL-STD-105B.

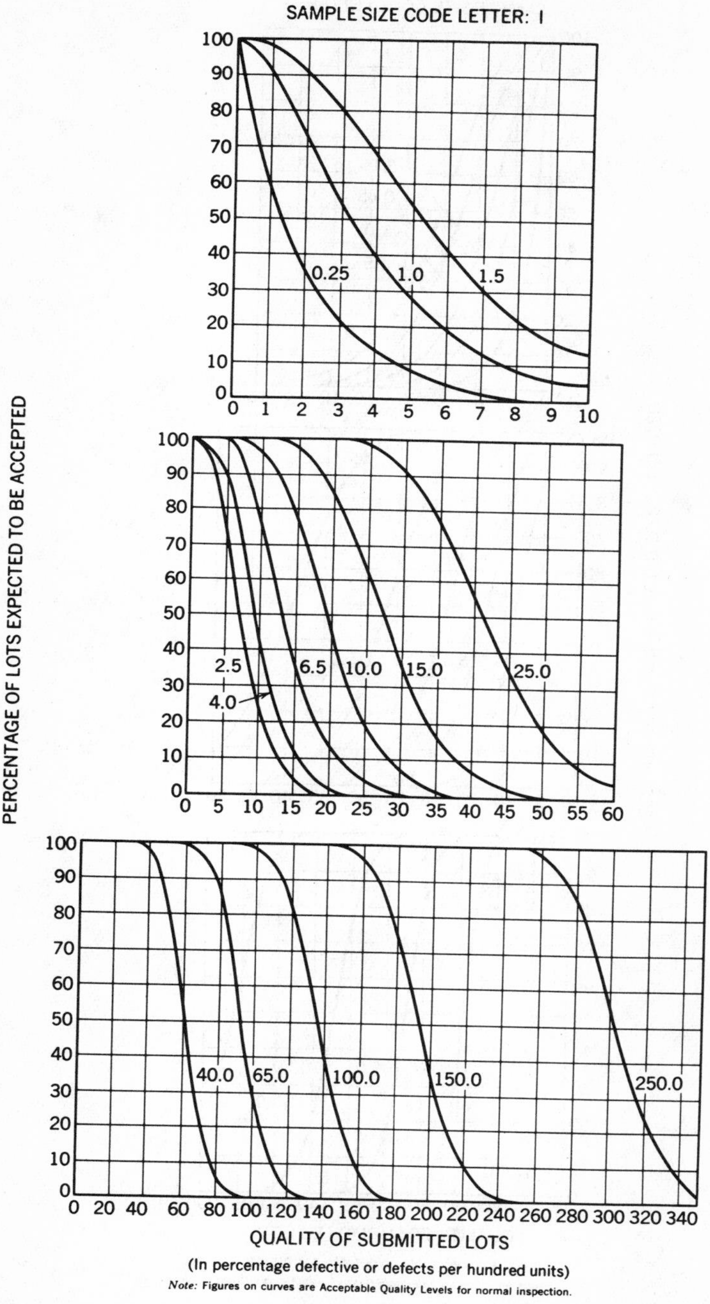

Fig. 13.6. (cont.) OC curves from MIL-STD-105B.

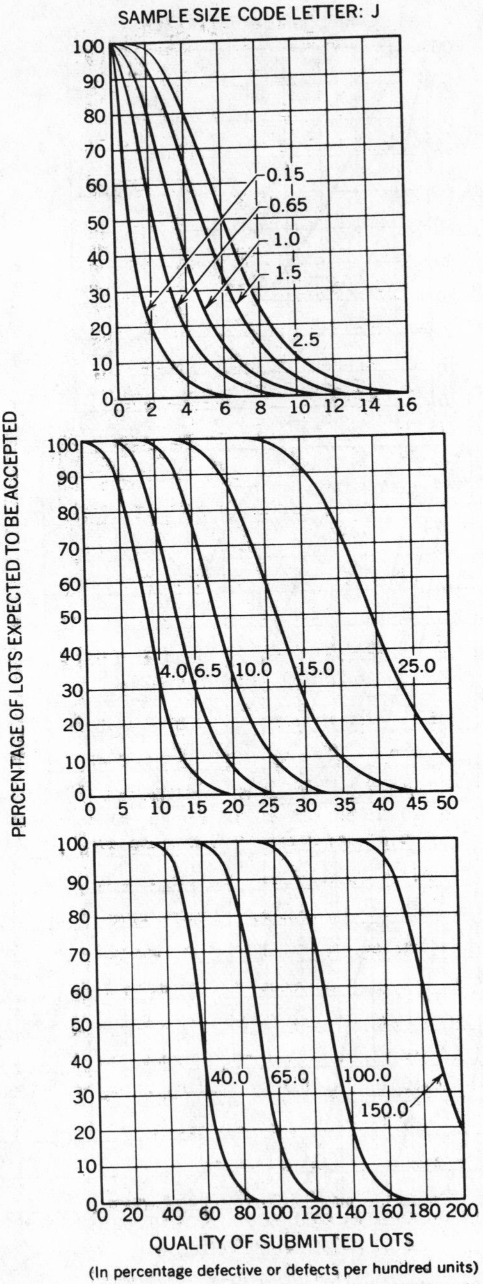

Note: Figures on curves are Acceptable Quality Levels for normal inspection.

Fig. 13.6. (cont.) OC curves from MIL-STD-105B.

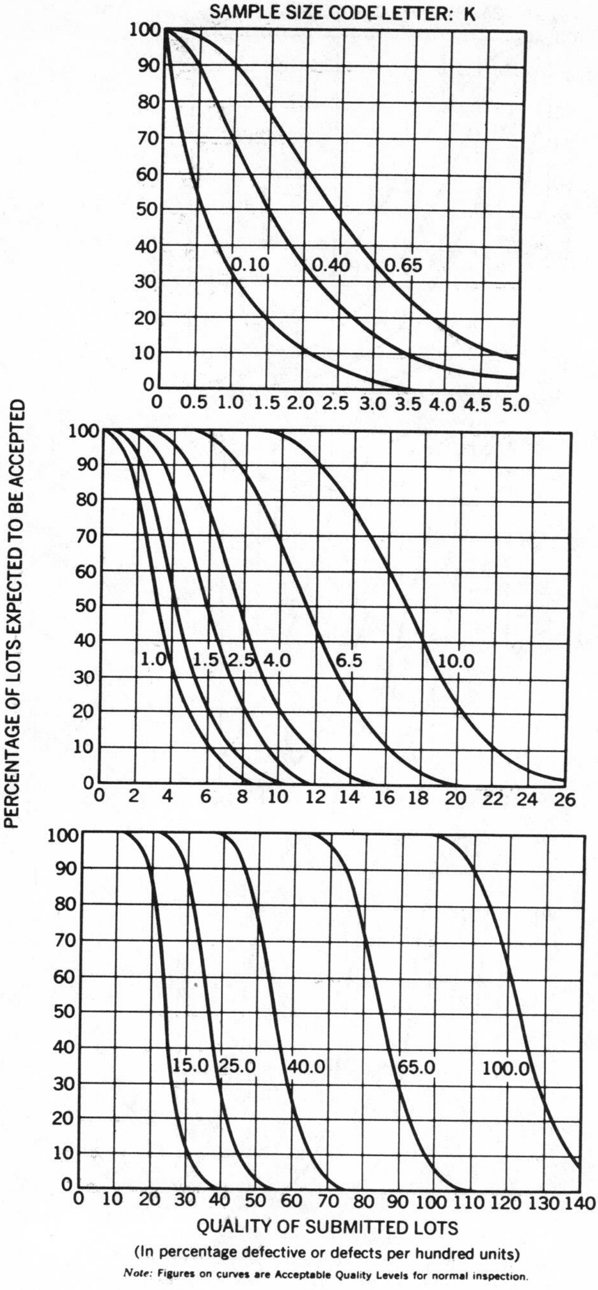

Fig. 13.6. (cont.) OC curves from MIL-STD-105B.

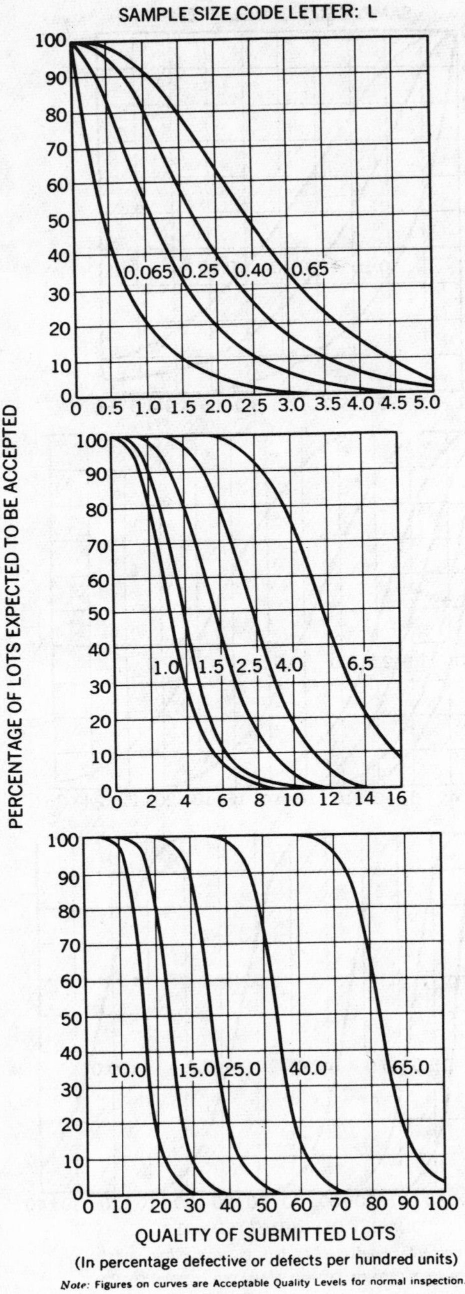

Fig. 13.6. (cont.) OC curves from MIL-STD-105B.

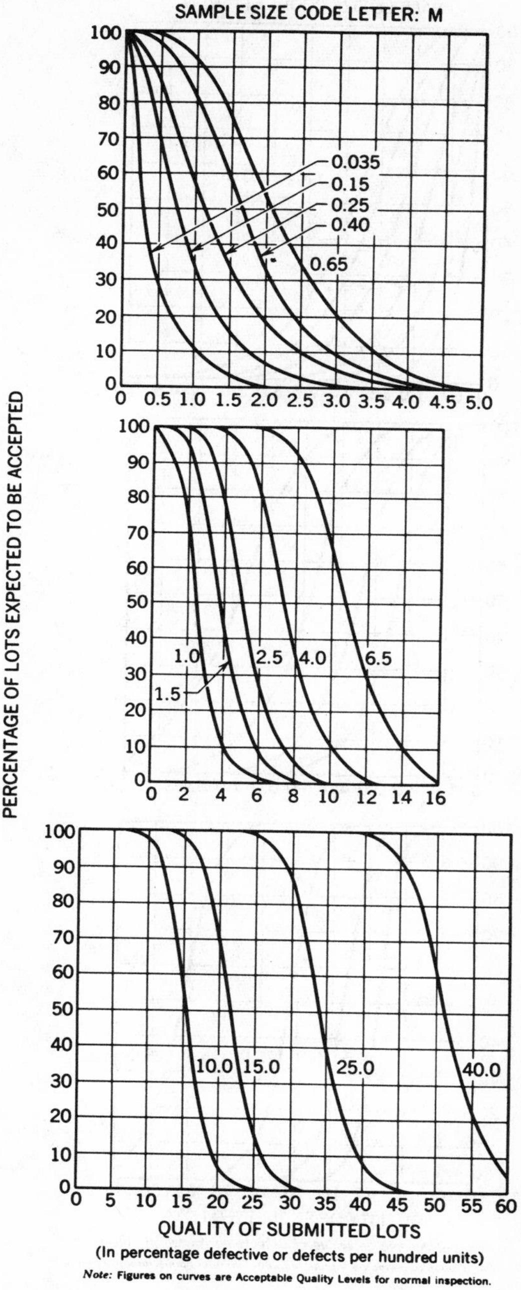

Fig. 13.6. (cont.) OC curves from MIL-STD-105B.

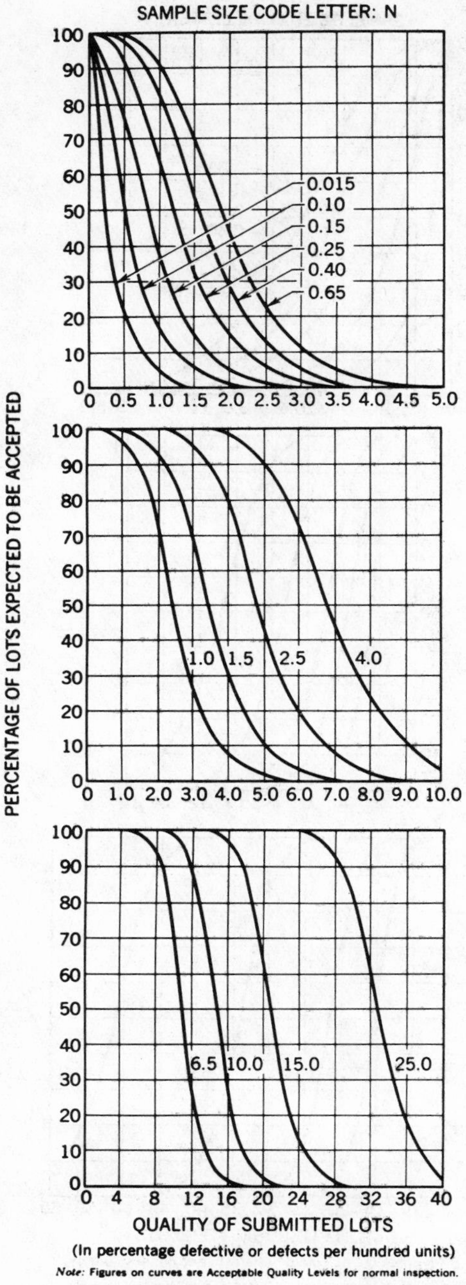

Fig. 13.6. (cont.) OC curves from MIL STD-105B.

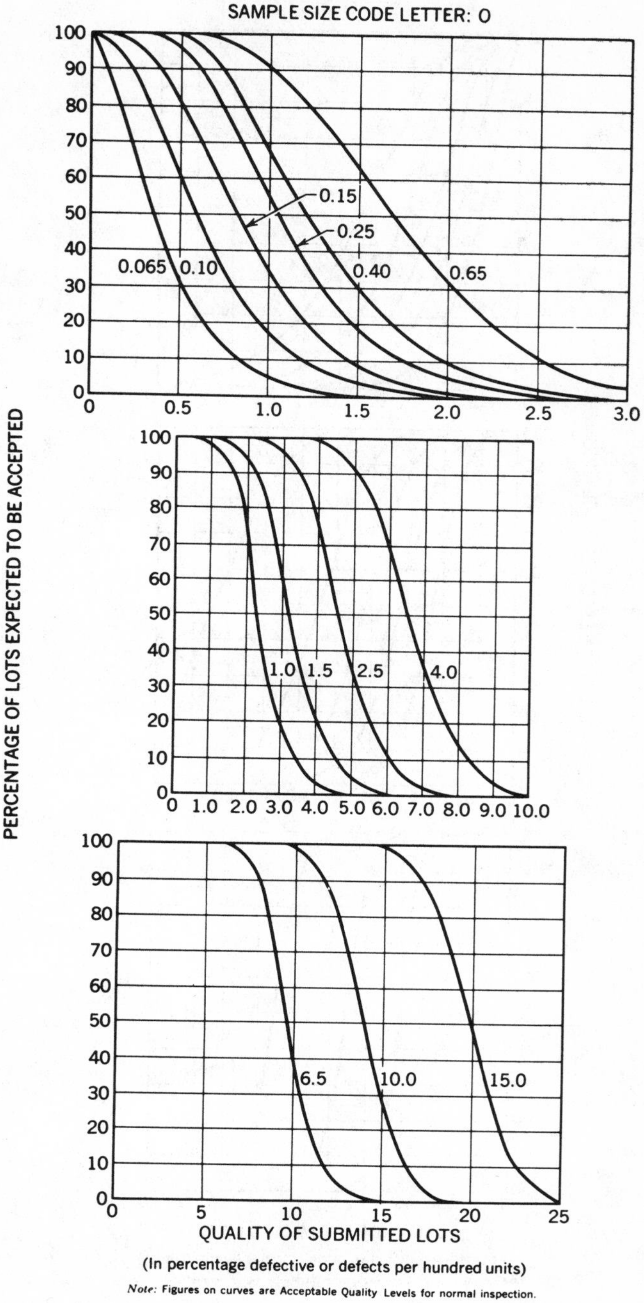

Fig. 13.6. (cont.) OC curves from MIL-STD-105B.

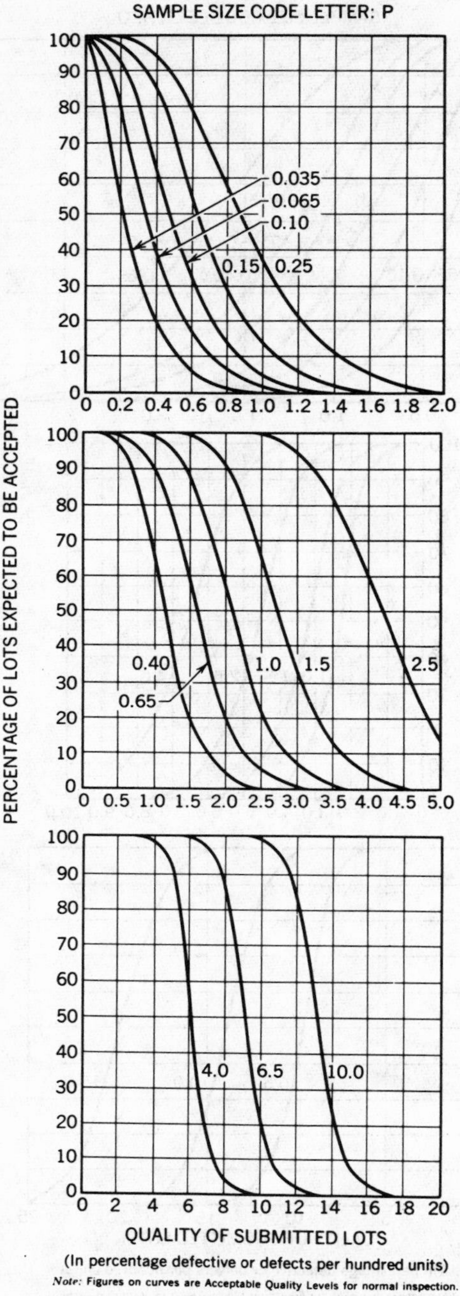

Fig. 13.6. (cont.) OC curves from MIL-STD-105B.

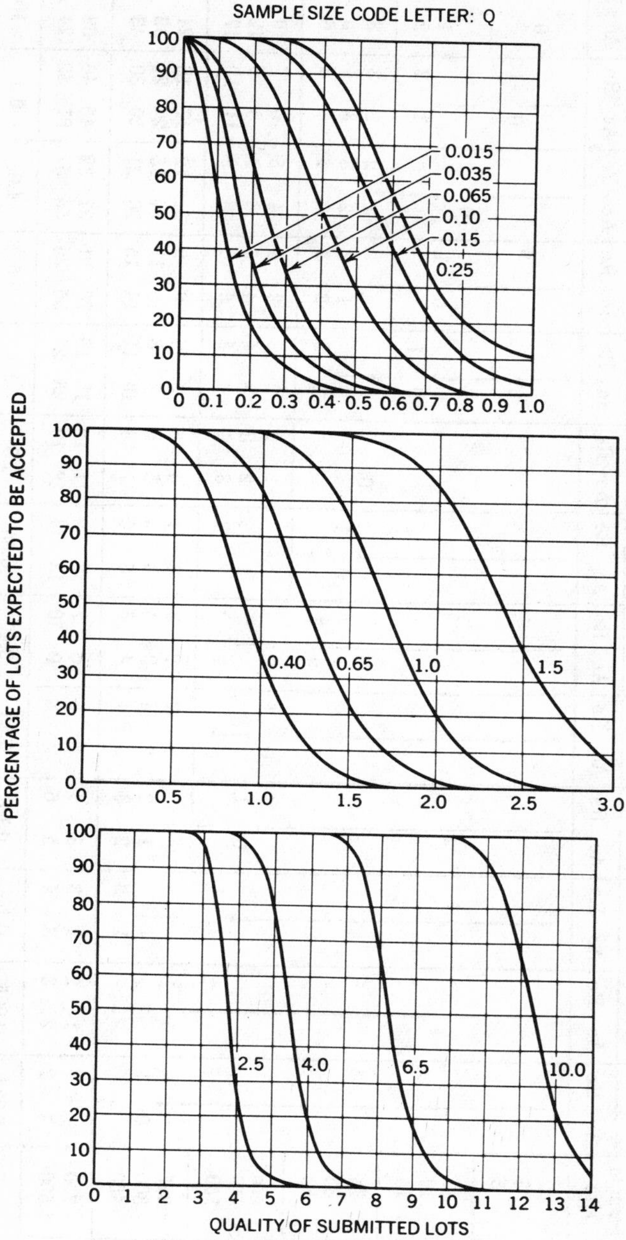

Note: Figures on curves are Acceptable Quality Levels for normal inspection.

Fig. 13.6. (cont.) OC curves from MIL-STD-105B.

Table 13.9. Master Table for Normal and Tightened Inspection
(Single sampling; Table IV A of MIL-STD-105B)

Sample size code letter	Sample size	Acceptable quality levels (normal inspection)																									
		0.015		0.035		0.065		0.10		0.15		0.25		0.40		0.65		1.0		1.5		2.5		4.0		6.5	
		Ac	Re	Ac	Re	Ac	Re	Ac	Re	Ac	Re	Ac	Re	Ac	Re	Ac	Re	Ac	Re	Ac	Re	Ac	Re	Ac	Re	Ac	Re
A	2	↓		↓		↓		↓		↓		↓		↓		↓		↓		↓		↓		↓		↓	
B	3	↓		↓		↓		↓		↓		↓		↓		↓		↓		↓		↓		↓		0	1
C	5	↓		↓		↓		↓		↓		↓		↓		↓		↓		↓		↓		0	1	↑	
D	7	↓		↓		↓		↓		↓		↓		↓		↓		↓		↓		0	1	↑		↓	
E	10	↓		↓		↓		↓		↓		↓		↓		↓		↓		0	1	↑		↓		1	2
F	15	↓		↓		↓		↓		↓		↓		↓		↓		0	1	↑		↓		1	2	2	3
G	25	↓		↓		↓		↓		↓		↓		↓		0	1	↑		↓		1	2	2	3	3	4
H	35	↓		↓		↓		↓		↓		↓		0	1	↑		↓		1	2	2	3	3	4	5	6
I	50	↓		↓		↓		↓		↓		0	1	↑		↓		1	2	2	3	3	4	4	5	6	7
J	75	↓		↓		↓		↓		0	1	↑		↓		1	2	2	3	3	4	4	5	6	7	9	10
K	110	↓		↓		↓		0	1	↑		↓		1	2	2	3	3	4	4	5	6	7	8	9	12	13
L	150	↓		↓		0	1	↑		↓		1	2	2	3	3	4	4	5	5	6	8	9	11	12	17	18
M	225	↓		0	1	↑		↓		1	2	2	3	3	4	4	5	5	6	8	9	11	12	17	18	24	25
N	300	0	1	↑		↓		1	2	2	3	3	4	4	5	5	6	7	8	10	11	14	15	20	21	32	33
O	450	↑		↓		1	2	2	3	3	4	4	5	5	6	7	8	10	11	14	15	20	21	29	30	43	44
P	750	↓		1	2	2	3	3	4	4	5	6	7	8	9	11	12	15	16	20	21	31	32	45	46	68	69
Q	1,500	1	2	2	3	3	4	5	6	7	8	9	10	13	14	18	19	25	26	35	36	56	57	81	82	124	125
		0.035		0.065		0.10		0.15		0.25		0.40		0.65		1.0		1.5		2.5		4.0		6.5		10.0	
		Acceptable quality levels (tightened inspection)																									

Table 13.9. (cont.) Master Table for Normal and Tightened Inspection
(Single sampling; Table IV A of MIL-STD-105B)

Sample size code letter	Sample size	Acceptable quality levels (normal inspection)																					
		10.0		15.0		25.0		40.0		65.0		100.0		150.0		250.0		400.0		650.0		1000.0	
		Ac	Re	Ac	Re	Ac	Re	Ac	Re	Ac	Re	Ac	Re	Ac	Re	Ac	Re	Ac	Re	Ac	Re	Ac	Re
A	2	0	1	↓		↓		1	2	2	3	3	4	5	6	8	9	12	13	19	20	28	29
B	3	↑		↓		1	2	2	3	3	4	5	6	8	9	12	13	18	19	28	29	41	42
C	5	↓		1	2	2	3	3	4	5	6	8	9	12	13	19	20	29	30	44	45	65	66
D	7	1	2	2	3	3	4	5	6	7	8	11	12	17	18	26	27	39	40	60	61	89	90
E	10	2	3	3	4	5	6	7	8	10	11	15	16	23	24	36	37	54	55	83	84	123	124
F	15	3	4	4	5	7	8	10	11	15	16	22	23	33	34	51	52	78	79	121	122	178	179
G	25	5	6	7	8	11	12	16	17	24	25	35	36	51	52	80	81	124	125	192	193	↑	
H	35	7	8	10	11	15	16	22	23	33	34	48	49	69	70	110	111	168	169	↑		↑	
I	50	9	10	13	14	20	21	30	31	46	47	67	68	96	97	151	152	↑		↑		↑	
J	75	13	14	19	20	29	30	43	44	66	67	96	97	138	139	↑		↑		↑		↑	
K	110	18	19	26	27	40	41	60	61	93	94	135	136	↑		↑		↑		↑		↑	
L	150	24	25	34	35	53	54	80	81	123	124	↑		↑		↑		↑		↑		↑	
M	225	34	35	48	49	76	77	115	116	↑		↑		↑		↑		↑		↑		↑	
N	300	44	45	63	64	98	99	↑		↑		↑		↑		↑		↑		↑		↑	
O	450	62	63	89	90	↑		↑		↑		↑		↑		↑		↑		↑		↑	
P	750	98	99	↑		↑		↑		↑		↑		↑		↑		↑		↑		↑	
Q	1,500	184	185	↑		↑		↑		↑		↑		↑		↑		↑		↑		↑	
		15.0		25.0		40.0		65.0		100.0		150.0		250.0		400.0		650.0		1000.0		X	
		Acceptable quality levels (tightened inspection)																					

↓ = Use first sampling plan below arrow. When sample size equals or exceeds lot size, do 100% inspection. ↑ = Use first sampling plan above arrow. Ac = Acceptance number. Re = Rejection number. Tightened sampling plans are not provided for AQL = 0.015.

Table 13.10. Master Table for Normal and Tightened Inspection
(Double sampling; Table IV B of MIL-STD-105B)

Sample size code letter	Sample	Sample size	Cumulative sample size	Acceptance Quality Levels (normal inspection)								
				0.015	0.035	0.065	0.10	0.15	0.25	0.40	0.65	1.0
				Ac Re	Ac Re	Ac Re	Ac Re	Ac Re	Ac Re	Ac Re	Ac Re	Ac Re
A B C	No double sampling plans for these sample size code letters: Use single sampling.											
D	First	5	5	⇩	⇩	⇩	⇩	⇩	⇩	⇩	⇩	⇩
	Second	10	15									
E	First	7	7	⇩	⇩	⇩	⇩	⇩	⇩	⇩	⇩	⇩
	Second	14	21									
F	First	10	10	⇩	⇩	⇩	⇩	⇩	⇩	⇩	⇩	*
	Second	20	30									
G	First	15	15	⇩	⇩	⇩	⇩	⇩	⇩	⇩	*	⇧
	Second	30	45									
H	First	25	25	⇩	⇩	⇩	⇩	⇩	⇩	*	⇧	⇩
	Second	50	75									
I	First	35	35	⇩	⇩	⇩	⇩	⇩	*	⇧	⇩	0 3
	Second	70	105									2 3
J	First	50	50	⇩	⇩	⇩	⇩	*	⇧	⇩	0 3	1 4
	Second	100	150								2 3	3 4
K	First	75	75	⇩	⇩	⇩	*	⇧	⇩	0 3	1 3	1 6
	Second	150	225							2 3	2 3	5 6
L	First	100	100	⇩	⇩	*	⇧	⇩	0 3	1 4	1 6	2 6
	Second	200	300						2 3	3 4	5 6	5 6
M	First	150	150	⇩	*	⇧	⇩	0 3	1 3	2 5	2 7	3 8
	Second	300	450					2 3	2 3	4 5	6 7	7 8
N	First	200	200	*	⇧	⇩	0 3	1 3	1 6	2 7	3 8	4 10
	Second	400	600				2 3	2 3	5 6	6 7	7 8	9 10
O	First	300	300	⇧	⇩	0 3	1 3	1 6	2 7	3 9	4 11	6 17
	Second	600	900			2 3	2 3	5 6	6 7	8 9	10 11	16 17
P	First	500	500	⇩	0 3	1 4	1 6	2 7	3 10	5 13	6 22	9 25
	Second	1000	1500		2 3	3 4	5 6	6 7	9 10	12 13	21 22	24 25
Q	First	1000	1000	0 3	1 4	1 6	2 9	4 13	5 17	7 26	11 33	15 47
	Second	2000	3000	2 3	3 4	5 6	8 9	12 13	16 17	25 26	32 33	46 47
				0.035	0.065	0.10	0.15	0.25	0.40	0.65	1.0	1.5
				Acceptance Quality Levels (tightened inspection)								

⇩ = Use first sampling below arrow. When sample size equals or exceeds lot size, do 100 percent inspection.

⇧ = Use first sampling plan above arrow.

* = Use corresponding single sampling plan, table IV–A.
Ac = Acceptance number.
Re = Rejection number.

Tightened sampling plans are not provided for AQL

Table 13.10. (cont.) Master Table for Normal and Tightened Inspection
(Double sampling; Table IV B of MIL-STD-105B)

Sample size code letter	Sample	Sample size	Cumulative sample size	Acceptable Quality Levels (normal inspection)							
				2.5		4.0		6.5		10.0	
				Ac	Re	Ac	Re	Ac	Re	Ac	Re
A B C	No double sampling plans for these sample size code letters: Use single sampling.										
D	First	5	5	*		*		⇩		1	3
	Second	10	15							2	3
E	First	7	7	⇧		⇩		0	3	1	5
	Second	14	21					2	3	4	5
F	First	10	10	⇩		0	3	1	4	2	6
	Second	20	30			2	3	3	4	5	6
G	First	15	15	0	3	1	4	1	5	3	7
	Second	30	45	2	3	3	4	4	5	6	7
H	First	25	25	1	4	2	5	3	7	5	11
	Second	50	75	3	4	4	5	6	7	10	11
I	First	35	35	1	5	2	7	3	12	6	15
	Second	70	105	4	5	6	7	11	12	14	15
J	First	50	50	2	7	3	10	5	15	8	21
	Second	100	150	6	7	9	10	14	15	20	21
K	First	75	75	4	9	5	12	7	20	12	29
	Second	150	225	8	9	11	12	19	20	28	29
L	First	100	100	5	12	7	17	10	31	14	49
	Second	200	300	11	12	16	17	30	31	48	49
M	First	150	150	7	19	11	29	15	47	21	65
	Second	300	450	18	19	28	29	46	47	64	65
N	First	200	200	9	25	12	36	18	67	27	89
	Second	400	600	24	25	35	36	66	67	88	89
O	First	300	300	12	36	18	55	26	88	38	123
	Second	600	900	35	36	54	55	87	88	122	123
P	First	500	500	18	65	27	89	43	131	62	191
	Second	1000	1500	64	65	88	89	130	131	190	191
Q	First	1000	1000	34	113	50	160	79	243	119	348
	Second	2000	3000	112	113	159	160	242	243	347	348
				4.0		6.5		10.0		X	
				Acceptable Quality Levels (tightened inspection)							

⇩ = Use first sampling below arrow. When sample size equals or exceeds lot size, do 100 percent inspection.

⇧ = Use first sampling plan above arrow.

* = Use corresponding single sampling plan in table IV–A.
Ac = Acceptance number.
Re = Rejection number.

Tightened sampling plans are not provided for AQL: 0.015.

Table 13.11. Master Table for Normal and Tightened Inspection
(Multiple sampling; Table IV C of MIL-STD-105B)

Sample size code letter	Sample	Sample size	Cumulative sample size	0.015	0.035	0.065	0.10	0.15	0.25	0.40	0.65	1.0	1.5	2.5	4.0	6.5
				Acceptable Quality Levels (normal inspection)												
				Ac Re	Ac Re	Ac Re	Ac Re	Ac Re	Ac Re	Ac Re	Ac Re	Ac Re	Ac Re	Ac Re	Ac Re	Ac Re
A B C	No multiple sampling plan for these sample size code letters. Use single sampling.															
D	First	3	3	↓	↓	↓	↓	↓	↓	↓	↓	↓	↓	*	*	↓
	Second	3	6													
	Third	3	9													
	Fourth	3	12													
E	First	4	4										*	↑	↓	# 2
	Second	4	8													0 2
	Third	4	12													1 3
	Fourth	4	16													1 3
	Fifth	4	20													3 4
F	First	5	5									*	↑	↓	# 2	# 2
	Second	5	10												0 2	1 3
	Third	5	15												0 3	1 4
	Fourth	5	20												1 3	1 4
	Fifth	5	25												2 3	3 4
G	First	7	7								*	↑	↓	# 1	# 2	# 2
	Second	7	14											# 2	0 3	1 3
	Third	7	21											0 2	1 3	2 4
	Fourth	7	28											1 3	2 4	3 5
	Fifth	7	35											2 4	3 5	4 6
	Sixth	7	42											3 4	4 5	5 6
H	First	10	10							*	↑	↓	# 2	# 2	# 3	0 4
	Second	10	20										# 2	0 3	0 3	1 5
	Third	10	30										0 2	1 3	1 4	3 7
	Fourth	10	40										0 3	1 4	2 6	4 8
	Fifth	10	50										1 3	2 4	3 7	5 9
	Sixth	10	60										2 4	2 4	4 7	7 10
	Seventh	10	70										3 4	4 5	6 7	10 11
I	First	14	14						*	↑	↓	# 2	# 2	# 2	# 3	0 4
	Second	14	28									0 2	0 3	0 3	1 4	2 5
	Third	14	42									0 2	1 3	1 4	2 4	4 8
	Fourth	14	56									0 2	1 3	3 5	3 6	5 9
	Fifth	14	70									1 3	2 4	3 5	4 6	7 10
	Sixth	14	84									1 3	3 5	3 5	5 7	10 12
	Seventh	14	98									2 3	4 5	4 5	6 7	11 12
				0.035	0.065	0.10	0.15	0.25	0.40	0.65	1.0	1.5	2.5	4.0	6.5	10.0
				Acceptable Quality Levels (tightened inspection)												

↓ = Use first sampling plan below arrow. When sample size equals or exceeds lot size, do 100 percent inspection.

↑ = Use first sampling plan above arrow.

-¦-¦→ = Use corresponding double sampling plan.
* = Use corresponding single sampling plan in table IV–A.
# = Acceptance not permitted at this sample size.
Ac = Acceptance number.
Re = Rejection number.

Tightened sampling plans are not provided for AQ[illegible]

Table 13.11. (cont.) Master Table for Normal and Tightened Inspection

(Multiple sampling; Table IV C of MIL-STD-105B)

Sample	Sample size	Cumulative sample size	0.015	0.035	0.065	0.10	0.15	0.25	0.40	0.65	1.0	1.5	2.5	4.0	6.5	10.0
			Acceptable Quality Levels (normal inspection)													
			Ac Re	Ac Re	Ac Re	Ac Re	Ac Re	Ac Re	Ac Re	Ac Re	Ac Re	Ac Re	Ac Re	Ac Re	Ac Re	Ac Re
First	20	20	↓	↓	↓	↓	*	↑	↓	# 2	# 2	# 3	# 3	0 4	1 5	1 6
Second	20	40								# 2	0 3	0 4	1 4	1 5	3 7	5 10
Third	20	60								0 2	1 3	1 4	2 5	3 6	5 10	9 14
Fourth	20	80								1 3	2 4	2 5	3 6	5 8	7 12	12 17
Fifth	20	100								1 3	2 4	4 6	5 7	8 10	9 13	15 20
Sixth	20	120								1 3	2 4	4 6	6 8	9 11	12 16	19 23
Seventh	20	140								2 3	3 4	5 6	7 8	10 11	16 17	22 23
First	30	30	↓	↓	↓	*	↑	↓	# 2	# 2	# 3	# 3	0 4	0 4	1 6	2 8
Second	30	60							0 2	0 2	0 3	1 4	2 5	3 7	5 9	8 13
Third	30	90							0 2	1 3	1 4	2 5	3 8	5 9	8 13	12 18
Fourth	30	120							0 2	2 4	2 6	4 6	4 9	7 11	11 16	17 22
Fifth	30	150							1 3	3 5	3 6	5 7	6 10	9 13	13 19	21 27
Sixth	30	180							1 3	3 5	5 7	6 8	8 12	12 15	16 22	27 32
Seventh	30	210							2 3	4 5	6 7	7 8	11 12	14 15	21 22	35 36
First	40	40	↓	↓	*	↑	↓	# 2	# 2	# 3	# 3	# 4	0 4	1 6	2 8	4 10
Second	40	80						# 2	0 3	0 3	1 4	1 5	2 7	4 8	6 12	10 16
Third	40	120						0 2	1 3	1 4	2 4	3 6	5 9	7 11	11 16	16 23
Fourth	40	160						0 3	3 5	3 5	4 6	4 7	7 11	10 14	16 21	22 29
Fifth	40	200						1 3	3 5	4 6	5 7	6 8	9 13	13 17	22 26	28 35
Sixth	40	240						2 4	3 5	6 8	6 8	6 8	11 15	16 21	28 31	36 42
Seventh	40	280						4 5	4 5	8 9	8 9	9 10	15 16	20 21	30 31	44 45
First	50	50	↓	*	↑	↓	# 2	# 2	# 2	# 3	# 3	0 4	0 5	2 7	3 10	5 12
Second	50	100					# 2	0 2	0 4	0 4	1 4	2 6	3 8	4 11	8 15	12 20
Third	50	150					0 2	0 3	2 4	1 4	2 6	3 8	5 11	8 15	13 19	19 28
Fourth	50	200					0 3	1 3	2 5	3 6	3 7	5 10	8 13	12 19	18 24	26 35
Fifth	50	250					1 3	2 4	3 6	4 7	4 8	7 12	10 15	16 22	23 30	33 42
Sixth	50	300					1 3	2 4	4 7	5 8	5 9	9 14	13 18	21 26	28 35	40 49
Seventh	50	350					2 4	2 4	5 7	6 8	7 10	11 16	15 20	25 29	33 40	48 56
Eighth	50	400					3 4	3 4	6 7	7 8	9 10	15 16	19 20	28 29	40 41	56 57
First	75	75	*	↑	↓	# 2	# 2	# 3	# 3	# 4	0 4	0 5	1 7	2 9	4 13	8 16
Second	75	150				# 2	0 2	0 3	1 4	1 5	2 5	3 8	4 10	7 13	11 21	18 27
Third	75	225				0 2	0 4	1 4	1 5	3 5	3 7	6 10	8 14	13 18	19 29	28 37
Fourth	75	300				0 3	2 4	2 5	2 6	4 7	5 9	8 13	12 18	18 23	27 37	39 48
Fifth	75	375				1 4	2 5	3 6	3 6	5 8	7 10	11 16	15 21	23 27	35 45	50 59
Sixth	75	450				2 4	2 5	4 6	5 8	7 9	9 11	13 18	18 25	28 32	43 53	61 69
Seventh	75	525				3 4	4 5	7 8	7 8	9 10	10 11	18 19	24 25	34 35	55 56	74 75
			0.035	0.065	0.10	0.15	0.25	0.40	0.65	1.0	1.5	2.5	4.0	6.5	10.0	X
			Acceptable Quality Levels (tightened inspection)													

Jse first sampling plan below arrow. When sample size equals or exceeds lot size, do 100 percent inspection.

'se first sampling plan above arrow.

Jse corresponding single sampling plan in table IV-A.
Acceptance not permitted at this sample size.
Acceptance number.
Rejection number.

Tightened sampling plans are not provided for AQL: 0.015.

Table 13.11. (cont.) Master Table for Normal and Tightened Inspection
(Multiple sampling; Table IV C of MIL-STD-105B)

Sample size code letter	Sample	Sample size	Cumulative sample size	Acceptable Quality Levels (normal inspection)																	
				0.015		0.035		0.065		0.10		0.15		0.25		0.40		0.65		1.0	
				Ac	Re	Ac	Re	Ac	Re	Ac	Re	Ac	Re	Ac	Re	Ac	Re	Ac	Re	Ac	Re
O	First	100	100	↑		↓		#	2	#	2	#	3	#	3	#	3	#	4	0	5
	Second	100	200					#	2	0	3	0	3	0	4	1	5	1	5	2	7
	Third	100	300					0	2	0	3	0	4	1	4	1	6	3	7	5	9
	Fourth	100	400					0	3	1	3	1	4	3	6	3	7	5	9	7	11
	Fifth	100	500					1	3	1	4	2	5	4	7	4	8	7	11	10	14
	Sixth	100	600					1	3	1	4	3	6	5	8	5	9	9	12	12	16
	Seventh	100	700					1	3	2	4	4	7	6	8	7	10	11	13	15	18
	Eighth	100	800					2	3	3	4	6	7	7	8	9	10	12	13	18	19
P	First	150	150	↓		#	2	#	2	#	2	#	3	#	3	0	4	0	5	0	6
	Second	150	300			#	2	#	2	0	3	0	4	1	5	1	5	2	7	4	9
	Third	150	450			0	2	0	3	0	4	1	4	2	6	2	7	4	9	7	12
	Fourth	150	600			0	2	0	3	0	4	2	5	3	7	4	9	7	12	10	15
	Fifth	150	750			0	2	1	4	1	4	3	6	4	9	6	11	9	14	13	18
	Sixth	150	900			0	3	1	5	2	5	4	7	6	10	8	13	12	16	16	21
	Seventh	150	1050			0	3	2	5	3	5	5	8	8	11	10	14	15	18	19	23
	Eighth	150	1200			1	3	3	5	4	6	6	8	9	11	12	15	17	20	22	26
	Ninth	150	1350			2	3	4	5	5	6	7	8	11	12	15	16	19	20	26	27
Q	First	300	300	#	2	#	2	#	3	#	3	#	4	#	5	0	6	1	7	2	9
	Second	300	600	#	2	#	2	#	3	0	4	1	5	1	6	2	9	5	11	7	14
	Third	300	900	0	2	0	3	0	4	1	5	3	6	3	8	4	11	9	15	12	18
	Fourth	300	1200	0	3	1	4	1	4	3	6	4	8	6	9	7	14	12	18	18	23
	Fifth	300	1500	0	3	1	5	1	5	4	7	6	9	8	11	10	17	15	21	23	28
	Sixth	300	1800	0	3	2	5	2	6	5	8	7	11	10	13	13	20	18	24	28	34
	Seventh	300	2100	0	3	2	5	3	7	7	9	9	13	13	16	15	23	22	27	33	39
	Eighth	300	2400	1	3	3	5	3	7	8	10	11	13	16	18	18	24	24	28	38	44
	Ninth	300	2700	2	3	4	5	6	7	9	10	13	14	19	20	23	24	29	30	46	47
				0.035		0.065		0.10		0.15		0.25		0.40		0.65		1.0		1.5	
				Acceptable Quality Levels (tightened inspection)																	

Sample size code letter	Sample	Sample size	Cumulative sample size	Acceptable Quality Levels (normal inspection)										
				1.5		2.5		4.0		6.5		10.0		
				Ac	Re	Ac	Re	Ac	Re	Ac	Re	Ac	Re	
O	First	100	100	0	7	1	8	3	11	6	15	9	19	
	Second	100	200	4	9	5	13	9	17	15	24	22	33	
	Third	100	300	6	13	10	18	16	23	24	33	37	47	
	Fourth	100	400	9	16	14	22	24	29	33	44	52	62	
	Fifth	100	500	12	19	18	26	30	35	42	53	66	74	
	Sixth	100	600	15	22	23	31	37	42	51	62	80	88	
	Seventh	100	700	19	25	27	35	43	47	62	72	94	102	
	Eighth	100	800	24	25	34	35	46	47	74	75	108	109	
P	First	150	150	1	8	2	11	6	15	10	21	15	29	
	Second	150	300	5	11	8	17	14	24	22	34	34	49	
	Third	150	450	10	15	14	23	24	35	37	48	54	69	
	Fourth	150	600	14	19	20	29	32	44	50	61	73	88	Notes: See the previous page.
	Fifth	150	750	17	24	26	36	40	52	62	74	92	108	
	Sixth	150	900	21	28	32	42	48	60	74	88	110	128	
	Seventh	150	1050	26	31	38	48	56	68	87	101	127	145	
	Eighth	150	1200	31	37	44	54	64	73	100	114	144	162	
	Ninth	150	1350	36	37	54	55	72	73	113	114	161	162	
Q	First	300	300	4	13	5	18	12	26	20	36	31	50	
	Second	300	600	11	18	16	30	28	41	43	60	67	85	
	Third	300	900	18	25	27	41	43	57	67	84	104	120	
	Fourth	300	1200	24	32	38	52	60	73	93	108	139	156	
	Fifth	300	1500	30	41	49	63	75	88	119	132	175	192	
	Sixth	300	1800	38	48	60	74	90	104	140	156	209	227	
	Seventh	300	2100	44	55	71	85	105	119	163	179	244	262	
	Eighth	300	2400	52	59	82	96	120	134	187	203	279	296	
	Ninth	300	2700	59	60	100	101	135	136	210	211	313	314	
				2.5		4.0		6.5		10.0				
				Acceptable Quality Levels (tightened inspection)										

13.2.7.1 *Computing the* OC *curve of a single sampling plan.* In Section 13.2.1.3 a method for computing the OC curve of a single sampling plan with sample size n and acceptance number c was presented. The OC curve may also be constructed from Table 13.13 by dividing each entry in the row for the given c by the sample size. The fraction defective p', for which the probability of acceptance is shown in the column heading, is obtained.

As an example, the OC curve for $n = 75$ and $c = 4$ will be found. Dividing the numbers in the row $c = 4$ by 75 in Table 13.13, we find the following 13 points on the OC curve:

Probability of Acceptance	*Fraction Defective*	*Probability of Acceptance*	*Fraction Defective*
0.995	$\frac{1.078}{75} = 0.01437$	0.250	$\frac{6.274}{75} = 0.08365$
0.990	$\frac{1.279}{75} = 0.01705$	0.100	$\frac{7.994}{75} = 0.1066$
0.975	$\frac{1.623}{75} = 0.02164$	0.050	$\frac{9.154}{75} = 0.1221$
0.950	$\frac{1.970}{75} = 0.02627$	0.025	$\frac{10.242}{75} = 0.1372$
0.900	$\frac{2.433}{75} = 0.03244$	0.010	$\frac{11.605}{75} = 0.1547$
0.750	$\frac{3.369}{75} = 0.04492$	0.005	$\frac{12.594}{75} = 0.1679$
0.500	$\frac{4.671}{75} = 0.06228$		

13.2.7.2 *Finding a sampling plan whose* OC *curve passes through two points.* As pointed out in Section 13.2.1.2, it is often useful to specify two points on an OC curve, p_1' and p_2' such that $L(p_1') = 1 - \alpha$ and $L(p_2') = \beta$. Here p_1' is usually denoted as acceptable quality (quality we want to accept), and p_2' usually represents unacceptable quality (quality we want to reject). Hence α and β are usually taken as small numbers, 0.01, 0.05, or 0.10. For these values, sampling plans may be found in Table 13.14. To construct a plan for a given p_1', α and p_2', β, calculate the ratio p_2'/p_1' and find the entry in the appropriate α, β column which is equal to or just less than the desired ratio. The acceptance number is read off directly and the sample size is determined by dividing np_1' by p_1'.

As an example, the sample size and acceptance number corresponding to $p_1' = 0.02$, $p_2' = 0.04$, $\alpha = 0.05$, and $\beta = 0.05$ will be found as follows:

$$\frac{p_2'}{p_1'} = \frac{0.04}{0.02} = 2.$$

The value in Table 13.14 just less than 2 is 1.999. Hence,

$$c = 22, \; n = 15.719/0.02 = 786.$$

Table 13.12. Master Table for Reduced Inspection
(Single sampling only; Table V of MIL-STD-105B)

Sample size code letter	Sample size	Acceptable quality levels 0.015 Ac	0.015 Re	0.035 Ac	0.035 Re	0.065 Ac	0.065 Re	0.10 Ac	0.10 Re	0.15 Ac	0.15 Re	0.25 Ac	0.25 Re	0.40 Ac	0.40 Re	0.65 Ac	0.65 Re	1.0 Ac	1.0 Re	1.5 Ac	1.5 Re	2.5 Ac	2.5 Re	4.0 Ac	4.0 Re
A, B, C, D	2	Reduced inspection not available for AQL: 0.015		↓		↓		↓		↓		↓		↓		↓		↓		0	1	↓		↓	
E	2																			0	1				
F	3															↓		0	1	↑		↓		1	2
G	5													↓		0	1	↑		↓		1	2	1	2
H	7											↓		0	1	↑		↓		1	2	1	2	2	3
I	10									↓		0	1	↑		↓		1	2	1	2	2	3	2	3
J	15							↓		0	1	↑		↓		1	2	1	2	2	3	2	3	3	4
K	22					↓		0	1	↑		↓		1	2	1	2	2	3	2	3	3	4	4	5
L	30			↓		0	1	↑		↓		1	2	1	2	2	3	2	3	3	4	4	5	5	6
M	45			0	1	↑		↓		1	2	1	2	2	3	2	3	3	4	4	5	5	6	7	8
N	60			↑		↓		1	2	1	2	2	3	2	3	3	4	4	5	5	6	6	7	9	10
O	90			↓		1	2	1	2	2	3	2	3	3	4	4	5	5	6	6	7	9	10	11	12
P	150			1	2	1	2	2	3	2	3	3	4	4	5	5	6	7	8	9	10	11	12	14	15
Q	300			1	2	2	3	3	4	4	5	5	6	6	7	8	9	11	12	13	14	16	17	21	22

Table 13.12. (cont.) Master Table for Reduced Inspection
(Single sampling only; Table V of MIL-STD-105B)

Sample size code letter	Sample size	Acceptable quality levels																							
		6.5		10.0		15.0		25.0		40.0		65.0		100.0		150.0		250.0		400.0		650.0		1000.0	
		Ac	Re	Ac	Re	Ac	Re	Ac	Re	Ac	Re	Ac	Re	Ac	Re	Ac	Re	Ac	Re	Ac	Re	Ac	Re	Ac	Re
A, B, C, D	1					1	2	1	2	1	2	2	3	4	5	5	6	8	9	12	13	15	16	19	20
E	2	1	2	1	2	2	3	3	4	3	4	4	5	7	8	10	11	13	14	17	18	22	23	29	30
F	3	1	2	2	3	2	3	3	4	5	6	7	8	10	11	13	14	17	18	21	22	29	30	37	38
G	5	2	3	3	4	3	4	5	6	7	8	10	11	13	14	17	18	22	23	29	30	39	40	↑	
H	7	3	4	3	4	5	6	7	8	9	10	13	14	16	17	20	21	27	28	36	37	↑			
I	10	3	4	4	5	6	7	9	10	12	13	16	17	19	20	24	25	33	34	↑					
J	15	4	5	6	7	8	9	12	13	15	16	19	20	24	25	31	32	↑							
K	22	5	6	8	9	11	12	14	15	18	19	23	24	31	32	↑									
L	30	7	8	11	12	12	13	16	17	21	22	29	30	↑											
M	45	10	11	13	14	15	16	20	21	27	28	↑													
N	60	12	13	15	16	18	19	24	25	↑															
O	90	14	15	18	19	23	24	↑																	
P	150	18	19	23	24	↑																			
Q	300	28	29	37	38																				

↓ = Use first sampling plan below arrow. When sample size equals or exceeds lot size, do 100% inspection. ↑ = Use first sampling plan above arrow. Ac=Acceptance number. Re=Rejection number.

Table 13.13. Values of np'_1 for Which the Probability of Acceptance of c or Fewer Defectives in a Sample of n is $P(A)$

[To find the fraction defective p, corresponding to a probability of acceptance $P(A)$ in a single sampling plan with sample size n and acceptance number c, divide by n the entry in the row for the given c and the column for the given $P(A)$].

c	$P(A) =$ 0.995	$P(A) =$ 0.990	$P(A) =$ 0.975	$P(A) =$ 0.950	$P(A) =$ 0.900	$P(A) =$ 0.750	$P(A) =$ 0.500	$P(A) =$ 0.250	$P(A) =$ 0.100	$P(A) =$ 0.050	$P(A) =$ 0.025	$P(A) =$ 0.010	$P(A) =$ 0.005
0	0.00501	0.0101	0.0253	0.0513	0.105	0.288	0.693	1.386	2.303	2.996	3.689	4.605	5.298
1	0.103	0.149	0.242	0.355	0.532	0.961	1.678	2.693	3.890	4.744	5.572	6.638	7.430
2	0.338	0.436	0.619	0.818	1.102	1.727	2.674	3.920	5.322	6.296	7.224	8.406	9.274
3	0.672	0.823	1.090	1.366	1.745	2.535	3.672	5.109	6.681	7.754	8.768	10.045	10.978
4	1.078	1.279	1.623	1.970	2.433	3.369	4.671	6.274	7.994	9.154	10.242	11.605	12.594
5	1.537	1.785	2.202	2.613	3.152	4.219	5.670	7.423	9.275	10.513	11.668	13.108	14.150
6	2.037	2.330	2.814	3.286	3.895	5.083	6.670	8.558	10.532	11.842	13.060	14.571	15.660
7	2.571	2.906	3.454	3.981	4.656	5.956	7.669	9.684	11.771	13.148	14.422	16.000	17.134
8	3.132	3.507	4.115	4.695	5.432	6.838	8.669	10.802	12.995	14.434	15.763	17.403	18.578
9	3.717	4.130	4.795	5.426	6.221	7.726	9.669	11.914	14.206	15.705	17.085	18.783	19.998
10	4.321	4.771	5.491	6.169	7.021	8.620	10.668	13.020	15.407	16.962	18.390	20.145	21.398
11	4.943	5.428	6.201	6.924	7.829	9.519	11.668	14.121	16.598	18.208	19.682	21.490	22.779
12	5.580	6.099	6.922	7.690	8.646	10.422	12.668	15.217	17.782	19.442	20.962	22.821	24.145
13	6.231	6.782	7.654	8.464	9.470	11.329	13.668	16.310	18.958	20.668	22.230	24.139	25.496
14	6.893	7.477	8.396	9.246	10.300	12.239	14.668	17.400	20.128	21.886	23.490	25.446	26.836
15	7.566	8.181	9.144	10.035	11.135	13.152	15.668	18.486	21.292	23.098	24.741	26.743	28.166
16	8.249	8.895	9.902	10.831	11.976	14.068	16.668	19.570	22.452	24.302	25.984	28.031	29.484
17	8.942	9.616	10.666	11.633	12.822	14.986	17.668	20.652	23.606	25.500	27.220	29.310	30.792
18	9.644	10.346	11.438	12.442	13.672	15.907	18.668	21.731	24.756	26.692	28.448	30.581	32.092
19	10.353	11.082	12.216	13.254	14.525	16.830	19.668	22.808	25.902	27.879	29.671	31.845	33.383
20	11.069	11.825	12.999	14.072	15.383	17.755	20.668	23.883	27.045	29.062	30.888	33.103	34.668
21	11.791	12.574	13.787	14.894	16.244	18.682	21.668	24.956	28.184	30.241	32.102	34.355	35.947
22	12.520	13.329	14.580	15.719	17.108	19.610	22.668	26.028	29.320	31.416	33.309	35.601	37.219
23	13.255	14.088	15.377	16.548	17.975	20.540	23.668	27.098	30.453	32.586	34.512	36.841	38.485
24	13.995	14.853	16.178	17.382	18.844	21.471	24.668	28.167	31.584	33.752	35.710	38.077	39.745
25	14.740	15.623	16.984	18.218	19.717	22.404	25.667	29.234	32.711	34.916	36.905	39.308	41.000
26	15.490	16.397	17.793	19.058	20.592	23.338	26.667	30.300	33.836	36.077	38.096	40.535	42.252
27	16.245	17.175	18.606	19.900	21.469	24.273	27.667	31.365	34.959	37.234	39.284	41.757	43.497

28	17.004	17.957	19.422	20.746	22.348	25.209	28.667	32.428	36.080	38.389	40.468	42.975	44.738
29	17.767	18.742	20.241	21.594	23.229	26.147	29.667	33.491	37.198	39.541	41.649	44.190	45.976
30	18.534	19.532	21.063	22.444	24.113	27.086	30.667	34.552	38.315	40.690	42.827	45.401	47.210
31	19.305	20.324	21.888	23.298	24.998	28.025	31.667	35.613	39.430	41.838	44.002	46.609	48.440
32	20.079	21.120	22.716	24.152	25.885	28.966	32.667	36.672	40.543	42.982	45.174	47.813	49.666
33	20.856	21.919	23.546	25.010	26.774	29.907	33.667	37.731	41.654	44.125	46.344	49.015	50.888
34	21.638	22.721	24.379	25.870	27.664	30.849	34.667	38.788	42.764	45.266	47.512	50.213	52.108
35	22.422	23.525	25.214	26.731	28.556	31.792	35.667	39.845	43.872	46.404	48.676	51.409	53.324
36	23.208	24.333	26.052	27.594	29.450	32.736	36.667	40.901	44.978	47.540	49.840	52.601	54.538
37	23.998	25.143	26.891	28.460	30.345	33.681	37.667	41.957	46.083	48.676	51.000	53.791	55.748
38	24.791	25.955	27.733	29.327	31.241	34.626	38.667	43.011	47.187	49.808	52.158	54.979	56.956
39	25.586	26.770	28.576	30.196	32.139	35.572	39.667	44.065	48.289	50.940	53.314	56.164	58.160
40	26.384	27.587	29.422	31.066	33.038	36.519	40.667	45.118	49.390	52.069	54.469	57.347	59.363
41	27.184	28.406	30.270	31.938	33.938	37.466	41.667	46.171	50.490	53.197	55.622	58.528	60.563
42	27.986	29.228	31.120	32.812	34.839	38.414	42.667	47.223	51.589	54.324	56.772	59.717	61.761
43	28.791	30.051	31.970	33.686	35.742	39.363	43.667	48.274	52.686	55.449	57.921	60.884	62.956
44	29.598	30.877	32.824	34.563	36.646	40.312	44.667	49.325	53.782	56.572	59.068	62.059	64.150
45	30.408	31.704	33.678	35.441	37.550	41.262	45.667	50.375	54.878	57.695	60.214	63.231	65.340
46	31.219	32.534	34.534	36.320	38.456	42.212	46.667	51.425	55.972	58.816	61.358	64.402	66.529
47	32.032	33.365	35.392	37.200	39.363	43.163	47.667	52.474	57.065	59.936	62.500	65.571	67.716
48	32.848	34.198	36.250	38.082	40.270	44.115	48.667	53.522	58.158	61.054	63.641	66.738	68.901
49	33.664	35.032	37.111	38.965	41.179	45.067	49.667	54.571	59.249	62.171	64.780	67.903	70.084

[1] Reprinted by permission from J. M. Cameron, "Tables for Constructing and Computing the Operating Characteristics of Single-Sampling Plans," *Industrial Quality Control*, July 1952, p. 39.

Table 13.14. Values of np'_1 and c for Constructing Single Sampling Plans Whose OC Curve Is Required to Pass through the Two Points $(p'_1, 1-\alpha)$ and (p'_2, β)

(Here p'_1 is the fraction defective for which the risk of rejection is to be α, and p'_2 is the fraction defective for which the risk of acceptance is to be β. To construct the plan, find the tabular value of p'_2/p'_1 in the column for the given α and β which is equal to or just less than the value of the ratio. The sample size is found by dividing the np'_1 corresponding to the selected ratio by p'_1. The acceptance number is the value of c corresponding to the selected value of the ratio.)

c	Values of p_2/p_1 for: $\alpha = 0.05$, $\beta = 0.10$	$\alpha = 0.05$, $\beta = 0.05$	$\alpha = 0.05$, $\beta = 0.01$	np_1	c	Values of p_2/p_1 for: $\alpha = 0.01$, $\beta = 0.10$	$\alpha = 0.01$, $\beta = 0.05$	$\alpha = 0.01$, $\beta = 0.01$	np_1
0	44.890	58.404	89.781	0.052	0	229.105	298.073	458.210	0.010
1	10.946	13.349	18.681	0.355	1	26.184	31.933	44.686	0.149
2	6.509	7.699	10.280	0.818	2	12.206	14.439	19.278	0.436
3	4.890	5.675	7.352	1.366	3	8.115	9.418	12.202	0.823
4	4.057	4.646	5.890	1.970	4	6.249	7.156	9.072	1.279
5	3.549	4.023	5.017	2.613	5	5.195	5.889	7.343	1.785
6	3.206	3.604	4.435	3.286	6	4.520	5.082	6.253	2.330
7	2.957	3.303	4.019	3.981	7	4.050	4.524	5.506	2.906
8	2.768	3.074	3.707	4.695	8	3.705	4.115	4.962	3.507
9	2.618	2.895	3.462	5.426	9	3.440	3.803	4.548	4.130
10	2.497	2.750	3.265	6.169	10	3.229	3.555	4.222	4.771
11	2.397	2.630	3.104	6.924	11	3.058	3.354	3.959	5.428
12	2.312	2.528	2.968	7.690	12	2.915	3.188	3.742	6.099
13	2.240	2.442	2.852	8.464	13	2.795	3.047	3.559	6.782
14	2.177	2.367	2.752	9.246	14	2.692	2.927	3.403	7.477
15	2.122	2.302	2.665	10.035	15	2.603	2.823	3.269	8.181
16	2.073	2.244	2.588	10.831	16	2.524	2.732	3.151	8.895
17	2.029	2.192	2.520	11.633	17	2.455	2.652	3.048	9.616
18	1.990	2.145	2.458	12.442	18	2.393	2.580	2.956	10.346
19	1.954	2.103	2.403	13.254	19	2.337	2.516	2.874	11.082

20	1.922	2.065	2.352	14.072	20	2.287	2.458	2.799	11.825
21	1.892	2.030	2.307	14.894	21	2.241	2.405	2.733	12.574
22	1.865	1.999	2.265	15.719	22	2.200	2.357	2.671	13.329
23	1.840	1.969	2.226	16.548	23	2.162	2.313	2.615	14.088
24	1.817	1.942	2.191	17.382	24	2.126	2.272	2.564	14.853
25	1.795	1.917	2.158	18.218	25	2.094	2.235	2.516	15.623
26	1.775	1.893	2.127	19.058	26	2.064	2.200	2.472	16.397
27	1.757	1.871	2.098	19.900	27	2.035	2.168	2.431	17.175
28	1.739	1.850	2.071	20.746	28	2.009	2.138	2.393	17.957
29	1.723	1.831	2.046	21.594	29	1.985	2.110	2.358	18.742
30	1.707	1.813	2.023	22.444	30	1.962	2.083	2.324	19.532
31	1.692	1.796	2.001	23.298	31	1.940	2.059	2.293	20.324
32	1.679	1.780	1.980	24.152	32	1.920	2.035	2.264	21.120
33	1.665	1.764	1.960	25.010	33	1.900	2.013	2.236	21.919
34	1.653	1.750	1.941	25.870	34	1.882	1.992	2.210	22.721
35	1.641	1.736	1.923	26.731	35	1.865	1.973	2.185	23.525
36	1.630	1.723	1.906	27.594	36	1.848	1.954	2.162	24.333
37	1.619	1.710	1.890	28.460	37	1.833	1.936	2.139	25.143
38	1.609	1.698	1.875	29.327	38	1.818	1.920	2.118	25.955
39	1.599	1.687	1.860	30.196	39	1.804	1.903	2.098	26.770
40	1.590	1.676	1.846	31.066	40	1.790	1.887	2.079	27.587
41	1.581	1.666	1.833	31.938	41	1.777	1.873	2.060	28.406
42	1.572	1.656	1.820	32.812	42	1.765	1.859	2.043	29.228
43	1.564	1.646	1.807	33.686	43	1.753	1.845	2.026	30.051
44	1.556	1.637	1.796	34.563	44	1.742	1.832	2.010	30.877
45	1.548	1.628	1.784	35.441	45	1.731	1.820	1.994	31.704
46	1.541	1.619	1.773	36.320	46	1.720	1.808	1.980	32.534
47	1.534	1.611	1.763	37.200	47	1.710	1.796	1.965	33.365
48	1.527	1.603	1.752	38.082	48	1.701	1.785	1.952	34.198
49	1.521	1.596	1.743	38.965	49	1.691	1.775	1.938	35.032

[1] Reprinted by permission from J. M. Cameron, "Tables for Constructing and Computing the Operating Characteristics of Single-Sampling Plans," *Industrial Quality Control*, July 1952, p. 39.

13.2.7.3 *Design of item-by-item sequential plans.* It is not possible to give simple formulas for the acceptance and rejection numbers in double or multiple sampling plans; however, simple formulas do exist for item-by-item sequential plans, i.e., plans in which the decision to accept, reject, or continue sampling is made after each observation. The plan can be represented graphically as a pair of parallel lines (Figure 13.7). The plan is determined by the acceptance line $a_m = -h_1 + ms$ and the rejection line $r_m = h_2 + ms$. To calculate h_1, h_2, and s, we introduce the quantities

$$b = \ln\frac{1-\alpha}{\beta}; \quad a = \ln\frac{1-\beta}{\alpha}; \quad g_1 = \ln\frac{p_2'}{p_1'}; \quad g_2 = \ln\frac{1-p_1'}{1-p_2'} .$$

Then
$$h_1 = \frac{b}{g_1 + g_2}; \quad h_2 = \frac{a}{g_1 + g_2}; \quad s = \frac{g_2}{g_1 + g_2} .$$

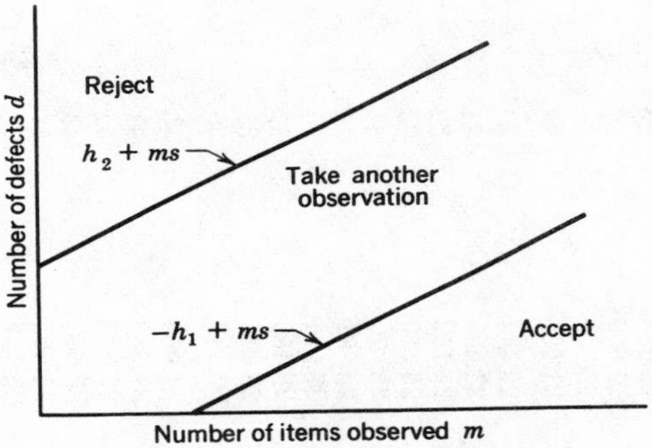

Fig. 13.7. Graphic procedure for item-by-item sequential sampling plan.

As an example, suppose we want $p_1' = 0.01$, $p_2' = 0.10$, $\alpha = 0.05$, $\beta = 0.20$; we calculate

$$b = \ln\frac{0.95}{0.20} = 1.55815; \quad a = \ln\frac{0.80}{0.05} = 2.77259$$

$$g_1 = \ln\frac{0.10}{0.01} = 2.30258; \quad g_2 = \ln\frac{0.99}{0.90} = 0.09531$$

$$h_1 = 0.649796; \quad h_2 = 1.156407; \quad s = 0.039747$$

and get the equations

$$a_m = -0.649796 + 0.039747(m); \quad r_m = 1.156407 + 0.039747(m) .$$

Usually the sequential plan is applied in tabular rather than graphical form; a_m and r_m are computed for every n and listed in Table 13.15. Note that a_m is rounded down to the nearest integer and r_m is rounded up to the nearest integer.

Simple formulas may be found for the probability of acceptance and the average sample number (ASN) from five values of the fraction defective including the two points defining the plan.

$$p' = 0; \qquad L(p') = 1; \qquad \bar{n}_0 = \frac{h_1}{s} .$$

$$p' = p_1'; \qquad L(p') = 1 - \alpha; \qquad \bar{n}_{p_1'} = \frac{(1-\alpha)h_1 - \alpha h_2}{s - p_1'} .$$

$$p' = s; \qquad L(p') = \frac{h_2}{h_1 + h_2}; \qquad \bar{n}_s = \frac{h_1 h_2}{s(1-s)} .$$

$$p' = p_2'; \qquad L(p') = \beta; \qquad \bar{n}_{p_2'} = \frac{(1-\beta)h_2 - \beta h_1}{p_2' - s} .$$

$$p' = 1; \qquad L(p') = 0; \qquad \bar{n}_1 = \frac{h_2}{1-s} .$$

In the example above,

$$p' = 0; \qquad L(0) = 1; \qquad \bar{n}_0 = \frac{0.649796}{0.039747} = 16.348 .$$

$$p' = 0.01; \qquad L(0.01) = 0.95;$$

$$\bar{n}_{0.01} = \frac{(0.95)(0.649796) - (0.05)(1.156407)}{0.029747} = 18.808 .$$

$$p' = 0.039747; \qquad L(0.039747) = \frac{1.156407}{0.649796 + 1.156407} = 0.64024;$$

$$\bar{n}_{(0.039747)} = \frac{(0.649796)(1.156407)}{(0.039747)(0.960253)} = 19.688 .$$

$$p' = 0.10; \qquad L(0.10) = 0.20;$$

$$\bar{n}_{0.10} = \frac{0.80(1.156407) - 0.20(0.649796)}{0.10 - 0.039747} = 13.197 .$$

$$p' = 1.00; \qquad L(1.00) = 0; \qquad \bar{n}_{1.00} = \frac{1.156407}{0.960253} = 1.2043 .$$

Usually, these five points may be used to make an adequate sketch of OC and ASN curves such as the ones in Figures 13.8 and 13.9.

Table 13.15. Tabular Procedure for Item-by-Item Sequential Sampling Plan

Number of Observations	Acceptance Number	Rejection Number	Observed Number of Defects
m	a_m	r_m	d
1	...	...	
2	...	...	
3	...	2	
⋮	⋮	⋮	
17	0	2	
⋮	⋮	⋮	
22	0	3	
⋮	⋮	⋮	
42	0	3	
⋮	⋮	⋮	
47	1	4	
⋮	⋮	⋮	
67	2	4	
⋮	⋮	⋮	
72	2	5	
⋮	⋮	⋮	
97	3	6	
⋮	⋮	⋮	
117	4	6	
⋮	⋮	⋮	
122	4	7	
⋮	⋮	⋮	
143	5	7	
⋮	⋮	⋮	
148	5	8	

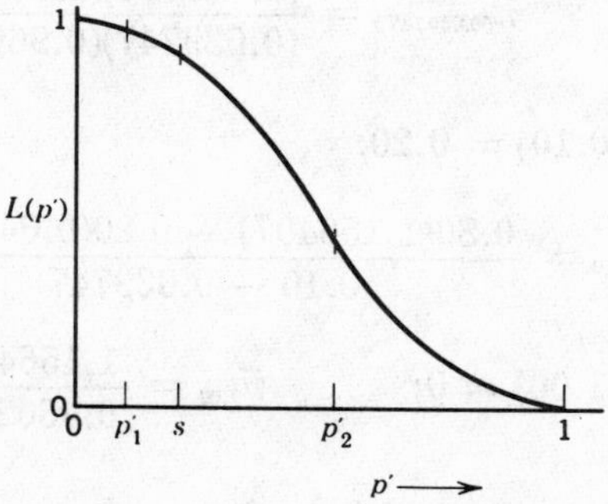

Fig. 13.8. OC curve of sequential plan sketched from five points.

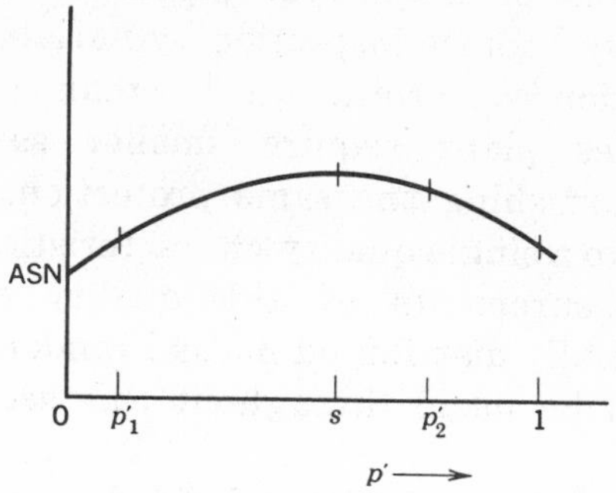

Fig. 13.9. ASN curve of sequential plan sketched from five points.

If more points are needed, they may be found by substituting values between 1 and ∞ for x in the equations

$$L(p') = \frac{x^{h_1+h_2} - x^{h_1}}{x^{h_1+h_2} - 1}; \qquad p' = \frac{x^s - 1}{x - 1}.$$

The point $[p', L(p')]$ has the conjugate $[p'_c, L(p'_c)]$ given by

$$L(p'_c) = \frac{L(p')}{x^{h_1}}; \qquad p'_c = p'(x^{1-s}) .$$

The expected number of observations is

$$\bar{n}_{p'} = \frac{L(p')(h_1 + h_2) - h_2}{s - p'} .$$

13.3 Lot-by-Lot Sampling Inspection by Variables

13.3.1 Introduction

In lot-by-lot acceptance sampling by attributes, each item of a sample drawn from a lot of manufactured items is classified simply as defective or non-defective. In single sampling, a random sample is drawn from the lot and the lot is either accepted or rejected depending solely upon the number of defectives in the sample; i.e., the lot is accepted if in a sample of size n, the number of defectives, d, does not exceed some preassigned constant, c, or equivalently, if the estimate of the fraction defective, $p = d/n$, does not exceed $M = c/n$. Inspection procedures by variables are based on the outcomes of a quality characteristic, measured on a continuous scale, and the decision to accept or to reject the lot is a function of these measurements (as opposed to the number of defectives). Variables inspection is applicable whenever the testing of individual items involves measurement on a continuous scale, and the form of the distribution is known. Since inspection by variables makes greater use of the information concerning the lot than does inspection by attributes, variables plans require smaller sample sizes than attributes plans furnishing the same protection. Variables sampling plans pertain to a single quality characteristic, and it is usually assumed that measurements of this quality characteristic are independent, identically distributed normal random variables. Such an assumption will be made throughout this section on variables inspection.

Sampling inspection by variables is divided into three categories: known standard deviation plans, unknown standard deviation plans based upon the sample standard deviation, and unknown standard deviation plans based upon the average range. Known standard deviation plans are functions of the sample mean and the known standard deviation. Unknown standard deviation plans based upon the sample standard deviation are functions of the sample mean and the sample standard deviation. Unknown standard deviation plans based upon the average range are functions of the sample mean and the average range in sub-samples.

13.3.2 General inspection criteria

Associated with each inspection characteristic are the design specifications. If only an upper specification limit, U, is given, the item is considered defective if its measurement exceeds U; the percent defective in the entire lot, p'_U is represented by the shaded area

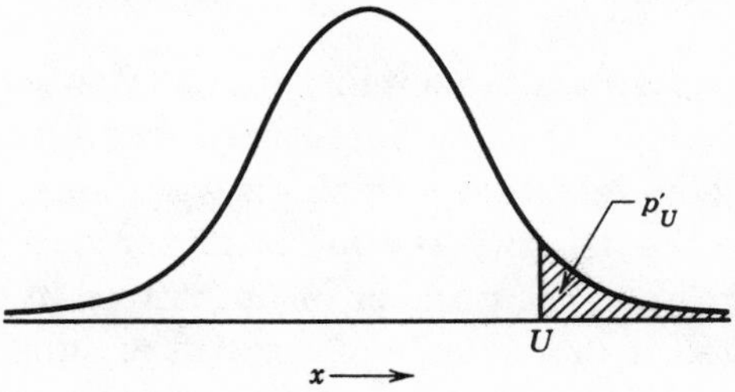

Fig. 13.10. The shaded area represents the fraction defective in the lot, given an upper specification limit.

in Figure 13.10. If only a lower specification limit, L, is given, the item is considered defective if its measurement is smaller than L; the percent defective in the entire lot, p'_L, is represented by the shaded area in Figure 13.11. Whenever both upper and lower

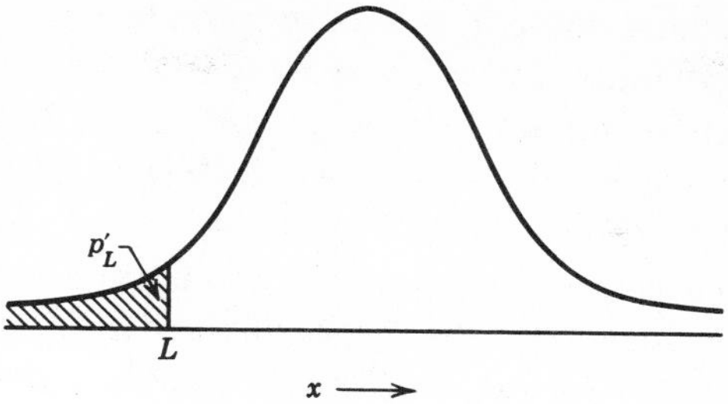

Fig. 13.11. The shaded area represents the fraction defective in the lot, given a lower specification limit.

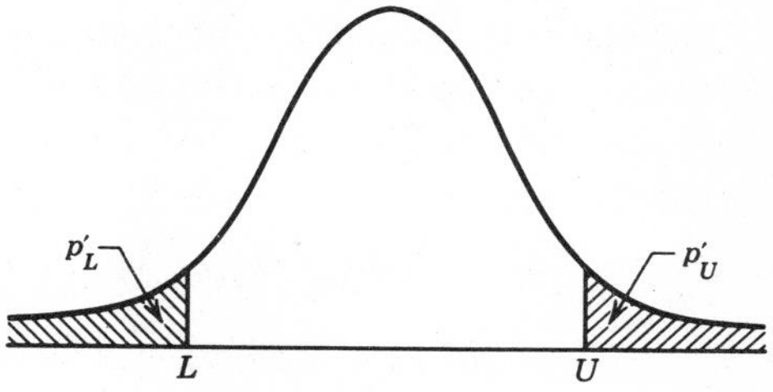

Fig. 13.12. The shaded areas represent the fraction defective in the lot, given both upper and lower specification limits.

limits are specified, the item is considered defective if its measurement either exceeds U or is smaller than L; the percent defective in the entire lot, $p'_U + p'_L$, is represented by the shaded area in Figure 13.12. If the mathematical model is realistic, i.e., the assumption of a normal distribution is valid, the ratio of the total number of defective units in the lot to the lot size should approximate the area outside of the specification limits.

If the percent defective of a submitted lot is known, i.e., p'_L and/or p'_U are given (whichever is appropriate), sampling inspection is unnecessary to determine whether or not the lot is to be accepted. If the percent defective is sufficiently small, the lot is accepted; otherwise, it is rejected. Since such knowledge about the incoming quality is rare, a logical procedure is to *estimate* the percent defective from a sample, and to accept or reject the lot on the basis of this estimate. A sampling plan is then described as consisting of the sample size n; a method for estimating the percent defective with the estimate being denoted by p_U, p_L, or $p_U + p_L$, whichever is appropriate; and a maximum allowable percent defective, M. If only an upper specification limit, U, is given, the estimate of the percentage above this limit, p_U, is obtained from the sample of size n. If

$p_U \leqq M$, the lot is accepted. If only a lower specification limit, L, is given, the estimate of the percentage below this limit, p_L is obtained from the sample of size n. If $p_L \leqq M$, the lot is accepted. If a double specification limit is given, both p_U and p_L are computed. If $p_U + p_L \leqq M$, the lot is accepted.

It is evident that these acceptance procedures are similar to the procedures used in attributes inspection, with p_L and/or p_U playing the role of d/n as the estimate of the incoming quality and M playing the role of c/n as the acceptance criterion. The OC curve is still a function of the incoming quality, p', and depends upon the choice of n and M.[1]

As in attributes inspection, increasing the sample size, while holding M fixed, increases the steepness of the OC curve. Similarly, increasing M, while holding the sample size fixed, essentially shifts the OC curve to the right.

13.3.3 Estimates of the Percent Defective

13.3.3.1 *Estimate of the percent defective when the standard deviation is unknown but estimated by the sample standard deviation.* As indicated in Chapter VIII, a "good" estimate of a parameter satisfies certain requirements. If the property of unbiasedness is considered important, the "best" estimate of the percent defective in the lot is the minimum variance unbiased estimate; i.e., it is the most efficient estimate among all unbiased estimates. For an upper specification limit, U, the minimum variance unbiased estimate is a function of $Q_U = \dfrac{U - \bar{x}}{s}$ and is denoted by p_U, where $\bar{x}$ is the sample mean and s is the sample standard deviation, i.e. $s=\sqrt{\dfrac{\sum(x_i-\bar{x})^2}{n-1}}$. For a lower specification limit, L, the minimum variance unbiased estimate is a function of $Q_L = \dfrac{\bar{x} - L}{s}$ and is denoted by p_L. The form of the functions, p_U and p_L, although rather complicated, is the same, thereby facilitating tabulation.[2] Such a tabulation is

[1] p' denotes the incoming quality. If there is only an upper specification limit given, the symbol p'_U coincides with p'. If there is only a lower specification limit given, the symbol p'_L is interpreted as p'. If there are both upper and lower specification limits, p' is equivalent to $p'_U + p'_L$.

[2] These estimates are described in detail in an article by G. J. Lieberman and G. J. Resnikoff "Sampling Plans For Inspection By Variables," *Journal of the American Statistical Association*, June, 1955, Vol. 50, pp. 457–516.

given in Table 13.16 for sample sizes of 3, 4, 5, 7, 10, 15, 20, 25, 30, 35, 40, 50, 75, 100, and 200. Therefore, in order to obtain the quantity p_U or p_L, Table 13.16 is entered with the sample size and the quality index Q_U or the quality index Q_L, whichever is appropriate; the required estimate, in percent, is read from the table. For two-sided specification limits, the minimum variance unbiased estimate of the percent defective is given by $p_U + p_L$ where these quantities are read from Table 13.16 as just described.

For an upper specification limit, the lot is accepted whenever $p_U \leqq M$, where M is a preassigned constant. Alternatively, it can be shown that $p_U \leqq M$ if, and only if, $\frac{U - \bar{x}}{s} \geqq k$, where k is a constant related to M. Hence, an equivalent procedure is to accept the lot if $\frac{U - \bar{x}}{s} \geqq k$.

For a lower specification limit, the lot is accepted whenever $p_L \leqq M$. Alternatively, it can be shown that $p_L \leqq M$ if, and only if, $\frac{\bar{x} - L}{s} \geqq k$. Hence, an equivalent procedure is to accept the lot if $\frac{\bar{x} - L}{s} \geqq k$.

Finally, for two-sided specification limits the lot is accepted whenever $p_U + p_L \leqq M$. There is no equivalent procedure in terms of k, but an approximation is given by accepting the lot if the following three criteria are *all* met:

(1) $\frac{U - \bar{x}}{s} \geqq k$;

(2) $\frac{\bar{x} - L}{s} \geqq k$;

(3) $s \leqq$ Maximum Standard Deviation (MSD), where MSD is also a constant depending upon M.

13.3.3.2 *Estimate of the percent defective when the standard deviation is unknown but estimated by the average range.* The estimate of the sample standard deviation σ' based on the average range has the advantage of computational ease over the estimate based on the sample standard deviation. However, it is known that the average range estimate is not as efficient as the sample standard deviation estimate. Moreover, there is no minimum variance unbiased estimate of p' which is a function of $\bar{x}$ and $\overline{R}$. A statistic has been found whose distribution is approximately equivalent to

Table 13.16. Estimating the Lot Percent Defective Using Standard Deviation Method*
(Table B–5 of MIL-STD-414)

Q_L or Q_U	Sample Size															
	3	4	5	7	10	15	20	25	30	35	40	50	75	100	150	
0	50.00	50.00	50.00	50.00	50.00	50.00	50.00	50.00	50.00	50.00	50.00	50.00	50.00	50.00	50.00	5
.1	47.24	46.67	46.44	46.26	46.16	46.10	46.08	46.06	46.05	46.05	46.04	46.04	46.03	46.03	46.02	4
.2	44.46	43.33	42.90	42.54	42.35	42.24	42.19	42.16	42.15	42.13	42.13	42.11	42.10	42.09	42.08	4
.3	41.63	40.00	39.37	38.87	38.60	38.44	38.37	38.33	38.31	38.29	38.28	38.27	38.25	38.24	38.22	3
.31	41.35	39.67	39.02	38.50	38.23	38.06	37.99	37.95	37.93	37.91	37.90	37.89	37.87	37.86	37.84	3
.32	41.06	39.33	38.67	38.14	37.86	37.69	37.62	37.58	37.55	37.54	37.52	37.51	37.49	37.48	37.46	3
.33	40.77	39.00	38.32	37.78	37.49	37.31	37.24	37.20	37.18	37.16	37.15	37.13	37.11	37.10	37.09	3
.34	40.49	38.67	37.97	37.42	37.12	36.94	36.87	36.83	36.80	36.78	36.77	36.75	36.73	36.72	36.71	3
.35	40.20	38.33	37.62	37.06	36.75	36.57	36.49	36.45	36.43	36.41	36.40	36.38	36.36	36.35	36.33	3
.36	39.91	38.00	37.28	36.69	36.38	36.20	36.12	36.08	36.05	36.04	36.02	36.01	35.98	35.97	35.96	3
.37	39.62	37.67	36.93	36.33	36.02	35.83	35.75	35.71	35.68	35.66	35.65	35.63	35.61	35.60	35.59	3
.38	39.33	37.33	36.58	35.98	35.65	35.46	35.38	35.34	35.31	35.29	35.28	35.26	35.24	35.23	35.22	3
.39	39.03	37.00	36.23	35.62	35.29	35.10	35.01	34.97	34.94	34.93	34.91	34.89	34.87	34.86	34.85	3
.40	38.74	36.67	35.88	35.26	34.93	34.73	34.65	34.60	34.58	34.56	34.54	34.53	34.50	34.49	34.48	3
.41	38.45	36.33	35.54	34.90	34.57	34.37	34.28	34.24	34.21	34.19	34.18	34.16	34.13	34.12	34.11	3
.42	38.15	36.00	35.19	34.55	34.21	34.00	33.92	33.87	33.85	33.83	33.81	33.79	33.77	33.76	33.74	3
.43	37.85	35.67	34.85	34.19	33.85	33.64	33.56	33.51	33.48	33.46	33.45	33.43	33.40	33.39	33.38	3
.44	37.56	35.33	34.50	33.84	33.49	33.28	33.20	33.15	33.12	33.10	33.09	33.07	33.04	33.03	33.02	3
.45	37.26	35.00	34.16	33.49	33.13	32.92	32.84	32.79	32.76	32.74	32.73	32.71	32.68	32.67	32.66	3
.46	36.96	34.67	33.81	33.13	32.78	32.57	32.48	32.43	32.40	32.38	32.37	32.35	32.32	32.31	32.30	3
.47	36.66	34.33	33.47	32.78	32.42	32.21	32.12	32.07	32.04	32.02	32.01	31.99	31.96	31.95	31.94	3
.48	36.35	34.00	33.12	32.43	32.07	31.85	31.77	31.72	31.69	31.67	31.65	31.63	31.61	31.60	31.58	3
.49	36.05	33.67	32.78	32.08	31.72	31.50	31.41	31.36	31.33	31.31	31.30	31.28	31.25	31.24	31.23	3
.50	35.75	33.33	32.44	31.74	31.37	31.15	31.06	31.01	30.98	30.96	30.95	30.93	30.90	30.89	30.87	3
.51	35.44	33.00	32.10	31.39	31.02	30.80	30.71	30.66	30.63	30.61	30.60	30.57	30.55	30.54	30.52	3
.52	35.13	32.67	31.76	31.04	30.67	30.45	30.36	30.31	30.28	30.26	30.25	30.23	30.20	30.19	30.17	3
.53	34.82	32.33	31.42	30.70	30.32	30.10	30.01	29.96	29.93	29.91	29.90	29.88	29.85	29.84	29.83	2
.54	34.51	32.00	31.08	30.36	29.98	29.76	29.67	29.62	29.59	29.57	29.55	29.53	29.51	29.49	29.48	2
.55	34.20	31.67	30.74	30.01	29.64	29.41	29.32	29.27	29.24	29.22	29.21	29.19	29.16	29.15	29.14	2
.56	33.88	31.33	30.40	29.67	29.29	29.07	28.98	28.93	28.90	28.88	28.87	28.85	28.82	28.81	28.79	2
.57	33.57	31.00	30.06	29.33	28.95	28.73	28.64	28.59	28.56	28.54	28.53	28.51	28.48	28.47	28.45	2
.58	33.25	30.67	29.73	28.99	28.61	28.39	28.30	28.25	28.22	28.20	28.19	28.17	28.14	28.13	28.12	2
.59	32.93	30.33	29.39	28.66	28.28	28.05	27.96	27.92	27.89	27.87	27.85	27.83	27.81	27.79	27.78	2
.60	32.61	30.00	29.05	28.32	27.94	27.72	27.63	27.58	27.55	27.53	27.52	27.50	27.47	27.46	27.45	2
.61	32.28	29.67	28.72	27.98	27.60	27.39	27.30	27.25	27.22	27.20	27.18	27.16	27.14	27.13	27.11	2
.62	31.96	29.33	28.39	27.65	27.27	27.05	26.96	26.92	26.89	26.87	26.85	27.83	26.81	26.80	26.78	2
.63	31.63	29.00	28.05	27.32	26.94	26.72	26.63	26.59	26.56	26.54	26.52	26.50	26.48	26.47	26.45	2
.64	31.30	28.67	27.72	26.99	26.61	26.39	26.31	26.26	26.23	26.21	26.20	26.18	26.15	26.14	26.13	2
.65	30.97	28.33	27.39	26.66	26.28	26.07	25.98	25.93	25.90	25.88	25.87	25.85	25.83	25.82	25.80	2
.66	30.63	28.00	27.06	26.33	25.96	25.74	25.66	25.61	25.58	25.56	25.55	25.53	25.51	25.49	25.48	2
.67	30.30	27.67	26.73	26.00	25.63	25.42	25.33	25.29	25.26	25.24	25.23	25.21	25.19	25.17	25.16	2
.68	29.96	27.33	26.40	25.68	25.31	25.10	25.01	24.97	24.94	24.92	24.91	24.89	24.87	24.86	24.84	2
.69	29.61	27.00	26.07	25.35	24.99	24.78	24.70	24.65	24.62	24.60	24.59	24.57	24.55	24.54	24.53	2
.70	29.27	26.67	25.74	25.03	24.67	24.46	24.38	24.33	24.31	24.29	24.28	24.26	24.24	24.23	24.21	2
.71	28.92	26.33	25.41	24.71	24.35	24.15	24.06	24.02	23.99	23.98	23.96	23.95	23.92	23.91	23.90	2
.72	28.57	26.00	25.09	24.39	24.03	23.83	23.75	23.71	23.68	23.67	23.65	23.64	23.61	23.60	23.59	2
.73	28.22	25.67	24.76	24.07	23.72	23.52	23.44	23.40	23.37	23.36	23.34	23.33	23.31	23.30	23.29	2
.74	27.86	25.33	24.44	23.75	23.41	23.21	23.13	23.09	23.07	23.05	23.04	23.02	23.00	22.99	22.98	2

* Values tabulated are read in percent.

Table 13.16. (cont.) **Estimating the Lot Percent Defective Using Standard Deviation Method***
(Table B–5 of MIL-STD-414)

Sample Size															
3	4	5	7	10	15	20	25	30	35	40	50	75	100	150	200
27.50	25.00	24.11	23.44	23.10	22.90	22.83	22.79	22.76	22.75	22.73	22.72	22.70	22.69	22.68	22.67
27.13	24.67	23.79	23.12	22.79	22.60	22.52	22.48	22.46	22.44	22.43	22.42	22.40	22.39	22.38	22.37
26.77	24.33	23.47	22.81	22.48	22.30	22.22	22.18	22.16	22.14	22.13	22.12	22.10	22.09	22.08	22.08
26.39	24.00	23.15	22.50	22.18	21.99	21.92	21.89	21.86	21.85	21.84	21.82	21.80	21.79	21.78	21.78
26.02	23.67	22.83	22.19	21.87	21.70	21.63	21.59	21.57	21.55	21.54	21.53	21.51	21.50	21.49	21.49
25.64	23.33	22.51	21.88	21.57	21.40	21.33	21.29	21.27	21.26	21.25	21.23	21.22	21.21	21.20	21.20
25.25	23.00	22.19	21.58	21.27	21.10	21.04	21.00	20.98	20.97	20.96	20.94	20.93	20.92	20.91	20.91
24.86	22.67	21.87	21.27	20.98	20.81	20.75	20.71	20.69	20.68	20.67	20.65	20.64	20.63	20.62	20.62
24.47	22.33	21.56	20.97	20.68	20.52	20.46	20.42	20.40	20.39	20.38	20.37	20.35	20.35	20.34	20.34
24.07	22.00	21.24	20.67	20.39	20.23	20.17	20.14	20.12	20.11	20.10	20.09	20.07	20.06	20.06	20.05
23.67	21.67	20.93	20.37	20.10	19.94	19.89	19.86	19.84	19.82	19.82	19.80	19.79	19.78	19.78	19.77
23.26	21.33	20.62	20.07	19.81	19.66	19.60	19.57	19.56	19.54	19.54	19.53	19.51	19.51	19.50	19.50
22.84	21.00	20.31	19.78	19.52	19.38	19.32	19.30	19.28	19.27	19.26	19.25	19.24	19.23	19.22	19.22
22.62	20.67	20.00	19.48	19.23	19.10	19.04	19.02	19.00	18.99	18.98	18.98	18.96	18.96	18.95	18.65
21.99	20.33	19.69	19.19	18.95	18.82	18.77	18.74	18.73	18.72	18.71	18.70	18.69	18.69	18.68	18.68
21.55	20.00	19.38	18.90	18.67	18.54	18.50	18.47	18.46	18.45	18.44	18.43	18.42	18.42	18.41	18.41
21.11	19.67	19.07	18.61	18.39	18.27	18.22	18.20	18.19	18.18	18.17	18.17	18.16	18.15	18.15	18.15
20.66	19.33	18.77	18.33	18.11	18.00	17.96	17.94	17.92	17.92	17.91	17.90	17.89	17.89	17.88	17.88
20.20	19.00	18.46	18.04	17.84	17.73	17.69	17.67	17.66	17.65	17.65	17.64	17.63	17.63	17.62	17.62
19.74	18.67	18.16	17.76	17.57	17.46	17.43	17.41	17.40	17.39	17.39	17.38	17.37	17.37	17.36	17.36
19.25	18.33	17.86	17.48	17.29	17.20	17.17	17.15	17.14	17.13	17.13	17.12	17.12	17.11	17.11	17.11
18.76	18.00	17.56	17.20	17.03	16.94	16.91	16.89	16.88	16.88	16.87	16.87	16.86	16.86	16.86	16.85
18.25	17.67	17.25	16.92	16.76	16.68	16.65	16.63	16.63	16.62	16.62	16.61	16.61	16.61	16.60	16.60
17.74	17.33	16.96	16.65	16.49	16.42	16.39	16.38	16.37	16.37	16.37	16.36	16.36	16.36	16.36	16.36
17.21	17.00	16.66	16.37	16.23	16.16	16.14	16.13	16.12	16.12	16.12	16.12	16.11	16.11	16.11	16.11
16.67	16.67	16.36	16.10	15.97	15.91	15.89	15.88	15.88	15.87	15.87	15.87	15.87	15.87	15.87	15.87
16.11	16.33	16.07	15.83	15.72	15.66	15.64	15.63	15.63	15.63	15.63	15.63	15.62	15.62	15.62	15.62
15.53	16.00	15.78	15.56	15.46	15.41	15.40	15.39	15.39	15.39	15.39	15.38	15.38	15.38	15.38	15.38
14.93	15.67	15.48	15.30	15.21	15.17	15.15	15.15	15.15	15.15	15.15	15.15	15.15	15.15	15.15	15.15
14.31	15.33	15.19	15.03	14.96	14.92	14.91	14.91	14.91	14.91	14.91	14.91	14.91	14.91	14.91	14.91
13.66	15.00	14.91	14.77	14.71	14.66	14.67	14.67	14.67	14.67	14.68	14.68	14.68	14.68	14.68	14.68
12.98	14.67	14.62	14.51	14.46	14.44	14.44	14.44	14.44	14.44	14.44	14.45	14.45	14.45	14.45	14.45
12.27	14.33	14.33	14.26	14.22	14.20	14.20	14.21	14.21	14.21	14.21	14.22	14.22	14.22	14.22	14.23
11.51	14.00	14.05	14.00	13.97	13.97	13.97	13.98	13.91	13.98	13.99	13.99	13.99	14.00	14.00	14.00
10.71	13.67	13.76	13.75	13.73	13.74	13.74	13.75	13.75	13.76	13.76	13.77	13.77	13.77	13.78	13.78
9.84	13.33	13.48	13.49	13.50	13.51	13.52	13.52	13.53	13.54	13.54	13.54	13.55	13.55	13.56	13.56
8.89	13.00	13.20	13.25	13.26	13.28	13.29	13.30	13.31	13.31	13.32	13.32	13.33	13.34	13.34	13.34
7.82	12.67	12.93	13.00	13.03	13.05	13.07	13.08	13.09	13.10	13.10	13.11	13.12	13.12	13.12	13.13
6.60	12.33	12.65	12.75	12.80	12.83	12.85	12.86	12.87	12.88	12.89	12.89	12.90	12.91	12.91	12.92
5.08	12.00	12.37	12.51	12.57	12.61	12.63	12.65	12.66	12.67	12.67	12.68	12.69	12.70	12.70	12.70
0.29	11.67	12.10	12.27	12.34	12.39	12.42	12.44	12.45	12.46	12.46	12.47	12.48	12.49	12.49	12.50
0.00	11.33	11.83	12.03	12.12	12.18	12.21	12.22	12.24	12.25	12.25	12.26	12.28	12.28	12.29	12.29
0.00	11.00	11.56	11.79	11.90	11.96	12.00	12.02	12.03	12.04	12.05	12.06	12.07	12.08	12.08	12.09
0.00	10.67	11.29	11.56	11.68	11.75	11.79	11.81	11.82	11.84	11.84	11.85	11.87	11.88	11.88	11.89
0.00	10.33	11.02	11.33	11.46	11.54	11.58	11.61	11.62	11.63	11.64	11.65	11.67	11.68	11.69	11.69

* Values tabulated are read in percent.

Table 13.16. (cont.) **Estimating the Lot Percent Defective Using Standard Deviation Method***
(Table B–5 of MIL-STD-414)

Q_L or Q_U	Sample Size														
	3	4	5	7	10	15	20	25	30	35	40	50	75	100	150
1.20	0.00	10.00	10.76	11.10	11.24	11.34	11.38	11.41	11.42	11.43	11.44	11.46	11.47	11.48	11.49
1.21	0.00	9.67	10.50	10.87	11.03	11.13	11.18	11.21	11.22	11.24	11.25	11.26	11.28	11.29	11.30
1.22	0.00	9.33	10.23	10.65	10.82	10.93	10.98	11.01	11.03	11.04	11.05	11.07	11.09	11.09	11.10
1.23	0.00	9.00	9.97	10.42	10.61	10.73	10.78	10.81	10.84	10.85	10.86	10.88	10.90	10.91	10.91
1.24	0.00	8.67	9.72	10.20	10.41	10.53	10.59	10.62	10.64	10.66	10.67	10.69	10.71	10.72	10.73
1.25	0.00	8.33	9.46	9.98	10.21	10.34	10.40	10.43	10.46	10.47	10.48	10.50	10.52	10.53	10.54
1.26	0.00	8.00	9.21	9.77	10.00	10.15	10.21	10.25	10.27	10.29	10.30	10.32	10.34	10.35	10.36
1.27	0.00	7.67	8.96	9.55	9.81	9.96	10.02	10.06	10.09	10.10	10.12	10.13	10.16	10.17	10.18
1.28	0.00	7.33	8.71	9.34	9.61	9.77	9.84	9.88	9.90	9.92	9.94	9.95	9.98	9.99	10.00
1.29	0.00	7.00	8.46	9.13	9.42	9.58	9.65	9.70	9.72	9.74	9.76	9.78	9.80	9.82	9.83
1.30	0.00	6.67	8.21	8.93	9.22	9.40	9.48	9.52	9.55	9.57	9.58	9.60	9.63	9.64	9.65
1.31	0.00	6.33	7.97	8.72	9.03	9.22	9.30	9.34	9.37	9.39	9.41	9.43	9.46	9.47	9.48
1.32	0.00	6.00	7.73	8.52	8.85	9.04	9.12	9.17	9.20	9.22	9.24	9.26	9.29	9.30	9.31
1.33	0.00	5.67	7.49	8.32	8.66	8.86	8.95	9.00	9.03	9.05	9.07	9.09	9.12	9.13	9.15
1.34	0.00	5.33	7.25	8.12	8.48	8.69	8.78	8.83	8.86	8.88	8.90	8.92	8.95	8.97	8.98
1.35	0.00	5.00	7.02	7.92	8.30	8.52	8.61	8.66	8.69	8.72	8.74	8.76	8.79	8.81	8.82
1.36	0.00	4.67	6.79	7.73	8.12	8.35	8.44	8.50	8.53	8.55	8.57	8.60	8.63	8.65	8.66
1.37	0.00	4.33	6.56	7.54	7.95	8.18	8.28	8.33	8.37	8.39	8.41	8.44	8.47	8.49	8.50
1.38	0.00	4.00	6.33	7.35	7.77	8.01	8.12	8.17	8.21	8.24	8.25	8.28	8.31	8.33	8.35
1.39	0.00	3.67	6.10	7.17	7.60	7.85	7.96	8.01	8.05	8.08	8.10	8.12	8.16	8.18	8.19
1.40	0.00	3.33	5.88	6.98	7.44	7.69	7.80	7.86	7.90	7.92	7.94	7.97	8.01	8.02	8.04
1.41	0.00	3.00	5.66	6.80	7.27	7.53	7.64	7.70	7.74	7.77	7.79	7.82	7.86	7.87	7.89
1.42	0.00	2.67	5.44	6.62	7.10	7.37	7.49	7.55	7.59	7.62	7.64	7.67	7.71	7.73	7.74
1.43	0.00	2.33	5.23	6.45	6.94	7.22	7.34	7.40	7.44	7.47	7.50	7.52	7.56	7.58	7.60
1.44	0.00	2.00	5.01	6.27	6.78	7.07	7.19	7.26	7.30	7.33	7.35	7.38	7.42	7.44	7.46
1.45	0.00	1.67	4.81	6.10	6.63	6.92	7.04	7.11	7.15	7.18	7.21	7.24	7.28	7.30	7.31
1.46	0.00	1.33	4.60	5.93	6.47	6.77	6.90	6.97	7.01	7.04	7.07	7.10	7.14	7.16	7.18
1.47	0.00	1.00	4.39	5.77	6.32	6.63	6.75	6.83	6.87	6.90	6.93	6.96	7.00	7.02	7.04
1.48	0.00	0.67	4.19	5.60	6.17	6.48	6.61	6.69	6.73	6.77	6.79	6.82	6.86	6.88	6.90
1.49	0.00	0.33	3.99	5.44	6.02	6.34	6.48	6.55	6.60	6.63	6.65	6.69	6.73	6.75	6.77
1.50	0.00	0.00	3.80	5.28	5.87	6.20	6.34	6.41	6.46	6.50	6.52	6.55	6.60	6.62	6.64
1.51	0.00	0.00	3.61	5.13	5.73	6.06	6.20	6.28	6.33	6.36	6.39	6.42	6.47	6.49	6.51
1.52	0.00	0.00	3.42	4.97	5.59	5.93	6.07	6.15	6.20	6.23	6.26	6.29	6.34	6.36	6.38
1.53	0.00	0.00	3.23	4.82	5.45	5.80	5.94	6.02	6.07	6.11	6.13	6.17	6.21	6.24	6.26
1.54	0.00	0.00	3.05	4.67	5.31	5.67	5.81	5.89	5.95	5.98	6.01	6.04	6.09	6.11	6.13
1.55	0.00	0.00	2.87	4.52	5.18	5.54	5.69	5.77	5.82	5.86	5.88	5.92	5.97	5.99	6.01
1.56	0.00	0.00	2.69	4.38	5.05	5.41	5.56	5.65	5.70	5.74	5.76	5.80	5.85	5.87	5.89
1.57	0.00	0.00	2.52	4.24	4.92	5.29	5.44	5.53	5.58	5.62	5.64	5.68	5.73	5.75	5.78
1.58	0.00	0.00	2.35	4.10	4.79	5.16	5.32	5.41	5.46	5.50	5.53	5.56	5.61	5.64	5.66
1.59	0.00	0.00	2.19	3.96	4.66	5.04	5.20	5.29	5.34	5.38	5.41	5.45	5.50	5.52	5.54
1.60	0.00	0.00	2.03	3.83	4.54	4.92	5.09	5.17	5.23	5.27	5.30	5.33	5.38	5.41	5.43
1.61	0.00	0.00	1.87	3.69	4.41	4.81	4.97	5.06	5.12	5.16	5.18	5.22	5.27	5.30	5.32
1.62	0.00	0.00	1.72	3.57	4.30	4.69	4.86	4.95	5.01	5.04	5.07	5.11	5.16	5.19	5.21
1.63	0.00	0.00	1.57	3.44	4.18	4.58	4.75	4.84	4.90	4.94	4.97	5.01	5.06	5.08	5.11
1.64	0.00	0.00	1.42	3.31	4.06	4.47	4.64	4.73	4.79	4.83	4.86	4.90	4.95	4.98	5.00

* Values tabulated are read in percent.

Table 13.16. (cont.) Estimating the Lot Percent Defective Using Standard Deviation Method*
(Table B–5 of MIL-STD-414)

)L r 'U	Sample Size															
	3	4	5	7	10	15	20	25	30	35	40	50	75	100	150	200
65	0.00	0.00	1.28	3.19	3.95	4.36	4.53	4.62	4.68	4.72	4.75	4.79	4.85	4.87	4.90	4.91
66	0.00	0.00	1.15	3.07	3.84	4.25	4.43	4.52	4.58	4.62	4.65	4.69	4.74	4.77	4.80	4.81
67	0.00	0.00	1.02	2.95	3.73	4.15	4.32	4.42	4.48	4.52	4.55	4.59	4.64	4.67	4.70	4.71
68	0.00	0.00	0.89	2.84	3.62	4.05	4.22	4.32	4.38	4.42	4.45	4.49	4.55	4.57	4.60	4.61
69	0.00	0.00	0.77	2.73	3.52	3.94	4.12	4.22	4.28	4.32	4.35	4.39	4.45	4.47	4.50	4.51
70	0.00	0.00	0.66	2.62	3.41	3.84	4.02	4.12	4.18	4.22	4.25	4.30	4.35	4.38	4.41	4.42
71	0.00	0.00	0.55	2.51	3.31	3.75	3.93	4.02	4.09	4.13	4.16	4.20	4.26	4.29	4.31	4.32
72	0.00	0.00	0.45	2.41	3.21	3.65	3.83	3.93	3.99	4.04	4.07	4.11	4.17	4.19	4.22	4.23
73	0.00	0.00	0.36	2.30	3.11	3.56	3.74	3.84	3.90	3.94	3.98	4.02	4.08	4.10	4.13	4.14
74	0.00	0.00	0.27	2.20	3.02	3.46	3.65	3.75	3.81	3.85	3.89	3.93	3.99	4.01	4.04	4.05
75	0.00	0.00	0.19	2.11	2.93	3.37	3.56	3.66	3.72	3.77	3.80	3.84	3.90	3.93	3.95	3.97
76	0.00	0.00	0.12	2.01	2.83	3.28	3.47	3.57	3.63	3.68	3.71	3.76	3.81	3.84	3.87	3.88
77	0.00	0.00	0.06	1.92	2.74	3.20	3.38	3.48	3.55	3.59	3.63	3.67	3.73	3.76	3.78	3.80
78	0.00	0.00	0.02	1.83	2.66	3.11	3.30	3.40	3.47	3.51	3.54	3.59	3.64	3.67	3.70	3.71
79	0.00	0.00	0.00	1.74	2.57	3.03	3.21	3.32	3.38	3.43	3.46	3.51	3.56	3.59	3.62	3.63
80	0.00	0.00	0.00	1.65	2.49	2.94	3.13	3.24	3.30	3.35	3.38	3.43	3.48	3.51	3.54	3.55
81	0.00	0.00	0.00	1.57	2.40	2.86	3.05	3.16	3.22	3.27	3.30	3.35	3.40	3.43	3.46	3.47
82	0.00	0.00	0.00	1.49	2.32	2.79	2.98	3.08	3.15	3.19	3.22	3.27	3.33	3.36	3.38	3.40
83	0.00	0.00	0.00	1.41	2.25	2.71	2.90	3.00	3.07	3.11	3.15	3.19	3.25	3.28	3.31	3.32
84	0.00	0.00	0.00	1.34	2.17	2.63	2.82	2.93	2.99	3.04	3.07	3.12	3.18	3.21	3.23	3.25
85	0.00	0.00	0.00	1.26	2.09	2.56	2.75	2.85	2.92	2.97	3.00	3.05	3.10	3.13	3.16	3.17
86	0.00	0.00	0.00	1.19	2.02	2.48	2.68	2.78	2.85	2.89	2.93	2.97	3.03	3.06	3.09	3.10
87	0.00	0.00	0.00	1.12	1.95	2.41	2.61	2.71	2.78	2.82	2.86	2.90	2.96	2.99	3.02	3.03
88	0.00	0.00	0.00	1.06	1.88	2.34	2.54	2.64	2.71	2.75	2.79	2.83	2.89	2.92	2.95	2.96
89	0.00	0.00	0.00	0.99	1.81	2.28	2.47	2.57	2.64	2.69	2.72	2.77	2.83	2.85	2.88	2.90
90	0.00	0.00	0.00	0.93	1.75	2.21	2.40	2.51	2.57	2.62	2.65	2.70	2.76	2.79	2.82	2.83
91	0.00	0.00	0.00	0.87	1.68	2.14	2.34	2.44	2.51	2.56	2.59	2.63	2.69	2.72	2.75	2.77
92	0.00	0.00	0.00	0.81	1.62	2.08	2.27	2.38	2.45	2.49	2.52	2.57	2.63	2.66	2.69	2.70
93	0.00	0.00	0.00	0.76	1.56	2.02	2.21	2.32	2.38	2.43	2.46	2.51	2.57	2.60	2.62	2.64
94	0.00	0.00	0.00	0.70	1.50	1.96	2.15	2.25	2.32	2.37	2.40	2.45	2.51	2.54	2.56	2.58
95	0.00	0.00	0.00	0.65	1.44	1.90	2.09	2.19	2.26	2.31	2.34	2.39	2.45	2.48	2.50	2.52
96	0.00	0.00	0.00	0.60	1.38	1.84	2.03	2.14	2.20	2.25	2.28	2.33	2.39	2.42	2.44	2.46
97	0.00	0.00	0.00	0.56	1.33	1.78	1.97	2.08	2.14	2.19	2.22	2.27	2.33	2.36	2.39	2.40
98	0.00	0.00	0.00	0.51	1.27	1.73	1.92	2.02	2.09	2.13	2.17	2.21	2.27	2.30	2.33	2.34
99	0.00	0.00	0.00	0.47	1.22	1.67	1.86	1.97	2.03	2.08	2.11	2.16	2.22	2.25	2.27	2.29
00	0.00	0.00	0.00	0.43	1.17	1.62	1.81	1.91	1.98	2.03	2.06	2.10	2.16	2.19	2.22	2.23
01	0.00	0.00	0.00	0.39	1.12	1.57	1.76	1.86	1.93	1.97	2.01	2.05	2.11	2.14	2.17	2.18
02	0.00	0.00	0.00	0.36	1.07	1.52	1.71	1.81	1.87	1.92	1.95	2.00	2.06	2.09	2.11	2.13
03	0.00	0.00	0.00	0.32	1.03	1.47	1.66	1.76	1.82	1.87	1.90	1.95	2.01	2.04	2.06	2.08
04	0.00	0.00	0.00	0.29	0.98	1.42	1.61	1.71	1.77	1.82	1.85	1.90	1.96	1.99	2.01	2.03
05	0.00	0.00	0.00	0.26	0.94	1.37	1.56	1.66	1.73	1.77	1.80	1.85	1.91	1.94	1.96	1.98
06	0.00	0.00	0.00	0.23	0.90	1.33	1.51	1.61	1.68	1.72	1.76	1.80	1.86	1.89	1.92	1.93
07	0.00	0.00	0.00	0.21	0.86	1.28	1.47	1.57	1.63	1.68	1.71	1.76	1.81	1.84	1.87	1.88
08	0.00	0.00	0.00	0.18	0.82	1.24	1.42	1.52	1.59	1.63	1.66	1.71	1.77	1.79	1.82	1.84
09	0.00	0.00	0.00	0.16	0.78	1.20	1.38	1.48	1.54	1.59	1.62	1.66	1.72	1.75	1.78	1.79

* Values tabulated are read in percent.

Table 13.16. (cont.) Estimating the Lot Percent Defective Using Standard Deviation Method*
(Table B–5 of MIL-STD-414)

Q_L or Q_U	Sample Size														
	3	4	5	7	10	15	20	25	30	35	40	50	75	100	150	2
2.10	0.00	0.00	0.00	0.14	0.74	1.16	1.34	1.44	1.50	1.54	1.58	1.62	1.68	1.71	1.73	1.
2.11	0.00	0.00	0.00	0.12	0.71	1.12	1.30	1.39	1.46	1.50	1.53	1.58	1.63	1.66	1.69	1.
2.12	0.00	0.00	0.00	0.10	0.67	1.08	1.26	1.35	1.42	1.46	1.49	1.54	1.59	1.62	1.65	1.
2.13	0.00	0.00	0.00	0.08	0.64	1.04	1.22	1.31	1.38	1.42	1.45	1.50	1.55	1.58	1.61	1.
2.14	0.00	0.00	0.00	0.07	0.61	1.00	1.18	1.28	1.34	1.38	1.41	1.46	1.51	1.54	1.57	1.
2.15	0.00	0.00	0.00	0.06	0.58	0.97	1.14	1.24	1.30	1.34	1.37	1.42	1.47	1.50	1.53	1.
2.16	0.00	0.00	0.00	0.05	0.55	0.93	1.10	1.20	1.26	1.30	1.34	1.38	1.43	1.46	1.49	1.
2.17	0.00	0.00	0.00	0.04	0.52	0.90	1.07	1.16	1.22	1.27	1.30	1.34	1.40	1.42	1.45	1.
2.18	0.00	0.00	0.00	0.03	0.49	0.87	1.03	1.13	1.19	1.23	1.26	1.30	1.36	1.39	1.41	1.
2.19	0.00	0.00	0.00	0.02	0.46	0.83	1.00	1.09	1.15	1.20	1.23	1.27	1.32	1.35	1.38	1.
2.20	0.000	0.000	0.000	0.015	0.437	0.803	0.968	1.061	1.120	1.161	1.192	1.233	1.287	1.314	1.340	1.
2.21	0.000	0.000	0.000	0.010	0.413	0.772	0.936	1.028	1.087	1.128	1.158	1.199	1.253	1.279	1.305	1.
2.22	0.000	0.000	0.000	0.006	0.389	0.743	0.905	0.996	1.054	1.095	1.125	1.166	1.219	1.245	1.271	1.
2.23	0.000	0.000	0.000	0.003	0.366	0.715	0.875	0.965	1.023	1.063	1.093	1.134	1.186	1.212	1.238	1.
2.24	0.000	0.000	0.000	0.002	0.345	0.687	0.845	0.935	0.992	1.032	1.061	1.102	1.154	1.180	1.205	1.
2.25	0.000	0.000	0.000	0.001	0.324	0.660	0.816	0.905	0.962	1.002	1.031	1.071	1.123	1.148	1.173	1.
2.26	0.000	0.000	0.000	0.000	0.304	0.634	0.789	0.876	0.933	0.972	1.001	1.041	1.092	1.117	1.142	1.
2.27	0.000	0.000	0.000	0.000	0.285	0.609	0.762	0.848	0.904	0.943	0.972	1.011	1.062	1.087	1.112	1.
2.28	0.000	0.000	0.000	0.000	0.267	0.585	0.735	0.821	0.876	0.915	0.943	0.982	1.033	1.058	1.082	1.
2.29	0.000	0.000	0.000	0.000	0.250	0.561	0.710	0.794	0.849	0.887	0.915	0.954	1.004	1.029	1.053	1.
2.30	0.000	0.000	0.000	0.000	0.233	0.538	0.685	0.769	0.823	0.861	0.888	0.927	0.977	1.001	1.025	1.
2.31	0.000	0.000	0.000	0.000	0.218	0.516	0.661	0.743	0.797	0.834	0.862	0.900	0.949	0.974	0.997	1.
2.32	0.000	0.000	0.000	0.000	0.203	0.495	0.637	0.719	0.772	0.809	0.836	0.874	0.923	0.947	0.971	0.
2.33	0.000	0.000	0.000	0.000	0.189	0.474	0.614	0.695	0.748	0.784	0.811	0.848	0.897	0.921	0.944	0.
2.34	0.000	0.000	0.000	0.000	0.175	0.454	0.592	0.672	0.724	0.760	0.787	0.824	0.872	0.895	0.915	0.
2.35	0.000	0.000	0.000	0.000	0.163	0.435	0.571	0.650	0.701	0.736	0.763	0.799	0.847	0.870	0.893	0.
2.36	0.000	0.000	0.000	0.000	0.151	0.416	0.550	0.628	0.678	0.714	0.740	0.776	0.823	0.846	0.869	0.
2.37	0.000	0.000	0.000	0.000	0.139	0.398	0.530	0.606	0.656	0.691	0.717	0.753	0.799	0.822	0.845	0.
2.38	0.000	0.000	0.000	0.000	0.128	0.381	0.510	0.586	0.635	0.670	0.695	0.730	0.777	0.799	0.822	0.
2.39	0.000	0.000	0.000	0.000	0.118	0.364	0.491	0.566	0.614	0.648	0.674	0.709	0.754	0.777	0.799	0.
2.40	0.000	0.000	0.000	0.000	0.109	0.348	0.473	0.546	0.594	0.628	0.653	0.687	0.732	0.755	0.777	0.
2.41	0.000	0.000	0.000	0.000	0.100	0.332	0.455	0.527	0.575	0.608	0.633	0.667	0.711	0.733	0.755	0.
2.42	0.000	0.000	0.000	0.000	0.091	0.317	0.437	0.509	0.555	0.588	0.613	0.646	0.691	0.712	0.734	0.
2.43	0.000	0.000	0.000	0.000	0.083	0.302	0.421	0.491	0.537	0.569	0.593	0.627	0.670	0.692	0.713	0.
2.44	0.000	0.000	0.000	0.000	0.076	0.288	0.404	0.474	0.519	0.551	0.575	0.608	0.651	0.672	0.693	0.
2.45	0.000	0.000	0.000	0.000	0.069	0.275	0.389	0.457	0.501	0.533	0.556	0.589	0.632	0.653	0.673	0.
2.46	0.000	0.000	0.000	0.000	0.063	0.262	0.373	0.440	0.484	0.516	0.539	0.571	0.613	0.634	0.654	0.
2.47	0.000	0.000	0.000	0.000	0.057	0.249	0.339	0.425	0.468	0.499	0.521	0.553	0.595	0.615	0.635	0.
2.48	0.000	0.000	0.000	0.000	0.051	0.237	0.344	0.409	0.452	0.482	0.505	0.536	0.577	0.597	0.617	0.
2.49	0.000	0.000	0.000	0.000	0.046	0.226	0.331	0.394	0.436	0.466	0.488	0.519	0.560	0.580	0.600	0.
2.50	0.000	0.000	0.000	0.000	0.041	0.214	0.317	0.380	0.421	0.451	0.473	0.503	0.543	0.563	0.582	0.
2.51	0.000	0.000	0.000	0.000	0.037	0.204	0.304	0.366	0.407	0.436	0.457	0.487	0.527	0.546	0.565	0.
2.52	0.000	0.000	0.000	0.000	0.033	0.193	0.292	0.352	0.392	0.421	0.442	0.472	0.511	0.530	0.549	0.
2.53	0.000	0.000	0.000	0.000	0.029	0.184	0.280	0.339	0.379	0.407	0.428	0.457	0.495	0.514	0.533	0.
2.54	0.000	0.000	0.000	0.000	0.026	0.174	0.268	0.326	0.365	0.393	0.413	0.442	0.480	0.499	0.517	0.

* Values tabulated are read in percent.

Table 13.16. (cont.) Estimating the Lot Percent Defective Using Standard Deviation Method*

(Table B–5 of MIL-STD-414)

	Sample Size															
	3	4	5	7	10	15	20	25	30	35	40	50	75	100	150	200
55	0.000	0.000	0.000	0.000	0.023	0.165	0.257	0.314	0.352	0.379	0.400	0.428	0.465	0.484	0.502	0.511
56	0.000	0.000	0.000	0.000	0.020	0.156	0.246	0.302	0.340	0.366	0.386	0.414	0.451	0.469	0.487	0.496
57	0.000	0.000	0.000	0.000	0.017	0.148	0.236	0.291	0.327	0.354	0.373	0.401	0.437	0.455	0.473	0.482
58	0.000	0.000	0.000	0.000	0.015	0.140	0.226	0.279	0.316	0.341	0.361	0.388	0.424	0.441	0.459	0.468
59	0.000	0.000	0.000	0.000	0.013	0.133	0.216	0.269	0.304	0.330	0.349	0.375	0.410	0.428	0.445	0.454
60	0.000	0.000	0.000	0.000	0.011	0.125	0.207	0.258	0.293	0.318	0.337	0.363	0.398	0.415	0.432	0.441
61	0.000	0.000	0.000	0.000	0.009	0.118	0.198	0.248	0.282	0.307	0.325	0.351	0.385	0.402	0.419	0.428
62	0.000	0.000	0.000	0.000	0.008	0.112	0.189	0.238	0.272	0.296	0.314	0.339	0.373	0.390	0.406	0.415
63	0.000	0.000	0.000	0.000	0.007	0.105	0.181	0.229	0.262	0.285	0.303	0.328	0.361	0.378	0.394	0.402
64	0.000	0.000	0.000	0.000	0.005	0.099	0.172	0.220	0.252	0.275	0.293	0.317	0.350	0.366	0.382	0.390
65	0.000	0.000	0.000	0.000	0.005	0.094	0.165	0.211	0.243	0.265	0.282	0.307	0.339	0.355	0.371	0.379
66	0.000	0.000	0.000	0.000	0.004	0.088	0.157	0.202	0.233	0.256	0.273	0.296	0.328	0.344	0.359	0.367
67	0.000	0.000	0.000	0.000	0.003	0.083	0.150	0.194	0.224	0.246	0.263	0.286	0.317	0.333	0.348	0.356
68	0.000	0.000	0.000	0.000	0.002	0.078	0.143	0.186	0.216	0.237	0.254	0.277	0.307	0.322	0.338	0.345
69	0.000	0.000	0.000	0.000	0.002	0.073	0.136	0.179	0.208	0.229	0.245	0.267	0.297	0.312	0.327	0.335
70	0.000	0.000	0.000	0.000	0.001	0.069	0.130	0.171	0.200	0.220	0.236	0.258	0.288	0.302	0.317	0.325
71	0.000	0.000	0.000	0.000	0.001	0.064	0.124	0.164	0.192	0.212	0.227	0.249	0.278	0.293	0.307	0.315
72	0.000	0.000	0.000	0.000	0.000	0.060	0.118	0.157	0.184	0.204	0.219	0.241	0.269	0.283	0.298	0.305
73	0.000	0.000	0.000	0.000	0.000	0.057	0.112	0.151	0.177	0.197	0.211	0.232	0.260	0.274	0.288	0.296
74	0.000	0.000	0.000	0.000	0.000	0.053	0.107	0.144	0.170	0.189	0.204	0.224	0.252	0.266	0.279	0.286
75	0.000	0.000	0.000	0.000	0.000	0.049	0.102	0.138	0.163	0.182	0.196	0.216	0.243	0.257	0.271	0.277
76	0.000	0.000	0.000	0.000	0.000	0.046	0.087	0.132	0.157	0.175	0.189	0.209	0.235	0.249	0.262	0.269
77	0.000	0.000	0.000	0.000	0.000	0.043	0.092	0.126	0.151	0.168	0.182	0.201	0.227	0.241	0.254	0.260
78	0.000	0.000	0.000	0.000	0.000	0.040	0.087	0.121	0.145	0.162	0.175	0.194	0.220	0.233	0.246	0.252
79	0.000	0.000	0.000	0.000	0.000	0.037	0.083	0.115	0.139	0.156	0.169	0.187	0.212	0.225	0.238	0.244
80	0.000	0.000	0.000	0.000	0.000	0.035	0.079	0.110	0.133	0.150	0.162	0.181	0.205	0.218	0.230	0.237
81	0.000	0.000	0.000	0.000	0.000	0.032	0.075	0.105	0.128	0.144	0.156	0.174	0.198	0.211	0.223	0.229
82	0.000	0.000	0.000	0.000	0.000	0.030	0.071	0.101	0.122	0.138	0.150	0.168	0.192	0.204	0.216	0.222
83	0.000	0.000	0.000	0.000	0.000	0.028	0.067	0.096	0.117	0.133	0.145	0.162	0.185	0.197	0.209	0.215
84	0.000	0.000	0.000	0.000	0.000	0.026	0.064	0.092	0.112	0.128	0.139	0.156	0.179	0.190	0.202	0.208
85	0.000	0.000	0.000	0.000	0.000	0.024	0.060	0.088	0.108	0.122	0.134	0.150	0.173	0.184	0.195	0.201
86	0.000	0.000	0.000	0.000	0.000	0.022	0.057	0.084	0.103	0.118	0.129	0.145	0.167	0.178	0.189	0.195
87	0.000	0.000	0.000	0.000	0.000	0.020	0.054	0.080	0.099	0.113	0.124	0.139	0.161	0.172	0.183	0.188
88	0.000	0.000	0.000	0.000	0.000	0.019	0.051	0.076	0.094	0.108	0.119	0.134	0.155	0.166	0.177	0.182
89	0.000	0.000	0.000	0.000	0.000	0.017	0.048	0.073	0.090	0.104	0.114	0.129	0.150	0.160	0.171	0.176
90	0.000	0.000	0.000	0.000	0.000	0.016	0.046	0.069	0.087	0.100	0.110	0.125	0.145	0.155	0.165	0.171
91	0.000	0.000	0.000	0.000	0.000	0.015	0.043	0.066	0.083	0.096	0.106	0.120	0.140	0.150	0.160	0.165
92	0.000	0.000	0.000	0.000	0.000	0.013	0.041	0.063	0.079	0.092	0.101	0.115	0.135	0.145	0.155	0.160
93	0.000	0.000	0.000	0.000	0.000	0.012	0.038	0.060	0.076	0.088	0.097	0.111	0.130	0.140	0.149	0.154
94	0.000	0.000	0.000	0.000	0.000	0.011	0.036	0.057	0.072	0.084	0.093	0.107	0.125	0.135	0.144	0.149
95	0.000	0.000	0.000	0.000	0.000	0.010	0.034	0.054	0.069	0.081	0.090	0.103	0.121	0.130	0.140	0.144
96	0.000	0.000	0.000	0.000	0.000	0.009	0.032	0.051	0.066	0.077	0.086	0.099	0.117	0.126	0.135	0.140
97	0.000	0.000	0.000	0.000	0.000	0.009	0.030	0.049	0.063	0.074	0.083	0.095	0.112	0.121	0.130	0.135
98	0.000	0.000	0.000	0.000	0.000	0.008	0.028	0.046	0.060	0.071	0.079	0.091	0.108	0.117	0.126	0.130
99	0.000	0.000	0.000	0.000	0.000	0.007	0.027	0.044	0.057	0.068	0.076	0.088	0.104	0.113	0.122	0.126

* Values tabulated are read in percent.

Table 13.16. (cont.) Estimating the Lot Percent Defective Using Standard Deviation Method*
(Table B–5 of MIL-STD-414)

Q_L or Q_U	Sample Size														
	3	4	5	7	10	15	20	25	30	35	40	50	75	100	150	
3.00	0.000	0.000	0.000	0.000	0.000	0.006	0.025	0.042	0.055	0.065	0.073	0.084	0.101	0.109	0.118	0
3.01	0.000	0.000	0.000	0.000	0.000	0.006	0.024	0.040	0.052	0.062	0.070	0.081	0.097	0.105	0.114	0.
3.02	0.000	0.000	0.000	0.000	0.000	0.005	0.022	0.038	0.050	0.059	0.067	0.078	0.093	0.101	0.110	0
3.03	0.000	0.000	0.000	0.000	0.000	0.005	0.021	0.036	0.048	0.057	0.064	0.075	0.090	0.098	0.106	0.
3.04	0.000	0.000	0.000	0.000	0.000	0.004	0.019	0.034	0.045	0.054	0.061	0.072	0.087	0.094	0.102	0.
3.05	0.000	0.000	0.000	0.000	0.000	0.004	0.018	0.032	0.043	0.052	0.059	0.069	0.083	0.091	0.099	0
3.06	0.000	0.000	0.000	0.000	0.000	0.003	0.017	0.030	0.041	0.050	0.056	0.066	0.080	0.088	0.095	0.
3.07	0.000	0.000	0.000	0.000	0.000	0.003	0.016	0.029	0.039	0.047	0.054	0.064	0.077	0.085	0.092	0
3.08	0.000	0.000	0.000	0.000	0.000	0.003	0.015	0.027	0.037	0.045	0.052	0.061	0.074	0.081	0.089	0.
3.09	0.000	0.000	0.000	0.000	0.000	0.002	0.014	0.026	0.036	0.043	0.049	0.059	0.072	0.079	0.086	0.
3.10	0.000	0.000	0.000	0.000	0.000	0.002	0.013	0.024	0.034	0.041	0.047	0.056	0.069	0.076	0.083	0.
3.11	0.000	0.000	0.000	0.000	0.000	0.002	0.012	0.023	0.032	0.039	0.045	0.054	0.066	0.073	0.080	0.
3.12	0.000	0.000	0.000	0.000	0.000	0.002	0.011	0.022	0.031	0.038	0.043	0.052	0.064	0.070	0.077	0
3.13	0.000	0.000	0.000	0.000	0.000	0.002	0.011	0.021	0.029	0.036	0.041	0.050	0.061	0.068	0.044	0.
3.14	0.000	0.000	0.000	0.000	0.000	0.001	0.010	0.019	0.028	0.034	0.040	0.048	0.059	0.065	0.071	0.
3.15	0.000	0.000	0.000	0.000	0.000	0.001	0.009	0.018	0.026	0.033	0.038	0.046	0.057	0.063	0.069	0.
3.16	0.000	0.000	0.000	0.000	0.000	0.001	0.009	0.017	0.025	0.031	0.036	0.044	0.055	0.060	0.066	0.
3.17	0.000	0.000	0.000	0.000	0.000	0.001	0.008	0.016	0.024	0.030	0.035	0.042	0.053	0.058	0.064	0.
3.18	0.000	0.000	0.000	0.000	0.000	0.001	0.007	0.015	0.022	0.028	0.033	0.040	0.050	0.056	0.062	0.
3.19	0.000	0.000	0.000	0.000	0.000	0.001	0.007	0.015	0.021	0.027	0.032	0.038	0.049	0.054	0.059	0
3.20	0.000	0.000	0.000	0.000	0.000	0.001	0.006	0.014	0.020	0.026	0.030	0.037	0.047	0.052	0.057	0.
3.21	0.000	0.000	0.000	0.000	0.000	0.000	0.006	0.013	0.019	0.024	0.029	0.035	0.045	0.050	0.055	0
3.22	0.000	0.000	0.000	0.000	0.000	0.000	0.005	0.012	0.018	0.023	0.027	0.034	0.043	0.048	0.053	0.
3.23	0.000	0.000	0.000	0.000	0.000	0.000	0.005	0.011	0.017	0.022	0.026	0.032	0.041	0.046	0.051	0.
3.24	0.000	0.000	0.000	0.000	0.000	0.000	0.005	0.011	0.016	0.021	0.025	0.031	0.040	0.044	0.049	0.
3.25	0.000	0.000	0.000	0.000	0.000	0.000	0.004	0.010	0.015	0.020	0.024	0.030	0.038	0.043	0.048	0.
3.26	0.000	0.000	0.000	0.000	0.000	0.000	0.004	0.009	0.015	0.019	0.023	0.028	0.037	0.041	0.046	0.
3.27	0.000	0.000	0.000	0.000	0.000	0.000	0.004	0.009	0.014	0.019	0.022	0.027	0.035	0.040	0.044	0.
3.28	0.000	0.000	0.000	0.000	0.000	0.000	0.003	0.008	0.013	0.017	0.021	0.026	0.034	0.038	0.042	0.
3.29	0.000	0.000	0.000	0.000	0.000	0.000	0.003	0.008	0.012	0.016	0.020	0.025	0.032	0.037	0.041	0
3.30	0.000	0.000	0.000	0.000	0.000	0.000	0.003	0.007	0.012	0.015	0.019	0.024	0.031	0.035	0.039	0
3.31	0.000	0.000	0.000	0.000	0.000	0.000	0.003	0.007	0.011	0.015	0.018	0.023	0.030	0.034	0.038	0.
3.32	0.000	0.000	0.000	0.000	0.000	0.000	0.002	0.006	0.010	0.014	0.017	0.022	0.029	0.032	0.036	0.
3.33	0.000	0.000	0.000	0.000	0.000	0.000	0.002	0.006	0.010	0.013	0.016	0.021	0.027	0.031	0.035	0
3.34	0.000	0.000	0.000	0.000	0.000	0.000	0.002	0.006	0.009	0.013	0.015	0.020	0.026	0.030	0.034	0.
3.35	0.000	0.000	0.000	0.000	0.000	0.000	0.002	0.005	0.009	0.012	0.015	0.019	0.025	0.029	0.032	0.
3.36	0.000	0.000	0.000	0.000	0.000	0.000	0.002	0.005	0.008	0.011	0.014	0.018	0.024	0.028	0.031	0.
3.37	0.000	0.000	0.000	0.000	0.000	0.000	0.002	0.005	0.008	0.011	0.013	0.017	0.023	0.026	0.030	0.
3.38	0.000	0.000	0.000	0.000	0.000	0.000	0.001	0.004	0.007	0.010	0.013	0.016	0.022	0.025	0.029	0.
3.39	0.000	0.000	0.000	0.000	0.000	0.000	0.001	0.004	0.007	0.010	0.012	0.016	0.021	0.024	0.028	0.
3.40	0.000	0.000	0.000	0.000	0.000	0.000	0.001	0.004	0.007	0.009	0.011	0.015	0.020	0.023	0.027	0.
3.41	0.000	0.000	0.000	0.000	0.000	0.000	0.001	0.003	0.006	0.009	0.011	0.014	0.020	0.022	0.026	0.
3.42	0.000	0.000	0.000	0.000	0.000	0.000	0.001	0.003	0.006	0.008	0.010	0.014	0.019	0.022	0.025	0.
3.43	0.000	0.000	0.000	0.000	0.000	0.000	0.001	0.003	0.005	0.008	0.010	0.013	0.018	0.021	0.024	0.
3.44	0.000	0.000	0.000	0.000	0.000	0.000	0.001	0.003	0.005	0.007	0.009	0.012	0.017	0.020	0.023	0.

* Values tabulated are read in percent.

Table 13.16. (cont.) Estimating the Lot Percent Defective Using Standard Deviation Method*

(Table B-5 of MIL-STD-414)

	Sample Size															
	3	4	5	7	10	15	20	25	30	35	40	50	75	100	150	200
5	0.000	0.000	0.000	0.000	0.000	0.000	0.001	0.003	0.005	0.007	0.009	0.012	0.016	0.019	0.022	0.023
6	0.000	0.000	0.000	0.000	0.000	0.000	0.001	0.002	0.005	0.007	0.008	0.011	0.016	0.018	0.021	0.022
7	0.000	0.000	0.000	0.000	0.000	0.000	0.001	0.002	0.004	0.006	0.008	0.011	0.015	0.017	0.020	0.022
8	0.000	0.000	0.000	0.000	0.000	0.000	0.001	0.002	0.004	0.006	0.007	0.010	0.014	0.017	0.019	0.021
9	0.000	0.000	0.000	0.000	0.000	0.000	0.000	0.002	0.004	0.005	0.007	0.010	0.014	0.016	0.019	0.020
0	0.000	0.000	0.000	0.000	0.000	0.000	0.000	0.002	0.003	0.005	0.007	0.009	0.013	0.015	0.018	0.019
1	0.000	0.000	0.000	0.000	0.000	0.000	0.000	0.002	0.003	0.005	0.006	0.009	0.013	0.015	0.017	0.018
2	0.000	0.000	0.000	0.000	0.000	0.000	0.000	0.002	0.003	0.005	0.006	0.008	0.012	0.014	0.017	0.018
3	0.000	0.000	0.000	0.000	0.000	0.000	0.000	0.001	0.003	0.004	0.006	0.008	0.012	0.014	0.016	0.017
4	0.000	0.000	0.000	0.000	0.000	0.000	0.000	0.001	0.003	0.004	0.005	0.008	0.011	0.013	0.015	0.016
5	0.000	0.000	0.000	0.000	0.000	0.000	0.000	0.001	0.003	0.004	0.005	0.007	0.011	0.012	0.015	0.016
6	0.000	0.000	0.000	0.000	0.000	0.000	0.000	0.001	0.002	0.004	0.005	0.007	0.010	0.012	0.014	0.015
7	0.000	0.000	0.000	0.000	0.000	0.000	0.000	0.001	0.002	0.003	0.005	0.006	0.010	0.011	0.013	0.014
8	0.000	0.000	0.000	0.000	0.000	0.000	0.000	0.001	0.002	0.003	0.004	0.006	0.009	0.011	0.013	0.014
9	0.000	0.000	0.000	0.000	0.000	0.000	0.000	0.001	0.002	0.003	0.004	0.006	0.009	0.010	0.012	0.013
0	0.000	0.000	0.000	0.000	0.000	0.000	0.000	0.001	0.002	0.003	0.004	0.006	0.008	0.010	0.012	0.013
1	0.000	0.000	0.000	0.000	0.000	0.000	0.000	0.001	0.002	0.003	0.004	0.005	0.008	0.010	0.011	0.012
2	0.000	0.000	0.000	0.000	0.000	0.000	0.000	0.001	0.002	0.003	0.003	0.005	0.008	0.009	0.011	0.012
3	0.000	0.000	0.000	0.000	0.000	0.000	0.000	0.001	0.001	0.002	0.003	0.005	0.007	0.009	0.010	0.011
4	0.000	0.000	0.000	0.000	0.000	0.000	0.000	0.001	0.001	0.002	0.003	0.004	0.007	0.008	0.010	0.011
5	0.000	0.000	0.000	0.000	0.000	0.000	0.000	0.001	0.001	0.002	0.003	0.004	0.007	0.008	0.010	0.010
6	0.000	0.000	0.000	0.000	0.000	0.000	0.000	0.000	0.001	0.002	0.003	0.004	0.006	0.008	0.009	0.010
7	0.000	0.000	0.000	0.000	0.000	0.000	0.000	0.000	0.001	0.002	0.003	0.004	0.006	0.007	0.009	0.010
8	0.000	0.000	0.000	0.000	0.000	0.000	0.000	0.000	0.001	0.002	0.002	0.004	0.006	0.007	0.008	0.009
9	0.000	0.000	0.000	0.000	0.000	0.000	0.000	0.000	0.001	0.002	0.002	0.003	0.005	0.007	0.008	0.009
0	0.000	0.000	0.000	0.000	0.000	0.000	0.000	0.000	0.001	0.002	0.002	0.003	0.005	0.006	0.008	0.008
1	0.000	0.000	0.000	0.000	0.000	0.000	0.000	0.000	0.001	0.001	0.002	0.003	0.005	0.006	0.007	0.008
2	0.000	0.000	0.000	0.000	0.000	0.000	0.000	0.000	0.001	0.001	0.002	0.003	0.005	0.006	0.007	0.008
3	0.000	0.000	0.000	0.000	0.000	0.000	0.000	0.000	0.001	0.001	0.002	0.003	0.005	0.006	0.007	0.007
4	0.000	0.000	0.000	0.000	0.000	0.000	0.000	0.000	0.001	0.001	0.002	0.003	0.004	0.005	0.007	0.007
5	0.000	0.000	0.000	0.000	0.000	0.000	0.000	0.000	0.001	0.001	0.002	0.002	0.004	0.005	0.006	0.007
6	0.000	0.000	0.000	0.000	0.000	0.000	0.000	0.000	0.001	0.001	0.001	0.002	0.004	0.005	0.006	0.007
7	0.000	0.000	0.000	0.000	0.000	0.000	0.000	0.000	0.001	0.001	0.001	0.002	0.004	0.005	0.006	0.006
8	0.000	0.000	0.000	0.000	0.000	0.000	0.000	0.000	0.000	0.001	0.001	0.002	0.004	0.004	0.005	0.006
9	0.000	0.000	0.000	0.000	0.000	0.000	0.000	0.000	0.000	0.001	0.001	0.002	0.003	0.004	0.005	0.006
0	0.000	0.000	0.000	0.000	0.000	0.000	0.000	0.000	0.000	0.001	0.001	0.002	0.003	0.004	0.005	0.006
1	0.000	0.000	0.000	0.000	0.000	0.000	0.000	0.000	0.000	0.001	0.001	0.002	0.003	0.004	0.005	0.005
2	0.000	0.000	0.000	0.000	0.000	0.000	0.000	0.000	0.000	0.001	0.001	0.002	0.003	0.004	0.005	0.005
3	0.000	0.000	0.000	0.000	0.000	0.000	0.000	0.000	0.000	0.001	0.001	0.002	0.003	0.004	0.004	0.005
4	0.000	0.000	0.000	0.000	0.000	0.000	0.000	0.000	0.000	0.001	0.001	0.001	0.003	0.003	0.004	0.005
5	0.000	0.000	0.000	0.000	0.000	0.000	0.000	0.000	0.000	0.001	0.001	0.001	0.002	0.003	0.004	0.004
6	0.000	0.000	0.000	0.000	0.000	0.000	0.000	0.000	0.000	0.000	0.001	0.001	0.002	0.003	0.004	0.004
7	0.000	0.000	0.000	0.000	0.000	0.000	0.000	0.000	0.000	0.000	0.001	0.001	0.002	0.003	0.004	0.004
8	0.000	0.000	0.000	0.000	0.000	0.000	0.000	0.000	0.000	0.000	0.001	0.001	0.002	0.003	0.004	0.004
9	0.000	0.000	0.000	0.000	0.000	0.000	0.000	0.000	0.000	0.000	0.001	0.001	0.002	0.003	0.003	0.004
0	0.000	0.000	0.000	0.000	0.000	0.000	0.000	0.000	0.000	0.000	0.001	0.001	0.002	0.003	0.003	0.004

* Values tabulated are read in percent.

Table 13.17. Estimating the Lot Percent Defective Using Range Method*
(Table C–5 of MIL-STD-414)

Q_L or Q_U	Sample Size														
	3	4	5	7	10	15	25	30	35	40	50	60	85	115	175
0	50.00	50.00	50.00	50.00	50.00	50.00	50.00	50.00	50.00	50.00	50.00	50.00	50.00	50.00	50.00
.1	47.24	46.67	46.44	46.29	46.20	46.13	46.08	46.07	46.06	46.05	46.05	46.04	46.03	46.03	46.02
.2	44.46	43.33	42.90	42.60	42.42	42.29	42.19	42.17	42.16	42.15	42.13	42.12	42.10	42.10	42.08
.3	41.63	40.00	39.37	38.95	38.70	38.51	38.38	38.34	38.32	38.31	38.28	38.27	38.26	38.24	38.23
.31	41.35	39.67	39.02	38.59	38.33	38.14	38.00	37.96	37.94	37.93	37.90	37.89	37.88	37.86	37.85
.32	41.06	39.33	38.67	38.23	37.96	37.77	37.63	37.59	37.57	37.55	37.53	37.51	37.50	37.48	37.47
.33	40.77	39.00	38.32	37.87	37.60	37.39	37.25	37.21	37.19	37.18	37.15	37.14	37.12	37.11	37.09
.34	40.49	38.67	37.97	37.51	37.23	37.02	36.88	36.84	36.82	36.80	36.77	36.76	36.74	36.73	36.71
.35	40.20	38.33	37.62	37.15	36.87	36.65	36.50	36.46	36.44	36.43	36.40	36.39	36.37	36.36	36.34
.36	39.91	38.00	37.28	36.79	36.50	36.29	36.13	36.09	36.07	36.05	36.03	36.01	35.99	35.97	35.96
.37	39.62	37.67	36.93	36.43	36.14	35.92	35.76	35.72	35.70	35.68	35.65	35.64	35.62	35.61	35.59
.38	39.33	37.33	36.58	36.07	35.78	35.55	35.39	35.35	35.33	35.31	35.28	35.27	35.25	35.24	33.22
.39	39.03	37.00	36.23	35.72	35.41	35.19	35.02	34.98	34.96	34.94	34.92	34.90	34.88	34.87	34.85
.40	38.74	36.67	35.88	35.36	35.05	34.82	34.66	34.62	34.59	34.58	34.55	34.53	34.51	34.49	34.48
.41	38.45	36.33	35.54	35.01	34.69	34.46	34.29	34.25	34.23	34.21	34.18	34.17	34.14	34.12	34.11
.42	38.15	36.00	35.19	34.65	34.33	34.10	33.93	33.89	33.86	33.85	33.82	33.80	33.78	33.77	33.75
.43	37.85	35.67	34.85	34.30	33.98	33.74	33.57	33.53	33.50	33.48	33.45	33.44	33.41	33.39	33.38
.44	37.56	35.33	34.50	33.95	33.62	33.38	33.21	33.17	33.14	33.12	33.09	33.08	33.05	33.03	33.02
.45	37.26	35.00	34.16	33.60	33.27	33.02	32.85	32.81	32.78	32.76	32.73	32.72	32.69	32.67	32.66
.46	36.96	34.67	33.81	33.24	32.91	32.66	32.49	32.45	32.42	32.40	32.37	32.36	32.33	32.31	32.30
.47	36.66	34.33	33.47	32.89	32.56	32.31	32.13	32.09	32.06	32.04	32.01	32.00	31.97	31.95	31.94
.48	36.35	34.00	33.12	32.55	32.21	31.96	31.78	31.74	31.71	31.69	31.66	31.64	31.62	31.61	31.59
.49	36.05	33.67	32.78	32.20	31.86	31.60	31.42	31.38	31.35	31.33	31.30	31.29	31.26	31.24	31.23
.50	35.75	33.33	32.44	31.85	31.51	31.25	31.07	31.03	31.00	30.98	30.95	30.94	30.91	30.89	30.88
.51	35.44	33.00	32.10	31.51	31.16	30.90	30.72	30.68	30.65	30.63	30.60	30.59	30.55	30.55	30.53
.52	35.13	32.67	31.76	31.16	30.81	30.55	30.37	30.33	30.30	30.28	30.25	30.24	30.21	30.19	30.18
.53	34.82	32.33	31.42	30.82	30.46	30.21	30.02	29.98	29.95	29.93	29.90	29.89	29.86	29.84	29.83
.54	34.51	32.00	31.08	30.47	30.12	29.86	29.68	29.64	29.61	29.59	29.56	29.54	29.52	29.50	29.48
.55	34.20	31.67	30.74	30.13	29.78	29.52	29.33	29.29	29.26	29.24	29.21	29.20	29.17	29.15	29.14
.56	33.88	31.33	30.40	29.79	29.44	29.18	28.99	28.95	28.92	28.90	28.87	28.86	28.83	28.81	28.80
.57	33.57	31.00	30.06	29.45	29.09	28.83	28.65	28.61	28.58	28.56	28.53	28.52	28.49	28.47	28.46
.58	33.25	30.67	29.73	29.11	28.76	28.50	28.31	28.27	28.24	28.22	28.19	28.18	28.15	28.13	28.12
.59	32.93	30.33	29.39	28.77	28.42	28.16	27.97	27.93	27.91	27.89	27.86	27.84	27.82	27.80	27.78
.60	32.61	30.00	29.05	28.44	28.08	27.82	27.64	27.60	27.57	27.55	27.52	27.51	27.48	27.46	27.45
.61	32.28	29.67	28.72	28.10	27.75	27.49	27.31	27.27	27.24	27.22	27.19	27.17	27.15	27.14	27.12
.62	31.96	29.33	28.39	27.77	27.41	27.16	26.97	26.93	26.91	26.89	26.86	26.84	26.82	26.81	26.79
.63	31.63	29.00	28.05	27.44	27.08	26.82	26.64	26.60	26.58	26.56	26.53	26.51	26.49	26.48	26.46
.64	31.30	28.67	27.72	27.11	26.75	26.50	26.32	26.28	26.25	26.23	26.20	26.19	26.16	26.14	26.13
.65	30.97	28.33	27.39	26.78	26.42	26.17	25.99	25.95	25.92	25.90	25.87	25.86	25.84	25.83	25.81
.66	30.63	28.00	27.06	26.45	26.10	25.84	25.67	25.63	25.60	25.58	25.55	25.54	25.52	25.50	25.48
.67	30.30	27.67	26.73	26.12	25.77	25.52	25.34	25.30	25.28	25.26	25.23	25.22	25.20	25.18	25.16
.68	29.96	27.33	26.40	25.79	25.45	25.20	25.02	24.98	24.96	24.94	24.91	24.90	24.88	24.87	24.85
.69	29.61	27.00	26.07	25.47	25.12	24.88	24.71	24.67	24.64	24.62	24.59	24.58	24.56	24.55	24.53
.70	29.27	26.67	25.74	25.14	24.80	24.56	24.39	24.35	24.32	24.31	24.28	24.27	24.25	24.24	24.22
.71	28.92	26.33	25.41	24.82	24.48	24.24	24.07	24.03	24.01	23.99	23.97	23.95	23.93	23.91	23.90
.72	28.57	26.00	25.09	24.50	24.17	23.93	23.76	23.72	23.70	23.68	23.66	23.64	23.62	23.60	23.59
.73	28.22	25.67	24.76	24.18	23.85	23.61	23.45	23.41	23.39	23.37	23.35	23.33	23.32	23.30	23.29
.74	27.86	25.33	24.44	23.86	23.54	23.30	23.14	23.10	23.08	23.07	23.04	23.03	23.01	23.00	22.98

* Values tabulated are read in percent.

Table 13.17. (cont.) Estimating the Lot Percent Defective Using Range Method*
(Table C–5 of MIL-STD-414)

Q_L or Q_U	Sample Size															
	3	4	5	7	10	15	25	30	35	40	50	60	85	115	175	230
.75	27.50	25.00	24.11	23.55	23.22	22.99	22.84	22.80	22.78	22.76	22.74	22.72	22.71	22.69	22.68	22.68
.76	27.13	24.67	23.79	23.23	22.91	22.69	22.53	22.49	22.47	22.46	22.43	22.42	22.41	22.39	22.38	22.38
.77	26.77	24.33	23.47	22.92	22.60	22.38	22.23	22.19	22.17	22.16	22.13	22.12	22.11	22.09	22.08	22.08
.78	26.39	24.00	23.15	22.60	22.30	22.08	21.93	21.90	21.88	21.86	21.85	21.83	21.81	21.80	21.78	21.78
.79	26.02	23.67	22.83	22.29	21.99	21.78	21.64	21.60	21.58	21.57	21.54	21.53	21.52	21.50	21.49	21.49
.80	25.64	23.33	22.51	21.98	21.69	21.48	21.34	21.30	21.28	21.27	21.26	21.24	21.22	21.22	21.20	21.20
.81	25.25	23.00	22.19	21.68	21.39	21.18	21.04	21.01	20.99	20.98	20.97	20.95	20.93	20.93	20.91	20.91
.82	24.86	22.67	21.87	21.37	21.09	20.89	20.75	20.72	20.70	20.69	20.68	20.66	20.64	20.64	20.62	20.62
.83	24.47	22.33	21.56	21.06	20.79	20.59	20.46	20.43	20.41	20.40	20.38	20.37	20.36	20.35	20.34	20.34
.84	24.07	22.00	21.24	20.76	20.49	20.30	20.17	20.15	20.13	20.12	20.10	20.09	20.08	20.06	20.06	20.06
.85	23.67	21.67	20.93	20.46	20.20	20.01	19.89	19.87	19.85	19.84	19.82	19.81	19.79	19.79	19.78	19.78
.86	23.26	21.33	20.62	20.16	19.90	19.73	19.60	19.58	19.57	19.56	19.54	19.54	19.52	19.51	19.50	19.50
.87	22.84	21.00	20.31	19.86	19.61	19.44	19.32	19.31	19.29	19.28	19.26	19.25	19.24	19.24	19.22	19.22
.88	22.42	20.67	20.00	19.57	19.33	19.16	19.04	19.03	19.01	19.00	18.98	18.98	18.97	18.96	18.95	18.95
.89	21.99	20.33	19.69	19.27	19.04	18.88	18.77	18.75	18.74	18.73	18.71	18.70	18.69	18.69	18.68	18.68
.90	21.55	20.00	19.38	18.98	18.75	18.60	18.50	18.48	18.47	18.46	18.44	18.43	18.42	18.42	18.41	18.41
.91	21.11	19.67	19.07	18.69	18.47	18.32	18.22	18.21	18.20	18.19	18.17	18.17	18.17	18.16	18.15	18.15
.92	20.66	19.33	18.77	18.40	18.19	18.05	17.96	17.95	17.93	17.92	17.92	17.90	17.89	17.89	17.88	17.88
.93	20.20	19.00	18.46	18.11	17.91	17.78	17.69	17.68	17.67	17.66	17.65	17.65	17.63	17.63	17.62	17.62
.94	19.74	18.67	18.16	17.82	17.64	17.51	17.43	17.42	17.41	17.40	17.39	17.39	17.37	17.37	17.36	17.36
.95	19.25	18.33	17.86	17.54	17.36	17.24	17.17	17.16	17.15	17.14	17.13	17.13	17.12	17.12	17.11	17.11
.96	18.76	18.00	17.56	17.26	17.09	16.98	16.91	16.90	16.89	16.88	16.88	16.87	16.86	16.86	16.86	16.86
.97	18.25	17.67	17.25	16.97	16.82	16.71	16.65	16.64	16.63	16.63	16.62	16.62	16.61	16.61	16.60	16.60
.98	17.74	17.33	16.96	16.70	16.55	16.45	16.39	16.38	16.38	16.37	16.37	16.37	16.36	16.36	16.36	16.36
.99	17.21	17.00	16.66	16.42	16.28	16.19	16.14	16.13	16.13	16.12	16.12	16.12	16.11	16.11	16.11	16.11
1.00	16.67	16.67	16.36	16.14	16.02	15.94	15.89	15.88	15.88	15.88	15.87	15.87	15.87	15.87	15.87	15.87
1.01	16.11	16.33	16.07	15.87	15.76	15.68	15.64	15.63	15.63	15.63	15.63	15.63	15.62	15.62	15.62	15.62
1.02	15.53	16.00	15.78	15.60	15.50	15.43	15.40	15.39	15.39	15.39	15.39	15.39	15.38	15.38	15.38	15.38
1.03	14.93	15.67	15.48	15.33	15.24	15.18	15.15	15.15	15.15	15.15	15.15	15.15	15.15	15.15	15.15	15.15
1.04	14.31	15.33	15.19	15.06	14.98	14.94	14.91	14.91	14.91	14.91	14.91	14.91	14.91	14.91	14.91	14.91
1.05	13.66	15.00	14.91	14.79	14.73	14.69	14.67	14.67	14.67	14.67	14.67	14.68	14.68	14.68	14.68	14.68
1.06	12.98	14.67	14.62	14.53	14.48	14.45	14.44	14.44	14.44	14.44	14.44	14.44	14.45	14.45	14.45	14.45
1.07	12.27	14.33	14.33	14.27	14.23	14.21	14.20	14.21	14.21	14.21	14.21	14.21	14.22	14.22	14.22	14.22
1.08	11.51	14.00	14.05	14.01	13.98	13.97	13.97	13.98	13.98	13.98	13.98	13.99	13.99	13.99	14.00	14.00
1.09	10.71	13.67	13.76	13.75	13.74	13.73	13.74	13.75	13.75	13.75	13.76	13.76	13.77	13.77	13.78	13.78
1.10	9.84	13.33	13.48	13.50	13.49	13.50	13.52	13.52	13.52	13.53	13.54	13.54	13.55	13.55	13.56	13.56
1.11	8.89	13.00	13.20	13.24	13.25	13.27	13.29	13.30	13.30	13.31	13.31	13.32	13.32	13.33	13.34	13.34
1.12	7.82	12.67	12.93	12.99	13.02	13.04	13.07	13.08	13.08	13.09	13.10	13.10	13.12	13.12	13.12	13.12
1.13	6.60	12.33	12.65	12.74	12.78	12.81	12.85	12.86	12.86	12.87	12.89	12.89	12.89	12.90	12.91	12.91
1.14	5.00	12.00	12.37	12.49	12.55	12.59	12.63	12.64	12.65	12.66	12.67	12.67	12.69	12.69	12.70	12.70
1.15	0.29	11.67	12.10	12.25	12.31	12.37	12.42	12.43	12.44	12.45	12.46	12.46	12.48	12.48	12.49	12.49
1.16	0.00	11.33	11.83	12.00	12.08	12.15	12.21	12.22	12.23	12.24	12.25	12.25	12.27	12.28	12.29	12.29
1.17	0.00	11.00	11.56	11.76	11.86	11.93	12.00	12.01	12.02	12.03	12.05	12.06	12.07	12.07	12.08	12.08
1.18	0.00	10.67	11.29	11.52	11.63	11.71	11.79	11.80	11.81	11.82	11.84	11.84	11.86	11.88	11.88	11.88
1.19	0.00	10.33	11.02	11.29	11.41	11.50	11.58	11.60	11.61	11.62	11.64	11.65	11.66	11.68	11.69	11.69

* Values tabulated are read in percent.

Table 13.17. (cont.) Estimating the Lot Percent Defective Using Range Method*
(Table C–5 of MIL-STD-414)

Q_L or Q_U	Sample Size														
	3	4	5	7	10	15	25	30	35	40	50	60	85	115	175	230
1.20	0.00	10.00	10.76	11.05	11.19	11.29	11.38	11.40	11.41	11.42	11.43	11.45	11.47	11.47	11.49	11.4
1.21	0.00	9.67	10.50	10.82	10.97	11.08	11.18	11.20	11.21	11.22	11.25	11.26	11.27	11.29	11.30	11.3
1.22	0.00	9.33	10.23	10.59	10.76	10.88	10.98	11.00	11.02	11.03	11.04	11.06	11.08	11.09	11.10	11.1
1.23	0.00	9.00	9.97	10.36	10.54	10.67	10.78	10.80	10.82	10.84	10.85	10.87	10.89	10.90	10.91	10.9
1.24	0.00	8.67	9.72	10.13	10.33	10.47	10.58	10.61	10.63	10.64	10.67	10.68	10.70	10.71	10.73	10.7
1.25	0.00	8.33	9.46	9.91	10.12	10.27	10.39	10.42	10.44	10.46	10.47	10.49	10.51	10.52	10.54	10.5
1.26	0.00	8.00	9.21	9.69	9.92	10.08	10.20	10.24	10.26	10.27	10.30	10.31	10.33	10.34	10.36	10.3
1.27	0.00	7.67	8.96	9.47	9.71	9.88	10.01	10.05	10.07	10.09	10.11	10.13	10.15	10.17	10.18	10.1
1.28	0.00	7.33	8.71	9.25	9.51	9.69	9.83	9.87	9.89	9.90	9.93	9.95	9.97	9.99	10.00	10.0
1.29	0.00	7.00	8.46	9.04	9.31	9.50	9.64	9.68	9.71	9.72	9.75	9.77	9.79	9.81	9.83	9.8
1.30	0.00	6.67	8.21	8.86	9.11	9.32	9.47	9.51	9.53	9.55	9.58	9.59	9.62	9.64	9.65	9.6
1.31	0.00	6.33	7.97	8.62	8.92	9.13	9.29	9.33	9.35	9.37	9.40	9.42	9.45	9.47	9.48	9.4
1.32	0.00	6.00	7.73	8.41	8.73	8.95	9.11	9.15	9.18	9.20	9.23	9.25	9.28	9.30	9.31	9.3
1.33	0.00	5.67	7.49	8.20	8.54	8.77	8.94	8.98	9.01	9.03	9.06	9.08	9.11	9.13	9.14	9.1
1.34	0.00	5.33	7.25	8.00	8.35	8.59	8.77	8.81	8.84	8.86	8.89	8.91	8.94	8.96	8.98	8.9
1.35	0.00	5.00	7.02	7.80	8.16	8.41	8.60	8.64	8.67	8.69	8.73	8.75	8.78	8.80	8.82	8.8
1.36	0.00	4.67	6.79	7.60	7.98	8.24	8.43	8.48	8.51	8.53	8.56	8.59	8.62	8.64	8.66	8.6
1.37	0.00	4.33	6.56	7.40	7.80	8.07	8.27	8.31	8.34	8.37	8.40	8.43	8.46	8.48	8.50	8.5
1.38	0.00	4.00	6.33	7.21	7.62	7.90	8.11	8.15	8.18	8.21	8.25	8.26	8.30	8.32	8.34	8.
1.39	0.00	3.67	6.10	7.02	7.45	7.73	7.95	7.99	8.02	8.05	8.09	8.11	8.14	8.17	8.19	8.1
1.40	0.00	3.33	5.88	6.83	7.27	7.57	7.79	7.84	7.88	7.90	7.93	7.96	8.00	8.02	8.03	8.
1.41	0.00	3.00	5.66	6.65	7.10	7.41	7.63	7.68	7.71	7.74	7.78	7.81	7.85	7.87	7.88	7.
1.42	0.00	2.67	5.44	6.46	6.93	7.25	7.48	7.53	7.56	7.59	7.63	7.66	7.70	7.72	7.74	7.
1.43	0.00	2.33	5.23	6.28	6.76	7.09	7.33	7.38	7.41	7.44	7.49	7.51	7.54	7.57	7.59	7.6
1.44	0.00	2.00	5.01	6.10	6.60	6.93	7.18	7.24	7.28	7.30	7.34	7.37	7.41	7.43	7.45	7.
1.45	0.00	1.67	4.81	5.93	6.44	6.78	7.03	7.09	7.13	7.15	7.20	7.23	7.27	7.29	7.30	7.
1.46	0.00	1.33	4.60	5.75	6.28	6.63	6.89	6.95	6.99	7.01	7.06	7.09	7.13	7.15	7.17	7.
1.47	0.00	1.00	4.39	5.58	6.12	6.48	6.74	6.80	6.85	6.87	6.92	6.95	6.99	7.01	7.03	7.
1.48	0.00	0.67	4.19	5.41	5.96	6.34	6.60	6.66	6.71	6.73	6.78	6.81	6.85	6.87	6.89	6.
1.49	0.00	0.33	3.99	5.24	5.81	6.19	6.47	6.53	6.57	6.60	6.64	6.67	6.72	6.74	6.76	6.
1.50	0.00	0.00	3.80	5.08	5.66	6.05	6.33	6.39	6.43	6.46	6.51	6.54	6.58	6.61	6.63	6.
1.51	0.00	0.00	3.61	4.92	5.51	5.91	6.19	6.25	6.30	6.33	6.38	6.41	6.45	6.48	6.50	6.
1.52	0.00	0.00	3.42	4.76	5.37	5.77	6.06	6.12	6.17	6.20	6.25	6.28	6.32	6.35	6.37	6.
1.53	0.00	0.00	3.23	4.60	5.22	5.64	5.93	5.99	6.04	6.07	6.12	6.15	6.20	6.22	6.25	6.
1.54	0.00	0.00	3.05	4.45	5.08	5.50	5.80	5.86	5.91	5.95	6.00	6.03	6.07	6.10	6.12	6.
1.55	0.00	0.00	2.87	4.30	4.94	5.37	5.68	5.74	5.79	5.82	5.87	5.90	5.95	5.98	6.00	6.
1.56	0.00	0.00	2.69	4.15	4.81	5.24	5.55	5.62	5.67	5.70	5.75	5.78	5.83	5.86	5.88	5.
1.57	0.00	0.00	2.52	4.01	4.67	5.11	5.43	5.50	5.55	5.58	5.63	5.66	5.71	5.74	5.77	5.
1.58	0.00	0.00	2.35	3.86	4.54	4.99	5.31	5.38	5.43	5.46	5.52	5.55	5.59	5.62	5.65	5.
1.59	0.00	0.00	2.19	3.72	4.41	4.86	5.19	5.26	5.31	5.34	5.40	5.43	5.48	5.51	5.53	5.
1.60	0.00	0.00	2.03	3.58	4.28	4.74	5.08	5.14	5.19	5.23	5.29	5.32	5.36	5.39	5.42	5.
1.61	0.00	0.00	1.87	3.45	4.16	4.62	4.96	5.03	5.08	5.12	5.17	5.20	5.25	5.28	5.31	5.
1.62	0.00	0.00	1.72	3.31	4.03	4.51	4.85	4.92	4.97	5.01	5.06	5.09	5.14	5.17	5.20	5.
1.63	0.00	0.00	1.57	3.18	3.91	4.39	4.74	4.81	4.86	4.90	4.96	4.99	5.04	5.07	5.10	5.
1.64	0.00	0.00	1.42	3.06	3.79	4.28	4.63	4.70	4.75	4.79	4.85	4.88	4.93	4.96	4.99	5.

* Values tabulated are read in percent.

Table 13.17. (cont.) Estimating the Lot Percent Defective Using Range Method*
(Table C–5 of MIL-STD-414)

L r U	Sample Size															
	3	4	5	7	10	15	25	30	35	40	50	60	85	115	175	230
65	0.00	0.00	1.28	2.93	3.68	4.17	4.52	4.59	4.64	6.68	4.74	4.77	4.83	4.86	4.89	4.91
66	0.00	0.00	1.15	2.81	3.56	4.06	4.41	4.49	4.54	6.58	4.64	4.67	4.72	4.75	4.79	4.81
67	0.00	0.00	1.02	2.69	3.45	3.95	4.31	4.39	4.44	4.48	4.54	4.57	4.62	4.65	4.69	4.71
68	0.00	0.00	0.89	2.57	3.34	3.85	4.21	4.29	4.34	4.38	4.44	4.47	4.53	4.56	4.59	4.61
69	0.00	0.00	0.77	2.46	3.23	3.74	4.10	4.19	4.24	4.28	4.34	4.37	4.43	4.46	4.49	4.51
70	0.00	0.00	0.66	2.35	3.13	3.64	4.00	4.09	4.14	4.18	4.24	4.28	4.33	4.36	4.40	4.42
71	0.00	0.00	0.55	2.24	3.02	3.54	3.92	3.99	4.05	4.09	4.15	4.18	4.24	4.27	4.30	4.31
72	0.00	0.00	0.45	2.13	2.92	3.45	3.82	3.90	3.95	3.99	4.06	4.09	4.15	4.18	4.21	4.23
73	0.00	0.00	0.36	2.03	2.82	3.35	3.73	3.81	3.86	3.90	3.96	4.00	4.06	4.09	4.12	4.14
74	0.00	0.00	0.27	1.93	2.73	3.26	3.63	3.72	3.77	3.81	3.87	3.91	3.97	4.00	4.03	4.05
75	0.00	0.00	0.19	1.83	2.63	3.16	3.54	3.63	3.68	3.72	3.79	3.82	3.88	3.91	3.94	3.96
76	0.00	0.00	0.12	1.73	2.54	3.07	3.45	3.54	3.59	3.63	3.70	3.74	3.79	3.82	3.86	3.88
77	0.00	0.00	0.06	1.64	2.45	2.99	3.37	3.45	3.51	3.55	3.61	3.65	3.71	3.74	3.77	3.79
78	0.00	0.00	0.02	1.55	2.36	2.90	3.28	3.37	3.43	3.47	3.53	3.57	3.62	3.65	3.69	3.71
79	0.00	0.00	0.00	1.46	2.27	2.81	3.20	3.28	3.34	3.38	3.45	3.49	3.54	3.57	3.61	3.63
80	0.00	0.00	0.00	1.38	2.19	2.73	3.11	3.20	3.26	3.30	3.37	3.41	3.46	3.49	3.53	3.55
81	0.00	0.00	0.00	1.29	2.10	2.65	3.03	3.12	3.18	3.22	3.29	3.33	3.38	3.41	3.45	3.47
82	0.00	0.00	0.00	1.21	2.02	2.57	2.96	3.05	3.11	3.15	3.21	3.25	3.31	3.34	3.37	3.39
83	0.00	0.00	0.00	1.14	1.94	2.49	2.88	2.97	3.03	3.07	3.13	3.17	3.23	3.26	3.30	3.32
84	0.00	0.00	0.00	1.06	1.87	2.42	2.80	2.89	2.95	2.99	3.06	3.10	3.16	3.19	3.22	3.24
85	0.00	0.00	0.00	0.99	1.79	2.34	2.73	2.82	2.88	2.92	2.99	3.03	3.08	3.11	3.15	3.17
86	0.00	0.00	0.00	0.92	1.72	2.27	2.66	2.75	2.81	2.85	2.91	2.95	3.01	3.04	3.08	3.10
87	0.00	0.00	0.00	0.86	1.65	2.20	2.59	2.68	2.74	2.78	2.84	2.88	2.94	2.97	3.01	3.03
88	0.00	0.00	0.00	0.79	1.58	2.13	2.52	2.61	2.67	2.71	2.77	2.81	2.87	2.90	2.94	2.96
89	0.00	0.00	0.00	0.73	1.51	2.06	2.45	2.56	2.60	2.64	2.71	2.75	2.81	2.84	2.87	2.89
90	0.00	0.00	0.00	0.67	1.45	1.99	2.38	2.47	2.53	2.57	2.64	2.68	2.74	2.77	2.81	2.83
91	0.00	0.00	0.00	0.62	1.38	1.93	2.32	2.41	2.47	2.51	2.58	2.61	2.67	2.70	2.74	2.76
92	0.00	0.00	0.00	0.56	1.32	1.86	2.25	2.34	2.41	2.45	2.51	2.55	2.61	2.64	2.68	2.70
93	0.00	0.00	0.00	0.51	1.26	1.80	2.19	2.28	2.34	2.38	2.45	2.49	2.55	2.58	2.61	2.63
94	0.00	0.00	0.00	0.46	1.20	1.74	2.13	2.22	2.28	2.32	2.39	2.43	2.49	2.52	2.55	2.57
95	0.00	0.00	0.00	0.42	1.15	1.68	2.07	2.16	2.22	2.26	2.33	2.37	2.43	2.46	2.49	2.51
96	0.00	0.00	0.00	0.37	1.09	1.62	2.01	2.10	2.16	2.20	2.27	2.31	2.37	2.40	2.43	2.45
97	0.00	0.00	0.00	0.33	1.04	1.57	1.95	2.04	2.10	2.14	2.21	2.25	2.31	2.34	2.38	2.40
98	0.00	0.00	0.00	0.30	0.99	1.51	1.90	1.99	2.05	2.09	2.15	2.19	2.25	2.28	2.32	2.34
99	0.00	0.00	0.00	0.26	0.94	1.46	1.84	1.93	1.99	2.03	2.10	2.14	2.20	2.23	2.26	2.28
00	0.00	0.00	0.00	0.23	0.89	1.41	1.79	1.88	1.94	1.98	2.05	2.08	2.14	2.17	2.21	2.23
01	0.00	0.00	0.00	0.20	0.84	1.36	1.74	1.83	1.89	1.93	1.99	2.03	2.09	2.12	2.16	2.18
02	0.00	0.00	0.00	0.17	0.80	1.31	1.69	1.78	1.83	1.87	1.94	1.98	2.04	2.07	2.10	2.12
03	0.00	0.00	0.00	0.14	0.75	1.26	1.64	1.73	1.78	1.82	1.89	1.93	1.99	2.02	2.05	2.07
04	0.00	0.00	0.00	0.12	0.71	1.21	1.59	1.68	1.73	1.77	1.84	1.88	1.94	1.97	2.00	2.02
05	0.00	0.00	0.00	0.10	0.67	1.17	1.54	1.63	1.69	1.73	1.79	1.83	1.89	1.92	1.95	1.97
06	0.00	0.00	0.00	0.08	0.63	1.12	1.49	1.58	1.64	1.68	1.74	1.78	1.84	1.87	1.91	1.93
07	0.00	0.00	0.00	0.06	0.60	1.08	1.45	1.54	1.59	1.63	1.70	1.74	1.79	1.82	1.86	1.88
08	0.00	0.00	0.00	0.05	0.56	1.04	1.40	1.49	1.55	1.59	1.65	1.69	1.75	1.78	1.81	1.83
09	0.00	0.00	0.00	0.03	0.53	1.00	1.36	1.45	1.50	1.54	1.61	1.64	1.70	1.73	1.77	1.79

* Values tabulated are read in percent.

Table 13.17. (cont.) Estimating the Lot Percent Defective Using Range Method*
(Table C–5 of MIL-STD-414)

Q_L or Q_U	Sample Size															
	3	4	5	7	10	15	25	30	35	40	50	60	85	115	175	2
2.10	0.00	0.00	0.00	0.02	0.49	0.96	1.32	1.41	1.46	1.50	1.56	1.60	1.66	1.69	1.72	1.
2.11	0.00	0.00	0.00	0.01	0.46	0.92	1.28	1.36	1.42	1.46	1.52	1.56	1.61	1.64	1.68	1.
2.12	0.00	0.00	0.00	0.00	0.43	0.88	1.24	1.32	1.38	1.42	1.48	1.52	1.57	1.60	1.64	1.
2.13	0.00	0.00	0.00	0.00	0.40	0.85	1.20	1.28	1.34	1.38	1.44	1.48	1.53	1.56	1.60	1.
2.14	0.00	0.00	0.00	0.00	0.38	0.81	1.16	1.25	1.30	1.34	1.40	1.44	1.49	1.52	1.56	1.
2.15	0.00	0.00	0.00	0.00	0.35	0.78	1.13	1.21	1.26	1.30	1.36	1.40	1.45	1.48	1.52	1.
2.16	0.00	0.00	0.00	0.00	0.32	0.75	1.09	1.17	1.22	1.26	1.32	1.36	1.41	1.44	1.48	1.
2.17	0.00	0.00	0.00	0.00	0.30	0.71	1.06	1.13	1.18	1.22	1.29	1.32	1.38	1.41	1.44	1.
2.18	0.00	0.00	0.00	0.00	0.28	0.68	1.02	1.10	1.15	1.19	1.25	1.28	1.34	1.37	1.40	1.
2.19	0.00	0.00	0.00	0.00	0.26	0.65	0.99	1.06	1.11	1.15	1.22	1.25	1.30	1.33	1.37	1.
2.20	0.000	0.000	0.000	0.000	0.236	0.625	0.954	1.030	1.083	1.122	1.178	1.214	1.267	1.299	1.330	1.
2.21	0.000	0.000	0.000	0.000	0.217	0.597	0.922	0.997	1.050	1.089	1.144	1.180	1.233	1.265	1.295	1.
2.22	0.000	0.000	0.000	0.000	0.199	0.570	0.891	0.966	1.018	1.056	1.111	1.147	1.199	1.231	1.261	1.
2.23	0.000	0.000	0.000	0.000	0.182	0.544	0.861	0.935	0.986	1.025	1.079	1.115	1.167	1.197	1.228	1.
2.24	0.000	0.000	0.000	0.000	0.166	0.519	0.831	0.905	0.956	0.994	1.048	1.083	1.135	1.165	1.195	1.
2.25	0.000	0.000	0.000	0.000	0.150	0.495	0.802	0.875	0.926	0.964	1.018	1.052	1.104	1.134	1.163	1.
2.26	0.000	0.000	0.000	0.000	0.136	0.471	0.775	0.847	0.897	0.935	0.987	1.022	1.073	1.103	1.132	1.
2.27	0.000	0.000	0.000	0.000	0.123	0.449	0.748	0.819	0.869	0.906	0.958	0.993	1.043	1.073	1.103	1.
2.28	0.000	0.000	0.000	0.000	0.111	0.427	0.722	0.792	0.841	0.878	0.930	0.964	1.014	1.044	1.073	1.
2.29	0.000	0.000	0.000	0.000	0.099	0.406	0.697	0.766	0.814	0.851	0.902	0.936	0.986	1.015	1.044	1.
2.30	0.000	0.000	0.000	0.000	0.089	0.386	0.672	0.741	0.789	0.825	0.875	0.909	0.959	0.988	1.016	1.
2.31	0.000	0.000	0.000	0.000	0.079	0.367	0.648	0.716	0.763	0.799	0.849	0.882	0.931	0.960	0.988	1.
2.32	0.000	0.000	0.000	0.000	0.070	0.348	0.624	0.691	0.739	0.774	0.823	0.856	0.905	0.934	0.962	0.
2.33	0.000	0.000	0.000	0.000	0.061	0.330	0.601	0.668	0.715	0.750	0.798	0.831	0.879	0.908	0.935	0.
2.34	0.000	0.000	0.000	0.000	0.054	0.313	0.579	0.645	0.691	0.726	0.774	0.807	0.854	0.882	0.909	0.
2.35	0.000	0.000	0.000	0.000	0.047	0.296	0.558	0.623	0.669	0.703	0.750	0.782	0.829	0.857	0.884	0.
2.36	0.000	0.000	0.000	0.000	0.040	0.280	0.538	0.602	0.646	0.680	0.728	0.759	0.806	0.833	0.860	0.
2.37	0.000	0.000	0.000	0.000	0.035	0.265	0.518	0.580	0.624	0.658	0.705	0.736	0.782	0.809	0.836	0.
2.38	0.000	0.000	0.000	0.000	0.029	0.250	0.498	0.560	0.604	0.637	0.683	0.714	0.759	0.787	0.813	0.
2.39	0.000	0.000	0.000	0.000	0.025	0.236	0.479	0.541	0.584	0.616	0.662	0.693	0.737	0.764	0.791	0.
2.40	0.000	0.000	0.000	0.000	0.021	0.223	0.461	0.521	0.564	0.596	0.641	0.671	0.715	0.742	0.769	0.
2.41	0.000	0.000	0.000	0.000	0.017	0.210	0.443	0.503	0.545	0.577	0.621	0.651	0.695	0.721	0.747	0.
2.42	0.000	0.000	0.000	0.000	0.014	0.198	0.426	0.485	0.526	0.557	0.601	0.631	0.674	0.701	0.726	0.
2.43	0.000	0.000	0.000	0.000	0.011	0.186	0.410	0.467	0.508	0.539	0.582	0.611	0.654	0.679	0.705	0.
2.44	0.000	0.000	0.000	0.000	0.009	0.175	0.393	0.450	0.491	0.521	0.564	0.593	0.635	0.660	0.685	0.
2.45	0.000	0.000	0.000	0.000	0.007	0.165	0.378	0.434	0.473	0.503	0.545	0.573	0.616	0.641	0.665	0.
2.46	0.000	0.000	0.000	0.000	0.005	0.154	0.362	0.417	0.456	0.486	0.528	0.556	0.597	0.622	0.646	0.
2.47	0.000	0.000	0.000	0.000	0.004	0.145	0.348	0.403	0.441	0.470	0.511	0.538	0.579	0.604	0.627	0.
2.48	0.000	0.000	0.000	0.000	0.003	0.136	0.333	0.387	0.425	0.454	0.494	0.522	0.562	0.586	0.609	0.
2.49	0.000	0.000	0.000	0.000	0.002	0.127	0.321	0.372	0.409	0.438	0.478	0.504	0.545	0.569	0.593	0.
2.50	0.000	0.000	0.000	0.000	0.001	0.118	0.307	0.358	0.395	0.423	0.463	0.489	0.528	0.552	0.575	0.
2.51	0.000	0.000	0.000	0.000	0.001	0.111	0.294	0.345	0.381	0.409	0.447	0.473	0.512	0.536	0.558	0.
2.52	0.000	0.000	0.000	0.000	0.000	0.103	0.282	0.331	0.367	0.394	0.432	0.458	0.497	0.519	0.542	0.
2.53	0.000	0.000	0.000	0.000	0.000	0.096	0.270	0.319	0.354	0.381	0.418	0.444	0.481	0.503	0.526	0.
2.54	0.000	0.000	0.000	0.000	0.000	0.089	0.258	0.306	0.340	0.367	0.404	0.428	0.466	0.488	0.510	0.

* Values tabulated are read in percent.

Table 13.17. (cont.) Estimating the Lot Percent Defective Using Range Method*
(Table C–5 of MIL-STD-414)

L r U	Sample Size															
	3	4	5	7	10	15	25	30	35	40	50	60	85	115	175	230
55	0.000	0.000	0.000	0.000	0.000	0.083	0.247	0.294	0.328	0.354	0.390	0.415	0.451	0.473	0.495	0.506
56	0.000	0.000	0.000	0.000	0.000	0.077	0.237	0.283	0.316	0.341	0.377	0.401	0.437	0.459	0.480	0.491
57	0.000	0.000	0.000	0.000	0.000	0.071	0.227	0.272	0.304	0.328	0.364	0.388	0.424	0.445	0.466	0.477
58	0.000	0.000	0.000	0.000	0.000	0.066	0.217	0.261	0.292	0.317	0.352	0.376	0.411	0.432	0.452	0.463
59	0.000	0.000	0.000	0.000	0.000	0.061	0.207	0.251	0.282	0.305	0.340	0.363	0.397	0.418	0.439	0.449
60	0.000	0.000	0.000	0.000	0.000	0.056	0.198	0.240	0.271	0.294	0.328	0.351	0.385	0.406	0.426	0.436
61	0.000	0.000	0.000	0.000	0.000	0.052	0.189	0.231	0.260	0.283	0.317	0.339	0.372	0.393	0.413	0.423
62	0.000	0.000	0.000	0.000	0.000	0.048	0.181	0.221	0.250	0.273	0.306	0.327	0.360	0.381	0.400	0.410
63	0.000	0.000	0.000	0.000	0.000	0.044	0.173	0.212	0.241	0.263	0.295	0.316	0.349	0.368	0.388	0.398
64	0.000	0.000	0.000	0.000	0.000	0.040	0.164	0.203	0.232	0.253	0.285	0.306	0.338	0.357	0.376	0.386
65	0.000	0.000	0.000	0.000	0.000	0.037	0.157	0.195	0.223	0.244	0.274	0.295	0.327	0.346	0.365	0.375
66	0.000	0.000	0.000	0.000	0.000	0.034	0.149	0.186	0.213	0.234	0.265	0.285	0.316	0.335	0.353	0.363
67	0.000	0.000	0.000	0.000	0.000	0.031	0.143	0.179	0.205	0.225	0.255	0.275	0.305	0.324	0.342	0.352
68	0.000	0.000	0.000	0.000	0.000	0.028	0.136	0.171	0.197	0.217	0.246	0.266	0.296	0.314	0.332	0.342
69	0.000	0.000	0.000	0.000	0.000	0.025	0.129	0.164	0.190	0.209	0.238	0.257	0.286	0.304	0.321	0.331
70	0.000	0.000	0.000	0.000	0.000	0.023	0.123	0.156	0.182	0.201	0.228	0.248	0.277	0.295	0.311	0.321
71	0.000	0.000	0.000	0.000	0.000	0.021	0.117	0.150	0.174	0.193	0.220	0.239	0.267	0.285	0.302	0.311
72	0.000	0.000	0.000	0.000	0.000	0.019	0.111	0.143	0.167	0.185	0.212	0.231	0.259	0.275	0.292	0.301
73	0.000	0.000	0.000	0.000	0.000	0.017	0.106	0.137	0.160	0.178	0.205	0.222	0.250	0.266	0.283	0.292
74	0.000	0.000	0.000	0.000	0.000	0.015	0.101	0.131	0.153	0.171	0.197	0.215	0.241	0.258	0.274	0.282
75	0.000	0.000	0.000	0.000	0.000	0.014	0.096	0.125	0.147	0.164	0.189	0.207	0.233	0.248	0.266	0.274
76	0.000	0.000	0.000	0.000	0.000	0.012	0.091	0.120	0.141	0.158	0.182	0.200	0.225	0.241	0.257	0.265
77	0.000	0.000	0.000	0.000	0.000	0.011	0.086	0.114	0.135	0.152	0.175	0.192	0.217	0.232	0.249	0.257
78	0.000	0.000	0.000	0.000	0.000	0.010	0.081	0.109	0.130	0.146	0.169	0.185	0.210	0.226	0.241	0.249
79	0.000	0.000	0.000	0.000	0.000	0.008	0.077	0.103	0.124	0.140	0.163	0.179	0.202	0.218	0.233	0.241
80	0.000	0.000	0.000	0.000	0.000	0.007	0.074	0.099	0.118	0.134	0.156	0.172	0.196	0.210	0.225	0.233
81	0.000	0.000	0.000	0.000	0.000	0.007	0.070	0.094	0.113	0.129	0.150	0.165	0.189	0.204	0.218	0.226
82	0.000	0.000	0.000	0.000	0.000	0.006	0.066	0.090	0.109	0.123	0.144	0.159	0.183	0.194	0.211	0.219
83	0.000	0.000	0.000	0.000	0.000	0.005	0.062	0.085	0.103	0.118	0.139	0.154	0.176	0.190	0.204	0.212
84	0.000	0.000	0.000	0.000	0.000	0.004	0.059	0.082	0.099	0.113	0.134	0.148	0.170	0.184	0.197	0.205
85	0.000	0.000	0.000	0.000	0.000	0.004	0.055	0.078	0.095	0.109	0.128	0.143	0.164	0.178	0.191	0.198
86	0.000	0.000	0.000	0.000	0.000	0.003	0.053	0.074	0.091	0.104	0.124	0.137	0.159	0.172	0.185	0.192
87	0.000	0.000	0.000	0.000	0.000	0.003	0.050	0.070	0.087	0.100	0.119	0.132	0.152	0.166	0.179	0.185
88	0.000	0.000	0.000	0.000	0.000	0.002	0.047	0.067	0.082	0.095	0.114	0.127	0.147	0.160	0.173	0.179
89	0.000	0.000	0.000	0.000	0.000	0.002	0.044	0.064	0.079	0.091	0.109	0.122	0.142	0.155	0.167	0.173
90	0.000	0.000	0.000	0.000	0.000	0.002	0.042	0.061	0.075	0.088	0.105	0.117	0.138	0.149	0.161	0.168
91	0.000	0.000	0.000	0.000	0.000	0.001	0.039	0.057	0.072	0.084	0.101	0.112	0.132	0.145	0.156	0.162
92	0.000	0.000	0.000	0.000	0.000	0.001	0.037	0.055	0.069	0.080	0.097	0.107	0.127	0.140	0.151	0.157
93	0.000	0.000	0.000	0.000	0.000	0.001	0.035	0.052	0.066	0.077	0.093	0.104	0.123	0.134	0.146	0.151
94	0.000	0.000	0.000	0.000	0.000	0.001	0.033	0.049	0.062	0.073	0.089	0.100	0.118	0.129	0.141	0.146
95	0.000	0.000	0.000	0.000	0.000	0.001	0.031	0.047	0.059	0.070	0.086	0.096	0.114	0.125	0.136	0.142
96	0.000	0.000	0.000	0.000	0.000	0.001	0.029	0.044	0.056	0.067	0.082	0.092	0.110	0.121	0.132	0.137
97	0.000	0.000	0.000	0.000	0.000	0.000	0.027	0.042	0.054	0.064	0.079	0.088	0.105	0.116	0.127	0.132
98	0.000	0.000	0.000	0.000	0.000	0.000	0.025	0.039	0.051	0.061	0.075	0.085	0.101	0.112	0.123	0.128
99	0.000	0.000	0.000	0.000	0.000	0.000	0.024	0.038	0.049	0.058	0.072	0.082	0.098	0.108	0.119	0.124

* Values tabulated are read in percent.

Table 13.17. (cont.) Estimating the Lot Percent Defective Using Range Method*
(Table C-5 of MIL-STD-414)

Q_L or Q_U	Sample Size														
	3	4	5	7	10	15	25	30	35	40	50	60	85	115	175	2
3.00	0.000	0.000	0.000	0.000	0.000	0.000	0.022	0.036	0.047	0.056	0.069	0.078	0.094	0.105	0.115	0.
3.01	0.000	0.000	0.000	0.000	0.000	0.000	0.022	0.034	0.044	0.053	0.066	0.075	0.091	0.101	0.111	0.
3.02	0.000	0.000	0.000	0.000	0.000	0.000	0.020	0.032	0.042	0.050	0.063	0.072	0.087	0.097	0.107	0.
3.03	0.000	0.000	0.000	0.000	0.000	0.000	0.019	0.030	0.040	0.048	0.061	0.069	0.084	0.094	0.103	0.
3.04	0.000	0.000	0.000	0.000	0.000	0.000	0.017	0.028	0.038	0.045	0.058	0.066	0.081	0.090	0.099	0.
3.05	0.000	0.000	0.000	0.000	0.000	0.000	0.016	0.027	0.036	0.043	0.056	0.064	0.078	0.086	0.096	0.
3.06	0.000	0.000	0.000	0.000	0.000	0.000	0.015	0.025	0.034	0.041	0.053	0.061	0.075	0.083	0.092	0.
3.07	0.000	0.000	0.000	0.000	0.000	0.000	0.014	0.024	0.032	0.039	0.051	0.059	0.072	0.080	0.089	0.
3.08	0.000	0.000	0.000	0.000	0.000	0.000	0.013	0.022	0.030	0.037	0.049	0.056	0.069	0.077	0.086	0.
3.09	0.000	0.000	0.000	0.000	0.000	0.000	0.012	0.021	0.029	0.036	0.046	0.054	0.067	0.075	0.083	0.
3.10	0.000	0.000	0.000	0.000	0.000	0.000	0.011	0.020	0.027	0.034	0.044	0.051	0.064	0.072	0.080	0.
3.11	0.000	0.000	0.000	0.000	0.000	0.000	0.011	0.019	0.026	0.032	0.042	0.050	0.061	0.069	0.077	0.
3.12	0.000	0.000	0.000	0.000	0.000	0.000	0.010	0.018	0.025	0.031	0.041	0.048	0.060	0.067	0.074	0.
3.13	0.000	0.000	0.000	0.000	0.000	0.000	0.010	0.017	0.024	0.029	0.039	0.046	0.057	0.064	0.072	0.
3.14	0.000	0.000	0.000	0.000	0.000	0.000	0.009	0.015	0.023	0.028	0.037	0.044	0.055	0.062	0.069	0.
3.15	0.000	0.000	0.000	0.000	0.000	0.000	0.008	0.014	0.021	0.026	0.036	0.042	0.053	0.060	0.067	0.
3.16	0.000	0.000	0.000	0.000	0.000	0.000	0.008	0.014	0.020	0.025	0.034	0.040	0.051	0.057	0.064	0.
3.17	0.000	0.000	0.000	0.000	0.000	0.000	0.007	0.013	0.019	0.024	0.033	0.038	0.049	0.056	0.062	0.
3.18	0.000	0.000	0.000	0.000	0.000	0.000	0.006	0.012	0.017	0.022	0.031	0.036	0.046	0.053	0.060	0.
3.19	0.000	0.000	0.000	0.000	0.000	0.000	0.006	0.012	0.017	0.021	0.030	0.034	0.044	0.052	0.057	0.
3.20	0.000	0.000	0.000	0.000	0.000	0.000	0.005	0.011	0.016	0.020	0.028	0.033	0.043	0.049	0.055	0.
3.21	0.000	0.000	0.000	0.000	0.000	0.000	0.005	0.010	0.015	0.019	0.027	0.032	0.041	0.047	0.053	0.
3.22	0.000	0.000	0.000	0.000	0.000	0.000	0.004	0.009	0.014	0.018	0.025	0.031	0.040	0.045	0.051	0.
3.23	0.000	0.000	0.000	0.000	0.000	0.000	0.004	0.009	0.013	0.017	0.024	0.029	0.037	0.043	0.049	0.
3.24	0.000	0.000	0.000	0.000	0.000	0.000	0.004	0.009	0.013	0.016	0.023	0.028	0.037	0.042	0.047	0.
3.25	0.000	0.000	0.000	0.000	0.000	0.000	0.003	0.008	0.012	0.015	0.022	0.027	0.035	0.040	0.046	0.
3.26	0.000	0.000	0.000	0.000	0.000	0.000	0.003	0.007	0.011	0.015	0.021	0.025	0.033	0.039	0.044	0.
3.27	0.000	0.000	0.000	0.000	0.000	0.000	0.003	0.007	0.011	0.014	0.021	0.024	0.032	0.037	0.042	0.
3.28	0.000	0.000	0.000	0.000	0.000	0.000	0.003	0.006	0.010	0.013	0.019	0.023	0.031	0.036	0.040	0.
3.29	0.000	0.000	0.000	0.000	0.000	0.000	0.003	0.006	0.009	0.012	0.018	0.023	0.029	0.034	0.039	0.
3.30	0.000	0.000	0.000	0.000	0.000	0.000	0.003	0.005	0.009	0.012	0.017	0.021	0.028	0.033	0.037	0.
3.31	0.000	0.000	0.000	0.000	0.000	0.000	0.003	0.005	0.008	0.011	0.017	0.021	0.027	0.032	0.036	0.
3.32	0.000	0.000	0.000	0.000	0.000	0.000	0.002	0.004	0.007	0.010	0.016	0.020	0.026	0.030	0.034	0.
3.33	0.000	0.000	0.000	0.000	0.000	0.000	0.002	0.004	0.007	0.010	0.015	0.019	0.025	0.029	0.033	0.
3.34	0.000	0.000	0.000	0.000	0.000	0.000	0.002	0.004	0.007	0.009	0.014	0.018	0.024	0.028	0.032	0.
3.35	0.000	0.000	0.000	0.000	0.000	0.000	0.002	0.004	0.006	0.009	0.014	0.017	0.023	0.027	0.031	0.
3.36	0.000	0.000	0.000	0.000	0.000	0.000	0.002	0.004	0.006	0.008	0.013	0.016	0.022	0.026	0.030	0.
3.37	0.000	0.000	0.000	0.000	0.000	0.000	0.002	0.004	0.006	0.008	0.012	0.015	0.021	0.024	0.028	0.
3.38	0.000	0.000	0.000	0.000	0.000	0.000	0.001	0.003	0.005	0.007	0.012	0.014	0.019	0.024	0.027	0.
3.39	0.000	0.000	0.000	0.000	0.000	0.000	0.001	0.003	0.005	0.007	0.011	0.014	0.019	0.022	0.027	0.
3.40	0.000	0.000	0.000	0.000	0.000	0.000	0.001	0.003	0.005	0.007	0.010	0.013	0.018	0.021	0.026	0.
3.41	0.000	0.000	0.000	0.000	0.000	0.000	0.001	0.002	0.004	0.006	0.010	0.012	0.018	0.021	0.025	0.
3.42	0.000	0.000	0.000	0.000	0.000	0.000	0.001	0.002	0.004	0.006	0.009	0.012	0.017	0.020	0.024	0.
3.43	0.000	0.000	0.000	0.000	0.000	0.000	0.001	0.002	0.004	0.005	0.009	0.011	0.016	0.019	0.023	0.
3.44	0.000	0.000	0.000	0.000	0.000	0.000	0.001	0.002	0.004	0.005	0.008	0.011	0.015	0.018	0.022	0.

* Values tabulated are read in percent.

Table 13.17. (cont.) Estimating the Lot Percent Defective Using Range Method*
(Table C–5 of MIL-STD-414)

	Sample Size															
	3	4	5	7	10	15	25	30	35	40	50	60	85	115	175	230
5	0.000	0.000	0.000	0.000	0.000	0.000	0.001	0.002	0.004	0.005	0.008	0.011	0.014	0.017	0.021	0.023
6	0·000	0.000	0.000	0.000	0.000	0.000	0.001	0.002	0.003	0.005	0.008	0.010	0.014	0.017	0.020	0.022
7	0.000	0.000	0.000	0.000	0.000	0.000	0.001	0.002	0.003	0.004	0.007	0.010	0.014	0.016	0.019	0.021
8	0.000	0.000	0.000	0.000	0.000	0.000	0.001	0.002	0.003	0.004	0.007	0.009	0.013	0.015	0.018	0.020
9	0.000	0.000	0.000	0.000	0.000	0.000	0.000	0.001	0.003	0.004	0.006	0.009	0.012	0.015	0.018	0.020
0	0.000	0.000	0.000	0.000	0.000	0.000	0.000	0.001	0.002	0.003	0.006	0.008	0.012	0.014	0.017	0.019
1	0.000	0.000	0.000	0.000	0.000	0.000	0.000	0.001	0.002	0.003	0.006	0.008	0.011	0.014	0.016	0.018
2	0.000	0.000	0.000	0.000	0.000	0.000	0.000	0.001	0.002	0.003	0.006	0.007	0.010	0.013	0.016	0.017
3	0.000	0.000	0.000	0.000	0.000	0.000	0.000	0.001	0.002	0.003	0.005	0.007	0.010	0.013	0.015	0.016
4	0.000	0.000	0.000	0.000	0.000	0.000	0.000	0.001	0.002	0.003	0.004	0.007	0.010	0.012	0.014	0.015
5	0.000	0.000	0.000	0.000	0.000	0.000	0.000	0.001	0.002	0.003	0.004	0.006	0.009	0.012	0.014	0.015
6	0.000	0.000	0.000	0.000	0.000	0.000	0.000	0.001	0.001	0.002	0.004	0.006	0.009	0.011	0.013	0.014
7	0.000	0.000	0.000	0.000	0.000	0.000	0.000	0.001	0.001	0.002	0.004	0.005	0.008	0.011	0.012	0.013
8	0.000	0.000	0.000	0.000	0.000	0.000	0.000	0.001	0.001	0.002	0.003	0.005	0.008	0.010	0.012	0.013
9	0.000	0.000	0.000	0.000	0.000	0.000	0.000	0.001	0.001	0.002	0.003	0.005	0.008	0.010	0.011	0.012
0	0.000	0.000	0.000	0.000	0.000	0.000	0.000	0.001	0.001	0.002	0.003	0.005	0.007	0.009	0.011	0.012
1	0.000	0.000	0.000	0.000	0.000	0.000	0.000	0.001	0.001	0.002	0.003	0.004	0.007	0.009	0.011	0.011
2	0.000	0.000	0.000	0.000	0.000	0.000	00000	0.001	0.001	0.002	0.003	0.004	0.007	0.009	0.010	0.011
3	0.000	0.000	0.000	0.000	0.000	0.000	00000	0.001	0.001	0.001	0.003	0.004	0.006	0.008	0.010	0.010
4	0.000	0.000	0.000	0.000	0.000	0.000	00000	0.001	0.001	0.001	0.003	0.003	0.006	0.008	0.009	0.010
5	0.000	0.000	0.000	0.000	0.000	0.000	0.000	0.001	0.001	0.001	0.003	0.003	0.006	0.008	0.009	0.010
6	0.000	0.000	0.000	0.000	0.000	0.000	0.000	0.000	0.000	0.001	0.003	0.003	0.005	0.007	0.009	0.009
7	0.000	0.000	0.000	0.000	0.000	0.000	0.000	0.000	0.000	0.001	0.003	0.003	0.005	0.007	0.008	0.009
8	0.000	0.000	0.000	0.000	0.000	0.000	0.000	0.000	0.000	0.001	0.002	0.003	0.005	0.006	0.008	0.008
9	0.000	0.000	0.000	0.000	0.000	0.000	0.000	0.000	0.000	0.001	0.002	0.002	0.004	0.006	0.008	0.008
0	0.000	0.000	0.000	0.000	0.000	0.000	0.000	0.000	0.000	0.001	0.002	0.002	0.004	0.006	0.007	0.008
1	0.000	0.000	0.000	0.000	0.000	0.000	0.000	0.000	0.000	0.001	0.001	0.002	0.004	0.006	0.007	0.007
2	0.000	0.000	0.000	0.000	0.000	0.000	0.000	0.000	0.000	0.001	0.001	0.002	0.004	0.006	0.007	0.007
3	0.000	0.000	0.000	0.000	0.000	0.000	0.000	0.000	0.000	0.001	0.001	0.002	0.004	0.006	0.007	0.007
4	0.000	0.000	0.000	0.000	0.000	0.000	0.000	0.000	0.000	0.001	0.001	0.002	0.004	0.005	0.006	0.007
5	0.000	0.000	0.000	0.000	0.000	0.000	0.000	0.000	0.000	0.001	0.001	0.002	0.003	0.005	0.006	0.006
6	0.000	0.000	0.000	0.000	0.000	0.000	0.000	0.000	0.000	0.001	0.001	0.002	0.003	0.005	0.006	0.006
7	0.000	0.000	0.000	0.000	0.000	0.000	0.000	0.000	0.000	0.001	0.001	0.002	0.003	0.005	0.006	0.006
8	0.000	0.000	0.000	0.000	0.000	0.000	0.000	0.000	0.000	0.000	0.001	0.002	0.003	0.004	0.005	0.005
9	0.000	0.000	0.000	0.000	0.000	0.000	0.000	0.000	0.000	0.000	0.001	0.002	0.003	0.003	0.005	0.005
0	0.000	0.000	0.000	0.000	0.000	0.000	0.000	0.000	0.000	0.000	0.001	0.002	0.003	0.003	0.005	0.005
1	0.000	0.000	0.000	0.000	0.000	0.000	0.000	0.000	0.000	0.000	0.001	0.002	0.003	0.003	0.005	0.005
2	0.000	0.000	0.000	0.000	0.000	0.000	0.000	0.000	0.000	0.000	0.001	0.002	0.003	0.003	0.005	0.005
3	0.000	0.000	0.000	0.000	0.000	0.000	0.000	0.000	0.000	0.000	0.001	0.002	0.003	0.003	0.004	0.004
4	0.000	0.000	0.000	0.000	0.000	0.000	0.000	0.000	0.000	0.000	0.001	0.001	0.002	0.003	0.004	0.004
5	0.000	0.000	0.000	0.000	0.000	0.000	0.000	0.000	0.000	0.000	0.001	0.001	0.001	0.002	0.004	0.004
6	0.000	0.000	0.000	0.000	0.000	0.000	0.000	0.000	0.000	0.000	0.000	0.001	0.001	0.002	0.004	0.004
7	0.000	0.000	0.000	0.000	0.000	0.000	0.000	0.000	0.000	0.000	0.000	0.001	0.001	0.002	0.004	0.004
8	0.000	0.000	0.000	0.000	0.000	0.000	0.000	0.000	0.000	0.000	0.000	0.001	0.001	0.002	0.004	0.004
9	0.000	0.000	0.000	0.000	0.000	0.000	0.000	0.000	0.000	0.000	0.000	0.001	0.001	0.002	0.003	0.003
0	0.000	0.000	0.000	0.000	0.000	0.000	0.000	0.000	0.000	0.000	0.000	0.001	0.001	0.002	0.003	0.003

* Values tabulated are read in percent.

the distribution of the minimum variance unbiased estimate of p' based on $\bar{x}$ and s, but at the expense of an increased sample size. For an upper specification limit, U, this estimate is a function of $Q_U = (U - \bar{x})c/\overline{R}$ and is denoted by p_U, where $\bar{x}$ is the sample mean, $\overline{R}$ is the average range of sub-group ranges (each subgroup usually consists of five measurements), and c is a constant depending upon the sample size. For a lower specification limit, L, the estimate is a function of $Q_L = (\bar{x} - L)c/\overline{R}$ and is denoted by p_L. The form of the functions, p_U and p_L, although rather complicated, is the same, thereby facilitating tabulation. Such a tabulation is given in Table 13.17 for sample sizes of 3, 4, 5, 7, 10, 15, 25, 30, 35, 40, 50, 60, 85, 115, 175, and 230. For sample sizes of 3, 4, 5, and 7, $\overline{R}$ is taken to be the range of the sample. The factor c corresponding to these sample sizes is given in Table 13.18. In order to obtain the quantities p_U and p_L Table 13.17 is entered with the sample size and the quality index Q_U or the quality index Q_L, whichever is appropriate; the required estimate, in percent, is read from the table. For two-sided specification limits, the estimate of the percent defective is given by $p_U + p_L$ where these quantities are read from Table 13.17 as just described.

Table 13.18. Table of *c* Factors

Sample Size	*c* Factor	Sample Size	*c* Factor
3	1.910	35	2.349
4	2.234	40	2.346
5	2.474	50	2.342
7	2.830	60	2.339
10	2.405	85	2.335
15	2.379	115	2.333
25	2.358	175	2.333
30	2.353	230	2.333

For an upper specification limit, the lot is accepted whenever $p_U \leqq M$, where M is a preassigned constant. Alternatively, it can be shown that $p_U \leqq M$ if, and only if, $\dfrac{U - \bar{x}}{\overline{R}} \geqq k$, where k is a constant related to M. Hence, an equivalent procedure is to accept the lot if $\dfrac{U - \bar{x}}{\overline{R}} \geqq k$.

For a lower specification limit, the lot is accepted whenever

$p_L \leqq M$. Alternatively, it can be shown that $p_L \leqq M$ if, and only if, $\frac{\bar{x} - L}{\bar{R}} \geqq k$. Hence, an equivalent procedure is to accept the lot if

$$\frac{\bar{x} - L}{\bar{R}} \geqq k.$$

Finally, for a two-sided specification limit, the lot is accepted whenever $p_U + p_L \leqq M$. There is no equivalent procedure in terms of k, but an approximation is given by accepting the lot if the following three criteria are *all* met.

(1) $\frac{U - \bar{x}}{\bar{R}} \geqq k$;

(2) $\frac{\bar{x} - L}{\bar{R}} \geqq k$;

(3) $\bar{R} \leqq$ Maximum Average Range (MAR), where MAR is also a constant depending upon M.

OC curves based on average range plans can be constructed which match OC curves based on sample standard deviation plans by suitably varying the sample size n and the maximum allowable percent defective M (or n and k if this procedure is used). However, the resulting sample size using the range plans will always be greater than or equal to the sample size using the sample standard deviation plans. If a sample standard deviation plan requires a small n, the comparable range plan requires a not too different sample size; this difference increases as the sample size of the sample standard deviation plan increases.

13.3.3.3 *Estimate of the percent defective when the standard deviation is known.* When the standard deviation, σ', of the normal distribution is known, there exists a minimum variance unbiased estimate of p'. For an upper specification limit, U, the minimum variance unbiased estimate is a function of $Q_U = \frac{(U - \bar{x})}{\sigma'}\sqrt{\frac{n}{n-1}}$ and is given by p_U, where

$$p_U = \int_{\sqrt{\frac{n}{n-1}}\left(\frac{U-\bar{x}}{\sigma'}\right)}^{\infty} \frac{1}{\sqrt{2\pi}} e^{-z^2/2}\, dz.$$

It is evident that p_U is just the area of the standardized normal distribution above Q_U, and can be evaluated using Appendix Table 1 (table of the percentage points of the normal distribution). For a lower specification limit, L, the minimum variance unbiased estimate is a function of $Q_L = \left(\frac{\bar{x} - L}{\sigma'}\right)\sqrt{\frac{n}{n-1}}$ and is given by p_L, where

$$p_L = \int_{-\infty}^{-\sqrt{\frac{n}{n-1}}\left(\frac{\bar{x}-L}{\sigma'}\right)} \frac{1}{\sqrt{2\pi}} e^{-z^2/2}\, dz.$$

This integral is equivalent to the same integral between the limits Q_L and ∞ so that p_L is just the area of the standardized normal distribution above Q_L, and also can be evaluated using Appendix Table 1. In order to obtain the quantities p_U and p_L, Appendix Table 1 is entered with the quality index Q_U or the quality index Q_L, whichever is appropriate. The required estimate is read from the table and converted to readings in percent by multiplying by one hundred. For two-sided specification limits, the minimum variance unbiased estimate of the percent defective is given by $p_U + p_L$.

For an upper specification limit, the lot is accepted whenever $p_U \leqq M$, where M is a preassigned constant. Alternatively, it is evident from the expression for p_U that $p_U \leqq M$ if, and only if, $\frac{U - \bar{x}}{\sigma'} \geqq k$, where k is a constant given by

$$k = \sqrt{\frac{n-1}{n}} K_M$$

and K_M is the M percentage point of the normal distribution. Hence, an equivalent procedure is to accept the lot if

$$\frac{U - \bar{x}}{\sigma'} \geqq k.$$

For a lower specification limit, the lot is accepted whenever $p_L \leqq M$. Alternatively, it is evident from the expression for p_L that $p_L \leqq M$ if, and only if, $\frac{\bar{x} - L}{\sigma'} \geqq k$, where k is defined as above. Hence, an equivalent procedure is to accept the lot if

$$\frac{\bar{x} - L}{\sigma'} \geqq k.$$

Finally, for a two-sided specification limit, the lot is accepted whenever $p_U + p_L \leqq M$.

OC curves based on known standard deviation procedures can be constructed which match OC curves for unknown standard deviation procedures by suitably varying the sample size n and the maximum allowable percent defective M (or n and k if this procedure is used). The resulting sample size using the known standard deviation procedure will always be smaller than the sample size using the unknown standard deviation procedure.

13.3.4 Comparison of Variables Procedures with M and k.

It was indicated that single specification limit procedures based on estimates of the percent defective are equivalent to single specification limit procedures involving k. Any advantages or disadvantages of one over the other are independent of OC curve arguments since they both result in the same OC curve, provided M and k are chosen properly. The estimation procedure has several advantages over the k procedure. Like attribute plans, variables plans based on the estimation procedure can involve measurement of lot quality by the percent defective. The k procedure involves measurement of lot quality by the average and variability of the measurements. Thus, with the estimation procedure it is unnecessary to shift to attributes sampling whenever percent defective is the appropriate measure. Secondly, the estimation procedure has intuitive appeal because it is a logical one, whereas relating a measurement of lot quality based on the average and variability of the measurements to a practical criterion is difficult. In other words, rejecting a lot because the estimated percent defective is too large is understandable whereas rejecting a lot because $\frac{U - \bar{x}}{s}$ is too large is difficult to explain. Finally, the estimation procedure leaves the vendor and the consumer with an estimate of lot quality whether the lot is accepted or rejected. On the other hand the k procedure has the advantage of requiring one less step in carrying through the procedure. It does not require a table look-up for the estimate of the lot quality before deciding on acceptance or rejection.

All variables sampling procedures for two-sided specification limits suffer from the fact that the probability of accepting a submitted lot with given percent defective p' does not depend on p' alone but on the division of p' into two components, the percent lying above the upper specification limit, p'_U, and the percent lying below the lower specification limit, p'_L. For this reason a two-sided procedure does not have a unique OC curve but rather a band of curves, each curve within the band representing a possible division of p'. Using the double specification procedure based on the estimate of the lot quality, this band is so narrow as to be, for all practical purposes, a single curve. Since the OC curve for the single specification limit is contained within this narrow band (corresponding to all the defectives outside one limit and none outside the other), it is used as the OC curve for the double specification procedure. The k procedure, with its associated MSD or MAR, is just an attempt to approximate this estimation procedure.

13.3.5 THE MILITARY STANDARD FOR INSPECTION BY VARIABLES, MIL-STD-414

13.3.5.1 *Introduction.* In recent years there has been a recognition of the importance of inspection by variables for percent defective as a test procedure to evaluate product quality. Both government and industry prepared sets of standard sampling plans designed for their own needs. The use of sampling inspection by variables increased to such an extent that the need to prepare a military standard, to serve as an alternative to Military Standard 105B for inspection by attributes, became apparent. In the latter part of 1957, the first military standard for inspection by variables, MIL-STD-414 was issued.

The variables standard is divided into four sections, namely:
Section A—General Description of Sampling Plans;
Section B—Variability Unknown—Standard Deviation Method;
Section C—Variability Unknown—Range Method;
Section D—Variability Known.

Section A will always be used in conjunction with the other sections since it provides general concepts and definitions needed for sampling inspection by variables. Each of the Sections B, C, and D consists of three parts:

(1) Sampling Plans for the Single Specification Limit Case;
(2) Sampling Plans for the Double Specification Limit Case;
(3) Procedures for Estimation of Process Average and Criteria for Tightened and Reduced Inspection.

For the single specification limit case, the acceptability criterion is given in two forms, namely, the k procedure (form 1), e.g., accept the lot if $\frac{U - \bar{x}}{s} \geqq k$; and the estimation procedure based upon M (form 2), e.g., accept the lot if $p_U \leqq M$. Either of the forms may be used since they require the same sample size and have identical OC curves. Form 2 is required for the estimation of the process average, if such an estimate is desired. The estimation procedure (form 2) is required for double specification limits.

13.3.5.2 *Section A—General description of sampling plans.* The variables standard has many features which are similar to MIL-STD-105B—inspection by attributes. Defects are grouped into three classes, critical, major, and minor with the definition of each identical with that in MIL-STD-105B. Like the attributes standard, the sampling plans are indexed by AQL, inspection level, and lot size. The definition of the AQL is similar to that found in MIL-STD-105B. The acceptable quality level is a nominal value expressed in terms of percent defective specified for a single quality characteristic. The AQL values given, in percent, are

0.04	0.40	4.0
0.065	0.65	6.5
0.10	1.0	10.0
0.15	1.5	15.0
0.25	2.5	

These values include almost all those given in MIL-STD-105B for percent defective. The probabilities of acceptance of submitted lots having these AQL range from about 0.89 for the smallest sample sizes to about 0.99 for the largest sample sizes. The particular AQL value to be used for a single quality characteristic of a given product must be specified. In the case of a double specification limit, either an AQL value is specified for the total percent defective outside both upper and lower specification limits or two AQL values are specified, one for the upper limit and another for the lower limit.

Sample sizes are designated by code letters B to Q. The sample size code letter depends on the inspection level and the lot size. There are five inspection levels (as opposed to three in MIL-STD-105B), I, II, III, IV, V. Unless otherwise specified inspection level IV is used. The sample size code letter applicable to the specified inspection level and for lots of given size is obtained from Table 13.19 (Table A–2 of MIL-STD-414).

After the AQL, inspection level, and sample code letter are chosen, the appropriate sampling plan is chosen from a master table in Section B, C, or D of the standard, whichever is appropriate. It is interesting to note that the variables standard states that "unless otherwise specified, unknown variability, standard deviation method sampling plans and the acceptability criterion of form 2 (for the single specification limit case) shall be used." This refers to Section B and the estimation procedure.

Table 13.19. Sample Size Code Letters[1]
(Table A-2 of MIL-STD-414)

Lot Size	Inspection Levels				
	I	II	III	IV	V
3 to 8	B	B	B	B	C
9 to 15	B	B	B	B	D
16 to 25	B	B	B	C	E
26 to 40	B	B	B	D	F
41 to 65	B	B	C	E	G
66 to 110	B	B	D	F	H
111 to 180	B	C	E	G	I
181 to 300	B	D	F	H	J
301 to 500	C	E	G	I	K
501 to 800	D	F	H	J	L
801 to 1,300	E	G	I	K	L
1,301 to 3,200	F	H	J	L	M
3,201 to 8,000	G	I	L	M	N
8,001 to 22,000	H	J	M	N	O
22,001 to 110,000	I	K	N	O	P
110,001 to 550,000	I	K	O	P	Q
550,001 and over	I	K	P	Q	Q

Section A also contains all the OC curves. These OC curves represent the plans in Section B exactly. The OC curves of the corresponding plans in Sections C and D were matched as well as possible and are extremely close in most cases. Any discrepancies are due to the requirement of a constant sample size (multiple of five) for a code letter for the range plans, and integer values of the sample size for the known standard deviation plans. Hence, for a given lot size, inspection level, and AQL, the user may select a plan from any of Sections B, C, or D, and be assured of the same probability of accepting or rejecting material for any given quality. These OC curves are reproduced in Figure 13.13.

13.3.5.3 *Section B—Variability unknown—standard deviation method.* When there is a single specification limit and form 1 is used (k procedure), the appropriate master sampling table is selected as follows:[2]

[1] Sample size code letters given in body of table are applicable when the indicated inspection levels are to be used.

[2] The k procedure cannot be used for double specification limits, even though a table for obtaining maximum standard deviation (MSD) factors is included. The MSD serves as a guide, but lots cannot be accepted or rejected using this criterion.

For normal or tightened inspection, Table 13.20 (Table B-1 of MIL-STD-414).

For reduced inspection, Table 13.21 (Table B-2 of MIL-STD-414).

The sample size, n, and acceptance criterion, k, are read from these tables. The lot is accepted if $\frac{U - \bar{x}}{s} \geqq k$ or $\frac{\bar{x} - L}{s} \geqq k$, whichever is appropriate.

When there is a single specification limit or a double specification limit (with the AQL applying to both limits) and form 2 is used (estimation procedure), the appropriate master sampling table is selected as follows:

For normal or tightened inspection, Table 13.22 (Table B–3 of MIL-STD-414)

For reduced inspection, Table 13.23 (Table B–4 of MIL-STD-414).

The sample size, n, and the acceptance criterion, M, are read from these tables. The quality indices $Q_U = \frac{U - \bar{x}}{s}$ and/or $Q_L = \frac{\bar{x} - L}{s}$ are computed. Table 13.16 (Table B–5 of MIL-STD-414) is entered with n and Q_U and/or Q_L, whichever is appropriate ; the estimates p_U and/or p_L, in percent, are read from the table. For an upper specification limit, the lot is accepted if $p_U \leqq M$; for a lower specification limit, the lot is accepted if $p_L \leqq M$; and for both upper and lower specification limits, the lot is accepted if $p_U + p_L \leqq M$. Procedures are also given for two-sided specification limits when one AQL value is specified for the upper limit and another for the lower limit.

As in MIL-STD-105B, the variables standard provides for normal, tightened, and reduced inspection. The decision as to which type of sampling to choose usually depends upon the estimate of the process average. The estimated process average is the arithmetic mean of the estimated lot defective computed from the sampling inspection results of the preceding ten lots or as many as may be otherwise designated. This estimated process average is denoted by $\bar{p}_U$ if an upper specification limit is given, $\bar{p}_L$ if a lower specification limit is given, and $\bar{p} = \overline{p_L + p_U}$ if double specification limits are given.

The variables standard calls for tightened inspection to be instituted when the estimated process average computed from the preceding ten lots (or other such number of lots designated) is greater than the AQL, and when more than a certain number, T, of these lots have estimates of the percent defective exceeding the AQL. The T values are given in Table 13.24 (Table B–6 of MIL-STD-414) for the process average computed from 5, 10, or 15 lots. Normal inspection is reinstated if the estimated process average of lots under tightened inspection is equal to or less than the AQL.

SAMPLE SIZE CODE LETTER: B

(Curves for sampling plans based on range method and known variability are essentially equivalent)

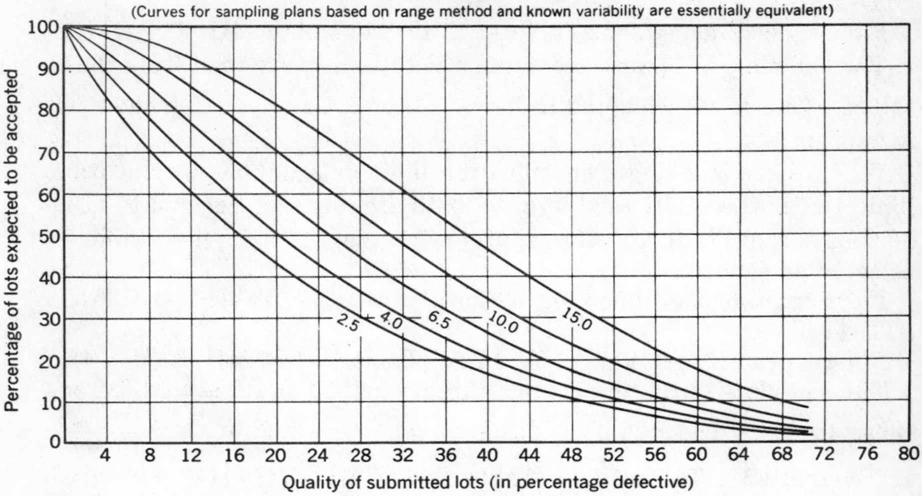

Note: Figures on curves are Acceptable Quality Levels for normal inspection.

SAMPLE SIZE CODE LETTER: C

(Curves for sampling plans based on range method and known variability are essentially equivalent)

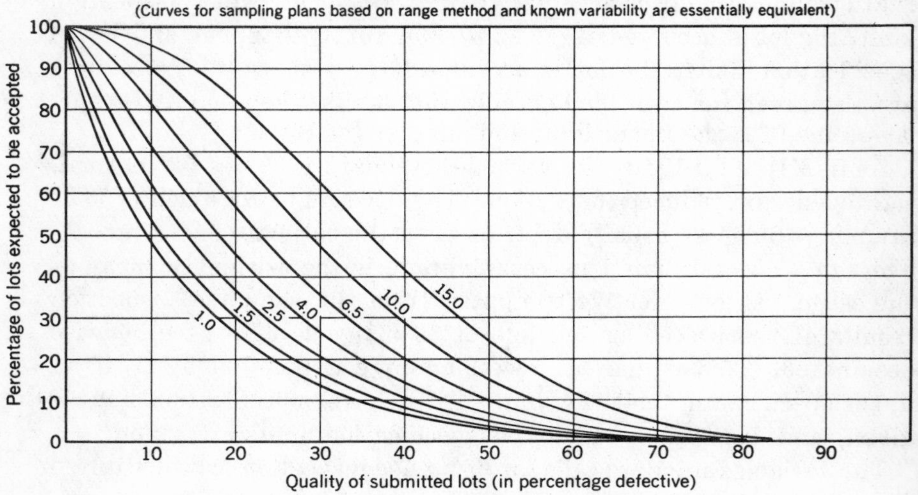

Note: Figures on curves are Acceptable Quality Levels for normal inspection.

Fig. 13.13. OC curves from MIL-STD-414.

SAMPLE SIZE CODE LETTER: D

(Curves for sampling plans based on range method and known variability are essentially equivalent)

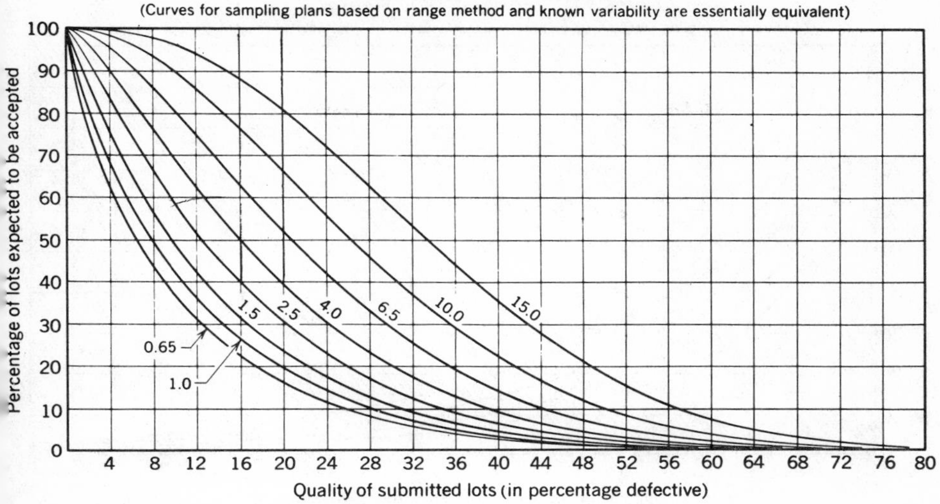

Note: Figures on curves are Acceptable Quality Levels for normal inspection.

SAMPLE SIZE CODE LETTER: E

(Curves for sampling plans based on range method and known variability are essentially equivalent)

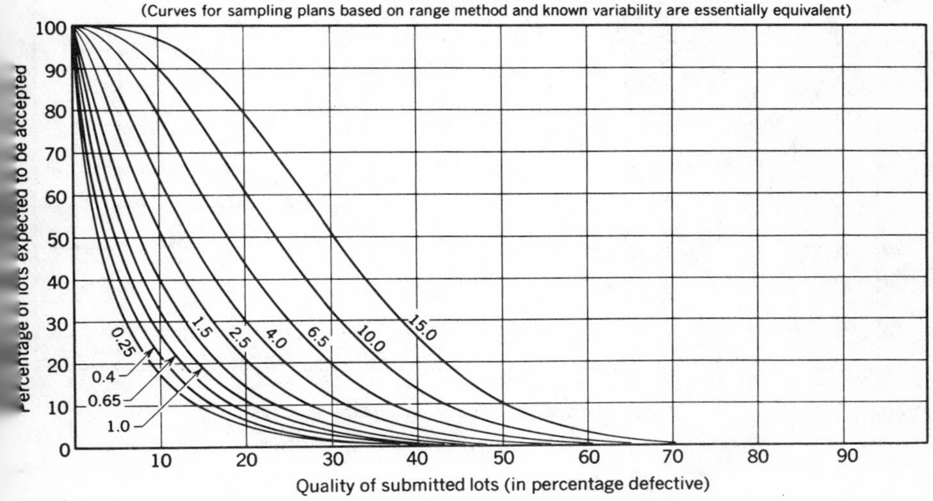

Note: Figures on curves are Acceptable Quality Levels for normal inspection.

Fig. 13.13. (cont.) OC curves from MIL-STD-414.

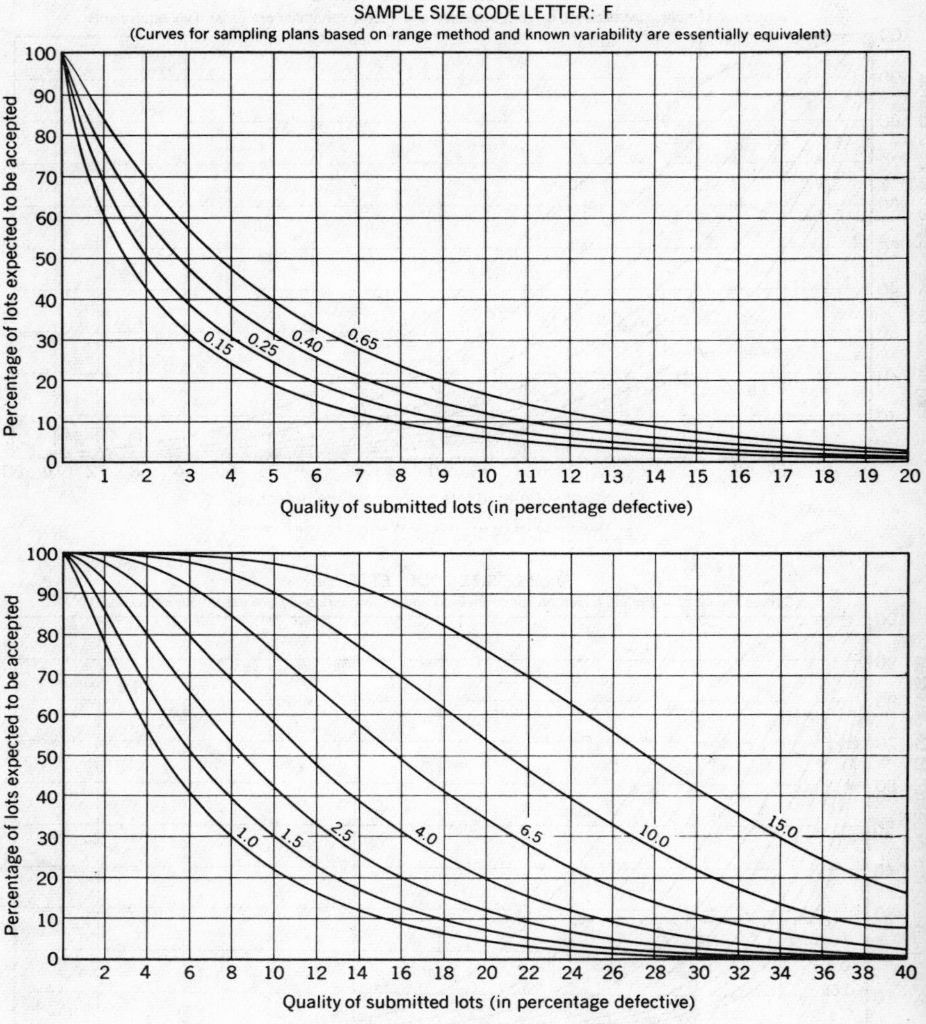

Fig. 13.13. (cont.) OC curves from MIL-STD-414.

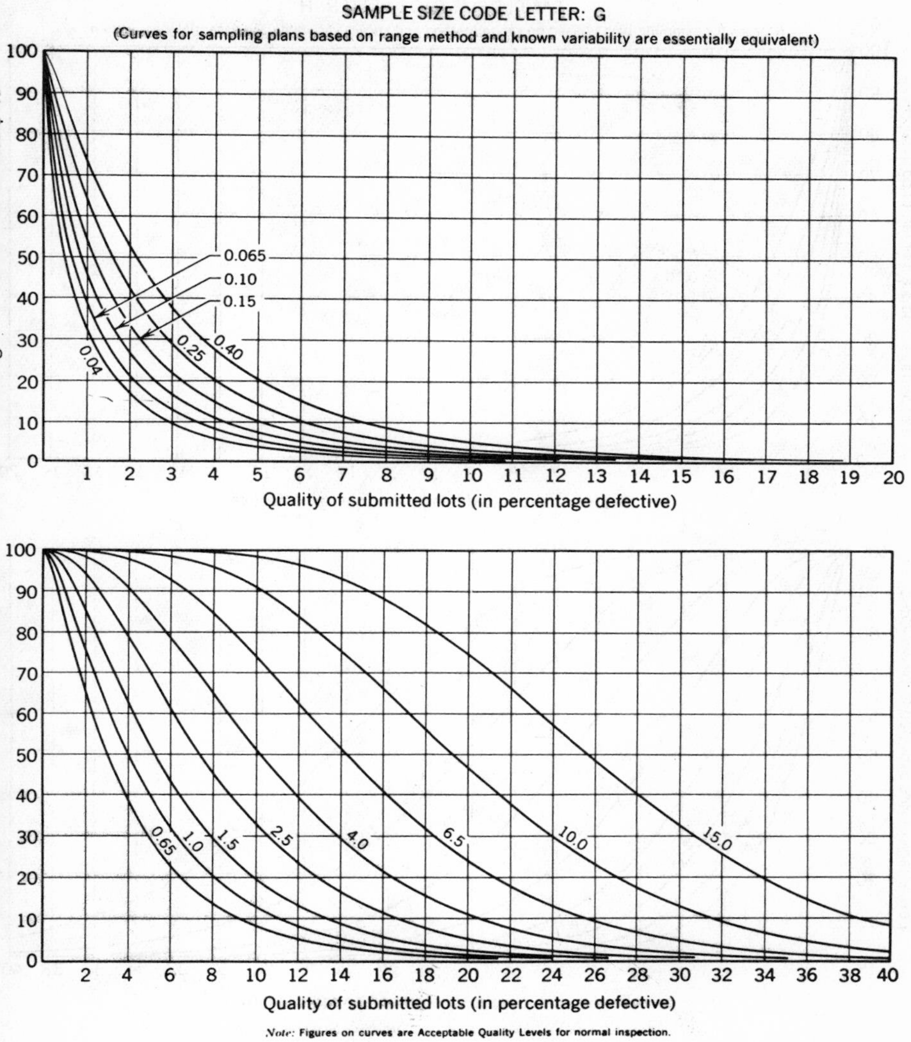

Fig. 13.13. (cont.) OC curves from MIL-STD-414.

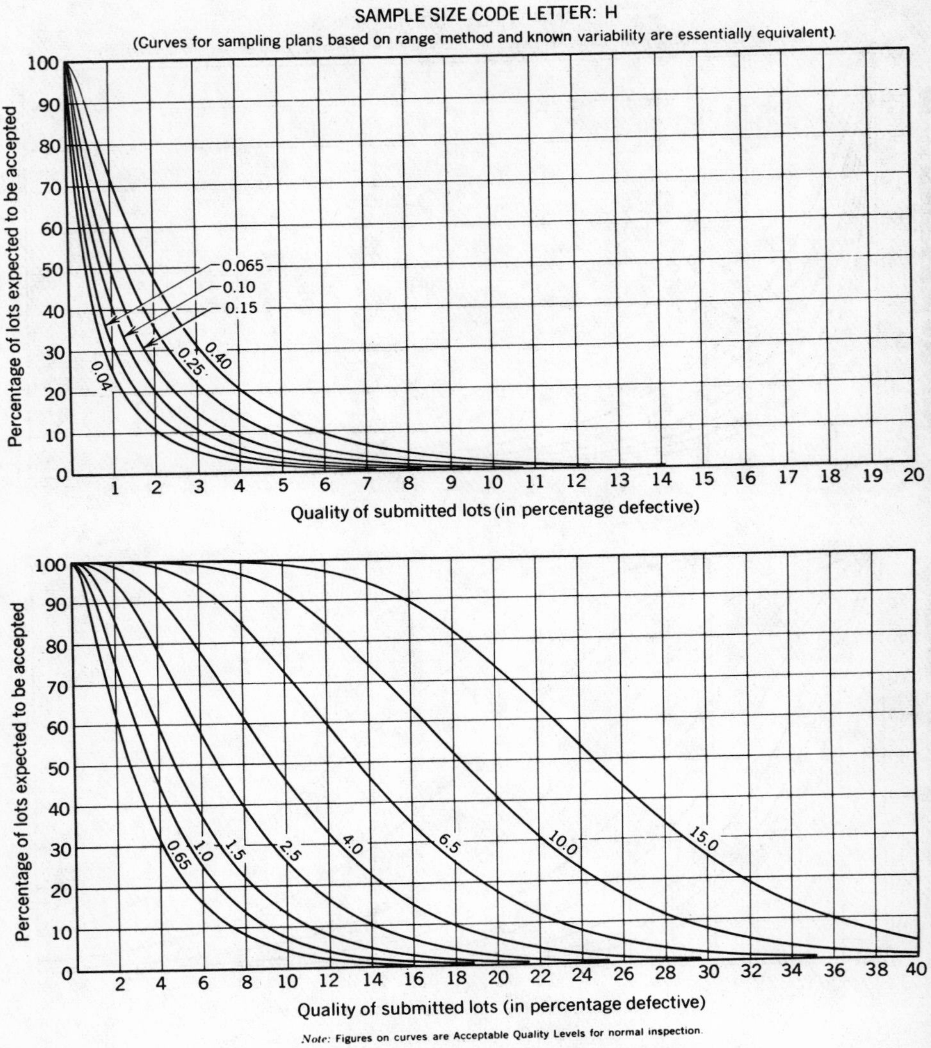

Note: Figures on curves are Acceptable Quality Levels for normal inspection.

Fig. 13.13. (cont.) OC curves from MIL-STD-414.

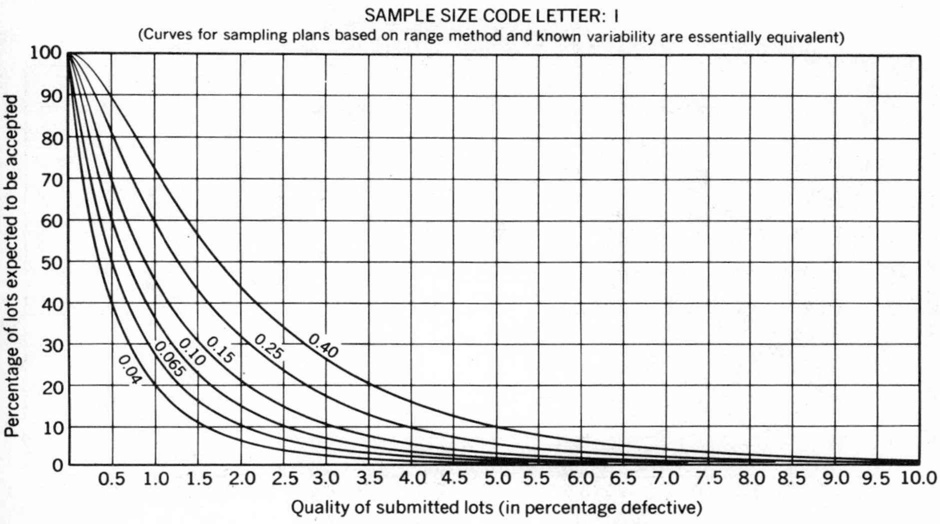

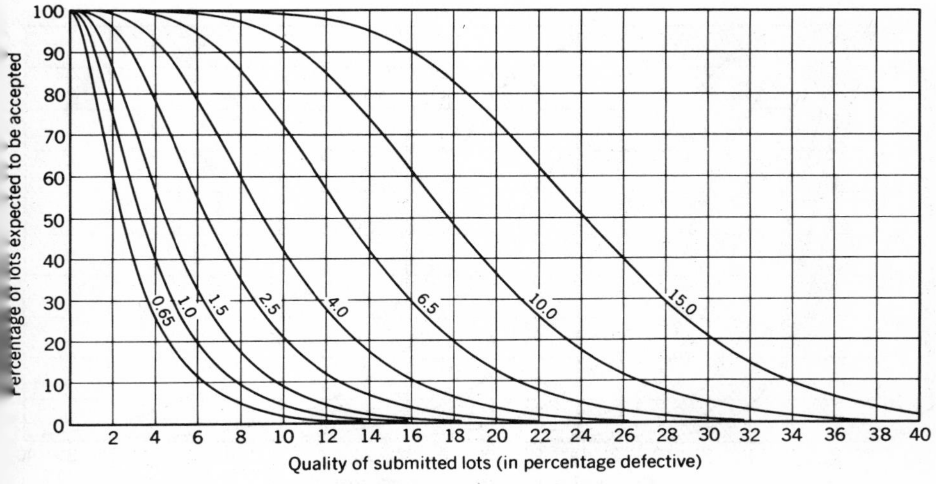

Note: Figures on curves are Acceptable Quality Levels for normal inspection.

Fig. 13.13. (cont.) OC curves from MIL-STD-414.

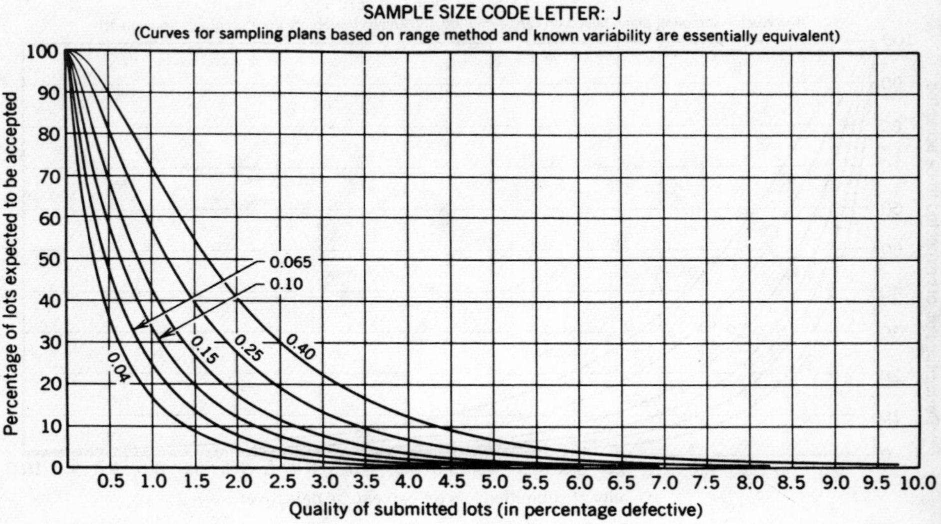

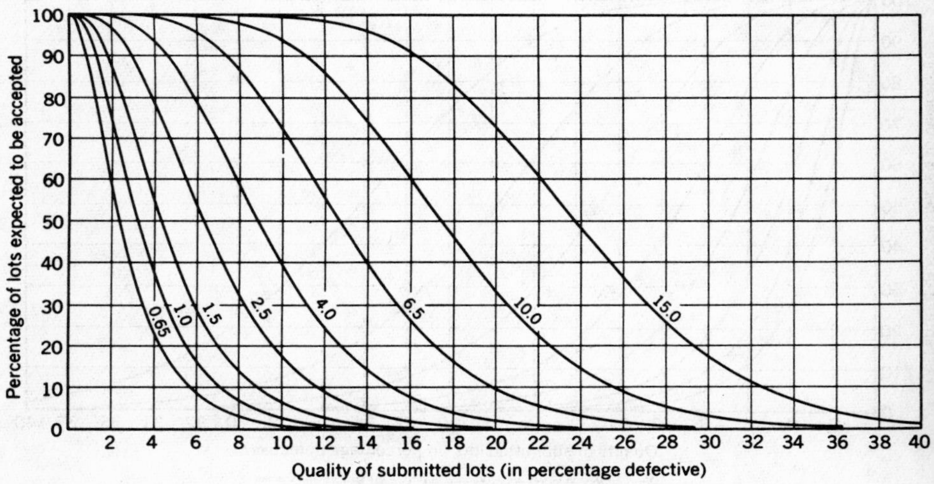

Note: Figures on curves are Acceptable Quality Levels for normal inspection.

Fig. 13.13. (cont.) OC curves from MIL-STD-414.

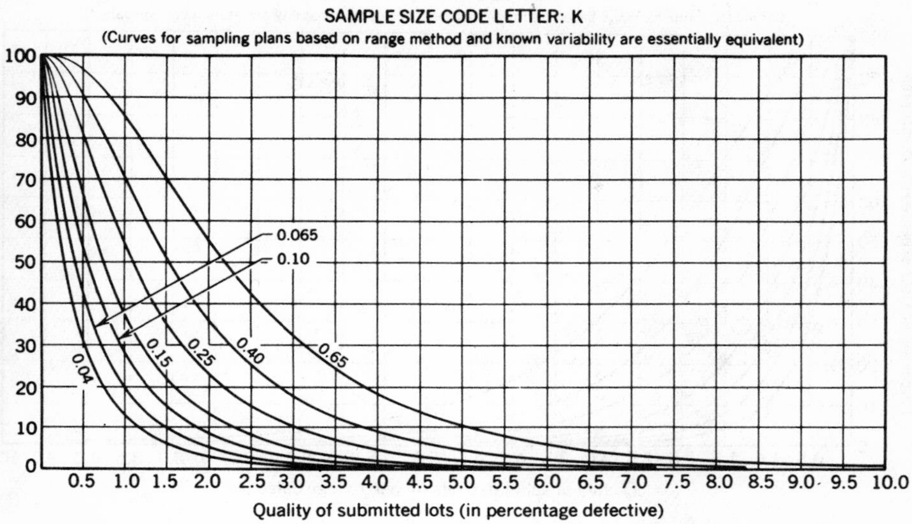

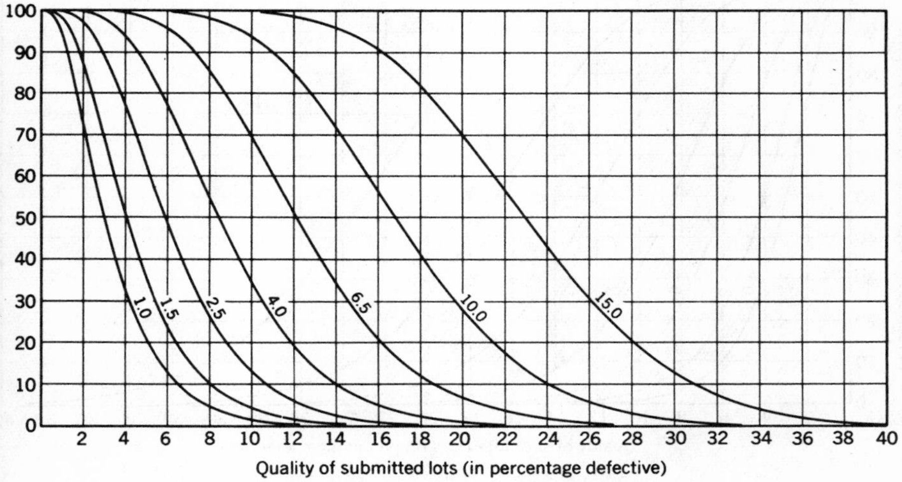

Note: Figures on curves are Acceptable Quality Levels for normal inspection.

Fig. 13.13. (cont.) OC curves from MIL-STD-414.

SAMPLE SIZE CODE LETTER: L

(Curves for sampling plans based on range method and known variability are essentially equivalent)

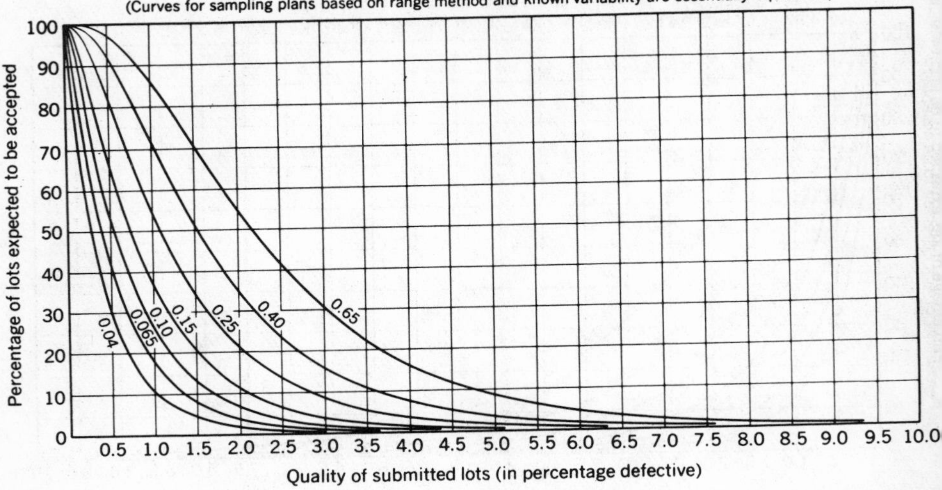

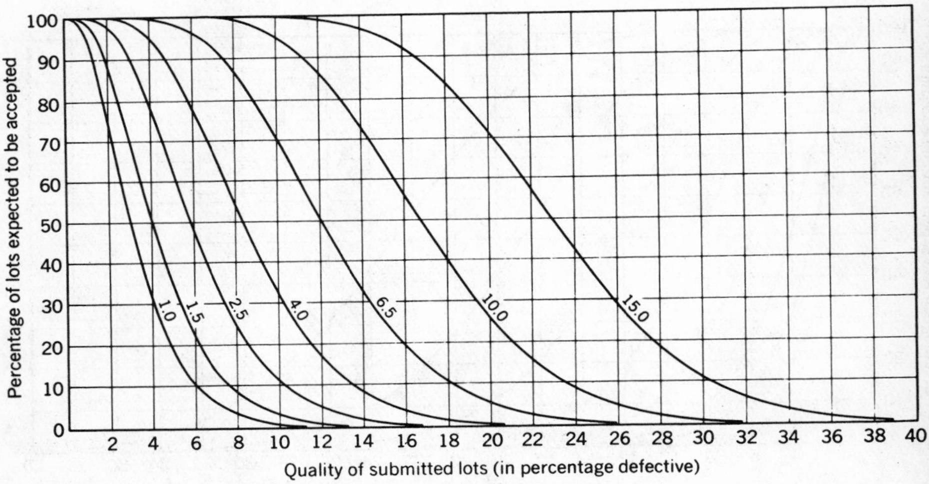

Note: Figures on curves are Acceptable Quality Levels for normal inspection

Fig. 13.13. (cont.) OC curves from MIL-STD-414.

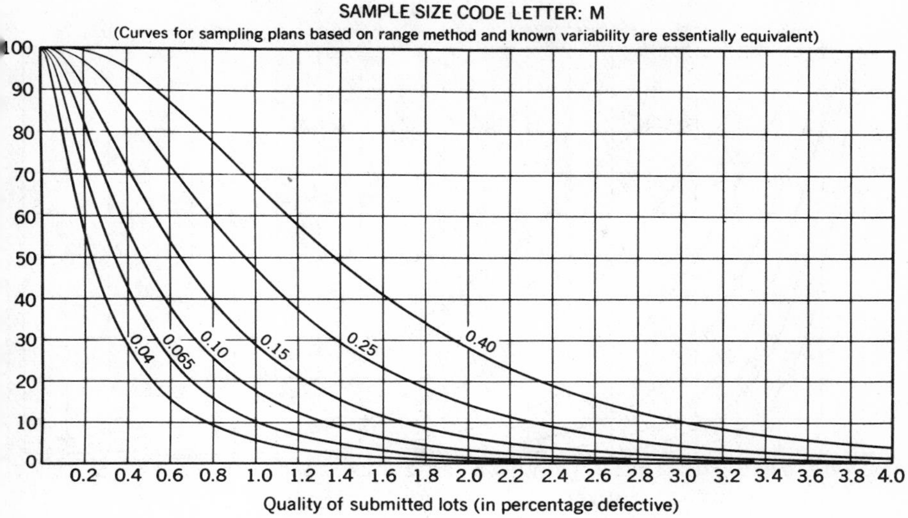

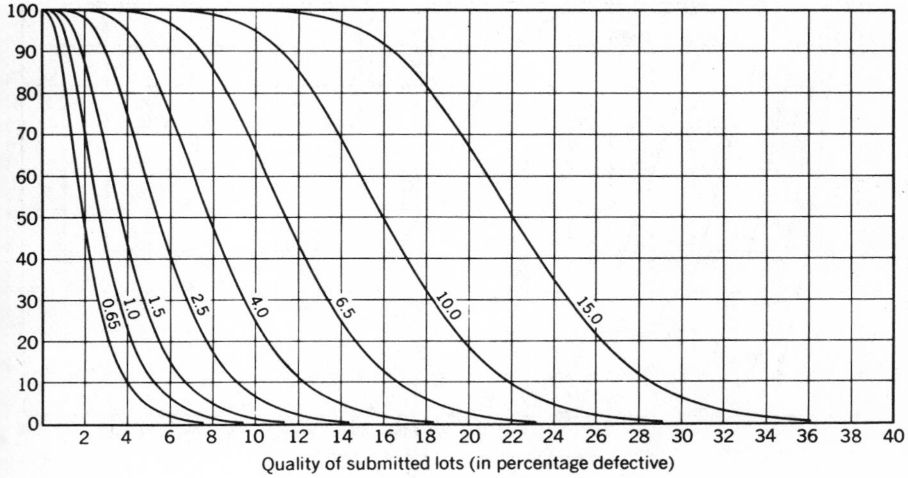

Note: Figures on curves are Acceptable Quality Levels for normal inspection.

Fig. 13.13. (cont.) OC curves from MIL-STD-414.

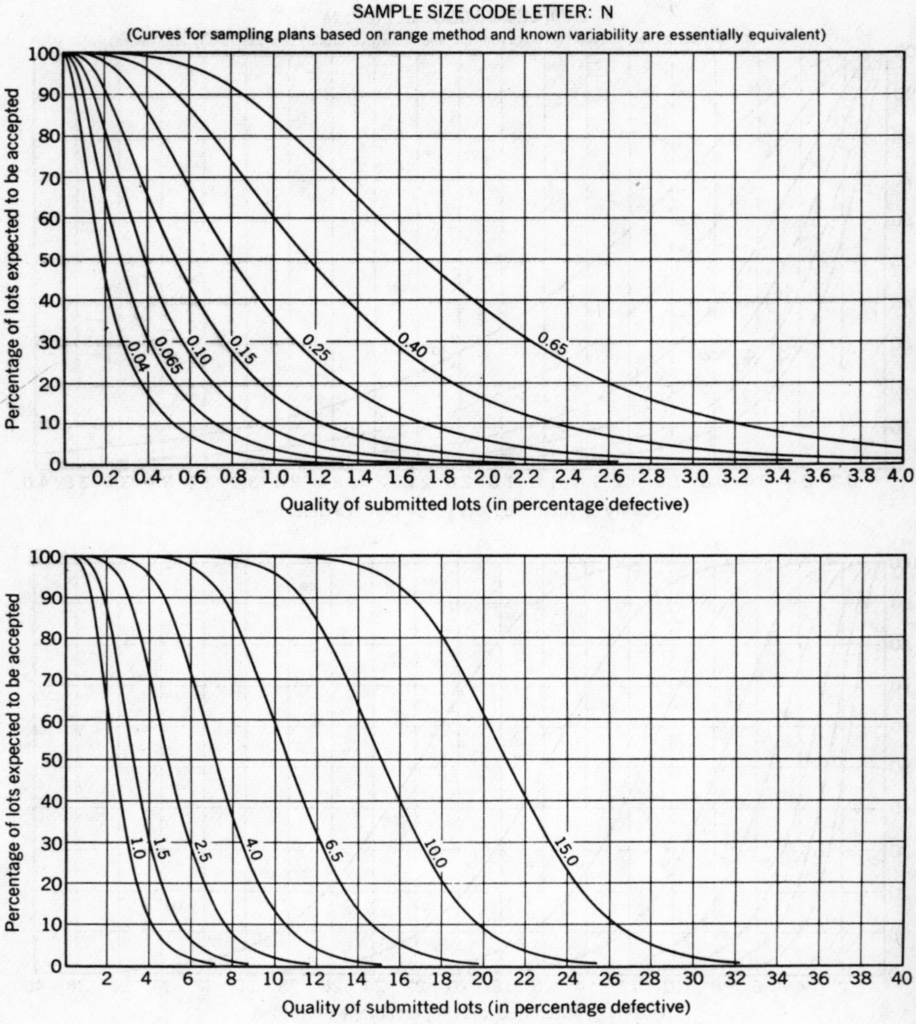

Note: Figures on curves are Acceptable Quality Levels for normal inspection.

Fig. 13.13. (cont.) OC curves from MIL-STD-414.

SAMPLE SIZE CODE LETTER: O
(Curves for sampling plans based on range method and known variability are essentially equivalent)

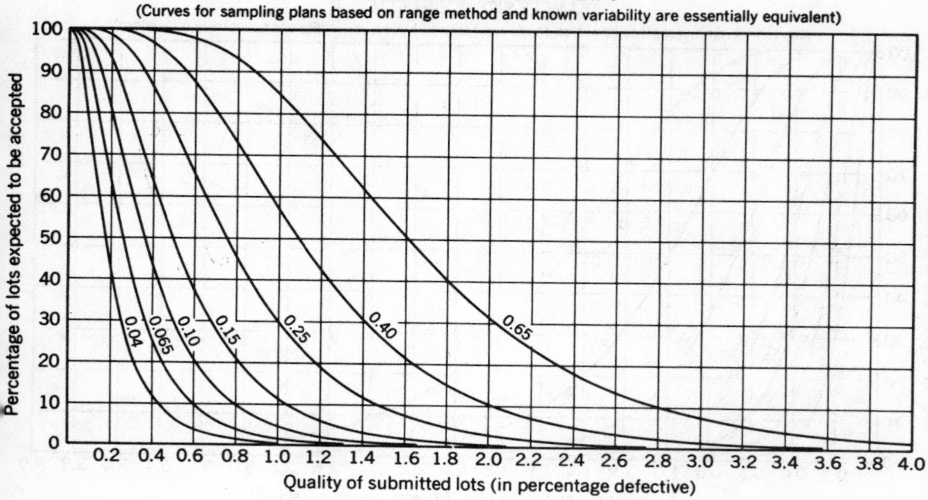

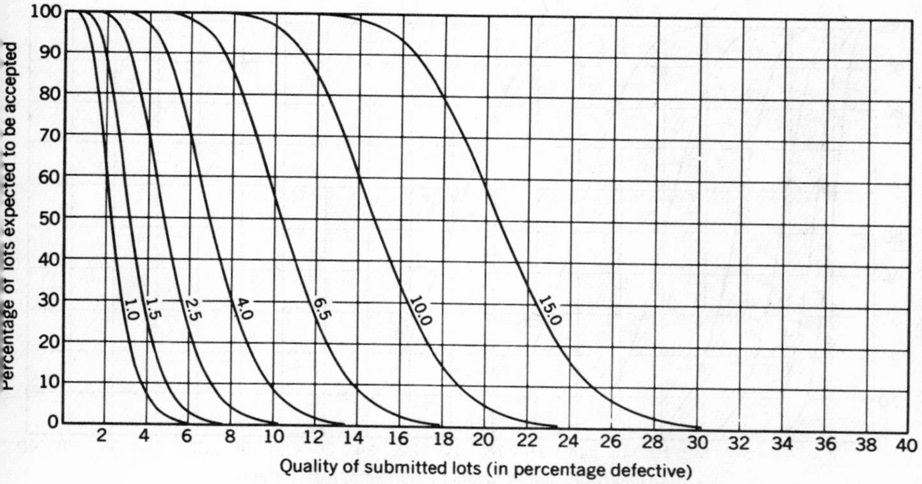

Note: Figures on curves are Acceptable Quality Levels for normal inspection.

Fig. 13.13. (cont.) OC curves from MIL-STD-414.

SAMPLE SIZE CODE LETTER: P

(Curves for sampling plans based on range method and known variability are essentially equivalent)

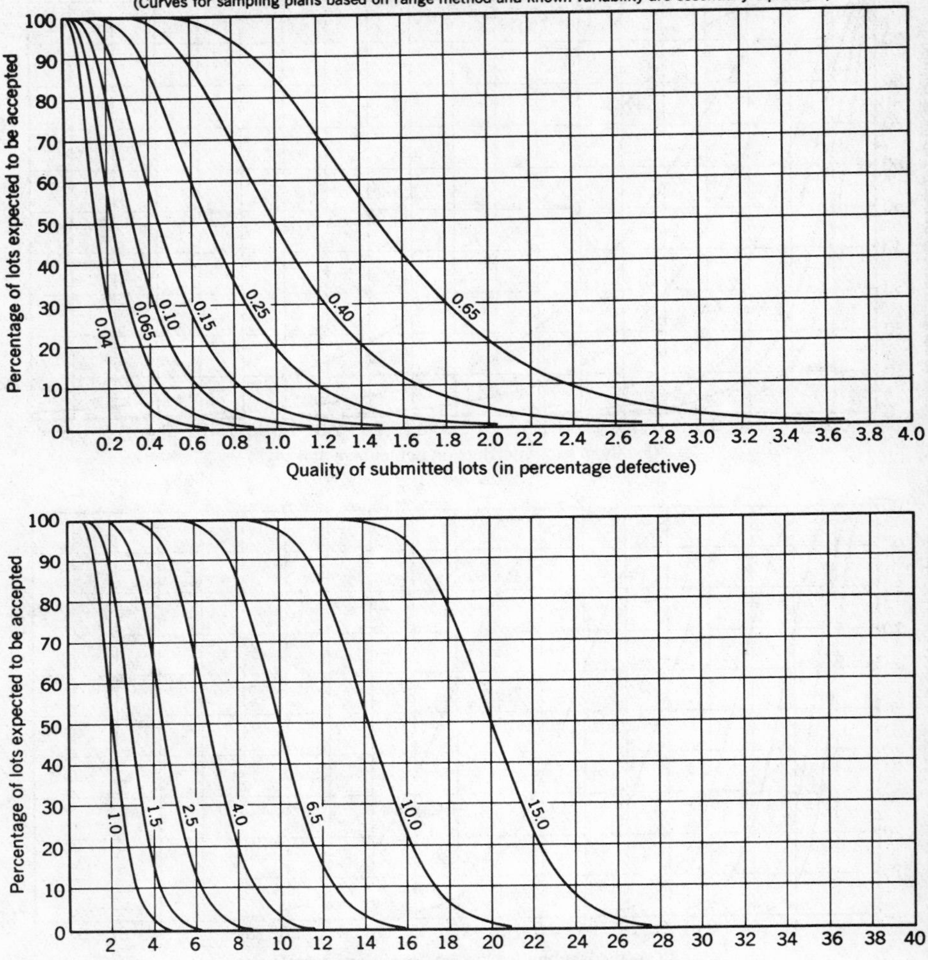

Note: Figures on curves are Acceptable Quality Levels for normal inspection.

Fig. 13.13. (cont.) OC curves from MIL-STD-414.

SAMPLE SIZE CODE LETTER: Q

(Curves for sampling plans based on range method and known variability are essentially equivalent)

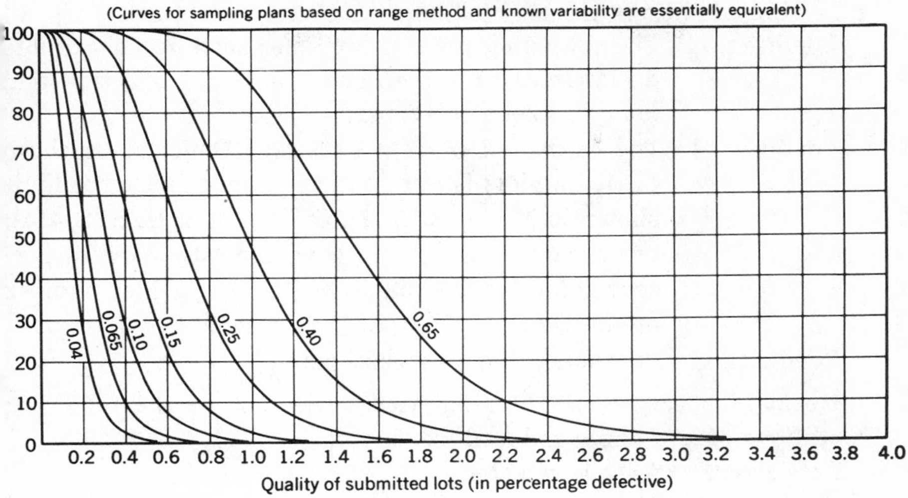

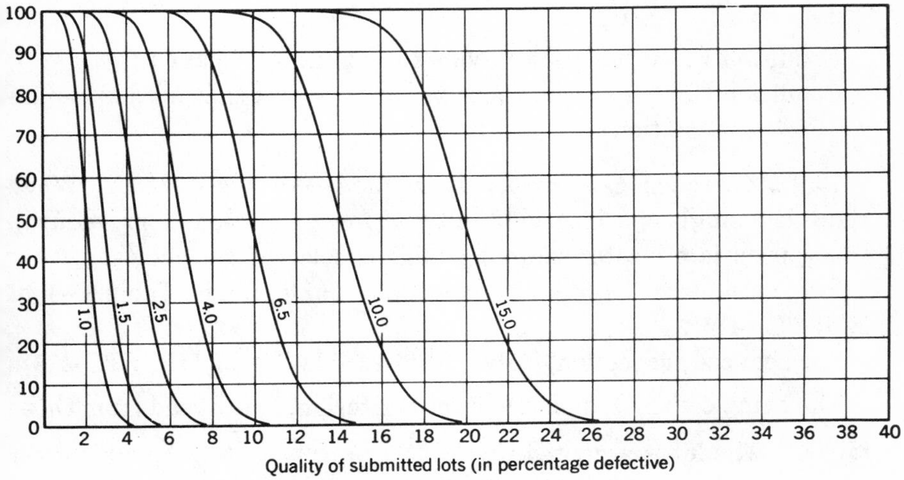

Note: Figures on curves are Acceptable Quality Levels for normal inspection.

Fig. 13.13. (cont.) OC curves from MIL-STD-414.

Reduced inspection may be instituted provided that all of the following conditions are satisfied:

Condition A. The preceding ten lots (or other such number of lots designated) have been under normal inspection and none has been rejected.

Condition B. The estimated percent defective for each of these preceding lots is less than the applicable lower limit shown in Table 13.25 (Table B–7 of MIL-STD-414); or for certain sampling plans, the estimated lot percent defective is equal to zero for a specified number of consecutive lots (see Table 13.25).

Condition C. Production is at a steady rate.

Normal inspection shall be reinstated if any one of the following conditions occurs under reduced inspection.

Condition D. A lot is rejected.

Condition E. The estimated process average is greater than the AQL.

Condition F. Production becomes irregular or delayed.

Condition G. Other conditions as may warrant that normal inspection be reinstated.

13.3.5.4 *Section C—Variability unknown—range method.* When there is a single specification limit and form 1 is used (k procedure), the appropriate master sampling table is selected as follows:

For normal or tightened inspection, Table 13.26 (Table C–1 of MIL-STD-414).

For reduced inspection, Table 13.27 (Table C–2 of MIL-STD-414).

The sample size, n, and acceptance criterion, k, are read from these tables. The lot is accepted if $\frac{U - \bar{x}}{\bar{R}} \geq k$ or $\frac{\bar{x} - L}{\bar{R}} \geq k$, whichever is appropriate.

When there is a single specification limit or a double specification limit (with the AQL applying to both limits) and form 2 is used (estimation procedure), the appropriate master sampling table is selected as follows:

For normal or tightened inspection, Table 13.28 (Table C–3 of MIL-STD-414).

For reduced inspection, Table 13.29 (Table C–4 of MIL-STD-414).

The sample size, n, and the acceptance criterion, M, are read from these tables. The quality indices $Q_U = (U - \bar{x})c/\bar{R}$ and/or $Q_L = (\bar{x} - L)c/\bar{R}$ are computed. c is obtained from Table 13.18. c is also given in the Master Tables 13.28 and 13.29. Table 13.17 (Table C-5 of MIL-STD-414) is entered with n and Q_U and/or Q_L, whichever is appropriate; the estimates p_U and/or p_L in percent, are read from the table. For an upper specification limit, the lot is accepted if $p_U \leqq M$; for a lower specification limit, the lot is accepted if $p_L \leqq M$; and for both upper and lower specification limits, the lot is accepted if $p_U + p_L \leqq M$.

The estimated process average is obtained as explained in Section 13.3.5.3. Criteria for tightened and reduced inspection are identical with the criteria given in Section 13.3.5.3 except that Table 13.24 (Table B-6 of MIL-STD-414) is replaced by Table 13.30 (Table C-6 of MIL-STD-414) and Table 13.25 (Table B-7 of MIL-STD-414) is replaced by Table 13.31 (Table C-7 of MIL-STD-414).

13.3.5.5 *Section D—Variability known.* This section assumes that the standard deviation, σ', of the normal distribution is known.[1]

When there is a single specification limit and form 1 is used (k procedure), the appropriate master sampling table is selected as follows:

For normal or tightened inspection, Table 13.32 (Table D-1 of MIL-STD-414).

For reduced inspection, Table 13.33 (Table D-2 of MIL-STD-414).

The sample size, n, and acceptance criterion, k, are read from these tables. The lot is accepted if $\dfrac{U - \bar{x}}{\sigma'} \geqq k$ or $\dfrac{\bar{x} - L}{\sigma'} \geqq k$, whichever is appropriate.

When there is a single specification limit or a double specification limit (with the AQL applying to both limits) and form 2 is used (estimation procedure), the appropriate master sampling table is selected as follows:

For normal or tightened inspection, Table 13.34 (Table D-3 of MIL-STD-414).

For reduced inspection, Table 13.35 (Table D-4 of MIL-STD-414).

The sample size, n, and the acceptance criterion, M, are read from these tables. The quality indices $Q_U = \dfrac{(U - \bar{x})}{\sigma'}\sqrt{\dfrac{n}{n-1}}$ and/or $Q_L =$

[1] The known standard deviation is denoted by σ in MIL-STD-414. σ' will be used in this section so that the notation conforms to the notation of the American Society for Quality Control.

$\frac{(\bar{x} - L)}{\sigma'}\sqrt{\frac{n}{n-1}}$ are computed. The quantity $\sqrt{\frac{n}{n-1}}$ is denoted by v and is given in the master table. Appendix Table 1 is entered with Q_U and/or Q_L, whichever is appropriate; the estimates p_U and/or p_L are read from the table and converted to readings in percent by multiplying by one hundred. For an upper specification limit, the lot is accepted if $p_U \leqq M$; for a lower specification limit, the lot is accepted if $p_L \leqq M$; and for both upper and lower specification limits, the lot is accepted if $p_U + p_L \leqq M$.

The estimated process average is obtained as explained in Section 13.3.5.3. Criteria for tightened and reduced inspection are identical with the criteria given in Section 13.3.5.3 except that Table 13.24 (Table B–6 of MIL-STD-414) is replaced by Table 13.36 (Table D–6 of MIL-STD-414) and Table 13.25 (Table B–7 of MIL-STD-414) is replaced by Table 13.37 (Table D–7 of MIL-STD-414).

13.3.5.6 *Example using* MIL-STD-414. The following example of a two-sided specification limits procedure, when the variability is unknown and the standard method is used, is given in MIL-STD-414.

EXAMPLE OF CALCULATIONS. VARIABILITY UNKNOWN — STANDARD DEVIATION METHOD (*One* AQL *Value for Both Upper and Lower Specification Limit Combined*)

The minimum temperature of operation for a certain device is specified as 180° F. The maximum temperature is 209° F. A lot of 40 items is submitted for inspection. Inspection Level IV, normal inspection, with AQL = 1 % is to be used. From the appropriate tables it is seen that a sample of size 5 is required. Suppose the measurements obtained are as follows:

197°, 188°, 184°, 205°, and 201°,

and compliance with the acceptability criterion is to be determined.

The lot meets the acceptability criterion, since $p = p_U + p_L$ is less than M.

Line	*Information Needed*	*Value Obtained*	*Explanation*
1	Sample size: n	5	
2	Sum of measurements: $\sum x$	975	
3	Sum of squared measurements: $\sum x^2$	190,435	
4	Correction factor (CF): $(\sum x)^2/n$	190,125	$(975)^2/5$
5	Corrected sum of squares (SS): $\sum x^2$ − CF	310	190,435—190,125
6	Variance (V): $SS/(n-1)$	77.5	310/4
7	Estimate of lot standard deviation (s): $\sqrt{V}$	8.81	$\sqrt{77.5}$
8	Sample mean ($\bar{x}$): $\sum x/n$	195	975/5

9	Upper specification limit: U	209	
10	Lower specification limit: L	180	
11	Quality index: $Q_U = (U - \bar{x})/s$	1.59	(209–195)/8.81
12	Quality index: $Q_L = (\bar{x} - L)/s$	1.70	(195–180)/8.81
13	Est. of lot % defective: above $U(p_U)$	2.19%	See Table 13.16
14	Est. of lot % defective: below $L(p_L)$	0.66%	See Table 13.16
15	Total est. % defective: in lot (p): $p = p_U + p_L$	2.85%	2.19% + .66%
16	Max. allowable % defective: (M)	3.32%	See Table 13.22
17	Acceptability Criterion: Compare $p = p_U + p_L$ with M	2.85% < 3.32%.	

13.4 Continuous Sampling Inspection[1]

13.4.1 Introduction

The purpose of this section on continuous sampling is to present the continuous sampling plans most commonly used in industry. These plans are used where the formation of inspection lots for lot-by-lot acceptance may be impractical or artificial, often the case for conveyor line production. The inspection is carried out by alternate sequences of consecutive item inspection (often called the 100% inspection) and sequences of production which are not inspected or from which sample items are inspected. In the plans discussed in this paper, each item inspected is classified as defective or non-defective. Unless otherwise noted, it is assumed that all defective items found when sampling or when on 100% inspection are replaced by non-defective items. Procedures which use variables quality measures have not been derived.

13.4.2 Dodge continuous sampling plans

The simplest continuous sampling plan is the one proposed by Dodge in his pioneer paper in 1943.[2] This procedure (called CSP–1) is as follows: At the outset of inspection, inspect 100% of the units consecutively as produced and continue such inspection until i units in succession are found clear of defects. When this happens discontinue 100% inspection and inspect only a fraction f of the units, selecting one unit at random from each segment of $1/f$ items. If a single defective is found revert immediately to 100% inspection of succeeding units and continue until again i units in succession are found clear of defects.

The objective of this plan and a wide class of generalizations of it

[1] Most of the material in this section is taken from a paper given by the authors at the meeting of the International Statistical Institute in Stockholm in 1957. The title of the paper is "Recent Development in Continuous Sampling."

[2] H. F. Dodge, "A Sampling Inspection Plan For Continuous Production," *Annals of Mathematical Statistics*, 1943, Vol. 14, pp. 264–279.

Table 13.20. Master Table for Normal and Tightened Inspection for Plans Based on Variability Unknown, Standard Deviation Method
(Single Specification Limit—Form 1; Table B-1 of MIL-STD-414)

Sample size code letter	Sample size	Acceptable quality levels (normal inspection)													
		0.04	0.065	0.10	0.15	0.25	0.40	0.65	1.00	1.50	2.50	4.00	6.50	10.00	15.00
		k	*k*	*k*	*k*	*k*	*k*	*k*	*k*	*k*	*k*	*k*	*k*	*k*	*k*
B	3	↓	↓	↓	↓	↓	↓	↓	↓	↓	1.12	0.958	0.765	0.566	0.341
C	4	↓	↓	↓	↓	↓	↓	↓	1.45	1.34	1.17	1.01	0.814	0.617	0.393
D	5	↓	↓	↓	↓	↓	↓	1.65	1.53	1.40	1.24	1.07	0.874	0.675	0.455
E	7	↓	↓	↓	↓	2.00	1.88	1.75	1.62	1.50	1.33	1.15	0.955	0.755	0.536
F	10	↓	↓	↓	2.24	2.11	1.98	1.84	1.72	1.58	1.41	1.23	1.03	0.828	0.611
G	15	2.64	2.53	2.42	2.32	2.20	2.06	1.91	1.79	1.65	1.47	1.30	1.09	0.886	0.664
H	20	2.69	2.58	2.47	2.36	2.24	2.11	1.96	1.82	1.69	1.51	1.33	1.12	0.917	0.695
I	25	2.72	2.61	2.50	2.40	2.26	2.14	1.98	1.85	1.72	1.53	1.35	1.14	0.936	0.712
J	30	2.73	2.61	2.51	2.41	2.28	2.15	2.00	1.86	1.73	1.55	1.36	1.15	0.946	0.723
K	35	2.77	2.65	2.54	2.45	2.31	2.18	2.03	1.89	1.76	1.57	1.39	1.18	0.969	0.745
L	40	2.77	2.66	2.55	2.44	2.31	2.18	2.03	1.89	1.76	1.58	1.39	1.18	0.971	0.746
M	50	2.83	2.71	2.60	2.50	2.35	2.22	2.08	1.93	1.80	1.61	1.42	1.21	1.00	0.774
N	75	2.90	2.77	2.66	2.55	2.41	2.27	2.12	1.98	1.84	1.65	1.46	1.24	1.03	0.804
O	100	2.92	2.80	2.69	2.58	2.43	2.29	2.14	2.00	1.86	1.67	1.48	1.26	1.05	0.819
P	150	2.96	2.84	2.73	2.61	2.47	2.33	2.18	2.03	1.89	1.70	1.51	1.29	1.07	0.841
Q	200	2.97	2.85	2.73	2.62	2.47	2.33	2.18	2.04	1.89	1.70	1.51	1.29	1.07	0.845
		0.065	0.10	0.15	0.25	0.40	0.65	1.00	1.50	2.50	4.00	6.50	10.00	15.00	
		Acceptable quality levels (tightened inspection)													

All AQL values are in percent defective.
Use first sampling plan below arrow, that is, both sample size as well as *k* value. When sample size equals or exceeds lot size, every item in the lot must be inspected.

Table 13.21. Master Table for Reduced Inspection for Plans Based on Variability Unknown, Standard Deviation Method
(Single Specification Limit—Form 1; Table B-2 of MIL-STD-414)

Sample size code letter	Sample size	Acceptable quality levels 0.04	0.065	0.10	0.15	0.25	0.40	0.65	1.00	1.50	2.50	4.00	6.50	10.00
		k	k	k	k	k	k	k	k	k	k	k	k	k
B	3	↓	↓	↓	↓	↓	↓	↓	↓	1.12	0.958	0.765	0.566	0.341
C	3	↓	↓	↓	↓	↓	↓	↓	↓	1.12	0.958	0.765	0.566	0.341
D	3	↓	↓	↓	↓	↓	↓	↓	↓	1.12	0.958	0.765	0.566	0.341
E	3	↓	↓	↓	↓	↓	↓	↓	↓	1.12	0.958	0.765	0.566	0.341
F	4	↓	↓	↓	↓	↓	↓	1.45	1.34	1.17	1.01	0.814	0.617	0.393
G	5	↓	↓	↓	↓	↓	1.65	1.53	1.40	1.24	1.07	0.874	0.675	0.455
H	7	↓	↓	↓	2.00	1.88	1.75	1.62	1.50	1.33	1.15	0.955	0.755	0.536
I	10	↓	↓	2.24	2.11	1.98	1.84	1.72	1.58	1.41	1.23	1.03	0.828	0.611
J	10	↓	↓	2.24	2.11	1.98	1.84	1.72	1.58	1.41	1.23	1.03	0.828	0.611
K	15	2.53	2.42	2.32	2.20	2.06	1.91	1.79	1.65	1.47	1.30	1.09	0.886	0.664
L	20	2.58	2.47	2.36	2.24	2.11	1.96	1.82	1.69	1.51	1.33	1.12	0.917	0.695
M	20	2.58	2.47	2.36	2.24	2.11	1.96	1.82	1.69	1.51	1.33	1.12	0.917	0.695
N	25	2.61	2.50	2.40	2.26	2.14	1.98	1.85	1.72	1.53	1.35	1.14	0.936	0.712
O	30	2.61	2.51	2.41	2.28	2.15	2.00	1.86	1.73	1.55	1.36	1.15	0.946	0.723
P	50	2.71	2.60	2.50	2.35	2.22	2.08	1.93	1.80	1.61	1.42	1.21	1.00	0.774
Q	75	2.77	2.66	2.55	2.41	2.27	2.12	1.98	1.84	1.65	1.46	1.24	1.03	0.804

All AQL values are in percent defective.

Use first sampling plan below arrow, that is, both sample size as well as k value. When sample size equals or exceeds lot size, every item in the lot must be inspected.

Table 13.22. Master Table for Normal and Tightened Inspection for Plans Based on Variability Unknown, Standard Deviation Method
(Double Specification Limit and Form 2—Single Specification Limit; Table B–3 of MIL-STD-414)

Sample size code letter	Sample size	Acceptable quality levels (normal inspection)													
		0.04	0.065	0.10	0.15	0.25	0.40	0.65	1.00	1.50	2.50	4.00	6.50	10.00	15.00
		M	*M*	*M*	*M*	*M*	*M*	*M*	*M*	*M*	*M*	*M*	*M*	*M*	*M*
B	3	↓	↓	↓	↓	↓	↓	↓	↓	↓	7.59	18.86	26.94	33.69	40.47
C	4	↓	↓	↓	↓	↓	↓	↓	1.53	5.50	10.92	16.45	22.86	29.45	36.90
D	5	↓	↓	↓	↓	↓	↓	1.33	3.32	5.83	9.80	14.39	20.19	26.56	33.99
E	7	↓	↓	↓	↓	0.422	1.06	2.14	3.55	5.35	8.40	12.20	17.35	23.29	30.50
F	10	↓	↓	↓	0.349	0.716	1.30	2.17	3.26	4.77	7.29	10.54	15.17	20.74	27.57
G	15	0.099	0.186	0.312	0.503	0.818	1.31	2.11	3.05	4.31	6.56	9.46	13.71	18.94	25.61
H	20	0.135	0.228	0.365	0.544	0.846	1.29	2.05	2.95	4.09	6.17	8.92	12.99	18.03	24.53
I	25	0.155	0.250	0.380	0.551	0.877	1.29	2.00	2.86	3.97	5.97	8.63	12.57	17.51	23.97
J	30	0.179	0.280	0.413	0.581	0.879	1.29	1.98	2.83	3.91	5.86	8.47	12.36	17.24	23.58
K	35	0.170	0.264	0.388	0.535	0.847	1.23	1.87	2.68	3.70	5.57	8.10	11.87	16.65	22.91
L	40	0.179	0.275	0.401	0.566	0.873	1.26	1.88	2.71	3.72	5.58	8.09	11.85	16.61	22.86
M	50	0.163	0.250	0.363	0.503	0.789	1.17	1.71	2.49	3.45	5.20	7.61	11.23	15.87	22.00
N	75	0.147	0.228	0.330	0.467	0.720	1.07	1.60	2.29	3.20	4.87	7.15	10.63	15.13	21.11
O	100	0.145	0.220	0.317	0.447	0.689	1.02	1.53	2.20	3.07	4.69	6.91	10.32	14.75	20.66
P	150	0.134	0.203	0.293	0.413	0.638	0.949	1.43	2.05	2.89	4.43	6.57	9.88	14.20	20.02
Q	200	0.135	0.204	0.294	0.414	0.637	0.945	1.42	2.04	2.87	4.40	6.53	9.81	14.12	19.92
		0.065	0.10	0.15	0.25	0.40	0.65	1.00	1.50	2.50	4.00	6.50	10.00	15.00	
		Acceptable quality levels (tightened inspection)													

All AQL and table values are in percent defective.
Use first sampling plan below arrow, that is, both sample size as well as *M* value. When sample size equals or exceeds lot size, every item in the lot must be inspected.

Table 13.23. Master Table for Reduced Inspection for Plans Based on Variability Unknown, Standard Deviation Method
(Double Specification Limit and Form 2—Single Specification Limit; Table B-4 of MIL-STD-414)

Sample size code letter	Sample size	Acceptable quality levels												
		0.04	0.065	0.10	0.15	0.25	0.40	0.65	1.00	1.50	2.50	4.00	6.50	10.00
		M	*M*	*M*	*M*	*M*	*M*	*M*	*M*	*M*	*M*	*M*	*M*	*M*
B	3									7.59	18.86	26.94	33.69	40.47
C	3									7.59	18.86	26.94	33.69	40.47
D	3									7.59	18.86	26.94	33.69	40.47
E	3							↓	↓	7.59	18.86	26.94	33.69	40.47
F	4						↓	1.53	5.50	10.92	16.45	22.86	29.45	36.90
G	5				↓	↓	1.33	3.32	5.83	9.80	14.39	20.19	26.56	33.99
H	7			↓	0.422	1.06	2.14	3.55	5.35	8.40	12.20	17.35	23.29	30.50
I	10			0.349	0.716	1.30	2.17	3.26	4.77	7.29	10.54	15.17	20.74	27.57
J	10	↓	↓	0.349	0.716	1.30	2.17	3.26	4.77	7.29	10.54	15.17	20.74	27.57
K	15	0.186	0.312	0.503	0.818	1.31	2.11	3.05	4.31	6.56	9.46	13.71	18.94	25.61
L	20	0.228	0.365	0.544	0.846	1.29	2.05	2.95	4.09	6.17	8.92	12.99	18.03	24.53
M	20	0.228	0.365	0.544	0.846	1.29	2.05	2.95	4.09	6.17	8.92	12.99	18.03	24.53
N	25	0.250	0.380	0.551	0.877	1.29	2.00	2.86	3.97	5.97	8.63	12.57	17.51	23.97
O	30	0.280	0.413	0.581	0.879	1.29	1.98	2.83	3.91	5.86	8.47	12.36	17.24	23.58
P	50	0.250	0.363	0.503	0.789	1.17	1.71	2.49	3.45	5.20	7.61	11.23	15.87	22.00
Q	75	0.228	0.330	0.467	0.720	1.07	1.60	2.29	3.20	4.87	7.15	10.63	15.13	21.11

All AQL and table values are in percent defective.

Use first sampling plan below arrow, that is, both sample size as well as *M* value. When sample size equals or exceeds lot size, every item in the lot must be inspected.

Table 13.24. Values of *T* for Tightened Inspection, Standard Deviation Method
(Table B-6 of MIL-STD-414)

Code letter	Acceptable quality level														Number of lots
	.04	.065	.10	.15	.25	.40	.65	1.0	1.5	2.5	4.0	6.5	10.0	15.0	
B	*	*	*	*	*	*	*	*	*	2 4 5	3 5 6	4 6 8	4 7 9	4 8 11	5 10 15
C	*	*	*	*	*	*	*	2 3 5	2 4 6	3 5 7	3 6 8	4 7 9	4 7 10	4 8 11	5 10 15
D	*	*	*	*	*	*	2 4 5	3 4 6	3 5 7	3 6 8	4 6 9	4 7 10	4 8 11	4 8 11	5 10 15
E	*	*	*	*	2 4 5	3 4 6	3 5 6	3 5 7	4 6 8	4 6 9	4 7 9	4 7 10	4 8 11	4 8 11	5 10 15
F	*	*	*	3 4 6	3 5 6	3 5 7	3 6 8	4 6 8	4 6 9	4 7 9	4 7 10	4 8 11	4 8 11	4 8 11	5 10 15
G	3 4 6	3 5 6	3 5 6	3 5 7	3 6 7	4 6 8	4 6 9	4 7 9	4 7 9	4 7 10	4 7 10	4 8 11	4 8 11	4 8 11	5 10 15
H	3 5 6	3 5 7	3 5 7	3 6 8	4 6 8	4 6 9	4 7 9	4 7 9	4 7 10	4 7 10	4 8 11	4 8 11	4 8 11	4 8 11	5 10 15
I	3 5 7	3 6 7	4 6 8	4 6 8	4 6 9	4 7 9	4 7 9	4 7 10	4 7 10	4 7 10	4 8 11	4 8 11	4 8 11	4 8 11	5 10 15
J	3 6 8	4 6 8	4 6 8	4 6 9	4 7 9	4 7 9	4 7 10	4 7 10	4 7 10	4 8 11	4 8 11	4 8 11	4 8 11	4 8 11	5 10 15
K	4 6 8	4 6 8	4 6 9	4 6 9	4 7 9	4 7 9	4 7 10	4 7 10	4 8 10	4 8 11	4 8 11	4 8 11	4 8 11	4 8 11	5 10 15
L	4 6 8	4 6 9	4 6 9	4 7 9	4 7 9	4 7 10	4 7 10	4 7 10	4 8 10	4 8 11	4 8 11	4 8 11	4 8 11	4 8 11	5 10 15
M	4 6 9	4 7 9	4 7 9	4 7 9	4 7 10	4 7 10	4 7 10	4 7 10	4 8 11	4 8 11	4 8 11	4 8 11	4 8 11	4 8 11	5 10 15
N	4 7 9	4 7 9	4 7 10	4 7 10	4 7 10	4 7 10	4 8 11	4 8 11	4 8 11	4 8 11	4 8 11	4 8 11	4 8 11	4 8 11	5 10 15
O	4 7 10	4 7 10	4 7 10	4 7 10	4 7 10	4 8 11	4 8 11	4 8 11	4 8 11	4 8 11	4 8 11	4 8 11	4 8 11	4 8 11	5 10 15
P	4 7 10	4 7 10	4 7 10	4 8 10	4 8 11	4 8 11	4 8 11	4 8 11	4 8 11	4 8 11	4 8 11	4 8 11	4 8 11	4 8 12	5 10 15
Q	4 7 10	4 8 11	4 8 11	4 8 11	4 8 11	4 8 11	4 8 11	4 8 11	4 8 11	4 8 11	4 8 11	4 8 11	4 8 11	4 8 12	5 10 15

* There are no sampling plans provided in this Standard for these code letters and AQL values.

The top figure in each block refers to the preceding 5 lots, the middle figure to the preceding 10 lots and the bottom figure to the preceding 15 lots.

Tightened inspection is required when the number of lots with estimates of percent defective above the AQL from the preceding 5, 10, or 15 lots is greater than the given value of *T* in the table, and the process average from these lots exceeds the AQL.

All estimates of the lot percent defective are obtained from Table 13.16 (Table B-5 of MIL-STD-414).

Table 13.25. Limits of Estimated Lot Percent Defective for Reduced Inspection, Standard Deviation Method

(Table B–7 of MIL-STD-414)

Code letter	Acceptable quality level														Number of lots
	.04	.065	.10	.15	.25	.40	.65	1.0	1.5	2.5	4.0	6.5	10.0	15.0	
B	*	*	*	*	*	*	*	*	*	[40]**	[30]**	[20]**	[12]**	[8]**	
C	*	*	*	*	*	*	*	[35]**	[25]**	↓ 0	↓ .57	[8]**	[6]**	0 ↑	5 10 15
D	*	*	*	*	*	*	[35]**	[25]**	[20]**	↓ .77	↓ .53 3.95	↓ 4.40 ↑	.74 9.96 ↑	6.06 ↑	5 10 15
E	*	*	*	*	[25]**	[20]**	↓ .02	↓ 0 .25	↓ .10 .88	↓ .88 2.49	.13 2.65 ↑	1.38 5.96 ↑	4.24 10.00 ↑	9.09 ↑	5 10 15
F	*	*	*	↓ .002	↓ .001 .029	↓ .016 .123	↓ .101 .369	.003 .317 .81	.044 .74 1.50	.306 1.80 ↑	1.05 3.56 ↑	2.81 6.50 ↑	5.79 10.00 ↑	10.47 ↑	5 10 15
G	↓ .003	↓ .002 .010	↓ .006 .028	↓ .018 .062	.002 .057 .151	.011 .143 .315	.047 .330 .626	.136 .643 1.00	.323 1.14 ↑	.84 2.23 ↑	1.84 3.94 ↑	3.80 6.50 ↑	6.86 10.00 ↑	11.52 15.00 ↑	5 10 15
H	↓ .004 .013	↓ .010 .029	.002 .023 .058	.005 .048 .105	.017 .111 .215	.048 .225 .396	.123 .445 ↑	.266 .785 ↑	.521 1.31 ↑	1.14 2.40 ↑	2.24 4.00 ↑	4.29 6.50 ↑	7.40 10.00 ↑	12.07 15.00 ↑	5 10 15
I	.001 .009 .021	.002 .020 .043	.006 .039 .077	.014 .071 .133	.037 .146 .248	.083 .274 .40	.185 .509 ↑	.360 .863 ↑	.653 1.39 ↑	1.33 2.48 ↑	2.49 4.00 ↑	4.59 6.50 ↑	7.74 10.00 ↑	12.43 15.00 ↑	5 10 15
J	.002 .013 .027	.005 .027 .052	.012 .050 .089	.023 .087 .146	.054 .169 .25	.113 .306 ↑	.233 .550 ↑	.431 .909 ↑	.750 1.44 ↑	1.47 2.50 ↑	2.66 4.00 ↑	4.81 6.50 ↑	7.98 10.00 ↑	12.69 15.00 ↑	5 10 15
K	.004 .017 .032	.008 .033 .058	.017 .059 .097	.032 .099 .15	.069 .186 ↑	.137 .328 ↑	.270 .577 ↑	.483 .940 ↑	.821 1.47 ↑	1.57 2.50 ↑	2.79 4.00 ↑	4.96 6.50 ↑	8.15 10.00 ↑	12.88 15.00 ↑	5 10 15
L	.005 .020 .035	.011 .038 .063	.022 .065 .10	.040 .108 .15	.082 .199 ↑	.157 .343 ↑	.300 .596 ↑	.525 .961 ↑	.876 1.49 ↑	1.64 2.50 ↑	2.88 4.00 ↑	5.08 6.50 ↑	8.29 10.00 ↑	13.03 15.00 ↑	5 10 15
M	.008 .025 .040	.016 .045 .065	.030 .075 .10	.052 .120 ↑	.102 .215 ↑	.187 .364 ↑	.345 .621 ↑	.587 .989 ↑	.959 1.50 ↑	1.76 2.50 ↑	3.03 4.00 ↑	5.27 6.50 ↑	8.50 10.00 ↑	13.25 15.00 ↑	5 10 15
N	.014 .031 .04	.026 .054 ↑	.044 .087 ↑	.072 .136 ↑	.134 .236 ↑	.235 .389 ↑	.414 .650 ↑	.681 1.00 ↑	1.082 1.50 ↑	1.92 2.50 ↓	3.24 4.00 ↑	5.52 6.50 ↑	8.81 10.00 ↑	13.60 15.00 ↑	5 10 15
O	.018 .034 ↑	.032 .058 ↑	.053 .093 ↑	.085 .143 ↑	.153 .245 ↑	.261 .400 ↑	.453 .65 ↑	.733 1.00 ↑	1.149 1.50 ↑	2.01 2.50 ↑	3.36 4.00 ↑	5.67 6.50 ↑	8.98 10.00 ↑	13.80 15.00 ↑	5 10 15
P	.023 .038 ↑	.039 .064 ↑	.064 .100 ↑	.101 .15 ↑	.177 .25 ↑	.296 .40 ↑	.501 .65 ↑	.799 1.00 ↑	1.237 1.50 ↑	2.13 2.50 ↑	3.52 4.00 ↑	5.87 6.50 ↑	9.22 10.00 ↑	14.07 15.00 ↑	5 10 15
Q	.025 .040 ↑	.044 .065 ↑	.069 .10 ↑	.108 .15 ↑	.188 .25 ↑	.312 .40 ↑	.525 .65 ↑	.830 1.00 ↑	1.276 1.50 ↑	2.19 2.50 ↑	3.59 4.00 ↑	5.96 6.50 ↑	9.32 10.00 ↑	14.19 15.00 ↑	5 10 15

* There are no sampling plans provided in this Standard for these code letters and AQL values.

↑↓ Use the first figure in direction of arrow and corresponding number of lots. In each block the top figure refers to the preceding 5 lots, the middle figure to the preceding 10 lots, and the bottom figure to the preceding 15 lots.

Reduced inspection may be instituted when every estimated lot percent defective from the preceding 5, 10, or 15 lots is below the figure given in the table; reduced inspection for sampling plans marked (**) in the table requires that the estimated lot percent defective is equal to zero for the number of consecutive lots indicated in brackets. In addition, all other conditions for reduced inspection must be satisfied.

All estimates of the lot percent defective are obtained from Table 13.16 (Table B–5 of MIL–STD–414).

Table 13.26. Master Table for Normal and Tightened Inspection for Plans Based on Variability Unknown, Range Method
(Single Specification Limit—Form 1; Table C–1 of MIL-STD-414)

Sample size code letter	Sample size	Acceptable quality levels (normal inspection)													
		0.04	0.065	0.10	0.15	0.25	0.40	0.65	1.00	1.50	2.50	4.00	6.50	10.00	15.00
		k	*k*	*k*	*k*	*k*	*k*	*k*	*k*	*k*	*k*	*k*	*k*	*k*	*k*
B	3	↓	↓	↓	↓	↓	↓	↓	↓	↓	0.587	0.502	0.401	0.296	0.178
C	4	↓	↓	↓	↓	↓	↓	↓	0.651	0.598	0.525	0.450	0.364	0.276	0.176
D	5	↓	↓	↓	↓	↓	↓	0.663	0.614	0.565	0.498	0.431	0.352	0.272	0.184
E	7	↓	↓	↓	↓	0.702	0.659	0.613	0.569	0.525	0.465	0.405	0.336	0.266	0.189
F	10	↓	↓	↓	0.916	0.863	0.811	0.755	0.703	0.650	0.579	0.507	0.424	0.341	0.252
G	15	1.09	1.04	0.999	0.958	0.903	0.850	0.792	0.738	0.684	0.610	0.536	0.452	0.368	0.276
H	25	1.14	1.10	1.05	1.01	0.951	0.896	0.835	0.779	0.723	0.647	0.571	0.484	0.398	0.305
I	30	1.15	1.10	1.06	1.02	0.959	0.904	0.843	0.787	0.730	0.654	0.577	0.490	0.403	0.310
J	35	1.16	1.11	1.07	1.02	0.964	0.908	0.848	0.791	0.734	0.658	0.581	0.494	0.406	0.313
K	40	1.18	1.13	1.08	1.04	0.978	0.921	0.860	0.803	0.746	0.668	0.591	0.503	0.415	0.321
L	50	1.19	1.14	1.09	1.05	0.988	0.931	0.893	0.812	0.754	0.676	0.598	0.510	0.421	0.327
M	60	1.21	1.16	1.11	1.06	1.00	0.948	0.885	0.826	0.768	0.689	0.610	0.521	0.432	0.336
N	85	1.23	1.17	1.13	1.08	1.02	0.962	0.899	0.839	0.780	0.701	0.621	0.530	0.441	0.345
O	115	1.24	1.19	1.14	1.09	1.03	0.975	0.911	0.851	0.791	0.711	0.631	0.539	0.449	0.353
P	175	1.26	1.21	1.16	1.11	1.05	0.994	0.929	0.868	0.807	0.726	0.644	0.552	0.460	0.363
Q	230	1.27	1.21	1.16	1.12	1.06	0.996	0.931	0.870	0.809	0.728	0.646	0.553	0.462	0.364
		0.065	0.10	0.15	0.25	0.40	0.65	1.00	1.50	2.50	4.00	6.50	10.00	15.00	
		Acceptable quality levels (tightened inspection)													

All AQL values are in percent defective.
Use first sampling plan below arrow, that is, both sample size as well as *k* value. When sample size equals or exceeds lot size, every item in the lot must be inspected.

Table 13.27. Master Table for Reduced Inspection for Plans Based on Variability Unknown, Range Method
(Single Specification Limit—Form 1; Table C–2 of MIL-STD-414)

Sample size code letter	Sample size	Acceptable quality levels												
		0.04	0.065	0.10	0.15	0.25	0.40	0.65	1.00	1.50	2.50	4.00	6.50	10.00
		k	k	k	k	k	k	k	k	k	k	k	k	k
B	3	↓	↓	↓	↓	↓	↓	↓	↓	0.587	0.502	0.401	0.296	0.178
C	3	↓	↓	↓	↓	↓	↓	↓	↓	0.587	0.502	0.401	0.296	0.178
D	3	↓	↓	↓	↓	↓	↓	↓	↓	0.587	0.502	0.401	0.296	0.178
E	3	↓	↓	↓	↓	↓	↓	↓	↓	0.587	0.502	0.401	0.296	0.178
F	4	↓	↓	↓	↓	↓	↓	0.651	0.598	0.525	0.451	0.364	0.276	0.176
G	5	↓	↓	↓	↓	↓	0.666	0.617	0.567	0.500	0.432	0.353	0.273	0.184
H	7	↓	↓	↓	0.702	0.659	0.613	0.569	0.525	0.465	0.405	0.336	0.266	0.189
I	10	↓	↓	0.916	0.863	0.811	0.755	0.703	0.650	0.579	0.507	0.424	0.341	0.252
J	10	↓	↓	0.916	0.863	0.811	0.755	0.703	0.650	0.579	0.507	0.424	0.341	0.252
K	15	1.04	0.999	0.958	0.903	0.850	0.792	0.738	0.684	0.610	0.536	0.452	0.368	0.276
L	25	1.10	1.05	1.01	0.951	0.896	0.835	0.779	0.723	0.647	0.571	0.484	0.398	0.305
M	25	1.10	1.05	1.01	0.951	0.896	0.835	0.779	0.723	0.647	0.571	0.484	0.398	0.305
N	30	1.10	1.06	1.02	0.959	0.904	0.843	0.787	0.730	0.654	0.577	0.490	0.403	0.310
O	35	1.11	1.07	1.02	0.964	0.908	0.848	0.791	0.734	0.658	0.581	0.494	0.406	0.313
P	60	1.16	1.11	1.06	1.00	0.948	0.885	0.826	0.768	0.689	0.610	0.521	0.432	0.336
Q	85	1.17	1.13	1.08	1.02	0.962	0.899	0.839	0.780	0.701	0.621	0.530	0.441	0.345

All AQL values are in percent defective.

Use first sampling plan below arrow, that is, both sample size as well as k value. When sample size equals or exceeds lot size, every item in the lot must be inspected.

Table 13.28. Master Table for Normal and Tightened Inspection for Plans Based on Variability Unknown, Range Method (Double Specification Limit and Form 2—Single Specification Limit; Table C–3 of MIL-STD-414)

Sample size code letter	Sample size	c factor	Acceptable quality levels (normal inspection)													
			0.04	0.065	0.10	0.15	0.25	0.40	0.65	1.00	1.50	2.50	4.00	6.50	10.00	15.00
			M	*M*	*M*	*M*	*M*	*M*	*M*	*M*	*M*	*M*	*M*	*M*	*M*	*M*
B	3	1.910	↓	↓	↓	↓	↓	↓	↓	↓	↓	7.59	18.86	26.94	33.69	40.47
C	4	2.234	↓	↓	↓	↓	↓	↓	↓	1.53	5.50	10.92	16.45	22.86	29.45	36.90
D	5	2.474	↓	↓	↓	↓	↓	↓	1.42	3.44	5.93	9.90	14.47	20.27	26.59	33.95
E	7	2.830	↓	↓	↓	↓	0.28	0.89	1.99	3.46	5.32	8.47	12.35	17.54	23.50	30.66
F	10	2.405	↓	↓	↓	0.23	0.58	1.14	2.05	3.23	4.77	7.42	10.79	15.49	21.06	27.90
G	15	2.379	0.061	0.136	0.253	0.430	0.786	1.30	2.10	3.11	4.44	6.76	9.76	14.09	19.30	25.92
H	25	2.358	0.125	0.214	0.336	0.506	0.827	1.27	1.95	2.82	3.96	5.98	8.65	12.59	17.48	23.79
I	30	2.353	0.147	0.240	0.366	0.537	0.856	1.29	1.96	2.81	3.92	5.88	8.50	12.36	17.19	23.42
J	35	2.349	0.165	0.261	0.391	0.564	0.883	1.33	1.98	2.82	3.90	5.85	8.42	12.24	17.03	23.21
K	40	2.346	0.160	0.252	0.375	0.539	0.842	1.25	1.88	2.69	3.73	5.61	8.11	11.84	16.55	22.38
L	50	2.342	0.169	0.261	0.381	0.542	0.838	1.25	1.60	2.63	3.64	5.47	7.91	11.57	16.20	22.26
M	60	2.339	0.158	0.244	0.356	0.504	0.781	1.16	1.74	2.47	3.44	5.17	7.54	11.10	15.64	21.63
N	85	2.335	0.156	0.242	0.350	0.493	0.755	1.12	1.67	2.37	3.30	4.97	7.27	10.73	15.17	21.05
O	115	2.333	0.153	0.230	0.333	0.468	0.718	1.06	1.58	2.25	3.14	4.76	6.99	10.37	14.74	20.57
P	175	2.333	0.139	0.210	0.303	0.427	0.655	0.972	1.46	2.08	2.93	4.47	6.60	9.89	14.15	19.88
Q	230	2.333	0.142	0.215	0.308	0.432	0.661	0.976	1.47	2.08	2.92	4.46	6.57	9.84	14.10	19.82
			0.065	0.10	0.15	0.25	0.40	0.65	1.00	1.50	2.50	4.00	6.50	10.00	15.00	
			Acceptable quality level (tightened inspection)													

All AQL and table values are in percent defective.

Use first sampling plan below arrow, that is, both sample size as well as *M* value. When sample size equals or exceeds lot size, every item in the lot must be inspected.

Table 13.29. Master Table for Reduced Inspection for Plans Based on Variability Unknown, Range Method
(Double Specification Limit and Form 2—Single Specification Limit; Table C-4 of MIL-STD-414)

Sample size code letter	Sample size	c factor	Acceptable quality levels												
			0.04	0.065	0.10	0.15	0.25	0.40	0.65	1.00	1.50	2.50	4.00	6.50	10.00
			M	*M*	*M*	*M*	*M*	*M*	*M*	*M*	*M*	*M*	*M*	*M*	*M*
B	3	1.910	↓	↓	↓	↓	↓	↓	↓	↓	7.59	18.86	26.94	33.69	40.47
C	3	1.910	↓	↓	↓	↓	↓	↓	↓	↓	7.59	18.86	26.94	33.69	40.47
D	3	1.910	↓	↓	↓	↓	↓	↓	↓	↓	7.59	18.86	26.94	33.69	40.47
E	3	1.910	↓	↓	↓	↓	↓	↓	↓	↓	7.59	18.86	26.94	33.69	40.47
F	4	2.234	↓	↓	↓	↓	↓	↓	1.53	5.50	10.92	16.45	22.86	29.45	36.90
G	5	2.474	↓	↓	↓	↓	↓	1.42	3.44	5.93	9.90	14.47	20.27	26.59	33.95
H	7	2.830	↓	↓	↓	0.28	0.89	1.99	3.46	5.32	8.47	12.35	17.54	23.50	30.66
I	10	2.405	↓	↓	0.23	0.58	1.14	2.05	3.23	4.77	7.42	10.79	15.49	21.06	27.90
J	10	2.405	↓	↓	0.23	0.58	1.14	2.05	3.23	4.77	7.42	10.79	15.49	21.06	27.90
K	15	2.379	0.136	0.253	0.430	0.786	1.30	2.10	3.11	4.44	6.76	9.76	14.09	19.30	25.92
L	25	2.358	0.214	0.336	0.506	0.827	1.27	1.95	2.82	3.96	5.98	8.65	12.59	17.48	23.79
M	25	2.358	0.214	0.336	0.506	0.827	1.27	1.95	2.82	3.96	5.98	8.65	12.59	17.48	23.79
N	30	2.353	0.240	0.366	0.537	0.856	1.29	1.96	2.81	3.92	5.88	8.50	12.36	17.19	23.42
O	35	2.349	0.261	0.381	0.542	0.838	1.25	1.60	2.63	3.64	5.47	7.91	11.57	16.20	22.26
P	60	2.339	0.244	0.356	0.504	0.781	1.16	1.74	2.47	3.44	5.17	7.54	11.10	15.64	21.63
Q	85	2.335	0.242	0.350	0.493	0.755	1.12	1.67	2.37	3.30	4.97	7.27	10.73	15.17	21.05

All AQL and table values are in percent defective.

Use first sampling plan below arrow, that is, both sample size as well as *M* value. When sample size equals or exceeds lot size, every item in the lot must be inspected.

Table 13.30. Values of *T* for Tightened Inspection, Range Method
(Table C–6 of MIL-STD-414)

Code letter	Acceptable quality levels														Number of lots
	.04	.065	.10	.15	.25	.40	.65	1.0	1.5	2.5	4.0	6.5	10.0	15.0	
B	*	*	*	*	*	*	*	*	*	2 4 5	3 5 6	4 6 8	4 7 9	4 8 11	5 10 15
C	*	*	*	*	*	*	*	2 3 4	2 4 5	3 5 6	3 6 7	4 7 9	4 7 10	4 8 11	5 10 15
D	*	*	*	*	*	*	2 3 4	2 4 5	3 4 6	3 5 7	4 6 8	4 7 9	4 7 10	4 8 11	5 10 15
E	*	*	*	*	2 3 4	2 4 5	3 4 6	3 5 6	3 5 7	4 6 8	4 7 9	4 7 10	4 7 10	4 8 11	5 10 15
F	*	*	*	2 4 5	3 4 5	3 5 6	3 5 7	3 5 7	4 6 8	4 6 9	4 7 9	4 7 10	4 8 11	4 8 11	5 10 15
G	2 4 5	2 4 5	3 4 6	3 5 6	3 5 7	3 5 7	4 6 8	4 6 8	4 6 9	4 7 9	4 7 10	4 8 11	4 8 11	4 8 11	5 10 15
H	3 5 6	3 5 7	3 5 7	3 6 7	4 6 8	4 6 8	4 7 9	4 7 9	4 7 10	4 7 10	4 7 11	4 8 11	4 8 11	4 8 11	5 10 15
I	3 5 7	3 5 7	3 6 7	4 6 8	4 6 8	4 6 9	4 7 9	4 7 9	4 7 10	4 7 10	4 8 11	4 8 11	4 8 11	4 8 11	5 10 15
J	3 5 7	3 6 7	4 6 8	4 6 8	4 6 9	4 7 9	4 7 9	4 7 10	4 7 10	4 7 10	4 8 11	4 8 11	4 8 11	4 8 11	5 10 15
K	3 6 7	4 6 8	4 6 8	4 6 8	4 7 9	4 7 9	4 7 10	4 7 10	4 7 10	4 8 11	4 8 11	4 8 11	4 8 11	4 8 11	5 10 15
L	4 6 8	4 6 8	4 6 9	4 6 9	4 7 9	4 7 9	4 7 10	4 7 10	4 7 10	4 8 11	4 8 11	4 8 11	4 8 11	4 8 11	5 10 15
M	4 6 8	4 6 9	4 7 9	4 7 9	4 7 9	4 7 10	4 7 10	4 7 10	4 8 11	4 8 11	4 8 11	4 8 11	4 8 11	4 8 11	5 10 15
N	4 7 9	4 7 9	4 7 9	4 7 10	4 7 10	4 7 10	4 7 10	4 8 11	4 8 11	4 8 11	4 8 11	4 8 11	4 8 11	4 8 11	5 10 15
O	4 7 9	4 7 10	4 7 10	4 7 10	4 7 10	4 7 10	4 8 11	4 8 11	4 8 11	4 8 11	4 8 11	4 8 11	4 8 11	4 8 11	5 10 15
P	4 7 10	4 7 10	4 7 10	4 7 10	4 7 10	4 8 11	4 8 11	4 8 11	4 8 11	4 8 11	4 8 11	4 8 11	4 8 11	4 8 12	5 10 15
Q	4 7 10	4 7 10	4 7 10	4 8 10	4 8 11	4 8 11	4 8 11	4 8 11	4 8 11	4 8 11	4 8 11	4 8 11	4 8 11	4 8 12	5 10 15

* There are no sampling plans provided in this Standard for these code letters and AQL values.

The top figure in each block refers to the preceding 5 lots, the middle figure to the preceding 10 lots and the bottom figure to the preceding 15 lots.

Tightened inspection is required when the number of lots with estimates of percent defective above the AQL from the preceding 5, 10, or 15 lots is greater than the given value of *T* in the table, and the process average from these lots exceeds the AQL.

All estimates of the lot percent defective are obtained from Table 13.17 (Table C–5 of MIL–STD–414).

Table 13.31. Limits of Estimated Lot Percent Defective for Reduced Inspection, Range Method
(Table C–7 of MIL-STD-414)

Code letter	Acceptable quality levels														Number of lots
	.04	.065	.10	.15	.25	.40	.65	1.0	1.5	2.5	4.0	6.5	10.0	15.0	
B	*	*	*	*	*	*	*	*	*	[40]**	[30]**	[20]**	[12]**	[8]**	
C	*	*	*	*	*	*	*	[35]**	[25]**	↓ 0	↓ 0 .57	[8]**	[6]**	0 ↑	5 10 15
D	*	*	*	*	*	*	[45]**	[30]**	[20]**	↓ 0	↓ 0 2.34	↓ 2.73 6.50	0 8.10 ↑	4.15 15.00 ↑	5 10 15
E	*	*	*	*	[40]**	[30]**	[20]**	↓ .01	↓ 0 .43	↓ .35 1.84	0 1.84 4.00	.50 5.02 ↑	2.93 9.65 ↑	7.71 15.00 ↑	5 10 15
F	*	*	*	[20]**	↓ 0	↓ .020	↓ .008 .158	↓ .104 .50	↓ .40 1.14	.061 1.32 2.50	.53 3.01 ↑	2.04 6.06 ↑	4.92 10.00 ↑	9.66 15.00 ↑	5 10 15
G	[20]**	↓ .001	↓ 0 .005	↓ .002 .020	↓ .015 .074	0 .060 .199	.006 .192 .466	.040 .449 .90	.148 .90 1.50	.536 1.94 ↑	1.41 3.63 ↑	3.27 6.50 ↑	6.30 10.00 ↑	11.01 15.00 ↑	5 10 15
H	↓ .003 .011	0 .009 .025	.002 .020 .052	.004 .042 .096	.014 .101 .199	.042 .209 .374	.112 .422 .65	.248 .755 ↑	.498 1.26 ↑	1.12 2.34 ↑	2.20 4.00 ↑	4.27 6.50 ↑	7.40 10.00 ↑	12.13 15.00 ↑	5 10 15
I	.001 .006 .017	.002 .015 .037	.004 .032 .067	.010 .061 .118	.028 .130 .230	.069 .252 .40	.162 .478 ↑	.326 .822 ↑	.608 1.34 ↑	1.27 2.42 ↑	2.42 4.00 ↑	4.52 6.50 ↑	7.68 10.00 ↑	12.43 15.00 ↑	5 10 15
J	.001 .010 .022	.004 .021 .044	.007 .042 .079	.017 .075 .133	.042 .151 .248	.094 .281 .40	.202 .516 ↑	.386 .867 ↑	.691 1.39 ↑	1.39 2·47 ↑	2.57 4.00 ↑	4.71 6.50 ↑	7.91 10.00 ↑	12.65 15.00 ↑	5 10 15
K	.002 .013 .026	.005 .027 .050	.012 .049 .088	.024 .087 .144	.056 .167 .25	.114 .302 ↑	.235 .544 ↑	.435 .899 ↑	.758 1.43 ↑	1.48 2.50 ↑	2.69 4.00 ↑	4.86 6.50 ↑	8.06 10.00 ↑	12.82 15.00 ↑	5 10 15
L	.004 .018 .033	.010 .036 .059	.020 .062 .099	.036 .102 .15	.076 .190 ↑	.148 .332 ↑	.288 .581 ↑	.509 .942 ↑	.857 1.47 ↑	1.62 2.50 ↑	2.86 4.00 ↑	5.07 6.50 ↑	8.31 10.00 ↑	13.09 15.00 ↑	5 10 15
M	.007 .023 .036	.014 .041 .064	.026 .069 .10	.046 .112 .15	.092 .206 ↑	.174 .352 ↑	.326 .604 ↑	.562 .968 ↑	.927 1.50 ↑	1.72 2.50 ↑	2.99 4.00 ↑	5.22 6.50 ↑	8.48 10.00 ↑	13.27 15.00 ↑	5 10 15
N	.012 .028 .042	.022 .051 .065	.038 .082 ↑	.064 .129 ↑	.122 .226 ↑	.216 .378 ↑	.389 .636 ↑	.648 1.00 ↑	1.041 1.50 ↑	1.87 2.50 ↑	3.19 4.00 ↑	5.46 6.50 ↑	8.76 10.00 ↑	13.57 15.00 ↑	5 10 15
O	.015 .033 .040	.029 .056 ↑	.048 .089 ↑	.078 .139 ↑	.144 .238 ↑	.246 .393 ↑	.434 .65 ↑	.709 1.00 ↑	1.119 1.50 ↑	1.98 2.50 ↑	3.32 4.00 ↑	5.63 6.50 ↑	8.95 10.00 ↑	13.79 15.00 ↑	5 10 15
P	.021 .036 ↑	.036 .061 ↑	.059 .095 ↑	.093 .146 ↑	.166 .248 ↑	.280 .40 ↑	.480 .65 ↑	.771 1.00 ↑	1.199 1.50 ↑	2.08 2.50 ↑	3.46 4.00 ↑	5.80 6.50 ↑	9.15 10.00 ↑	14.02 15.00 ↑	5 10 15
Q	.024 .038 ↑	.040 .063 ↑	.065 .099 ↑	.103 .149 ↑	.179 .25 ↑	.300 .40 ↑	.507 .65 ↑	.808 1.00 ↑	1.248 1.50 ↑	2.15 2.50 ↑	3.54 4.00 ↑	5.90 6.50 ↑	9.27 10.00 ↑	14.15 15.00 ↑	5 10 15

* There are no sampling plans provided in this Standard for these code letters and AQL values.

Use the first figure in direction of arrow and corresponding number of lots. In each block the top figure refers to the preceding 5 lots, the middle figure to the preceding 10 lots, and the bottom figure to the preceding 15 lots.

Reduced inspection may be instituted when every estimated lot percent defective from the preceding 5, 10, or 15 lots is below the figure given in the table; reduced inspection for sampling plans marked (**) in the table requires that the estimated lot percent defective is equal to zero for the number of consecutive lots indicated in brackets. In addition, all other conditions for reduced inspection, must be satisfied.

All estimates of the lot percent defective are obtained from Table 13.17 (Table C–5 of MIL–STD–414).

Table 13.32. Master Table for Normal and Tightened Inspection for Plans Based on Variability Known
(Single Specification Limit—Form 1; Table D-1 of MIL-STD-414)

Code letter	Acceptable quality levels (normal inspection)													
	0.04		0.065		0.10		0.15		0.25		0.40		0.65	
	n	k	n	k	n	k	n	k	n	k	n	k	n	k
B	↓		↓		↓		↓		↓		↓		↓	
C	↓		↓		↓		↓		↓		↓		↓	
D	↓		↓		↓		↓		↓		↓		2	1.58
E	↓		↓		↓		↓		2	1.94	2	1.81	3	1.69
F	↓		↓		↓		3	2.19	3	2.07	3	1.91	4	1.80
G	3	2.58	3	2.49	4	2.39	4	2.30	4	2.14	5	2.05	5	1.88
H	4	2.65	4	2.55	5	2.46	5	2.34	6	2.23	6	2.08	7	1.95
I	5	2.69	6	2.59	6	2.49	6	2.37	7	2.25	8	2.13	8	1.96
J	6	2.72	6	2.58	7	2.50	7	2.38	8	2.26	9	2.13	10	1.99
K	7	2.77	7	2.63	8	2.54	9	2.45	9	2.29	10	2.16	11	2.01
L	8	2.77	8	2.64	9	2.54	10	2.45	11	2.31	12	2.18	13	2.03
M	10	2.83	11	2.72	11	2.59	12	2.49	13	2.35	14	2.21	16	2.07
N	14	2.88	15	2.77	16	2.65	17	2.54	19	2.41	21	2.27	23	2.12
O	19	2.92	20	2.80	22	2.69	23	2.57	25	2.43	27	2.29	30	2.14
P	27	2.96	30	2.84	31	2.72	34	2.62	37	2.47	40	2.33	44	2.17
Q	37	2.97	40	2.85	42	2.73	45	2.62	49	2.48	54	2.34	59	2.18
	0.065		0.10		0.15		0.25		0.40		0.65		1.00	
	Acceptable quality levels (tightened inspection)													

All AQL values are in percent defective.

Use first sampling plan below arrow, that is, both sample size as well as k value. When sample size equals or exceeds lot size, every item in the lot must be inspected.

Table 13.32. (cont.) Master Table for Normal and Tightened Inspection for Plans Based on Variability Known
(Single Specification Limit—Form 1; Table D–1 of MIL-STD-414)

Code letter	Acceptable quality levels (normal inspection)													
	1.00		1.50		2.50		4.00		6.50		10.00		15.00	
	n	k	n	k	n	k	n	k	n	k	n	k	n	k
B	↓		↓		↓		↓		↓		↓		↓	
C	2	1.36	2	1.25	2	1.09	2	0.936	3	0.755	3	0.573	4	0.344
D	2	1.42	2	1.33	3	1.17	3	1.01	3	0.825	4	0.641	4	0.429
E	3	1.56	3	1.44	4	1.28	4	1.11	5	0.919	5	0.728	6	0.515
F	4	1.69	4	1.53	5	1.39	5	1.20	6	0.991	7	0.797	8	0.584
G	6	1.78	6	1.62	7	1.45	8	1.28	9	1.07	11	0.877	12	0.649
H	7	1.80	8	1.68	9	1.49	10	1.31	12	1.11	14	0.906	16	0.685
I	9	1.83	10	1.70	11	1.51	13	1.34	15	1.13	17	0.924	20	0.706
J	11	1.86	12	1.72	13	1.53	15	1.35	18	1.15	21	0.942	24	0.719
K	12	1.88	14	1.75	15	1.56	18	1.38	20	1.17	24	0.964	27	0.737
L	14	1.89	15	1.75	18	1.57	20	1.38	23	1.17	27	0.965	31	0.741
M	17	1.93	19	1.79	22	1.61	25	1.42	29	1.21	33	0.995	38	0.770
N	25	1.97	28	1.84	32	1.65	36	1.46	42	1.24	49	1.03	56	0.803
O	33	2.00	36	1.86	42	1.67	48	1.48	55	1.26	64	1.05	75	0.819
P	49	2.03	54	1.89	61	1.69	70	1.51	82	1.29	95	1.07	111	0.841
Q	65	2.04	71	1.89	81	1.70	93	1.51	109	1.29	127	1.07	147	0.845
	1.50		2.50		4.00		6.50		10.00		15.00			
	Acceptable quality levels (tightened inspection)													

All AQL values are in percent defective.

Use first sampling plan below arrow, that is, both sample size as well as k value. When sample size equals or exceeds lot size, every item in the lot must be inspected.

Table 13.33. Master Table for Reduced Inspection for Plans Based on Variability Known

(Single Specification Limit—Form 1; Table D–2 of MIL-STD-414)

Code letter	Acceptable quality levels													
	0.04		0.065		0.10		0.15		0.25		0.40		0.65	
	n	k	n	k	n	k	n	k	n	k	n	k	n	k
B	↓		↓		↓		↓		↓		↓		↓	
C	↓		↓		↓		↓		↓		↓		↓	
D	↓		↓		↓		↓		↓		↓		↓	
E	↓		↓		↓		↓		↓		↓		↓	
F	↓		↓		↓		↓		↓		↓		2	1.36
G	↓		↓		↓		↓		↓		2	1.58	2	1.42
H	↓		↓		↓		2	1.94	2	1.81	3	1.69	3	1.56
I	↓		↓		3	2.19	3	2.07	3	1.91	4	1.80	4	1.69
J	↓		↓		3	2.19	3	2.07	3	1.91	4	1.80	4	1.69
K	3	2.49	4	2.39	4	2.30	4	2.14	5	2.05	5	1.88	6	1.78
L	4	2.55	5	2.46	5	2.34	6	2.23	6	2.08	7	1.95	7	1.80
M	4	2.55	5	2.46	5	2.34	6	2.23	6	2.08	7	1.95	7	1.80
N	6	2.59	6	2.49	6	2.37	7	2.25	8	2.13	8	1.96	9	1.83
O	6	2.58	7	2.50	7	2.38	8	2.26	9	2.13	10	1.99	10	1.86
P	11	2.72	11	2.59	12	2.49	13	2.35	14	2.21	16	2.07	17	1.93
Q	15	2.77	16	2.65	17	2.54	19	2.41	21	2.27	23	2.12	25	1.97

All AQL values are in percent defective.

Use first sampling plan below arrow, that is, both sample size as well as k value. When sample size equals or exceeds lot size, every item in the lot must be inspected.

Table 13.33. (cont.) Master Table for Reduced Inspection for Plans Based on Variability Known
(Single Specification Limit—Form 1; Table D–2 of MIL-STD-414)

Code letter	1.00		1.50		2.50		4.0		6.5		10.00	
	n	k	n	k	n	k	n	k	n	k	n	k
B	↓		↓		↓		↓		↓		↓	
C	↓		↓		↓		↓		↓		↓	
D	↓		↓		↓		↓		↓		↓	
E	↓		↓		↓		↓		↓		↓	
F	2	1.25	2	1.09	2	0.936	3	0.755	3	0.573	4	0.344
G	2	1.33	3	1.17	3	1.01	3	0.825	4	0.641	4	0.429
H	3	1.44	4	1.28	4	1.11	5	0.919	5	0.728	6	0.515
I	4	1.53	5	1.39	5	1.20	6	0.991	7	0.797	8	0.584
J	4	1.53	5	1.39	5	1.20	6	0.991	7	0.797	8	0.584
K	6	1.62	7	1.45	8	1.28	9	1.07	11	0.877	12	0.649
L	8	1.68	9	1.49	10	1.31	12	1.11	14	0.906	16	0.685
M	8	1.68	9	1.49	10	1.31	12	1.11	14	0.906	16	0.685
N	10	1.70	11	1.51	13	1.34	15	1.13	17	0.924	20	0.706
O	12	1.72	13	1.53	15	1.35	18	1.15	21	0.942	24	0.719
P	19	1.79	22	1.61	25	1.42	29	1.21	33	0.995	38	0.770
Q	28	1.84	32	1.65	36	1.46	42	1.24	49	1.03	56	0.803

All AQL values are in percent defective.

Use first sampling plan below arrow, that is, both sample size as well as k value. When sample size equals or exceeds lot size, every item in the lot must be inspected.

Table 13.34. Master Table for Normal and Tightened Inspection for Plans Based on Variability Known
(Double Specification Limit and Form 2—Single Specification Limit; Table D-3 of MIL-STD-414)

Code letter	Acceptable quality levels (normal inspection)																				
	0.04			0.065			0.10			0.15			0.25			0.40			0.65		
	n	*M*	*v*	*n*	*M*	*v*	*n*	*M*	*v*	*n*	*M*	*v*	*n*	*M*	*v*	*n*	*M*	*v*	*n*	*M*	*v*
B		↓			↓			↓			↓			↓			↓			↓	
C		↓			↓			↓			↓			↓			↓			↓	
D		↓			↓			↓			↓			↓			↓		2	1.28	1.414
E		↓			↓			↓			↓		2	0.310	1.414	2	0.510	1.414	3	1.94	1.225
F		↓			↓			↓		3	0.369	1.225	3	0.568	1.225	3	0.959	1.225	4	1.88	1.155
G	3	0.079	1.225	3	0.114	1.225	4	0.290	1.155	4	0.399	1.155	4	0.681	1.155	5	1.09	1.118	5	1.76	1.118
H	4	0.111	1.155	4	0.161	1.155	5	0.296	1.118	5	0.445	1.118	6	0.721	1.095	6	1.14	1.095	7	1.75	1.080
I	5	0.130	1.118	6	0.230	1.095	6	0.321	1.095	6	0.478	1.095	7	0.756	1.080	8	1.14	1.069	8	1.80	1.069
J	6	0.145	1.095	6	0.234	1.095	7	0.343	1.080	7	0.507	1.080	8	0.791	1.069	9	1.18	1.061	10	1.79	1.054
K	7	0.141	1.080	7	0.226	1.080	8	0.330	1.069	9	0.469	1.061	9	0.760	1.061	10	1.14	1.054	11	1.73	1.049
L	8	0.153	1.069	8	0.243	1.069	9	0.351	1.061	10	0.494	1.054	11	0.768	1.049	12	1.15	1.045	13	1.74	1.041
M	10	0.141	1.054	11	0.217	1.049	11	0.326	1.049	12	0.461	1.045	13	0.721	1.041	14	1.08	1.038	16	1.62	1.033
N	14	0.138	1.038	15	0.211	1.035	16	0.308	1.033	17	0.438	1.031	19	0.673	1.027	11	1.00	1.025	23	1.51	1.023
O	19	0.134	1.027	20	0.207	1.026	22	0.296	1.024	23	0.423	1.023	25	0.655	1.021	27	0.980	1.019	30	1.47	1.017
P	27	0.129	1.019	30	0.193	1.017	31	0.283	1.017	34	0.397	1.015	37	0.615	1.014	40	0.921	1.013	44	1.39	1.012
Q	37	0.130	1.014	40	0.196	1.013	42	0.285	1.012	45	0.402	1.011	49	0.620	1.010	54	0.920	1.009	59	1.39	1.009
	0.065			0.10			0.15			0.25			0.40			0.65			1.00		
	Acceptable quality levels (tightened inspection)																				

Table 13.34. (cont.) Master Table for Normal and Tightened Inspection for Plans Based on Variability Known
(Double Specification Limit and Form 2—Single Specification Limit; Table D-3 of MIL-STD-414)

Code letter	Acceptable quality levels (normal inspection)																				
	1.00			1.50			2.50			4.00			6.50			10.00			15.00		
	n	*M*	*v*	*n*	*M*	*v*	*n*	*M*	*v*	*n*	*M*	*v*	*n*	*M*	*v*	*n*	*M*	*v*	*n*	*M*	*v*
B		↓			↓			↓			↓			↓			↓			↓	
C	2	2.73	1.414	2	3.90	1.414	2	6.11	1.414	2	9.27	1.414	3	17.74	1.225	3	24.22	1.225	4	33.67	1.225
D	2	2.23	1.414	2	3.00	1.414	3	7.56	1.225	3	10.79	1.225	3	15.60	1.225	4	22.97	1.155	4	31.01	1.155
E	3	2.76	1.225	3	3.85	1.225	4	6.99	1.155	4	9.97	1.155	5	15.21	1.118	5	20.80	1.118	6	28.64	1.095
F	4	2.58	1.155	4	3.87	1.155	5	6.05	1.118	5	8.92	1.118	6	13.89	1.095	7	19.46	1.080	8	26.64	1.069
G	6	2.57	1.095	6	3.77	1.095	7	5.83	1.080	8	8.62	1.069	9	12.88	1.061	11	17.88	1.049	12	24.88	1.045
H	7	2.62	1.080	8	3.68	1.069	9	5.68	1.061	10	8.43	1.054	12	12.35	1.045	14	17.36	1.038	16	23.96	1.033
I	9	2.59	1.061	10	3.63	1.054	11	5.60	1.049	13	8.13	1.041	15	12.04	1.035	17	17.05	1.031	20	23.43	1.026
J	11	2.57	1.049	12	3.61	1.045	13	5.58	1.041	15	8.13	1.035	18	11.88	1.029	21	16.71	1.025	24	23.13	1.022
K	12	2.49	1.045	14	3.43	1.038	15	5.34	1.035	18	7.72	1.029	20	11.57	1.026	24	16.23	1.022	27	22.63	1.019
L	14	2.51	1.038	15	3.54	1.035	18	5.29	1.029	20	7.80	1.026	23	11.56	1.023	27	16.27	1.019	31	22.57	1.017
M	17	2.35	1.031	19	3.28	1.027	22	4.98	1.024	25	7.34	1.021	29	10.93	1.018	33	15.61	1.016	38	21.77	1.013
N	25	2.19	1.021	28	3.05	1.018	32	4.68	1.016	36	6.95	1.014	42	10.40	1.012	49	14.87	1.010	56	20.90	1.009
O	33	2.12	1.016	36	2.99	1.014	42	4.55	1.012	48	6.75	1.011	55	10.17	1.009	64	14.58	1.008	75	20.48	1.007
P	49	2.00	1.010	54	2.82	1.009	61	4.35	1.008	70	6.48	1.007	82	9.76	1.006	95	14.09	1.005	111	19.90	1.005
Q	65	2.00	1.008	71	2.82	1.007	81	4.34	1.006	93	6.46	1.005	109	9.73	1.005	127	14.02	1.004	147	19.84	1.003
	1.50			2.50			4.00			6.50			10.00			15.00					
	Acceptable quality levels (tightened inspection)																				

All AQL and table values are in percent defective.

Use first sampling plan below arrow, that is, both sample size as well as *M* value. When sample size equals or exceeds lot size, every item in the lot must be inspected.

Table 13.35. Master Table for Reduced Inspection for Plans Based on Variability Known
(Double Specification Limit and Form 2—Single Specification Limit; Table D–4 of MIL-STD-414)

Code letter	Acceptable quality levels																				
	0.04			0.065			0.10			0.15			0.25			0.40			0.65		
	n	*M*	*v*	*n*	*M*	*v*	*n*	*M*	*v*	*n*	*M*	*v*	*n*	*M*	*v*	*n*	*M*	*v*	*n*	*M*	*v*
B		↓			↓			↓			↓			↓			↓			↓	
C		↓			↓			↓			↓			↓			↓			↓	
D		↓			↓			↓			↓			↓			↓			↓	
E		↓			↓			↓			↓			↓			↓			↓	
F		↓			↓			↓			↓			↓			↓		2	2.73	1.414
G		↓			↓			↓			↓			↓		2	1.28	1.414	2	2.23	1.414
H		↓			↓			↓		2	0.310	1.414	2	0.510	1.414	3	1.94	1.225	3	2.76	1.225
I		↓			↓		3	0.369	1.225	3	0.568	1.225	3	0.959	1.225	4	1.88	1.155	4	2.58	1.155
J		↓			↓		3	0.369	1.225	3	0.568	1.225	3	0.959	1.225	4	1.88	1.155	4	2.58	1.155
K	3	0.11	1.225	4	0.290	1.155	4	0.399	1.155	4	0.681	1.155	5	1.09	1.118	5	1.76	1.118	6	2.57	1.095
L	4	0.16	1.155	5	0.296	1.118	5	0.445	1.118	6	0.721	1.095	6	1.14	1.095	7	1.75	1.080	7	2.62	1.080
M	4	0.161	1.155	5	0.296	1.118	5	0.445	1.118	6	0.721	1.095	6	1.14	1.095	7	1.75	1.080	7	2.62	1.080
N	6	0.230	1.095	6	0.321	1.095	6	0.478	1.095	7	0.756	1.080	8	1.14	1.069	8	1.80	1.069	9	2.59	1.061
O	6	0.234	1.095	7	0.343	1.080	7	0.507	1.080	8	0.791	1.069	9	1.18	1.061	10	1.79	1.054	11	2.57	1.049
P	11	0.217	1.049	11	0.326	1.049	12	0.461	1.045	13	0.721	1.041	14	1.08	1.038	16	1.62	1.033	17	2.35	1.031
Q	15	0.211	1.035	16	0.308	1.033	17	0.438	1.031	19	0.673	1.027	21	1.00	1.025	23	1.51	1.023	25	2.19	1.021

Table 13.35. (cont.) Master Table for Reduced Inspection for Plans Based on Variability Known
(Double Specification Limit and Form 2—Single Specification Limit; Table D–4 of MIL-STD-414)

Code letter	Acceptable quality levels																	
	1.00			1.50			2.50			4.0			6.5			10.00		
	n	*M*	*v*	*n*	*M*	*v*	*n*	*M*	*v*	*n*	*M*	*v*	*n*	*M*	*v*	*n*	*M*	*v*
B		↓			↓			↓			↓			↓			↓	
C		↓			↓			↓			↓			↓			↓	
D		↓			↓			↓			↓			↓			↓	
E		↓			↓			↓			↓			↓			↓	
F	2	3.90	1.414	2	6.11	1.414	2	9.27	1.414	3	17.74	1.225	3	24.22	1.225	4	33.67	1.225
G	2	3.00	1.414	3	7.56	1.225	3	10.79	1.225	3	15.60	1.225	4	22.97	1.155	4	31.01	1.155
H	3	3.85	1.225	4	6.99	1.155	4	9.97	1.155	5	15.21	1.118	5	20.80	1.118	6	28.64	1.095
I	4	3.87	1.155	5	6.05	1.118	5	8.92	1.118	6	13.89	1.095	7	19.46	1.080	8	26.64	1.069
J	4	3.87	1.155	5	6.05	1.118	5	8.92	1.118	6	13.89	1.095	7	19.46	1.080	8	26.64	1.069
K	6	3.77	1.095	7	5.83	1.080	8	8.62	1.069	9	12.88	1.061	11	17.88	1.049	12	24.88	1.045
L	8	3.68	1.069	9	5.68	1.061	10	8.43	1.054	12	12.35	1.045	14	17.36	1.038	16	23.96	1.033
M	8	3.68	1.069	9	5.68	1.061	10	8.43	1.054	12	12.35	1.045	14	17.36	1.038	16	23.96	1.033
N	10	3.63	1.054	11	5.60	1.049	13	8.13	1.041	15	12.04	1.035	17	17.05	1.031	20	23.43	1.026
O	12	3.61	1.045	13	5.58	1.041	15	8.13	1.035	18	11.88	1.029	21	16.71	1.025	24	23.13	1.022
P	19	3.28	1.027	22	4.98	1.024	25	7.34	1.021	29	10.93	1.018	33	15.61	1.016	38	21.77	1.013
Q	28	3.05	1.018	32	4.68	1.016	36	6.95	1.014	42	10.40	1.012	49	14.87	1.010	56	20.90	1.009

All AQL and table values are in percent defective.

Use first sampling plan below arrow, that is, both sample size as well as *M* value. When sample size equals or exceeds lot size, every item in the lot must be inspected.

Table 13.36. Values of T for Tightened Inspection, Variability Known

(Table D–6 of MIL-STD-414)

Code letter	Acceptable quality levels														Number of lots
	.04	.065	.10	.15	.25	.40	.65	1.0	1.5	2.5	4.0	6.5	10.0	15.0	
B	*	*	*	*	*	*	*	*	*	*	*	*	*	*	5
															10
															15
C	*	*	*	*	*	*	*	3	3	3	3	4	4	4	5
								5	5	5	6	7	7	7	10
								6	6	7	7	9	9	10	15
D	*	*	*	*	*	*	3	3	3	4	4	4	4	4	5
							4	5	5	6	6	7	7	7	10
							6	6	6	8	9	9	10	10	15
E	*	*	*	*	2	3	3	3	3	4	4	4	4	4	5
					4	4	5	6	6	6	7	7	7	8	10
					5	6	7	7	8	9	9	10	10	11	15
F	*	*	*	3	3	3	4	4	4	4	4	4	4	4	5
				5	5	5	6	6	6	7	7	7	7	8	10
				6	7	7	8	8	8	9	9	10	10	11	15
G	3	3	3	3	3	4	4	4	4	4	4	4	4	4	5
	4	4	5	5	6	6	6	7	7	7	7	7	8	8	10
	6	6	7	7	7	8	8	9	9	10	10	10	11	11	15
H	3	3	3	3	4	4	4	4	4	4	4	4	4	4	5
	5	5	6	6	6	6	7	7	7	7	7	8	8	8	10
	6	7	7	8	8	9	9	9	10	10	10	11	11	11	15
I	3	4	4	4	4	4	4	4	4	4	4	4	4	4	5
	5	6	6	6	6	7	7	7	7	7	7	8	8	8	10
	7	8	8	8	9	9	9	9	10	10	10	11	11	11	15
J	3	4	4	4	4	4	4	4	4	4	4	4	4	4	5
	6	6	6	6	6	7	7	7	7	7	8	8	8	8	10
	8	8	8	8	9	9	9	10	10	10	11	11	11	11	15
K	4	4	4	4	4	4	4	4	4	4	4	4	4	4	5
	6	6	6	6	7	7	7	7	7	7	8	8	8	8	10
	8	8	9	9	9	9	10	10	10	10	11	11	11	11	15
L	4	4	4	4	4	4	4	4	4	4	4	4	4	4	5
	6	6	6	7	7	7	7	7	7	8	8	8	8	8	10
	8	9	9	9	9	10	10	10	10	11	11	11	11	11	15
M	4	4	4	4	4	4	4	4	4	4	4	4	4	4	5
	6	7	7	7	7	7	7	7	7	8	8	8	8	8	10
	9	9	9	9	10	10	10	10	10	11	11	11	11	11	15
N	4	4	4	4	4	4	4	4	4	4	4	4	4	4	5
	7	7	7	7	7	7	7	8	8	8	8	8	8	8	10
	9	9	10	10	10	10	10	11	11	11	11	11	11	11	15
O	4	4	4	4	4	4	4	4	4	4	4	4	4	4	5
	7	7	7	7	7	7	8	8	8	8	8	8	8	8	10
	10	10	10	10	10	10	11	11	11	11	11	11	11	11	15
P	4	4	4	4	4	4	4	4	4	4	4	4	4	4	5
	7	7	7	7	8	8	8	8	8	8	8	8	8	8	10
	10	10	10	10	11	11	11	11	11	11	11	11	11	11	15
Q	4	4	4	4	4	4	4	4	4	4	4	4	4	4	5
	7	7	8	8	8	8	8	8	8	8	8	8	8	8	10
	10	10	11	11	11	11	11	11	11	11	11	11	11	11	15

* There are no sampling plans provided in this Standard for these code letters and AQL values.

The top figure in each block refers to the preceding 5 lots, the middle figure to the preceding 10 lots and the bottom figure to the preceding 15 lots.

Tightened inspection required when the number of lots with estimates of percent defective above the AQL from the preceding 5, 10, or 15 lots is greater than the given value of T in the table, and the process average from these lots exceeds the AQL.

All estimates of the lot percent defective are obtained from Appendix Table 1.

Table 13.37. Limits of Estimated Lot Percent Defective for Reduced Inspection, Variability Known

(Table D-7 of MIL-STD-414)

Code letter	Acceptable quality level														Number of lots
	.04	.065	.10	.15	.25	.40	.65	1.0	1.5	2.5	4.0	6.5	10.0	15.0	
B	*	*	*	*	*	*	*	*	*	*	*	*	*	*	
C	*	*	*	*	*	*	*	.011	.027	.077	.205	1.645	3.226	7.714	5
								.109	.222	.542	1.217	4.496	7.912	14.291	10
								.209	.558	1.253	2.592	6.50	↑	↑	15
D	*	*	*	*	*	*	.005	.011	.027	.369	.769	1.645	4.386	7.714	5
							.050	.109	.222	1.248	2.354	4.496	8.845	14.291	10
							.144	.290	.558	2.145	3.850	6.50	↑	↑	15
E	*	*	*	*	.001	.002	.045	.088	.166	.637	1.225	2.937	5.154	9.479	5
					.009	.021	.197	.357	.622	1.643	2.924	5.697	9.330	15.00	10
					.029	.064	.384	.669	1.124	2.50	4.00	↑	↑	↑	15
F	*	*	*	.005	.010	.021	.098	.178	.313	.846	1.560	3.325	6.114	10.436	5
				.025	.052	.100	.309	.528	.874	1.880	3.250	5.958	9.806	15.00	10
				.056	.110	.204	.522	.867	1.394	2.50	↑	↑	↑	↑	15
G	.001	.001	.007	.013	.026	.078	.147	.322	.533	1.136	2.166	4.045	7.093	11.478	5
	.004	.008	.029	.049	.093	.217	.385	.718	1.139	2.141	3.698	6.342	10.00	15.00	10
	.010	.018	.055	.090	.167	.347	.602	1.00	1.50	↑	↑	↑	↑	↑	15
H	.002	.004	.013	.022	.057	.103	.223	.375	.677	1.326	2.403	4.453	7.502	12.054	5
	.009	.017	.041	.067	.147	.252	.478	.773	1.270	2.277	3.831	6.50	10.00	15.00	10
	.018	.033	.071	.114	.227	.382	.65	1.00	↑	↑	↑	↑	↑	↑	15
I	.004	.011	.018	.030	.070	.142	.252	.457	.778	1.461	2.643	4.719	7.786	12.427	5
	.014	.031	.051	.081	.164	.298	.508	.847	1.346	2.359	3.942	6.50	10.00	15.00	10
	.025	.051	.082	.129	.244	.40	.65	↑	↑	↑	↑	↑	↑	↑	15
J	.006	.011	.023	.038	.082	.158	.298	.516	.853	1.562	2.758	4.909	8.055	12.693	5
	.018	.031	.058	.092	.177	.313	.549	.892	1.394	2.412	3.987	6.50	10.00	15.00	10
	.030	.051	.090	.140	.25	.40	↑	↑	↑	↑	↑	↑	↑	↑	15
K	.008	.014	.028	.051	.091	.171	.317	.540	.910	1.641	2.891	5.009	8.205	12.848	5
	.021	.036	.064	.108	.188	.326	.564	.908	1.427	2.449	4.00	6.50	10.00	15.00	10
	.033	.056	.096	.15	.25	.40	↑	↑	↑	↑	↑	↑	↑	↑	15
L	.009	.017	.032	.056	.107	.193	.348	.581	.934	1.732	2.960	5.131	8.328	13.017	5
	.023	.040	.069	.113	.203	.344	.586	.934	1.440	2.486	4.00	6.50	10.00	15.00	10
	.036	.060	.100	.15	.25	↑	↑	↑	↑	↑	↑	↑	↑	↑	15
M	.012	.023	.038	.064	.120	.211	.383	.627	1.010	1.821	3.093	5.310	8.516	13.238	5
	.027	.048	.076	.121	.214	.357	.608	.959	1.475	2.50	4.00	6.50	10.00	15.00	10
	.039	.065	.10	.15	↑	↑	↑	↑	↑	↑	↑	↑	↑	↑	15
N	.017	.030	.049	.080	.146	.251	.435	.705	1.113	1.959	3.272	5.546	8.822	13.588	5
	.032	.054	.086	.134	.232	.382	.635	.994	1.50	2.50	4.00	6.50	10.00	15.00	10
	.04	↑	↑	↑	↑	↑	↑	↑	↑	↑	↑	↑	↑	↑	15
O	.020	.035	.058	.091	.161	.272	.467	.750	1.168	2.041	3.386	5.685	8.990	13.801	5
	.035	.058	.092	.141	.241	.392	.648	1.00	1.50	2.50	4.00	6.50	10.00	15.00	10
	↑	↑	↑	↑	↑	↑	↑	↑	↑	↑	↑	↑	↑	↑	15
P	.024	.041	.066	.103	.180	.299	.505	.803	1.239	2.132	3.509	5.852	9.192	14.034	5
	.037	.062	.097	.147	.249	.40	.65	1.00	1.50	2.50	4.00	6.50	10.00	15.00	10
	↑	↑	↑	↑	↑	↑	↑	↑	↑	↑	↑	↑	↑	↑	15
Q	.027	.045	.071	.110	.191	.316	.528	.834	1.278	2.188	3.583	5.949	9.312	14.173	5
	.039	.064	.099	.150	.25	.40	.65	1.00	1.50	2.50	4.00	6.50	10.00	15.00	10
	↑	↑	↑	↑	↑	↑	↑	↑	↑	↑	↑	↑	↑	↑	15

* There are no sampling plans provided in this Standard for these code letters and AQL values.

↑ Use the first figure in direction of arrow and corresponding number of lots.

In each block the top figure refers to the preceding 5 lots, the middle figure to the preceding 10 lots, and the bottom figure to the preceding 15 lots.

Reduced inspection may be instituted when every estimated lot percent defective from the preceding 5, 10, or 15 lots is below the figure given in the table.

In addition, all other conditions for reduced inspection must be satisfied.

All estimates of the lot percent defective are obtained from Appendix Table 1.

is to provide assurance that the long run percentage of defective units in the accepted product will be held within a prescribed limit, the average outgoing quality limit or AOQL. Evaluation of the statistical properties of the plan has been made under the assumption of control—qualities of the items are mutually independent binomial random variables with constant parameter p'. Its statistical properties may be described by an average outgoing quality (AOQ) curve and by an average fraction inspected (AFI) curve. The statistical properties of most of the plans discussed in this section are described in the same way; AFI and AOQ curves are computed under the assumption of control and plans are classified by AOQL and sampling rate. The AOQL is defined as the worst average outgoing quality that will result from using a continuous sampling acceptance plan over a long run period, regardless of the presented quality. The average outgoing quality (AOQ) will usually be better than the AOQL. The AOQL to be used for a given item is determined by the risk of accepting some defective items that the consumer is willing to take.

In his paper, Dodge presented equations and charts for determining the AOQL as functions of the parameters f and i, under the assumption that the process is in a state of statistical control. Figure 13.14 presents the necessary chart for the selection of the appropriate plan.[1]

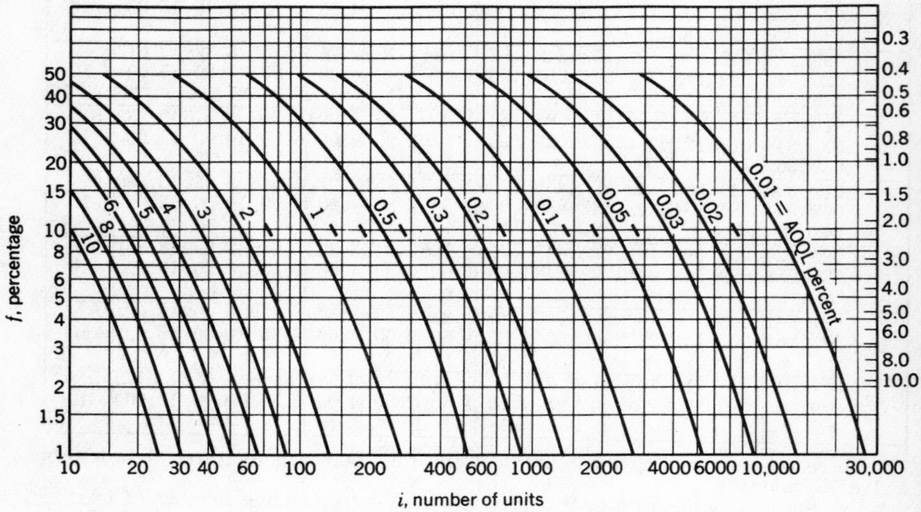

Fig. 13.14. Curves for determining values of f and i for a given value of AOQL in Dodge's CSP-1 plan for continuous production.

[1] If defective units are removed but not replaced, i should be increased by one to maintain the prescribed AOQL while holding f fixed.

Clearly an abrupt change between 100% inspection and partial inspection may sometimes be unnecessary. In the first place, even a production process which is at a satisfactory quality level will produce a certain number of defectives and eventually one of the sampled items will be defective. Further, this abrupt change may lead to hardships in personnel assignments and, hence, in administration. In a very complicated and expensive item, such as an aircraft engine, this transition may require major readjustments. In a later paper published in 1951, Dodge and Torrey propose two modifications of the plan (CSP–2 and CSP–3) which delay the beginning of 100% inspection and also add some protection against spotty quality.[1] In the CSP–2 plan inspection is begun as in the CSP–1 plan by inspecting 100% until i successive good items are found, after which partial inspection is introduced; the plan reverts to 100% inspection not on the basis of a single defective, but whenever two defectives occur spaced less than k units apart. The CSP–3 plan is similar to the CSP–2 plan, except that after a defective is found the next four units are inspected.

13.4.3 Multi-level sampling plans

A plan which allows for smooth transition between sampling and 100% inspection, which requires 100% inspection only when quality is quite inferior, and which allows for the inspection to continue to reduce when quality is definitely good, is a multi-level sampling plan. This plan allows for any number of sampling levels subject to the provisions that transitions can occur only between adjacent levels. The origin of this plan is somewhat obscure. It seems to have been used by the Air Force for the inspection of aircraft engines and was proposed by several people including Joseph Greenwood of the Navy Bureau of Aeronautics.[2] A particular multi-level plan has recently been discussed in considerable detail by Lieberman and Solomon.[3] Their plan is called MLP and is as follows.

As with the Dodge plan, inspect 100% of the units consecutively as produced and continue until i units in succession are found clear

[1] H. F. Dodge and M. N. Torrey, "Additional Continuous Sampling Plans," *Industrial Quality Control*, 1951, Vol. 7, pp. 7–12.

[2] J. A. Greenwood, "A Continuous Sampling Plan and Its Operating Characteristics," Bureau of Aeronautics, Navy Department, Washington, D.C., unpublished memorandum.

[3] G. J. Lieberman and H. Solomon, "Multi-Level Continuous Sampling Plans," *Annals of Mathematical Statistics*, 1955, Vol. 26, pp. 686–704.

of defects. When i units are found clear of defects, discontinue 100% inspection and inspect only a fraction f. If the next i units inspected are non-defective then proceed to sampling at rate f^2. If a defective is found, revert to inspection at the next lowest level, etc. This plan can be used with any number of levels, but from two to four levels seem to be of the greatest practical interest. The first Dodge plan, CSP–1, is easily recognized as a special case containing only one sampling level. Curves of constant AOQL are given in Figure 13.15 as a function of f and i for one-level and infinite level plans. Figures 13.16, 13.17, and 13.18 present curves of constant AOQL for two-, three-, and four-level plans respectively.

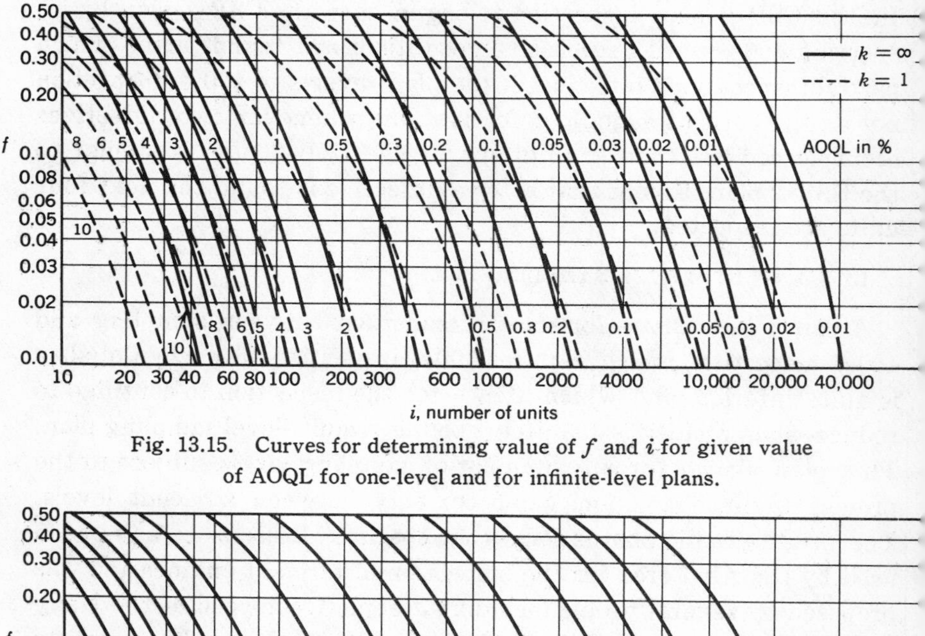

Fig. 13.15. Curves for determining value of f and i for given value of AOQL for one-level and for infinite-level plans.

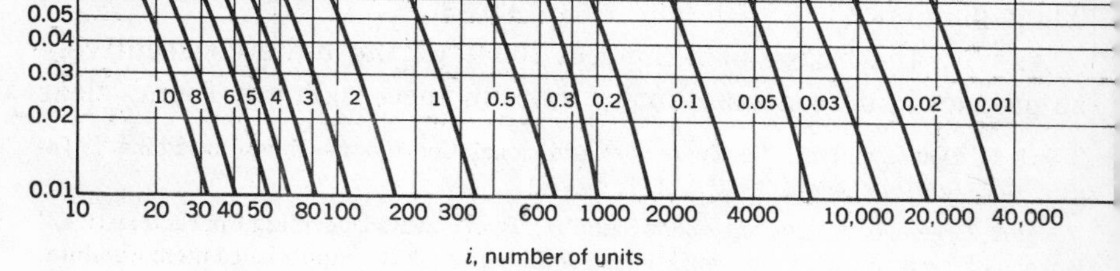

Fig. 13.16. Curves for determining value of f and i for a given value of AOQL for two-level plans.

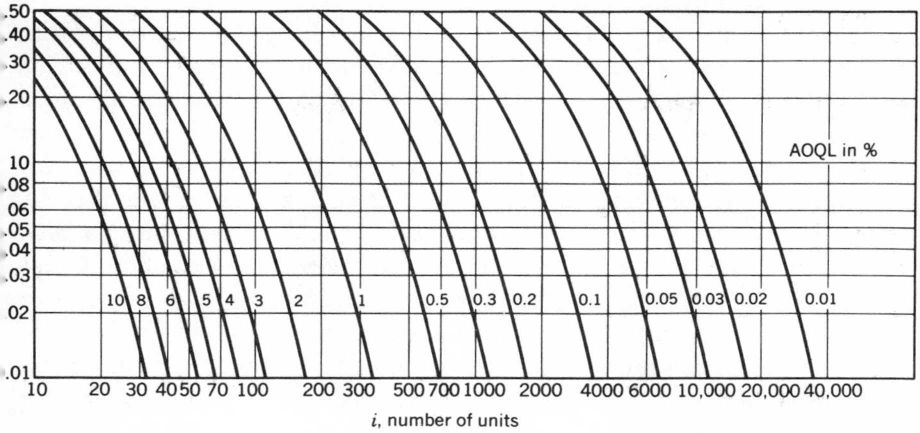

Fig. 13.17. Curves for determining values of f and i for a given value of AOQL for three-level plans.

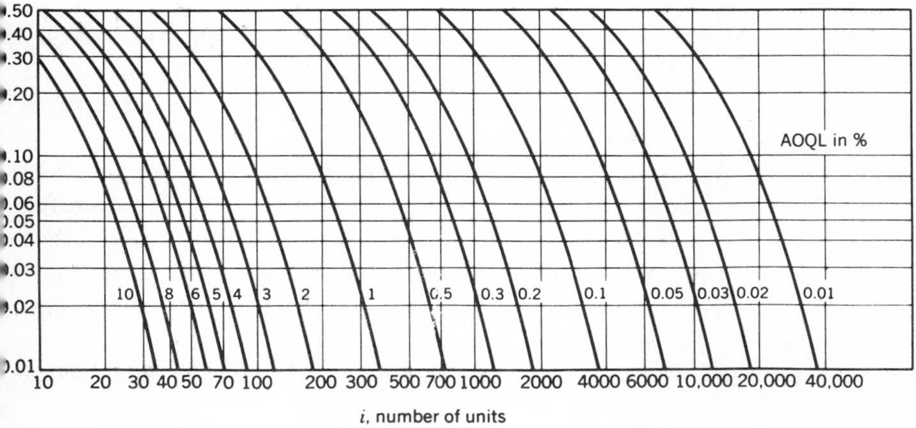

Fig. 13.18. Curves for determining values of f and i for a given value of AOQL for four-level plans.

The paper of Lieberman and Solomon seems to have stimulated a great interest in these multi-level plans. The original Lieberman-Solomon plan suggested that sampling be reduced to the next lower level if a defective is found. That is, if sampling were at the rate f^k it would be reduced to the rate f^{k-1}. In a subsequent paper,[1] various modifications were considered, schemes which permit reversion to sampling at any rate when a defective is found or even back

[1] C. Derman, S. Littauer, and H. Solomon, "Tightened Multi-level Continuous Sampling Plans," *Annals of Mathematical Statistics*, 1957, Vol. 28, pp. 395–404.

to 100% inspection.

One feature of all these plans is the requirement that, while sampling, one item is to be selected for inspection from a block of $1/f^j$ items. This type of sampling is called block sampling and assumes that each block is passed whether the sampled unit is defective or not. The AOQ function is derived under this assumption. If the item is found to be defective and the block contains a large number of items, the inspection department might indeed be reluctant to pass the remainder of the block without inspection, since finding a defective might be the result of a sudden shift in the process due to machine breakdown or a similar occurrence. Even if the block is small, as in the production of complex items, inspection might be initiated on one of the early items of the block before all of the items are through the production line. Knowledge that one of the items has already been chosen for inspection and that the remainder are to be passed without inspection might prejudice the quality of the remaining items in the block.

An alternative scheme is to inspect every item with probability f and pass every item without inspection with probability $1 - f$. If the decision whether or not to inspect is made by coin tossing or some other random device after the item is off the production line, the processors will be unable to determine beforehand whether the item is to be inspected. Resnikoff[1] showed that the AOQ function for MLP with this modified type of sampling is identical with that obtained by Lieberman and Solomon for MLP with block sampling.

Another problem discussed by Resnikoff is that of truncation. When probability sampling is used instead of block sampling, there is a possibility of too long a string of uninspected items. With block sampling the inspection department is assured of having a "look" at the process within a specified production interval; with probability sampling no such assurance is possible. A natural method of overcoming the possibility of passing too many items without inspection is to truncate the sampling in some manner. For example, this might be done as follows:

Inspect each item with probability f^j when sampling at the jth level. It is mandatory, however, to inspect the $2/f^j$ item if $2/f^j - 1$ consecutive items have been passed without inspection. After inspecting the $2/f^j$ item resume the probability sampling.

[1] G. J. Resnikoff, "Some Modifications of the Lieberman-Solomon Multi-Level Continuous Sampling Plan, MLP," *Technical Report No. 26*, Applied Mathematics and Statistics Laboratory, Stanford University, February 8, 1957.

Some numerical calculation reveals that for almost any reasonable truncation scheme the effect of truncation is negligible. The use of the AOQL contours already published by Lieberman and Solomon and Dodge is sufficiently close to those which might be computed with truncation, for practical purposes.

One of the most useful plans of this general type has already been adopted by the Air Force.[1] As in the Lieberman-Solomon plans, at the outset 100% of the product is inspected until a specified number of consecutive items, i, is found free of defects. At that time, the inspection is reduced to a fraction, f, of the items presented. If i successively inspected items are found free of defects, the inspection rate is reduced to a new fraction, f^2, of the items presented. This procedure is continued through the several levels for the plan. However, if at any level a defective item is found before i successively inspected items are passed, the plan provides a special procedure to determine if sampling should continue at the present rate or revert back to the previous sampling rate.

The special procedure, after a defective item has been found, consists of inspecting the next four consecutive items. If no defectives are found in the four consecutive items, sampling is resumed at the same rate as when the defective was found. If no more defectives are found before a total (including the four consecutive items) of i successively inspected items are determined to be acceptable, the sampling rate is reduced to the next smaller fraction (the next level). If a second defective is found, either in the four consecutive items or after resuming the sampling rate, revert to the former level (next larger sampling rate) and inspect four consecutive items, repeating the procedure just described. Thus, the sampling rate may be reduced or increased by specific steps as the quality of the presented items varies up and down. At the same time, the plans provide a check to differentiate between the random occurrence of a defective, which can occur even though the overall quality level is acceptable, and inferior quality or spotty quality.

13.4.4 The Dodge CSP-1 plan without control

Up to this point the formulas for AOQ were derived on the assumption that the process is in control—a mathematical model for many processes which is rarely realized exactly. Several workers in the field have turned their attention to this problem. One approach

[1] Multi-Level Continuous Sampling Acceptance Plans for Attributes, *AMC Manual* 74–23, Headquarters AMC, Wright Patterson AFB, Ohio, September 14, 1956.

is to examine the common procedures to see how they are affected by relaxing this condition.

In 1953, Lieberman[1] showed that the Dodge procedure guarantees an AOQL whether or not the process is in a state of statistical control. In fact, without the assumption of control, and for a given f and i,

$$\text{AOQL} = \frac{(1/f) - 1}{(1/f) + i}.$$

Naturally this value of the AOQL is higher than the value given by the Dodge result for a fixed f and i. This is not to imply that the Dodge result is not useful. The AOQL is itself an upper bound, rarely achieved. The AOQ may be much less than the AOQL. Consequently, the AOQ for a process which is not in control may be less than the Dodge AOQL. In fact, this AOQL is achieved only for the pathological process which produces all defective items during partial inspection, and produces all non-defective items during 100% inspection. However, the result does point out that if the process behaves irregularly, the standard plans can lead to trouble.

13.4.5 Wald-Wolfowitz continuous sampling plans

Another approach has been to derive plans specifically designed to prescribe an AOQL. A major result in the field was obtained by Wald and Wolfowitz[2] in 1945 when they proposed a class of what might be called one-sided sequential plans. A particular type of this general class of plans is as follows:

Begin with partial inspection, choosing one at random from each successive segment of $1/f$ items. Let d_n be the total number of defectives found after the nth stage of partial inspection. Continue partial inspection as long as $d_n < h + sn$, where h and s are fixed constants.

If for some n, $d_n \geqq h + sn$, terminate partial inspection and inspect h/s segments 100%. Repeat the procedure. Except for a slight approximation the AOQL for this plan is equal to $(1 - f)s$.

This one-sided sequential inspection plan is optimum from the point of view of cost of inspection (at least in the case of controlled production); it is not a plan which can be recommended for cases in which the excessive variability of the outgoing quality in finite batches of

[1] G. J. Lieberman, "A Note on Dodge's Continuous Sampling Plan," *Annals of Mathematical Statistics*, 1953, Vol. 24, pp. 480-484.

[2] A. Wald and J. Wolfowitz, "Sampling Inspection Plans for Continuous Production Which Insure a Prescribed Limit on the Outgoing Quality," *Annals of Mathematical Statistics,* 1945, Vol. 16, pp. 30-49.

the product is an important factor. This becomes apparent when we consider the fact that in order for any material to be inspected 100%, the point

$$(d_j, j),$$

which represents the total number of defectives found after the jth stage of partial inspection, must reach or exceed the line $y = h + sj$ for some sample size j. But if a few defective items are found during a long stretch of partial inspection, the point can wander so far away from the line that a great deal of unacceptable material can pass by before this fact is noted and the quality of the product improved by the 100% inspection. This situation can be remedied to some extent by employing a two-sided sequential inspection plan.

A two-sided sequential inspection plan is defined by two lines

$$y = h_2 + sj$$

and

$$y = h_1 + sj$$

where h_1 and h_2 are positive constants and j represents the number of items inspected at the jth stage of partial inspection. In this plan, partial inspection continues as long as the point

$$(d_j, j)$$

lies between the two lines above. Partial inspection terminates when, for some $j = n$, either

$$d_n \leqq -h_1 + sn$$

or

$$d_n \geqq h_2 + sn\ .$$

In the former case no 100% inspection is called for and the inspection procedure is simply repeated on new material. In the latter case, the inspection is resumed on new material only after h_2/s segments, that is, h_2/fs items, have been inspected 100%.

Except for the action taken, the inspection procedure described above is identical with the sequential method of lot-by-lot acceptance inspection. Consequently, all the sequential theory can be employed in studying the plan.

13.4.6 GIRSHICK CONTINUOUS SAMPLING PLAN

One other type of sequential plan was presented recently by M. A. Girshick.[1] It is defined by the three constants m, N, and f. The

[1] M. A. Girshick, "A Sequential Inspection Plan for Quality Control," *Technical Report No. 16*, Applied Mathematics and Statistics Laboratory, Stanford University, 1954.

plan operates as follows. The units of product in the production sequence are divided into segments of size $1/f$ items. The items are inspected in sequence and the number of defectives found, as well as the number of items examined, are cumulated. This procedure is continued until the cumulate number of defectives reaches m. At this point, the size of the sample n is compared with the integer N. If $n \geqq N$, the product which has passed through inspection is considered acceptable and the inspection procedure is repeated on the new incoming product. If, on the other hand, $n < N$, the following actions are taken: (a) the next $N - n$ segments $[(N - n)1/f$ units$]$ are inspected 100% and (b) after that, the inspection procedure is repeated. This procedure always guarantees, whether the process is in control or not, that the AOQL cannot exceed $(1 - f)m/N$.

Girshick also presented various operating characteristics of this plan, under control, such as the probability of inspection terminating with acceptance ($n \geqq N$), and discussed the biased and unbiased estimates of the process average.

13.4.7 Plans which provide for termination of production

A criticism of all these continuous sampling plans, particularly those mentioned in the last section, is that they emphasize doing enough inspection to bring quality down to the AOQL but do not provide automatic penalties for poor quality. If the inspection is performed by a consumer or purchaser, he may do a very large amount of inspection to insure quality. Sometimes this may be avoided by administrative action. Dodge's original paper says:

"The inspection plan is most effective in practice if it is administered in such a way as to provide an incentive to clear up causes of trouble promptly. Such an incentive may be had by imposing a penalty on the operating or manufacturing department when defects are encountered. Normally no such penalty is imposed if both the sampling inspection and the 100% inspection are performed by this same person or group of persons; then the two costs merge. The inspector then merely serves as an agency for screening defects when quality goes bad. It is accordingly recommended that sampling inspection and 100% inspection operations be treated as two separate functions." With this possibility in mind, Army Ordnance[1] brought out a very comprehensive set of Dodge plans to use as a

[1] "Procedures and Tables for Continuous Sampling by Attributes," *Ordnance Inspection Handbook*, ORD-M608-11, August, 1954.

standard procedure for continuous sampling. In these plans, the government inspector performs all the sequences of partial inspection and the manufacturer is required to provide a screening crew to perform the 100% inspection. The government, in some cases, does 100% inspection but will charge the contractors for it.

The government inspector may verify the 100% inspection by inspecting the screened material, and, of course, take fairly drastic action if any errors are found.

Another attack on this problem has been made by Rosenblatt and Weingarten of Navy Ordnance,[1] who modify the Dodge plan by limiting the number of 100% inspection sequences an inspector is allowed to perform. In addition to f and i, their plan specifies a maximum allowable number of sequences of consecutive item inspection beyond the first. If this number is exceeded at any time during a day's production, all inspection is stopped and the remaining items on the line are not accepted (although any product that has passed the inspector is accepted). The inspector informs the manufacturer as to which defects have occurred. He will not begin inspection again until the manufacturer locates the source of difficulty and gives him assurance that he has removed the cause.

A feature of the Navy Ordnance plan is that it comes close to satisfying the assumption of control. A state of control is rarely achieved over a long period of time. However, over a short period of time, like a day, the process may be in a state of control. The operating characteristics of this plan are calculated for a day's production.

This provision for shutting down the acceptance line seems to be desirable in a continuous sampling plan. That is, the plan should check on the product, do enough inspection to make quality good if there are any minor fluctuations, and if they are major to provide a basis for shutting down the line and looking for corrective action. The original Dodge plan and the Army Ordnance plan rely largely on the fact that if a product is generally bad or if it deteriorates, a large amount of the product will have to be inspected. However, the Navy Ordnance plan has the advantage that specific criteria for making this decision are provided. Other plans providing for termination of inspection have been introduced using various mathema-

[1] H. Rosenblatt and H. Weingarten, "Sampling Procedures and Tables for Inspection on a Moving Line," *Continuous Sampling Plans*, NAVORD-Std 81.

tical models for the production process.[1] However, there has been little industrial use of these procedures.

PROBLEMS

1. What is the probability of accepting a lot whose incoming quality is 10% defective, using a sample of size 5 and an acceptance number of 0? What is the probability using an acceptance number of 1?

2. What is the probability of accepting a lot whose incoming quality is 8% defective using a sample of size 4 and acceptance numbers of 0, 1, 2, and 3?

3. A lot containing 4,000 pieces is submitted for acceptance by a vendor. The consumer knows that this vendor's previous lots have averaged 3.0% defective. If it is assumed that this lot is not significantly different from previous lots; i.e., it contains 3.0% defective, (a) what is the probability of the lot containing exactly three defectives by the binomial approximation? (Set up the equation.) (b) What is the probability of the sample containing two or more defectives by the Poisson approximation?

4. A product in lots is submitted for inspection.

$$n_1=50 \qquad c_1=0 \qquad n_2=60 \qquad c_2=3$$

Assuming the lot is very large compared to the sample sizes, determine the probability of acceptance if the submitted quality is 4% defective. Use the Poisson approximation.

5. A process is in control from which the $\overline{X}'=100$ and $\sigma'=4$. This quality characteristic is going to be accepted by an attributes sampling acceptance plan. The purchaser has indicated that it will accept practically all the lots which are 2% defective or less. The specification is 100 ± 10 for this characteristic. (a) What percent defective is being submitted? (b) If the producer were running a p chart as a quality assurance check on the in-process quality control function, what should the limits for this p chart be? Why? (c) Using the binomial approximation, determine the expression for the probability of a lot being accepted by the consumer if he is using the single sampling plan $n=50$, $c=1$. (d) Using the Poisson table select a sampling plan that will accept 2% defective lots 95% of

[1] The following papers deal with this type of continuous sampling plan: M. A. Girshick and H. Rubin, "A Bayes Approach to a Quality Control Model," *Annals of Mathematical Statistics*, 1952, Vol. 23, pp. 114–125. Geoffrey Gregory, "An Economic Approach to the Choice of Continuous Sampling Plans," *Technical Report No. 30*, Applied Mathematics and Statistics Laboratory, Stanford University, Sept. 20, 1956. I. R. Savage, "A Three Decision Continuous Sampling Plan for Attributes," *Technical Report No. 20*, Applied Mathematics and Statistics Laboratory, Stanford University, 1955. I. R. Savage, "Statistical Production Models and Sampling Plans," *Technical Report No. 29*, Applied Mathematics and Statistics Laboratory, Stanford University, Aug. 8, 1956.

the time. Use $c=3$.

6. The following double sampling plan is used in the acceptance inspection of a product:

First sample size..........	200	Second sample size........	400
Acceptance number, first sample	2	Cumulative sample size....	600
Rejection number, first sample	5	Acceptance number, combined samples	4
		Rejection number, combined samples	5

(a) If a lot that is 1.5% defective is submitted to this plan, what is the probability that it will be accepted? You may assume the Poisson distribution will give an answer of satisfactory accuracy. (b) If a great many lots that are 1.5% defective are submitted to this plan, what will be the average amount of inspection per lot? Consider only the inspection done by the receiver who does not screen rejected lots.

7. When complex items which are costly are tested using a destructive test, a common acceptance procedure calls for taking a sample of size 1. If the item is good, the lot is accepted. If it is bad, a second sample of size 1 is obtained. If this item is also defective, the lot is rejected. If it is good, the lot is accepted. (a) Write down the expression for the OC curve for this plan assuming that the submitted lot is large enough so that the selection of a defective article on the first sample does not make any appreciable change in the probability of getting a defective article on the second sample. (b) A single sampling procedure having the same OC curve as found in (a) is as follows: Take a sample of size $n=2$. If the number of defectives is less than or equal to one, the lot is accepted. Otherwise, the lot is rejected. Prove this statement. (c) Compare the "average sample number" for the single and double sampling plans above.

8. A single test on a fuse costs more than the whole lot of 1,000. Alternative tests not proving satisfactory, it was decreed that not more than three tests of a fuse in actual performance could be made per lot. The following plan was prepared. Test two. If both function, accept the lot. If both fail, reject. Otherwise, test another. If it fails, reject; or if it functions, accept. (a) Assuming the binomial distribution is applicable, write down the expression for the OC curve. Express the final result in terms of q' only. (b) Suppose the single sampling plan $n=3$, $c=1$ were used instead of the one described above. Write down the expression for the OC curve. Express the final result in terms of q' only. (c) Compare the "average sample number" of the plans described in (a) and (b).

9. (a) Determine from the Dodge-Romig Single Sampling AOQL Table the single sampling plan for an AOQL of 2%, a process average of 1.5%, and lot size of 700. (b) What will be the total percentage of inspection considering both sampling inspection and screening inspection of rejected lots if the incoming lots are 1.5% defective? (c) What is the average outgoing quality (AOQ) if the incoming lots are 1.5% defective?

10. Determine the average outgoing quality for the single sampling plan $N=100$, $n=10$, $c=0$, when submitted quality is 4% defective. Use the OC curve from MIL-STD-105B for finding the probability of acceptance.

11. A manufacturer is using a Dodge-Romig 5% LTPD single sampling plan on his own product. The lot size is 900 units and his process average is running at 1%. The government is purchasing this product and is using normal single sampling under MIL-STD-105B with an AQL of 6.5%. (a) What is the probability that a 6.5% defective lot will pass the manufacturer's sampling inspection and be rejected by the government? (b) Suppose the first ten lots submitted by the vendor are exactly 8% defective. This is somewhat worse than the AQL of 6.5%. What is the approximate probability that the process average estimated from the first ten samples will turn out to be bad enough to force a shift to tightened inspection?

12. Single sampling with an AQL of 1% is being used with acceptance criteria determined by MIL-STD-105B. The sample size code letter is *F*. (a) In the samples from the first 20 lots zero defectives have been found and no lots rejected. Does this qualify for reduced inspection? If not, what is the least number of additional lots that must be inspected before qualification is possible? (b) Suppose the first 20 lots submitted are each exactly 2% defective. What is the approximate probability that the process average estimated from the first 20 samples will turn out to be bad enough to force a shift to tightened inspection?

13. (a) In acceptance sampling under MIL-STD-105B, single sampling is to be used with inspection level I, an AQL of 2.5% and a lot size of 200. What are the acceptance criteria under normal inspection? (b) What is the highest number of defective items that may be found in the samples from the first ten lots that will permit qualification for reduced inspection?

14. Billions of dollars worth of material are purchased under MIL-STD-105B. For any given lot, the probability of accepting a lot whose quality is at the next higher AQL is very good. Yet, most of the submitted product is at a quality level better than the AQL. Explain how MIL-STD-105B applies pressure to the producer to submit quality better than the AQL.

15. (a) In acceptance sampling under MIL-STD-105B single sampling is to be used with inspection level II, an AQL of 4%, and a lot size of 10,000. What are acceptance criteria under (1) normal, (2) tightened, and (3) reduced inspection? (b) Use the Poisson table to compute the approximate probability of acceptance of a 4% defective lot under normal, tightened, and reduced inspection. (c) The first ten lots are sampled under normal inspection. The number of defectives found are 6, 15, 9, 8, 3, 10, 11, 7, 16, and 11, respectively. Make the necessary calculations to determine whether or not a shift to tightened inspection is required. (d) If the incoming quality was 6%, what was the probability that the process average estimated from the first ten samples would turn out to be bad enough to force a shift to tightened inspection?

16. A product is submitted for acceptance inspection in lots of 25,000. Inspection is to be done in accordance with MIL-STD-105B single sampling. The AQL specified is 1%. Normal inspection at inspection level II is to be used. What is the probability of accepting a lot if the product is submitted at the AQL? Suppose the process average were estimated from ten lots. What is the highest process average permitting continuance of normal inspection?

17. A product is submitted for acceptance inspection by attributes in lots of 10,000. Inspection is to be done in accordance with MIL-STD-105B single sampling. The AQL specified is 0.4%. Normal inspection at inspection level II is to be used. What is the probability of accepting a lot if the product is submitted at the AQL (use Poisson approximation to calculate the probability)? Suppose the process average was estimated from ten lots. What is the highest process average permitting continuance of normal inspection? If the incoming quality of the ten lots is at the next higher AQL, what is the approximate probability that the process average estimated from the first ten samples will turn out to be bad enough to force a shift to tightened inspection?

18. In normal single sampling under MIL-STD-105B, the AQL is 2.5% and the lot size is 200. Inspection level II is to be used. Of the samples from the first 20 lots only 6 defectives have been found and no lots rejected. (a) Does this qualify for reduced inspection? (b) If not, what is the least number of additional lots that must be inspected before qualification is possible? (c) If the process average is actually at 1%, what is the probability that reduced inspection will be instituted in the minimum time found in part (b)?

19. Solve Problem 16 using double sampling. For process average estimation only the results of the first sample are used.

20. Solve Problem 17 using double sampling. For process average estimation only the results of the first sample are used.

21. Find a single sample attribute plan with the following properties (use Table 13.14): $p_1'=0.025$, $\alpha=0.05$; $p_2'=0.121$, $\beta=0.10$. Sketch the OC curve of the plan.

22. Suppose we have a single sampling plan with $p_1'=0.01$, $\alpha=0.01$, and $p_2'=0.0195$. Find n if $\beta=0.01$, $\beta=0.05$, and $\beta=0.10$.

23. Find a single sampling plan with $p_1'=0.04$, $p_2'=0.10$, $\alpha=0.05$, $\beta=0.05$. Sketch the OC curve of the plan.

24. Find the boundaries for the graphical procedure of the item-by-item sequential plan for the problem above. Sketch the sequential OC curve from five points. How does this OC curve compare with the OC curve of the single sampling plan determined by the same p_1', α and p_2', β?

25. Find a single sampling plan with properties $p_1'=0.01$, $\alpha=0.05$, $p_2'=0.08$, $\beta=0.05$ from Table 13.14. Find a plan in MIL-STD-105B which closely approximates this plan. Compare the two OC curves. Read the OC curve for the MIL-STD-105A plan from the graph.

26. Suppose the following two points on the OC curve are specified:

$$(p_1'=0.04,\ \alpha=0.05) \text{ and } (p_2'=0.08,\ \beta=0.10).$$

Thus,

$$(1) \qquad P(d \leqq c)=0.95, \quad \text{where } p'=0.04$$

$$(2) \qquad P(d \leqq c)=0.10, \quad \text{where } p'=0.08\ .$$

If we assume that d is normally distributed with mean np' and standard deviation $\sqrt{np'q'}$, then

$$(3) \qquad P(d \leqq c)=P\left(\frac{d-np'}{\sqrt{np'q'}} \leqq \frac{c-np'}{\sqrt{np'q'}}\right)$$

$$=P\left(X \leqq \frac{c-np'}{\sqrt{np'q'}}\right),$$

where X is normally distributed with mean 0 and standard deviation 1, Using the knowledge about equations (1), (2), and (3), two simultaneous equations can be set up which can be solved for the two unknowns n and c, thereby determining a sampling plan passing through two specified points. Find the values of n and c.

27. The minimum value of the shear strength in pounds of spot welds is 300 pounds. A lot containing 800 such welds is submitted for inspection under MIL-STD-414. Inspection level IV, normal inspection with AQL=1%, is to be used. The variability is unknown and the standard deviation method is to be used. (a) Using form 2, what are the acceptance criteria? (b) If $\overline{X}=351.4$ and $s=22.3$, would the lot be accepted? (c) What are the criteria for reduced and tightened inspection based upon ten lots?

28. The maximum specification for a certain device is 6.5. A lot containing 750 items is submitted for inspection under MIL-STD-414. Inspection level II, normal inspection with AQL=4%, is to be used. The variability is unknown and the standard deviation method is to be used. (a) Using form 2, what are the acceptance criteria? (b) Suppose the data are as follows: 5.00, 5.88, 3.86, 4.13, 4.01, 3.92, 2.92, 5.82, 6.20, and 4.90. Should the lot be accepted? (c) What are the acceptance criteria if form 1 is used?

29. The specification limits for a certain type of resistance is specified as 75±5 ohms. A lot containing 300 items is submitted for inspection under MIL-STD-414. Inspection level III, normal inspection with AQL=4%, is to be used. The variability is unknown and the standard deviation method is to be used. (a) Using form 2, what are the acceptance criteria? (b) Suppose the data are as follows: 74.3, 75.1, 71.9, 72.7, 75.8, 77.0, 77.3, 70.1, 73.8, and 75.0. Determine whether or not to accept the lot.

30. The minimum requirement on the content of a particular type of can is 80 ounces. A lot containing 700 cans is submitted for inspection under MIL-STD-414. Inspection level III, tightened inspection with an AQL=1.5%, is to be used. The variability is unknown and the standard deviation method is to be used. (a) Using form 2, what are the acceptance criteria? (b) Suppose the data are follows: 81.4, 81.0, 81.0, 80.2, 82.1, 79.8, 80.3, 80.5, 80.8, 80.6, 80.3, 81.0, 80.5, 81.0, 82.1, 81.4, 80.6, 81.9, 80.8, and 80.7. Would you accept the lot? (c) What are the criteria for tightened and reduced inspection based upon 15 lots?

31. The temperature specifications of a certain device are given as 195 degrees $\pm$ 15 degrees. A lot containing 5,000 items is submitted for inspection under MIL-STD-414. Inspection level V, normal inspection with an AQL=0.65%, is to be used. The variability is unknown and the standard deviation method is to be used. (a) What are the acceptance criteria? (b) If $\bar{x}=194.8$ degrees and $s=5.1$ degrees, would the lot be accepted?

32. Solve Problem 27 assuming that the variability is unknown, but using the average range method. Assume that $\bar{R}=49.7$ in part (b).

33. Solve Problem 28 assuming that the variability is unknown, but using the average range method. Assume that the observations presented are in order reading from left to right.

34. Solve Problem 29 assuming that the variability is unknown, but using the average range method. Assume that the observations presented are in order reading from left to right.

35. Can you solve Problem 30 using the average range method with the data given? If not, assume that the average of the additional observations equals the average of the 20 observations given in part (b) of Problem 30. Also assume that the ranges of any additional subgroups equal the average range of the four subgroups of five given in part (b) of Problem 30.

36. Solve Problem 31 assuming that the variability is unknown, but using the average range method. Assume that $\bar{R}=11.7$ in part (b).

37. Solve Problem 27 assuming that the variability is known and equal to $\sigma'=22$.

38. Solve Problem 28 assuming that the variability is known and equal to $\sigma'=1.07$. Use only the required number of observations in the order that they are presented reading from left to right.

39. Solve Problem 29 assuming that the variability is known and equal to $\sigma'=1.25$. Use only the required number of observations in the order that they are presented reading from left to right.

40. Solve Problem 30 assuming that the variability is known and equal to $\sigma'=0.80$. Use only the required number of observations in the order that they are presented reading from left to right.

41. Solve Problem 31 assuming that the variability is known and equal to $\sigma'=5$ degrees.

42. In Problem 38 verify that the value of k used in form 1 equals $K_M\sqrt{\frac{n-1}{n}}$.

43. In Problem 38 calculate exactly the probability of accepting the lot when the incoming quality is at the AQL.

44. In Problem 40 calculate exactly the probability of accepting the lot when the incoming quality is at the AQL.

45. (a) Assuming the process is in control and the CSP-1 procedure is used, what is the necessary value of i if $f=\frac{1}{10}$ and the desired AOQL=4%? (b) If control is not assumed, what is the required value of i?

46. (a) Assuming the process is in control and the CSP-1 procedure is used, what is the necessary value of i if $f=\frac{1}{20}$ and the desired AOQL=1%? (b) If control is not assumed, what is the required value of i?

47. In Dodge's CSP-1 continuous sampling plan, it is desired to apply inspection to one piece out of every ten and to maintain an AOQL of 2%. (a) What should be the value of i assuming the process is in control? (b) What should be the value of i assuming the process is not in control? (c) If the process is in control at $p'=2\%$ and sampling inspection is in effect, what is the probability of returning to 100% inspection on the next item looked at?

48. Assuming the process is in control and the MLP procedure is used with two levels, what is the necessary value of i if $f=\frac{1}{10}$ and the desired AOQL=4%? If the process is in control at $p'=4\%$ and sampling inspection is in effect at level III, what is the probability of returning to 100% inspection after the next two items are looked at?

49. Assuming the process is in control and the MLP procedure is used with three levels, what is the necessary value of i if $f=\frac{1}{10}$ and the desired AOQL=2%? What is the value of i if $f=\frac{1}{2}, \frac{1}{3}$?

50. Assuming the process is in control and the MLP procedure is used with four levels, what is the necessary value of i if $f=\frac{1}{2}$ and the desired AOQL=1%?

51. Using the Girshick continuous sampling procedure, what is the required value of N corresponding to $m=2$, $f=\frac{1}{10}$ and AOQL=2%?

52. Using the Girshick continuous sampling procedure, what is the AOQL corresponding to $m=2$, $f=\frac{1}{10}$, and $N=100$?

APPENDIX

Table 1. Areas under the Normal Curve from K_α to ∞*

$$\int_{K_\alpha}^{\infty} \frac{1}{\sqrt{2\pi}} e^{-x^2/2}\, dx = \alpha$$

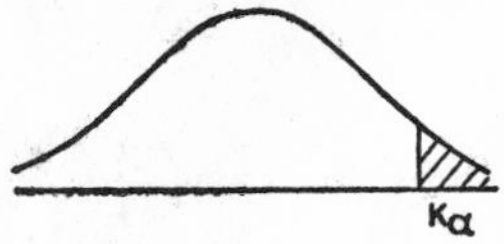

K_α	.00	.01	.02	.03	.04	.05	.06	.07	.08	.09
0.0	.5000	.4960	.4920	.4880	.4840	.4801	.4761	.4721	.4681	.4641
0.1	.4602	.4562	.4522	.4483	.4443	.4404	.4364	.4325	.4286	.4247
0.2	.4207	.4168	.4129	.4090	.4052	.4013	.3974	.3936	.3897	.3859
0.3	.3821	.3783	.3745	.3707	.3669	.3632	.3594	.3557	.3520	.3483
0.4	.3446	.3409	.3372	.3336	.3300	.3264	.3228	.3192	.3156	.3121
0.5	.3085	.3050	.3015	.2981	.2946	.2912	.2877	.2843	.2810	.2776
0.6	.2743	.2709	.2676	.2643	.2611	.2578	.2546	.2514	.2483	.2451
0.7	.2420	.2389	.2358	.2327	.2296	.2266	.2236	.2206	.2177	.2148
0.8	.2119	.2090	.2061	.2033	.2005	.1977	.1949	.1922	.1894	.1867
0.9	.1841	.1814	.1788	.1762	.1736	.1711	.1685	.1660	.1635	.1611
1.0	.1587	.1562	.1539	.1515	.1492	.1469	.1446	.1423	.1401	.1379
1.1	.1357	.1335	.1314	.1292	.1271	.1251	.1230	.1210	.1190	.1170
1.2	.1151	.1131	.1112	.1093	.1075	.1056	.1038	.1020	.1003	.0985
1.3	.0968	.0951	.0934	.0918	.0901	.0885	.0869	.0853	.0838	.0823
1.4	.0808	.0793	.0778	.0764	.0749	.0735	.0721	.0708	.0694	.0681
1.5	.0668	.0655	.0643	.0630	.0618	.0606	.0594	.0582	.0571	.0559
1.6	.0548	.0537	.0526	.0516	.0505	.0495	.0485	.0475	.0465	.0455
1.7	.0446	.0436	.0427	.0418	.0409	.0401	.0392	.0384	.0375	.0367
1.8	.0359	.0351	.0344	.0336	.0329	.0322	.0314	.0307	.0301	.0294
1.9	.0287	.0281	.0274	.0268	.0262	.0256	.0250	.0244	.0239	.0233
2.0	.0228	.0222	.0217	.0212	.0207	.0202	.0197	.0192	.0188	.0183
2.1	.0179	.0174	.0170	.0166	.0162	.0158	.0154	.0150	.0146	.0143
2.2	.0139	.0136	.0132	.0129	.0125	.0122	.0119	.0116	.0113	.0110
2.3	.0107	.0104	.0102	.00990	.00964	.00939	.00914	.00889	.00866	.00842
2.4	.00820	.00798	.00776	.00755	.00734	.00714	.00695	.00676	.00657	.00639
2.5	.00621	.00604	.00587	.00570	.00554	.00539	.00523	.00508	.00494	.00480
2.6	.00466	.00453	.00440	.00427	.00415	.00402	.00391	.00379	.00368	.00357
2.7	.00347	.00336	.00326	.00317	.00307	.00298	.00289	.00280	.00272	.00264
2.8	.00256	.00248	.00240	.00233	.00226	.00219	.00212	.00205	.00199	.00193
2.9	.00187	.00181	.00175	.00169	.00164	.00159	.00154	.00149	.00144	.00139

K_α	.0	.1	.2	.3	.4	.5	.6	.7	.8	.9
3	.00135	$.0^{3}968$	$.0^{3}687$	$.0^{3}483$	$.0^{3}337$	$.0^{3}233$	$.0^{3}159$	$.0^{3}108$	$.0^{4}723$	$.0^{4}481$
4	$.0^{4}317$	$.0^{4}207$	$.0^{4}133$	$.0^{5}854$	$.0^{5}541$	$.0^{5}340$	$.0^{5}211$	$.0^{5}130$	$.0^{6}793$	$.0^{6}479$
5	$.0^{6}287$	$.0^{6}170$	$.0^{7}996$	$.0^{7}579$	$.0^{7}333$	$.0^{7}190$	$.0^{7}107$	$.0^{8}599$	$.0^{8}332$	$.0^{8}182$
6	$.0^{9}987$	$.0^{9}530$	$.0^{9}282$	$.0^{9}149$	$.0^{10}777$	$.0^{10}402$	$.0^{10}206$	$.0^{10}104$	$.0^{11}523$	$.0^{11}260$

* Reprinted by permission from Frederick E. Croxton, *Elementary Statistics with Applications in Medicine*, Prentice-Hall, Inc., Englewood Cliffs, N. J., 1953, p. 323.

Table 2. Percentage Points of the χ^2 Distribution

Table of $\chi^2_{\alpha;\nu}$—the 100 α percentage point of the χ^2 distribution for ν degrees of freedom

ν \ α	.995	.99	.98	.975	.95	.90	.80	.75	.70	.50
1	$.0^4393$	$.0^3157$	$.0^3628$	$.0^3982$	.00393	.0158	.0642	.102	.148	.455
2	.0100	.0201	.0404	.0506	.103	.211	.446	.575	.713	1.386
3	.0717	.115	.185	.216	.352	.584	1.005	1.213	1.424	2.366
4	.207	.297	.429	.484	.711	1.064	1.649	1.923	2.195	3.357
5	.412	.554	.752	.831	1.145	1.610	2.343	2.675	3.000	4.351
6	.676	.872	1.134	1.237	1.635	2.204	3.070	3.455	3.828	5.348
7	.989	1.239	1.564	1.690	2.167	2.833	3.822	4.255	4.671	6.346
8	1.344	1.646	2.032	2.180	2.733	3.490	4.594	5.071	5.527	7.344
9	1.735	2.088	2.532	2.700	3.325	4.168	5.380	5.899	6.393	8.343
10	2.156	2.558	3.059	3.247	3.940	4.865	6.179	6.737	7.267	9.342
11	2.603	3.053	3.609	3.816	4.575	5.578	6.989	7.584	8.148	10.341
12	3.074	3.571	4.178	4.404	5.226	6.304	7.807	8.438	9.034	11.340
13	3.565	4.107	4.765	5.009	5.892	7.042	8.634	9.299	9.926	12.340
14	4.075	4.660	5.368	5.629	6.571	7.790	9.467	10.165	10.821	13.339
15	4.601	5.229	5.985	6.262	7.261	8.547	10.307	11.036	11.721	14.339
16	5.142	5.812	6.614	6.908	7.962	9.312	11.152	11.912	12.624	15.338
17	5.697	6.408	7.255	7.564	8.672	10.085	12.002	12.792	13.531	16.338
18	6.265	7.015	7.906	8.231	9.390	10.865	12.857	13.675	14.440	17.338
19	6.844	7.633	8.567	8.907	10.117	11.651	13.716	14.562	15.352	18.338
20	7.434	8.260	9.237	9.591	10.851	12.443	14.578	15.452	16.266	19.337
21	8.034	8.897	9.915	10.283	11.591	13.240	15.445	16.344	17.182	20.337
22	8.643	9.542	10.600	10.982	12.338	14.041	16.314	17.240	18.101	21.337
23	9.260	10.196	11.293	11.688	13.091	14.848	17.187	18.137	19.021	22.337
24	9.886	10.856	11.992	12.401	13.848	15.659	18.062	19.037	19.943	23.337
25	10.520	11.524	12.697	13.120	14.611	16.473	18.940	19.939	20.867	24.337
26	11.160	12.198	13.409	13.844	15.379	17.292	19.820	20.843	21.792	25.336
27	11.808	12.879	14.125	14.573	16.151	18.114	20.703	21.749	22.719	26.336
28	12.461	13.565	14.847	15.308	16.928	18.939	21.588	22.657	23.647	27.336
29	13.121	14.256	15.574	16.047	17.708	19.768	22.475	23.567	24.577	28.336
30	13.787	14.953	16.306	16.791	18.493	20.599	23.364	24.478	25.508	29.336

For values of $\nu > 30$, approximate values for χ^2 may be obtained from the expression $\nu\left[1 \pm \frac{2}{9\nu} \pm \frac{x}{\sigma}\sqrt{\frac{2}{9\nu}}\right]^3$, where $\frac{x}{\sigma}$ is the normal deviate cutting off the tails of a normal distribution. If $\frac{x}{\sigma}$ is taken at the 0.02 level, so that 0.01 of the normal distribution is in each tail, the expression yields χ^2 at the 0.99 and 0.01 points. For very large values of ν, it is sufficiently accurate to compute $\sqrt{2\chi^2}$, the distribution of which is

Table 2. (cont.) Percentage Points of the χ^2 Distribution

Table of $\chi^2_{\alpha;\nu}$—the 100 α percentage point of the χ^2 distribution for ν degrees of freedom

.30	.25	.20	.10	.05	.025	.02	.01	.005	.001	α / ν
1.074	1.323	1.642	2.706	3.841	5.024	5.412	6.635	7.879	10.827	1
2.408	2.773	3.219	4.605	5.991	7.378	7.824	9.210	10.597	13.815	2
3.665	4.108	4.642	6.251	7.815	9.348	9.837	11.345	12.838	16.268	3
4.878	5.385	5.989	7.779	9.488	11.143	11.668	13.277	14.860	18.465	4
6.064	6.626	7.289	9.236	11.070	12.832	13.388	15.086	16.750	20.517	5
7.231	7.841	8.558	10.645	12.592	14.449	15.033	16.812	18.548	22.457	6
8.383	9.037	9.803	12.017	14.067	16.013	16.622	18.475	20.278	24.322	7
9.524	10.219	11.030	13.362	15.507	17.535	18.168	20.090	21.955	26.125	8
10.656	11.389	12.242	14.684	16.919	19.023	19.679	21.666	23.589	27.877	9
11.781	12.549	13.442	15.987	18.307	20.483	21.161	23.209	25.188	29.588	10
12.899	13.701	14.631	17.275	19.675	21.920	22.618	24.725	26.757	31.264	11
14.011	14.845	15.812	18.549	21.026	23.337	24.054	26.217	28.300	32.909	12
15.119	15.984	16.985	19.812	22.362	24.736	25.472	27.688	29.819	34.528	13
16.222	17.117	18.151	21.064	23.685	26.119	26.873	29.141	31.319	36.123	14
17.322	18.245	19.311	22.307	24.996	27.488	28.259	30.578	32.801	37.697	15
18.418	19.369	20.465	23.542	26.296	28.845	29.633	32.000	34.267	39.252	16
19.511	20.489	21.615	24.769	27.587	30.191	30.995	33.409	35.718	40.790	17
20.601	21.605	22.760	25.989	28.869	31.526	32.346	34.805	37.156	42.312	18
21.689	22.718	23.900	27.204	30.144	32.852	33.687	36.191	38.582	43.820	19
22.775	23.828	25.038	28.412	31.410	34.170	35.020	37.566	39.997	45.315	20
23.858	24.935	26.171	29.615	32.671	35.479	36.343	38.932	41.401	46.797	21
24.939	26.039	27.301	30.813	33.924	36.781	37.659	40.289	42.796	48.268	22
26.018	27.141	28.429	32.007	35.172	38.076	38.968	41.638	44.181	49.728	23
27.096	28.241	29.553	33.196	36.415	39.364	40.270	42.980	45.558	51.179	24
28.172	29.339	30.675	34.382	37.652	40.646	41.566	44.314	46.928	52.620	25
29.246	30.434	31.795	35.563	38.885	41.923	42.856	45.642	48.290	54.052	26
30.319	31.528	32.912	36.741	40.113	43.194	44.140	46.963	49.645	55.476	27
31.391	32.620	34.027	37.916	41.337	44.461	45.419	48.278	50.993	56.893	28
32.461	33.711	35.139	39.087	42.557	45.722	46.693	49.588	52.336	58.302	29
33.530	34.800	36.250	40.256	43.773	46.979	47.962	50.892	53.672	59.703	30

approximately normal around a mean of $\sqrt{2\nu - 1}$ and with a standard deviation of 1.

This table is taken by consent from R. A. Fisher and F. Yates, *Statistical Tables for Biological, Agricultural, and Medical Research*, Oliver and Boyd, Edinburgh, and from Catherine M. Thomson, "Table of Percentage Points of the χ^2 Distribution," *Biometrika* Vol. 32, Part II, October 1941, pp. 187–191. Reproduced in Croxton *Elementary Statistics with Applications in Medicine*, Appendix VI, pp. 328–329.

Table 3. Percentage Points of the t Distribution*

Table of $t_{\alpha;\nu}$—the 100 α percentage point of the t distribution for ν degrees of freedom

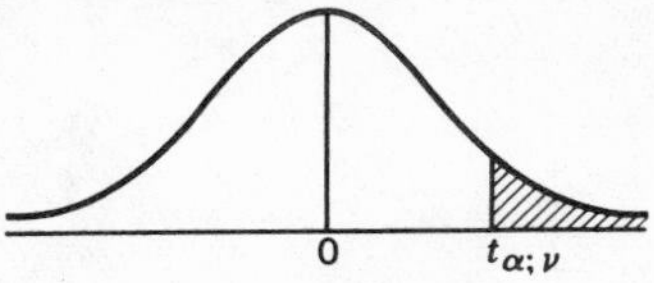

ν \ α	.40	.30	.20	.10	.050	.025	.010	.005	.001	.0005
1	.325	.727	1.376	3.078	6.314	12.71	31.82	63.66	318.3	636.6
2	.289	.617	1.061	1.886	2.920	4.303	6.965	9.925	22.33	31.60
3	.277	.584	.978	1.638	2.353	3.182	4.541	5.841	10.22	12.94
4	.271	.569	.941	1.533	2.132	2.776	3.747	4.604	7.173	8.610
5	.267	.559	.920	1.476	2.015	2.571	3.365	4.032	5.893	6.859
6	.265	.553	.906	1.440	1.943	2.447	3.143	3.707	5.208	5.959
7	.263	.549	.896	1.415	1.895	2.365	2.998	3.499	4.785	5.405
8	.262	.546	.889	1.397	1.860	2.306	2.896	3.355	4.501	5.041
9	.261	.543	.883	1.383	1.833	2.262	2.821	3.250	4.297	4.781
10	.260	.542	.879	1.372	1.812	2.228	2.764	3.169	4.144	4.587
11	.260	.540	.876	1.363	1.796	2.201	2.718	3.106	4.025	4.437
12	.259	.539	.873	1.356	1.782	2.179	2.681	3.055	3.930	4.318
13	.259	.538	.870	1.350	1.771	2.160	2.650	3.012	3.852	4.221
14	.258	.537	.868	1.345	1.761	2.145	2.624	2.977	3.787	4.140
15	.258	.536	.866	1.341	1.753	2.131	2.602	2.947	3.733	4.073
16	.258	.535	.865	1.337	1.746	2.120	2.583	2.921	3.686	4.015
17	.257	.534	.863	1.333	1.740	2.110	2.567	2.898	3.646	3.965
18	.257	.534	.862	1.330	1.734	2.101	2.552	2.878	3.611	3.922
19	.257	.533	.861	1.328	1.729	2.093	2.539	2.861	3.579	3.883
20	.257	.533	.860	1.325	1.725	2.086	2.528	2.845	3.552	3.850
21	.257	.532	.859	1.323	1.721	2.080	2.518	2.831	3.527	3.819
22	.256	.532	.858	1.321	1.717	2.074	2.508	2.819	3.505	3.792
23	.256	.532	.858	1.319	1.714	2.069	2.500	2.807	3.485	3.767
24	.256	.531	.857	1.318	1.711	2.064	2.492	2.797	3.467	3.745
25	.256	.531	.856	1.316	1.708	2.060	2.485	2.787	3.450	3.725
26	.256	.531	.856	1.315	1.706	2.056	2.479	2.779	3.435	3.707
27	.256	.531	.855	1.314	1.703	2.052	2.473	2.771	3.421	3.690
28	.256	.530	.855	1.313	1.701	2.048	2.467	2.763	3.408	3.674
29	.256	.530	.854	1.311	1.699	2.045	2.462	2.756	3.396	3.659
30	.256	.530	.854	1.310	1.697	2.042	2.457	2.750	3.385	3.646
40	.255	.529	.851	1.303	1.684	2.021	2.423	2.704	3.307	3.551
50	.255	.528	.849	1.298	1.676	2.009	2.403	2.678	3.262	3.495
60	.254	.527	.848	1.296	1.671	2.000	2.390	2.660	3.232	3.460
80	.254	.527	.846	1.292	1.664	1.990	2.374	2.639	3.195	3.415
100	.254	.526	.845	1.290	1.660	1.984	2.365	2.626	3.174	3.389
200	.254	.525	.843	1.286	1.653	1.972	2.345	2.601	3.131	3.339
500	.253	.525	.842	1.283	1.648	1.965	2.334	2.586	3.106	3.310
∞	.253	.524	.842	1.282	1.645	1.960	2.326	2.576	3.090	3.291

* This table is reproduced from *Statistical Tables and Formulas* by A. Hald, published by John Wiley & Sons, Inc., New York, the greater part of which has been reproduced from Table III of R. A. Fisher and F. Yates, *Statistical Tables*, Oliver & Boyd, Edinburgh, by permission of the authors and publishers.

Table 4. 10 Percentage Points of the F Distribution*

Table of $F_{0.10;\nu_1,\nu_2}$

Degrees of freedom for the denominator (ν_2)	Degrees of freedom for the numerator (ν_1)																	
	1	2	3	4	5	6	7	8	9	10	15	20	30	50	100	200	500	∞
1	39.9	49.5	53.6	55.8	57.2	58.2	58.9	59.4	59.9	60.2	61.2	61.7	62.3	62.7	63.0	63.2	63.3	63.3
2	8.53	9.00	9.16	9.24	9.29	9.33	9.35	9.37	9.38	9.39	9.42	9.44	9.46	9.47	9.48	9.49	9.49	9.49
3	5.54	5.46	5.39	5.34	5.31	5.28	5.27	5.25	5.24	5.23	5.20	5.18	5.17	5.15	5.14	5.14	5.14	5.13
4	4.54	4.32	4.19	4.11	4.05	4.01	3.98	3.95	3.94	3.92	3.87	3.84	3.82	3.80	3.78	3.77	3.76	3.76
5	4.06	3.78	3.62	3.52	3.45	3.40	3.37	3.34	3.32	3.30	3.24	3.21	3.17	3.15	3.13	3.12	3.11	3.10
6	3.78	3.46	3.29	3.18	3.11	3.05	3.01	2.98	2.96	2.94	2.87	2.84	2.80	2.77	2.75	2.73	2.73	2.72
7	3.59	3.26	3.07	2.96	2.88	2.83	2.78	2.75	2.72	2.70	2.63	2.59	2.56	2.52	2.50	2.48	2.48	2.47
8	3.46	3.11	2.92	2.81	2.73	2.67	2.62	2.59	2.56	2.54	2.46	2.42	2.38	2.35	2.32	2.31	2.30	2.29
9	3.36	3.01	2.81	2.69	2.61	2.55	2.51	2.47	2.44	2.42	2.34	2.30	2.25	2.22	2.19	2.17	2.17	2.16
10	3.28	2.92	2.73	2.61	2.52	2.46	2.41	2.38	2.35	2.32	2.24	2.20	2.16	2.12	2.09	2.07	2.06	2.06
11	3.23	2.86	2.66	2.54	2.45	2.39	2.34	2.30	2.27	2.25	2.17	2.12	2.08	2.04	2.00	1.99	1.98	1.97
12	3.18	2.81	2.61	2.48	2.39	2.33	2.28	2.24	2.21	2.19	2.10	2.06	2.01	1.97	1.94	1.92	1.91	1.90
13	3.14	2.76	2.56	2.43	2.35	2.28	2.23	2.20	2.16	2.14	2.05	2.01	1.96	1.92	1.88	1.86	1.85	1.85
14	3.10	2.73	2.52	2.39	2.31	2.24	2.19	2.15	2.12	2.10	2.01	1.96	1.91	1.87	1.83	1.82	1.80	1.80
15	3.07	2.70	2.49	2.36	2.27	2.21	2.16	2.12	2.09	2.06	1.97	1.92	1.87	1.83	1.79	1.77	1.76	1.76
16	3.05	2.67	2.46	2.33	2.24	2.18	2.13	2.09	2.06	2.03	1.94	1.89	1.84	1.79	1.76	1.74	1.73	1.72
17	3.03	2.64	2.44	2.31	2.22	2.15	2.10	2.06	2.03	2.00	1.91	1.86	1.81	1.76	1.73	1.71	1.69	1.69
18	3.01	2.62	2.42	2.29	2.20	2.13	2.08	2.04	2.00	1.98	1.89	1.84	1.78	1.74	1.70	1.68	1.67	1.66
19	2.99	2.61	2.40	2.27	2.18	2.11	2.06	2.02	1.98	1.96	1.86	1.81	1.76	1.71	1.67	1.65	1.64	1.63
20	2.97	2.59	2.38	2.25	2.16	2.09	2.04	2.00	1.96	1.94	1.84	1.79	1.74	1.69	1.65	1.63	1.62	1.61
22	2.95	2.56	2.35	2.22	2.13	2.06	2.01	1.97	1.93	1.90	1.81	1.76	1.70	1.65	1.61	1.59	1.58	1.57
24	2.93	2.54	2.33	2.19	2.10	2.04	1.98	1.94	1.91	1.88	1.78	1.73	1.67	1.62	1.58	1.56	1.54	1.53
26	2.91	2.52	2.31	2.17	2.08	2.01	1.96	1.92	1.88	1.86	1.76	1.71	1.65	1.59	1.55	1.53	1.51	1.50
28	2.89	2.50	2.29	2.16	2.06	2.00	1.94	1.90	1.87	1.84	1.74	1.69	1.63	1.57	1.53	1.50	1.49	1.48
30	2.88	2.49	2.28	2.14	2.05	1.98	1.93	1.88	1.85	1.82	1.72	1.67	1.61	1.55	1.51	1.48	1.47	1.46
40	2.84	2.44	2.23	2.09	2.00	1.93	1.87	1.83	1.79	1.76	1.66	1.61	1.54	1.48	1.43	1.41	1.39	1.38
50	2.81	2.41	2.20	2.06	1.97	1.90	1.84	1.80	1.76	1.73	1.63	1.57	1.50	1.44	1.39	1.36	1.34	1.33
60	2.79	2.39	2.18	2.04	1.95	1.87	1.82	1.77	1.74	1.71	1.60	1.54	1.48	1.41	1.36	1.33	1.31	1.29
80	2.77	2.37	2.15	2.02	1.92	1.85	1.79	1.75	1.71	1.68	1.57	1.51	1.44	1.38	1.32	1.28	1.26	1.24
100	2.76	2.36	2.14	2.00	1.91	1.83	1.78	1.73	1.70	1.66	1.56	1.49	1.42	1.35	1.29	1.26	1.23	1.21
200	2.73	2.33	2.11	1.97	1.88	1.80	1.75	1.70	1.66	1.63	1.52	1.46	1.38	1.31	1.24	1.20	1.17	1.14
500	2.72	2.31	2.10	1.96	1.86	1.79	1.73	1.68	1.64	1.61	1.50	1.44	1.36	1.28	1.21	1.16	1.12	1.09
∞	2.71	2.30	2.08	1.94	1.85	1.77	1.72	1.67	1.63	1.60	1.49	1.42	1.34	1.26	1.18	1.13	1.08	1.00

Example: $P\{F_{.10;8,20} < 2.00\} = 90\%$.

$F_{.90;\nu_1,\nu_2} = 1/F_{.10;\nu_2,\nu_1}$. **Example:** $F_{.90;8,20} = 1/F_{.10;20,8} = 1/2.42 = 0.413$.

Approximate formula for ν_1 and ν_2 larger than 30: $\log_{10} F_{.10;\nu_1,\nu_2} \simeq \dfrac{1.1131}{\sqrt{h} - 0.77} - 0.527\left(\dfrac{1}{\nu_1} - \dfrac{1}{\nu_2}\right)$,

where $\dfrac{1}{h} = \dfrac{1}{2}\left(\dfrac{1}{\nu_1} + \dfrac{1}{\nu_2}\right)$.

* This table is abridged from A. Hald, *Statistical Tables and Formulas*, John Wiley and Sons, Inc., New York, a major part of which has been abridged from "Tables of Percentage Points of the Inverted Beta (F) Distribution," computed by M. Merrington and C. M. Thompson, *Biometrika* Vol. 33, 1943, pp. 73–88, by permission of the proprietors, or reproduced from Table V of R. A. Fisher and F. Yates, *Statistical Tables*, Oliver and Boyd, Edinburgh, by permission of the authors and the publishers.

Table 4. (cont.) 5 Percentage Points of the F Distribution

Table of $F_{0.05;\nu_1,\nu_2}$

Degrees of freedom for the numerator (ν_1) across; degrees of freedom for the denominator (ν_2) down.

ν_2	1	2	3	4	5	6	7	8	9	10	11	12	13	14	15	16	17	18
1	161	200	216	225	230	234	237	239	241	242	243	244	245	245	246	246	247	247
2	18.5	19.0	19.2	19.2	19.3	19.3	19.4	19.4	19.4	19.4	19.4	19.4	19.4	19.4	19.4	19.4	19.4	19.4
3	10.1	9.55	9.28	9.12	9.01	8.94	8.89	8.85	8.81	8.79	8.76	8.74	8.73	8.71	8.70	8.69	8.68	8.67
4	7.71	6.94	6.59	6.39	6.26	6.16	6.09	6.04	6.00	5.96	5.94	5.91	5.89	5.87	5.86	5.84	5.83	5.82
5	6.61	5.79	5.41	5.19	5.05	4.95	4.88	4.82	4.77	4.74	4.70	4.68	4.66	4.64	4.62	4.60	4.59	4.58
6	5.99	5.14	4.76	4.53	4.39	4.28	4.21	4.15	4.10	4.06	4.03	4.00	3.98	3.96	3.94	3.92	3.91	3.90
7	5.59	4.74	4.35	4.12	3.97	3.87	3.79	3.73	3.68	3.64	3.60	3.57	3.55	3.53	3.51	3.49	3.48	3.47
8	5.32	4.46	4.07	3.84	3.69	3.58	3.50	3.44	3.39	3.35	3.31	3.28	3.26	3.24	3.22	3.20	3.19	3.17
9	5.12	4.26	3.86	3.63	3.48	3.37	3.29	3.23	3.18	3.14	3.10	3.07	3.05	3.03	3.01	2.99	2.97	2.96
10	4.96	4.10	3.71	3.48	3.33	3.22	3.14	3.07	3.02	2.98	2.94	2.91	2.89	2.86	2.85	2.83	2.81	2.80
11	4.84	3.98	3.59	3.36	3.20	3.09	3.01	2.95	2.90	2.85	2.82	2.79	2.76	2.74	2.72	2.70	2.69	2.67
12	4.75	3.89	3.49	3.26	3.11	3.00	2.91	2.85	2.80	2.75	2.72	2.69	2.66	2.64	2.62	2.60	2.58	2.57
13	4.67	3.81	3.41	3.18	3.03	2.92	2.83	2.77	2.71	2.67	2.63	2.60	2.58	2.55	2.53	2.51	2.50	2.48
14	4.60	3.74	3.34	3.11	2.96	2.85	2.76	2.70	2.65	2.60	2.57	2.53	2.51	2.48	2.46	2.44	2.43	2.41
15	4.54	3.68	3.29	3.06	2.90	2.79	2.71	2.64	2.59	2.54	2.51	2.48	2.45	2.42	2.40	2.38	2.37	2.35
16	4.49	3.63	3.24	3.01	2.85	2.74	2.66	2.59	2.54	2.49	2.46	2.42	2.40	2.37	2.35	2.33	2.32	2.30
17	4.45	3.59	3.20	2.96	2.81	2.70	2.61	2.55	2.49	2.45	2.41	2.38	2.35	2.33	2.31	2.29	2.27	2.26
18	4.41	3.55	3.16	2.93	2.77	2.66	2.58	2.51	2.46	2.41	2.37	2.34	2.31	2.29	2.27	2.25	2.23	2.22
19	4.38	3.52	3.13	2.90	2.74	2.63	2.54	2.48	2.42	2.38	2.34	2.31	2.28	2.26	2.23	2.21	2.20	2.18
20	4.35	3.49	3.10	2.87	2.71	2.60	2.51	2.45	2.39	2.35	2.31	2.28	2.25	2.22	2.20	2.18	2.17	2.15
21	4.32	3.47	3.07	2.84	2.68	2.57	2.49	2.42	2.37	2.32	2.28	2.25	2.22	2.20	2.18	2.16	2.14	2.12
22	4.30	3.44	3.05	2.82	2.66	2.55	2.46	2.40	2.34	2.30	2.26	2.23	2.20	2.17	2.15	2.13	2.11	2.10
23	4.28	3.42	3.03	2.80	2.64	2.53	2.44	2.37	2.32	2.27	2.23	2.20	2.18	2.15	2.13	2.11	2.09	2.07
24	4.26	3.40	3.01	2.78	2.62	2.51	2.42	2.36	2.30	2.25	2.21	2.18	2.15	2.13	2.11	2.09	2.07	2.05
25	4.24	3.39	2.99	2.76	2.60	2.49	2.40	2.34	2.28	2.24	2.20	2.16	2.14	2.11	2.09	2.07	2.05	2.04
26	4.23	3.37	2.98	2.74	2.59	2.47	2.39	2.32	2.27	2.22	2.18	2.15	2.12	2.09	2.07	2.05	2.03	2.02
27	4.21	3.35	2.96	2.73	2.57	2.46	2.37	2.31	2.25	2.20	2.17	2.13	2.10	2.08	2.06	2.04	2.02	2.00
28	4.20	3.34	2.95	2.71	2.56	2.45	2.36	2.29	2.24	2.19	2.15	2.12	2.09	2.06	2.04	2.02	2.00	1.99
29	4.18	3.33	2.93	2.70	2.55	2.43	2.35	2.28	2.22	2.18	2.14	2.10	2.08	2.05	2.03	2.01	1.99	1.97
30	4.17	3.32	2.92	2.69	2.53	2.42	2.33	2.27	2.21	2.16	2.13	2.09	2.06	2.04	2.01	1.99	1.98	1.96
32	4.15	3.29	2.90	2.67	2.51	2.40	2.31	2.24	2.19	2.14	2.10	2.07	2.04	2.01	1.99	1.97	1.95	1.94
34	4.13	3.28	2.88	2.65	2.49	2.38	2.29	2.23	2.17	2.12	2.08	2.05	2.02	1.99	1.97	1.95	1.93	1.92
36	4.11	3.26	2.87	2.63	2.48	2.36	2.28	2.21	2.15	2.11	2.07	2.03	2.00	1.98	1.95	1.93	1.92	1.90
38	4.10	3.24	2.85	2.62	2.46	2.35	2.26	2.19	2.14	2.09	2.05	2.02	1.99	1.96	1.94	1.92	1.90	1.88
40	4.08	3.23	2.84	2.61	2.45	2.34	2.25	2.18	2.12	2.08	2.04	2.00	1.97	1.95	1.92	1.90	1.89	1.87
42	4.07	3.22	2.83	2.59	2.44	2.32	2.24	2.17	2.11	2.06	2.03	1.99	1.96	1.93	1.91	1.89	1.87	1.86
44	4.06	3.21	2.82	2.58	2.43	2.31	2.23	2.16	2.10	2.05	2.01	1.98	1.95	1.92	1.90	1.88	1.86	1.84
46	4.05	3.20	2.81	2.57	2.42	2.30	2.22	2.15	2.09	2.04	2.00	1.97	1.94	1.91	1.89	1.87	1.85	1.83
48	4.04	3.19	2.80	2.57	2.41	2.29	2.21	2.14	2.08	2.03	1.99	1.96	1.93	1.90	1.88	1.86	1.84	1.82
50	4.03	3.18	2.79	2.56	2.40	2.29	2.20	2.13	2.07	2.03	1.99	1.95	1.92	1.89	1.87	1.85	1.83	1.81
55	4.02	3.16	2.77	2.54	2.38	2.27	2.18	2.11	2.06	2.01	1.97	1.93	1.90	1.88	1.85	1.83	1.81	1.79
60	4.00	3.15	2.76	2.53	2.37	2.25	2.17	2.10	2.04	1.99	1.95	1.92	1.89	1.86	1.84	1.82	1.80	1.78
65	3.99	3.14	2.75	2.51	2.36	2.24	2.15	2.08	2.03	1.98	1.94	1.90	1.87	1.85	1.82	1.80	1.78	1.76
70	3.98	3.13	2.74	2.50	2.35	2.23	2.14	2.07	2.02	1.97	1.93	1.89	1.86	1.84	1.81	1.79	1.77	1.75
80	3.96	3.11	2.72	2.49	2.33	2.21	2.13	2.06	2.00	1.95	1.91	1.88	1.84	1.82	1.79	1.77	1.75	1.73
90	3.95	3.10	2.71	2.47	2.32	2.20	2.11	2.04	1.99	1.94	1.90	1.86	1.83	1.80	1.78	1.76	1.74	1.72
100	3.94	3.09	2.70	2.46	2.31	2.19	2.10	2.03	1.97	1.93	1.89	1.85	1.82	1.79	1.77	1.75	1.73	1.71
125	3.92	3.07	2.68	2.44	2.29	2.17	2.08	2.01	1.96	1.91	1.87	1.83	1.80	1.77	1.75	1.72	1.70	1.69
150	3.90	3.06	2.66	2.43	2.27	2.16	2.07	2.00	1.94	1.89	1.85	1.82	1.79	1.76	1.73	1.71	1.69	1.67
200	3.89	3.04	2.65	2.42	2.26	2.14	2.06	1.98	1.93	1.88	1.84	1.80	1.77	1.74	1.72	1.69	1.67	1.66
300	3.87	3.03	2.63	2.40	2.24	2.13	2.04	1.97	1.91	1.86	1.82	1.78	1.75	1.72	1.70	1.68	1.66	1.64
500	3.86	3.01	2.62	2.39	2.23	2.12	2.03	1.96	1.90	1.85	1.81	1.77	1.74	1.71	1.69	1.66	1.64	1.62
1000	3.85	3.00	2.61	2.38	2.22	2.11	2.02	1.95	1.89	1.84	1.80	1.76	1.73	1.70	1.68	1.65	1.63	1.61
∞	3.84	3.00	2.60	2.37	2.21	2.10	2.01	1.94	1.88	1.83	1.79	1.75	1.72	1.69	1.67	1.64	1.62	1.60

Example: $P\{F_{.05;8,20} < 2.45\} = 95\%$.

$F_{.95;\nu_1,\nu_2} = 1/F_{.05;\nu_2,\nu_1}$. Example: $F_{.95;8,20} = 1/F_{.05;20,8} = 1/3.15 = 0.317$.

Table 4. (cont.) 5 Percentage Points of the F Distribution

Table of $F_{0.05;\nu_1,\nu_2}$

Degrees of freedom for the numerator (ν_1)

Degrees of freedom for the denominator (ν_2)	19	20	22	24	26	28	30	35	40	45	50	60	80	100	200	500	∞
1	248	248	249	249	249	250	250	251	251	251	252	252	252	253	254	254	254
2	19.4	19.4	19.5	19.5	19.5	19.5	19.5	19.5	19.5	19.5	19.5	19.5	19.5	19.5	19.5	19.5	19.5
3	8.67	8.66	8.65	8.64	8.63	8.62	8.62	8.60	8.59	8.59	8.58	8.57	8.56	8.55	8.54	8.53	8.53
4	5.81	5.80	5.79	5.77	5.76	5.75	5.75	5.73	5.72	5.71	5.70	5.69	5.67	5.66	5.65	5.64	5.63
5	4.57	4.56	4.54	4.53	4.52	4.50	4.50	4.48	4.46	4.45	4.44	4.43	4.41	4.41	4.39	4.37	4.37
6	3.88	3.87	3.86	3.84	3.83	3.82	3.81	3.79	3.77	3.76	3.75	3.74	3.72	3.71	3.69	3.68	3.67
7	3.46	3.44	3.43	3.41	3.40	3.39	3.38	3.36	3.34	3.33	3.32	3.30	3.29	3.27	3.25	3.24	3.23
8	3.16	3.15	3.13	3.12	3.10	3.09	3.08	3.06	3.04	3.03	3.02	3.01	2.99	2.97	2.95	2.94	2.93
9	2.95	2.94	2.92	2.90	2.89	2.87	2.86	2.84	2.83	2.81	2.80	2.79	2.77	2.76	2.73	2.72	2.71
10	2.78	2.77	2.75	2.74	2.72	2.71	2.70	2.68	2.66	2.65	2.64	2.62	2.60	2.59	2.56	2.55	2.54
11	2.66	2.65	2.63	2.61	2.59	2.58	2.57	2.55	2.53	2.52	2.51	2.49	2.47	2.46	2.43	2.42	2.40
12	2.56	2.54	2.52	2.51	2.49	2.48	2.47	2.44	2.43	2.41	2.40	2.38	2.36	2.35	2.32	2.31	2.30
13	2.47	2.46	2.44	2.42	2.41	2.39	2.38	2.36	2.34	2.33	2.31	2.30	2.27	2.26	2.23	2.22	2.21
14	2.40	2.39	2.37	2.35	2.33	2.32	2.31	2.28	2.27	2.25	2.24	2.22	2.20	2.19	2.16	2.14	2.13
15	2.34	2.33	2.31	2.29	2.27	2.26	2.25	2.22	2.20	2.19	2.18	2.16	2.14	2.12	2.10	2.08	2.07
16	2.29	2.28	2.25	2.24	2.22	2.21	2.19	2.17	2.15	2.14	2.12	2.11	2.08	2.07	2.04	2.02	2.01
17	2.24	2.23	2.21	2.19	2.17	2.16	2.15	2.12	2.10	2.09	2.08	2.06	2.03	2.02	1.99	1.97	1.96
18	2.20	2.19	2.17	2.15	2.13	2.12	2.11	2.08	2.06	2.05	2.04	2.02	1.99	1.98	1.95	1.93	1.92
19	2.17	2.16	2.13	2.11	2.10	2.08	2.07	2.05	2.03	2.01	2.00	1.98	1.96	1.94	1.91	1.89	1.88
20	2.14	2.12	2.10	2.08	2.07	2.05	2.04	2.01	1.99	1.98	1.97	1.95	1.92	1.91	1.88	1.86	1.84
21	2.11	2.10	2.07	2.05	2.04	2.02	2.01	1.98	1.96	1.95	1.94	1.92	1.89	1.88	1.84	1.82	1.81
22	2.08	2.07	2.05	2.03	2.01	2.00	1.98	1.96	1.94	1.92	1.91	1.89	1.86	1.85	1.82	1.80	1.78
23	2.06	2.05	2.02	2.00	1.99	1.97	1.96	1.93	1.91	1.90	1.88	1.86	1.84	1.82	1.79	1.77	1.76
24	2.04	2.03	2.00	1.98	1.97	1.95	1.94	1.91	1.89	1.88	1.86	1.84	1.82	1.80	1.77	1.75	1.73
25	2.02	2.01	1.98	1.96	1.95	1.93	1.92	1.89	1.87	1.86	1.84	1.82	1.80	1.78	1.75	1.73	1.71
26	2.00	1.99	1.97	1.95	1.93	1.91	1.90	1.87	1.85	1.84	1.82	1.80	1.78	1.76	1.73	1.71	1.69
27	1.99	1.97	1.95	1.93	1.91	1.90	1.88	1.86	1.84	1.82	1.81	1.79	1.76	1.74	1.71	1.69	1.67
28	1.97	1.96	1.93	1.91	1.90	1.88	1.87	1.84	1.82	1.80	1.79	1.77	1.74	1.73	1.69	1.67	1.65
29	1.96	1.94	1.92	1.90	1.88	1.87	1.85	1.83	1.81	1.79	1.77	1.75	1.73	1.71	1.67	1.65	1.64
30	1.95	1.93	1.91	1.89	1.87	1.85	1.84	1.81	1.79	1.77	1.76	1.74	1.71	1.70	1.66	1.64	1.62
32	1.92	1.91	1.88	1.86	1.85	1.83	1.82	1.79	1.77	1.75	1.74	1.71	1.69	1.67	1.63	1.61	1.59
34	1.90	1.89	1.86	1.84	1.82	1.80	1.80	1.77	1.75	1.73	1.71	1.69	1.66	1.65	1.61	1.59	1.57
36	1.88	1.87	1.85	1.82	1.81	1.79	1.78	1.75	1.73	1.71	1.69	1.67	1.64	1.62	1.59	1.56	1.55
38	1.87	1.85	1.83	1.81	1.79	1.77	1.76	1.73	1.71	1.69	1.68	1.65	1.62	1.61	1.57	1.54	1.53
40	1.85	1.84	1.81	1.79	1.77	1.76	1.74	1.72	1.69	1.67	1.66	1.64	1.61	1.59	1.55	1.53	1.51
42	1.84	1.83	1.80	1.78	1.76	1.74	1.73	1.70	1.68	1.66	1.65	1.62	1.59	1.57	1.53	1.51	1.49
44	1.83	1.81	1.79	1.77	1.75	1.73	1.72	1.69	1.67	1.65	1.63	1.61	1.58	1.56	1.52	1.49	1.48
46	1.82	1.80	1.78	1.76	1.74	1.72	1.71	1.68	1.65	1.64	1.62	1.60	1.57	1.55	1.51	1.48	1.46
48	1.81	1.79	1.77	1.75	1.73	1.71	1.70	1.67	1.64	1.62	1.61	1.59	1.56	1.54	1.49	1.47	1.45
50	1.80	1.78	1.76	1.74	1.72	1.70	1.69	1.66	1.63	1.61	1.60	1.58	1.54	1.52	1.48	1.46	1.44
55	1.78	1.76	1.74	1.72	1.70	1.68	1.67	1.64	1.61	1.59	1.58	1.55	1.52	1.50	1.46	1.43	1.41
60	1.76	1.75	1.72	1.70	1.68	1.66	1.65	1.62	1.59	1.57	1.56	1.53	1.50	1.48	1.44	1.41	1.39
65	1.75	1.73	1.71	1.69	1.67	1.65	1.63	1.60	1.58	1.56	1.54	1.52	1.49	1.46	1.42	1.39	1.37
70	1.74	1.72	1.70	1.67	1.65	1.64	1.62	1.59	1.57	1.55	1.53	1.50	1.47	1.45	1.40	1.37	1.35
80	1.72	1.70	1.68	1.65	1.63	1.62	1.60	1.57	1.54	1.52	1.51	1.48	1.45	1.43	1.38	1.35	1.32
90	1.70	1.69	1.66	1.64	1.62	1.60	1.59	1.55	1.53	1.51	1.49	1.46	1.43	1.41	1.36	1.32	1.30
100	1.69	1.68	1.65	1.63	1.61	1.59	1.57	1.54	1.52	1.49	1.48	1.45	1.41	1.39	1.34	1.31	1.28
125	1.67	1.65	1.63	1.60	1.58	1.57	1.55	1.52	1.49	1.47	1.45	1.42	1.39	1.36	1.31	1.27	1.25
150	1.66	1.64	1.61	1.59	1.57	1.55	1.53	1.50	1.48	1.45	1.44	1.41	1.37	1.34	1.29	1.25	1.22
200	1.64	1.62	1.60	1.57	1.55	1.53	1.52	1.48	1.46	1.43	1.41	1.39	1.35	1.32	1.26	1.22	1.19
300	1.62	1.61	1.58	1.55	1.53	1.51	1.50	1.46	1.43	1.41	1.39	1.36	1.32	1.30	1.23	1.19	1.15
500	1.61	1.59	1.56	1.54	1.52	1.50	1.48	1.45	1.42	1.40	1.38	1.34	1.30	1.28	1.21	1.16	1.11
1000	1.60	1.58	1.55	1.53	1.51	1.49	1.47	1.44	1.41	1.38	1.36	1.33	1.29	1.26	1.19	1.13	1.08
∞	1.59	1.57	1.54	1.52	1.50	1.48	1.46	1.42	1.39	1.37	1.35	1.32	1.27	1.24	1.17	1.11	1.00

Approximate formula for ν_1 and ν_2 larger than 30:

$$\log_{10} F_{.05;\nu_1,\nu_2} \simeq \frac{1.4287}{\sqrt{h} - 0.95} - 0.681\left(\frac{1}{\nu_1} - \frac{1}{\nu_2}\right), \text{ where } \frac{1}{h} = \frac{1}{2}\left(\frac{1}{\nu_1} + \frac{1}{\nu_2}\right).$$

Table 4. (cont.) 2.5 Percentage Points of the F Distribution

Table of $F_{0.025;\nu_1,\nu_2}$

Degrees of freedom for the denominator (ν_2)	Degrees of freedom for the numerator (ν_1) 1	2	3	4	5	6	7	8	9	10	11	12	13	14	15	16	17	18
1	648	800	864	900	922	937	948	957	963	969	973	977	980	983	985	987	989	990
2	38.5	39.0	39.2	39.2	39.3	39.3	39.4	39.4	39.4	39.4	39.4	39.4	39.4	39.4	39.4	39.4	39.4	39.4
3	17.4	16.0	15.4	15.1	14.9	14.7	14.6	14.5	14.5	14.4	14.4	14.3	14.3	14.3	14.3	14.2	14.2	14.2
4	12.2	10.6	9.98	9.60	9.36	9.20	9.07	8.98	8.90	8.84	8.79	8.75	8.72	8.69	8.66	8.64	8.62	8.60
5	10.0	8.43	7.76	7.39	7.15	6.98	6.85	6.76	6.68	6.62	6.57	6.52	6.49	6.46	6.43	6.41	6.39	6.37
6	8.81	7.26	6.60	6.23	5.99	5.82	5.70	5.60	5.52	5.46	5.41	5.37	5.33	5.30	5.27	5.25	5.23	5.21
7	8.07	6.54	5.89	5.52	5.29	5.12	4.99	4.90	4.82	4.76	4.71	4.67	4.63	4.60	4.57	4.54	4.52	4.50
8	7.57	6.06	5.42	5.05	4.82	4.65	4.53	4.43	4.36	4.30	4.24	4.20	4.16	4.13	4.10	4.08	4.05	4.03
9	7.21	5.71	5.08	4.72	4.48	4.32	4.20	4.10	4.03	3.96	3.91	3.87	3.83	3.80	3.77	3.74	3.72	3.70
10	6.94	5.46	4.83	4.47	4.24	4.07	3.95	3.85	3.78	3.72	3.66	3.62	3.58	3.55	3.52	3.50	3.47	3.45
11	6.72	5.26	4.63	4.28	4.04	3.88	3.76	3.66	3.59	3.53	3.47	3.43	3.39	3.36	3.33	3.30	3.28	3.26
12	6.55	5.10	4.47	4.12	3.89	3.73	3.61	3.51	3.44	3.37	3.32	3.28	3.24	3.21	3.18	3.15	3.13	3.11
13	6.41	4.97	4.35	4.00	3.77	3.60	3.48	3.39	3.31	3.25	3.20	3.15	3.12	3.08	3.05	3.03	3.00	2.98
14	6.30	4.86	4.24	3.89	3.66	3.50	3.38	3.29	3.21	3.15	3.09	3.05	3.01	2.98	2.95	2.92	2.90	2.88
15	6.20	4.76	4.15	3.80	3.58	3.41	3.29	3.20	3.12	3.06	3.01	2.96	2.92	2.89	2.86	2.84	2.81	2.79
16	6.12	4.69	4.08	3.73	3.50	3.34	3.22	3.12	3.05	2.99	2.93	2.89	2.85	2.82	2.79	2.76	2.74	2.72
17	6.04	4.62	4.01	3.66	3.44	3.28	3.16	3.06	2.98	2.92	2.87	2.82	2.79	2.75	2.72	2.70	2.67	2.65
18	5.98	4.56	3.95	3.61	3.38	3.22	3.10	3.01	2.93	2.87	2.81	2.77	2.73	2.70	2.67	2.64	2.62	2.60
19	5.92	4.51	3.90	3.56	3.33	3.17	3.05	2.96	2.88	2.82	2.76	2.72	2.68	2.65	2.62	2.59	2.57	2.55
20	5.87	4.46	3.86	3.51	3.29	3.13	3.01	2.91	2.84	2.77	2.72	2.68	2.64	2.60	2.57	2.55	2.52	2.50
21	5.83	4.42	3.82	3.48	3.25	3.09	2.97	2.87	2.80	2.73	2.68	2.64	2.60	2.56	2.53	2.51	2.48	2.46
22	5.79	4.38	3.78	3.44	3.22	3.05	2.93	2.84	2.76	2.70	2.65	2.60	2.56	2.53	2.50	2.47	2.45	2.43
23	5.75	4.35	3.75	3.41	3.18	3.02	2.90	2.81	2.73	2.67	2.62	2.57	2.53	2.50	2.47	2.44	2.42	2.39
24	5.72	4.32	3.72	3.38	3.15	2.99	2.87	2.78	2.70	2.64	2.59	2.54	2.50	2.47	2.44	2.41	2.39	2.36
25	5.69	4.29	3.69	3.35	3.13	2.97	2.85	2.75	2.68	2.61	2.56	2.51	2.48	2.44	2.41	2.38	2.36	2.34
26	5.66	4.27	3.67	3.33	3.10	2.94	2.82	2.73	2.65	2.59	2.54	2.49	2.45	2.42	2.39	2.36	2.34	2.31
27	5.63	4.24	3.65	3.31	3.08	2.92	2.80	2.71	2.63	2.57	2.51	2.47	2.43	2.39	2.36	2.34	2.31	2.29
28	5.61	4.22	3.63	3.29	3.06	2.90	2.78	2.69	2.61	2.55	2.49	2.45	2.41	2.37	2.34	2.32	2.29	2.27
29	5.59	4.20	3.61	3.27	3.04	2.88	2.76	2.67	2.59	2.53	2.48	2.43	2.39	2.36	2.32	2.30	2.27	2.25
30	5.57	4.18	3.59	3.25	3.03	2.87	2.75	2.65	2.57	2.51	2.46	2.41	2.37	2.34	2.31	2.28	2.26	2.23
32	5.53	4.15	3.56	3.22	3.00	2.84	2.72	2.62	2.54	2.48	2.43	2.38	2.34	2.31	2.28	2.25	2.22	2.20
34	5.50	4.12	3.53	3.19	2.97	2.81	2.69	2.59	2.52	2.45	2.40	2.35	2.31	2.28	2.25	2.22	2.19	2.17
36	5.47	4.09	3.51	3.17	2.94	2.79	2.66	2.57	2.49	2.43	2.37	2.33	2.29	2.25	2.22	2.20	2.17	2.15
38	5.45	4.07	3.48	3.15	2.92	2.76	2.64	2.55	2.47	2.41	2.35	2.31	2.27	2.23	2.20	2.17	2.15	2.13
40	5.42	4.05	3.46	3.13	2.90	2.74	2.62	2.53	2.45	2.39	2.33	2.29	2.25	2.21	2.18	2.15	2.13	2.11
42	5.40	4.03	3.45	3.11	2.89	2.73	2.61	2.51	2.44	2.37	2.32	2.27	2.23	2.20	2.16	2.14	2.11	2.09
44	5.39	4.02	3.43	3.09	2.87	2.71	2.59	2.50	2.42	2.36	2.30	2.26	2.21	2.18	2.15	2.12	2.10	2.07
46	5.37	4.00	3.42	3.08	2.86	2.70	2.58	2.48	2.41	2.34	2.29	2.24	2.20	2.17	2.13	2.11	2.08	2.06
48	5.35	3.99	3.40	3.07	2.84	2.69	2.57	2.47	2.39	2.33	2.27	2.23	2.19	2.15	2.12	2.09	2.07	2.05
50	5.34	3.98	3.39	3.06	2.83	2.67	2.55	2.46	2.38	2.32	2.26	2.22	2.18	2.14	2.11	2.08	2.06	2.03
55	5.31	3.95	3.36	3.03	2.81	2.65	2.53	2.43	2.36	2.29	2.24	2.19	2.15	2.11	2.08	2.05	2.03	2.01
60	5.29	3.93	3.34	3.01	2.79	2.63	2.51	2.41	2.33	2.27	2.22	2.17	2.13	2.09	2.06	2.03	2.01	1.98
65	5.27	3.91	3.32	2.99	2.77	2.61	2.49	2.39	2.32	2.25	2.20	2.15	2.11	2.07	2.04	2.01	1.99	1.97
70	5.25	3.89	3.31	2.98	2.75	2.60	2.48	2.38	2.30	2.24	2.18	2.14	2.10	2.06	2.03	2.00	1.97	1.95
80	5.22	3.86	3.28	2.95	2.73	2.57	2.45	2.36	2.28	2.21	2.16	2.11	2.07	2.03	2.00	1.97	1.95	1.93
90	5.20	3.84	3.27	2.93	2.71	2.55	2.43	2.34	2.26	2.19	2.14	2.09	2.05	2.02	1.98	1.95	1.93	1.91
100	5.18	3.83	3.25	2.92	2.70	2.54	2.42	2.32	2.24	2.18	2.12	2.08	2.04	2.00	1.97	1.94	1.91	1.89
125	5.15	3.80	3.22	2.89	2.67	2.51	2.39	2.30	2.22	2.15	2.10	2.05	2.01	1.97	1.94	1.91	1.89	1.86
150	5.13	3.78	3.20	2.87	2.65	2.49	2.37	2.28	2.20	2.13	2.08	2.03	1.99	1.95	1.92	1.89	1.87	1.84
200	5.10	3.76	3.18	2.85	2.63	2.47	2.35	2.26	2.18	2.11	2.06	2.01	1.97	1.93	1.90	1.87	1.84	1.82
300	5.08	3.74	3.16	2.83	2.61	2.45	2.33	2.23	2.16	2.09	2.04	1.99	1.95	1.91	1.88	1.85	1.82	1.80
500	5.05	3.72	3.14	2.81	2.59	2.43	2.31	2.22	2.14	2.07	2.02	1.97	1.93	1.89	1.86	1.83	1.80	1.78
1000	5.04	3.70	3.13	2.80	2.58	2.42	2.30	2.20	2.13	2.06	2.01	1.96	1.92	1.88	1.85	1.82	1.79	1.77
∞	5.02	3.69	3.12	2.79	2.57	2.41	2.29	2.19	2.11	2.05	1.99	1.94	1.90	1.87	1.83	1.80	1.78	1.75

Example: $P\{F_{.025;8,20} < 2.91\} = 97.5\%$.

$F_{.975;\nu_1,\nu_2} = 1/F_{.025;\nu_2,\nu_1}$. Example: $F_{.975;8,29} = 1/F_{.025;29,8} = 1/4.00 = 0.250$.

Table 4. (cont.) 2.5 Percentage Points of the F Distribution

Table of $F_{0.025;\nu_1,\nu_2}$

Degrees of freedom for the numerator (ν_1)

19	20	22	24	26	28	30	35	40	45	50	60	80	100	200	500	∞	Degrees of freedom for the denominator (ν_2)
992	993	995	997	999	1000	1001	1004	1006	1007	1008	1010	1012	1013	1016	1017	1018	1
39.4	39.4	39.5	39.5	39.5	39.5	39.5	39.5	39.5	39.5	39.5	39.5	39.5	39.5	39.5	39.5	39.5	2
14.2	14.2	14.1	14.1	14.1	14.1	14.1	14.1	14.0	14.0	14.0	14.0	14.0	14.0	13.9	13.9	13.9	3
8.58	8.56	8.53	8.51	8.49	8.48	8.46	8.44	8.41	8.39	8.38	8.36	8.33	8.32	8.29	8.27	8.26	4
6.35	6.33	6.30	6.28	6.26	6.24	6.23	6.20	6.18	6.16	6.14	6.12	6.10	6.08	6.05	6.03	6.02	5
5.19	5.17	5.14	5.12	5.10	5.08	5.07	5.04	5.01	4.99	4.98	4.96	4.93	4.92	4.88	4.86	4.85	6
4.48	4.47	4.44	4.42	4.39	4.38	4.36	4.33	4.31	4.29	4.28	4.25	4.23	4.21	4.18	4.16	4.14	7
4.02	4.00	3.97	3.95	3.93	3.91	3.89	3.86	3.84	3.82	3.81	3.78	3.76	3.74	3.70	3.68	3.67	8
3.68	3.67	3.64	3.61	3.59	3.58	3.56	3.53	3.51	3.49	3.47	3.45	3.42	3.40	3.37	3.35	3.33	9
3.44	3.42	3.39	3.37	3.34	3.33	3.31	3.28	3.26	3.24	3.22	3.20	3.17	3.15	3.12	3.09	3.08	10
3.24	3.23	3.20	3.17	3.15	3.13	3.12	3.09	3.06	3.04	3.03	3.00	2.97	2.96	2.92	2.90	2.88	11
3.09	3.07	3.04	3.02	3.00	2.98	2.96	2.93	2.91	2.89	2.87	2.85	2.82	2.80	2.76	2.74	2.72	12
2.96	2.95	2.92	2.89	2.87	2.85	2.84	2.80	2.78	2.76	2.74	2.72	2.69	2.67	2.63	2.61	2.60	13
2.86	2.84	2.81	2.79	2.77	2.75	2.73	2.70	2.67	2.65	2.64	2.61	2.58	2.56	2.53	2.50	2.49	14
2.77	2.76	2.73	2.70	2.68	2.66	2.64	2.61	2.58	2.56	2.55	2.52	2.49	2.47	2.44	2.41	2.40	15
2.70	2.68	2.65	2.63	2.60	2.58	2.57	2.53	2.51	2.49	2.47	2.45	2.42	2.40	2.36	2.33	2.32	16
2.63	2.62	2.59	2.56	2.54	2.52	2.50	2.47	2.44	2.42	2.41	2.38	2.35	2.33	2.29	2.26	2.25	17
2.58	2.56	2.53	2.50	2.48	2.46	2.44	2.41	2.38	2.36	2.35	2.32	2.29	2.27	2.23	2.20	2.19	18
2.53	2.51	2.48	2.45	2.43	2.41	2.39	2.36	2.33	2.31	2.30	2.27	2.24	2.22	2.18	2.15	2.13	19
2.48	2.46	2.43	2.41	2.39	2.37	2.35	2.31	2.29	2.27	2.25	2.22	2.19	2.17	2.13	2.10	2.09	20
2.44	2.42	2.39	2.37	2.34	2.33	2.31	2.27	2.25	2.23	2.21	2.18	2.15	2.13	2.09	2.06	2.04	21
2.41	2.39	2.36	2.33	2.31	2.29	2.27	2.24	2.21	2.19	2.17	2.14	2.11	2.09	2.05	2.02	2.00	22
2.37	2.36	2.33	2.30	2.28	2.26	2.24	2.20	2.18	2.15	2.14	2.11	2.08	2.06	2.01	1.99	1.97	23
2.35	2.33	2.30	2.27	2.25	2.23	2.21	2.17	2.15	2.12	2.11	2.08	2.05	2.02	1.98	1.95	1.94	24
2.32	2.30	2.27	2.24	2.22	2.20	2.18	2.15	2.12	2.10	2.08	2.05	2.02	2.00	1.95	1.92	1.91	25
2.29	2.28	2.24	2.22	2.19	2.17	2.16	2.12	2.09	2.07	2.05	2.03	1.99	1.97	1.92	1.90	1.88	26
2.27	2.25	2.22	2.19	2.17	2.15	2.13	2.10	2.07	2.05	2.03	2.00	1.97	1.94	1.90	1.87	1.85	27
2.25	2.23	2.20	2.17	2.15	2.13	2.11	2.08	2.05	2.03	2.01	1.98	1.94	1.92	1.88	1.85	1.83	28
2.23	2.21	2.18	2.15	2.13	2.11	2.09	2.06	2.03	2.01	1.99	1.96	1.92	1.90	1.86	1.83	1.81	29
2.21	2.20	2.16	2.14	2.11	2.09	2.07	2.04	2.01	1.99	1.97	1.94	1.90	1.88	1.84	1.81	1.79	30
2.18	2.16	2.13	2.10	2.08	2.06	2.04	2.00	1.98	1.95	1.93	1.91	1.87	1.85	1.80	1.77	1.75	32
2.15	2.13	2.10	2.07	2.05	2.03	2.01	1.97	1.95	1.92	1.90	1.88	1.84	1.82	1.77	1.74	1.72	34
2.13	2.11	2.08	2.05	2.03	2.00	1.99	1.95	1.92	1.90	1.88	1.85	1.81	1.79	1.74	1.71	1.69	36
2.11	2.09	2.05	2.03	2.00	1.98	1.96	1.93	1.90	1.87	1.85	1.82	1.79	1.76	1.71	1.68	1.66	38
2.09	2.07	2.03	2.01	1.98	1.96	1.94	1.90	1.88	1.85	1.83	1.80	1.76	1.74	1.69	1.66	1.64	40
2.07	2.05	2.02	1.99	1.96	1.94	1.92	1.89	1.86	1.83	1.81	1.78	1.74	1.72	1.67	1.64	1.62	42
2.05	2.03	2.00	1.97	1.95	1.93	1.91	1.87	1.84	1.82	1.80	1.77	1.73	1.70	1.65	1.62	1.60	44
2.04	2.02	1.99	1.96	1.93	1.91	1.89	1.85	1.82	1.80	1.78	1.75	1.71	1.69	1.63	1.60	1.58	46
2.02	2.01	1.97	1.94	1.92	1.90	1.88	1.84	1.81	1.79	1.77	1.73	1.69	1.67	1.62	1.58	1.56	48
2.01	1.99	1.96	1.93	1.91	1.88	1.87	1.83	1.80	1.77	1.75	1.72	1.68	1.66	1.60	1.57	1.55	50
1.99	1.97	1.93	1.90	1.88	1.86	1.84	1.80	1.77	1.74	1.72	1.69	1.65	1.62	1.57	1.54	1.51	55
1.96	1.94	1.91	1.88	1.86	1.83	1.82	1.78	1.74	1.72	1.70	1.67	1.62	1.60	1.54	1.51	1.48	60
1.95	1.93	1.89	1.86	1.84	1.82	1.80	1.76	1.72	1.70	1.68	1.65	1.60	1.58	1.52	1.48	1.46	65
1.93	1.91	1.88	1.85	1.82	1.80	1.78	1.74	1.71	1.68	1.66	1.63	1.58	1.56	1.50	1.46	1.44	70
1.90	1.88	1.85	1.82	1.79	1.77	1.75	1.71	1.68	1.65	1.63	1.60	1.55	1.53	1.47	1.43	1.40	80
1.88	1.86	1.83	1.80	1.77	1.75	1.73	1.69	1.66	1.63	1.61	1.58	1.53	1.50	1.44	1.40	1.37	90
1.87	1.85	1.81	1.78	1.76	1.74	1.71	1.67	1.64	1.61	1.59	1.56	1.51	1.48	1.42	1.38	1.35	100
1.84	1.82	1.79	1.75	1.73	1.71	1.68	1.64	1.61	1.58	1.56	1.52	1.48	1.45	1.38	1.34	1.30	125
1.82	1.80	1.77	1.74	1.71	1.69	1.67	1.62	1.59	1.56	1.54	1.50	1.45	1.42	1.35	1.31	1.27	150
1.80	1.78	1.74	1.71	1.68	1.66	1.64	1.60	1.56	1.53	1.51	1.47	1.42	1.39	1.32	1.27	1.23	200
1.77	1.75	1.72	1.69	1.66	1.64	1.62	1.57	1.54	1.51	1.48	1.45	1.39	1.36	1.28	1.23	1.18	300
1.76	1.74	1.70	1.67	1.64	1.62	1.60	1.55	1.51	1.49	1.46	1.42	1.37	1.34	1.25	1.19	1.14	500
1.74	1.72	1.69	1.65	1.63	1.60	1.58	1.54	1.50	1.47	1.44	1.41	1.35	1.32	1.23	1.16	1.09	1000
1.73	1.71	1.67	1.64	1.61	1.59	1.57	1.52	1.48	1.45	1.43	1.39	1.33	1.30	1.21	1.13	1.00	∞

Approximate formula for ν_1 and ν_2 larger than 30: $\log_{10} F_{.025;\nu_1,\nu_2} \simeq \frac{1.7023}{\sqrt{h} - 1.14} - 0.846\left(\frac{1}{\nu_1} - \frac{1}{\nu_2}\right)$, where $\frac{1}{h} = \frac{1}{2}\left(\frac{1}{\nu_1} + \frac{1}{\nu_2}\right)$.

Table 4. (cont.) 1 Percentage Points of the F Distribution

Table of $F_{0.01;\nu_1,\nu_2}$

Degrees of freedom for the denominator (ν_2)	Degrees of freedom for the numerator (ν_1)																	
	1	2	3	4	5	6	7	8	9	10	11	12	13	14	15	16	17	18
	Multiply the numbers of the first row (ν_2 = 1) by 10.																	
1	*405*	*500*	*540*	*563*	*576*	*586*	*593*	*598*	*602*	*606*	*608*	*611*	*613*	*614*	*616*	*617*	*618*	*619*
2	98.5	99.0	99.2	99.2	99.3	99.3	99.4	99.4	99.4	99.4	99.4	99.4	99.4	99.4	99.4	99.4	99.4	99.4
3	34.1	30.8	29.5	28.7	28.2	27.9	27.7	27.5	27.3	27.2	27.1	27.1	27.0	26.9	26.9	26.8	26.8	26.8
4	21.2	18.0	16.7	16.0	15.5	15.2	15.0	14.8	14.7	14.5	14.4	14.4	14.3	14.2	14.2	14.2	14.1	14.1
5	16.3	13.3	12.1	11.4	11.0	10.7	10.5	10.3	10.2	10.1	9.96	9.89	9.82	9.77	9.72	9.68	9.64	9.61
6	13.7	10.9	9.78	9.15	8.75	8.47	8.26	8.10	7.98	7.87	7.79	7.72	7.66	7.60	7.56	7.52	7.48	7.45
7	12.2	9.55	8.45	7.85	7.46	7.19	6.99	6.84	6.72	6.62	6.54	6.47	6.41	6.36	6.31	6.27	6.24	6.21
8	11.3	8.65	7.59	7.01	6.63	6.37	6.18	6.03	5.91	5.81	5.73	5.67	5.61	5.56	5.52	5.48	5.44	5.41
9	10.6	8.02	6.99	6.42	6.06	5.80	5.61	5.47	5.35	5.26	5.18	5.11	5.05	5.00	4.96	4.92	4.89	4.86
10	10.0	7.56	6.55	5.99	5.64	5.39	5.20	5.06	4.94	4.85	4.77	4.71	4.65	4.60	4.56	4.52	4.49	4.46
11	9.65	7.21	6.22	5.67	5.32	5.07	4.89	4.74	4.63	4.54	4.46	4.40	4.34	4.29	4.25	4.21	4.18	4.15
12	9.33	6.93	5.95	5.41	5.06	4.82	4.64	4.50	4.39	4.30	4.22	4.16	4.10	4.05	4.01	3.97	3.94	3.91
13	9.07	6.70	5.74	5.21	4.86	4.62	4.44	4.30	4.19	4.10	4.02	3.96	3.91	3.86	3.82	3.78	3.75	3.72
14	8.86	6.51	5.56	5.04	4.70	4.46	4.28	4.14	4.03	3.94	3.86	3.80	3.75	3.70	3.66	3.62	3.59	3.56
15	8.68	6.36	5.42	4.89	4.56	4.32	4.14	4.00	3.89	3.80	3.73	3.67	3.61	3.56	3.52	3.49	3.45	3.42
16	8.53	6.23	5.29	4.77	4.44	4.20	4.03	3.89	3.78	3.69	3.62	3.55	3.50	3.45	3.41	3.37	3.34	3.31
17	8.40	6.11	5.18	4.67	4.34	4.10	3.93	3.79	3.68	3.59	3.52	3.46	3.40	3.35	3.31	3.27	3.24	3.21
18	8.29	6.01	5.09	4.58	4.25	4.01	3.84	3.71	3.60	3.51	3.43	3.37	3.32	3.27	3.23	3.19	3.16	3.13
19	8.18	5.93	5.01	4.50	4.17	3.94	3.77	3.63	3.52	3.43	3.36	3.30	3.24	3.19	3.15	3.12	3.08	3.05
20	8.10	5.85	4.94	4.43	4.10	3.87	3.70	3.56	3.46	3.37	3.29	3.23	3.18	3.13	3.09	3.05	3.02	2.99
21	8.02	5.78	4.87	4.37	4.04	3.81	3.64	3.51	3.40	3.31	3.24	3.17	3.12	3.07	3.03	2.99	2.96	2.93
22	7.95	5.72	4.82	4.31	3.99	3.76	3.59	3.45	3.35	3.26	3.18	3.12	3.07	3.02	2.98	2.94	2.91	2.88
23	7.88	5.66	4.76	4.26	3.94	3.71	3.54	3.41	3.30	3.21	3.14	3.07	3.02	2.97	2.93	2.89	2.86	2.83
24	7.82	5.61	4.72	4.22	3.90	3.67	3.50	3.36	3.26	3.17	3.09	3.03	2.98	2.93	2.89	2.85	2.82	2.79
25	7.77	5.57	4.68	4.18	3.86	3.63	3.46	3.32	3.22	3.13	3.06	2.99	2.94	2.89	2.85	2.81	2.78	2.75
26	7.72	5.53	4.64	4.14	3.82	3.59	3.42	3.29	3.18	3.09	3.02	2.96	2.90	2.86	2.82	2.78	2.74	2.72
27	7.68	5.49	4.60	4.11	3.78	3.56	3.39	3.26	3.15	3.06	2.99	2.93	2.87	2.82	2.78	2.75	2.71	2.68
28	7.64	5.45	4.57	4.07	3.75	3.53	3.36	3.23	3.12	3.03	2.96	2.90	2.84	2.79	2.75	2.72	2.68	2.65
29	7.60	5.42	4.54	4.04	3.73	3.50	3.33	3.20	3.09	3.00	2.93	2.87	2.81	2.77	2.73	2.69	2.66	2.63
30	7.56	5.39	4.51	4.02	3.70	3.47	3.30	3.17	3.07	2.98	2.91	2.84	2.79	2.74	2.70	2.66	2.63	2.60
32	7.50	5.34	4.46	3.97	3.65	3.43	3.26	3.13	3.02	2.93	2.86	2.80	2.74	2.70	2.66	2.62	2.58	2.55
34	7.44	5.29	4.42	3.93	3.61	3.39	3.22	3.09	2.98	2.89	2.82	2.76	2.70	2.66	2.62	2.58	2.55	2.51
36	7.40	5.25	4.38	3.89	3.57	3.35	3.18	3.05	2.95	2.86	2.79	2.72	2.67	2.62	2.58	2.54	2.51	2.48
38	7.35	5.21	4.34	3.86	3.54	3.32	3.15	3.02	2.92	2.83	2.75	2.69	2.64	2.59	2.55	2.51	2.48	2.45
40	7.31	5.18	4.31	3.83	3.51	3.29	3.12	2.99	2.89	2.80	2.73	2.66	2.61	2.56	2.52	2.48	2.45	2.42
42	7.28	5.15	4.29	3.80	3.49	3.27	3.10	2.97	2.86	2.78	2.70	2.64	2.59	2.54	2.50	2.46	2.43	2.40
44	7.25	5.12	4.26	3.78	3.47	3.24	3.08	2.95	2.84	2.75	2.68	2.62	2.56	2.52	2.47	2.44	2.40	2.37
46	7.22	5.10	4.24	3.76	3.44	3.22	3.06	2.93	2.82	2.73	2.66	2.60	2.54	2.50	2.45	2.42	2.38	2.35
48	7.19	5.08	4.22	3.74	3.43	3.20	3.04	2.91	2.80	2.72	2.64	2.58	2.53	2.48	2.44	2.40	2.37	2.33
50	7.17	5.06	4.20	3.72	3.41	3.19	3.02	2.89	2.79	2.70	2.63	2.56	2.51	2.46	2.42	2.38	2.35	2.32
55	7.12	5.01	4.16	3.68	3.37	3.15	2.98	2.85	2.75	2.66	2.59	2.53	2.47	2.42	2.38	2.34	2.31	2.28
60	7.08	4.98	4.13	3.65	3.34	3.12	2.95	2.82	2.72	2.63	2.56	2.50	2.44	2.39	2.35	2.31	2.28	2.25
65	7.04	4.95	4.10	3.62	3.31	3.09	2.93	2.80	2.69	2.61	2.53	2.47	2.42	2.37	2.33	2.29	2.26	2.23
70	7.01	4.92	4.08	3.60	3.29	3.07	2.91	2.78	2.67	2.59	2.51	2.45	2.40	2.35	2.31	2.27	2.23	2.20
80	6.96	4.88	4.04	3.56	3.26	3.04	2.87	2.74	2.64	2.55	2.48	2.42	2.36	2.31	2.27	2.23	2.20	2.17
90	6.93	4.85	4.01	3.54	3.23	3.01	2.84	2.72	2.61	2.52	2.45	2.39	2.33	2.29	2.24	2.21	2.17	2.14
100	6.90	4.82	3.98	3.51	3.21	2.99	2.82	2.69	2.59	2.50	2.43	2.37	2.31	2.26	2.22	2.19	2.15	2.12
125	6.84	4.78	3.94	3.47	3.17	2.95	2.79	2.66	2.55	2.47	2.39	2.33	2.28	2.23	2.19	2.15	2.11	2.08
150	6.81	4.75	3.92	3.45	3.14	2.92	2.76	2.63	2.53	2.44	2.37	2.31	2.25	2.20	2.16	2.12	2.09	2.06
200	6.76	4.71	3.88	3.41	3.11	2.89	2.73	2.60	2.50	2.41	2.34	2.27	2.22	2.17	2.13	2.09	2.06	2.02
300	6.72	4.68	3.85	3.38	3.08	2.86	2.70	2.57	2.47	2.38	2.31	2.24	2.19	2.14	2.10	2.06	2.03	1.99
500	6.69	4.65	3.82	3.36	3.05	2.84	2.68	2.55	2.44	2.36	2.28	2.22	2.17	2.12	2.07	2.04	2.00	1.97
1000	6.66	4.63	3.80	3.34	3.04	2.82	2.66	2.53	2.43	2.34	2.27	2.20	2.15	2.10	2.06	2.02	1.98	1.95
∞	6.63	4.61	3.78	3.32	3.02	2.80	2.64	2.51	2.41	2.32	2.25	2.18	2.13	2.08	2.04	2.00	1.97	1.93

Example: $P\{F_{.01;8,20} < 3.56\} = 99\%$.

$F_{.99;\nu_1,\nu_2} = 1/F_{.01;\nu_2,\nu_1}$. Example: $F_{.99;8,20} = 1/F_{.01;20,8} = 1/5.36 = 0.187$.

Table 4. (cont.) 1 Percentage Points of the F Distribution

Table of $F_{0.01;\nu_1,\nu_2}$

Degrees of freedom for the numerator (ν_1)

Multiply the numbers of the first row ($\nu_2 = 1$) by 10.

19	20	22	24	26	28	30	35	40	45	50	60	80	100	200	500	∞	Degrees of freedom for the denominator (ν_2)
620	*621*	*622*	*623*	*624*	*625*	*626*	*628*	*629*	*630*	*630*	*631*	*633*	*633*	*635*	*636*	*637*	1
99.4	99.4	99.5	99.5	99.5	99.5	99.5	99.5	99.5	99.5	99.5	99.5	99.5	99.5	99.5	99.5	99.5	2
26.7	26.7	26.6	26.6	26.6	26.5	26.5	26.5	26.4	26.4	26.4	26.3	26.3	26.2	26.2	26.1	26.1	3
14.0	14.0	14.0	13.9	13.9	13.9	13.8	13.8	13.7	13.7	13.7	13.7	13.6	13.6	13.5	13.5	13.5	4
9.58	9.55	9.51	9.47	9.43	9.40	9.38	9.33	9.29	9.26	9.24	9.20	9.16	9.13	9.08	9.04	9.02	5
7.42	7.40	7.35	7.31	7.28	7.25	7.23	7.18	7.14	7.11	7.09	7.06	7.01	6.99	6.93	6.90	6.88	6
6.18	6.16	6.11	6.07	6.04	6.02	5.99	5.94	5.91	5.88	5.86	5.82	5.78	5.75	5.70	5.67	5.65	7
5.38	5.36	5.32	5.28	5.25	5.22	5.20	5.15	5.12	5.09	5.07	5.03	4.99	4.96	4.91	4.88	4.86	8
4.83	4.81	4.77	4.73	4.70	4.67	4.65	4.60	4.57	4.54	4.52	4.48	4.44	4.42	4.36	4.33	4.31	9
4.43	4.41	4.36	4.33	4.30	4.27	4.25	4.20	4.17	4.14	4.12	4.08	4.04	4.01	3.96	3.93	3.91	10
4.12	4.10	4.06	4.02	3.99	3.96	3.94	3.89	3.86	3.83	3.81	3.78	3.73	3.71	3.66	3.62	3.60	11
3.88	3.86	3.82	3.78	3.75	3.72	3.70	3.65	3.62	3.59	3.57	3.54	3.49	3.47	3.41	3.38	3.36	12
3.69	3.66	3.62	3.59	3.56	3.53	3.51	3.46	3.43	3.40	3.38	3.34	3.30	3.27	3.22	3.19	3.17	13
3.53	3.51	3.46	3.43	3.40	3.37	3.35	3.30	3.27	3.24	3.22	3.18	3.14	3.11	3.06	3.03	3.00	14
3.40	3.37	3.33	3.29	3.26	3.24	3.21	3.17	3.13	3.10	3.08	3.05	3.00	2.98	2.92	2.89	2.87	15
3.28	3.26	3.22	3.18	3.15	3.12	3.10	3.05	3.02	2.99	2.97	2.93	2.89	2.86	2.81	2.78	2.75	16
3.18	3.16	3.12	3.08	3.05	3.03	3.00	2.96	2.92	2.89	2.87	2.83	2.79	2.76	2.71	2.68	2.65	17
3.10	3.08	3.03	3.00	2.97	2.94	2.92	2.87	2.84	2.81	2.78	2.75	2.70	2.68	2.62	2.59	2.57	18
3.03	3.00	2.96	2.92	2.89	2.87	2.84	2.80	2.76	2.73	2.71	2.67	2.63	2.60	2.55	2.51	2.49	19
2.96	2.94	2.90	2.86	2.83	2.80	2.78	2.73	2.69	2.67	2.64	2.61	2.56	2.54	2.48	2.44	2.42	20
2.90	2.88	2.84	2.80	2.77	2.74	2.72	2.67	2.64	2.61	2.58	2.55	2.50	2.48	2.42	2.38	2.36	21
2.85	2.83	2.78	2.75	2.72	2.69	2.67	2.62	2.58	2.55	2.53	2.50	2.45	2.42	2.36	2.33	2.31	22
2.80	2.78	2.74	2.70	2.67	2.64	2.62	2.57	2.54	2.51	2.48	2.45	2.40	2.37	2.32	2.28	2.26	23
2.76	2.74	2.70	2.66	2.63	2.60	2.58	2.53	2.49	2.46	2.44	2.40	2.36	2.33	2.27	2.24	2.21	24
2.72	2.70	2.66	2.62	2.59	2.56	2.54	2.49	2.45	2.42	2.40	2.36	2.32	2.29	2.23	2.19	2.17	25
2.69	2.66	2.62	2.58	2.55	2.53	2.50	2.45	2.42	2.39	2.36	2.33	2.28	2.25	2.19	2.16	2.13	26
2.66	2.63	2.59	2.55	2.52	2.49	2.47	2.42	2.38	2.35	2.33	2.29	2.25	2.22	2.16	2.12	2.10	27
2.63	2.60	2.56	2.52	2.49	2.46	2.44	2.39	2.35	2.32	2.30	2.26	2.22	2.19	2.13	2.09	2.06	28
2.60	2.57	2.53	2.49	2.46	2.44	2.41	2.36	2.33	2.30	2.27	2.23	2.19	2.16	2.10	2.06	2.03	29
2.57	2.55	2.51	2.47	2.44	2.41	2.39	2.34	2.30	2.27	2.25	2.21	2.16	2.13	2.07	2.03	2.01	30
2.53	2.50	2.46	2.42	2.39	2.36	2.34	2.29	2.25	2.22	2.20	2.16	2.11	2.08	2.02	1.98	1.96	32
2.49	2.46	2.42	2.38	2.35	2.32	2.30	2.25	2.21	2.18	2.16	2.12	2.07	2.04	1.98	1.94	1.91	34
2.45	2.43	2.38	2.35	2.32	2.29	2.26	2.21	2.17	2.14	2.12	2.08	2.03	2.00	1.94	1.90	1.87	36
2.42	2.40	2.35	2.32	2.28	2.26	2.23	2.18	2.14	2.11	2.09	2.05	2.00	1.97	1.90	1.86	1.84	38
2.39	2.37	2.33	2.29	2.26	2.23	2.20	2.15	2.11	2.08	2.06	2.02	1.97	1.94	1.87	1.83	1.80	40
2.37	2.34	2.30	2.26	2.23	2.20	2.18	2.13	2.09	2.06	2.03	1.99	1.94	1.91	1.85	1.80	1.78	42
2.35	2.32	2.28	2.24	2.21	2.18	2.15	2.10	2.06	2.03	2.01	1.97	1.92	1.89	1.82	1.78	1.75	44
2.33	2.30	2.26	2.22	2.19	2.16	2.13	2.08	2.04	2.01	1.99	1.95	1.90	1.86	1.80	1.75	1.73	46
2.31	2.28	2.24	2.20	2.17	2.14	2.12	2.06	2.02	1.99	1.97	1.93	1.88	1.84	1.78	1.73	1.70	48
2.29	2.27	2.22	2.18	2.15	2.12	2.10	2.05	2.01	1.97	1.95	1.91	1.86	1.82	1.76	1.71	1.68	50
2.25	2.23	2.18	2.15	2.11	2.08	2.06	2.01	1.97	1.93	1.91	1.87	1.81	1.78	1.71	1.67	1.64	55
2.22	2.20	2.15	2.12	2.08	2.05	2.03	1.98	1.94	1.90	1.88	1.84	1.78	1.75	1.68	1.63	1.60	60
2.20	2.17	2.13	2.09	2.06	2.03	2.00	1.95	1.91	1.88	1.85	1.81	1.75	1.72	1.65	1.60	1.57	65
2.18	2.15	2.11	2.07	2.03	2.01	1.98	1.93	1.89	1.85	1.83	1.78	1.73	1.70	1.62	1.57	1.54	70
2.14	2.12	2.07	2.03	2.00	1.97	1.94	1.89	1.85	1.81	1.79	1.75	1.69	1.66	1.58	1.53	1.49	80
2.11	2.09	2.04	2.00	1.97	1.94	1.92	1.86	1.82	1.79	1.76	1.72	1.66	1.62	1.54	1.49	1.46	90
2.09	2.07	2.02	1.98	1.94	1.92	1.89	1.84	1.80	1.76	1.73	1.69	1.63	1.60	1.52	1.47	1.43	100
2.05	2.03	1.98	1.94	1.91	1.88	1.85	1.80	1.76	1.72	1.69	1.65	1.59	1.55	1.47	1.41	1.37	125
2.03	2.00	1.96	1.92	1.88	1.85	1.83	1.77	1.73	1.69	1.66	1.62	1.56	1.52	1.43	1.38	1.33	150
2.00	1.97	1.93	1.89	1.85	1.82	1.79	1.74	1.69	1.66	1.63	1.58	1.52	1.48	1.39	1.33	1.28	200
1.97	1.94	1.89	1.85	1.82	1.79	1.76	1.71	1.66	1.62	1.59	1.55	1.48	1.44	1.35	1.28	1.22	300
1.94	1.92	1.87	1.83	1.79	1.76	1.74	1.68	1.63	1.60	1.56	1.52	1.45	1.41	1.31	1.23	1.16	500
1.92	1.90	1.85	1.81	1.77	1.74	1.72	1.66	1.61	1.57	1.54	1.50	1.43	1.38	1.28	1.19	1.11	1000
1.90	1.88	1.83	1.79	1.76	1.72	1.70	1.64	1.59	1.55	1.52	1.47	1.40	1.36	1.25	1.15	1.00	∞

Approximate formula for ν_1 and ν_2 larger than 30: $\log_{10} F_{.01;\nu_1,\nu_2} \simeq \dfrac{2.0206}{\sqrt{h} - 1.40} - 1.073\left(\dfrac{1}{\nu_1} - \dfrac{1}{\nu_2}\right)$, where $\dfrac{1}{h} = \dfrac{1}{2}\left(\dfrac{1}{\nu_1} + \dfrac{1}{\nu_2}\right)$.

Table 4. (cont.) 0.5 Percentage Points of the F Distribution

Table of $F_{0.005;\nu_1,\nu_2}$

Degrees of freedom for the numerator (ν_1) across the top; degrees of freedom for the denominator (ν_2) down the side.

Multiply the numbers of the first row ($\nu_2 = 1$) by 100.

ν_2	1	2	3	4	5	6	7	8	9	10	11	12	13	14	15	16	17	18
1	*162*	*200*	*216*	*225*	*231*	*234*	*237*	*239*	*241*	*242*	*243*	*244*	*245*	*246*	*246*	*247*	*247*	*248*
2	198	199	199	199	199	199	199	199	199	199	199	199	199	199	199	199	199	199
3	55.6	49.8	47.5	46.2	45.4	44.8	44.4	44.1	43.9	43.7	43.5	43.4	43.3	43.2	43.1	43.0	42.9	42.9
4	31.3	26.3	24.3	23.2	22.5	22.0	21.6	21.4	21.1	21.0	20.8	20.7	20.6	20.5	20.4	20.4	20.3	20.3
5	22.8	18.3	16.5	15.6	14.9	14.5	14.2	14.0	13.8	13.6	13.5	13.4	13.3	13.2	13.1	13.1	13.0	13.0
6	18.6	14.5	12.9	12.0	11.5	11.1	10.8	10.6	10.4	10.2	10.1	10.0	9.95	9.88	9.81	9.76	9.71	9.66
7	16.2	12.4	10.9	10.0	9.52	9.16	8.89	8.68	8.51	8.38	8.27	8.18	8.10	8.03	7.97	7.93	7.87	7.83
8	14.7	11.0	9.60	8.81	8.30	7.95	7.69	7.50	7.34	7.21	7.10	7.01	6.94	6.87	6.81	6.76	6.72	6.68
9	13.6	10.1	8.72	7.96	7.47	7.13	6.88	6.69	6.54	6.42	6.31	6.23	6.15	6.09	6.03	5.98	5.94	5.90
10	12.8	9.43	8.08	7.34	6.87	6.54	6.30	6.12	5.97	5.85	5.75	5.66	5.59	5.53	5.47	5.42	5.38	5.34
11	12.2	8.91	7.60	6.88	6.42	6.10	5.86	5.68	5.54	5.42	5.32	5.24	5.16	5.10	5.05	5.00	4.96	4.92
12	11.8	8.51	7.23	6.52	6.07	5.76	5.52	5.35	5.20	5.09	4.99	4.91	4.84	4.77	4.72	4.67	4.63	4.59
13	11.4	8.19	6.93	6.23	5.79	5.48	5.25	5.08	4.94	4.82	4.72	4.64	4.57	4.51	4.46	4.41	4.37	4.33
14	11.1	7.92	6.68	6.00	5.56	5.26	5.03	4.86	4.72	4.60	4.51	4.43	4.36	4.30	4.25	4.20	4.16	4.12
15	10.8	7.70	6.48	5.80	5.37	5.07	4.85	4.67	4.54	4.42	4.33	4.25	4.18	4.12	4.07	4.02	3.98	3.95
16	10.6	7.51	6.30	5.64	5.21	4.91	4.69	4.52	4.38	4.27	4.18	4.10	4.03	3.97	3.92	3.87	3.83	3.80
17	10.4	7.35	6.16	5.50	5.07	4.78	4.56	4.39	4.25	4.14	4.05	3.97	3.90	3.84	3.79	3.75	3.71	3.67
18	10.2	7.21	6.03	5.37	4.96	4.66	4.44	4.28	4.14	4.03	3.94	3.86	3.79	3.73	3.68	3.64	3.60	3.56
19	10.1	7.09	5.92	5.27	4.85	4.56	4.34	4.18	4.04	3.93	3.84	3.76	3.70	3.64	3.59	3.54	3.50	3.46
20	9.94	6.99	5.82	5.17	4.76	4.47	4.26	4.09	3.96	3.85	3.76	3.68	3.61	3.55	3.50	3.46	3.42	3.38
21	9.83	6.89	5.73	5.09	4.68	4.39	4.18	4.01	3.88	3.77	3.68	3.60	3.54	3.48	3.43	3.38	3.34	3.31
22	9.73	6.81	5.65	5.02	4.61	4.32	4.11	3.94	3.81	3.70	3.61	3.54	3.47	3.41	3.36	3.31	3.27	3.24
23	9.63	6.73	5.58	4.95	4.54	4.26	4.05	3.88	3.75	3.64	3.55	3.47	3.41	3.35	3.30	3.25	3.21	3.18
24	9.55	6.66	5.52	4.89	4.49	4.20	3.99	3.83	3.69	3.59	3.50	3.42	3.35	3.30	3.25	3.20	3.16	3.12
25	9.48	6.60	5.46	4.84	4.43	4.15	3.94	3.78	3.64	3.54	3.45	3.37	3.30	3.25	3.20	3.15	3.11	3.08
26	9.41	6.54	5.41	4.79	4.38	4.10	3.89	3.73	3.60	3.49	3.40	3.33	3.26	3.20	3.15	3.11	3.07	3.03
27	9.34	6.49	5.36	4.74	4.34	4.06	3.85	3.69	3.56	3.45	3.36	3.28	3.22	3.16	3.11	3.07	3.03	2.99
28	9.28	6.44	5.32	4.70	4.30	4.02	3.81	3.65	3.52	3.41	3.32	3.25	3.18	3.12	3.07	3.03	2.99	2.95
29	9.23	6.40	5.28	4.66	4.26	3.98	3.77	3.61	3.48	3.38	3.29	3.21	3.15	3.09	3.04	2.99	2.95	2.92
30	9.18	6.35	5.24	4.62	4.23	3.95	3.74	3.58	3.45	3.34	3.25	3.18	3.11	3.06	3.01	2.96	2.92	2.89
32	9.09	6.28	5.17	4.56	4.17	3.89	3.68	3.52	3.39	3.29	3.20	3.12	3.06	3.00	2.95	2.90	2.86	2.83
34	9.01	6.22	5.11	4.50	4.11	3.84	3.63	3.47	3.34	3.24	3.15	3.07	3.01	2.95	2.90	2.85	2.81	2.78
36	8.94	6.16	5.06	4.46	4.06	3.79	3.58	3.42	3.30	3.19	3.10	3.03	2.96	2.90	2.85	2.81	2.77	2.73
38	8.88	6.11	5.02	4.41	4.02	3.75	3.54	3.39	3.25	3.15	3.06	2.99	2.92	2.87	2.82	2.77	2.73	2.70
40	8.83	6.07	4.98	4.37	3.99	3.71	3.51	3.35	3.22	3.12	3.03	2.95	2.89	2.83	2.78	2.74	2.70	2.66
42	8.78	6.03	4.94	4.34	3.95	3.68	3.48	3.32	3.19	3.09	3.00	2.92	2.86	2.80	2.75	2.71	2.67	2.63
44	8.74	5.99	4.91	4.31	3.92	3.65	3.45	3.29	3.16	3.06	2.97	2.89	2.83	2.77	2.72	2.68	2.64	2.60
46	8.70	5.96	4.88	4.28	3.90	3.62	3.42	3.26	3.14	3.03	2.94	2.87	2.80	2.75	2.70	2.65	2.61	2.58
48	8.66	5.93	4.85	4.25	3.87	3.60	3.40	3.24	3.11	3.01	2.92	2.85	2.78	2.72	2.67	2.63	2.59	2.55
50	8.63	5.90	4.83	4.23	3.85	3.58	3.38	3.22	3.09	2.99	2.90	2.82	2.76	2.70	2.65	2.61	2.57	2.53
55	8.55	5.84	4.77	4.18	3.80	3.53	3.33	3.17	3.05	2.94	2.85	2.78	2.71	2.66	2.61	2.56	2.52	2.49
60	8.49	5.80	4.73	4.14	3.76	3.49	3.29	3.13	3.01	2.90	2.82	2.74	2.68	2.62	2.57	2.53	2.49	2.45
65	8.44	5.75	4.68	4.11	3.73	3.46	3.26	3.10	2.98	2.87	2.79	2.71	2.65	2.59	2.54	2.49	2.45	2.42
70	8.40	5.72	4.65	4.08	3.70	3.43	3.23	3.08	2.95	2.85	2.76	2.68	2.62	2.56	2.51	2.47	2.43	2.39
80	8.33	5.67	4.61	4.03	3.65	3.39	3.19	3.03	2.91	2.80	2.72	2.64	2.58	2.52	2.47	2.43	2.39	2.35
90	8.28	5.62	4.57	3.99	3.62	3.35	3.15	3.00	2.87	2.77	2.68	2.61	2.54	2.49	2.44	2.39	2.35	2.32
100	8.24	5.59	4.54	3.96	3.59	3.33	3.13	2.97	2.85	2.74	2.66	2.58	2.52	2.46	2.41	2.37	2.33	2.29
125	8.17	5.53	4.49	3.91	3.54	3.28	3.08	2.93	2.80	2.70	2.61	2.54	2.47	2.42	2.37	2.32	2.28	2.24
150	8.12	5.49	4.45	3.88	3.51	3.25	3.05	2.89	2.77	2.67	2.58	2.51	2.44	2.38	2.33	2.29	2.25	2.21
200	8.06	5.44	4.41	3.84	3.47	3.21	3.01	2.85	2.73	2.63	2.54	2.47	2.40	2.35	2.30	2.25	2.21	2.18
300	8.00	5.39	4.37	3.80	3.43	3.17	2.97	2.81	2.69	2.59	2.51	2.43	2.37	2.31	2.26	2.21	2.17	2.14
500	7.95	5.36	4.33	3.76	3.40	3.14	2.94	2.79	2.66	2.56	2.48	2.40	2.34	2.28	2.23	2.19	2.14	2.11
1000	7.92	5.33	4.31	3.74	3.37	3.11	2.92	2.77	2.64	2.54	2.45	2.38	2.32	2.26	2.21	2.16	2.12	2.09
∞	7.88	5.30	4.28	3.72	3.35	3.09	2.90	2.74	2.62	2.52	2.43	2.36	2.29	2.24	2.19	2.14	2.10	2.06

Example: $P\{F_{.005;8,20} < 4.09\} = 99.5\%$.

$F_{.995;\nu_1,\nu_2} = 1/F_{.005;\nu_2,\nu_1}$. Example: $F_{.995;8,20} = 1/F_{.005;20,8} = 1/6.61 = 0.151$.

Table of $F_{0.005;\nu_1,\nu_2}$

Degrees of freedom for the numerator (ν_1)

Multiply the numbers of the first row (ν_2 = 1) by 100.

Degrees of freedom for the denominator (ν_2)	19	20	22	24	26	28	30	35	40	45	50	60	80	100	200	500	∞
1	*248*	*248*	*249*	*249*	*250*	*250*	*250*	*251*	*251*	*252*	*252*	*253*	*253*	*253*	*254*	*254*	*255*
2	199	199	199	199	199	199	199	199	199	199	199	199	199	199	199	200	200
3	42.8	42.8	42.7	42.6	42.6	42.5	42.5	42.4	42.3	42.3	42.2	42.1	42.1	42.0	41.9	41.9	41.8
4	20.2	20.2	20.1	20.0	20.0	19.9	19.9	19.8	19.8	19.7	19.7	19.6	19.5	19.5	19.4	19.4	19.3
5	12.9	12.9	12.8	12.8	12.7	12.7	12.7	12.6	12.5	12.5	12.5	12.4	12.3	12.3	12.2	12.2	12.1
6	9.62	9.59	9.53	9.47	9.43	9.39	9.36	9.29	9.24	9.20	9.17	9.12	9.06	9.03	8.95	8.91	8.88
7	7.79	7.75	7.69	7.64	7.60	7.57	7.53	7.47	7.42	7.38	7.35	7.31	7.25	7.22	7.15	7.10	7.08
8	6.64	6.61	6.55	6.50	6.46	6.43	6.40	6.33	6.29	6.25	6.22	6.18	6.12	6.09	6.02	5.98	5.95
9	5.86	5.83	5.78	5.73	5.69	5.65	5.62	5.56	5.52	5.48	5.45	5.41	5.36	5.32	5.26	5.21	5.19
10	5.30	5.27	5.22	5.17	5.13	5.10	5.07	5.01	4.97	4.93	4.90	4.86	4.80	4.77	4.71	4.67	4.64
11	4.89	4.86	4.80	4.76	4.72	4.68	4.65	4.60	4.55	4.52	4.49	4.44	4.39	4.36	4.29	4.25	4.23
12	4.56	4.53	4.48	4.43	4.39	4.36	4.33	4.27	4.23	4.19	4.17	4.12	4.07	4.04	3.97	3.93	3.90
13	4.30	4.27	4.22	4.17	4.13	4.10	4.07	4.01	3.97	3.94	3.91	3.87	3.81	3.78	3.71	3.67	3.65
14	4.09	4.06	4.01	3.96	3.92	3.89	3.86	3.80	3.76	3.73	3.70	3.66	3.60	3.57	3.50	3.46	3.44
15	3.91	3.88	3.83	3.79	3.75	3.72	3.69	3.63	3.58	3.55	3.52	3.48	3.43	3.39	3.33	3.29	3.26
16	3.76	3.73	3.68	3.64	3.60	3.57	3.54	3.48	3.44	3.40	3.37	3.33	3.28	3.25	3.18	3.14	3.11
17	3.64	3.61	3.56	3.51	3.47	3.44	3.41	3.35	3.31	3.28	3.25	3.21	3.15	3.12	3.05	3.01	2.98
18	3.53	3.50	3.45	3.40	3.36	3.33	3.30	3.25	3.20	3.17	3.14	3.10	3.04	3.01	2.94	2.90	2.87
19	3.43	3.40	3.35	3.31	3.27	3.24	3.21	3.15	3.11	3.07	3.04	3.00	2.95	2.91	2.85	2.80	2.78
20	3.35	3.32	3.27	3.22	3.18	3.15	3.12	3.07	3.02	2.99	2.96	2.92	2.86	2.83	2.76	2.72	2.69
21	3.27	3.24	3.19	3.15	3.11	3.08	3.05	2.99	2.95	2.91	2.88	2.84	2.78	2.75	2.68	2.64	2.61
22	3.20	3.18	3.12	3.08	3.04	3.01	2.98	2.92	2.88	2.84	2.82	2.77	2.72	2.69	2.62	2.57	2.55
23	3.15	3.12	3.06	3.02	2.98	2.95	2.92	2.86	2.82	2.78	2.76	2.71	2.66	2.62	2.56	2.51	2.48
24	3.09	3.06	3.01	2.97	2.93	2.90	2.87	2.81	2.77	2.73	2.70	2.66	2.60	2.57	2.50	2.46	2.43
25	3.04	3.01	2.96	2.92	2.88	2.85	2.82	2.76	2.72	2.68	2.65	2.61	2.55	2.52	2.45	2.41	2.38
26	3.00	2.97	2.92	2.87	2.83	2.80	2.77	2.72	2.67	2.64	2.61	2.56	2.51	2.47	2.40	2.36	2.33
27	2.96	2.93	2.88	2.83	2.79	2.76	2.73	2.67	2.63	2.59	2.57	2.52	2.47	2.43	2.36	2.32	2.29
28	2.92	2.89	2.84	2.79	2.76	2.72	2.69	2.64	2.59	2.56	2.53	2.48	2.43	2.39	2.32	2.28	2.25
29	2.88	2.86	2.80	2.76	2.72	2.69	2.66	2.60	2.56	2.52	2.49	2.45	2.39	2.36	2.28	2.24	2.21
30	2.85	2.82	2.77	2.73	2.69	2.66	2.63	2.57	2.52	2.49	2.46	2.42	2.36	2.32	2.25	2.21	2.18
32	2.80	2.77	2.71	2.67	2.63	2.60	2.57	2.51	2.47	2.43	2.40	2.36	2.30	2.26	2.19	2.15	2.11
34	2.75	2.72	2.66	2.62	2.58	2.55	2.52	2.46	2.42	2.38	2.35	2.30	2.25	2.21	2.14	2.09	2.06
36	2.70	2.67	2.62	2.58	2.54	2.50	2.48	2.42	2.37	2.33	2.30	2.26	2.20	2.17	2.09	2.04	2.01
38	2.66	2.63	2.58	2.54	2.50	2.47	2.44	2.38	2.33	2.29	2.27	2.22	2.16	2.12	2.05	2.00	1.97
40	2.63	2.60	2.55	2.50	2.46	2.43	2.40	2.34	2.30	2.26	2.23	2.18	2.12	2.09	2.01	1.96	1.93
42	2.60	2.57	2.52	2.47	2.43	2.40	2.37	2.31	2.26	2.23	2.20	2.15	2.09	2.06	1.98	1.93	1.90
44	2.57	2.54	2.49	2.44	2.40	2.37	2.34	2.28	2.24	2.20	2.17	2.12	2.06	2.03	1.95	1.90	1.87
46	2.54	2.51	2.46	2.42	2.38	2.34	2.32	2.26	2.21	2.17	2.14	2.10	2.04	2.00	1.92	1.87	1.84
48	2.52	2.49	2.44	2.39	2.36	2.32	2.29	2.23	2.19	2.15	2.12	2.07	2.01	1.97	1.89	1.84	1.81
50	2.50	2.47	2.42	2.37	2.33	2.30	2.27	2.21	2.16	2.13	2.10	2.05	1.99	1.95	1.87	1.82	1.79
55	2.45	2.42	2.37	2.33	2.29	2.26	2.23	2.16	2.12	2.08	2.05	2.00	1.94	1.90	1.82	1.77	1.73
60	2.42	2.39	2.33	2.29	2.25	2.22	2.19	2.13	2.08	2.04	2.01	1.96	1.90	1.86	1.78	1.73	1.69
65	2.39	2.36	2.30	2.26	2.22	2.19	2.16	2.09	2.05	2.01	1.98	1.93	1.87	1.83	1.74	1.69	1.65
70	2.36	2.33	2.28	2.23	2.19	2.16	2.13	2.07	2.02	1.98	1.95	1.90	1.84	1.80	1.71	1.66	1.62
80	2.32	2.29	2.23	2.19	2.15	2.11	2.08	2.02	1.97	1.93	1.90	1.85	1.79	1.75	1.66	1.60	1.56
90	2.28	2.25	2.20	2.15	2.12	2.08	2.05	1.99	1.94	1.90	1.87	1.82	1.75	1.71	1.62	1.56	1.52
100	2.26	2.23	2.17	2.13	2.09	2.05	2.02	1.96	1.91	1.87	1.84	1.79	1.72	1.68	1.59	1.53	1.49
125	2.21	2.18	2.13	2.08	2.04	2.01	1.98	1.91	1.86	1.82	1.79	1.74	1.67	1.63	1.53	1.47	1.42
150	2.18	2.15	2.10	2.05	2.01	1.98	1.94	1.88	1.83	1.79	1.76	1.70	1.63	1.59	1.49	1.42	1.37
200	2.14	2.11	2.06	2.01	1.97	1.94	1.91	1.84	1.79	1.75	1.71	1.66	1.59	1.54	1.44	1.37	1.31
300	2.10	2.07	2.02	1.97	1.93	1.90	1.87	1.80	1.75	1.71	1.67	1.61	1.54	1.50	1.39	1.31	1.25
500	2.07	2.04	1.99	1.94	1.90	1.87	1.84	1.77	1.72	1.67	1.64	1.58	1.51	1.46	1.35	1.26	1.18
1000	2.05	2.02	1.97	1.92	1.88	1.84	1.81	1.75	1.69	1.65	1.61	1.56	1.48	1.43	1.31	1.22	1.13
∞	2.03	2.00	1.95	1.90	1.86	1.82	1.79	1.72	1.67	1.63	1.59	1.53	1.45	1.40	1.28	1.17	1.00

Approximate formula for ν_1 and ν_2 larger than 30: $\log_{10} F_{.005;\nu_1,\nu_2} \simeq \dfrac{2.2373}{\sqrt{h} - 1.61} - 1.250\left(\dfrac{1}{\nu_1} - \dfrac{1}{\nu_2}\right)$, where $\dfrac{1}{h} = \dfrac{1}{2}\left(\dfrac{1}{\nu_1} + \dfrac{1}{\nu_2}\right)$.

Table 4. (cont.) 0.1 Percentage Points of the F Distribution

Table of $F_{0.001;\nu_1,\nu_2}$

Degrees of freedom for the numerator (ν_1)

Degrees of freedom for the denominator (ν_2)	1	2	3	4	5	6	7	8	9	10	15	20	30	50	100	200	500	∞
	Multiply the numbers of the first row ($\nu_2 = 1$) by 1000.																	
1	*405*	*500*	*540*	*562*	*576*	*586*	*593*	*598*	*602*	*606*	*616*	*621*	*626*	*630*	*633*	*635*	*636*	*637*
2	998	999	999	999	999	999	999	999	999	999	999	999	999	999	999	999	999	999
3	168	148	141	137	135	133	132	131	130	129	127	126	125	125	124	124	124	124
4	74.1	61.2	56.2	53.4	51.7	50.5	49.7	49.0	48.5	48.0	46.8	46.1	45.4	44.9	44.5	44.3	44.1	44.0
5	47.0	36.6	33.2	31.1	29.8	28.8	28.2	27.6	27.2	26.9	25.9	25.4	24.9	24.4	24.1	23.9	23.8	23.8
6	35.5	27.0	23.7	21.9	20.8	20.0	19.5	19.0	18.7	18.4	17.6	17.1	16.7	16.3	16.0	15.9	15.8	15.8
7	29.2	21.7	18.8	17.2	16.2	15.5	15.0	14.6	14.3	14.1	13.3	12.9	12.5	12.2	11.9	11.8	11.7	11.7
8	25.4	18.5	15.8	14.4	13.5	12.9	12.4	12.0	11.8	11.5	10.8	10.5	10.1	9.80	9.57	9.46	9.39	9.34
9	22.9	16.4	13.9	12.6	11.7	11.1	10.7	10.4	10.1	9.89	9.24	8.90	8.55	8.26	8.04	7.93	7.86	7.81
10	21.0	14.9	12.6	11.3	10.5	9.92	9.52	9.20	8.96	8.75	8.13	7.80	7.47	7.19	6.98	6.87	6.81	6.76
11	19.7	13.8	11.6	10.4	9.58	9.05	8.66	8.35	8.12	7.92	7.32	7.01	6.68	6.41	6.21	6.10	6.04	6.00
12	18.6	13.0	10.8	9.63	8.89	8.38	8.00	7.71	7.48	7.29	6.71	6.40	6.09	5.83	5.63	5.52	5.46	5.42
13	17.8	12.3	10.2	9.07	8.35	7.86	7.49	7.21	6.98	6.80	6.23	5.93	5.62	5.37	5.17	5.07	5.01	4.97
14	17.1	11.8	9.73	8.62	7.92	7.43	7.08	6.80	6.58	6.40	5.85	5.56	5.25	5.00	4.80	4.70	4.64	4.60
15	16.6	11.3	9.34	8.25	7.57	7.09	6.74	6.47	6.26	6.08	5.53	5.25	4.95	4.70	4.51	4.41	4.35	4.31
16	16.1	11.0	9.00	7.94	7.27	6.81	6.46	6.19	5.98	5.81	5.27	4.99	4.70	4.45	4.26	4.16	4.10	4.06
17	15.7	10.7	8.73	7.68	7.02	6.56	6.22	5.96	5.75	5.58	5.05	4.78	4.48	4.24	4.05	3.95	3.89	3.85
18	15.4	10.4	8.49	7.46	6.81	6.35	6.02	5.76	5.56	5.39	4.87	4.59	4.30	4.06	3.87	3.77	3.71	3.67
19	15.1	10.2	8.28	7.26	6.61	6.18	5.84	5.59	5.39	5.22	4.70	4.43	4.14	3.90	3.71	3.61	3.55	3.51
20	14.8	9.95	8.10	7.10	6.46	6.02	5.69	5.44	5.24	5.08	4.56	4.29	4.01	3.77	3.58	3.48	3.42	3.38
22	14.4	9.61	7.80	6.81	6.19	5.76	5.44	5.19	4.99	4.83	4.32	4.06	3.77	3.53	3.34	3.25	3.19	3.15
24	14.0	9.34	7.55	6.59	5.98	5.55	5.23	4.99	4.80	4.64	4.14	3.87	3.59	3.35	3.16	3.07	3.01	2.97
26	13.7	9.12	7.36	6.41	5.80	5.38	5.07	4.83	4.64	4.48	3.99	3.72	3.45	3.20	3.01	2.92	2.86	2.82
28	13.5	8.93	7.19	6.25	5.66	5.24	4.93	4.69	4.50	4.35	3.86	3.60	3.32	3.08	2.89	2.79	2.73	2.70
30	13.3	8.77	7.05	6.12	5.53	5.12	4.82	4.58	4.39	4.24	3.75	3.49	3.22	2.98	2.79	2.69	2.63	2.59
40	12.6	8.25	6.60	5.70	5.13	4.73	4.43	4.21	4.02	3.87	3.40	3.15	2.87	2.64	2.44	2.34	2.28	2.23
50	12.2	7.95	6.34	5.46	4.90	4.51	4.22	4.00	3.82	3.67	3.20	2.95	2.68	2.44	2.24	2.14	2.07	2.03
60	12.0	7.76	6.17	5.31	4.76	4.37	4.09	3.87	3.69	3.54	3.08	2.83	2.56	2.31	2.11	2.01	1.93	1.89
80	11.7	7.54	5.97	5.13	4.58	4.21	3.92	3.70	3.53	3.39	2.93	2.68	2.40	2.16	1.95	1.84	1.77	1.72
100	11.5	7.41	5.85	5.01	4.48	4.11	3.83	3.61	3.44	3.30	2.84	2.59	2.32	2.07	1.87	1.75	1.68	1.62
200	11.2	7.15	5.64	4.81	4.29	3.92	3.65	3.43	3.26	3.12	2.67	2.42	2.15	1.90	1.68	1.55	1.46	1.39
500	11.0	7.01	5.51	4.69	4.18	3.82	3.54	3.33	3.16	3.02	2.58	2.33	2.05	1.80	1.57	1.43	1.32	1.23
∞	10.8	6.91	5.42	4.62	4.10	3.74	3.47	3.27	3.10	2.96	2.51	2.27	1.99	1.73	1.49	1.34	1.21	1.00

Example: $P\{F_{.001;8,20} < 5.44\} = 99.9\%$.

$F_{.999;\nu_1,\nu_2} = 1/F_{.001;\nu_2,\nu_1}$. Example: $F_{.999;8,20} = 1/F_{.001;20,8} = 1/10.5 = 0.095$.

Approximate formula for ν_1 and ν_2 larger than 30: $\log_{10} F_{.001;\nu_1,\nu_2} \simeq \dfrac{2.6841}{\sqrt{h} - 2.09} - 1.672\left(\dfrac{1}{\nu_1} - \dfrac{1}{\nu_2}\right)$, where $\dfrac{1}{h} = \dfrac{1}{2}\left(\dfrac{1}{\nu_1} + \dfrac{1}{\nu_2}\right)$.

Index

A

B

C

D

L

M

P

Q

R

T

U

V

W